A Publication of the
National Research Council of Canada
Monograph Publishing Program

THE QUEST FOR SUSTAINABLE INTERNATIONAL FISHERIES

Regional Efforts To Implement The 1995 United Nations Fish Stocks Agreement:

An Overview For The May 2006 Review Conference

EVELYNE MELTZER
evelyne.meltzer@gmail.com

with the assistance of Susanna D. Fuller

NRC·CNRC

NRC Research Press
Ottawa, 2009

www.cabi.org

ISBN 978-0-660-19724-1
NRC No. 46857
Catalogue No.: NR16-98/2008E

Published exclusively in Canada and non-exclusively in other territories of the world (excluding Europe, Asia, Africa, the Middle East, and South America) by NRC Research Press as 978-0-660-19724-1
NRC No. 46857

Published exclusively in Europe, Asia, Africa, the Middle East, and South America, and non-exclusively in other territories of the world (excluding Australia, Canada, and New Zealand) by CABI Publishing, a division of CAB International, with the ISBN 978-1-84593-583-2.

CABI Publishing, CAB International, Wallingford, Oxon OX10 8DE UK

CABI Head Office
Nosworthy Way
Wallingford
Oxfordshire OX10 8DE
UK

Tel: +44 (0)1491 832111
Fax: +44 (0)1491 833508
E-mail: cabi@cabi.org
Website: www.cabi.org

CABI North American Office
875 Massachusetts Avenue
7th Floor
Cambridge, MA 02139
USA

Tel: +1 617 395 4056
Fax: +1 617 354 6875
E-mail: cabi-nao@cabi.org

This monograph is also distributed in Australia and New Zealand by CSIRO Publishing, 150 Oxford Street, Collingwood, Victoria 3066, Australia.

Library and Archives Canada Cataloguing in Publication

Meltzer, Evelyne
The quest for sustainable international fisheries : regional efforts to implement the 1995 United Nations Fish Stock Agreement : an overview for the May 2006 review / Evelyne Meltzer with the assistance of Susanna D. Fuller.

Includes bibliographical references.
Issued by: National Research Council Canada.
ISBN 978-0-660-19724-1

1. Fishery management, International. 2. Sustainable fisheries --International cooperation. 3. Fishes--Conservation--International cooperation. 4. United Nations Convention on the Law of the Sea (1982). Protocols, etc., 1995 Aug. 4. I. Fuller, Susanna II. National Research Council Canada III. Title.

QH75.M43 2009 333.95'616 C2008-980321-3

NRC Monograph Publishing Program

Inquiries: Monograph Publishing Program, NRC Research Press, National Research Council of Canada, Ottawa, Ontario K1A 0R6, Canada. Web site: http://pubs.nrc-cnrc.gc.ca

Correct citation for this publication: Meltzer, E. 2009. The quest for sustainable international fisheries: regional efforts to implement the 1995 United Nations Fish Stock Agreement: an overview for the May 2006 review NRC Research Press, Ottawa, Ontario, Canada. 448pp.

In loving memory of

Garfield Theodore Meltzer

1922–2008

Contents

Table List xii

Figure List xii

Preface xv

Selected Acronym List xxv

Summary xxxi

Part I: Legal, Institutional, and Biological Context ***1***

Chapter 1 Introduction **3**

Chapter 2 Overview of International Legal Instruments Pertaining to Straddling Fish Stocks and Highly Migratory Fish Stocks **11**

2.1 Introduction 11
2.2 The international legal framework: hard law, soft law, and customary international law 11
2.3 Progressive development of international fisheries law — post-LOSC 14
2.4 Key developments in international fisheries law 17
2.4.1 Law of the Sea Convention (1982) 17
2.4.2 1992 United Nations Conference on Environment and Development 18
2.4.3 FAO Compliance Agreement (1993) 18
2.4.4 FAO Code of Conduct for Responsible Fisheries (1995) 19
2.4.5 United Nations Fish Stocks Agreement (UNFSA) (1995) 21
2.4.6 FAO International Plans of Action (1999, 2001) 22
2.4.6.1 International Plan of Action for Conservation and Management of Sharks (IPOA-Sharks) 22
2.4.6.2 International Plan of Action for Reducing Incidental Catch of Seabirds in Longline Fisheries (IPOA-Seabirds) 22
2.4.6.3 International Plan of Action for the Management of Fishing Capacity (IPOA - Fishing Capacity) 23
2.4.6.4 International Plan of Action to Prevent, Deter, and Eliminate Illegal, Unreported and Unregulated Fishing (IPOA-IUU) 23
2.4.7 Reykjavik Declaration (2001) 23
2.4.8 World Summit on Sustainable Development Plan of Implementation (2002) 24
2.4.9 FAO Strategy for Improving Information on Status and Trends of Capture Fisheries (2003) 25
2.4.10 United Nations General Assembly 25
2.4.10.1 United Nations Open-ended Informal Consultative Process on Oceans and the Law of the Sea (2000) 26
2.4.10.2 UNFSA Informal Consultation of States Parties (2001) 26
2.4.11 FAO Ministerial Declarations 26
2.4.11.1 Rome Consensus on World Fisheries (1995) 26
2.4.11.2 The Rome Declaration on the Implementation of the Code of Conduct for Responsible Fisheries (1999) 26

2.4.12 The Rome Declaration on Illegal, Unreported and Unregulated Fishing (2005) 27
2.4.13 Ministerial Declaration of the St. John's Conference on the Governance of High Seas Fisheries and the UN Fish Agreement (2005) 27
2.4.14 Related international legal instruments 27
2.4.14.1 International Convention for the Regulation of Whaling (1946) 27
2.4.14.2 The Convention on International Trade in Endangered Species of Wild Fauna and Flora (1973) 28
2.4.14.3 Convention on the Conservation of Migratory Species of Wild Animals (1979) 28
2.4.14.4 Convention on Biological Diversity (1992) 29
2.4.14.5 The World Trade Organization (1994) 29
2.4.14.6 Agreement on the Conservation of Albatrosses and Petrels (2001) 30
2.5 Summary and conclusion 30
2.5.1 The legal framework for sustainable high seas fishing: not just words 31

Chapter 3 Juridical Status of Fish Stocks 41
3.1 Introduction 41
3.2 Regional fishery bodies with competence for straddling fish stocks 41
3.3 Straddling fish stocks 42
3.4 Highly migratory fish stocks 45
3.5 Discrete high seas stocks 45
3.6 Deepwater species 46
3.7 Transboundary, joint, and shared stocks 46
3.8 Conclusion 47

Chapter 4 International Fisheries Governance 53
4.1 Introduction 53
4.2 New model for global good governance 53
4.2.1 Beyond the tragedy of the commons: collaborative regulation of the commons 54
4.2.2 Regional governance and regime building 54
4.3 The evolution of regional fishery organizations 55
4.4 The typology of RFBs 56
4.5 RFMOs as vehicles of good governance 57
4.5.1 Strengthening and modernizing existing pre-UNFSA RFMOs 58
4.5.1.1 IATTC 59
4.5.1.2 ICCAT 59
4.5.1.3 NAFO 60
4.5.1.4 NEAFC 60
4.5.2 Creation of new RFMOs 60
4.5.2.1 SEAFO 61
4.5.2.2 WCPFC 61
4.5.2.3 SIOFA 61
4.5.2.4 Galapagos Agreement 61
4.5.2.5 South Pacific RFMO 62
4.5.3 IUU fishing 62

4.6 Membership in RFMOs 63
4.6.1 Regulating access to the high seas fisheries resources 63
4.6.2 Real interest 63
4.6.3 The "free-rider" problem 64
4.6.4 Efforts to increase membership or participation 64
4.6.4.1 Fishing entities 65
4.6.4.2 Vessel registers: positive and negative lists 66
4.6.4.3 Catch and trade documentation schemes 66
4.6.5 Implications of increased membership 66
4.7 Conclusion 67

Chapter 5 State of the Fisheries 77
5.1 Introduction 77
5.2 Status of straddling fish stocks and highly migratory fish stocks 78
5.2.1 Global overview 78
5.2.2 UNFSA provisions for stock assessment 80
5.2.3 Stock assessment in practice 80
5.2.4 RFMO variability on stock assessments 81
5.3 What overfishing means 81
5.3.1 Concepts 81
5.3.2 MSY 82
5.3.2.1 Overview 82
5.3.2.2 Defining MSY 83
5.3.3 Shortcomings of MSY 84
5.3.4 MSY: from target to limit 84
5.3.5 Use of MSY as a limit by RFMOs 85
5.4 Status of major stocks 86
5.4.1 Straddling fish stocks 86
5.4.1.1 Cod in Northwest Atlantic Ocean 86
5.4.1.2 Shrimp in Northwest Atlantic Ocean 87
5.4.1.3 Greenland halibut in Northwest Atlantic Ocean 87
5.4.1.4. Blue whiting in Northeast Atlantic Ocean 87
5.4.1.5 Patagonian toothfish in Southern Ocean 88
5.4.2 Highly migratory fish stocks 88
5.4.2.1 Overview 88
5.4.2.2 Southern bluefin tuna 91
5.4.2.3 Bigeye tuna in Eastern Pacific Ocean 93
5.4.2.4 Western Atlantic bluefin tuna 93
5.4.2.5 Yellowfin tuna, skipjack tuna, and bigeye tuna in Indian Ocean 93
5.4.3 Stock recovery 94
5.4.3.1 Overview within RFMOs 94
5.4.3.2 Atlanto-Scandian (Norwegian) spring-spawning herring 94
5.4.4 Unregulated species 95
5.4.5 New and exploratory fisheries 95

5.5 Beyond optimum economic yield 96
5.5.1 Overcapacity and overfishing 96
5.5.2 Subsidies and overcapacity 96
5.5.2.1 IATTC 97
5.5.2.2 ICCAT 98
5.5.2.3 IOTC 98
5.5.2.4 CCSBT 98
5.5.3 Dealing with subsidies 98
5.6 Beyond MSY and excess capacity 99
5.6.1 The spectre of the single vessel 100
5.6.2 Species at risk 100
5.6.3 RFMOs' response to setting nonsustainable TACs: harvest control or decision rules 102
5.7 Conclusion 102

Part II: Selected Themes Pertaining To RFMO Regulation Of Straddling Fish Stocks And Highly Migratory Fish Stocks 113

Chapter 6 The Precautionary Approach to Fisheries 115
6.1 Introduction 115
6.2 International legal underpinnings 116
6.3 Defining the precautionary approach 116
6.3.1 Reference points and decision rules 117
6.3.2 Precautionary management strategies and plans 119
6.4 Efforts to adopt and implement the PA to fisheries 120
6.4.1 Pre-UNFSA RFMOs 121
6.4.1.1 CCAMLR 121
6.4.1.2 IATTC 124
6.4.1.3 ICCAT 124
6.4.1.4 GFCM 125
6.4.1.5 NEAFC 125
6.4.1.6 NAFO 126
6.4.1.7 Bering Sea Donut Hole 127
6.4.1.8 CCSBT 127
6.4.2 Post-UNFSA RFMOs 129
6.4.2.1 IOTC 129
6.4.2.2 WCPFC 129
6.4.2.3 SEAFO 129
6.4.3 RFMOs–RFMAs not yet in force 129
6.4.3.1 SIOFA 129
6.5 Challenges and opportunities for the implementation of the PA 130
6.6 Conclusion 133

Chapter 7 The Ecosystem Approach to Fisheries 143
7.1 Introduction 143

7.2 International legal underpinnings ... 143
7.3 Terminology associated with the ecosystem approach ... 145
7.4 Efforts to adopt and elaborate EAF ... 147
7.4.1 Increased data needs for understanding marine ecosystem impacts ... 148
7.4.2 Increased uncertainty and the precautionary approach ... 148
7.4.3 Cooperation, compatibility, and transparency ... 149
7.4.4 Precautionary, rule-based decision making and efficient dispute settlement ... 149
7.4.5 Reducing fishing pressure through capacity reduction ... 150
7.4.6. Ecosystem indicators and objectives ... 150
7.5 Implementing EAF ... 150
7.6 RFMO implementation of EAF ... 151
7.6.1 CCAMLR ... 154
7.6.1.1 Seabird bycatch ... 156
7.6.1.2 Bycatch and impact on other species ... 157
7.6.1.3 Pollution and marine debris ... 158
7.6.1.4 Data needs ... 158
7.6.2 IATTC ... 159
7.6.2.1 Tuna–dolphin interactions ... 159
7.6.2.2 IATTC bycatch initiatives ... 160
7.6.2.3 IATTC's current ecosystem activities ... 162
7.6.3 ICCAT ... 162
7.6.4 GFCM ... 164
7.6.5 NAFO ... 165
7.6.6 NEAFC ... 166
7.6.7 IOTC ... 167
7.6.8 CCSBT ... 167
7.6.9 Central Bering Sea Donut Hole Arrangement ... 168
7.6.10 SEAFO ... 168
7.6.11 WCPFC ... 169
7.6.12 SIOFA ... 170
7.7 Practical challenges to EAF implementation ... 170
7.8 Conclusion ... 172

Chapter 8 Decision-Making ... 187
8.1 Introduction ... 187
8.2 State sovereignty and RFMO prescriptive authority ... 187
8.3 Efforts to overcome misuse of State sovereignty ... 189
8.4 Effective decision making ... 190
8.4.1 Institutional structure ... 190
8.4.2 Quorum ... 190
8.4.3 The vote-based decision-making process ... 191
8.4.3.1 Unanimous voting ... 191
8.4.4 Consensus-based decision making ... 191
8.4.4.1 What is consensus? ... 191

8.4.4.2 Advantages of consensus-based decision making 192
8.4.4.3 Shortcomings of consensus-based decision making 192
8.4.4.4 Efforts to address the shortcomings of consensus-based decision making 192
8.4.5. Objection and opt-out procedures 193
8.4.5.1 Addressing abuse of objection procedures 193
8.5. Factors guiding RFMO decisions 195
8.5.1 Science-based and rule-based decision making 195
8.5.2. Science, precaution, and the ecosystem approach 195
8.5.3 Institutional structure to support science-based decision making 196
8.5.4 Decision rules 197
8.5.5 Allocation criteria and fishing possibilities 198
8.5.5.1 NAFO 199
8.5.5.2 ICCAT 200
8.5.5.3 NEAFC 200
8.5.6 Transparency 200
8.5.7 Compatibility 201
8.5.7.1 The 1982 LOSC 202
8.5.7.2 UNFSA 203
8.5.7.3 Pre-UNFSA treaties for the conservation and management of straddling fish stocks 203
8.5.7.4 Post-UNFSA treaties for the conservation and management of straddling fish stocks and highly migratory fish stocks 205
8.5.7.5 Interpreting compatibility 206
8.5.7.6 Inter-RFMO cooperation and compatibility 207
8.6 Dispute settlement 208
8.7 Conclusion 209

Chapter 9 The MCS and Enforcement Regime 219
9.1 Introduction 219
9.2 International legal framework for MCS and enforcement 220
9.3 What are MCS and enforcement mechanisms? 222
9.3.1 MCS 222
9.3.1.1 Monitoring 222
9.3.1.2 Control 222
9.3.1.3 Surveillance 222
9.3.2 Enforcement 223
9.4 Implementation of MCS and enforcement measures in RFMOs 223
9.4.1 Fishing vessel registry 223
9.4.1.1 Negative lists 223
9.4.1.2 Positive lists 224
9.4.1.3 Challenges 224
9.4.2 Data collection and reporting 225
9.4.3 Trade information and market-related measures 226
9.4.4 Regulation of transshipment 228

9.4.5 Observer programs ... 229
9.4.5.1 Role of the observer ... 229
9.4.5.2 Types of observer programs ... 229
9.4.5.3 Observer coverage levels ... 230
9.4.5.4 Impact of observer program ... 230
9.4.6 Aerial surveillance ... 231
9.4.7 Satellite-based vessel monitoring systems ... 231
9.4.7.1 Various schemes ... 232
9.4.7.2 Vessel detection systems ... 234
9.4.8 At-sea boarding and inspection ... 234
9.4.8.1 Reciprocal boarding and inspection ... 235
9.4.8.2 Articles 21 and 22 procedures ... 235
9.4.8.3 At-sea boarding and inspection schemes ... 236
9.4.8.3.1 NAFO ... 236
9.4.8.3.2 CCAMLR ... 240
9.4.9 Issues arising from UNFSA Articles 21–22 ... 241
9.4.9.1 Obligation to adopt a boarding and inspection scheme ... 241
9.4.9.2 The question of consistency ... 242
9.4.9.3 Use of force ... 244
9.4.9.4 Alternative mechanisms for the purposes of Article 21(15) ... 245
9.4.9.5 Other issues of application ... 246
9.4.10 Port-State control schemes ... 247
9.4.10.1 ICCAT ... 247
9.4.10.2 IOTC ... 247
9.4.10.3 Other RFMOs ... 247
9.4.10.4 Challenges to implementing port-State controls ... 248
9.4.11 Enforcement measures ... 248
9.5 Conclusion ... 252

Afterword ... 261

Part III: Matrices and Maps ... 267

Matrices and Maps of the RFMOs ... 269

Appendix I – UNFSA ... 307

Appendix 2 – FAO Compliance Agreement ... 347

Appendix 3 – FAO Code Of Conduct For Responsible Fisheries ... 359

Selected Bibliography ... 409

Index ... 435

Table List

Chapter 1
Table 1.1: Parties to UNFSA and FAO Compliance Agreement (as of May 1, 2006) ... 4
Table 1.2: State Participation in RFMOs (as of May 1, 2006) ... 6

Chapter 5
Table 5.1: Status of straddling fish stocks managed by RFMOs (as of March 2006) ... 89
Table 5.2: Status of highly migratory fish stocks managed by RFMOs (as of March 2006) ... 92

Chapter 9
Table 9.1. Comparison among Article 11 of UNFSA and provisions in the NAFO and NEAFC schemes ... 243

Afterword
Table 10.1. Parties to UNFSA and FAO Compliance Agreement (as of July 1, 2008) ... 266

Figure List

Summary
Figure S.1 Global Overview of RFMOs: Straddling Fish Stocks ... xliv
Figure S.2 Global Overview of RFMOs: Highly Migratory Fish Stocks ... xlv

Chapter 2
Figure 2.1 International fisheries law and related instruments pertaining to straddling fish stocks and highly migratory fish stocks ... 12

Chapter 3
Figure 3.1 Defining straddling fish stocks, highly migratory fish stocks, transboundary and discrete high seas stocks ... 43

Chapter 5
Figure 5.1 Global catch trends by FAO region ... 79
Figure 5.2 Surplus production model illustrating MEY, MSY and overexploitation ... 84
Figure 5.3 Cod Catches in the NAFO Convention Area ... 86
Figure 5.4 Shrimp catches in the NAFO Convention Area ... 86
Figure 5.5 Greenland halibut catches in the NAFO Convention Area ... 87
Figure 5.6 Blue whiting landings in the NEAFC Convention Area ... 87
Figure 5.7 Catches of Patagonian toothfish inside and outside the CCAMLR Area (as of 2000) ... 88
Figure 5.8 World cumulative tuna catches, by ocean ... 88
Figure 5.9 World catches of major tuna species ... 91
Figure 5.10 Catches of southern bluefin tuna ... 91
Figure 5.11 Catches of bigeye in the Eastern Pacific Ocean, by fishing gear ... 91
Figure 5.12 Bluefin tuna catches in the Western Atlantic Ocean by fishing gear ... 93
Figure 5.13 Catches of tuna in the Indian Ocean ... 93
Figure 5.14 Catches of Atlanto-Scandian (Norwegian) spring-spawning herring ... 94
Figure 5.15 Revenue-yield model illustrating MEY, MSY and implications of subsidies ... 95

Figure 5.16 Beyond Sustainable Fishing: comparing conventional fisheries management terminiology and those of species at risk ... 97
Figure 5.17 Differences between overfished and overfishing ... 101
Figure 5.18 Harvest control rule schema as developed by NAFO ... 101

Chapter 6

Figure 6.1 Precautionary approach reference points, as defined by ICES ... 118

Chapter 7

Figure 7.1 Simplified trophic food web in the Southern Ocean ... 145
Figure 7.2 Comparing conventional fisheries management and the EAF ... 152
Figure 7.3 Simplified food-web diagram of the pelagic ecosystem in the tropical Eastern Pacific Ocean ... 158
Figure 7.4 Reduction in dolphin mortality in Eastern Pacific Ocean? ... 161

Part III: Matrices and Maps

Matrices

NAFO ... 270
NEAFC ... 272
SEAFO ... 274
CCAMLR ... 276
BSDH ... 278
ICCAT ... 280
GFCM ... 282
IATTC ... 284
WCPFC ... 286
CCSBT ... 288
IOTC ... 290

Maps

NAFO ... 292
NEAFC ... 293
SEAFO ... 294
CCAMLR ... 295
BSDH ... 296
Sea of Okhotsk "Peanut Hole" ... 297
Barents Sea "Loop Hole" ... 298
South Tasman Rise Arrangement ... 299
SIOFA ... 300
ICCAT ... 301
GFCM ... 302
IATTC ... 303
WCPFC ... 304
CCSBT ... 305
IOTC ... 306

Preface

Genesis of the book

Updating 1994 journal article

In August of 2004, I received an e-mail from Greg Peacock, a colleague at the Department of Fisheries and Oceans (DFO) in Nova Scotia, asking to see me about a paper I had written. I wondered which paper. To my surprise Greg informed me that he had attended a meeting in the United States where Canada had agreed, at the suggestion of Ambassador David Balton, to commission an update of an article I wrote entitled "Global Overview of Straddling and Highly Migratory Fish Stocks: The Non-sustainable Nature of High Seas Fishing", published in the September 1994 edition of *Ocean Development and International Law* (ODIL, Vol. 25, pp. 255–344). Ambassador Balton had proposed that an updated study of each regional fisheries management organization (RFMO) would contribute to the preparatory efforts for the 2006 Review Conference of the United Nations Fish Stocks Agreement (UNFSA). In late October 2004, under the championship of David Bevan and Nadia Bouffard of DFO in Ottawa, together with the support of DFO's Maritimes Regional Director General, Neil Bellefontaine, I was assigned the project to research and write an updated version of my 1994 article that would present an impartial factual account of the RFMOs charged with regulating straddling fish stocks or highly migratory fish stocks as well as an overview of the initiatives to establish new RFMOs in marine areas where none existed. So began the project.

As I had had limited direct involvement in international fisheries matters in some years, little did I realize there had been so many new developments since the 1994 publication of my article and the 1995 adoption of UNFSA. RFMOs such as the North East Atlantic Fisheries Commission (NEAFC) and the Commission for the Conservation of Antarctic Marine Living Resources (CCAMLR) had become major players; new RFMOs such as the Commission for the Conservation of Southern Bluefin Tuna (CCSBT), the South East Atlantic Fisheries Organisation (SEAFO), and the Western and Central Pacific Fisheries Commission (WCPFC) had been established; the Inter-American Tropical Tuna Commission (IATTC), in an effort to modernize and strengthen its Commission, had adopted but had not ratified a new convention known as the Antigua Convention; the Galapagos Agreement, covering the Eastern Pacific, had been adopted but not ratified by the coastal States of the Permanent South Pacific Commission (CPPS); and another agreement establishing an organization called the Southern Indian Ocean Fisheries Arrangement (SIOFA) was being negotiated in the Southern Indian Ocean. It quickly became clear to me that including these new developments in the updated publication would entail writing a substantially larger monograph than my ODIL publication. In this information age of sound bites, executive summaries, and two-page briefing notes, I also realized that many people might prefer a condensed tabular presentation of the material that could be used as a handy reference to understand and compare the RFMOs. In December 2004, I prepared the first of many draft templates and proposed to DFO that I develop a set of matrices — one for each RFMO — that would accompany a narrative. As it turned out there was a subsequent request from DFO officials for a condensed yet comprehensive presentation of the information in preparation for the UNFSA Review Conference.

Matrices, maps, figures, compendium of international law, and annotated bibliography

In May 2005, a package of materials entitled "Global Overview of Straddling and Highly Migratory Fish Stocks" by Evelyne Meltzer was distributed to participants attending the May 2005 Conference *Governance of High Seas Fisheries and the UN Fish Agreement — Moving from Words to Action* in St. John's, Newfoundland and Labrador, Canada. These materials were also posted in English and French on the Conference's Web site (http://www.dfo-mpo.gc.ca/fgc-cgp/documents/meltzer_e.htm).

This global overview included a matrix and an updated coloured map for each RFMO and one regional fisheries management arrangement (the Bering Sea Donut Hole), as well as two global overview maps; key figures; tables depicting the international legal framework, participation in UNFSA and the FAO Compliance Agreement, the participation of States, regional economic integration organizations (REIOs), and fishing entities in the relevant RFMOs; and a long list of acronyms for a field of study that had become a thick alphabet soup for the uninitiated. These materials were well received by Conference participants.

The matrices were unconventional in that they were not footnoted and not accompanied by a bibliography. They were prepared based on interviews I conducted by phone and in person with experts throughout the world, as well as on a massive literature review that resulted in an annotated bibliography, the compilation of a compendium of international legal instruments relevant to international fisheries (available on CD), and short case studies prepared on each of the RFMOs. I was particularly fortunate to attend the March 2005 meeting of the FAO Committee on Fisheries (COFI) in Rome, where many individuals kindly agreed to be interviewed. Following the COFI meetings I revised the matrix template once again in response to the key issues identified. Together with a small team of part-time research assistants, I prepared the draft matrices. Each matrix was vetted by the respective RFMO as well as by a large number of scientists, academics, nongovernmental organizations (NGOs), FAO personnel, and government officials throughout the world. As there were often factual discrepancies, I would only include information if I had the original hard copy or if a fact from a secondary source could be corroborated by three or more people. The subjective information — found in the sections entitled Efforts to Strengthen RFMO and Challenges in each matrix — was based on my research.

The maps and matrices prepared for the May 2005 St. John's Conference were updated and published as "Global Overview of Straddling and Highly Migratory Fish Stocks: Maps and Charts Detailing Coverage and Implementation" in the November 2005 special double edition of the *International Journal of Marine and Coastal Law* (Vol. 20, No. 3, pp. 573–608).

The matrices, maps, figures, and tables were further updated to reflect ongoing initiatives and reforms and, together with an introductory note, formed a booklet I entitled "Post-UNFA Overview of Regional Fisheries Management Organizations (RFMOs): Regulation of Straddling Fish Stocks and Highly Migratory Fish Stocks". This booklet was intended to be a factual representation of the state of regional implementation of UNFSA and as such was submitted as a contribution by Canada to David Balton, Chair of the Informal Consultations of State Parties (ICSP-5) to UNFSA held in New York, March 20–25, 2006, in preparation for the May Review Conference, and distributed to all delegates and observers attending the meeting The matrices were then submitted by Canada to be part of the official documents for the UN Review Conference (see A/CONF.210/2006/10 of April 12, 2006).

These matrices were further updated for two posters (by E. Meltzer and S.D. Fuller) that I presented on the margin of the May 2006 Review Conference. A slightly modified May 2006 version of the matrices and maps is found in Part III of this book.

The summary

At the end of the March 2006 Preparatory Meeting in New York (ICSP-5), Canada offered to prepare a written summary overviewing the state of implementation of UNFSA for the May 2006 Review Conference. I was asked by the head of the Canadian delegation to prepare this summary, which would serve as a stand alone document for the Review Conference, as a companion piece for the March 2006 booklet, and as the summary for this monograph. The draft summary was circulated broadly within Canada and internationally (to RFMOs, experts, FAO) for review and comment. The final summary was submitted by Canada on March 31, 2006, as an official Conference document A/CONF.210/2006/10 of April 12, 2006. This document was subsequently reworked for inclusion as the summary in this book.

The monograph

In the wake of the March 2005 COFI meetings in Rome, the May 2005 St. John's Conference, and the Fourth Informal Consultations of States Parties to the Agreement (ICSP-4) held in New York, 31 May – 3 June 2005, where there was interest in independent performance reviews of RFMOs, individually and collectively, DFO decided to revamp the assignment. There was concern that a narrative on each RFMO and on the areas not under a formal governance regime, as originally envisaged in keeping with the format of my 1994 article, could be perceived as a performance assessment of RFMOs. It was also thought that the matrices, maps, and figures had fulfilled that part of the assignment to update the ODIL article. Instead, Nadia Bouffard suggested that a thematic approach would better serve preparatory efforts for the Review Conference and would assist those interested in advancing RFMO reform and modernization. The monograph was to be now a factual thematic overview to illustrate the state of implementation of UNFSA by RFMOs and provide an impartial depiction of regional governance of these international fisheries since the adoption of the Agreement. The key themes identified to be covered by the monograph were the precautionary approach (PA), the ecosystem approach (EA), monitoring, control, and surveillance (MCS), decision making, along with chapters providing the contextual background on the law, the governance regime, and the status of the stocks and associated issues. These were the elements considered critical to RFMO reform adopted in the May 2005 St. John's Declaration. I began working on what was to become this book in July 2005.

I was asked to submit the first complete draft by the end of October 2005 for initial review. Given the need to have a draft prepared for circulation and review as soon as possible, it was suggested that individual chapters be distributed when completed. The working title of the October 25, 2005, submission of the draft manuscript was "Global Overview of Straddling and Highly Migratory Fish Stocks".

Over the summer of 2005 it had been agreed that the research undertaken would be shared with the UN Division for Ocean Affairs and Law of the Sea (DOALOS) to inform the UN Secretary-General's Report. The October 2005 version of the manuscript was shared with Terje Løbach, Norwegian Legal Advisor, who was seconded to DOALOS to prepare the Secretary-General's Report to the Review Conference. Terje did a masterful job. Reference to the contribution of this earlier draft was generously included in the attribution in footnote 3 of the Secretary-General's Report (document A/CONF.210/2006/1, dated January 4, 2006). I was pleased that this first draft along with the matrices could serve as another of many contributions on the part of Canada in the efforts to prepare for the Review Conference.

Over the next several months many people in Canada and throughout the world commented on various versions of draft chapters, and a few brave souls graciously read and commented on successive drafts of the entire manuscript. This informal internal and external peer-review process sparked dialogues, conversations, and e-mail exchanges within Canada and globally, enabling the dissemination of information together with various perspectives and positions. I repeatedly revised, updated, and expanded the chapters based on the feedback I received and on my ongoing research. In an effort to be objective, different and sometimes contentious perspectives are included in this monograph. Drafts were submitted to DFO for review and distribution. The final pre-UNFSA Review Conference manuscript was submitted, as requested, to DFO March 31, 2006, with the editorial assistance of Martin Tsamenyi and Bill Edeson. This version of the manuscript was used internally by the Government of Canada to prepare for the May 2006 Review Conference, and drafts were circulated abroad for discussion purposes. The manuscript was submitted for consideration to NRC Research Press in Ottawa, Ontario, Canada. In the time leading up to the Review Conference further revisions were made to the text to reflect ongoing initiatives and reforms, the bibliography was completed, and the list of key words was expanded.

Publication

It had initially been DFO's plan to self publish this monograph. However, for a variety of reasons, the National Research Council of Canada (NRC) was considered the publisher of choice. NRC Research Press accepted the work for publication (subject to peer review) in early May 2006, prior to the UNFSA Review Conference. The Government of Canada organized a breakfast meeting, hosted by Lori Ridgeway with the assistance of Anne Frenette, on the margin of the Conference to announce that the book would be published by NRC Research Press and to enable me to present an overview of this overview.

Following the Review Conference the manuscript was further refined and was sent to NRC Research Press on June 8, 2006, for distribution to its reviewers. In February 2007, I received the comments from the reviewers. Chapter 5 was expanded and reworked with Susanna D. Fuller and submitted for review in July 2007. In October 2007 the revised manuscript was submitted to NRC Research Press for production. With the wonderful and patient assistance of Stephanie Haddad, Publication Officer at NRC Research Press — "my new best friend", who now knows a lot more about international law of the sea, RFMOs, and the challenges and opportunities confronting international fisheries — the manuscript has become a book.

The scope of the book

This book covers a broad range of activities, issues, and organizations related to straddling fish stocks and highly migratory fish stocks. It is intended to provide an overview of the management efforts by RFMOs to regulate these fisheries in accordance with UNFSA. To this end, the book provides a snapshot of the trends across a whole spectrum of practices in the management of international fisheries. While the emphasis is on the period following the coming into force of the Agreement, the book takes a somewhat historical view and endeavours to provide the context for the international legal regime, regional governance model, and the scientific use of maximum sustainable yield (MSY), and traces the development of the PA and the EA. I was asked by DFO, in providing the overview of the state of implementation of UNFSA by RFMOs, to select one or two RFMOs in each of the thematic chapters to highlight progress, reforms, or initiatives made since 1995. For example, Chapter 7: Ecosystem Approach highlights the efforts of CCAMLR and IATTC, whereas Chapter 9: The MCS and Enforcement Regime reports on some of the efforts introduced by the Northwest Atlantic Fisheries Organization (NAFO) and NEAFC. For this reason, within each thematic chapter, though the efforts of all RFMOs are discussed, certain RFMOs are featured. I have endeavoured to provide an accurate and balanced representation of the RFMOs' efforts to implement the key provisions of UNFSA.

While the book reviews a substantial body of work, it is not exhaustive. Other topics or themes such as trade barriers, subsidies, fisheries economics, overcapacity, and peaceful dispute settlement are worthy of much fuller examination than the time and space afforded in preparing and publishing this monograph.

The post-UNFSA period is one of continual change and improvements in the regulatory regime governing straddling fish stocks and highly migratory fish stocks, and the book captures many of these developments. It should be emphasized, as the title indicates, that this book provides an overview up to but not including the May 2006 Review Conference. To record the outcomes of the 2006 Review Conference and some of the many initiatives in the subsequent two years, I thought it would be helpful to include an Afterword. Terje Løbach and Nathalie Lavoie generously agreed to contribute this section.

The intended audiences and uses

The book was written to serve as a reference for those new to the subject as well as for those with years of experience in one or more RFMOs or with expertise in various disciplines such as law, biology, economics, and resource management. As such it is hoped this book will assist those who are not terribly familiar with international fisheries as well as those individuals who are experts in a particular thematic area or RFMO, but may not have as full an appreciation of other thematic issues or the experiences of other regional organizations.

Accordingly, the book is prepared so that it can accommodate a variety of readers and uses. The book can be read sequentially or read in parts or be used as a reference for looking up particular issues. For this reason each chapter is, for the most part, self-contained — acronyms are spelled out the first time they appear in each chapter and each chapter has endnotes. A summary is found at the beginning of the book. The matrices and maps found in Part III of this book can be used alone, can be examined as a group for comparison purposes, or can serve as a companion to the summary and individual chapters.

Quirks

Spelling can be quirky. The spelling in this book is for the most part what my publisher would consider Canadian, which has some American English forms and some British styles. So favour not favor, behaviour not behavior, honour not honor, program not programme, organization not organisation, realization not realisation. However when there is an official spelling of a word, I used it, for example, NAFO uses Organization, whereas SEAFO uses Organisation; ICCAT use "utilisation" and IATTC uses "utilization" in their respective constitutive agreements. Similarly with by-catch versus bycatch, and pre-agreed versus preagreed.

When I submitted the manuscript for copy-editing little did I realize that capitalization of certain words would become such a headache. Does one use uppercase for Resolution, Recommendation, Southern, Northern, Precautionary Approach, Ecosystem Approach, etc. It was agreed that if the Recommendation was binding as in the International Commission for the Conservation of Atlantic Tunas (ICCAT) and the General Fisheries Commission for the Mediterranean (GFCM), it would be capitalized. Similarly Resolutions, which are binding in the case of IATTC, are capitalized (recommendations are nonbinding and are therefore found in lowercase). Conservation Measures, which is the term used in CCAMLR for binding regulations, is found in uppercase. Everywhere else in the text conservation measures are generic, and therefore the term appears in lowercase. For others I conferred with and often deferred to my copy editor, Stephanie Haddad.

Acronyms can be tricky. There are many acronyms in this book, as there are many acronyms in the study of straddling fish stocks and highly migratory fish stocks. When I was a grad student, UNCLOS was used to refer to the United Nations Conference on the Law of the Sea, as in UNCLOS I, II, and III. The Law of the Sea Convention is abbreviated to LOSC. The United Nations Fish Stocks Agreement also has different short forms. This book uses UNFSA, as it is the acronym adopted by the United Nations. Some authors use the acronym RFMO/MA to refer to RFMOs and RFMAs. In this book I use the term RFMOs unless expressly noted otherwise to include RFMAs when referring to regional management frameworks responsible for straddling fish stocks and (or) highly migratory fish stocks. It was decided with NRC Press not to abbreviate the terms straddling fish stocks and highly migratory fish stocks.

In conducting the research for this project, the internet was extremely useful, though not always reliable, as I discovered mistakes made on one site or by one author sometimes became entrenched as fact. Most RFMOs have established Web sites, and over the course of preparing this book, there have been major improvements in the amount of information available on these sites as well as in their design. Similarly, FAO has brilliantly succeeded in coordinating and providing web-based information through FIRMS and FIGIS. The transparency of and timely access to information, particularly through electronic distribution and the World Wide Web, have continually improved in the last three years, altering the paradigm for awareness raising of issues and member activities, participation, inclusion, scrutiny, and accountability.

Concluding thoughts

Calls for action: few words, more words, many words

In the early 1990s when I began my research on high seas overfishing, particularly for straddling fish stocks for presentation at the 1991 and 1992 Preparatory Conferences for the UN Conference on Environment and Development (UNCED), there were comparatively few published articles or books on high seas fishing, and it was often difficult to obtain information on straddling fish stocks and highly migratory fish stocks. Though overfishing had been a preoccupying problem for several decades, the public was, in general, not aware of the difficulties on the high seas, and many academics, scientists, and bureaucrats thought the problems were overstated. Much of my research was done by telephone or in-person interviews and exchanges by fax. Information was sometimes not available in English, and many key documents and faxes had to be translated.

When I wrote and disseminated the *New York Declaration on High Seas Fishing: Protecting the Commons* in February 1992 as an NGO delegate (representing the United Nations Association of Canada) to the UNCED Preparatory Meetings in New York, many had not heard of high seas overfishing. Those who were aware of the issue thought the crisis was confined to the Northwest Atlantic for cod stocks. In fact, one of the NGO delegates, upon hearing about the straddling stocks problem, asked me what the stock market had to do with sustainable development! Of course that was another topic, and we instead discussed unregulated and excessive fishing, competing interests over a common resource, depleted fish stocks, and devastated coastal communities. The New York Declaration was subsequently signed by some 300 NGOs and 20 governments. It was then unanimously endorsed by the NGO plenary and submitted to Maurice Strong and the UNCED Secretariat in New York on March 31, 1992. The Declaration was translated into French, Portuguese, and Spanish and circulated to all governments in May 1992 along with the first draft of my monograph "Global Overview of Straddling Fish Stocks".

Following the Earth Summit, I was commissioned by the Government of Canada to expand my research to include international management of the tuna and tuna-like species, and I produced another paper entitled "Global Overview of Straddling and Highly Migratory Fish Stocks: Nonsustainable Nature of High Seas Fisheries". I was asked to present my findings as the keynote speaker at the *Meeting of Like-Minded Countries on High Seas Fisheries*, held in St. John's, Newfoundland and Labrador, Canada, in January 1993. It is this paper that was further developed and published in ODIL. Though it was a lengthy article, taking up almost the entire September 1994 issue of the journal, Jon Jacobson deemed it worthy of printing *in toto*.

The banner title from "Words to Action" for the May 2005 St John's Conference is apt because there have been many words written in the years following the 1995 adoption of UNFSA on the subject of high seas fishing. Indeed, as I embarked upon my research to update the ODIL article and then to research the themes for this book, I was surprised to find not only a large body of soft and hard law pertaining to international fisheries, but also a tremendous amount of literature published in the form of journal articles, books, technical reports, studies, and grey literature as well as a myriad of Web sites on the subject of straddling fish stocks and highly migratory fish stocks in general, and on UNFSA, specific RFMOs, or specific issues such as the PA, the EA, and illegal, unregulated, and unreported (IUU) fishing. This impressive body of written material is a reflection of the growing interest by academics, scientists, intergovernmental organizations (IGOs), NGOs, and civil society in the health and sustainable use of the oceans and responsible deployment of fishing vessels since 1995. It is also a response to the continued conflicts, excessive use and misuse, illegal fishing, declining fish stocks, and enormous technological issues that allow overcapacity and threaten environmental degradation.

Time for action

The Chinese word for "crisis" is composed of two characters: one signifying "danger" and the other "opportunity". It is this perspective of crisis that I believe summarizes best the current situation with respect to international fisheries, particularly those related to straddling fish stocks and highly migratory fish stocks. Excessive and indiscriminate exploitation of fishery resources is no longer a sectoral issue but is considered a global environmental problem — a problem that has reached critical proportions, exacerbated by marine pollution, habitat degradation, and significant perturbations due to climate change. The danger is not exaggerated. Fish were once thought to be inexhaustible and perennially resilient animals. With the industrialization of fishing operations capable of fishing in ever deeper water, continuous technological advances, and continued failure to institute, manage, and enforce sustainable fishing practices, large-scale commercial fisheries have often resulted in serious and sometimes irreparable harm to target and incidentally caught species. Though once unimaginable, some fish stocks are now listed as species at risk.

Following the 1992 Earth Summit's call for a Conference on High Seas Fishing and subsequent repeated calls for action, the international community responded by negotiating, adopting, and bringing into force UNFSA. The Agreement provides the blueprint for a transparent, precautionary, scientific, and ecosystem-based governance framework to responsibly regulate fishing for straddling fish stocks and highly migratory fish stocks together with dependent, associated, and other nontarget species. This legally binding, global framework reinforces multilateral cooperation and regional good governance, strengthening compliance and enforcement of RFMO management strategies. It constitutes a progressive structure of rights and duties and international minimum standards, and an institutional construct with the potential to secure the sustainable development of these international fisheries. This composite regime created by UNFSA and its supporting agreements, however, is set within the authority and constraints under which the various RFMOs, flag States, and coastal States operate.

While there has been progress and ongoing reform to varying degrees by the RFMOs examined in this book, there is still a discernible gap between the formal legal–institutional developments and the necessary action required to restore and achieve the long-term sustainability of straddling fish stocks and highly migratory fish stocks. If the international community is to successfully arrest and reverse this fisheries crisis there is an urgent and immediate need for this overarching framework to be rigorously implemented and enforced at the regional and national levels throughout the world. Full cooperation, accelerated reform, meaningful control, effective enforcement, and compliance both within and beyond areas under national jurisdiction are critical to meet conservation targets such as those set by the World Summit on Sustainable Development. Equally imperative is the need to further engage and renew the commitment of the "fishing industry" to fish responsibly, in keeping with the FAO Code of Conduct and other international hard and soft law instruments. The industry must take a leadership role in abating unauthorized fishing and in developing an incentive structure that will promote widespread compliance. Indeed, it is in the fishing industry's self-interest to ensure the long-term health and sustainability of the species upon which it depends. This is a central axiom of sustainable development.

Time is of the essence. The window of opportunity for achieving long-term conservation, restoration, and sustainable development of these fisheries is quickly closing. Sustainable international fisheries are not an elusive quest: but a practical, critical, and attainable goal.

> "Human progress is neither automatic nor inevitable. We are faced now with the fact that tomorrow is today. We are confronted with the fierce urgency of now. In this unfolding conundrum of life and history there is such a thing as being too late…We may cry out desperately for time to pause in her passage, but time is deaf to every plea and rushes on. Over the bleached bones and jumbled residues of numerous civilizations are written the pathetic words: Too late". *(Martin Luther King Jr., "Where Do We Go From Here: Chaos or Community" 1967, speaking in another context*

eloquently captures the looming danger and need to act, and act quickly).

Acknowledgements

DFO provided the impetus and support required to undertake the various components of this project, which all ultimately contributed to the research and writing of this book, and I express my sincere appreciation to a number of DFO employees for their individual support and contribution to this project whether administrative, technical, or substantive: Chris Allen, Jim Baird, Neil Bellefontaine, Guy Beaupre, Sophie Belisle, David Bevan, Nadia Bouffard, Bill Brodie, Willi Bruce, Lori Collins, Arthur Cosgrove, Carla Dale, Bill Doubleday, Gary Gammon, Anna Fiander, Anne Frennette, Rod Forbes, Terry Higgins, Blair Hodgson, Francis Kelly, Jim Jones, Sylvie Lapointe, Nathalie Lavoie, Jeff MacDonald, Bob O'Boyle, Candace Partridge, Greg Peacock, Julie Porter, Kym Purchase, Nadine Regimbal, Lori Ridgeway, Jake Rice, Denis Rivard, Gorazd Ruseski, Michael Sinclair, Paul Steele, Bob Steinbock, Robert Stephenson, Tim Surette, Norwood Whynot, Brian Wood, and Bobbi Zahra.

I am particularly indebted to Nadia Bouffard, who, at the suggestion of Ambassador David Balton, conceived and championed this project as part of a strategic vision for the Department as well as to advance the cause of sustainable international fisheries. Her commitment to improving and modernizing regional fisheries governance and overcoming systemic obstacles was a motivating force in persevering with what sometimes seemed like a Sisyphean endeavour. In November 2005, after I had submitted the complete first draft of the manuscript, she passed the torch for project oversight to Kym Purchase and Nathalie Lavoie, whose ongoing support, executive management, and commitment to the project were evident and greatly appreciated. Nathalie coordinated the review and distribution of the March 31, 2006 manuscript. She also generously stepped up to the plate in March 2008 to further update the August 2007 Afterword prepared by Terje Løbach. Merci beaucoup! I also wish to express my gratitude to Anne Frennette and Lori A. Ridgeway, who together with Kym and Nathalie coordinated the nature and scope of my involvement in the run up to the March 2006 Preparatory Meeting and the Review Conference, including the preparation and distribution at the United Nations in New York of the March 2006 booklet and the March 31st summary; the distribution of the chapters, matrices, maps, posters, etc. in Canada and abroad prior to and after the Review Conference; and the organization of the breakfast presentation of my work on the margin of the Review Conference in May 2006.

I am also indebted to Amos Donohue of the Canadian Department of Justice, and to Michael Shewchuk, Louis Simard, and Wendell Sanford of the Department of Foreign Affairs and International Trade for their ongoing support, generosity of spirit, and careful review of earlier draft chapters of the book.

I thank all those who kindly agreed to be interviewed, suggested contacts or resource materials, or provided comments and suggestions, including Greg Aikins, Don Aldous, Robin Allen, Alejandro Anganuzzi, Transform Aqorau, Alfonso Jalil Avarez, David Balton, Alicia Barcena, Jay Batongbacal, Kamil Bekyahev, Ron Berg, Luis Bombin, Alain Bonzon, Charlotte Breide, William Burke, Fuensanta Candela-Castillo, Robert Lihtorng Chen, Aldo Chircop, Biliana Cicin-Sain, Les Clark, Lewis Clifton, Jean Collins, John Croxall, Hasjim Djalal, David Doulman, Alexandra Edgar, Gudmundur Eiriksson, Steffan Ekwall, Sam Elsworth, Jose Jairo Escobar, Janet Farmer, Johanne Fischer, Serge Garcia, Michael Gardner, Matt Gianni, Kristina Gjerde, Alistair Graham, Jon Grant, Jan Pieter Groenhof, Ndiaga Gueye,Vesna Guzina, Arnor Halldorsson, Brian Hallman, Hashali Hamukuaya, John Herity, Cheryle Hislop, Kjartan Hoydal, Douglas Johnston, Terry Johnston, Bob Kennedy, Lee Kimball, Leonid Kokhovkin, Blaise Kuemlangan, Terje Løbach, Michael Lodge, Lois Loewen, Sebastian Losada, Loh-Lee Low, Indrani Lutchman, Brian MacDonald, Ronald St. John MacDonald, Alastair Macfarlane, Jennifer McDonald, Robin Mahon, Jacek Majkowski, Holger Martinsen, Moira McConnell, Ted McDorman, Driss Meski, Ed Miles, Denzil Miller, Erik Molenaar, Ransom Myers, Satya Nandan, Paul Niemeier, Carlos Palma, Kate Partridge, Daniel Pauly, Gonzalo Pereira Puchy, Jean-Francois Pulvenis, Rosemary Rayfuse, Victor Restrepo, Eric Reynolds, Len Rodwell, Phillip Saunders, Andrew Serdy, Ross Shotton, Cleo Small, Lisa Speer, Judith Swan, Dean Swanson, Feleti Teo, Adam Tugio, David Vanderwaag, Ann Kristin Westberg, Jane Willing, Earl Wiseman, Andrew Wright, Fernando Zegers Santa Cruz, and Jorge Zuzunaga.

I am especially thankful to the many staff and research assistants with whom I have worked in numerous settings since undertaking the project. Here, I acknowledge several individuals who participated in the research effort in various capacities for varying periods of time, whether for a few days, a few weeks, or a few months leading up to the May 2005 version of the matrices, maps, figures, individual RFMO case studies, the preliminary literature review, annotated bibliography, and compendium of legal instruments: Lowell Bautista, Tom Crist, Bradford Hatch, Cornelia (Nell) Den Heyer, Godwin Djokoto,

Susanna D. Fuller, Patricia Gonzalez, Leslie Kieley, Cheluchi Onymelukwe, Marie Palmer, Megan Sikaneta, Karen Traversy, and Karan Whiteman.

Following the July 2005 assignment to prepare this book, I engaged several individuals to assist with particular chapters. I am most grateful to research assistants Don Aldous, Jay L. Batongbacal, Godwin Djokoto, David Dzidzornu, Susanna D. Fuller, Leonie Renwrantz, and Hugh Williamson.

I am particularly grateful for the research assistance provided by Susanna D. Fuller. Because of her many commitments, her involvement was piecemeal and was repeatedly renewed on a part-time basis from her initial contract in February 2005, which, in the end, resulted in Susanna being the principal research assistant, through many of the iterations and redactions of this book, assisting me with consolidating the manuscript for submission in October 2005, again in March 2006, and yet again in May 2006; and in 2007 she assisted me with responding to the peer reviewers' comments and with responding to the copy editor's comments for one chapter and the matrices. She is the second author of this version of the matrices, second author of Chapter 5, and was the research assistant for Chapters 6 and 7. While several people have provided research assistance in the preparation of this book, all of whom are deserving and have been acknowledged in the first footnote of each chapter in keeping with protocol for law and social sciences, I wanted to highlight my appreciation for Susanna's overall contribution to this project, and therefore there is the special attribution on the front cover. Susanna is completing her doctorate in marine biology at Dalhousie University. The world will undoubtedly benefit from Susanna's tireless efforts and commitment to improving marine environmental conservation and protection.

I would also like to thank Susan Rolston for her assistance with consolidating the manuscript for submission in October 2005, for coordinating the response to the informal review process in early 2006, for providing some initial copy-editing services, and for assisting with the consolidation of the February 2006 draft. Susan is an expert in developing key words and indexes. I am indebted to her assistance in preparing the draft list of key words for the index, as it is a major task and one that is critical to assisting readers to effectively use a book such as this. I am also grateful to Susan for assisting me with responding to the copy editor's comments for two chapters.

Special thanks to Francis Kelly, Art Cosgrove, Rick Paine, Derek Sarty, and Norwood Whynot for graphic and design support for earlier drafts and components of the project completed prior to the May Review Conference, and to Patricia Gonzalez for carefully compiling the acronym list and proofreading earlier drafts of the monograph and the matrices. Patricia also prepared the master program for the annotated bibliography using ProCite.

I acknowledge and express my most sincere appreciation to David Pitcher of Spatial Metrics, Halifax, Nova Scotia, for his patience with the numerous revisions and iterations of the maps and for his excellent cartographic support services. I am cognizant that he did not realize that these services would require working with me for hours at a time, often burning the midnight oil, to produce the beautiful, functional maps for yet another deadline. I am proud of the maps and very pleased that David didn't pull out all his hair as I tried to determine how best to depict the RFMOs and reconcile the competing views about WCPFC, in particular. In the end, David also became a champion of the project and an advocate for responsible management of international fisheries. Thank you!

I would like to thank Rosemary Rayfuse, Professor of Law, University of New South Wales, Australia, for allowing herself to be sidetracked during an academic visit to Halifax in mid-September 2005 to work on Chapter 9 (MCS and Enforcement). Dr. Rayfuse is an expert in this field, widely published, and highly regarded. Her involvement was much appreciated, and I would not have met the October 25, 2005, deadline for submission of the first complete draft without her contribution.

In July 2007, Terje Løbach kindly agreed to contribute an Afterword updating the monograph to incorporate key initiatives and changes, covering the period from the May Review Conference to the end of June 2007. Terje is the Senior Legal Advisor, Directorate of Fisheries, Norway, and has participated for many years on the delegation to a number of RFMOs, was the lead in developing the FAO Model Port Scheme, was seconded to UNDOALOS to write the Secretary-General's Report for the UNFSA Review Conference, participated on the High Seas Task Force committee, and travels the world as an ambassador for the Agreement, working with developing coastal States on the domestic legislative and practical measures needed to become Party to UNFSA. To Terje for your input throughout this project and for coauthoring the Afterword — Tak!

In March 2008 it was clear that the book would not be published prior to July 2008, and therefore the reviewers had advised NRC Press that the book would benefit from an updated Afterword. Nathalie Lavoie was asked to update the Afterword. Natalie, a policy analyst and lawyer with DFO, has been involved with international fisheries and the project since 2005, had read and commented on earlier drafts of the manuscript, and was well placed to elaborate upon the version prepared by Terje. Nadia Bouffard and Anne Frenette reviewed the Afterword and provided Nathalie with their comments. Avec gratitude je veux reconnaître votre contribution considérable et votre support continuel. Merci!

There have been many other professional colleagues who have offered their thoughts and provided guidance along the way. As each chapter was written I sent it out for review. The informal review process was iterative and protracted. Some individuals reviewed and provided comments on one or more chapters; some reviewed the entire monograph. I am so very grateful to everyone who kindly and carefully read one or more redactions and offered their views. I found I learnt a great deal through the review process, considered competing perspectives and opinions, and was pleased that the review precipitated discussion and debate among individuals within Canada and internationally on many issues and may have served to clarify some issues prior to the Review Conference.

There is a long list of reviewers who generously provided comments on certain issues, particular chapters, or the whole of earlier drafts: Chris Allen, Robin Allen, Alejandro Anganuzzi, Transform Aqorau, David Balton, Alain Bonzon, Nadia Bouffard, William Burke, Fuensanta Candela-Castillo, John Croxall, Amos Donohue, Gudmundur Eiriksson, Johanne Fischer, Serge Garcia, Michael Gardner, Kristina Gjerde, Brian Hallman, Hashali Hamukuaya, Kjartan Hoydal, Jim Jones, Lee Kimball, Sylvie Lapointe, Nathalie Lavoie, Terje Løbach, Michael Lodge, Brian MacDonald, Ted McDorman, Don McRae, Driss Meski, Denzil Miller, Erik Molenaar, Bob O'Boyle, Scott Parsons, Greg Peacock, Julie Porter, Kym Purchase, Rosemary Rayfuse, Victor Restrepo, Jake Rice, Lori Ridgeway, Denis Rivard, Gorazd Ruseski, Wendell Sanford, Michael Shewchuk, Louis Simard, Michael Sinclair, Cleo Small, Bob Steinbock, Tim Surette, Dean Swanson, Brian Wood, and Andrew Wright.

Here, I also record special thanks to Professor William Edeson, Professorial Fellow at the Australian National Centre for Ocean Resources and Security (CORS), University of Wollongong, Australia, and his colleague Professor Martin Tsamenyi, Director of CORS, who provided technical inputs and editorial assistance to bring an earlier version of this study to completion for the March 31, 2006, deadline. Despite the intercontinental context, their busy schedules, and international travel itineraries, we managed to work very effectively. Their "fresh eyes", expertise, and experience were much appreciated.

I also acknowledge the Marine and Environmental Law Institute, Dalhousie Law School, for hosting me as a Visiting Scholar during part of the period over which this work was completed (November 2004 to September 2005). As well, I express my sincere gratitude to Gardner-Pinfold Consulting for accommodating the project from September 2005 to June 2006, and to Michael Gardner for providing tremendous moral support.

I extend my heartfelt thanks to P.B. Cavers and the Editorial Board at NRC Research Press for accepting this monograph for publication. I am especially appreciative of all the efforts undertaken by Dr. Art May, who as the Editorial Board designate reviewed, championed, and provided oversight to ensure this monograph became a published book. I would like to thank Rob Forrest at NRC Research Press, who coordinated the initial evaluation and acceptance process and oversaw the final stages of publication when he returned from his parental leave. To the anonymous NRC Research Press Reviewers, thank you. I also thank David Balton, Michael Gardner, Jim Jones, Ted Mcdorman, and Bob O'Boyle for further reviewing individual chapters as part of this process. Many thanks to Tatiana Branicki, graphics technician at NRC Research Press, to ALDI Design for page design and layout, and to Accurate Design and Communication Inc. for the cover design. I am particularly grateful to Suzanne Kettley, Head of the Monograph Program, and Stephanie Haddad, Publication Officer, for their patience, kindness, enthusiasm, and expertise in bringing this monograph to you. I learnt a great deal throughout this process and have truly developed an appreciation for what is involved in publishing a book. Merci milles fois pour toute votre gentillesse et aide.

A special note of appreciation to David Dzidzornu, Susanna D. Fuller, and Shona Godwin for proofreading the galleys.

Special tribute

I extend special thanks to Professor Bill Burke. It was my lucky day when I was accepted into the LL.M. program in Marine Affairs at the University of Washington in Seattle to study under the tutelage of the great Professor Burke, who independently and together with the wonderful Dr. Ed Miles has written many articles and books on the problems associated with high seas fishing. The

University of Washington graduate program and these two professors have been an academic touchstone for me since 1981. Bill is a great source of inspiration and motivation; I am profoundly grateful for his friendship and continued interest in my work and ideas.

I would also like to pay special tribute to Dr. Douglas Johnston, Dr. Ronald St. John MacDonald, and Dr. Ramsom Myers. These three men were all colossus in their fields and all passionate about creating a world where fisheries were managed responsibly and sustainably. I had the great privilege of knowing all three, and over the course of writing this book, I had several opportunities to discuss different ideas and concepts with each. All three have passed away since May 2006, leaving an impressive legacy of lives dedicated to scholarship and advocacy — all three had many interests, and all three endeavoured to improve international governance for the responsible use of the marine environment to achieve sustainable international fisheries.

To conclude a most heartfelt expression of immense gratitude to my husband for his understanding, good humour, and encouragement. To my wonderful children, this book is for you. I was pregnant in 1992 with you Michaela when I drafted the New York Declaration confident that the issue could be addressed effectively. You were on 16 airplanes before you were six months old as we travelled in the post-UNCED months for speaking engagements to discuss high seas fishing and sustainable development. Theo understood the issues well. By six years old you wondered why the world could not set aside part of the ocean for the fishes and the manatees. Thank you all for helping with the photocopying, printing, collating materials, picking up and returning books at the libraries, packing boxes for the repeated moves, surprising me with lunch and supper, and supporting me through what seemed like endless hours of reading, writing, and editing. To Al and Marieke, though you were busy pursuing your own careers in different parts of the world, your ongoing interest and support were always much appreciated. And last but certainly not least, a special thanks to my mom for her steadfast support.

While I continue to fear the spectre of an empty ocean, I strongly believe sustainable, robust international fisheries are an attainable goal. As Mahatma Gandhi said, "The difference between what we do and what we are capable of doing would solve most of the world's problems". The international community and the fishing industry are capable of collaborating, acting responsibly, and doing what is necessary to ensure the rehabilitation and conservation of straddling fish stocks and highly migratory fish stocks. UNFSA provides the blueprint. It is for this reason that I offer the following monograph for your consideration.

Disclaimer

While I am indebted to many people in the preparation of this book, I wish to emphasize that the views expressed and the information presented are solely those of the author (except where noted in Chapters 5 and 9 and the Afterword) and should in no way be construed to reflect the views or policies of the Government of Canada, NRC Research Press, or the Department of Fisheries and Oceans Canada. While I acknowledge the contribution of the many individuals noted above who kindly agreed to be interviewed; who offered comments and suggestions on the chapters; who reviewed the matrices, maps, and figures; or who provided research assistance, I take full responsibility for the information presented in this work and regret any errors or omissions.

Evelyne Meltzer
June 8, 2008, World Ocean Day

Selected Acronym List

Acronym	Name (RFMO)[1]
AC	Advisory Committee (IATTC)
ACAP	Agreement on the Conservation of Albatrosses and Petrels
ACCOBAMS	Agreement on the Conservation of Cetaceans of the Black Sea, Mediterranean Sea and Contiguous Atlantic Area (GFCM)
ACFM	Advisory Committee on Fisheries Management (ICES)
ADRIAMED	Scientific Cooperation to Support Responsible Fisheries in the Adriatic Sea (GFCM)
AIDCP	Agreement on the International Dolphin Conservation Program (IATTC)
AMSY	Average maximum sustainable yield
APEC	Asia Pacific Economic Cooperation
APFIC	Asia Pacific Fishery Commission
ASOC	Antarctic and Southern Ocean Coalition (CCAMLR)
ATS	Antarctic Treaty System (CCAMLR)
B&I	Boarding and inspection
BI-SWG	Biology Specialist Working Group (WCPFC)
B_{lim}	Biomass limit
B_{MSY}	Biomass of maximum sustainable yield
BRP	Biological reference point
BSDH	Central Bering Sea Donut Hole
BYP	Bluefin Year Program (ICCAT)
CAQ	Committee on Aquaculture
CBD	Convention on Biological Diversity
CCAMLR	Commission for the Conservation of Antarctic Marine Living Resources
CCAS	Convention for the Conservation of Antarctic Seals
CCM	Member, cooperating nonmember, and fishing entities (WCPFC)
CCSBT	Commission for the Conservation of Southern Bluefin Tuna
CDS	Catch documentation scheme (CCAMLR)
CECAF	Fishery Committee for the Eastern and Central Atlantic
CEMP	CCAMLR Ecosystem Monitoring Program (CCAMLR)
CEP	The Committee for Environmental Protection under the Environmental Protocol to the Antarctic Treaty (CCAMLR)
CIHEAM	Mediterranean Agriculture Institute (GFCM)
CIPS	Confédération internationale de la pêche sportive
CITES	Convention on International Trade in Endangered Species of Wild Fauna and Flora
CLCS	Commission on the Limits of the Continental Shelf
CMP-2	CCSBT Management Procedure – number two (CCSBT)
CNM	Cooperating nonmember
COFI	FAO Committee on Fisheries
COLTO	Coalition of Legal Toothfish Operators (CCAMLR)
COPEMED	Advice, Technical Support and Establishment of Cooperation Networks to Facilitate Coordination to Support Fisheries Management in the Western and Central Mediterranean (GFCM)

[1] To facilitate the use of these acronyms particularly with the matrices found in Part III of this book, the RFMO appears in parentheses following the name of certain acronyms.

Acronym	Name (RFMO)
CP	Contracting Party
CPC	Contracting Parties and Cooperating Non-Contracting Parties, Entities or Fishing Entities
CPPS	Permanent Commission for the South Pacific
CPUE	Catch per unit effort
CWP	Coordinating Working Party on Fishery Statistics
EA	Ecosystem approach
EAF	Ecosystem approach to fisheries
EAM	Environment and Aquaculture (GFCM)
EBFM	Ecosystem-based fisheries management
EB-SWG	Ecosystem and Bycatch Specialist Working Group (WCPFC)
EC	European Community
E-CDS	Electronic catch documentation scheme
EEZ	Exclusive economic zone
EFP	Experimental fishing program (Japan)
ENFO	NPAFC Committee on Enforcement
EPO	Eastern Pacific Ocean
ERSWG	Ecologically Related Species Working Group (CCSBT)
EU	European Union
F	Fishing mortality rate
FAD	Fish aggregating device
FAO	United Nations Food and Agriculture Organization
$F_{current}$	Current fishing mortality rate
FFA	Pacific Islands Forum Fisheries Agency (WCPFC)
FIGIS	FAO Fisheries Global Information System
FIRMS	FAO Fisheries Resources Monitoring System
FMC	Fisheries Monitoring Centre

Acronym	Name (RFMO)
F_{MSY}	Fishing mortality rate of maximum sustainable yield
FOC	Flags of convenience
FT-SWG	Fishing Technology Specialist Working Group (WCPFC)
GC	General Council (NAFO)
GFCM	General Fisheries Commission for the Mediterranean
GNAI	Greatest net annual increment
GYM	Generalized yield model
IAEA	International Atomic Energy Agency (IOTC)
IATTC	Inter-American Tropical Tuna Commission
ICCAT	International Commission for the Conservation of Atlantic Tunas
ICES	International Council for the Exploration of the Sea
ICJ	International Court of Justice
ICNAF	International Commission for the Northwest Atlantic Fisheries
ICRW	International Convention for the Regulation of Whaling
IDCP	International Dolphin Conservation Program
IGO	Intergovernmental organization
ILO	International Labour Organization
IMO	International Maritime Organization
IOC	Intergovernmental Oceanographic Commission
IOTC	Indian Ocean Tuna Commission
IOTTP	Indian Ocean Tuna Tagging Program (IOTC)
IPOA	International Plan of Action
IPOA-Capacity	International Plan of Action for the Management of Fishing Capacity
IPOA-IUU	International Plan of Action - Illegal, Unreported, and Unregulated Fishing

Acronym	Name (RFMO)
IPOA-Seabirds	International Plan of Action for Reducing Incidental Catch of Seabirds in Longline Fisheries
IPOA-Sharks	International Plan of Action for the Conservation and Management of Sharks
IRP	International Review Panel (IATTC)
ISC	Interim Scientific Committee for Tuna and Tuna-Like Species in the North Pacific Ocean (IATTC)
ITLOS	International Tribunal for the Law of the Sea
IUCN	World Conservation Union (International Union for the Conservation of Nature)
IUU	Illegal, unreported, and unregulated
IWC	International Whaling Commission
JAG	Joint Assessment Group (CCAMLR)
LOSC	United Nations Convention on the Law of the Sea
LSTLV	Large-scale tuna longline fishing vessel
MCS	Monitoring, control, and surveillance
MEDFISIS	Mediterranean Fishery Statistics Information System (GFCM)
Medisamak	Mediterranean Fishing Industry Association (GFCM)
MEDITS	Mediterranean International Trawl Survey Program (GFCM)
MEDLEM	Mediterranean Large Elasmobranchs Monitoring (GFCM)
MedSudMed	Assessment and Monitoring of the Fishery Resources and the Ecosystems in the Straits of Sicily (GFCM)
ME-SWG	Methods Specialist Working Group (WCPFC)
MMPA	Marine Mammal Protection Act
MOU	Memorandum of Understanding
MPA	Marine protected area
MSY	Maximum sustainable yield

Acronym	Name (RFMO)
NA	North Atlantic
NAF	North Atlantic Format (NAFO, NEAFC, CCAMLR)
NAFO	Northwest Atlantic Fisheries Organization
NARFMO	North Atlantic Regional Fisheries Management Organizations
NC	Northern Committee (WCPFC)
NCA	NAFO Convention Area
NCC	Non-Contracting Cooperating Party (Cooperating Non-Contracting Party, Entity or Fishing Entity)
NCP	Non-Contracting Party
NE	North East
NEAFC	North East Atlantic Fisheries Commission
NGO	Nongovernmental organization
NPAFC	North Pacific Anadromous Fish Commission
NPOA	National Plan of Action
NPOA-Seabirds	National Plan of Action for the Conservation of Seabirds
NRA	NAFO Regulatory Area
OECD	Organisation for Economic Co-operation and Development
OIS	CCAMLR System of Observation and Inspection
PA	Precautionary approach
PECCOE	Permanent Committee on Control and Enforcement (NEAFC)
PICES	North Pacific Marine Science Organization
Prep. Con.	Preparatory Conference
PWG	Permanent Working Group for the Improvement of ICCAT Statistics and Conservation Measures (ICCAT)
PWG	Provisional Working Group (SEAFO)

Acronym	Name (RFMO)
PWSAM	Permanent Working Group on Stock Assessment Methodology (GFCM)
RECOFI	Regional Commission for Fisheries
REIO	Regional economic integration organization
RFB	Regional fishery body
RFMA	Regional fisheries management arrangement
RFMO	Regional fisheries management organization
SA	South Atlantic
SAC	Scientific Advisory Committee (GFCM)
SAG	Stock Assessment Group (CCSBT)
SAR	Synthetic aperture radar
SBT	Southern bluefin tuna
SC	Sub-Committee (GFCM)
SC	Scientific Council (NAFO)
SC	Scientific Committee (CCAMLR, SEAFO, CCSBT, IOTC, WCPFC)
SCAF	Standing Committee on Administration and Finance (CCAMLR)
SCAR	Scientific Committee on Antarctic Research (CCAMLR)
SCESS	Sub-Committee on Economics and Social Sciences (GFCM)
SCIC	Standing Committee on Implementation and Compliance (CCAMLR)
SCM	Agreement on Subsidies and Countervailing Measures
SCMEE	Subcommittee on Marine Environment and Ecosystem (GFCM)
SCRS	Standing Committee on Research and Statistics (ICCAT)
SCSA	Sub-Committee on Stock Assessment (GFCM, ICCAT)
SCSI	Sub-Committee on Statistics and Information (GFCM)

Acronym	Name (RFMO)
SDP	Statistical documentation program (IATTC)
SDS	Statistical documentation scheme (IOTC)
SEAFO	South East Atlantic Fisheries Organisation
SELAM	Social Economic and Legal Aspect of Aquaculture (GFCM)
SG-ASAM	Sub-Group on Acoustic Survey and Analysis Methods (CCAMLR)
SIOFA	Southern Indian Ocean Fisheries Arrangement
SIPAM	Information System for the Promotion of Aquaculture (GFCM)
SPC	Secretariat of the Pacific Community (formerly the South Pacific Commission)
SPRFMO	South Pacific regional fisheries management organization
SPSA	Southeast Pacific Swordfish Arrangement
SRP	Scientific Research Program (CCSBT)
STACFAC	Standing Committee on Non-Contracting Party Fishing Activities (NAFO)
STACFAD	Standing Committee on Finance and Administration (NAFO, ICCAT)
STACFEN	Standing Committee on Fisheries and Environment (NAFO)
STACFIS	Standing Committee on Fisheries Science (NAFO)
STACPUB	Standing Committee on Publications (NAFO)
STACREC	Standing Committee on Research Coordination (NAFO)
STACTIC	Standing Committee on International Control (NAFO)
STC	Scientific and Technical Committee (BSDH)
STR	South Tasman Rise Arrangement

Acronym	Name (RFMO)
ST-SWG	Statistics Specialist Working Group (WCPFC)
SW	South West
SWG	Specialist Working Group (WCPFC)
SWIOFC	South West Indian Ocean Fisheries Commission
t	tonnes (= metric ton)
TAC	Total allowable catch
TCC	Technical and Compliance Committee (WCPFC)
TECAM	Technology and Aquaculture (GFCM)
TIS	Trade information scheme (CCSBT)
TOR	Terms of Reference
UN	United Nations
UNCED	United Nations Conference on Environment and Development
UNCLOS	United Nations Conference on the Law of the Sea
UNDOALOS	United Nations Division of Ocean Affairs and Law of the Sea
UNEP	United Nations Environment Programme
UNFSA	UN Fish Stocks Agreement (Agreement for the Implementation of the Provisions of the United Nations Convention of the Law of the Sea Relating to the Conservation and Management of Straddling Fish Stocks and Highly Migratory Fish Stocks)
UNGA	United Nations General Assembly
VMS	Vessel monitoring system (satellite-based)
WA	West Atlantic
WCPFC	Western and Central Pacific Fisheries Commission

Acronym	Name (RFMO)
WCPO	Western and Central Pacific Ocean
WECAFC	Western Central Atlantic Fisheries Commission
WG	Working Group (CCAMLR, GFCM, ICCAT, IATTC)
WG-Bycatch	Working Group on Bycatch (IATTC)
WGDEEP	Working Group on the Biology and Assessment of Deep-Sea Fisheries Resources (ICES) (WGDEEP)
WG-EMM	Working Group on Ecosystem Modeling and Management (CCAMLR)
WG-FSA	Working Group on Fish Stock Assessment (CCAMLR)
WG-IMAF	Working Group on Incidental Mortality Associated with Fishing (CCAMLR)
WG-IMALF	Working Group on Incidental Mortality Associated with Longline Fishing (CCAMLR, WG-IMAF after 1994)
WG-LRP	Working Group on Limit Reference Points (IATTC)
WG Reform	Working Group on NAFO Reform
WG-SA	Working Group on Stock Assessments (IATTC)
WP	Working Party (IOTC)
WPB	Working Party on Bycatch (IOTC)
WPM	Working Party on Methods (IOTC)
WPT	Working Party on Tagging (IOTC)
WPTMT	Working Party on Temperate Tunas (IOTC)
WPTT	Working Party on Tropical Tunas (IOTC)
WSSD	World Summit on Sustainable Development
WTO	World Trade Organization

Summary*

Background

The Agreement for the Implementation of the Provisions of the United Nations Convention on the Law of the Sea of 10 December 1982 relating to the Conservation and Management of Straddling Fish Stocks and Highly Migratory Fish Stocks (UNFSA)

UNFSA was negotiated over a three-year period and adopted in 1995 to strengthen certain provisions of the United Nations Convention on the Law of the Sea (LOSC) relating to straddling fish stocks and highly migratory fish stocks, particularly the provisions in Articles 63, 64, and 116–119, and to provide for the coherent and responsible conservation and management of these stocks throughout their geographic range. UNFSA entered into force in December 2001 following the deposit of the 30th instrument of ratification. The Agreement addresses concerns voiced during the 1992 United Nations Conference on the Environment and Development (UNCED) regarding the demise of key international fishery resources, the impact of fishing operations on ocean ecosystems, and the need for ensuring that all fishing is conducted responsibly in accordance with guidelines to achieve sustainable development.

UNFSA builds upon, reformulates, and refines LOSC to ensure the long-term conservation, sustainable use, and responsible management of straddling fish stocks and highly migratory fish stocks, and their associated and dependent species. As noted by the UN Legal Counsel in his March 2006 Statement to the Fifth round of Informal Consultations of States Parties to UNFSA (ICSP-5), UNFSA "is considered to be the most important legally binding global instrument for the conservation and management of fishery resources since the adoption of the [LOSC] in 1982".

The significance of UNFSA underscores the need for achieving its universal acceptance and effective implementation, as called for at the 2002 World Summit on Sustainable Development in the Johannesburg Plan of Implementation (WSSD-POI) and in recent UN General Assembly (UNGA) resolutions (60/31 and 59/25). As of May 2006, there were 149 Parties to LOSC, 57 to UNFSA, and 34 to the FAO Agreement to Promote Compliance with International Conservation and Management Measures by Fishing Vessels on the High Seas (FAO Compliance Agreement) (see Table 1.1). During the March 2006 ICSP-5, several States, including prominent high seas fishing nations Japan and the Republic of Korea (South Korea), voiced their intention to join UNFSA in 2006.

Together with the FAO Compliance Agreement and the FAO Code of Conduct for Responsible Fisheries 1995 (FAO Code of Conduct), UNFSA provides a set of norms and principles and international minimum standards for regulating fishing and related activities, with an emphasis on the duties of flag States. The responsibility of coastal States to sustainably manage these living marine resources within their exclusive economic zones (EEZs) is reinforced through the introduction and application of post-LOSC management principles and approaches such as the precautionary approach and the ecosystem approach to fisheries. These post-LOSC principles and standards apply throughout the range of straddling fish stocks and highly migratory fish stocks, both within areas under national jurisdiction and on the high seas. UNFSA

* An earlier version of this summary was prepared on behalf of the Government of Canada by Evelyne Meltzer and submitted 31 March 2006 as an official UN document for the May 2006 UNFSA Review Conference (see A/CONF.210/2006/10 of April 12, 2006). That summary also served as a companion to the document "Post-UNFA Overview of RFMO Regulation of Straddling Fish Stocks and Highly Migratory Fish Stocks" (containing figures, maps, and matrices) prepared by the author for the Fifth round of Informal Consultations of States Parties to the Agreement (ICSP-5) 20–24 March 2006, a preparatory meeting convened in New York at UN headquarters prior to the May Review Conference. The maps and matrices were updated where necessary for the Conference and are found in Part III of this book. The author reworked the summary prepared for the Review Conference for inclusion as the Summary of this book. Please note that this work provides an overview up to but not including the May 2006 Review Conference. For a brief update from May 2006 to June 2008 please see the Afterword found at the end of Part II kindly contributed by Terje Løbach and Nathalie Lavoie. The author gratefully acknowledges the review and input of many interested individuals, colleagues, and research assistants but wishes to emphasize that the views expressed and the information presented, unless stated otherwise, are solely those of the author and should not be construed in any way to reflect the policies or positions of the Government of Canada.

reaffirms the jurisdiction of flag States over their vessels fishing on the high seas and elaborates obligations and responsibilities that qualify the traditional freedom of fishing on the high seas. UNFSA also contains innovative provisions for action by nonflag States where noncompliance with conservation and management measures is suspected and the flag State fails to or is unable to act in a timely manner.

UNFSA reinforces the need for cooperative, compatible management between coastal States and flag States by strengthening the role of both regional fisheries management organizations (RFMOs) and regional fisheries management arrangements (RFMAs). Where there are gaps, new RFMOs or RFMAs should be established to conserve and manage the fisheries exploiting these fish stocks coherently throughout their geographic range.

Article 36 of UNFSA provides for the convening of a review conference by the Secretary General of the United Nations four years following the Agreement coming into force to assess the effectiveness of the Agreement in securing the conservation and management of straddling fish stocks and highly migratory fish stocks. Article 36(2) further enjoins the Conference to "review and assess the adequacy" of the provisions of UNFSA, and if necessary, "propose means of strengthening the substance and methods of implementation of those provisions in order better to address any continuing problems in the conservation and management of straddling fish stocks and highly migratory fish stocks". The Review Conference was held 22–26 May 2006 at the UN in New York. See Afterword for update.

Legal overview

LOSC, UNFSA, and the FAO Compliance Agreement are treaties, sometimes referred to as "hard law". UNFSA, the key legal instrument governing fisheries for straddling fish stocks and highly migratory fish stocks, was negotiated in parallel with the FAO Code of Conduct for Responsible Fisheries. The FAO Compliance Agreement entered into force two years after UNFSA in 2003.

The FAO Code of Conduct is a voluntary, non legally binding instrument and is thus often referred to as "soft law". However, as shown in Figure 2.1, the FAO Compliance Agreement, though a treaty, forms an integral and binding part of the FAO Code of Conduct. To supplement the FAO Code of Conduct (and UNFSA) there are four International Plans of Action (IPOAs) (to reduce bycatch of seabirds, to conserve and manage shark fisheries, to reduce excess fleet capacity, and to combat illegal, unreported, and unregulated (IUU) fishing), which were negotiated and adopted by FAO in the framework of the Code of Conduct. Further, a series of technical guidelines (including one on the precautionary approach, one on the ecosystem approach to fisheries, and another on IUU fishing) has been prepared by FAO to assist in the implementation of the FAO Code of Conduct and the IPOAs. The IPOAs have been given added impetus by the 2002 Johannesburg Political Declaration on Sustainable Development and WSSD-POI (United Nations 2003), which set tight timelines for their implementation.

There are now many international legal instruments and initiatives governing the conservation and management of fish stocks within and beyond areas of national jurisdiction (see Figure 2.1). Both hard and soft law instruments have been developed, particularly over the last decade, to support and reinforce the LOSC regime to conserve and manage straddling fish stocks and highly migratory fish stocks. These interrelated initiatives in international fisheries governance constitute an evolution in the new international law for fisheries. The question remains whether this new and continually evolving regime for sustainable fisheries will be embraced by coastal, port, and flag States alike; adopted by RFMOs; and implemented responsibly by the fishing industry. The challenge is to fully implement and comply with the international fisheries governance framework — moving from words to action.

Straddling fish stocks and highly migratory fish stocks

Straddling fish stocks or straddling stocks are described in LOSC Article 63(2) as "stock or stocks of associated species [occurring] both within the exclusive economic zone and in an area beyond and adjacent to the zone". While neither LOSC nor UNFSA provide a precise definition, these terms are now commonly understood to refer to fish stocks that straddle areas of national jurisdiction and the adjacent high seas. In other words, straddling stocks can be found both within the EEZ and in an area adjacent to the EEZ, *or* they can migrate between an EEZ and the high seas. Straddling fish stocks must be distinguished from transboundary fish stocks (sometimes referred to as shared or joint stocks), as described in Article 63(1) of LOSC, which occur within or migrate between the EEZs of adjacent coastal States (see Figure 3.1 for an illustration of straddling fish stocks). Highly migratory fish stocks are tuna and tuna-like species whose distribution and migration pattern cover great distances, moving through maritime areas under national jurisdiction and the high seas. While no definition is provided in LOSC, Article 64 uses the term highly migratory species to refer to species listed in Annex I. This annex lists 17 highly migratory species: albacore tuna, bluefin tuna, bigeye tuna, skipjack tuna, yellowfin tuna, blackfin tuna, little tuna, southern bluefin tuna, frigate mackerel, pomfrets, marlins, sailfishes, swordfish, sauries, dolphins, oceanic sharks, and cetaceans. (See Figure 3.1 for an illustration of highly migratory fish stocks). UNFSA, however, does not refer to LOSC Article 64 "highly migratory species" and instead confines the Agreement to highly migratory "fish" stocks (i.e., not cetaceans). While many species of marine living resources occur within areas under national jurisdiction and on the high seas, straddling fish stocks and highly migratory fish stocks do *not* include anadromous and catadromous species, marine mammals (included in LOSC Annex I), sedentary species (as defined by Article 77 of LOSC), or "purely" high seas or discrete high seas stocks, i.e., those species whose migration pattern does not include areas of national jurisdiction (Figure 3.1). However, it should be noted that some molluscs and crustaceans, while sedentary, are also regulated as straddling stocks by certain RFMOs. And while discrete high seas stocks are not straddling, the Commission for the Conservation of Antarctic Marine Living Resources (CCAMLR), the Northwest Atlantic Fisheries Organization (NAFO), the North East Atlantic Fisheries Commission (NEAFC), and the South East Atlantic Fisheries Organisation (SEAFO) (and as of May 2006 the not yet adopted Southern Indian Ocean Fisheries Arrangement (SIOFA)) have competence to regulate these purely high seas fish stocks as well (see Figure S.1).

Regional fishery bodies

Regional fishery organizations are generally referred to collectively as regional fishery bodies (RFBs). There are now over 30 RFBs worldwide. RFBs are diverse in scope and mandate and can generally be divided into three categories: (1) multilateral *management* entities (RFMOs or RFMAs) through which members directly establish management measures; (2) *advisory* bodies that provide members with scientific and management advice, including those that provide a regional coordination and development function; and (3) *scientific research* organizations that provide only scientific and data advice. To enhance inter-RFB cooperation and coordination, a meeting of RFBs was first held in 1999 under the auspices of FAO. This group, now known as the RFB Secretariats Network, meets biennially.

RFMOs

Nine RFMOs–RFMAs were established prior to UNFSA, and two RFMOs were established subsequent to UNFSA. The treaties establishing these RFMOs provide the areas of competence and the main objectives of the RFMO. This book primarily covers the activities of 10 RFMOs with competence over straddling fish stocks or highly migratory fish stocks (Figures S.1 and S.2).

The term RFMO refers to those RFBs that have management and enforcement responsibilities, a constitutive instrument, and a permanent Secretariat with regular organizational meetings. RFMOs are thus a geopolitical construct overlaid with operational or management functions in fisheries through agreement among interested States, regional economic integration organizations (REIOs), or fishing entities. Four basic elements characterize RFMOs: (1) a geographic area of application, or range as in the case of the Commission for the Conservation of Southern Bluefin Tuna (CCSBT), for a certain fish stock or stocks; (2) an institutional arrangement that establishes a body or structure for carrying out the management of the area; (3) an array of functions necessary for carrying out its management objectives; and (4) a constitutive instrument that sets out the terms of cooperation for the participating States, REIOs, or fishing entities.

RFMAs

An RFMA differs from an RFMO in that it is a cooperative fisheries management mechanism that does not establish a formal organization or body as such. Article 1(1)(d) of UNFSA defines an RFMA as a "cooperative mechanism established in accordance with the Convention and this Agreement by two or more States for the purpose, *inter alia*, of establishing conservation and management measures in a subregion or region for one or more straddling fish stocks or highly migratory fish stocks". This definition gives States considerable flexibility in devising a practical and effective management mechanism. Establishing a formal organization may be deemed unnecessary if a less-structured, effective mechanism is agreed upon. For instance, a practical, cost-effective regional arrangement can enable the Contracting Parties to rotate annual meetings and to collect data, with FAO or some other appropriate international organization serving as the repository. Examples of RFMAs include arrangements under the Central Bering Sea Donut Hole (BSDH) Convention; the not yet adopted SIOFA; the South Tasman Rise Agreement; arguably the 1999 Loop Hole Agreement among Iceland, Norway, and Russia; the bilateral fisheries agreements between Russia and China, Japan, Korea, and Poland over the "Peanut Hole" in the Sea of Okhotsk; and the 2001 Provisional Agreement for swordfish between the European Community (EC) and Chile; of these RFMAs, only the BSDH is further examined in the matrices (found in Part III of this book).

FAO RFMOs or RFMAs

RFMOs and RFMAs can also be established within the framework of FAO. In this respect, it is essential to distinguish between bodies that are established under Article VI of the FAO Constitution and those established under Article XIV. Article VI bodies have only recommendatory or advisory mandates and no decision-making or enforcement powers. Further, they are primarily funded by FAO, which also provides the secretariat. Bodies in this category include the Fishery Committee for the Eastern Central Atlantic (CECAF), the Western Central Atlantic Fishery Commission (WECAFC), and the South Western Indian Ocean Fisheries Commission (SWIOFC, established in November 2004). These are not considered RFMOs or RFMAs under UNFSA.

FAO Article XIV marine fisheries bodies, on the other hand, are bodies with an autonomous budget based on contributions from the members, and they have, subject to the specific provisions of the treaty in question, autonomous decision-making powers, including the power to adopt binding conservation and management measures. They are established under their own treaty regime adopted by the parties in accordance with Article XIV of the FAO Constitution and are registered with the UN. These include the General Fisheries Commission for the Mediterranean (GFCM), the Asia Pacific Fishery Commission (APFIC), the Indian Ocean Tuna Commission (IOTC), and the Regional Commission for Fisheries (RECOFI). GFCM and IOTC are the two RFMOs whose regulatory competence includes highly migratory fish stocks.

RFMOs as vehicles of good governance

A central element of the governance regime under UNFSA is international fisheries management through a network of RFMOs or RFMAs. RFMOs remain the primary mechanism through which coastal States, States fishing in the area, and others with a "real interest" in the fishery fulfill their duty under UNFSA to cooperate in the management of straddling fish stocks and highly migratory fish stocks. Since these stocks occur both within and beyond areas of national jurisdiction, a cooperative and compatible management strategy is required to avoid overfishing (e.g., through authorized access and precautionary total allowable catches (TACs) and quotas). Such bodies are intended to ensure that straddling fish stocks and highly migratory fish stocks, together with their associated and dependent fish and nonfish species, are sustainably managed throughout their geographic range.

Strengthening the RFMO network

UNFSA calls upon States to strengthen existing RFMOs and, pursuant to Article 8(5), to establish RFMOs or RFMAs for any high seas area where there is no regime for regulating straddling fish stocks or highly migratory fish stocks. With the entry into force of the Western and Central Pacific Fisheries Commission (WCPFC) in 2004, there is now global coverage of commercial tuna and tuna-like species. Six RFMOs have competence to manage tuna and tuna-like species (CCSBT, GFCM, the Inter-American Tropical Tuna Commission (IATTC), the International Commission for the Conservation of Atlantic Tunas (ICCAT), IOTC, and WCPFC) (Figure S.2). There is also the bilateral provisional arrangement between the EC and Chile for the management of swordfish in the Southeast Pacific. Two lesser valued commercial species, slender tuna (not a highly migratory fish stock under the LOSC Annex I) and the butterfly kingfish, occur in the Southern Ocean and are sometimes caught as bycatch in the southern bluefin fishery. There may be a need for broadening the competence of an RFMO to encompass these stocks.

There are four RFMOs and one RFMA with competence over straddling fish stocks: CCAMLR, NAFO, NEAFC, SEAFO, and the BSDH. Additionally there is the Peanut Hole in the Sea of Okhotsk, which is an UNFSA Article 16 enclave; the South Tasman Rise Agreement; and the Galapagos Agreement for the Southeast Pacific, which was negotiated, adopted, and signed in August 2000, but has not been ratified by the minimum number (four) of States and is not expected to come into force.

Post-UNFSA RFMOs

SEAFO, adopted in 2001, was the first post-UNFSA treaty to be ratified, entering into force in April 2003. WCPFC, the first post-UNFSA agreement to be adopted (2000), entered into force June 2004. The constitutive instruments of both these RFMOs are modelled on the provisions of UNFSA. Their operational aspects are still in their initial phases. Their challenges are therefore different in some ways from those of RFMOs established prior to UNFSA or even prior to LOSC (GFCM, IATTC, ICCAT, and NAFO).

Unregulated high seas areas for straddling fish stocks

High seas fishing for straddling fish stocks remains unregulated in parts of the world, namely in the Southern Indian Ocean, Southwest Atlantic, Northwest Pacific, and the South Pacific (see Figure S.1). Efforts have been made in the post-UNFSA period to address these gaps. The draft text of the agreement creating SIOFA has been finalized but the agreement as of May 2006 has not yet been adopted and signed. A plenipotentiary meeting is planned in Rome in mid-2006. The SIOFA agreement will likely be adopted at that time. Australia, Chile, and New Zealand have proposed the creation of a new RFMO to manage the non highly migratory fish stocks in the South Pacific Ocean. The first negotiating session for this initiative was held in Wellington, New Zealand, in February 2006. The second meeting to establish the South Pacific RFMO (SPRFMO) is to be held in Australia in November 2006. While this RFMO is under negotiation, the preliminary boundary for the proposed area may be seen in Figure S.1. There have also been calls by Japan, the Republic of Korea, Russia, and the United States to establish a regional management framework to govern international fisheries for straddling fish stocks and discrete high seas fish stocks in the North Pacific.

Reform and modernization of pre-UNFSA RFMOs

RFMOs established prior to UNFSA are, to varying degrees, incorporating the new principles and functions called for under the new regime. The challenge of RFMO reform or modernization has been taken up in various ways: IATTC negotiated a new constitutive agreement known as the Antigua Convention (key provisions are shown in italics in the IATTC matrix found in Part III) that will replace the existing IATTC Convention when it comes into force. The new agreement reflects modern developments in fisheries management, including application of the precautionary approach and compatibility principles as described in UNFSA and the FAO Code of Conduct. GFCM amended its agreement in 1997 in recognition of UNFSA and other international instruments to include the precautionary approach, membership of nonlittoral States, and an autonomous budget, and to restructure the Commission's subsidiary bodies. ICCAT has made efforts to increase participation in the Commission and improve criteria for the implementation of trade measures. At the November 2005 Annual Meeting, the parties agreed to create a regional observer program to monitor at-sea transshipment as well as launch a process to strengthen the Commission. These are among the steps taken by ICCAT to become more consistent with UNFSA and other relevant international instruments.

In 1996, NEAFC created the Working Group on the Future of NEAFC to examine the adequacy of the NEAFC Convention and to introduce measures and amend the Convention in keeping with the overarching framework provided by UNFSA. On the basis of the Working Group's recommendations, it was agreed at the November 2005 Annual Meeting to modernize the NEAFC Convention to provide a clearer mandate to pursue the ecosystem approach, the protection of biodiversity, and the precautionary approach. These amendments will be applied on a voluntary basis pending ratification. NEAFC Parties further agreed to conduct an independent performance review to ensure the alignment with UNFSA and related instruments. NAFO, after several years of incrementally adopting measures to align the organization with UNFSA, agreed at its September 2005 Annual Meeting to embark on a reform process. To this end, the Ad Hoc Working Group on NAFO Reform was established to evaluate and recommend changes to the NAFO Convention with respect to the decision-making process and to examine the current structure and operation of NAFO to, *inter alia*, strengthen its effectiveness in a manner that is consistent with UNFSA and the St. John's Declaration adopted at the St. John's, Newfoundland and Labrador, *Conference on High Seas Fisheries and the United Nations Fish Agreement* in May 2005. The first meeting of the Ad Hoc Working Group was held in Montréal in April 2006.

Participation

Members

UNFSA gives effect to the general duty to cooperate and to conserve fish stocks by calling on States, fishing entities, and REIOs to become members of RFMOs, or to agree to apply the conservation and management measures of such RFMOs regardless of whether they are members or not. The framework allows all States with a "real interest" in the fisheries concerned to become members or participants, and enjoins all States to comply with the management regime of the relevant RFMO or RFMA. While the term real interest is not defined in UNFSA, since 1995 there has been a concerted effort to encourage nonmember States, REIOs, and fishing entities to join RFMOs. As can be seen in Table 1.2, the major fishing nations have joined RFMOs, although membership by States known as "flags of noncompliance" is limited. Increased participation in and compliance with RFMOs will likely lead to better governance and effective regulation of straddling fish stocks and highly migratory fish stocks.

Cooperating Non-Contracting Parties and cooperating nonmembers

Fishing by vessels of nonmembers has long undermined the effectiveness of RFMOs in managing their regulatory areas. To achieve broad cooperation and participation and therefore compliance with the RFMO conservation regime as called for under UNFSA, most RFMOs created new categories for participation. Cooperating nonmember status typically allows for participation in meetings and quota allocations or fishing opportunities, but not in decision making. CCAMLR, CCSBT, GFCM, IATTC, ICCAT, IOTC, NEAFC, and WCPFC have all introduced differentiated participation status in recent years.

CCAMLR has introduced an additional subcategory of cooperating nonmember States, namely those States participating in CCAMLR's catch documentation scheme (CDS), which may be port States where catches of toothfish are landed or transshipped, rather than fishing nations. Recognizing the major role of such interstate arrangements in international fisheries management, several RFMOs now also include REIOs, such as the European Union (EU). The EU has concluded access arrangements with many countries and is a member of most RFMOs (CCAMLR, GFCM, ICCAT, IOTC, NAFO, NEAFC, SEAFO, WCPFC). Only the CCSBT and the IATTC do not permit REIOs to join as members. In practice the EU does participate informally in IATTC, and the Antigua Convention, not yet in force, will allow IATTC to accept REIO membership.

Fishing entities

Taiwan, (also known as Chinese Taipei or within the UN system as Taiwan Province of China), is precluded from becoming a member of RFMOs within the FAO framework (such as the GFCM and IOTC), as it is not recognized within the UN system as a State or separate political entity. Taiwan, however, has a large international fishing industry with a significant high seas fleet. The acceptance of the new concept of "fishing entity" with specified rights and obligations has permitted Taiwan to accede to the status of "member" and participate in WCPFC, as part of the Extended Commission in the CCSBT, and as a cooperating fishing entity in ICCAT and IATTC. The IATTC - Antigua Convention (not yet in force) provides for participation of the fishing entity as a member.

Increase in RFMO participation

In the post-UNFSA period, considerable pressure has been placed on fishing States, REIOs, and the fishing entity to participate in RFMOs. Limiting access to fishery resources only to RFMO members, or to those that agree to apply the conservation and management measures of such RFMOs, has resulted in a marked trend to join RFMOs. For example, as of 5 March 2006, 41 States and one REIO from five continents are members of ICCAT, 23 of whom joined since 1995 (see Table 1.2).

Transparency

Article 12 of UNFSA requires that States ensure transparent decision-making processes, and to this end, representatives of intergovernmental organizations (IGOs) and nongovernmental organizations (NGOs) should be allowed as observers in meetings of RFMOs. All RFMOs responsible for straddling fish stocks or highly migratory fish stocks have publicly accessible Web sites, with minutes of annual meetings and subsidiary bodies available online. Most RFMOs allow for NGOs to participate as observers at meetings, upon application and subject to applicable procedural rules, which can vary considerably among RFMOs.

Overview of RFMO regulation

Fisheries, precautionary approach, and ecosystem approach

Key species and gear

Each RFMO has jurisdiction over specific stocks of straddling or highly migratory fish stocks. RFMOs responsible for straddling fish stocks have competence over a variety of species that are particular to the relevant geographic area. Many species are demersal fish and small pelagics, with shrimp, squid, and crab composing the invertebrate fisheries. These stocks are generally harvested using bottom trawls, midwater trawls, bottom longlines, gill nets, purse seines, and traps. RFMOs responsible for highly migratory fish stocks generally refer to Annex 1 of LOSC for tuna and tuna-like stocks over which they have competence (WCPFC, however, goes beyond the LOSC Annex I species by defining highly migratory fish stocks to include "such other species of fish as the Commission may determine"). Large pelagic fish are generally captured using longlines and purse seines, with bait boats, gill nets, pole and line, trolling, hand-lining, and other artisanal gear making up smaller sectors of the fishery.

Stock status

The pre-UNFSA call for a new governance regime to conserve and manage straddling fish stocks and highly migratory fish stocks emerged from the 1992 Rio Declaration recognition that there was an urgent and immediate need to address the world's fisheries crisis. The history of fishing straddling fish stocks and highly migratory fish stocks is characterized by increased exploitation as a result of technological changes, increased catch efficiency, expansion into new waters, and the targeting of new species. The 2005 FAO review of the world fishery resources provides a valuable report on the past and current state of global fish stocks. In measured language, the review notes (at p. 3) that with an estimated marine catch of some 85 x 10^6 tonnes (t), the general trend "...suggests that the maximum long-term potential of the world marine capture fisheries has been reached, with some stocks and areas being overfished and some stocks not producing their full expected long-term potential". As stocks have declined, fisheries issues have been recast as part of the broad-scale environmental and biodiversity concerns about food security, ecological security, and ultimately human security. Determining stocks status is a critical obligation under UNFSA and key to understanding the relative state of fisheries and the corresponding need for effective conservation and management as provided in UNFSA and in other instruments.

Status of straddling fish stocks

The poor state of many straddling fish stocks remains a significant concern. Several stocks have not recovered despite moratoria on directed harvesting for over a decade. Currently NAFO has nine stocks under moratoria, six of which have been closed to directed fisheries for over 10 years; a tenth, Greenland halibut, is now under a rebuilding plan, and an eleventh stock, yellowtail flounder, is recovering. Alaskan pollock in the BSDH Convention Area has been under a moratorium since 1993. The 2005 pollock biomass in the area was estimated at 422 000 t, which is substantially below the biomass of 1.67×10^6 t required to trigger the reopening of the fishery. There is no direct fishery for pollock in the Sea of Okhotsk Peanut Hole. CCAMLR has several stocks that are depleted (the directed fishery for some species including Antarctic rock cod is prohibited, and there is no longer a fishery for crab, squid, or lightfish). Krill have been assessed in some areas. The status of Patagonian toothfish (commonly known as Chilean seabass) has been the subject of international concern because of large-scale IUU fishing. In NEAFC, blue whiting is considered to have full reproductive capacity, though catches in recent years have exceeded recommended TACs. Mackerel, which has been classified as having reduced reproductive potential, is considered at risk of being overfished. Atlanto-Scandian (Norwegian) spring-spawning herring is considered to have full reproductive capacity. Stock status is uncertain for redfish, Rockall haddock, and deep-sea species in the NEAFC Convention Area. SEAFO has yet to assess stocks under its competence. RFMOs responsible for straddling fish stocks also face the challenge of assessing and managing deep-sea straddling stocks such as orange roughy, which are typically long lived and slow growing and vulnerable to overexploitation.

Status of highly migratory fish stocks

Globally, tuna and tuna-like species are generally fully exploited or overexploited. Of the RFMOs responsible for highly migratory fish stocks, CCSBT faces the greatest challenges with regard to stock status, as southern bluefin tuna is estimated to have declined by 90% since 1960. Some Atlantic stocks under the competence of ICCAT are overfished or below target levels ($B < B_{MSY}$), including northern albacore, Western Atlantic bluefin (depleted), blue marlin, and white marlin. The latter three are currently under recovery plans. The status of Eastern Atlantic bluefin with regard to B_{MSY} is uncertain, and there is concern that this stock is being overexploited. There is also concern about the magnitude of stock mixing or overlap between the Western and Eastern Atlantic bluefin tuna populations, and that potential overfishing in the Eastern Atlantic is compromising the rebuilding efforts in the Western Atlantic. ICCAT's rebuilding plan for North Atlantic swordfish appears to have met with some success. In 2002, after four years into a 10-year recovery program, the stock had increased from 65% to 94% of its target stock size (B_{MSY}). Areas were closed to fishing, and spawning grounds were protected. The recovery plan is in place until 2008. South Atlantic albacore and blue sharks are above levels required to produce maximum sustainable yield (MSY). In 2005, GFCM and ICCAT developed data collection procedures for tuna farming to improve the estimation of amounts fished and size composition of the catches (i.e., to ensure that fattening operations are properly controlled to avoid overfishing, consistent with conservation and management measures). Swordfish in the Mediterranean are stable or declining. In the Eastern Pacific Ocean, several stocks under the competence of IATTC are above the biomass needed to produce MSY, including skipjack, southern albacore, striped marlin, and swordfish, with yellowfin biomass at MSY levels. Bigeye tuna is at its lowest levels of abundance since 1975, and therefore a concerted effort to rebuild the stock is imperative. The status of other pelagics, such as black marlin and sailfish, remains uncertain. Blue marlin is assessed to be near full exploitation albeit with some uncertainty. North Pacific albacore, for which management is being coordinated between IATTC and WCPFC, is considered to be fully exploited, although there is some uncertainty about the status of the stock. In the Western Pacific, in the area now regulated by WCPFC, skipjack and South Pacific albacore are above B_{MSY}, while bigeye and yellowfin are considered to be fully exploited. Fisheries for bigeye and yellowfin face expansion, and ensuring levels of harvest commensurate with the resource is a challenge for WCPFC and its members. Two tuna species of concern in the Indian Ocean under the competence of the IOTC are yellowfin and albacore. Further increases in effort beyond 2000 levels are not recommended for yellowfin, and there is currently no reliable assessment for albacore. Stock levels of skipjack in the Indian Ocean are seen as stable, as are those of swordfish, although overfishing ($F > F_{MSY}$) of swordfish is suspected.

In summary, assessments available from RFMOs and the FAO indicate that most stocks of tuna and tuna-like species are fully exploited, with some overfished, and a few depleted. Few opportunities to increase exploitation exist, except in some areas of the Pacific and possibly in the Indian Ocean, where increases in catches of skipjack tuna might be sustainable. However, if current fishing techniques are used, it is thought this can only be done at the expense of detrimental bycatches of other species, particularly yellowfin and bigeye tuna.

Stock recovery

One objective of Article 5 of UNFSA is the restoration of stocks. The commitments of the WSSD to restore fisheries to levels capable of producing MSY puts pressure on both national and international management bodies to develop strategies that will result in stock restoration. While a desirable outcome of fisheries management, restoring stocks has proven to be challenging, even when there is political will to reduce fishing effort. Most recovery and rebuilding plans implemented in the last decade have not yet met the rebuilding objectives for the fish stocks concerned. Fishing effort often continues through bycatch, and factors other than fishing affect stock structure and biomass. Environmental effects (e.g., North Atlantic Oscillation (NAO), El Niño, ocean acidification, and climate change) can have a dramatic impact on recruitment to the fishery. Habitat destruction and pollution are additional factors that impede stock recovery. Reducing a stock below a certain biomass effectively removes much of its resilience to withstand natural fluctuations in the environment, reinforcing the need for implementing risk-averse conservation measures.

Data and assessment challenges

UNFSA emphasizes the need for rigorous data reporting mechanisms that provide timely, accurate, and complete information. Data limitations circumscribe or restrict the predictive capacity and probative value of the best scientific evidence available, underscoring the need for applying the precautionary approach to fisheries. The inherent uncertainties in trying to count marine fauna and in determining an appropriate harvest limit, together with the uncertainties inherent in the data collection, increase the need for caution.

The standard reference in stock status is the biomass needed to produce MSY. While RFMOs may have different biological reference points, most refer to MSY as either a limit or a target, and hence the status of the stock is generally referred to as being below or above MSY, whether the variable is fishing mortality, biomass, or catch levels.

Precautionary approach to fisheries

One of the key innovations established under UNFSA is the precautionary approach (PA) to fisheries management (UNFSA Articles 6 and Annex I). The PA recognizes that fishing operations can pose a risk of serious or irreversible harm to fish stocks and the marine ecosystem. Accordingly, under the PA, where there are threats of serious or irreversible harm, lack of scientific certainty is not an acceptable reason for postponing or failing to take measures to prevent overexploitation and adverse impacts to the marine environment. As elaborated in the 1996 FAO Technical Guidelines for Responsible Fisheries *(Precautionary Approach to Capture Fisheries and Species Introductions)* the burden of proof is shifted to those opposing precautionary action to demonstrate that harvesting practices and fishing effort are within an acceptable level of harm, i.e., no serious or irreversible harm will result.

The PA takes into account, and allows for, the inherent uncertainty, error, and associated biological risk in predicting stock abundance, as well as the challenges associated with accurately assessing fishing mortality and effectively controlling fishing activity. Additionally, the PA recognizes the difficulty in determining the effects of the marine environment on the fish stocks and vice versa. To decrease uncertainty and error in stock assessment and improve the quota-allocation process, precautionary reference points and preagreed decision rules were established under UNFSA along with guidelines for prosecuting new and exploratory fisheries

To date, most RFMOs have examined how to adopt and implement precautionary reference points but have yet to operationalize the approach. Most RFMOs (with some notable exceptions), have reformulated MSY from a target management reference point to a limit, and some have introduced harvest control rules. In 1998 the International Council for the Exploration of the Sea (ICES) implemented the PA, setting reference points in a PA framework that was, in turn, used to provide advice to coastal States. Since 1999 ICES has used its PA framework to provide advice to NEAFC. CCAMLR was the first RFMO to introduce the PA and systematically uses a structured precautionary assessment procedure to determine catch limits.

If the objectives of UNFSA and WSSD are to be realized, the PA must be “applied widely” both within and beyond areas of national jurisdiction. This means that the PA will be perceived not solely as a scientific exercise to develop precautionary reference points, but as part of an overarching risk-averse, regulatory appraisal process, whereby managers err on the side of conservation in making management decisions.

Ecosystem approach to fisheries

Article 5 of UNFSA emphasizes the need to take ecosystem considerations into account in managing fisheries for straddling fish stocks and highly migratory fish stocks. UNFSA recognizes that fisheries can have adverse impacts on the marine ecosystem and that changes in the marine ecosystem can impact on fisheries (pollution, climate change, natural fluctuations and oscillations, and habitat degradation). The overexploitation of fish stocks raises environmental and biodiversity concerns, which have led to calls for certain species to be listed in the Convention on International Trade in Endangered Species (CITES) appendices and for the establishment of marine reserves. The need for avoiding overfishing, rebuilding fish stocks, and rehabilitating habitat spurred the adoption of the 2003 FAO Technical Guidelines for Responsible Fisheries *(Fisheries Management 2. The Ecosystem Approach to Fisheries)*, which provide a comprehensive framework that builds upon and moves beyond conventional fisheries management, known as the ecosystem approach to fisheries (EAF). A critical component of the EAF is the reduction of excess capacity to ensure that fishing effort is commensurate with a sustainable catch. This framework, articulated through the FAO guidelines on EAF, focuses on the need to control human activity through the wide application of the precautionary approach while increasing the knowledge base or human understanding of the marine ecosystem and the impact of fishing. While only CCAMLR has adopted a practical ecosystem approach, consistent with its mandate, other RFMOs have begun to include ecosystem considerations in their management decisions and to incrementally implement the EAF. Most RFMOs have developed bycatch mitigation measures (particularly those responsible for highly migratory fish stocks), thereby addressing the IPOA on seabirds, the IPOA on sharks, and sea turtle conservation concerns (CCSBT, IATTC, ICCAT, IOTC, WCPFC). NAFO introduced a ban on shark finning and bycatch restrictions on shark catches, including prohibition of discards. NEAFC banned the fishery for basking shark in 2006 and is temporarily prohibiting the use of gill nets, entangling nets, and trammel nets. RFMOs for straddling fish stocks have also closed areas to fishing (CCAMLR, NEAFC, NAFO, and GFCM in its role as an RFB for shared stocks). Increasingly, RFMOs are addressing ecosystem effects through working groups that consider both the ecosystem effects on fish stocks and fishing impacts on the ecosystem.

Decision making

To increase the effectiveness of RFMOs, efforts have been made to strengthen decision-making processes to increase the adoption of measures needed to regulate the fisheries; enhance the likelihood of implementation and compliance; and avoid and reduce the possibility of disputes. Several RFMOs have adopted science-based decisions, preagreed decision rules including harvest controls, and allocation criteria, all of which are intended to provide for a more predictable, robust, and transparent system accepted by all participants.

Consensus-based decision making

All RFMOs, with the notable exception of NEAFC, generally use consensus-based decision-making procedures even if their constitutive agreement contains voting procedures. While consensus can be an effective procedural device to avoid marginalizing a minority viewpoint and to enhance compliance through collective agreement, it can also result in a protracted decision-making process, often leading to delay or paralysis to the detriment of making timely, responsible, and effective management decisions for the sustainability of fish stocks. Some RFMOs have endeavoured to address these shortcomings in a manner that is consistent with UNFSA by establishing international minimum standards that serve as benchmarks for decisions. These standards act as a default position even for a provisional period until agreement is reached whether by consensus or an alternative mechanism to break a deadlock or resolve disputes. These standards are reinforced by preagreed decision rules, precautionary reference points, and criteria for setting TACs and determining quotas. Where consensus cannot be reached in a timely fashion, the process in some organizations can default to a voting procedure or a fast-track dispute settlement process.

Objection procedures

A central tension characterizing international fisheries organizations is the need for balancing State sovereignty with the obligation to cooperate within an RFMO. One mechanism adopted by RFMOs to achieve this balance is the objection procedure or opting-out clause, which allows a State to opt-out of conservation and management decisions to which it objects. However, objections often lead to contentious unilateral action that can undermine the RFMO conservation regime. In an effort to improve the effectiveness of the decision-making process, some

RFMOs (NEAFC and SEAFO) have introduced provisions stipulating conditions under which an objection can be made. These provisions require, for example, that reasons must be given for the objection, and that the objecting State must provide an acceptable post-opting-out plan that demonstrates that the conservation regime will not be undermined by unilateral action. CCSBT, IATTC (nor the Antigua Convention), and WCPFC do not have an objection procedure in their constitutive agreements.

Dispute settlement mechanisms

All of the major treaties on the law of the sea governing fisheries (LOSC, UNFSA, and FAO Compliance Agreement) make provision for the settlement of disputes. LOSC and UNFSA contain compulsory procedures entailing binding decisions. However, these treaties also emphasize the more fundamental obligation to settle disputes by peaceful means, such as negotiation, inquiry, mediation, conciliation, arbitration, judicial settlement, and resort to regional agencies or arrangements.

UNFSA contains a comprehensive system of dispute settlement linked to LOSC. The principal provisions relate to the obligation to settle disputes by peaceful means; the prevention of disputes by the adoption of efficient and expeditious decision-making procedures or the strengthening of existing procedures; the use of ad hoc expert panels for disputes of a technical nature; the compulsory and binding procedures for the settlement of disputes found in LOSC; the use of provisional measures; and certain limitations to protect the interests of coastal States in their EEZs as contained in Article 297(3) of LOSC.

The obligations in UNFSA (Article 28) to cooperate to prevent disputes, to agree on efficient and expeditious decision-making procedures within RFMOs, and to strengthen them as necessary are a strong acknowledgement of the need for adopting procedures to prevent disputes in the first place.

Not all RFMOs have procedures in their constitutive treaties for settlement of disputes. For several RFMOs established prior to the adoption of UNFSA in 1995, the pattern is mixed. Thus ICCAT and NAFO, for example, make no provision for compulsory dispute settlement, while CCSBT, GFCM, and IOTC contain clauses permitting resort to binding dispute settlement procedures. Of those RFMOs that have been updated since 1995 in the light of the provisions of UNFSA, IATTC makes limited provision for referring a dispute to an ad hoc expert panel with the mutual consent of the parties. NEAFC, however, has agreed to adopt amendments (not in force as of May 2006) that will establish a fast-track dispute settlement mechanism. The post-UNFSA RFMOs have included such provisions: both SEAFO and WCPFC incorporate the dispute settlement provisions of UNFSA. In addition, WCPFC's decision-making process (which includes making binding decisions) provides for the appointment of a conciliator or a review panel where consensus is not reached.

Assistance to developing States

Measures required by UNFSA and its implementation through RFMOs, in particular those regarding data collection and reporting; the practical application of the ecosystem approach; effective mechanisms for monitoring, control, surveillance, and enforcement; domestic legislative and management measures; and participation in international and regional meetings, require adequate financial and human resources. Developing States require financial assistance, technical support, and capacity to effectively participate in the contemporary international fisheries governance system. CCAMLR recently passed a resolution (24/XXIV) to assist developing States in keeping with Article 25 of UNFSA. Towards this end, the Part VII Fund has been established and is administered by FAO. As of March 2006, Canada, Iceland, Norway, and the United States made Part VII contributions. Additionally, there have been bilateral and multilateral efforts to promote the full participation of developing coastal and island States in RFMOs and to facilitate the realization of attendant benefits of sustainable development through the effective implementation of LOSC and UNFSA.

Monitoring, control, and surveillance

Data collecting and reporting

Article 10 of UNFSA elaborates the functions of RFMOs and provides for data collection and reporting to be used in stock assessment. States are to "obtain and evaluate scientific advice, review the status of the stocks and assess the impact of fishing on non-target and associated or dependent species" (Article 10(d)); "agree on standards for collection, reporting, verification and exchange of data on fisheries for the stocks" (Article 10(e)); "compile and disseminate accurate and complete statistical data, as described in Annex I, to ensure that the best scientific evidence is available" (Article 10(f)); and "promote and conduct scientific assessments of the stocks and relevant research and disseminate the results thereof" (Article 10(g)). Since the entry into force of UNFSA, all RFMOs have put into place measures to strengthen data collection and reporting. For example, ICCAT at its 2005 Annual Meeting adopted a measure placing the onus on members as well as Cooperating Non-Contracting Parties to demonstrate why reporting requirements were not met and provide plans for corrective action.

Observer programs

UNFSA Article 18(3)(g)(ii) requires the implementation of national, subregional, and regional observer programs. There are two functions of observer programs — the monitoring of compliance with conservation measures and the collection of scientific data. Most RFMOs responsible for straddling fish stocks, with the exception of NEAFC, have an observer program. CCAMLR's Scheme of International Scientific Observation (which is not compliance oriented) requires 100% coverage on all finfish vessels. Observers are present on some krill fishing vessels, but as of 2006, observer coverage is not mandatory. NAFO requires 100% observer coverage on all Contracting Party vessels fishing in the NAFO Regulatory Area and has introduced a pilot project on real-time electronic submission of observer data. The BSDH Convention requires 100% observer coverage. Observers from another party may be placed on a fishing vessel upon request. However, since the entry into force of the BSDH Convention there has been no commercial pollock fishery in the area. As of 1 January 2006, all vessels fishing in the SEAFO area are required to carry scientific observers.

RFMOs responsible for highly migratory fish stocks have limited observer programs. ICCAT requires that a minimum of 5% of large-scale longline vessels fishing for bigeye tuna carry observers, although coverage may be greater for some Contracting Parties. The 2005 ICCAT transshipment measures established a regional observer program to be implemented not later than 2007, requiring the presence of observers during at-sea transshipments involving large-scale tuna longline fishing vessels. As of May 2006, GFCM does not have an observer program, but adopted in February 2006 the ICCAT Recommendation on Transshipment, calling for the establishment of a GFCM regional observer program. AIDCP requires 100% coverage on large purse-seine vessels, with 70% of observers provided by the IATTC and the remainder supplied by Contracting Parties. The IATTC Resolutions call for the use of observers for data collection and other monitoring services in the Eastern Pacific Ocean. Both IOTC and CCSBT encourage members to implement national observer programs with suggested coverage at 10%. WCPFC provides for a regional observer program with implementation planned for 2006, in coordination with other national or regional programs.

VMS

The use of a satellite vessel monitoring system (VMS) is another tool that can improve the effectiveness of national and regional fisheries conservation measures. In terms of monitoring, control, and surveillance (MCS), VMS assists fisheries management authorities in data gathering (i.e., monitoring vessel movements and collecting catch data) to ensure compliance with management objectives. There is considerable potential for expanded use of this technology for MCS and enforcement. UNFSA, requires "the development and implementation of vessel monitoring systems, including, as appropriate, satellite transmitter systems, in accordance with any national programmes and those which have been subregionally, regionally or globally agreed among the States concerned" (Article 18 (3)(g)(iii)). VMS initiatives are being implemented in a number of coastal States and RFMOs, including CCAMLR, IATTC, ICCAT, NAFO, and NEAFC. As of 1 April 2006, all vessels authorized to fish in the SEAFO Convention Area are to comply with its VMS standards, with data transmitted to the flag State's fisheries monitoring centre (FMC) and then to the SEAFO Secretariat. Some of these RFMOs have had fully functional VMS for several years (frequency of position reports from fishing vessels are now transmitted every two hours in NAFO and NEAFC). WCPFC recently endorsed the establishment of a satellite VMS as an integral part of its MCS system.

Enforcement

RFMOs with competence for straddling fish stocks or highly migratory fish stocks have increasingly implemented enforcement mechanisms over the last decade, including at-sea and in-port inspections, measures to address transshipment, and the creation of "positive" and "negative" vessel lists. UNFSA Article 18(3)(h) requires flag States to take measures to regulate transshipment on the high seas to ensure that the effectiveness of conservation and management measures is not undermined. Several RFMOs have adopted regulations or prohibitions on transshipment and landing of transshipped fish: GFCM, IATTC (in respect of purse-seine vessels), ICCAT, IOTC, NAFO, and NEAFC. Article 27 of WCPFC Convention allows port States to prohibit landings and transshipments if catch was taken through noncompliance and provides that to the extent practical transshipment is to take place in port.

At-sea boarding and inspection

UNFSA Article 18(3)(g)(i) calls for the implementation of national inspection schemes and subregional and regional schemes for cooperation in enforcement pursuant to Articles 21 and 22, including requirements for such vessels to permit access by duly authorized inspectors from other States. The basic procedures for boarding and inspection are set out in Article 22. Most RFMOs and RFMAs responsible for straddling fish stocks have at-sea boarding and inspection schemes in force (CCAMLR, NAFO, and NEAFC). Both the BSDH Convention and the SEAFO Convention provide for an at-sea inspection scheme. To date, most RFMOs responsible for highly migratory fish stocks have not adopted at-sea boarding and inspection schemes; however, WCPFC agreed in December 2005 to apply UNFSA Articles 21 and 22 in the WCPFC Regulatory Area as of 19 June 2006. The boarding and inspection procedures will be further considered during the intersessional period and will not be used until finalized at the Third Annual Meeting of WCPFC.

In-port inspections and port-State measures

Fishing vessels seek port access for many reasons, including refuelling, provisioning, landing catch, transshipment, repairs, and in case of emergencies. Accordingly, port-State control is now a fundamental MCS obligation. Some RFMOs (including CCAMLR, ICCAT, IOTC, NAFO, NEAFC, and WCPFC) have either introduced or are in the process of introducing measures that will require member States to inspect IUU vessels in their ports and to prevent the landing or transshipment of unauthorized catch. The role of port States in catch and trade documentation schemes is an area for further development. Some of the challenges relate to jurisdiction (under international law, ports fall under the exclusive jurisdiction of the port State), implementation capacity of developing States, and the international trade law implications of port-State measures. The recently endorsed FAO Model Scheme on Port State Measures by the FAO Committee on Fisheries (COFI) provides a useful guide for RFMOs on how to address some of these issues.

RFMO registers

Pursuant to Article 18(3)(c) of UNFSA, flag States are to establish national records of vessels authorized to fish on the high seas. There is a need for ensuring that flag States effectively control vessels flying their flag, and that they refuse to register or grant licences to vessels over which they cannot exercise the necessary control. Further, there is a need for ensuring that appropriate sanctions are imposed for noncompliance. CCAMLR, CCSBT, GFCM, IATTC, ICCAT, IOTC, NAFO, NEAFC, and WCPFC have all established "lists of vessels authorized to fish in the area". These so-called "positive" vessel lists are often referred to as vessel records or registries. There is a presumption in these RFMOs that vessels fishing in their regulatory areas without authorization are "IUU" vessels. Many RFMOs have established "IUU negative" vessel lists. CCAMLR, IATTC, ICCAT, IOTC, NEAFC, and most recently, NAFO and GFCM have established IUU vessel lists. Sharing of such lists among RFMOs ensures that vessels known to have violated conservation and enforcement measures in one area cannot operate easily in another area.

Catch documentation and trade measures

Another new development in international fisheries law since the adoption of UNFSA is the use of multilaterally agreed nondiscriminatory trade measures to promote compliance with international conservation and management measures. Although it does not directly mention trade measures, Article 33(2) of UNFSA provides that "States Parties shall take measures consistent with this Agreement and international law to deter the activities of vessels flying the flag of non-parties which undermine the effective implementation of this Agreement". CCAMLR, CCSBT, IATTC, ICCAT, and IOTC have all implemented catch, statistical, or trade documentation schemes. ICCAT first introduced its statistical documentation program (SDP) in 1994 for Atlantic bluefin tuna. This was followed by a similar program for bigeye and swordfish in 2002. CCAMLR has a CDS for Patagonian toothfish. NAFO requires that all fish caught in its regulatory area be labelled as such.

The IPOA to combat IUU fishing expands on UNFSA by making specific provision for the use of multilateral trade measures by RFMOs to enforce regional conservation and management measures. Such measures may include the adoption of multilateral catch documentation and certification schemes, import and export controls, or prohibitions. Several RFMOs (CCAMLR, CCSBT, ICCAT, IOTC, NEAFC, and NAFO) have developed and implemented wide ranging trade-related measures to address IUU fishing. Measures include mandatory prohibition of imports of fish from nonmembers whose vessels undermine the effectiveness of agreed conservation measures (CCAMLR CDS); prohibition of imports from members who exceed their catch limits; prohibition of imports from nonmembers who fail to take corrective action; use of CDSs to determine the origin of fish and whether the fish is harvested in a manner consistent with the RFMOs' management measures; use of other statistical documentation programs that trace all fish and fish products that enter international trade; provision of catch documents to vessels including export (or reexport) validation by the exporting State; and marking or labeling of fish products to indicate species, production method (e.g., caught at sea, inland waters, or farmed), and the catch or production area.

Conclusion

For more than half a century the international community has endeavoured to reduce and avoid overfishing. The 1946 London Conference on Overfishing recognized that many commercial stocks were overexploited but participants were unable to reach agreement on the conservation and management measures needed to control fishing mortality. Many thought the problems were exaggerated and that fish were inexhaustible and perennially resilient. Since then there has been a steady decline in the state of many of the world's fisheries, despite repeated efforts to address this problem and warnings to the contrary from some scientists about the fragility of fish stocks. The excessive and indiscriminate exploitation of fishery resources has become a global environmental problem: a problem that has reached critical proportions, exacerbated by marine pollution, habitat degradation, and perturbations due to climate change.

The message of this book is the global imperative to arrest and redirect the trajectory set in motion by unsustainable patterns in the legal and illegal exploitation of living marine resources, particularly straddling fish stocks and highly migratory fish stocks. UNFSA provides the legally binding, overarching framework — a transparent, precautionary, scientific, and ecosystem-based governance structure to be implemented primarily through RFMOs. This composite regime, however, is set within the authority and constraints under which these RFMOs operate. Fundamental to the success of the progressive regime created by UNFSA and the supporting hard and soft law instruments is the rigorous implementation, meaningful control, effective enforcement, and widespread compliance by RFMOs and States both within EEZs and on the high seas and by the fishing industry. While there has been significant progress and ongoing reform to varying degrees within the RFMOs examined in this book, there is still a discernible gap between the formal legal–institutional developments and the action required to achieve long-term sustainability of fish stocks and nontarget species at restored levels both within and beyond areas under national jurisdiction. As a result, more than 10 years following the adoption of UNFSA and its subsequent entry into force, there is still a pressing need to improve regional and national management regimes and responsibly regulate international fisheries to stop overfishing and other nonsustainable fishing practices.

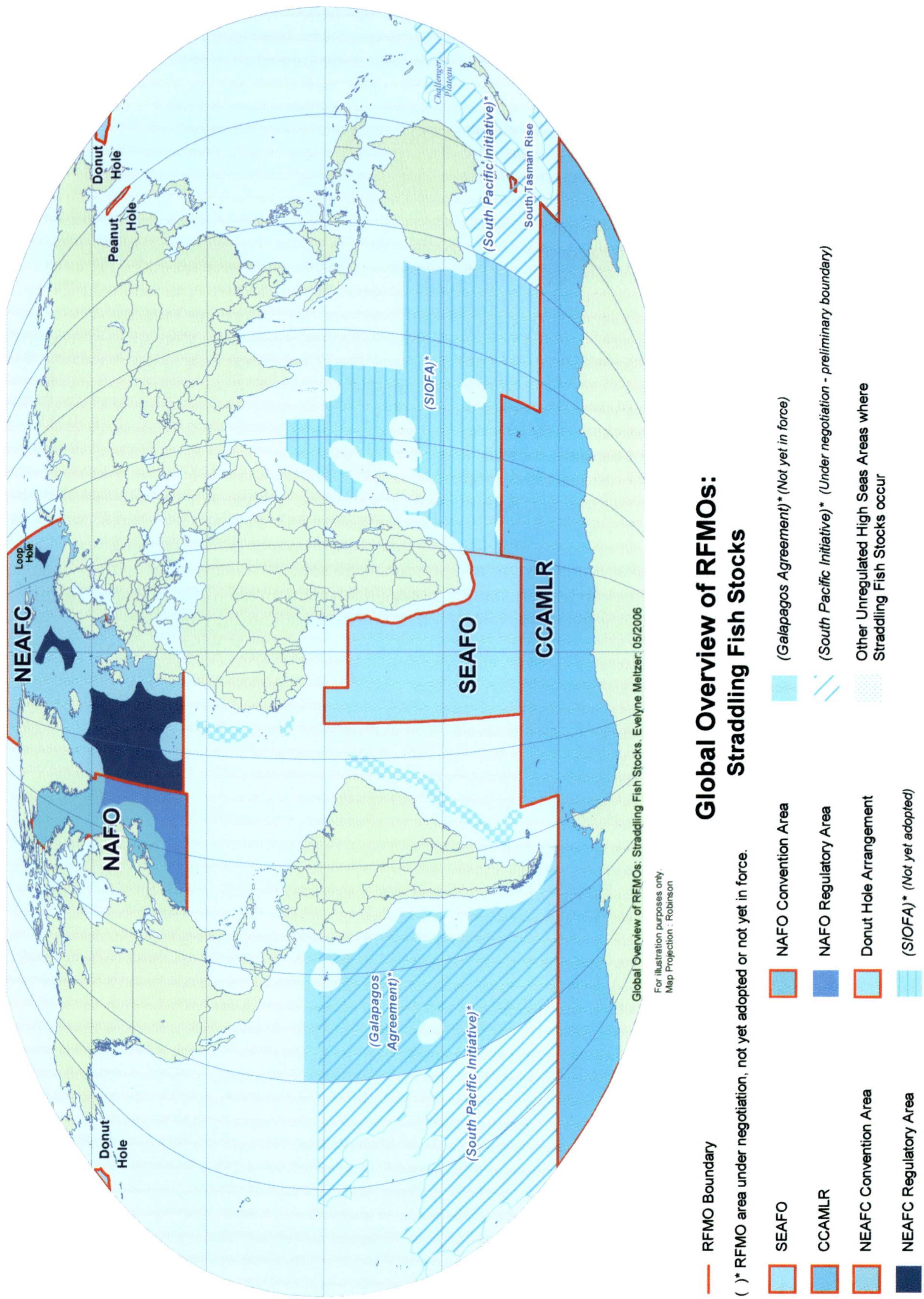
Global Overview of RFMOs:
Straddling Fish Stocks
NEAFC
NAFO
SEAFO
CCAMLR
Loop Hole
Peanut Hole
Donut Hole
Donut Hole
(SIOFA)*
(Galapagos Agreement)*
(South Pacific Initiative)*
(South Pacific Initiative)*
Challenger Plateau
South Tasman Rise
Global Overview of RFMOs: Straddling Fish Stocks. Evelyne Meltzer. 05/2006
For illustration purposes only.
Map Projection : Robinson
RFMO Boundary
( )* RFMO area under negotiation, not yet adopted or not yet in force.
SEAFO
CCAMLR
NEAFC Convention Area
NEAFC Regulatory Area
NAFO Convention Area
NAFO Regulatory Area
Donut Hole Arrangement
(SIOFA)* (Not yet adopted)
(Galapagos Agreement)* (Not yet in force)
(South Pacific Initiative)* (Under negotiation - preliminary boundary)
Other Unregulated High Seas Areas where Straddling Fish Stocks occur

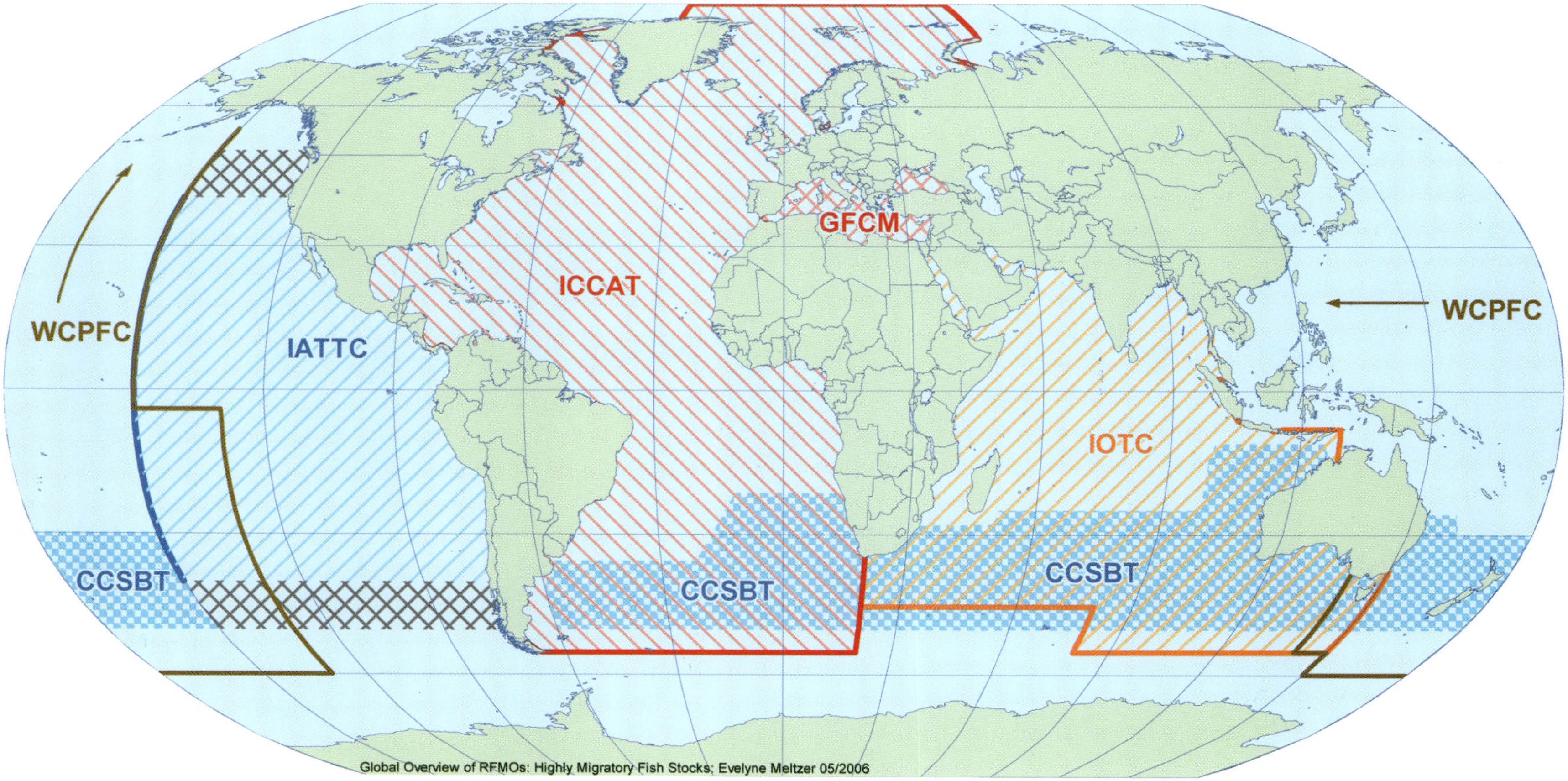

For illustration purposes only.
Map Projection : Robinson

Global Overview of RFMOs: Highly Migratory Fish Stocks (Tuna and Tuna-Like)

- IATTC
- Antigua Convention (not yet in force)
- GFCM
- CCSBT
- ICCAT
- IOTC
- WCPFC

WCPFC Note : Northern boundary and most of Western boundary for RFMO are not defined, and Area is not intended to include waters in South-East Asia which are not part of the Pacific Ocean; nor is it intended to include waters of the South China Sea.

Part I: Legal, Institutional, and Biological Context

Chapter 1
Introduction*

When the United Nations Law of the Sea Convention (LOSC) was adopted in 1982,[1] the high seas fisheries provisions (particularly Articles 63, 64, and 116 to 119) did not satisfactorily or conclusively address two juridical classifications of fish stocks, namely straddling fish stocks and highly migratory fish stocks. With the establishment of the exclusive economic zone (EEZ) regime under LOSC, maritime boundaries were reconfigured, resources were redistributed, and foreign fleets were displaced from waters previously considered international waters. Subsequently many commercial fish stocks were now solely within the EEZs of coastal States. Some fish stocks, however, such as cod, jack mackerel, pollock, orange roughy, and most of the tuna and tuna-like species still had migration patterns and geographic ranges encompassing both high seas and areas under national jurisdiction. These straddling fish stocks (stocks that straddle both the EEZ and the high seas) and highly migratory fish stocks (i.e., tuna and tuna-like species that typically cover several EEZs and vast areas of high seas) were vulnerable to overexploitation, as there was often no cooperative management framework capable of effectively regulating and conserving these international fisheries. Fisheries operations in international waters were often mismanaged, under-regulated, or not regulated. While the traditional freedom of fishing on the high seas had been qualified by conservation objectives and the need for cooperating found in LOSC, a lack of adequate control by flag States and lacuna in flag-State governance threatened the sustainability of international fisheries. Where regional management frameworks did exist, these cooperative efforts were often frustrated by competing interests or blatant disregard for the conservation measures introduced by the regional management organizations. Conflicts periodically arose with the adjacent coastal States, as the portion of the stocks occurring on the high seas was often fished indiscriminately or nonsustainably. Distant-water or high seas fishing nations countered that fisheries operations in domestic waters were not managed responsibly. By the 1992 United Nations Conference on Environment and Development (UNCED), these fish stocks experienced excessive fishing pressure and irresponsible fishing practices. Key straddling fish stocks and highly migratory fish stocks were significantly overfished, some to the point of commercial depletion, causing economic dislocation, food security issues, and environmental concerns. At the 1992 Earth Summit in Rio de Janeiro there was a call for a conference on high seas fishing to overcome this sustainable development problem.[2]

In 1995, after several negotiation sessions, the Agreement for the Implementation of the UN Convention of December 1982 relating to the Conservation and Management of Straddling Fish Stocks and Highly Migratory Fish Stocks was adopted.[3] This Agreement, also known as the UN Fish Stocks Agreement (UNFSA), the UN Fish Agreement (UNFA), the Straddling Stocks Agreement, the Fish Stocks Agreement (FSA), or the New York Agreement, entered into force in December 2001 with the deposit of the 30th instrument of ratification or accession. This book uses the acronym UNFSA, as adopted by the UN, when referring to the Agreement.

UNFSA builds upon, reformulates, and refines LOSC to ensure the long-term conservation, sustainable use, and responsible management of straddling fish stocks and highly migratory fish stocks, and their associated and dependent species. Moreover, UNFSA establishes an effective and enforceable regime to address the sustainability of these stocks, rebuild depleted fish stocks, protect marine biodiversity, protect habitat, and reduce overcapitalization to ensure the sustainable use of living marine resources both within areas under national jurisdiction and on the high seas.[4]

Together with the Agreement to Promote Compliance with International Conservation and Management Measures by Fishing Vessels on the High Seas (the FAO Compliance Agreement) and the FAO Code of Conduct for Responsible Fisheries (FAO Code of Conduct),[5] UNFSA provides a set of norms and principles and minimum international standards for regulating fishery resources and related activities, with an emphasis on the duties of

* The author expresses her appreciation to Professor Martin Tsamenyi for his comments and contribution to an early draft of this introductory chapter.

Table 1.1. Parties to UNFSA[1] and the FAO Compliance Agreement.[2]

State or Entity	UNFSA	Compliance Agreement
Albania		■
Angola		■
Argentina		■
Australia	■	■
Austria	■	*
Bahamas	■	
Barbados	■	■
Belgium	■	*
Belize	■	■
Benin		■
Brazil	■	
Canada	■	■
Cape Verde		■
Chile		■
Cook Islands	■	
Costa Rica	■	
Cyprus	■	■
Czech Republic		*
Denmark	■	*
Egypt		■
Estonia		*
EC[3]	■	■
Fiji	■	
Finland	■	*
France	■	*
Georgia		■
Germany	■	*
Ghana		■
Greece	■	*
Guinea	■	
Hungary		*
Iceland	■	
India	■	
Iran	■	
Ireland	■	*
Italy	■	*
Japan		■
Kiribati	■	
Korea		■
Kenya	■	
Latvia		*
Liberia	■	
Lithuania		*
Luxembourg	■	*
Madagascar		■
Maldives	■	
Malta	■	*
Marshall Islands	■	
Mauritius	■	■
Mexico		■
Micronesia (FSM)	■	
Monaco	■	
Morocco		■
Myanmar		■
Namibia	■	■
Nauru	■	
Netherlands	■	*
New Zealand	■	■
Norway	■	■
Papua New Guinea	■	
Peru		■
Poland	■	*
Portugal	■	*
Russian Federation	■	
St. Kitts & Nevis		■
St. Lucia	■	■
Samoa	■	
Senegal	■	
Seychelles	■	■
Slovakia		*
Slovenia		*
Solomon Islands	■	
South Africa	■	
Spain	■	*
Sri Lanka	■	
Sweden	■	■
Syria		■
Tanzania		■
Tonga	■	
Ukraine	■	
United Kingdom	■	*
USA	■	■
Uruguay	■	■
Total	**57**	**34 (57)***

[1] UNFSA: Agreement for the Implementation of the Provisions of the Convention Relating to the Conservation and Management of Straddling Fish Stocks and Highly Migratory Fish Stocks. Source: United Nations Oceans and Law of the Sea, Division of Ocean Affairs and Law of the Sea. http://www.un.org/Depts/los/reference_files/status2006.pdf, 1 May 2006.

[2] FAO Compliance Agreement to Promote Compliance with International Conservation and Management Measures by Fishing Vessels on the High Seas. Source: FAO http://www.fao.org/Legal/treaties/012s-e.htm, 1 May 2006.

[3] EC accepted FAO Compliance Agreement in 1996 on behalf of all EU member States, which now include Austria, Belgium, Cyprus, Czech Republic, Denmark, Estonia, Finland, France, Germany, Greece, Hungary, Ireland, Italy, Latvia, Lithuania, Luxembourg, Malta, Netherlands, Poland, Portugal, Slovakia, Slovenia, Spain, Sweden, and United Kingdom. Cyprus and Sweden accepted the FAO Compliance Agreement prior to joining the EC and are hence listed separately.

■ Indicates that the Parties have acceded or ratified these legal agreements.

* Indicates Parties to FAO Compliance Agreement under EC.

Evelyne Meltzer: 05/2006

flag States (see Table 1.1 for list of States party to UNFSA and to the FAO Compliance Agreement as of 1 May 2006). The responsibility of coastal States to manage the resources within their EEZs in a sustainable manner is reinforced through the introduction and application of post-LOSC management principles and approaches such as the precautionary approach and the ecosystem approach to fisheries. These post-LOSC principles and standards apply throughout the range of straddling fish stocks and highly migratory fish stocks, both within areas of national jurisdiction and on the high seas.

The FAO Compliance Agreement, the FAO Code of Conduct, and UNFSA reaffirm the jurisdiction of flag States over their vessels fishing on the high seas and elaborate flag-State obligations and responsibilities that qualify the traditional freedom of fishing on the high seas. There is general recognition that UNFSA has assisted "the ongoing process of clarifying the point that the freedom of fishing, as first articulated in conventional form in 1958, is a specifically qualified freedom".[6]

Additionally, UNFSA reinforces the need for cooperative, compatible management between coastal States and flag States by strengthening the roles of both regional fisheries management organizations (RFMOs) and regional fisheries management arrangements (RFMAs). Where there are gaps, new RFMOs or RFMAs should be established to conserve and manage these fish stocks coherently throughout their geographic range. This book generally uses the term RFMOs to include RFMAs when referring to fisheries management organizations responsible for these international fisheries.

With the creation of a network of RFMOs and membership requirements (see Table 1.2 for State participation in RFMOs), the adopted regime for straddling fish stocks and highly migratory fish stocks converts the open-access system to one of regulated participation with associated management measures, including catch restrictions. As a result, this regime endeavours to regulate access and control fishing in international waters.

The contemporary international law of fisheries enjoins all States with a "real interest" in the fishery to participate in and comply with the management regime of the relevant RFMO. Among these are the Bering Sea Donut Hole Arrangement (BSDH), the Commission for the Conservation of Antarctic Marine Living Resources (CCAMLR), the Commission for the Conservation of Southern Bluefin Tuna (CCSBT), the General Fisheries Commission for the Mediterranean (GFCM), the Inter-American Tropical Tuna Commission (IATTC), the International Commission for the Conservation of Atlantic Tunas (ICCAT), the Indian Ocean Tuna Commission (IOTC), the Northwest Atlantic Fisheries Organization (NAFO), the North East Atlantic Fisheries Commission (NEAFC), the South East Atlantic Fisheries Organisation (SEAFO), and the Western and Central Pacific Fisheries Commission (WCPFC).

There is now a large body of international legal instruments (legally binding and non legally binding) and initiatives governing the conservation and management of fisheries within and beyond areas of national jurisdiction. The FAO Code of Conduct, its four International Plans of Action (IPOAs), UNFSA, the FAO Compliance Agreement, and the 2002 World Summit on Sustainable Development (WSSD) Plan of Implementation,[7] together with the Ministerially-led High Seas Task Force on Illegal, Unreported and Unregulated Fishing, reflect the enormous efforts the international community has made to address fisheries conservation issues and enhance the coordinated, cooperative management of world fishery resources.[8]

The world community is increasingly concerned with the overexploitation of fishery resources, the impact of fishing operations on ocean ecosystems, and the need for ensuring that all fishing is conducted responsibly. There is much we do not know about ocean ecosystems and the serious or irreparable harm to marine resources that has resulted from mismanagement and a lack of regulations. UNFSA provides a precautionary, science-based, ecosystemic governance framework for the responsible regulation of straddling fish stocks and highly migratory fish stocks. Its significance in the progressive development of international fisheries law and in the recognition of the need for considering the broader environmental dimension in fishing operations cannot be underestimated. UNFSA reinforces and supplements the treaty-based regime to take into account the realities of international fisheries and the conservation imperative. The dominant interest of the international community is sustainable development of natural resources, which are of global concern and "in whose protection all states have a legal interest, and which all states have a duty to preserve from serious harm".[9] With UNFSA the world has moved beyond freedom of fishing on the high seas to a regime that is responsible, precautionary, and regulated to ensure intergenerational sustainability and biodiversity.

This monograph, produced to provide input into the UNFSA Review Conference of May 2006, provides a thematic overview of the extent to which RFMOs with competence over straddling or highly migratory fish stocks have adopted and implemented the international legal regime

Table 1.2. Participation in RFMOs for straddling fish stocks and highly migratory fish stocks.

RFMO Participation	ICCAT	GFCM	IATTC	WCPFC	IOTC	CCSBT	NAFO	NEAFC	SEAFO	BSDH	CCAMLR
Albania		▒									
Algeria	▒	▒									
Angola	▒								▒		
Argentina											▒
Australia				▒	▒	▒					▒
Barbados	▒										
Belgium											▒
Belize	▒										
Brazil	▒										▒
Bulgaria		▒					▒				█
Canada	▒		█	▒			▒	█			▨CDS
Cape Verde	▒										
Chile											▒
China	▒		█	▒	▒					▒	CDS
Comoros					▒						
Cook Islands				▒							█
Costa Rica			▒								
Côte d'Ivoire	▒										
Croatia	▒	▒									
Cuba							▒				
Cyprus		▒									
Denmark							▒	▒			
EC / EU	▒	▒	█	▒	▒		▒	▒	▒		▒
Ecuador			▒								
Egypt		▒									
El Salvador			▒								
Equatorial Guinea	▒										
Eritrea					▒						
Estonia								▒			
Fiji Islands				▒							
Finland											█
France	▒	▒	▒	▒	▒		▒				▒
Gabon	▒										
Germany											▒
Ghana	▒										
Greece		▒									█
Guatemala	▒		▒								
Guinea Conakry	▒				▒						
Guyana	█										
Honduras	▒		█								
Iceland	▒						▒	▒			
India					▒						▒
Indonesia				█	█						
Iran					▒						
Israel		▒									
Italy		▒									▒
Japan	▒	▒	▒	▒	▒	▒	▒	█		▒	▒
Kenya					▒						
Kiribati				▒							
Korea	▒		▒	▒	▒	▒	▒			▒	▒
Lebanon		▒									
Libya	▒	▒									
Madagascar					▒						
Malaysia					▒						
Malta		▒									
Marshall Islands				▒							
Mauritius					▒						▨CDS
Mexico	▒		▒								
Micronesia (FSM)				▒							
Monaco		▒									
Morocco	▒	▒									
Namibia	▒								▒		▒
Nauru				▒							
Netherlands	█										█
New Zealand				▒		▒		█			▒
Nicaragua	▒		▒								
Niue				▒							
Norway	▒						▒	▒	▒		▒
Oman					▒						
Pakistan					▒						
Palau				▒							
Panama	▒		▒								
Papua New Guinea				▒							
Peru			▒								▨CDS
Philippines	▒			▒	▒	█					
Poland								▒		▒	▒
Romania		▒									
Russian Federation	▒						▒	▒		▒	▒
Samoa				▒							
Sao Tomé and Principe	▒										
Senegal	▒										
Serbia and Montenegro		▒									
Seychelles					▒						CDS
Singapore											CDS
Slovenia		▒									
Solomon Islands				▒							
South Africa	▒				█						▒
Spain		▒	▒								▒
Sri Lanka					▒						
Sudan					▒						
Sweden											▒
Syria	▒	▒									
Taiwan (Chinese Taipei)[1]	█		█	▒ M		[M]					
Thailand					▒						
Tonga				▒							
Trinidad and Tobago	▒										
Tunisia	▒	▒									
Turkey	▒	▒									
Tuvalu				▒							
Ukraine							▒				▒
United Kingdom	▒				▒						▒
Uruguay	▒										▒
USA	▒		▒	█			▒			▒	▒
Vanuatu	▒		▒	▒	▒						█
Venezuela	▒		▒								
TOTAL members	**42**	**24**	**15**	**24**	**24**	**4**	**13**	**7**	**4**	**6**	**24**

[1] The Fishing Entity of Taiwan also known as Chinese Taipei (or within the UN system as Taiwan Province of China) is a member ▒ M of WCPFC and is a member [M] of CCSBT's Extended Commission.

▒ Contracting Parties

█ Cooperating States, Entities or Fishing Entity

CDS Nonmember participants in CCAMLR's catch documentation scheme (CDS)

▨CDS States party to CDS who have acceded to CCAMLR

Evelyne Meltzer: 05/2006

established by UNFSA and a range of other instruments in four thematic areas, namely (1) avoiding overfishing and restoring fish stocks through the precautionary approach; (2) introducing the ecosystem approach to fisheries management; (3) improving decision making; and (4) strengthening monitoring, control, and surveillance and enforcement. These four thematic areas are examined in Part II of this book. The nine substantive chapters are summarized below.

Chapter 2 lays the foundation for the study by providing an overview of the international legal framework pertaining to the conservation and management of straddling fish stocks and highly migratory fish stocks. From the time of Grotius (often referred to as the father of the law of the sea) in the early 1600s until the adoption of UNFSA, the concept of the freedom of the seas has dominated world fisheries. The concept is based on the natural law theory that the high seas is a global commons owned by all, to be used by all. Fisheries were considered inexhaustible, unbounded, and a common property resource, free of ownership until harvested. Governing the commons and common property resources is one of the fundamental challenges underlying high seas fishing. This theory, combined with the technological advances and nonsustainable fishing practices, has led to stock overexploitation and demands for new legal arrangements to regulate activities on the high seas and within areas under national jurisdiction. The chapter demonstrates that the international legal framework for fisheries is a complex, interrelated, and evolving system of "hard" law (legally binding) and "soft" law (non legally binding).

Contemporary international fisheries law for straddling fish stocks and highly migratory fish stocks encompasses three key legally binding global treaties that address the establishment of the new regime and minimum international standards for the conservation and responsible management of fish stocks, (i.e., LOSC, UNFSA, and the FAO Compliance Agreement). Additionally, an increasing number of international voluntary (soft law) instruments (in particular the FAO Code of Conduct); regional fisheries management instruments; and finally, global biodiversity, trade, and environmental instruments compose a suite of related instruments that despite being negotiated outside the international fisheries management framework provide a complementary set of principles, standards, and tools to supplement the fisheries-specific instruments in the quest for achieving sustainable, coherent, and cooperative fisheries management. The chapter provides an examination of the scope, content, and some of the interrelationships among these instruments.

Chapter 3 addresses the juridical classification of fish stocks, particularly straddling fish stocks and highly migratory fish stocks, under LOSC and UNFSA. The political and legal division of the oceans into various zones of jurisdiction and resulting maritime boundaries demarcate aquatic ecosystems and the resources they support into areas under national jurisdiction and the area within international waters known as the "high seas". The widespread adoption and establishment of EEZs or 200 nautical mile fisheries zones did not resolve the management of living marine resources whose ranges are not confined to the spatial and legal boundaries of one coastal State but also occur in the high seas. UNFSA was designed to address this problem, with particular focus on two juridical categories of fish stocks, namely straddling fish stocks and highly migratory fish stocks. The chapter begins with a discussion of the terminology associated with the juridical classification of fish stocks, the governance implications associated with defining these fish stocks, and some related issues. The chapter also discusses two other classifications of fish stocks: joint or shared stocks, now commonly referred to as transboundary stocks; and discrete or "purely" high seas fish stocks.

Chapter 4, which flows from the international legal framework discussed in Chapters 2 and 3, discusses the international governance of straddling fish stocks and highly migratory fish stocks. The governance framework for these two juridical classifications of fish stocks is based on the framework established by LOSC, further elaborated in UNFSA, and supported by the FAO Code of Conduct and the FAO Compliance Agreement, as well as by the host of soft law and other environmental instruments discussed in Chapter 2. UNFSA provided the opportunity to further strengthen and promote the regional approach to fisheries governance in high seas "common" areas of international interest. The institutional construct of RFMOs and the establishment of a network of RFMOs to responsibly regulate straddling fish stocks and highly migratory fish stocks in a compatible manner throughout their geographic range are effective mechanisms for overcoming the tragedy of the commons and ensuring coherent management. The chapter focuses on RFMOs as the "vehicle of good governance" for implementing the international conservation and management regime for straddling fish stocks and highly migratory fish stocks. It is principally through strengthening these RFMOs and the collaborative efforts of their Contracting Parties or members that the provisions of UNFSA can be implemented and improved fisheries governance, resource recovery, and sustainability realized. This chapter concludes with an overview of the concerted efforts to increase participation in

RFMOs in accordance with UNFSA and to control access for fishing stocks under the competence of RFMOs, together with a brief discussion of some of the issues associated with participation.

Chapter 5 provides an overview of fisheries reference points, particularly maximum sustainable yield (MSY) and related concepts. For centuries, the overexploitation of fish stocks was inconceivable not only because the supply of fish in the sea was considered inexhaustible but also because it was thought that a marked reduction in abundance would render commercial fishing uneconomical. For several decades the issue of overfishing has been at the centre of international fisheries management. The chapter discusses the terminology associated with overfishing and demonstrates that since the adoption of LOSC in 1982, the avoidance and reduction of overfishing and the obligation to maintain or restore populations of harvested species at levels that can produce MSY have become interrelated and fundamental objectives of the regime governing fisheries management. The chapter discusses the evolution of the concept of MSY, its shortcomings, and its formulation under UNFSA. The concept of maximum economic yield and the associated issues of overcapacity and subsidization are also briefly discussed. Despite international efforts, most recovery and rebuilding plans implemented in the last decade have not yet met the rebuilding objectives for the fish stocks concerned, mainly as a result of overfishing, although bycatch, environmental effects, habitat destruction, and pollution can also be contributing factors. The chapter also overviews the status of key commercial straddling fish stocks and highly migratory fish stocks regulated by RFMOs.

Chapter 6 provides analysis of the precautionary approach, a major innovation of the UNFSA regime in fisheries management. With the recognition of the high degree of uncertainty inherent in fisheries management, and in an effort to minimize the risk of serious or irreversible harm to fish stocks and the marine ecosystem associated with fishing operations, the precautionary approach to fisheries was introduced and embedded in UNFSA and the FAO Code of Conduct. It is a fundamental tenet of the precautionary approach that fishing operations can adversely and sometimes irreparably impact fish stocks and the marine ecosystem. Rigorous risk management, erring on the side of conservation, is therefore imperative in the quest for robust, sustainable fisheries and inter-generational equity. Under the approach, precaution commensurate to the risk should be applied to all aspects of fisheries management, corrective measures should be adopted within a reasonable time frame, conservation problems should be anticipated, and undesirable outcomes should be avoided where possible. If proposed or ongoing activities may result in serious or irreversible harm, unless proven otherwise, all reasonable efforts must be made to mitigate or prevent such undesirable consequences from occurring. In some cases the precautionary approach would preclude authorizing fishing operations or require the cessation of fishing.[10] The chapter begins with a brief outline of the history and definition of the precautionary approach, its specific applications in determining reference points and preagreed decision rules, as well as its wider application in fisheries management decision making. By way of illustration, efforts to apply the precautionary approach by RFMOs are discussed, followed by observations on some of the challenges in implementing the concept as well as on opportunities to integrate it into management of straddling fish stocks and highly migratory fish stocks. The central theme in Chapter 6 is that if the objectives of UNFSA are to be realized, the precautionary approach must be applied widely both within and beyond areas of national jurisdiction. This means that the approach will be perceived not solely as a scientific exercise for developing precautionary reference points, but as part of an overarching risk-averse, science-based regulatory appraisal process, whereby managers err on the side of conservation in making management decisions. As David Freestone observed, "The very survival of many capture fisheries may well depend on it".[11] The introduction and implementation of the precautionary approach signal a critical development in international fisheries law, incorporating the broader environmental dimension and conservation imperative in the management paradigm.

Chapter 7 is devoted to a discussion of the ecosystem approach to fisheries, which recognizes the dynamic, complex, and systemic interrelationships among living organisms and between these organisms and their physical environment, as well as the ecological influences on the variety and variability of living organisms and species. The chapter discusses the legal basis for the use of the ecosystem approach in fisheries management. As well, the associated terminology of ecosystem-based management and ecosystem-based fisheries management are discussed. This chapter assesses progress in the implementation of the ecosystem approach by RFMOs and some of the practical challenges to the implementation of the concept in international fisheries management.

Chapter 8 discusses aspects of the decision-making process associated with management of straddling fish stocks and highly migratory fish stocks by RFMOs. The issues addressed in the chapter include the nature of RFMO prescriptive authority; State sovereignty and flag-State responsibility; RFMO practices in decision making (voting, unanimity, consensus); RFMO opt-out provisions, including efforts made to qualify RFMO objection procedures; and decision-making processes adopted by RFMOs. The chapter then highlights some of the new principles introduced under UNFSA that qualify how decisions are made to ensure that fisheries are managed in a sustainable and coherent manner throughout the geographic range of the harvested species. The new principles discussed include the precautionary approach, the ecosystem approach to fisheries management, transparency, and compatibility. UNFSA recognizes that no matter how elaborate decision-making mechanisms may be, there is always the possibility of disputes and conflicts. Where disputes arise, a fair and timely process is required to reach agreement through peaceful means and to prevent impasses from compromising the conservation and management of target stocks and nontarget species. The chapter concludes with a brief discussion of peaceful settlement of disputes including compulsory and binding dispute resolution as provided for in UNFSA and LOSC, and how RFMOs have adopted and used these dispute settlement mechanisms.

Chapter 9, the final thematic chapter, is devoted to a discussion of monitoring, control, and surveillance (MCS) and enforcement, which are essential components of the international legal framework governing the sustainable management of straddling fish stocks and highly migratory fish stocks. The chapter begins with a discussion of the concepts of MCS and enforcement and the international legal framework supporting them. This is followed by an overview of the comprehensive, practical, and innovative measures provided for under UNFSA to address MSC and enforcement. The chapter examines the practice of RFMOs in implementing these measures. Practical MCS and enforcement measures discussed in the chapter include fishing vessel registries; data collection and reporting; trade-related measures; regulation of transshipment; observer programs; aerial surveillance; satellite-based vessel monitoring systems; vessel detection systems; at-sea boarding and inspection; requirements on the use of force; and port-State measures. The central message in Chapter 9 is that the objective of MCS measures is to ascertain what and how much is being caught, where, when, and by whom, to collect required scientific data, inform management decisions, and determine whether RFMO management measures are being undermined. To be effective, MCS must be accompanied by meaningful enforcement, including appropriate follow-up prosecution and punishment that serves as a deterrent and promotes widespread compliance. Noncompliance with RFMO conservation and management measures whether by authorized vessels or those of nonmembers poses a serious systemic threat to rebuilding depleted stocks and achieving sustainable international fisheries.

This publication overviews the state of implementation of UNFSA by the relevant RFMOs up to but not including the May 2006 Review Conference. The Afterword provides an update of key events from the 2006 Review Conference to May 2008.

This book is composed of three sections, with Part I containing those chapters that set the context, namely Chapters 1 to 5 (introduction, the international legal framework, the juridical classification of fish stocks, international governance of straddling fish stocks and highly migratory fish stocks, and a discussion of overfishing, overcapacity, and the status of some major fish stocks); Part II containing the thematic chapters, namely Chapters 6 to 9 (the precautionary approach; the ecosystem approach to fisheries; decision making in RFMOs; and MCS and enforcement). The Afterword completes Part II. Part III is made up of the matrices and maps. These materials in Part III provide the basis for a concise, detailed comparison of the scope, functions, and activities of the different RFMOs. The Summary found at the front of the book can serve as a companion to Part III or as an overview of the book.

Appended for easy reference are the full texts of UNFSA, the FAO Compliance Agreement, and the FAO Code of Conduct for Responsible Fisheries.

Endnotes

[1] *United Nations Convention on the Law of the Sea*, done at Montego Bay, Jamaica, 10 December 1982, entered into force 16 November 1994, 33 ILM 1309.

[2] For overview of problem associated with managing straddling fish stocks and highly migratory fish stocks prior to 1995 adoption of UNFSA please see E. Meltzer, "Global Overview of Straddling and Highly Migratory Fish Stocks: The Nonsustainable Nature of High Seas Fisheries" (1994) 25 *Ocean Development and International Law* 255–344.

[3] *Agreement for the Implementation of the Provisions of the United Nations Convention on the Law of the Sea of 10 December 1982 Relating to the Conservation and Management of Straddling Fish Stocks and Highly Migratory Fish Stocks*, done at New York, 4 December 1995, entered into force 11 December 2001, 2167 UNTS 88; (1995) 34 ILM 1542. For the purposes of this book, the acronym for this Agreement will be UNFSA. UNFSA aims to strengthen relevant provisions of LOSC, particularly Arts. 63, 64, and 116–119, which address the management and conservation of straddling fish stocks and highly migratory fish stocks. Together with the FAO Code of Conduct, UNFSA provides "a framework for the practical implementation of effective fisheries conservation and management. One of its main objectives is to progressively facilitate rational and long-term sustainable utilization of fisheries resources both on the high seas and in waters under national jurisdiction. It therefore accords high priority to ensuring effective co-operation between coastal States and high seas fishing States on a range of fundamental and technical issues". D. Miller and E. Molenaar, "The SEAFC Convention: A Comparative Analysis in a Developing Coastal State Perspective" in A. Chircop, S. Coffen-Smout, and M. McConnell, eds., *Ocean Yearbook 20* (Ardsley, N.Y.: Transnational Press, 2006) 305–375 at 305.

[4] As W. Burke notes in "Evolution in the Fisheries Provisions of UNCLOS" in N. Ando, E. McWhinney, R. Wolfrum, and B. Röben, eds., *Liber Amicorum Judge Shigeru Oda* (The Hague: Kluwer Law International, 2002) at 1355, "While in the last decade, most of the attention on the fisheries provisions of UNCLOS focused upon high seas fisheries, particularly in connection with straddling and highly migratory fish stocks, some developments directly concern coastal State authority over its EEZ fisheries".

[5] *Agreement to Promote Compliance with International Conservation and Management Measures by Fishing Vessels on the High Seas*, done at Rome, Italy, 24 November 1993, entered into force April 2003, (1994) 33 ILM 968. Also known as Flagging Agreement. Referred to as FAO Compliance Agreement. The FAO Compliance Agreement forms an integral part of the *FAO Code of Conduct for Responsible Fisheries* (Rome: FAO, 1995). See Part III of this book for reproduced version of text of these Agreements.

[6] D. Anderson, "The Straddling Stocks Agreement of 1995 — An Initial Assessment" (1996) 45 *International and Comparative Law Quarterly* 463–475 at 475.

[7] The 2002 Johannesburg Political Declaration on Sustainable Development and Plan of Implementation of the World Summit on Sustainable Development (WSSD-POI), United Nations 2003. Many of the WSSD–POI fisheries provisions echo the four FAO IPOAs that were adopted within the framework of the Code of Conduct prior to WSSD.

[8] See FAO, "Towards Ecosystem-based Fisheries Management" in M. Sinclair and G. Valdimarsson, eds., *Responsible Fisheries in the Marine Ecosystem* (Wallingford: CABI Publishing, 2003) 393-403 at 399. "The fisheries management context and framework have greatly improved through a range of initiatives at global, regional, and national levels. Overfishing and excess fleet capacities have been recognized generally (such as at UNCED and FAO) as being worldwide problems that call for socially acceptable and effective solutions. With the coming into force of the 1982 Convention on the Law of the Sea, the fisheries policy framework has become stronger and was reinforced by the adoption in 1995 of the FAO Code of Conduct for Responsible Fisheries. It [is] strengthened further by the coming into force of the 1993 FAO Compliance Agreement and the 1995 UN Fish Stocks Agreement. The Code has been complemented by a series of technical guidelines, including on fisheries management, on indicators for sustainable development of marine capture fisheries, and on the precautionary approach to capture fisheries and species introductions. Implementation of the Code will be also strengthened by the four International Plans of Action (IPOAs) recently adopted by FAO members: IPOA for the Management of Fishing Capacity; IPOA for the Conservation and Management of Sharks; IPOA for Reducing Incidental Catch of Seabirds in Longline Fisheries; and IPOA to Prevent, Deter and Eliminate Illegal, Unreported and Unregulated Fishing. The inherent existence of uncertainty and risk is formally recognized in all modern fisheries agreements…". These instruments will be further discussed in Chap. 2, which overviews the legal framework pertaining to straddling and highly migratory fish stocks.

[9] A. Boyle, "International Law and the Protection of the Global Atmosphere: Concepts, Categories and Principles" in R. Churchill and D. Freestone, eds., *International Law and Global Climate Change* (The Hague: Kluwer Law International, 1991) 7–19 at 19.

[10] S. Garcia, "Precautionary Approach to Fisheries Management", available at <http://www.oceansatlas.org/world_fisheries_and_aquaculture/html/govern/capture/precappr/default.htm> (accessed January 2008).

[11] D. Freestone, "Implementing Precaution Cautiously: The Precautionary Approach in the Straddling and Highly Migratory Fish Stocks Agreement" in E. Hey, ed., *Developments in International Fisheries Law* (The Hague: Kluwer Law International, 1999) 287–325 at 325.

Chapter 2

Overview of International Legal Instruments Pertaining to Straddling Fish Stocks and Highly Migratory Fish Stocks*

2.1 Introduction

The United Nations Fish Stocks Agreement (UNFSA), the Food and Agriculture Organization (FAO) Compliance Agreement, and the nonbinding FAO Code of Conduct for Responsible Fisheries are the principal international instruments concerning the conservation and management of straddling fish stocks and highly migratory fish stocks.These three instruments clarify, build upon, and implement the 1982 UN Convention on the Law of the Sea (LOSC)[1] to ensure a responsible, sustainable, and regulated fishery throughout the geographic range of these fish stocks. Additionally, many other binding instruments reinforce and influence the progressive development of the international legal regime for these two juridical classifications of fish.

The growing body of nonbinding, or soft law, instruments relating to high seas fisheries significantly informs and continues to drive the progressive development of contemporary international fisheries law. With the widespread acceptance of principles and practices agreed to through these nonbinding processes, soft law not only informs hard law but can also play a part in shaping customary international law. It is this legal framework of hard law and soft law that underpins the international governance of straddling fish stocks and highly migratory fish stocks. It is incumbent upon States within their maritime zones and within regional fisheries management organizations (RFMOs) to adopt and implement this regime to ensure the sustainable development and the responsible management of these fish stocks.

This chapter provides an overview of the international legal framework and related instruments pertaining to straddling fish stocks and highly migratory fish stocks and will outline developments in international fisheries law and international governance to promote sustainable development since the adoption of LOSC and in the post-UNFSA period up to the May 2006 Review Conference.

2.2 The international legal framework: hard law, soft law, and customary international law

International fisheries law, a subset of public international law,[2] is a complex and evolving system of both legally binding (hard law) and non legally binding instruments (soft law). Figure 2.1 provides an overview of the legal instruments affecting international fisheries management of straddling fish stocks and highly migratory fish stocks. Hard law refers to legally binding agreements that are typically embodied in international treaties and that require parties to behave in accord with their stipulations. They are the product of negotiations among States and, together with customary rules, are the basis of binding international law.[3]

In recent years, there has been an increasing use of non legally binding soft law,[4] which refers to nonbinding general norms or principles not readily enforceable through binding dispute resolution.[5] Soft law instruments take on many different forms, such as declarations of intergovernmental conferences, UN General Assembly (UNGA) resolutions, codes of conduct, guidelines, resolutions, and recommendations of international organizations.[6] Soft law instruments may be seen as both an alternative to and a part of the process of multilateral treaty making.[7] Soft law can be a precursor or catalyst of hard law, or reinforce and strengthen existing hard law instruments as well as prompt the formation of customary international law.

Soft law instruments may be used as mechanisms for authoritative interpretation or amplification of the terms of a treaty (e.g., through UNGA resolutions, recommendations and decisions of international organizations, and statements of conferences of parties)[8] or to provide the detailed rules and technical standards required for the implementation of treaties, e.g., the FAO Code of Conduct

* The author gratefully acknowledges the research assistance of Leonie Renwrantz, Jay L. Batongbacal, David Dzidzornu, and Lowell Bautista in preparing early drafts of this chapter.

Figure 2.1. International fisheries law and related instruments pertaining to straddling fish stocks and highly migratory fish stocks.

Hard Law

1946 International Convention for the Regulation of Whaling (ICRW), in force 1948*

1973 Convention on International Trade in Endangered Species of Wild Fauna and Flora (CITES), in force 1975

1979 Convention on the Conservation of Migratory Species of Wild Animals, in force 1983

1982 **Law of the Sea Convention (LOSC)**, in force 1994

1992 Convention on Biological Diversity (CBD), in force 1993

1993 **FAO Compliance Agreement**, adopted by FAO Conference; forms an integral part of 1995 FAO Code of Conduct, in force 2003

1994 Agreement Establishing the World Trade Organization (WTO), in force 1995

1995 **UN Fish Stocks Agreement (UNSFA)**, in force 2001

2001 Agreement on the Conservation of Albatrosses and Petrels (ACAP), in force 2004

Regional Fisheries Management Agreements

(establishing RFMOs and RFMAs)

Agreement for the Establishment of a General Fisheries Council for the Mediterranean, 1952 (GFCM) updated establishing General Fisheries Commission for the Mediterranean, 1997 (GFCM)

Convention for the Establishment of an Inter-American Tropical Tuna Commission, 1950, updated by Antigua Convention not in force (IATTC)

International Convention for the Conservation of Atlantic Tunas, 1969 (ICCAT)

Convention on Future Multilateral Cooperation in the Northwest Atlantic Fisheries, 1979 (NAFO)

Convention on Future Multilateral Co-Operation in North-East Atlantic Fisheries, 1982 (NEAFC)

Convention on the Conservation of Antarctic Marine Living Resources, 1982 (CCAMLR)

Convention for the Conservation of Southern Bluefin Tuna, 1994 (CCSBT)

Convention for the Conservation and Management of Pollock Resources in the Central Bering Sea, 1995 (Donut Hole)

Agreement for the Establishment of the Indian Ocean Tuna Commission, 1996 (IOTC)

Agreement between Iceland, Norway, and Russia Concerning Certain Aspects of Cooperation in the Area of Fisheries, 1999 (Loop Hole)

The Agreement on the International Dolphin Conservation Program, 1999, last amended 2005 (IDCP)

Arrangements between the Government of Australia and the Government of New Zealand for the Conservation and Management of Orange Roughy on the South Tasman Rise, 1998 and 2000 (STR)

Framework Agreement for the Conservation of the Living Marine Resources on the High Seas of the South Pacific (Galapagos Agreement), adopted 2000, not in force

The Provisional Agreement, 2001, (Southeast Pacific Swordfish Arrangement)

Convention for the Strengthening of the Inter-American Tropical Tuna Convention, signed 2003, not in force (IATTC - Antigua Convention)

Convention on the Conservation and Management of Fishery Resources in the South East Atlantic Ocean, 2003 (SEAFO)

Convention on the Conservation and Management of Highly Migratory Fish Stocks in the Western and Central Pacific Ocean, 2004 (WCPFC)

Southern Indian Ocean Fisheries Agreement, draft completed 2005, not yet formally adopted and signed (SIOFA)

South Pacific Commission, proposed at COFI March 2005, negotiating session February 2006

Soft Law

UNGA Resolutions, including: 46/215 Drift Net Moratorium; 49/118 Bycatch and Discards; 54/32 IUU; 58/14 and 59/L.23 UNFA Implementation (IUU, Ecosystem Approach, Precautionary Approach, fishing overcapacity, impacts of fishing on vulnerable marine ecosystems); 60/31Sustainable fisheries

1992 UNCED: Agenda 21; Rio Declaration on Environment and Development

1995 Rome Consensus on World Fisheries

1995 Jakarta Mandate (CBD)

1999 FAO Rome Declaration on Implementation of Code of Conduct

2001 Reykjavik Declaration on Responsible Fisheries in the Marine Ecosystem

2002 Johannesburg Declaration on Sustainable Development

2002 WSSD Plan of Implementation, setting targets: restore depleted stocks (2015); MPA-network (2012); Ecosystem Approach (2010); NPOA-Capacity (2005); and NPOA-IUU fishing (2004).

2005 Rome Declaration on Illegal, Unreported and Unregulated Fishing

2005 Ministerial Declaration of the St. John's Conference on the Governance of High Seas Fisheries and UN Fisheries Agreement

2005 APEC Joint Ministerial Declaration for Bali Plan of Action -Towards Healthy Oceans and Coasts (managing living resources sustainably)

FAO Code of Conduct for Responsible Fisheries, 1995

FAO Technical Guidelines

- Precautionary Approach, No. 2, 1996
- Fishing Operations, No. 1, 1996
- Fisheries Management, No. 4, 1997
- Sustainable Development Indicators, No. 8,1999
- IUU, No. 9, 2002
- Ecosystem Approach, No. 4, Suppl. 2, 2003

International Plans of Action

- IPOA-Seabirds, 1999
- IPOA-Sharks, 1999
- IPOA-Capacity, 1999
- IPOA-IUU, 2001

Strategy for Improving Information on Status and Trends in Capture Fisheries, 2003

- Fishery Resources Monitoring System (FIRMS)

* Year preceding name of instrument is date of adoption. Year following name of instrument is when it came into force.

for Responsible Fisheries.[9] The advantage of regulating international standards in this manner is that detailed rules may be easily amended or strengthened as the need arises.[10] Similarly, soft law can elaborate upon and further clarify standards found in hard law, such as in LOSC Article 119 (1)(a), which provides that States shall take into account "...any generally recommended international minimum standards...". LOSC also incorporates the recommendations and resolutions of the International Maritime Organization (IMO) and other international organizations under provisions requiring or permitting States to adopt "generally accepted [international] rules and standards established through the competent international organisation or general diplomatic conference".[11] Soft law instruments may themselves be supported and given binding force by treaties through incorporation by reference. For example, the Inter-American Tropical Tuna Commission (IATTC) used the FAO Code of Conduct as the key instrument of reference when negotiating the IATTC - Antigua Convention.[12]

Soft law is an increasingly effective mode of facilitating international dealings and strengthening cooperation to secure international agreement on global, regional, and transboundary concerns. Even though soft law instruments such as Agenda 21, the Rio Declaration, the FAO Code of Conduct, the World Summit on Sustainable Development (WSSD), and UNGA resolutions are not legally binding, they have been nonetheless arduously negotiated by the world community[13] in an effort to move in the direction of a more sustainable, responsible society. As Juda aptly states:

> While not binding, and thus unlike treaties in this important respect, soft law instruments are also the product of careful negotiations and may have substantial significance as a reflection of emerging international consensus and general statement of principles that may later be developed and incorporated into future treaties or emerge as norms of customary law over time. ...Such realities explain why states exert efforts to influence the nature and wording of nonbinding soft law instruments.

These voluntary instruments are also an increasingly effective means of achieving the compliance of States and international actors with the international framework to achieve sustainable and responsible fisheries.[14] Because of the nonbinding character of soft law instruments, some States are more inclined to cooperate in the development and conclusion of soft law instruments[15] and, as a result, are more likely to comply. While there are many factors that influence State practice, the extent to which soft law and arguably hard law are implemented depends largely on the political will of individual governments as they assess the relevance and relative benefits of the agreed laws to their perceived national interests. Pressure from citizens, industry, nongovernmental organizations (NGOs), intergovernmental organizations (IGOs), and other countries can also influence State practice and the degree of compliance.

In addition to treaties, rules of customary international law are binding on States and therefore considered hard law. Customary international law "... consists of the rules of law derived from the consistent conduct of States acting out of the belief that the law required them to act that way".[16] Customary international law can be discerned by a widespread repetition by States of similar international acts over time (State practice); acts must occur out of a sense of obligation (*opinio juris*); acts must be taken by a significant number of States and not be rejected by a significant number of States.[17] Both hard law and soft law may serve as evidence of customary international law. Nonbinding instruments can generate widespread and consistent State practice or provide evidence of *opinio juris* and the current state of the law. In conjunction with treaty norms, soft law instruments may lead to the emergence of a rule of customary international law.[18] Both hard law and soft law instruments can focus consensus on rules and principles and mobilize consistent, general response on the part of States.[19]

Under international treaty law, nonparties have neither duties nor rights under a treaty. When State practice transforms treaty law into customary international law, all States including nonparties are bound by the customary rules.[20] Therefore, there is the potential for the coexistence of customary international law and treaty law rules, both of which may be binding on a State.[21] The debate continues as to whether certain nonbinding norms and principles pertaining to international fisheries law have achieved the status of customary international law and are therefore binding on all States. For example, some authors suggest that the duty to cooperate in respect of high seas fisheries through the medium of RFMOs pursuant to Article 8(3) of UNFSA may already have evolved into customary international law.[22] Another principle is the obligation of all States to effectively control vessels flying their flag. The practices of both party and nonparty States indicate a growing acceptance of the flag-State responsibilities outlined in UNFSA and the FAO Compliance Agreement, regardless of whether the States are members of RFMOs.[23] Many would contend that the precautionary approach or the ecosystem approach to fisheries also have achieved the status of customary international law. Molenaar states that RFMOs "can by now quite safely rely on customary international law in applying concepts like the precautionary approach".[24]

Increasingly, the state of implementation and compliance with international instruments related to fisheries is monitored by international agencies. For example, since the adoption of LOSC in 1982, UNGA has held annual reviews based on annual comprehensive reports on the law of the sea prepared by the Secretary-General. Under an expanded mandate[25] the Secretary-General has been preparing annual reports on oceans and the law of the sea since 1996.[26] In addition to general information on the conservation and management of living marine resources contained in these annual reports, the Secretary-General prepares reports on particular issues relating to fisheries[27]and since 2004 has prepared an annual report on sustainable fisheries. As well, the Commission on Sustainable Development (CSD) monitors the implementation of Agenda 21 and reports annually to UNGA.[28] Members of FAO as well as RFMOs and some NGOs report biennially to the FAO Committee on Fisheries (COFI) on the implementation of the Code. In addition, the FAO FishCode Programme[29] assists in the implementation of the Code of Conduct through the provision of technical assistance.

The introduction of reporting and monitoring of State and RFMO implementation of these instruments facilitate and strengthen ongoing cooperation and assist international actors in meeting their international obligations. These processes, while imperfect and time consuming, reinforce the international framework and evolving global standards, as well as encourage participants to review and account for their performance on a continuing basis.

2.3 Progressive development of international fisheries law — post-LOSC

Public international law is not a static body of law, but a dynamic process evolving in response to global issues, changing socioeconomic and environmental conditions, and changing values. While the regime of the law of the sea was developed over centuries, dramatic changes have taken place in the last three decades in the international legal framework: We have seen the emergence of a new generation of fisheries agreements[30] on the use of coastal and oceans areas both within national jurisdiction and on the high seas. LOSC is one of the most dramatic illustrations of the progressive development of international law[31] and of the ongoing efforts of the international community to adopt measures for the sustainable and responsible use of the sea. Considered the "constitution of the oceans", LOSC codified aspects of the traditional law of the sea and introduced new concepts in international law pertaining to the conservation and management of living marine resources. LOSC will also serve as the international basis for future action.[32] With the advent of the exclusive economic zone (EEZ) and continental shelf regimes, the Convention radically altered and reconfigured maritime interests, maritime boundaries, and the economic distribution of marine resources. The rights, duties, and interests of coastal States over natural resources were significantly expanded through the creation of the EEZ extending to a maximum of 200 nautical miles from the baseline from which the territorial sea is measured, and on the continental shelf up to 200 nautical miles, and beyond in the case of coastal States with extended continental margins. This expanded area under national jurisdiction provided coastal States with the opportunity (as well as the challenge) to regulate ocean space and marine resource use in a more integrated fashion.

This partitioning of the oceans dramatically reduced the area of high seas subject to the *mare liberum* (freedom of the seas) doctrine espoused by Hugo Grotius in 1609[33] and excluded many foreign fishing fleets from areas historically fished as international waters.[34] This regime did not, however, resolve the fragmented jurisdiction and inchoate management of large marine ecosystems and associated living resources that straddle areas of the high seas and areas under national jurisdiction. The need for conserving and coherently managing straddling fish stocks and highly migratory fish stocks throughout their range was not adequately addressed by LOSC. The Convention did not reconcile the respective rights and obligations of coastal States and foreign fishing nations in a clear legal framework. Recognizing the biological unity of these fish stocks and the need for cooperative arrangements, LOSC requires parties to cooperate through "appropriate subregional or regional organizations".[35] With the extension of maritime jurisdiction, several RFMOs were established or reconstituted to manage straddling fish stocks in anticipation of or following the adoption of LOSC. The Northwest Atlantic Fisheries Organization (NAFO) was established in 1979, succeeding its antecedent organization, the International Convention for North Atlantic Fisheries (ICNAF). The North East Atlantic Fisheries Commission (NEAFC) was established to manage the fisheries in the Northeast Atlantic in 1982, and the Commission for the Conservation of Antarctic Marine Living Resources (CCAMLR) was also created in 1982 to oversee the conservation of living marine resources in the Southern Ocean. RFMOs with competence over highly migratory fish stocks in certain parts of the world had been established prior to LOSC (e.g., International Commission for the Conservation of Atlantic Tunas (ICCAT) established in 1969 and IATTC established in 1950). However, large parts of the high seas remained unregulated and without enforceable cooperative arrangements responsible for the conservation and management of straddling fish stocks and highly migratory fish stocks.

In the decade following the adoption of LOSC, industrialization of the oceans and freedom of fishing proceeded with few controls, and as a result, many stocks declined dramatically inside EEZs and, even more alarmingly, on the high seas. Fishing fleets became capable of harvesting and processing increasing amounts of fish at sea using sophisticated gear, navigation, and fish-finding devices. Global fishing capacity increased rapidly and exceeded the level of effort commensurate with the sustainable use of the resource. Fishing nations often did not cooperate, and nationals of some States reflagged their fishing vessels with flags of States not party to regional or global agreements to avoid conservation and management measures. Many States did not join RFMOs, asserting their freedom to fish to the detriment of the health of the stocks, and some States that participated in regional organizations chose to exercise their right to opt out of conservation and management decisions, including decisions on allocations, and thereby caused total catches of stocks to exceed the total allowable catch limits.

The international community responded with new attempts to reform and modernize the international fisheries regime through consistent and deliberate efforts in the negotiation of hard and soft law instruments pertaining to fisheries and the marine environment. The new global imperative was expressed in Agenda 21 and the Rio Declaration as sustainable development (i.e., the inextricable connection between environmental protection and economic development together with the concept of inter- and intragenerational equity) and was introduced through new concepts and principles, e.g., integrated management, food security, precaution, ecosystem considerations, biodiversity conservation, transparency and participation. This lexicon informed the discussions on the improved management of the high seas, straddling fish stocks and highly migratory fish stocks .

UNFSA was negotiated over three years (1993–1995), building upon and moving beyond LOSC and incorporating the principles, guidelines, and obligations negotiated in, *inter alia*, the Rio Declaration, the United Nations Conference on Environment and Development (UNCED) Agenda 21, the Convention on Biological Diversity, the FAO Compliance Agreement, and the FAO Code of Conduct.[36] As such, UNFSA provided significant clarification and development of the law pertaining to conservation and management of straddling fish stocks and highly migratory fish stocks. UNFSA provides innovative mechanisms for addressing the problems and issues associated with management of fisheries on the high seas, such as the reformulation of maximum sustainable yield (MSY)[37] from a target to a limit, the incorporation of the precautionary approach (PA)[38] and the ecosystem approach (EA)[39] to fisheries, and the strengthening and expansion of the rights of nonflag States to monitor and control high seas fishing activities to ensure compliance with the conservation and management regime of RFMOs.[40] Further, UNFSA has introduced the need for transparency in the activities of RFMOs. Through these innovations, UNFSA has created a governance regime for the sustainable development of straddling fish stocks and highly migratory fish stocks over the entire geographic range of these stocks, as many provisions apply to areas both within and beyond areas of national jurisdiction.

In the post-UNFSA period, there has been significant and progressive development of a large body of soft and hard law, at the global and regional levels, that informs, reinforces, and builds upon LOSC. These include the FAO International Plans of Action (IPOAs),[41] vessel registration requirements, trade restrictions (see, e.g., CITES),[42] the FAO technical guidelines on the elaboration of the EA and PA,[43] and the adoption of conventions establishing new RFMOs or reforming existing RFMOs (Antigua Convention,[44] South East Atlantic Fisheries Organisation (SEAFO),[45] Southern Indian Ocean Fisheries Arrangement (SIOFA),[46] WCFPC[47]). As Macdonald notes, LOSC

> stands as a constitution in its own right to the structure and process of continuing refinements such as those in the... [UNFSA. The LOSC] is basically a framework convention: it is to be filled in by literally hundreds of geographically or functionally sectoral agreements already in existence or yet to be created. Further progressive development, adjustment and crystallization of all aspects of the law of the sea and sustainable ocean management will continue under the benign guidance

of the overarching constitutional provisions referred to.[48]

UNFSA, together with the suite of legal instruments pertaining to fisheries and ocean resource use (trade, environmental and biodiversity protection, and sustainable development), is nested within this overarching legal framework and forms an integral part of the global system for high seas governance. It is this international legal regime, as well as the progressive development of contemporary fisheries law, that shapes and influences RFMOs in the management of straddling fish stocks and highly migratory fish stocks. Since there is an institutional disconnect between UNFSA and the RFMOs responsible for the conservation and management of high seas fisheries, certain aspects of the new international law of fisheries must be implemented by States Parties to UNFSA who are also members of RFMOs.[49] Anticipating that parties might be slow in implementing key provisions through RFMOs, the UNFSA drafters provided for default application of the boarding and inspection provisions in instances where an RFMO has failed to take UNFSA-consistent action within two years from the date of adoption (UNFSA Article 21.3). It is incumbent upon these member States[50] to actively implement the provisions of LOSC, UNFSA, the FAO Compliance Agreement, and the Code through the modernization, reform, and strengthening of existing RFMOs, for it is generally through these regional institutions that the new overarching regime negotiated and adopted in the UN is to be given effect. This is the prevailing governance paradigm within the international legal system. The international legal framework is given expression by sovereign States within areas under national jurisdiction and through the cooperation of flag States and nonflag States within RFMOs. Where regulatory gaps exist such as in the South Pacific Ocean and Southwest Atlantic Ocean, in keeping with LOSC and UNFSA, all efforts must be made to establish cooperative arrangements to ensure the long-term sustainable use of the living marine resources.

In addition to the need for strengthening and reforming RFMOs, the international community recognizes the need for improving the global legal regime to safeguard the resources for present and future generations. LOSC, along with the current legal framework, "does not adequately address a range of issues dealing with the high seas, including fisheries, marine environmental conservation, and emergent new technologies such as bioprospecting and underwater thermal energy. Related challenges include… [implementing more effective maritime monitoring, control, surveillance and enforcement in response to increasing frequency of] piracy, and terrorism, people-smuggling, illegal fishing, and ships who seek to fly multiple flags of registration in order to evade maritime laws".[51] Arguably, the most significant use and abuse of the high seas today is overfishing.[52]

The High Seas Task Force (HSTF) on Illegal, Unregulated and Unreported Fishing (IUU), which was established in 2003 and concluded in March 2006, examined the weaknesses in the current regime and endeavoured to address lacunae, encourage implementation, and promote compliance. The final report of the HSTF[53] provided a list of pragmatic, concrete recommendations. Prior to the Review Conference, States were forming coalitions to champion these various recommendations and to implement initiatives to strengthen the conservation and management of high seas living resources. The ministers at the May 2005 St. John's Conference on the Governance of High Seas Fisheries and the UN Fish Agreement[54] underscored in their declaration the duty of States to control the activities of their vessels in accordance with LOSC, UNFSA, and the FAO Compliance Agreement. There is a recognized urgent and immediate need for addressing IUU fishing both within and beyond areas of national jurisdiction through improved vessel monitoring systems (VMS), global registers, trade and certification programs, insurance requirements, etc. To address the pernicious problem of reflagging as of means of circumventing the duty to cooperate in the conservation and management of straddling fish stocks and highly migratory fish stocks, efforts are underway to define with greater precision the concept of "genuine link" between a flag State and vessels entitled to fly its flag, as set out in LOSC Article 91.

Another development since UNFSA is the increasing call for the establishment of marine protected areas (MPAs) within areas of national jurisdiction and on the high seas. There have been considerable efforts over the last decade to establish MPAs for the conservation and protection of aquatic biodiversity, critical habitats, or endangered species. Additionally, there are proposals for using MPAs and reserves for conservation of fisheries resources.[55] Their establishment is called for in the Convention on Biological Diversity (CBD),[56] the WSSD Plan of Implementation, and other international instruments of relevance to fisheries.[57] "Many countries are responding to these calls and progressively integrating MPAs into their fisheries management systems".[58] These and other initiatives for biodiversity conservation and fisheries management are part of the ongoing assessment, formulation, and progressive development of international fisheries law.

Two years after the 10th anniversary of the coming into force of LOSC, it is clear that the law of the sea remains in a state of flux. To promote a successful regime of effective supporting institutions, some commentators have opened the debate on whether there is a need for further reform to respond to existing and emerging challenges. Some argue for a new conception of oceans management,[59] calling for a fourth UN conference on the law of the sea.[60] However, most States and international actors call for the "'effective implementation and enforcement' of the [existing] regime and respect for the rule of law"[61] and, where necessary, the improvement of the regime through new subordinate instruments and State practice.[62]

2.4 Key developments in international fisheries law

As noted above, current international fisheries law is a blend of binding and nonbinding instruments and customary international law. The following is an overview of key international instruments, in chronological order, governing the conservation and management of straddling fish stocks and highly migratory fish stocks and incorporates international custom and hard and soft law instruments at both the global and regional levels. (Each of the headings in the next section has the name of the legal instrument followed by the date in brackets representing the year in which the instrument was adopted.)

2.4.1 Law of the Sea Convention (1982)

The United Nations Convention on the Law of the Sea (LOSC), which entered into force on 16 November 1994, comprises 320 articles and nine annexes that set forth the rights and obligations of States regarding their use of the ocean and its resources. "The Convention remains a shining example of international cooperation, diplomacy, and the role of international law in the regulation of international affairs and is considered to be one of the most complex and ultimately successful international diplomatic negotiations that took place in the 20th century".[63] It is based on the international law principle of cooperation and provides for compulsory and binding dispute settlement. LOSC is supplemented by the 1994 Deep Seabed Mining Agreement and UNFSA. Since the entry into force of LOSC, three international bodies have been established pursuant to its provisions: the International Seabed Authority, the International Tribunal for the Law of the Sea (ITLOS), and the Commission on the Limits of the Continental Shelf (CLCS).

LOSC is the first comprehensive, global international instrument to systematically codify the rights and obligations of coastal States and flag States in fisheries both within and beyond areas of national jurisdiction. It establishes the distribution of State rights and duties pertaining to fisheries, recognizing that coastal States exercise complete sovereignty over fisheries within internal waters and their territorial sea. The rights, duties, and interests of coastal States over natural resources in the EEZ and over the continental shelf were recognized and the right of distant water fishing States to fish on the high seas was qualified (as it had been in the 1958 High Seas Convention) by the principle of "reasonable regard"[64] or "due regard"[65] for the interests of other States and by the obligations to conserve high seas marine living resources and cooperate with other States in doing so (LOSC Articles 116–120). Over time, the traditional freedom of States to fish on the high seas has been subject to more detailed regulations and restrictions and to increased responsibilities of flag States. Further, it should be noted that many of the LOSC provisions on the EEZ (Articles 55–75) have evolved into customary international law and are therefore binding on States whether they are Parties to the Convention or not.[66] Where a fishery is not exploited or where projected surplus stocks are available, coastal States under LOSC are to provide access for foreign fishing interests. As a result many coastal States entered into fishery access agreements with distant-water fishing nations following extension of jurisdiction. At the same time, coastal States increased their fishing capacity to realize the potential of the resources within their EEZs. Over time the foreign fleets were displaced from these areas under national jurisdiction and increased their efforts in the high seas area adjacent and beyond the newly expanded zones under coastal State competence.

LOSC recognizes that fish know no political boundaries. Accordingly, to ensure the effective management of the ocean and its resources throughout all jurisdictions, LOSC requires the cooperation of States to coordinate and harmonize the exercise of their competing rights. Articles 63 and 64 set out the duty of States to cooperate directly or through appropriate regional or international organizations in the management of straddling fish stocks and highly migratory fish stocks. Articles 116–119, together with Articles 63(2) and 64, expressly qualify the freedom to fish on the high seas, imposing a duty on flag States and relevant coastal States to work out agreements on the measures needed for the conservation of straddling fish stocks, and to cooperate to ensure the conservation and optimum utilization of highly migratory fish stocks.

Since LOSC came into force, the international legal regime of fisheries has continued to evolve, especially with respect to the rights and duties of States relating to straddling fish stocks and highly migratory fish stocks. The principle of freedom of fishing on the high seas has steadily eroded with the advent of international agreements such as UNFSA, while the duty to cooperate and the principle of compatibility have been implemented through State practice in regional fisheries agreements. The need for States to apply compatible conservation and management measures both inside and outside areas of national jurisdiction (Article 61 LOSC, reinforced and elaborated by Article 7 UNFSA) is recognized by the international community and is increasingly being implemented through concerted State action.[67]

2.4.2 1992 United Nations Conference on Environment and Development

The United Nations Conference on Environment and Development (the "Earth Summit" or UNCED) was held in June 1992 in Rio de Janeiro, Brazil, following an extensive preparatory process. UNCED adopted the Rio Declaration on Environment and Development[68] and Agenda 21[69] — the global agenda for the management of the planet's environment into the 21st century. The UNCED process elaborated several concepts and governance principles relevant to international fisheries management, in particular the principles of sustainable development, integration, anticipatory[70] management, transparency, and precaution. Chapter 17 of Agenda 21 moved beyond the immediate obligations prescribed in LOSC to assert that the protection and sustainable development of the marine and coastal environment and its resources require new approaches to management and development.[71] Although the Rio Declaration and Agenda 21 are declaratory and not legally binding, these instruments have directed international policy by influencing subsequent treaties and soft law instruments on sustainable development of the oceans.

Program C of Chapter 17 is devoted to the use and conservation of high seas living resources. The provisions pertaining to high seas fishing, particularly those related to straddling and highly migratory fish stocks, were highly contentious.[72] Nonetheless, Chapter 17 (para. 17.45) acknowledges many of the problems associated with these fisheries:

> management of high seas fisheries, including the adoption, monitoring and enforcement of effective conservation measures, is inadequate in many areas and some resources are over-utilized. There are problems of unregulated fishing, overcapitalization, excessive fleet size, vessel reflagging to escape controls, insufficiently selective gear, unreliable data bases and lack of sufficient cooperation between States.[73]

At the 1992 Earth Summit, the United States led informal consultations on the issue of straddling fish stocks and highly migratory fish stocks, rather than allow lengthy and acrimonious discussions to continue between States. A compromise was achieved, and the resulting text for paragraph 17.52 reads: "States should convene an intergovernmental conference under UN auspices with a view to promoting effective implementation of the provisions of the Law of the Sea on straddling and highly migratory fish stocks". The European Community accepted this language once text was added to say that the conference would draw on scientific and technical studies by the FAO and be fully consistent with the provisions of LOSC, in particular the rights and obligations of coastal States and States fishing on the high seas.[74] This led to the convening of the Conference on High Seas Fishing, which eventually produced UNFSA (see further below).

2.4.3 FAO Compliance Agreement (1993)

The 1993 FAO Compliance Agreement entered into force on 24 April 2003. The purpose of this Agreement is to improve high seas fisheries management by committing flag States to specific obligations concerning the control of their fishing vessels on the high seas. It elaborates upon the provisions of Article 94 of LOSC (duties of flag States) and makes these obligations specifically applicable to fishing vessels on the high seas. The Agreement reemphasizes international cooperation and places the onus on flag States to exercise effective control over their fishing vessels and to address noncompliance of international conservation and management measures.

Article III on flag-State responsibility is the heart of the FAO Compliance Agreement, and it imposes three basic requirements upon each party:

- to ensure that its vessels do not engage in any activity that undermines the effectiveness of international fishery conservation and management measures, whether or not the flag State is a member of the regional fishery organization that adopted the measures;
- to prohibit its vessels from fishing on the high seas unless specifically authorized to do so;
- to deny such authority unless able to control the fishing activities of the vessel.[75]

Balton describes these three rules as "a new vision for high seas fisheries", as flag States "must now actively oversee the high seas fishing operations of their vessels. They must decide on a case-by-case basis whether to authorize any vessel to fish on the high seas. Most importantly, they may not permit any vessel to fish on the high seas at all unless they are able to prevent the vessel from undermining agreed conservation rules".[76]

The FAO Compliance Agreement requires each flag State to maintain a record of vessels entitled to fly its flag and authorized to fish on the high seas, and this information is to be provided to FAO. In addition to details about vessels and authorizations, the parties agree to cooperate and provide information about all fishing vessels that undermine the effectiveness of international measures.[77]

One principal legal obstacle to effective high seas fisheries management has been the ease of transferring a fishing vessel from one registry to another, a practice known as reflagging, which allows the vessel to escape restrictions agreed to by the former flag State. With the same objective, entrepreneurs often seek to register fishing vessels in States that are not party to international conservation and management measures. The FAO Compliance Agreement aims to discourage these manoeuvres by preventing parties from granting authorization to any vessel that has had its authorization to fish on the high seas suspended or withdrawn by its former flag State within the previous three years.[78]

There are some exceptions and limitations to this obligation, such as a bona fide change in the ownership of the vessel, or when the flag State determines that the grant of authorization would not "undermine the object and purposes of the Agreement" taking into account the circumstances in which the authorization was withdrawn.[79] According to some commentators, the latter formulation is too broad, and it will allow for exemptions of a large number of high seas fishing vessels. Another criticism relates to Article II(1), which states that the Agreement applies to "all fishing vessels that are used or intended for fishing on the high seas". This provision is narrowed substantially by an exemption in subparagraph 2 for vessels less than 24 m in length.[80]

To improve flag-State control over vessels, parties are required to take enforcement measures against vessels entitled to fly their flags.[81] Sanctions are to be sufficiently severe to deter noncompliance and may include depriving offenders of any benefits accruing from illegal activities, and refusal, suspension, or withdrawal of the authorization to fish on the high seas.[82]

The FAO Compliance Agreement complements UNFSA, and the use of a vessel registry as a mechanism for strengthening monitoring and control has been imported into UNFSA. The vessel registries, together with VMS, the emerging vessel detection systems, and the trade and catch documentation schemes of RFMOs, have considerably improved the governance regime.[83] Whether such an agreement will substantively contribute to improvement now depends in large measure on the willingness of flag States to ratify and implement the FAO Compliance Agreement. As of May 2006, most of the major fishing nations are party to this treaty, including Canada, the European Community, Japan, Mexico, Norway, South Korea, and the United States.

While Part V of UNFSA presents a condensed form of relevant sections of the FAO Compliance Agreement on the duties of flag States, there are a number of differences between the two instruments. UNFSA has specific provisions on flag-State inspection schemes and monitoring arrangements, while the FAO Compliance Agreement relies on the reports of violations being forwarded to the flag State for action, when a vessel is voluntarily in the port of a party other than the flag State.[84] Flag-State obligations to maintain a record of fishing vessels are somewhat stronger and more detailed under the FAO Compliance Agreement regarding ownership, operator changes, vessel specifications, areas of operation, gear, and details of the previous flag, as well as details of vessels scrapped, decommissioned, or that have had their licenses withdrawn.[85] The FAO promptly circulates the information to all parties and, subject to any restriction by a party, may also forward the information on request to any global, regional, or subregional fisheries organization. This highlights the need for developing a global register to monitor activity and effectively limit the "'shopping' around for new flags".[86]

2.4.4 FAO Code of Conduct for Responsible Fisheries (1995)

The need for a code of conduct was recognized by COFI at its 19th session in March 1991 prior to UNCED.[87] Uncontrolled or inadequately controlled exploitation within EEZs was "aggravated by the realization that unregulated fisheries on the high seas, in some cases involving straddling and highly migratory fish species, which occur within and outside EEZs, were becoming a matter of increasing concern".[88] FAO thought it necessary to have an overarching code or a set of guidelines establishing minimum standards and principles, as well as new concepts, as a means of achieving sustainable, responsible global fisheries.

The FAO Code of Conduct for Responsible Fisheries (Code of Conduct) assembles in a single nonbinding document all the undertakings contained in all major fishery instruments. It represents the "most complete and up-to-date expression of the principles of sustainable fisheries management and development".[89] The Code was unanimously adopted on 31 October 1995 by the FAO Conference after three years of development through working groups and technical consultations. The FAO Compliance Agreement, though legally binding and adopted prior to the completion of the Code, is an integral component of the Code of Conduct.[90]

The Code of Conduct expressly mirrors many of the provisions in UNFSA, as it was largely negotiated over three years in parallel with the UN Fish Stocks Conference (see Box 2.1).[91] Article 3.2(a) specifically states that the Code is to be interpreted and applied "in a manner consistent with the relevant provisions" of UNFSA.

There is general consistency between UNFSA and the Code's key principles in Article 6, the fisheries management provisions in Article 7, and Article 8 fishing operations, which sets out specific duties of States, flag States, and port States. Although it is directed to national governments and their national fisheries administrations, in keeping with UNFSA and other post-UNCED international fishery instruments, the Code also recognizes that RFMOs have a central role in ensuring the long-term conservation and sustainable use of stocks within and beyond areas of national jurisdiction.[92]

The Code of Conduct sets out minimum standards for achieving responsible fisheries (echoing Principle 19 of the Stockholm Declaration).[94] The avoidance and elimination of overfishing are some of the core objectives of the Code. A key principle underscores the duty to fish responsibly: "the right to fish carries with it the obligation to do so in a responsible manner so as to ensure effective conservation and management of the living aquatic resources".[95] The Code contains 12 articles and applies to all fisheries and all sectors associated with fishing. Article 1.2 of the Code gives it an extremely wide application:

> The Code is global in scope, and is directed toward members and non-members of FAO, fishing entities, subregional, regional and global organizations, whether governmental or non-governmental, and all persons concerned with the conservation of fishery resources and management and development of fisheries, such as fishers, those engaged in processing and marketing of fish and fishery products and other users of the aquatic environment in relation to fisheries.

Box 2.1. Development of the FAO Code of Conduct for Responsible Fisheries.

At the 1992 International Conference on Responsible Fishing in Cancun, Mexico, the FAO was requested to prepare an international code of conduct. The Declaration of Cancun endorsed by the Conference further developed the concept of responsible fisheries to encompass "the sustainable utilization of fisheries resources in harmony with the environment". This declaration then formed an important contribution to the UNCED Rio Summit in June 1992 and Agenda 21. The September 1992 FAO Technical Consultation on High Seas Fishing further recommended the elaboration of a code to address the contentious high seas fisheries issues, thereby augmenting the UNCED call for a high seas fishing conference. In November 1992, the FAO Council at its 102nd session recommended that priority be given to high seas issues in elaboration of the code, and the COFI March 1993 session requested that FAO "fast track" the proposals to prevent reflagging of fishing vessels that affect high seas conservation and management measures. By March 1995, the draft text of the Code had been reviewed by COFI, revised, and presented to the Ministerial Meeting on Fisheries following the COFI session. The FAO Council in June 1995[93] recommended that the final draft be completed following the expected conclusion of the August 1995 UN Fish Stocks Conference, allowing the drafters of the Code to harmonize the text linguistically and juridically, taking into account the final provisions of UNFSA. This three-year development process, which included FAO members, NGOs, and IGOs, culminated with the adoption by consensus of the Code of Conduct for Responsible Fisheries on 31 October 1995.

Article 6 outlines 19 general principles and international standards for responsible fishing and fisheries activities, all of which are consistent with UNFSA. These include

- applying the ecosystem approach;
- applying the precautionary approach;
- preventing overfishing and excess capacity;
- implementing management measures to ensure that fishing effort is commensurate with the productive capacity of the fishery resources and their sustainable use;
- taking measures to rehabilitate populations as far as possible, and when appropriate, developing and applying, to the extent practicable, selective and environmentally safe fishing gear and practices to maintain biodiversity;
- minimizing the catch of nontarget species, both fish and nonfish, and the impacts on associated and dependent species;
- protecting and rehabilitating critical habitats as far as possible and where necessary;
- using transparent and timely decision-making procedures;
- ensuring the effective participation of stakeholders and interested organizations.

As a nonbinding instrument, much more was included in the Code than could have been covered in a binding international agreement.[96] Under the aegis of the FAO, the Code has also been supplemented by a series of nonbinding technical guidelines;[97] four IPOAs,[98] and the 2003 FAO Strategy for Improving Information on Status and Trends of Capture Fisheries.[99]

The FAO considers the Code to be "a flagship instrument".[100] Although FAO is "committed to foster [the Code's] full and effective implementation",[101] COFI has not indicated priorities for implementation. Despite its limited resources, FAO is endeavouring to support the implementation of the Code *in toto*.[102] Further, given the symmetry between the Code and the WSSD Plan of Action as they pertain to fisheries, FAO is also actively promoting the WSSD Plan of Implementation by facilitating the implementation of the Code of Conduct.[103]

Though it is voluntary, the Code of Conduct enjoys wide support and has had a substantial impact on fisheries management at both national and international levels. Prior to UNFSA and the FAO Compliance Agreement coming into force, the Code was used to advance the conservation and management principles contained in UNFSA and to emphasize flag-State responsibilities. An illustration of the practical utility and importance of the Code is that it was considered the instrument of reference in reaching consensus on the Antigua Convention, a binding instrument that updates the IATTC constitutive agreement.[104]

2.4.5 United Nations Fish Stocks Agreement (UNFSA) (1995)

As noted previously, the 1992 UNCED Earth Summit called upon States to convene an international conference to address the problems of conservation and management of straddling fish stocks and highly migratory fish stocks on the high seas. This led to the eventual negotiation of the treaty, the full title of which is the Agreement for the Implementation of the Provisions of the United Nations Convention on the Law of the Sea of 10 December 1982 Relating to the Conservation and Management of Straddling Fish Stocks and Highly Migratory Fish Stocks (UNFSA). UNFSA is intended "to ensure the long-term conservation and sustainable use of straddling fish stocks and highly migratory fish stocks through effective implementation of the relevant provisions of the Convention".[105] It entered into force on 11 December 2001.

UNFSA became the new legal basis for conservation and management of straddling fish stocks and highly migratory fish stocks. It represents an evolution in the law of the sea, as it prescribes shifts from "unrestricted exploitation to increasing conservation, from freedom to responsibility, from confrontation to cooperation".[106] UNFSA provides innovative mechanisms for addressing the problems and issues associated with management of fisheries on the high seas, where freedom of fishing had resulted in inadequate regulation and deference to flag-State control over fishing vessels. As David Anderson states, UNFSA "will assist the ongoing process of clarifying the point that the freedom of fishing, as first articulated in conventional form in 1958, is a specifically qualified freedom".[107] UNFSA prescribes a conservation and management regime for straddling fish stocks and highly migratory fish stocks on the high seas, but also encourages States to manage stocks throughout their geographic range, regardless of whether they are found within or outside areas of national jurisdiction. UNFSA provides for the application of broad conservation principles and the precautionary approach within areas under national jurisdiction and establishes the principle of compatibility.[108]

Article 6.1 of UNFSA requires States to apply the precautionary approach (PA) "widely to conservation, management and exploitation of straddling fish stocks and highly migratory fish stocks in order to protect the living marine resources and preserve the marine environment".[109] States are obliged to improve decision making for fisheries conservation and management by acquiring and sharing the best scientific information available and by implementing improved techniques to address issues of risk and uncertainty. This approach is further detailed in Annex II of the Agreement, which provides guidelines for setting precautionary reference points in addition to recognizing MSY as a limit.[110]

By reformulating MSY as a limit within the precautionary approach, UNFSA addresses, in part, the shortcomings of MSY as a management target identified in the post-LOSC period.[111] UNFSA also prescribes a shift from single-species management, for which MSY was well suited, to multispecies management inherent in an ecosystem approach. While only certain fish stocks are now regulated, the UNFSA framework provides that associated and dependent species must be taken into account so that nontarget species are conserved, an ecosystem approach is adopted where possible, and critical habitat is protected. The issue of unregulated species is currently under examination, including straddling, highly migratory, and discrete high seas stocks.[112]

Most prominent in UNFSA is the identification of subregional and regional fisheries management organizations (RFMOs) as the appropriate institutional mechanism for States to carry out their duty to cooperate in the conservation and management of straddling fish stocks and highly migratory fish stocks.[113] States are required to become members of RFMOs, or at least agree to apply the RFMO's measures.[114] Articles 9 and 10 provide general guidelines for the establishment of RFMOs, their functions and responsibilities, their relationship with their members, and the rights and duties of RFMO members.

UNFSA also provides an elaborate system of compliance and enforcement that details flag-State responsibilities with regard to fishing vessels flying their flag.[115] The Agreement also provides for nonflag-State enforcement when the flag State is unable or unwilling to fulfill its duties in a timely fashion.[116] This is complemented by port-State control of vessels for the purpose of enforcing fisheries conservation and management measures.[117] This system of compliance and enforcement is designed to further promote and strengthen the effectiveness of RFMOs and their conservation and management measures.

2.4.6 FAO International Plans of Action (1999, 2001)

As part of the Code of Conduct, FAO has adopted four nonbinding instruments known as International Plans of Action (IPOAs) to address specific international fisheries problems. The first three IPOAs, which concern sharks, seabirds, and capacity, were adopted in 1999 by COFI and endorsed by the FAO Council in June 1999. The texts for these instruments were developed over the course of two intergovernmental meetings open to all FAO members. In 2001 FAO adopted a fourth IPOA on IUU fishing. Each of these instruments is discussed below.

2.4.6.1 International Plan of Action for Conservation and Management of Sharks (IPOA-Sharks)

The IPOA-Sharks[118] is a nonlegally binding instrument that applies to all States whose nationals engage in shark fisheries. This IPOA was developed in response to the increase in both target and incidental shark catches in many areas of the world and the negative consequences for certain shark species. States must conduct an assessment to determine whether there exists a problem with shark populations and, if so, determine the extent of the problem. Following this assessment States must adopt a National Plan of Action for the conservation and management of sharks, as well as procedures for national reviews and reporting.[119] IATTC, ICCAT, the Indian Ocean Tuna Commission (IOTC), and NAFO have encouraged member States to collect information about shark populations[120] and have developed regional databases for the purposes of stock assessment.

2.4.6.2 International Plan of Action for Reducing Incidental Catch of Seabirds in Longline Fisheries (IPOA-Seabirds)

The IPOA-Seabirds[121] is a nonbinding instrument with respect to incidental catch of seabirds in longline fisheries and applies to all States whose nationals engage in longline fisheries. This IPOA calls on States to conduct an assessment to determine whether there exists a seabird bycatch problem within their fisheries and, if so, determine the extent of the problem. States must also adopt a National Plan of Action for reducing seabird bycatch associated with longline fishing, as well as procedures for national reviews and reporting (Articles 11–21). The IPOA also provides a summary of mitigation measures.[122]

2.4.6.3 International Plan of Action for the Management of Fishing Capacity (IPOA - Fishing Capacity)

The IPOA - Fishing Capacity is a nonbinding instrument intended to bring global fishing capacity in line with fisheries resources. It applies to all States whose nationals engage in capture fisheries. This IPOA sets out mechanisms for achieving efficient, equitable, and transparent management of fishing capacity worldwide by 2005, including assessment and monitoring of fishing capacity; preparation and implementation of national strategies; and the exchange of information on issues related to management of fishing capacity.[123] It calls upon all countries to support the FAO in establishing an international record of fishing vessels operating on the high seas.[124] This IPOA is "far reaching" in its efforts to curb the well-known problem of "too many fishermen on too many vessels chasing too few fish".

The inability of States to meet the 2005 deadline is indicative of the complexity of the problem.[125] Certain forms of long-established fisheries subsidies contribute to overcapacity and overfishing and need to be eliminated. The cost of implementing transition can be high given the socioeconomic dislocation associated with reducing capacity and the need for introducing alternative livelihoods or compensation for withdrawing vessels or licences. Additionally, sophisticated newly constructed vessels equipped with improved technology for fish finding and fish harvesting and with improved transport systems are more efficient at distant-water harvesting and transport of fish. These technological developments compound the difficulty to reduce fishing capacity. The problem is further complicated by IUU fishing. Concerted political will is required to immediately effect the changes needed to address the issue of capacity.

RFMOs such as IATTC have adopted this IPOA and have made concerted efforts to reduce fleet capacity but have met with considerable resistance from some member States. This is a particularly worrisome situation, as capacity reduction is the primary means of conservation and restoration of fish stocks in many international fisheries (e.g., the tuna RFMOs). Excess capacity remains an urgent global concern, as it undermines the international community's commitments and ability to achieve sustainable fisheries.

2.4.6.4 International Plan of Action to Prevent, Deter, and Eliminate Illegal, Unreported and Unregulated Fishing (IPOA-IUU)

COFI introduced the IPOA-IUU in response to the alarming rate at which IUU fishing activities were increasing, the adverse impact on the marine ecosystem, the extent to which these activities were compromising management efforts, and the proliferation of vessels flying flags of convenience.[126] Adopted in 2002, the IPOA-IUU reiterates and elaborates upon the provisions in the Code of Conduct and UNFSA. The IPOA-IUU provides a comprehensive "toolbox" of measures that flag, coastal, and port States can take, both individually and collectively, to address the problem of IUU fishing.[127] These include internationally agreed market-related measures, measures to be implemented through RFMOs, as well as special requirements for developing States. Each State is to develop its own national plan of action to implement the IPOA. The principles underpinning this IPOA include the following:

- broad participation and coordination among States, as well as representatives from industry, fishing communities, and NGOs;
- the phasing in of action to implement the IPOA on the earliest possible timetable;
- the use of a comprehensive and integrated approach to address all impacts of IUU fishing;
- the conservation and long-term sustainable use of fish stocks and the protection of the marine environment;
- transparency;
- nondiscrimination in form or in fact against any State or its fishing vessels.[128]

Part IV is the heart of this IPOA. It deals with the implementation of measures to prevent, deter, and eliminate IUU fishing, which are covered under major headings concerning all-State responsibilities, flag-State responsibilities, coastal-State measures, port-State measures, internationally agreed market-related measures, research, and RFMOs.

There are separate parts on the special requirements of developing countries (Part V), on reporting, and on the role of FAO (Part VI).[129]

2.4.7 Reykjavik Declaration (2001)

FAO convened the 2001 Conference on Responsible Fisheries in the Marine Ecosystem in Reykjavik, Iceland, to assess progress on the implementation of the Code, with an emphasis on the measures pertaining to the conservation of ecosystems. The objectives of the Conference were to gather and review the best available knowledge on marine ecosystem issues, identify means by which ecosystem considerations could be included in capture-fisheries management, and identify future challenges and strategies. The Reykjavik Declaration[130] reinforces the need for sustainable and responsible marine fisheries through efforts to incorporate ecosystem considerations into fisheries management, not only from the standpoint of the impact of fisheries on the ecosystem, but also the impact of the ecosystem on fisheries.[131] The Declaration seeks to integrate ecosystem considerations into the work of RFMOs. It calls for States to improve cooperation between regional and international fisheries management organizations and "regional bodies in charge of managing and conserving the marine environment".[132] It emphasizes the need for strengthening international cooperation, through education and training, in the collection and processing of biological, oceanographic, ecological, and fisheries data required for the design, implementation, and upgrading of management strategies.

2.4.8 World Summit on Sustainable Development Plan of Implementation (2002)

The World Summit on Sustainable Development (WSSD) was held in Johannesburg, South Africa, in August 2002 to review progress made in the decade following UNCED. States negotiated and adopted two main documents: the Johannesburg Declaration on Sustainable Development[133] and the Plan of Implementation,[134] which were intended "to reinvigorate the global commitment to sustainable development". The Declaration specifically recognized the continuing depletion of fish stocks as one of the challenges to sustainable development. Chapter IV, paragraphs 30 to 36 of the Plan of Implementation deals with fisheries issues.

The Plan of Implementation is a strong endorsement of the ecosystem approach and of the need for protecting vulnerable areas not only within but also beyond areas of national jurisdiction. Many of the fisheries provisions in the Plan reinforce UNFSA, the FAO Compliance Agreement, the Code of Conduct, and the four FAO IPOAs. FAO's role in supporting national, regional, and international efforts towards implementation of WSSD targets "is an integral part of the Organization's efforts to promote the implementation of the Code of Conduct and, through it, other post-UNCED instruments".[135] Through its reporting requirements on implementing the Code of Conduct, FAO is in effect promoting and monitoring progress on the Plan of Implementation. RFMOs are aware of these targets; for example, for its 2004 Annual Meeting the NAFO Fisheries Commission requested that the NAFO Secretariat prepare a comparative review of the NAFO program in light of the WSSD commitments.

Outcomes identified by the Plan of Implementation that are necessary to achieve sustainable fisheries, and which may be most relevant to straddling fish stocks and highly migratory fish stocks are

- the restoration of depleted fish stocks to levels that can produce MSY by 2015;
- the development and implementation of national and regional plans of action to put into effect the IPOA-IUU by 2004;
- the development and implementation of national and regional plans of action to put into effect the IPOA-Capacity by 2005;
- the application of the ecosystem approach (to responsible fisheries) by 2010, noting the Reykjavik Declaration and Decision V/6 of the Conference of Parties to the Convention on Biological Diversity;[136]
- the establishment of "representative networks" of marine protected areas by 2012.

To achieve these outcomes, the Plan of Implementation

- exhorts States to ratify or accede to and effectively implement UNFSA, the FAO Compliance Agreement, and where appropriate, RFMOs or arrangements;
- exhorts States to implement the Code of Conduct together with the relevant IPOAs and FAO Technical Guidelines for Responsible Fisheries;
- encourages RFMOs and regional fisheries management arrangements (RFMAs) to "give due consideration to the rights, duties, and interests of coastal States and developing States when addressing the issue of allocation of share of fishery resources for straddling and highly migratory fish stocks" mindful of UNFSA;
- exhorts States to eliminate subsidies that contribute to IUU fishing and to overcapacity and to complete the efforts undertaken at the World Trade Organization (WTO) on fisheries subsidies.[137]

2.4.9 FAO Strategy for Improving Information on Status and Trends of Capture Fisheries (2003)

One of the fundamental principles underlying LOSC and UNFSA is that decisions should be based on the best scientific information available.[138] Fisheries management is data dependent, and data demands are ever increasing under the modern fisheries governance framework. Accurate, timely, and readily understandable information on the status and trends in fisheries, including socioeconomic aspects, is an essential component of all post-UNCED fisheries instruments.[139] The FAO Strategy for Improving Information on Status and Trends of Capture Fisheries (Strategy-STF)[140] is a voluntary instrument that applies to all States and entities. The Strategy provides a framework, strategy, and plan for "the improvement of knowledge and understanding of fishery status and trends as a basis for fisheries policy-making and management for the conservation and sustainable use of fishery resources within ecosystems".[141]

The Strategy-STF calls on States, directly or through participation in regional fishery bodies, to broaden the collection of information (including the effects of climatic, environmental, and socioeconomic factors)[142] and to address the issue of sustainable development indicators to support development of ecosystem-based fisheries management.[143] The Strategy-STF lists nine areas for action, including:

- capacity building in developing countries;
- expanding the scope of fisheries status and trend information to incorporate ecosystem considerations;
- compiling the global inventory of fisheries and fish stocks that has been established by the FAO called the Fisheries Global Information System (FIGIS);
- participating in and providing FIGIS with the best scientific information available;
- developing criteria for quality assurance, transparency, and security of data;
- monitoring data collection, analysis, and reporting systems.[144]

The Strategy-STF also outlines the role of RFMOs and arrangements, to the extent possible, in developing and adopting global standards and guidelines for information. States and RFMOs have extensive data collection, sharing, and analysis obligations to effect "sound policy making and responsible fisheries management".[145] Data collection and analysis, transparency, and other management obligations for UNFSA-related RFMOs may be strengthened by participation in the Strategy-STF and by partnering in FIGIS.

2.4.10 United Nations General Assembly

United Nations General Assembly (UNGA) resolutions are "formal expressions of the opinion or will of United Nations organs".[146] They can merely state the opinion of the General Assembly or set out a specific action to be taken. The General Assembly annually adopts two resolutions (one on oceans[147] and a second on fisheries) that raise issues related to the management of straddling and highly migratory fish stocks. In addition, other resolutions have been adopted on related issues such as large-scale pelagic driftnet fisheries (Resolution 46/215)[148] and fisheries bycatch and discards (Resolution 49/118).[149] The annual UNGA resolution on fisheries is a comprehensive initiative that reinforces new developments in international agreements and initiatives and reaffirms the international community's commitment to improving fisheries governance and sustainability.[150]

The 2005 resolution on sustainable fisheries (UNGA Resolution 60/31)[151] calls upon States to ratify and to accede to UNFSA (paras. 9 and 27) and the FAO Compliance Agreement (para. 29) and, in the interim, to consider applying them provisionally. Several references are made to supporting sustainable fisheries on the high seas, including the obligation of flag States to control fishing operations by vessels flying their flag (para. 2) and the application of the precautionary and ecosystem approaches (paras. 4, 63–82). The resolution emphasizes the role played by RMFOs in fisheries management (e.g., paras. 14–15, 26, 30, 41, 44, 54–62), including more recently established entities in the Southeast Atlantic (SEAFO, para. 18) and in the Western and Central Pacific (Western and Central Pacific Fisheries Commission (WCPFC), para. 19).

The resolution devotes 15 paragraphs to the pernicious problem of IUU fishing.[152] Establishing a "genuine link" (para. 38) between a flag State and a vessel and enhancing port-State controls (paras. 36 and 42) are cited as means of dealing with this issue. Paragraph 34 calls on "States not to permit vessels flying their flag to engage in fishing on the high seas or in areas under national jurisdiction of other States, unless duly authorized…without having effective control over their activities, including deterring the reflagging of vessels by their nationals". It further urges States and RFMOs to establish mandatory vessel monitoring, control, and surveillance systems for fishing vessels, including a comprehensive global record within FAO of fishing vessels (paras. 44–45). The resolution also calls upon States and relevant RFMOs to take measures with regard to fishing overcapacity (paras. 47–49), encourages capacity building in developing countries (paras. 83–88), and invites continued interagency cooperation within the UN system (paras. 89–91). In addition, the resolution reemphasizes the need for continued vigilance on issues that have been the subject of special UNGA resolutions, e.g., large-scale pelagic driftnet fishing (para. 50) and fisheries bycatch and discards (paras. 51–53). It is noteworthy that the resolution also encourages States to recognize that the general principles of UNFSA should apply to discrete fish stocks in the high seas (para. 12). This seems to indicate an emerging consensus among States according to which a number of the provisions of UNFSA not only apply to straddling fish stocks and highly migratory fish stocks but also to all marine capture fisheries, whether in areas under national jurisdiction or on the high seas.[153]

In 2005 the General Assembly established the Ad Hoc Open-ended Informal Working Group to study issues relating to the conservations and sustainable use of marine biological diversity beyond areas of national jurisdiction.[154] Its mandate includes: "Indicate, where appropriate, possible options and approaches to promote international cooperation and coordination for the conservation and sustainable use of marine biological diversity beyond areas of national jurisdiction".

2.4.10.1 United Nations Open-ended Informal Consultative Process on Oceans and the Law of the Sea (2000)

To enhance the effectiveness of the UN General Assembly's annual debate on oceans and law of the sea, the Commission on Sustainable Development (CSD) recommended that the General Assembly establish an open-ended, informal consultative process.[155] The General Assembly accepted this recommendation and in 2000 established what is now known as the United Nations Open-ended Informal Consultative Process on Oceans and the Law of the Sea (UNICPOLOS).[156] This new forum affords UN Member States the opportunity to discuss ocean affairs and law of the sea issues in much greater depth than has been previously possible in the annual General Assembly debate on the Secretary-General's Report on Oceans and the Law of the Sea.[157] This multistakeholder process considers the Secretary-General's annual reports and identifies specific issues where intergovernmental and interagency coordination and cooperation should be enhanced, for consideration by the General Assembly. The June 2005 meeting considered fisheries and its contribution to sustainable development.[158]

2.4.10.2 UNFSA Informal Consultation of States Parties (2001)

For the purpose, *inter alia,* of considering the subregional, regional, and global implementation of UNFSA, the General Assembly in 2001 established the Annual Informal Consultation of States Parties to the Agreement (ICSP).[159] The major focus in the June 2005 round of ICSP-4[160] and March 2006 ICSP-5 was preparation for the UNFSA Review Conference.

2.4.11 FAO Ministerial Declarations

2.4.11.1 Rome Consensus on World Fisheries (1995)

In 1995 the FAO Ministerial Conference on Fisheries, which was convened to review the state of world fisheries, noted that "the problems of overfishing in general, and overcapacity of industrial fishing fleets in particular, threaten the sustainability of the world's fisheries resources for present and future generations".[161] It was agreed that there was a need for continuing international cooperation and coordination to reestablish sustainability of world fisheries. Further, improved fisheries conservation and management and better protection of fisheries from harmful sea- and land-based activities are crucial to maintaining world fisheries resources and aquatic ecosystems.[162] The Rome Consensus on World Fisheries encourages the pursuit of an ecosystem approach to fisheries conservation and management.[163] It also affirmed FAO's mandate as the leading specialized UN agency for fisheries and commended FAO's efforts to bring the critical situation in fisheries to the attention of the international community.[164]

2.4.11.2 The Rome Declaration on the Implementation of the Code of Conduct for Responsible Fisheries (1999)

The Rome Declaration on the Implementation of the Code of Conduct for Responsible Fisheries was adopted by the FAO Ministerial Meeting on Fisheries in March 1999.[165] In addition to its call to implement the FAO Code of Conduct, the Declaration identifies several important issues in fisheries management including:

- implementation of UNFSA;
- use of the ecosystem approach to achieve sustainable fisheries;

- reduction of waste and destructive fishing practices;
- fisheries-related trade and environment issues;
- implementation of the IPOAs on sharks, seabirds, and capacity;
- development of the IPOA on IUU fishing.[166]

2.4.12 The Rome Declaration on Illegal, Unreported and Unregulated Fishing (2005)

The Rome Declaration on Illegal, Unreported and Unregulated Fishing was adopted by the FAO Ministerial Meeting on Fisheries in March 2005.[167] The Declaration calls on States to address the issue of IUU fishing through, *inter alia*:

- development and implementation of national and regional plans of action;
- adoption of internationally agreed market-related measures and catch certification schemes;
- information exchange on suspected IUU fishing, through FAO, RFMOs, and other relevant arrangements;
- strengthening and implementation of relevant monitoring, control, and surveillance measures;
- provision of assistance to developing countries to help them implement measures to address IUU fishing;
- adoption of appropriate measures to address fishing fleet overcapacity.

2.4.13 Ministerial Declaration of the St. John's Conference on the Governance of High Seas Fisheries and the UN Fish Agreement (2005)

The St. John's Ministerial Declaration was issued at the 2005 Conference on the Governance of High Seas Fisheries and the UN Fish Agreement — Moving from Words to Action held in St. John's, Newfoundland and Labrador, in May 2005.[168] The Declaration, supported by 19 fisheries ministers, signals a renewed commitment to reforms in fisheries conservation and management on the high seas.[169] It urges States to ratify international agreements, such as UNFSA, and to effectively implement these agreements in RFMOs. Specifically, the Declaration calls for States to review and strengthen RFMOs, provide a mechanism for addressing disputes, implement a decision-making process that incorporates the precautionary approach and ecosystem considerations to ensure the conservation of fish stocks, and strengthen monitoring and control mechanisms to improve detection and deter noncompliance.

The St. John's Ministerial Declaration marks a significant step ahead in facilitating the RFMO reform process required as part of UNFSA implementation.[170] Against the backdrop of the commitments undertaken in the Declaration, at its Annual Meeting held in September 2005, NAFO launched a review of the NAFO Convention, especially with a view to incorporating more integrated oceans management approaches, reforming decision-making processes, and examining the current structure of NAFO.[171] A similar reform process was launched by ICCAT at its Annual Meeting in November 2005, i.e., to reform ICCAT consistent with the St. John's Declaration but also with other recent political commitments, such as the FAO declaration on IUU fishing and the Bali Plan of Action.[172]

2.4.14 Related international legal instruments

2.4.14.1 International Convention for the Regulation of Whaling (1946)

In 1946 the International Convention for the Regulation of Whaling (ICRW) was signed by the representatives of 14 States.[173] The objective of the Convention is to ensure the conservation and development of whale stocks globally through the International Whaling Commission (IWC).[174] The mandate of the IWC is chiefly to review and revise, as necessary, the measures in the Schedule to the Convention[175] based on scientific findings[176] to regulate whaling.[177] These measures provide for the protection of certain species; designate specified areas as whale sanctuaries; set limits on the numbers and size of whales that may be taken; prescribe open and closed seasons and areas for whaling; and prohibit the capture of suckling calves and female whales accompanied by calves. Pursuant to the IWC decision in 1982[178] there has been a global moratorium on commercial whaling of all whale stocks since 1985.[179]

Whales are listed as highly migratory species under LOSC Annex I.[180] However, as whales are marine mammals and not fish, a distinction is made between highly migratory fish stocks and highly migratory species, and UNFSA deals only with fish stocks.

2.4.14.2 The Convention on International Trade in Endangered Species of Wild Fauna and Flora (1973)

The Convention on International Trade in Endangered Species of Wild Fauna and Flora (CITES)[181] entered into force in 1975. Recognizing that international trade directly contributes to the decimation of wild flora and fauna, CITES purports to regulate the trade in endangered species to prevent overexploitation. An import–export permit and certification system requires the movement of species protected by CITES to be subject to strict documentation and government control. The biennial Conference of the Parties oversees implementation of CITES and considers the inclusion of species on these lists.

The CITES "lists" include all species threatened with extinction that are affected by international trade (Appendix I species), species that may be threatened with extinction unless trade is strictly regulated, as well as associated species (Appendix II species). CITES prohibits international trade in Appendix I species and regulates international trade in Appendix II species by requiring extensive documentation of exports, import monitoring, and export quotas. Species that a party identifies as being subject to regulation within its jurisdiction for the purpose of preventing or restricting exploitation (Appendix III species) are subject to national rules and regulations.

Pressure from environmental groups to list certain straddling stocks and highly migratory fish stocks in CITES Appendix I or II has influenced States and RFMOs to adopt more rigorous conservation and management measures. For example, the intense discussion on the Atlantic bluefin tuna problem at the March 1992 Kyoto meeting of CITES led indirectly to the adoption of the ICCAT Bluefin Tuna Statistical Document Program in 1992[182] and, later, to prohibition of imports from Belize, Honduras, and Panama.[183]

Recognizing the complementary nature of many instruments and fora, States and NGOs have engaged in several institutions, including CITES, to resolve law of the sea challenges.[184] For example, in 2002, Australia endeavoured to effect change to secure international protection for the Patagonian toothfish through the competent RFMO (i.e., CCAMLR) as well as through CITES.[185]

CITES has generally become more prominent in the conservation of marine living resources in the post-UNFSA period because of the increasing number of formerly targeted or nontarget fish species being proposed for CITES listing under Appendix I. At the Conference of Parties in 2002, discussions and decisions involved several marine species, including minke and Bryde's whales, seahorses, sea cucumbers, whale sharks, basking sharks, dolphins, and green turtles.[186] The number of marine species proposed for CITES listing was significant enough to warrant proposals for mechanisms for cooperation and coordination between CITES and FAO and CITES and CCAMLR, and for further review of CITES criteria.[187] During the March 2005 COFI meeting, discussions regarding CITES were animated,[188] and FAO and CITES are as of May 2006 in the process of negotiating a memorandum of understanding to guide future collaboration.[189]

2.4.14.3 Convention on the Conservation of Migratory Species of Wild Animals (1979)

The 1979 Convention on the Conservation of Migratory Species of Wild Animals (Bonn Convention also known as CMS) has been in force since 1983.[190] Concluded under the aegis of the United Nations Environment Programme (UNEP), it establishes a framework for enhancing the conservation of rare and threatened "migratory species", which are defined as an "entire population or any geographically separate part of the population of any species or lower taxon of wild animals, a significant portion of whose members cyclically and predictably cross one or more national jurisdictional boundaries".[191] This includes terrestrial, marine, and avian migratory species and their habitats, throughout their range. Obligations are stipulated for "Range States", which refers to any State that exercises jurisdiction over any part of the range of the migratory species, or to a State whose flag vessels are engaged in taking the migratory species outside areas of national jurisdiction.[192] The Bonn Convention introduced the concept of the Range State in international law and provides a legal precedent for norms allowing the establishment of measures spanning the geographic range of a particular species.

The Bonn Convention established two lists. Appendix I lists species in danger of extinction. Stringent measures are imposed to protect these species. Appendix II lists species that are threatened and would benefit from international cooperation through the adoption of regional or global agreements for their conservation and management. A Conference of Parties is held triannually to review implementation of the Convention and to review and assess the conservation status of migratory species.[193] All parties are obliged to keep the Secretariat informed of species for which they may be considered to be Range States and to provide information on their flag vessels engaged in taking such species outside areas of national jurisdiction.[194] The Bonn Convention predates the 1982 LOSC and is without prejudice to the provisions of the latter.

2.4.14.4 Convention on Biological Diversity (1992)

The Convention on Biological Diversity (CBD) recognizes that the planet's biological diversity is being significantly reduced by human activity and that there is an urgent need to address the causes through the conservation of ecosystems and natural habitats and the recovery of viable populations of species in their natural surroundings.[195] The CBD also focuses on the sustainable use of the components of biological diversity. It calls for the fair and equitable sharing of benefits from the use of genetic resources, through appropriate funding, access to genetic resources, and transfer of relevant technologies, taking into account all relevant rights.[196] It includes both terrestrial and marine species.

The CBD declares that while States have the sovereign right to exploit their own natural resources pursuant to their own environmental policies, they have a corresponding responsibility for ensuring that activities subject to their jurisdiction or control do not cause damage to the environment of other States or to areas beyond the limits of national jurisdiction.[197] The Convention establishes a duty to cooperate, either directly or through competent international organizations, in respect of areas beyond national jurisdiction and on other matters of mutual interest for the conservation and sustainable use of biological diversity.[198] Since implementation of the CBD depends upon national action, the CBD only provides general guidance on best practices and does not provide for sanctions against countries that do not adhere to those practices.

The second Conference of Parties to the Convention on Biological Diversity in 1995 adopted the Ministerial Statement on the Implementation of the Convention on Biological Diversity, incorporating a program of action that has come to be known as the "Jakarta Mandate on Marine and Coastal Biodiversity". The Jakarta Mandate reflects a global consensus on the need for conservation and sustainable use of marine and coastal biodiversity,[199] and it urges Parties to the Convention to initiate immediate action to implement the decisions adopted on this issue.[200] The Jakarta Mandate focuses on five themes: integrated marine and coastal area management, the sustainable use of living resources, marine and coastal protected areas, mariculture, and alien species. Programme Element 2: Marine and Coastal Living Resources identifies measures for promoting ecosystem approaches to the conservation and sustainable use of marine and coastal living resource in areas within and beyond national jurisdiction. The Jakarta Mandate refers to the need for ecosystem-based approaches and the application of precaution. The work program on marine and coastal diversity adopted at the seventh Conference of the Parties is set out as an annex in Decision VII/5. In particular, Objective 2.1 suggests cooperative activities aimed at protecting biodiversity in marine areas beyond national jurisdiction. Decision VII/5 also addresses concerns over the impacts of destructive practices such as bottom trawling on vulnerable marine ecosystems in areas beyond national jurisdiction.[201] Parties are required to implement the CBD in accordance with State rights and obligations under LOSC.[202]

2.4.14.5 The World Trade Organization (1994)

Although it may not be an inherent part of international environmental or fisheries law, the World Trade Organization (WTO) impacts international fisheries management because of the links between international trade and fisheries and marine resources. Established in 1994, WTO is an intergovernmental body that negotiates and enforces trade rules and periodically convenes multilateral trade negotiations.[203]

WTO member States are required to grant equal treatment for like products (Article I); treat like imported products no less favourably than those produced domestically (Article III); and with few exceptions, not restrict the quantity of exports or imports (Article XI).[204] An exception provided in Article XX includes trade measures aimed at protecting human, animal, or plant life or measures "relating to the conservation of exhaustible natural resources". Increasingly, trade measures are being used as an effective means of enforcing the management strategies of RFMOs. Under UNFSA, RFMOs may take measures "consistent with this Agreement and international law" to deter activities of nonparty vessels that undermine the effectiveness of RFMO conservation and management measures.[205] Port States may also adopt regulations empowering the relevant national authorities to prohibit landings and transshipments when the catch has been taken in a manner that undermines the effectiveness of the management regime.[206] The CCAMLR catch documentation scheme (CDS) and the trade documentation schemes in the WCPFC, CCSBT, IATTC, ICCAT, and IOTC are considered to be compatible with the WTO requirements as long as they meet the transparency and nondiscrimination requirements.[207]

The WTO has a well-defined dispute settlement process to assist countries in resolving differences arising from the interpretation or application of the General Agreement on Trade and Tariff (GATT) 1994 and the other Uruguay Round Agreements.[208] Several countries have challenged trade embargoes on highly migratory fish stocks through GATT–WTO dispute settlement panels, for example, the Mexico – United States and the European Union (EU) – United States tuna–dolphin cases[209] and the EU–Chile swordfish case.[210] Decisions by GATT panels reflect a general preference for multilateral solutions to environmental problems, rather than unilateral trade action.

In 2001 at the Fourth WTO Ministerial Conference in Doha, WTO members agreed to launch negotiations to address how WTO rules apply to WTO members that are also parties to environmental agreements, and to clarify and improve WTO rules applicable to fisheries subsidies.[211] Members expressed particular concern about subsidies that lead to overcapacity and overfishing, and about trade restrictions and distortions arising from conservation and management measures taken by States.[212] Unfortunately, the 2003 Cancun meeting to assess progress on the Doha Development Agenda broke down because of marked divisions over all key areas under negotiation.[213] Perverse subsidies and their linkage to the problem of overcapacity and overfishing remain a key systemic barrier to achieving sustainable development.

The role of trade-related tools, such as certification and labeling, is growing in importance in the efforts of RFMOs to address the decline of certain fisheries and promote compliance with fisheries conservation and management regimes. A set of voluntary guidelines for ecolabelling[214] of fish products was adopted during the 26th session of COFI held 7–11 March 2005. Given the impact of these tools, considerable efforts have been made to ensure that trade-related environmental measures are consistent with WTO rules and form part of a coherent legal framework in support of sustainable development. To this end, further coordination and cooperation have been called for among international institutions, including RFMOs, FAO, UNEP, CITES, and CBD, to ensure that such measures do not create unnecessary obstacles to international trade, but enable fair competition and respect the sovereign rights of all States in the effort to promote implementation of and compliance with conservation and management regimes.[215]

2.4.14.6 Agreement on the Conservation of Albatrosses and Petrels (2001)

The 2001 Agreement on the Conservation of Albatrosses and Petrels (ACAP)[216] was developed under the auspices of the Convention on the Conservation of Migratory Species of Wild Animals (Bonn Convention). Its overall objective is to achieve and maintain a favourable conservation status for albatrosses and petrels.[217] ACAP is the first multilateral agreement seeking an integrated and holistic approach to promote the conservation of these birds, which are among the most threatened in the world.[218] ACAP entered into force on 1 February 2004. Prior to the May 2006 Review Conference 8 of the 11 signatories had ratified the Agreement.[219]

States Parties have a general obligation to take unilateral or multilateral conservation measures, at a minimum those listed in Article III and in the Action Plan included in Annex 2 of the ACAP. In implementing such measures, States shall widely apply the precautionary approach.[220] ACAP further stipulates that States shall cooperate to exchange data and other information, expertise, techniques, and knowledge, as well as to develop training and public awareness programs, thereby placing emphasis on multilateral rather than unilateral action.[221]

The decision-making body of the Agreement is the Meeting of the Parties,[222] which held its first session in Hobart, Australia, 10–12 November 2004. As called for by the ACAP,[223] the Meeting of the Parties established an Advisory Committee and a work program for this committee, which will provide expert advice and information to Parties to the ACAP, the Secretariat, and others.[224] The permanent Secretariat, which is still to be established,[225] will have its headquarters in Hobart. ACAP also establishes procedures for the settlement of disputes, including the possibility of referring disputes of a technical nature to a technical arbitration panel. It is noteworthy that in cases of conflict with LOSC, the Antarctic Treaty, or CCAMLR's Convention, the latter prevail.

2.5 Summary and conclusion

In creating "a new framework for the conduct of maritime affairs",[226] LOSC produced a dramatic change in the configuration of maritime areas and the economic distribution of marine resources. Approximately 10 years after LOSC was opened for signature, the international community, through UNCED, recognized the need for further development and adjustment of high seas fishing regimes. A consensus emerged on the need for strengthening the multilateral conservation and management of straddling fish stocks and highly migratory fish stocks through a more effective cooperative regime. COFI called for the development of new concepts that would lead to responsible and sustainable fisheries. The Code of Conduct for Responsible Fisheries, negotiated in parallel with UNFSA, serves as a globally recognized reference for minimum international standards and practices. The FAO Compliance Agreement, the Code of Conduct's legally binding component, emphasizes the flag State's duty to effectively control vessels entitled to fly its flag. After long and difficult discussions, UNFSA, a new regime for high seas fisheries, was adopted by consensus. Even though States that have not ratified either UNFSA or the FAO Compliance Agreement are not bound by their provisions, all States are bound by the customary international law obligation to cooperate in the conservation of marine living resources.

The legal regime for the conservation and management of straddling fish stocks and highly migratory fish stocks subjects fishing both inside and outside areas under national jurisdiction to international minimum standards. Underpinning this regime are the duties to cooperate and conserve, as enshrined in LOSC. Coastal and fishing States are increasingly being expected to fulfill these duties through cooperation in RFMOs. UNFSA imposes a duty on both RFMO members and nonmembers to cooperate in the conservation and management of these international fishing resources.[227] Access to the resource is typically subject to catch restrictions, quota allocations, reporting requirements, vessel registers, and trade measures.

2.5.1 The legal framework for sustainable high seas fishing: not just words

This complex and continually evolving legal framework governing ocean use and fisheries is "not just words" but is the normative order underpinning human behaviour, State relations, cooperative arrangements, and global standards. Since UNCED, new international legally binding agreements and voluntary soft law instruments have been negotiated, adopted, and increasingly implemented to prevent overfishing, rebuild fish stocks, reduce adverse impacts on the marine ecosystem, and protect marine biodiversity to ensure the sustainable development of fisheries.

Nonbinding instruments increasingly play an important role in the legal framework underpinning the conservation and management of marine living resources.[228] "They lend support and legitimacy to the overall management regime through the manner in which they were negotiated and adopted, as well as through restating and re-emphasizing the objectives and principles enshrined in binding instruments and vice versa".[229] Together, these hard and soft law instruments serve as the foundation for a new international regime for fishing straddling fish stocks and highly migratory fish stocks, considerably qualifying freedom of fishing and requiring sustainable, responsible, and precautionary fisheries management.

These instruments, so arduously negotiated, require States to fulfill their obligations within areas under national jurisdiction and on the high seas to ensure the good order of the oceans and the sustainability of the marine living resources for present and future generations. In this "Decade of Implementation" it is now time to put these words into action, in meeting rooms throughout the world, and on the water. It is understood that politicians and bureaucrats must implement the legal framework and make management decisions in keeping with contemporary international minimum standards and practices. It is also understood that fishing vessels must comply with the conservation measures. Ultimately it is the acceptance and compliance with conservation and management measures by vessels and the ability and willingness of States to responsibly regulate and control these vessels that will result in sustainable fishing. To effect the necessary systemic and behavioral changes, however, will require a broad-based societal push. With civil society's increasing concern for declining fish stocks and the adverse impacts to the marine ecosystem from fishing activities, there will undoubtedly be more scrutiny and pressure on States and RFMOs to be transparent and accountable. To this end, in keeping with the rapidly evolving notions of participatory democracy, it is critical that stakeholders from a broad representation of society (including the fishing industry, banks, insurers, restaurants, coastal communities, NGOs, and consumers) become engaged in the effort to achieve sustainable fisheries.

Endnotes

[1] *Agreement for the Implementation of the Provisions of the United Nations Convention on the Law of the Sea of 10 December 1982 Relating to the Conservation and Management of Straddling Fish Stocks and Highly Migratory Fish Stocks*, done at New York, 4 December 1995, entered into force 11 December 2001, 2167 UNTS 88; (1995) 34 ILM 1542. Also known as the Fish Stocks Agreement, the New York Agreement and UNFA. Referred to as UNFSA. *Agreement to Promote Compliance with International Conservation and Management Measures by Fishing Vessels on the High Seas*, done at Rome, Italy, 24 November 1993, entered into force April 2003, (1994) 33 ILM 968. Also known as the Flagging Agreement. Referred to as the FAO Compliance Agreement. The FAO Compliance Agreement forms an integral part of the *FAO Code of Conduct for Responsible Fisheries*, Rome: FAO 1995. See Part III of this book for reproduced version of text of these Agreements. *United Nations Convention on the Law of the Sea*, done at Montego Bay, Jamaica, 10 December 1982, entered into force 16 November 1994, 33 ILM 1309. Referred to as LOSC (Law of the Sea Convention) as distinct from the UN Conference on Law of the Sea (UNCLOS). The language in many international legal instruments and technical reports refers to the "conservation and management of fish stocks". The inherent uncertainty, environmental variability, and the dynamic nature of the marine ecosystem make it difficult, if not impossible, to manage fish stocks — the modern management objective is to responsibly regulate fisheries (i.e., manage human activity) so that the fish stocks are within sustainable levels of biomass while taking into account ecological interconnections and possible adverse impacts on the marine ecosystem. While recognizing that it is essentially the regulation of people involved in fishing operations, managing the fisheries by regulating the parameters for exploitation and protection of the living marine resources to conserve the fish stocks is the notion underlying the use of the terminology "conservation and management of fish stocks" used in this book.

[2] The purpose of international law is to establish a normative global order to guide the behaviour of States, their citizens, and other international actors; inform decision makers; and facilitate cooperative arrangements. Traditionally, public international law has been characterized as the system of rules freely accepted by sovereign States as legally binding upon their relations and interactions with one another. Rules, norms, and principles generally are thought to form the legal framework and conceptual foundation for international legal order, international systems, and the practices of States. Rules are generally considered legally binding and enforceable laws that carry the legal weight of a treaty such as LOSC, UNFSA, the FAO Compliance Agreement, Convention on Biodiversity (CBD), Convention for International Trade of Endangered Species (CITES), and the multilateral RFMO constitutive agreements. It is not uncommon for a rule to emerge first as a norm. Norms are customs, traditions, operational policies, or standards that serve as guides for human behaviour, influencing and informing decision making. Some norms become established practice or become moral principles and are then incorporated into a legal convention. This was the genesis of many rules now embodied in multilateral treaties, including LOSC. Some common law lawyers find the term "norm" vague, even though it is popular with civil law lawyers. The dictionary definition does not help much either. The generally accepted usage seems to mean standard, which is close to the Latin origin of norm.

[3] L. Juda, "Rio Plus Ten: The Evolution of International Marine Fisheries Governance" (2002) 33 *Ocean Development and International Law* 109–144 at 116. According to Art. 38 of the Statute of the International Court of Justice, there are three primary sources of international law: international treaties, custom, and general principles of law. Art. 38 of the Statute of the International Court of Justice enumerates the primary sources of international law as (1) international conventions, whether general or particular, establishing rules expressly negotiated by the Contracting Parties; (2) international custom, as evidence of a general practice accepted as law; (3) general principles of law recognized by civilized nations; (4) judicial decisions and the teachings of the most highly qualified publicists of the various nations, as subsidiary means for the determination of rules of law.

[4] Baxter identified the term "soft law" in 1980 when he noted three types of norms in international agreements that did not create enforceable legal obligations, but that were proliferating either as independent instruments or while within binding treaties and conventions: the *pactum de contrahendo*, the non-self-executing article of treaties requiring further agreements to give them effect, and the hortatory provisions. R. Baxter, "International Law in 'her Infinite Variety'" (1980) 29 *International and Comparative Law Quarterly* 549–566.

[5] A. Boyle, "Some Reflections on Treaties and Soft Law" (1999) 48 *International and Comparative Law Quarterly* 901–913 at 901–902.

[6] A. Boyle, at 901–902.

[7] A. Boyle, at 901–902.

[8] A. Boyle, at 905.

[9] A. Boyle, at 905.

[10] A. Boyle, at 905.

[11] A. Boyle, at 906.

[12] See, for instance Art. IV.1 of the Antigua Convention. *Convention for the Strengthening of the Inter-American Tropical Tuna Commission established by the 1949 Convention between the United States of America and the Republic of Costa Rica*, done at Washington, D.C., 14 November 2003, not yet in force (referred to here as the Antigua Convention). IATTC online <http://www.iattc.org/PDFFiles2/Antigua_Convention_Jun_2003.pdf> (accessed 21 October 2005).

[13] L. Juda, at 116.

[14] For a discussion on soft law, see R. Baxter, 549–566; D. Bodansky, "Customary (and Not So Customary) International Environmental Law" (1995) 3 *Indiana Journal of Global Legal Studies* 105–119; R. Bilder, "Beyond Compliance: Helping Nations Cooperate" in D. Shelton, ed., *Commitment and Compliance: The Role of Non-Binding Norms in the International Legal System* (New York: Oxford University Press, 2000) 65–73; M. Bothe, "Legal and Non-Legal Norms: A Meaningful Distinction in International Relations?" (1980) 11 *Netherlands Yearbook of International Law* 65–95; A. Boyle, 901–913; C. Chinkin, "The Challenge of Soft Law: Development and Change in International Law" (1989) 38 *International and Comparative Law Quarterly* 850–866; D. Dzidzornu, "Coastal State Obligations and Powers Respecting EEZ Environmental Protection Under Part XII of the UNCLOS: A Descriptive Analysis" (1997) 8 *Colorado Journal of International Environmental Law and Policy* 286–321 at 317–320; H. Hillgenberg, "A Fresh Look at Soft Law" (1999) 10 *European Journal of*

International Law 499–515; L. Juda, 109–144; J. Klabbers, "The Redundancy of Soft Law" (1996) 65 *Nordic Journal of International Law* 167–182; and "The Undesirability of Soft Law" (1998) 67 *Nordic Journal of International Law* 381–391; J.M. Macdonald, "Appreciating the Precautionary Principle as an Ethical Evolution in Ocean Management" (1995) 26 *Ocean Development and International Law* 255–286; D. Shelton, "Introduction: Law, Non-Law and the Problem of 'Soft Law'" in D. Shelton, ed., *Commitment and Compliance: The Role of Non-Binding Norms in the International Legal System* (New York: Oxford University Press, 2000) 1–18; F. Orrego Vicuña, "The Law Governing High Seas Fisheries: In Search of New Principles" in A. Chircop and M.L. McConnell, eds., *Ocean Yearbook 18* (Chicago: University of Chicago Press, 2004) 383–394.

[15] As noted by R. Bilder, "Beyond Compliance: Helping Nations Cooperate" at 66: "Because applicable norms are arrived at by largely voluntaristic and consensual means, a State's intention and decision about how willing or not willing it is to conform its behaviour to the norm typically will be reached and manifested, not as a separate decision reached after the norm is formulated, but in the very process of the norm's formation. Not only are the initial compliance decision and intention usually inextricably intertwined with the formation and nature of the norm, as publicly expressed, rather than independent of them, but the content, form, and contextual elements of the norm generally will reflect and signal the participating states' broadly shared intentions and expectations regarding compliance".

[16] S. Rosenne, *Practice and Methods of International Law* (London, New York: Ocean Publications, 1984) at 55.

[17] See discussion in S. Rosenne, at 55.

[18] A. Boyle, at 903 and 906, indicates this point is shown by International Court of Justice case law on the interplay between the UN Charter and resolutions of the UN General Assembly.

[19] A. Boyle, at 904.

[20] See Vienna Convention Arts. 34–38.

[21] This is confirmed by the decision of the International Court of Justice in the *Nicaragua* case, 1986 ICJ Rep. 14. See also R. Rayfuse, "To Our Children's Children's Children: From Promoting to Achieving Compliance in High Seas Fisheries" (2005) 20 *International Journal of Marine and Coastal Law* 509–532 at 511 note 11.

[22] E. Molenaar, "Global, Regional and Unilateral Approaches to Unregulated Deep-Sea Fisheries" in R. Shotton, ed., *Deep Sea 2003: Conference on the Governance and Management of Deep-sea Fisheries. Part I: Conference reports. Queenstown, New Zealand, 1–5 December 2003*, FAO Fisheries Proceedings No. 3/1. (Rome: FAO, 2005) 415–432 at 422; R. Rayfuse, at 525.

[23] R. Rayfuse, at 525–526.

[24] E. Molenaar, at 422.

[25] UN Doc. A/51/34 of 9 December 1996.

[26] The 2005 report of the Secretary-General on Oceans and the Law of the Sea dealt with issues relating to the conservation and sustainable use of marine biodiversity beyond areas of national jurisdiction, UN Doc. A/60/63/Add. 1 of 15 July 2005.

[27] Large-scale pelagic driftnet fishing; unauthorized fishing in zones of national jurisdiction; and fisheries bycatch and discards. Prior to 1996, fisheries issues were addressed in separate reports; in 1996, the Secretary-General began issuing a consolidated report on the issues, see UN Doc. A/51/404 of 25 September 1996. Moreover, the Secretary-General was requested to submit a report to assist the UNFSA Review Conference to implement its mandate, see UN Doc. A/59/25 of 17 November 2004, para. 17.

[28] See the report of the last CSD Session, UN Commission on Sustainable Development, *Report on the thirteenth session (30 April 2004 and 11–22 April 2005)*, UN Doc. E/CN.17/2005/12.

[29] In adopting the Code of Conduct, FAO member countries requested that the organization respond to the special requirements of developing countries through an Interregional Assistance Programme for its implementation. FishCode was thus established by the FAO Fisheries Department as a special program of global partnerships to promote responsible fisheries. See the FAO FishCode Web site <http://www.fao.org/figis/servlet/static?xml=fishcode_prog.xml&dom=org&xp_nav=1> (accessed 19 January 2005).

[30] See A. Rengifo, "Protection of Marine Biodiversity: A New Generation of Fisheries Agreements" (1997) 6/3 *Review of European Community and International Environmental Law* 313–321.

[31] "Building upon the original regimes established in the four 1958 Geneva Conventions, and developments in state practice that took place in the 1960s and 1970s, the LOS Convention completed the patchwork of maritime zones that are a cornerstone of the modern law of the sea". See D. Rothwell, "Building on the Strengths and Addressing the Challenges: The Role of Law of the Sea Institutions" (2004) 35 *Ocean Development and International Law* 131–156 at 132.

[32] R. Macdonald, "The Charter of the United Nations in Constitutional Perspective" (1999) 20 *The Australian Yearbook of International Law* 205–231 at 220 with further references.

[33] This is to be distinguished from John Selden's proposed *mare clausum*.

[34] For centuries, the area beyond the relatively narrow territorial sea belt was considered the high seas; this notion was later included in the 1958 Convention on the High Seas recognizing a three-mile territorial sea.

[35] LOSC Arts. 63(2) and 64(1).

[36] These and other instruments are discussed in detail later in this chapter.

[37] MSY is discussed in greater detail in Chap. 5.

[38] The PA is discussed in further detail in Chap. 6.

[39] The EA is discussed in further detail in Chap. 7.

[40] Monitoring, control, and surveillance (MCS) is discussed in further detail in Chap. 9.

[41] The FAO IPOAs (1999, 2001) are discussed further below in this chapter at 2.4.6.4.

[42] CITES is discussed further below in this chapter at 2.4.14.2.

[43] FAO, "Precautionary Approach to Capture Fisheries and Species Introductions", FAO Technical Guideline for Responsible Fisheries No. 2 (Rome: FAO, 1996); FAO, "Fisheries Management 2. The Ecosystem Approach to Fisheries", FAO Technical Guideline for Respon-

sible Fisheries No. 4 Suppl. 2 (Rome: FAO, 2003); FAO online <http://www.fao.org/fi/eims_search/advanced_s_result.asp?series=116&lang=en&sortorder=5&form_c=AND> (accessed 17 January 2006).

[44] *Convention for the Strengthening of the Inter-American Tropical Tuna Commission established by the 1949 Convention between the United States of America and the Republic of Costa Rica*, "Antigua Convention", done at Washington, D.C., 14 November 2003, not yet in force, IATTC online <http://www.iattc.org/PDFFiles2/Antigua_Convention_Jun_2003.pdf> (accessed 21 October 2005).

[45] *Convention on the Conservation and Management of Fishery Resources in the South-East Atlantic Ocean*, done at Windhoek, Namibia, 20 April 2001, entered into force 30 April 2003, SEAFO online <http://www.seafo.org/> (accessed 21 October 2005).

[46] *Southern Indian Ocean Fisheries Arrangement* (proposed).

[47] *Convention on the Conservation and Management of Highly Migratory Fish Stocks in the Western and Central Pacific Ocean*, done at Honolulu, Hawaii, 5 September 2000, entered into force 19 June 2004, 2004 ATS 15.

[48] R. Macdonald, "The Charter of the United Nations in Constitutional Perspective" at 220.

[49] Universal acceptance of LOSC and UNFSA and universality of participation are desirable. On the issue of participation in UNFSA, see M. Lodge and S. Nandan, "Some Suggestions Towards Better Implementation of the United Nations Agreement on Straddling Fish Stocks and Highly Migratory Fish Stocks of 1995" (2005) 20 *International Journal of Marine and Coastal Law* 345–379 at 352–354: "It is vitally important that all parties to the 1982 Convention become parties to UNF[S]A, so that, as originally intended, there is a seamless connection between the provisions of the Convention and the provisions of the implementing agreement".

[50] It could also be argued that the obligations of conservation and sustainable development devolve on all States in keeping with the evolving principle of solidarity. See R. Macdonald, "Solidarity in the Practice and Discourse of Public International Law" (1996) 8 *Pace International Law Review* 259–302. This is in keeping with the need for "mutual supportiveness" called for at UNCED to achieve sustainable development in advancing the common good. Furthermore, most provisions of LOSC are being recognized as having become part of international customary law. Given that UNFSA was negotiated in parallel with the non legally binding Code of Conduct, which enjoys widespread support and includes many of the same principles and provisions, there is evidence that some of these principles are also becoming part of customary international law and therefore binding on all States. See E. Brown, "The Law of the Sea at the Millenium" in H. Smith, ed., *The Oceans: Key Issues in Marine Affairs* (Dordrecht: Kluwer Academic Publishers, 2004) 25–62. See also E. Molenaar, at 9; R. Rayfuse, at 525–526.

[51] D. Rothwell, at 148.

[52] S. Kaye, "Implementing High Seas Biodiversity Conservation: Global Geopolitical Considerations" (2004) 28 *Marine Policy* 221–226 at 222.

[53] High Seas Task Force 2006, "Closing the Net: Stopping Illegal Fishing on the High Seas", Final Report of the Ministerially-led Task Force on IUU Fishing on the High Seas, Governments of Australia, Canada, Chile, Namibia, New Zealand, and the United Kingdom, WWF, IUCN, and the Earth Institute at Columbia University. For a brief account of the introduction of the acronym IUU at the international level (first in CCAMLR in 1997 and then in 1998 in ICCAT) see E. Molenaar, "Participation, Allocation and Unregulated Fishing: The Practice of Regional Fisheries Management Organisations" (2003) 18(4) *International Journal of Marine and Coastal Law* 457–480 at 459.

[54] *Conference on the Governance of High Seas Fisheries and the UN Fish Agreement*, St. John's, N.L., 5 May 2005. See online <http://www.dfo-mpo.gc.ca/fgc-cgp/declaration_e.htm> (accessed 7 January 2006).

[55] See "Marine Protected Areas (MPAs) and Fisheries", FAO COFI, 26th Session, Rome, Italy, 7–11 March 2005. See also IUCN, *Ten-Year High Seas Marine Protected Area Strategy: A Ten-year Strategy to Promote the Development of a Global Representative System of High Seas Marine Protected Area Networks*, as agreed by Marine Theme Participants at the Vth IUCN World Parks Congress, Durban, South Africa, 8–17 September 2003.

[56] CBD at Arts. 4(b), 6(a) and (b), and 22(2).

[57] A number of international instruments of relevance to fisheries refer to MPAs as essential tools for conserving marine resources and managing fisheries. These include the 1995 Jakarta Mandate on Marine and Coastal Biological Diversity within the framework of the CBD and the WSSD Plan of Implementation. The FAO Code of Conduct does not refer explicitly to MPAs but provides for, *inter alia*, protection and rehabilitation of critical habitats from all degradation. The related technical guidelines for fisheries management and the ecosystem approach to fisheries recognize that MPAs could play a critical role in ensuring sustainable fishing. See "Marine Protected Areas (MPAs) and Fisheries", FAO COFI, 26th Session, Rome, Italy, 7–11 March 2005, at para. 3.

[58] FAO COFI, 26th Session, Rome, Italy, 7–11 March 2005, at para. 3

[59] See G. Russ and D. Zeller, "From *Mare Liberum* to *Mare Reservarum*" (2003) 27 *Marine Policy* 75–78.

[60] E. Miles, "Preparing for UNCLOS IV", *Ocean Policy News* — Special Report July 1988; Y.K. Tyagi, "Towards UNCLOS IV?" (1993) 27/1 *Journal of World Trade* 143–154. See also the "RUNCLOS" project (ratification, review, and possible revision of the LOSC). The first public or semipublic discussion of the prospect of review (and potential revision) of the Convention and RUNCLOS took place at the SEAPOL Inter-Regional Conference, Bangkok, Thailand, 21–23 March 2001. The roundtable discussion was chaired by Professor Edward Miles. See SEAPOL online <http://www.seapol.org/NEWS%20LETTER%201.htm> (accessed 21 January 2006). Rothwell, however, holds the view that "there is not at present any serious international agenda for radical reform of the law or support for a Fourth UN Conference on the Law of the Sea". See D. Rothwell, at 149.

[61] See, generally, D. Rothwell, at 149.

[62] J. Norton Moore, "The Rule of Law in the Oceans" in M. Nordquist and J. Norton Moore, eds. *Security Flashpoints: Oil, Islands, Sea Access and Military Confrontation* (The Hague/Boston: Martinus Nijhoff Publishers, 1998) 467–475.

[63] D. Rothwell, at 131–132.

[64] LOSC Art. 87(2). Art. 87 of LOSC endorsed the concept that all States are free to fish on the high seas and vested the flag State with the responsibility for controlling the activities of its vessels. See D. Anderson, "The Straddling Stocks Agreement of 1995 — An Initial Assessment" (1996) 45 *International and Comparative Law Quarterly* 463–475.

[65] LOSC Art. 87(2).

[66] B. Kuemlangen, "Legal Implications and Options for Regional Cooperative Arrangements for the Management of Coastal and Oceanic Large Pelagic Resources for CARICOM States" in R. Mahon and P McConney, eds., *Management of Large Pelagic Fisheries in CARICOM Countries*, FAO Fisheries Technical Paper No. 464 (Rome: FAO, 2004) 5–14 at 7.

[67] See T. Løbach, *Background Paper — Workshop on the Vietnamese Accession to the 1995 UN Fish Stocks Agreement*, Hanoi, Vietnam, 22–23 February 2006, on file with the author.

[68] *Rio Declaration on Environment and Development*, Rio de Janeiro, Brazil, 14 June 1992; 31 ILM 874 referred to as the Rio Declaration.

[69] Report of the United Nations Conference on Environment and Development, UN Doc. A/CONF.151/26/REV.1, Vols. 1–III (1992) referred to as Agenda 21.

[70] Agenda 21 states "integrated in content and are precautionary and anticipatory in ambit". Report of the United Nations Conference on Environment and Development, UN Doc. A/CONF.151/26/REV.1, Vols. 1–III (1992).

[71] Chapter 17 of Agenda 21 addresses the "protection of the oceans, all kinds of seas, including enclosed seas and semi-enclosed seas, and coastal areas and the protection, rational use and development of their living resources".

[72] The issue of high seas fisheries proved to be among the most contentious during the extensive negotiations of Chapter 17 of Agenda 21. At the conclusion of PrepCom III, which was held in August 1991 in Geneva, straddling fish stocks and highly migratory fish stocks were one of five unresolved issues in the section on living marine resources. The issue proved so divisive that the Secretariat chose not to prepare text on high seas fisheries in its draft of Chapter 17 of Agenda 21 for the start of PrepCom IV, held in March 1992 in New York. For nearly four of the five weeks of PrepCom IV, a contact group on marine living resources (Programme Areas C and D of Chapter 17) met behind closed doors and attempted to reach consensus. By the conclusion of PrepCom IV, consensus remained elusive and the Plenary adopted the Oceans chapter with brackets around the paragraphs on straddling and highly migratory fish stocks.

[73] Agenda 21, at para. 17.45.

[74] International Institute for Sustainable Development (IISD), "A Summary of the Proceedings of the United Nations Conference on Environment and Development 3–14 June 1992", Earth Negotiations Bulletin No. ENB:02:13, Sec. II, Chap. 17 (Winnipeg, Man.: IISD, 1992). Available at <http://www.iisd.ca/vol02/0213017e.html>.

[75] Testimony of D. Balton, Deputy Assistant Secretary of State for Oceans and Fisheries, before the United States House Committee on Resources, Subcommittee on Fisheries Conservation, Wildlife and Oceans, 11 September 2003, US Congress online <http://resourcescommittee.house.gov/108cong/fish/2003sep11/balton.htm> (accessed 21 October 2005).

[76] Testimony of D. Balton, 11 September 2003.

[77] FAO Compliance Agreement Arts. V and VI, and in respect to nonparties Art. VIII.3.

[78] FAO Compliance Agreement Art. III.5.a.ii. The Forum Fisheries Agency (FFA) registry mechanism for monitoring and control was imported into the FAO Compliance Agreement. FFA had taken several steps regionally, independently of the FAO Compliance Agreement, to include violations by vessel operator. Noncompliance would result in removal from the registry, which would foreclose admission into the EEZ. This had significant economic consequences for fishing operations such as tuna, which are primarily found within the EEZs of the FFA member States.

[79] FAO Compliance Agreement Art. III.5.d.

[80] The FAO Compliance Agreement has qualified exemption provisions for fishing vessels less than 24 m in length. The State, however, if it has granted an exemption pursuant to Art. II(2) must take effective measures against such vessels if they have engaged in fishing practices that undermine international conservation measures, including halting fishing operations (Art. III).

[81] FAO Compliance Agreement Art. III.8.

[82] FAO Compliance Agreement Art. III.8.

[83] See C. Hedley, "Entry into force of the United Nations Fish Stocks Agreement: An Initial Assessment" (2001) 24 *International Fisheries Bulletin* at 10. Available at <http://www.intfish.net/ops/papers/7.htm>.

[84] FAO Compliance Agreement Arts. VI.1 and VI.2.

[85] FAO Compliance Agreement Arts. VI.1–VI.5.

[86] See G. Moore, "FAO — The Food and Agriculture Organisation of the United Nations Compliance Agreement" (1995) 10 *International Journal of Marine and Coastal Law* 412–425 at 414.

[87] FAO's COFI at its 19th session in March 1991, prior to UNCED, first identified the need. S. Garcia, personal communication, March 2005.

[88] FAO, *Code of Conduct for Responsible Fisheries* (Rome: FAO, 1995), at Preface, p. v.

[89] G. Moore, "Chapter 5: The Code of Conduct for Responsible Fisheries" in E. Hey, ed., *Developments in International Fisheries Law* (The Hague: Kluwer Law International, 1999) 85–105 at 85.

[90] Preamble, FAO Compliance Agreement and Art. 1 Code of Conduct.

[91] The FAO Council in June 1995 approved the work carried out by the Technical Committee revising the draft Code and endorsed its recommendation that another session be held in September 1995, allowing adequate time following the expected conclusion of the August 1995 UN Fish Stocks Conference for the drafters of the Code to harmonize the text linguistically and juridically, taking into account the outcome of UNFSA.

[92] See, in particular, Art. 7 of the Code of Conduct.

[93] The FAO Council in March 1995 issued the Rome Consensus on World Fisheries urging that "Governments and international organizations take prompt action to complete the International Code of Conduct for Responsible Fisheries with a view to submitting the final text

to the FAO Conference in October 1995", *The Rome Consensus on World Fisheries* adopted by the FAO Ministerial Conference on Fisheries, Rome, Italy, 14–15 March 1995, FAO Doc. AC441/E (1995), at para. 10.

[94] "Education in environmental matters, for the younger generation as well as adults, giving due consideration to the underprivileged, is essential in order to broaden the basis for an enlightened opinion and responsible conduct by individuals, enterprises and communities in protecting and improving the environment in its full human dimension. It is also essential that mass media of communications avoid contributing to the deterioration of the environment, but, on the contrary, disseminates information of an educational nature on the need to project and improve the environment in order to enable man to develop in every respect", see Principle 19 of the *Declaration of the United Nations Conference on the Environment*, Stockholm, 16 June 1972, (1972) 11 ILM 1416 (referred to here as the Stockholm Declaration).

[95] Code of Conduct Art. 6.1.

[96] W. Edeson, cited by W. Burke, "Evolution in the Fisheries Provisions of UNCLOS" in N. Ando, E. McWhinney, R. Wolfrum, and B. Röben, eds., *Liber Amicorum Judge Shigeru Oda*, Vol. 2 (The Hague: Kluwer Law International, 2002) 1355–1362 at 1361.

[97] To date there are 13 FAO Technical Guidelines for Responsible Fisheries, among others: Fishing Operations (No. 1 (1996)), Precautionary Approach to Capture Fisheries and Species Introductions (No. 2 (1996)), Fisheries Management (No. 4 (1997)), VMS (No. 1 Suppl. 1 (1998)), Indicators for Sustainable Development of Marine Capture Fisheries (No. 8 (1999)), Implementation of the IPOA to Deter, Prevent and Eliminate IUU Fishing (No. 9 (2002)), The Ecosystem Approach to Fisheries (No. 4 Suppl. 2 (2003)); FAO online <http://www.fao.org/fi/eims_search/advanced_s_result.asp?series=116&lang=en&sortorder=5&form_c=AND> (accessed 17 January 2006).

[98] The four FAO IPOAs are discussed further below in this chapter.

[99] FAO, *Strategy for Improving Information on Status and Trends of Capture Fisheries* (Rome: FAO, 2003), FAO online <http://www.fao.org/documents/show_cdr.asp?url_file=/DOCREP/006/Y4859T/Y4859T00.HTM> (accessed 17 January 2006).

[100] S. Garcia and D. Doulman, "FAO's Fisheries Programme and the Plan of Implementation of the World Summit on Sustainable Development" in S. Ebbin, A. Håkon Hoel, and A. Sydnes, eds., *A Sea Change: The Exclusive Economic Zone and Governance Institutions for Living Marine Resources* (Dordrecht: Springer, 2005) 169–193 at 170.

[101] S. Garcia and D. Doulman, at 170.

[102] S. Garcia and D. Doulman, at 170

[103] S. Garcia and D. Doulman, at 170.

[104] *Convention for the Strengthening of the Inter-American Tropical Tuna Commission established by the 1949 Convention between the United States of America and the Republic of Costa Rica*, "Antigua Convention", done at Washington, D.C., 14 November 2003, not yet in force, IATTC online <http://www.iattc.org/PDFFiles2/Antigua_Convention_Jun_2003.pdf> (accessed 21 October 2005).

[105] UNFSA Art 2.

[106] See Y. Fortier, "From Confrontation to Cooperation on the High Seas: Recent Developments in International Law Concerning the Conservation of Marine Resources" in N. Ando, E. McWhinney, R. Wolfrum, and B. Röben, eds., *Liber Amicorum Judge Shigeru Oda*, Vol. 2, 1377–1390 at 1378.

[107] D. Anderson, "The Straddling Stocks Agreement of 1995 — An Initial Assessment at 475.

[108] UNFSA Arts. 3, 5, 6, and 7.

[109] UNFSA Art. 6(1).

[110] These points are further elaborated upon in Chaps. 5 and 6.

[111] For a discussion of how MSY under LOSC morphed into a limit see D. Christie, "The Conservation and Management of Stocks Located Solely Within the Exclusive Economic Zone" in E. Hey, ed., *Developments in International Fisheries Law* (The Hague: Kluwer Law International, 1999) 395–419 at 402. See also Chap. 5 for a fuller discussion on the interpretation and reformulation of MSY.

[112] Some RFMOs, e.g., NAFO and SEAFO, have already included discrete stocks in the scope of their management mandate.

[113] UNFSA Art. 8.

[114] P. Birnie, "The Conservation and Management of Marine Mammals and Anadromous and Catadromous Species" in E. Hey, ed., *Developments in International Fisheries Law* (The Hague: Kluwer Law International, 1999) 357–393.

[115] See also the discussion concerning the FAO Compliance Agreement at 2.4.3 above.

[116] M. Lodge and S. Nandan, "Some Suggestions Towards Better Implementation of the United Nations Agreement on Straddling Fish Stocks and Highly Migratory Fish Stocks of 1995" (2005) 20 *International Journal of Marine and Coastal Law* 345–379 at 363.

[117] UNFSA Art. 23.

[118] This International Plan of Action for Conservation and Management of Sharks (IPOA-Sharks) has been developed through the meeting of the Technical Working Group on the Conservation and Management of Sharks in Tokyo from 23 to 27 April 1998 and the Consultation on Management of Fishing Capacity, Shark Fisheries and Incidental Catch of Seabirds in Longline Fisheries held in Rome from 26 to 30 October 1998 and its preparatory meeting held in Rome from 22 to 24 July 1998. See FAO, "Report of the FAO Technical Working Group on the Conservation and Management of Sharks", Tokyo, Japan, 23–27 April 1998, FAO Fisheries Report No. 583 (Rome: FAO, 1999). FAO, "Report of the Preparatory Meeting for the Consultation on the Management of Fishing Capacity, Shark Fisheries and Incidental Catch of Seabirds in Longline Fisheries", Rome, Italy, 22–24 July 1998, FAO Fisheries Report No. 584 (Rome: FAO, 1998). FAO, "Report of the Consultation on the Management of Fishing Capacity, Shark Fisheries and Incidental Catch of Seabirds in Longline Fisheries", Rome, Italy, 26–30 October 1998, FAO Fisheries Report No. 593 (Rome: FAO, 1998).

[119] IPOA-Sharks paras. 18, 21, 24, and 28.

[120] IATTC, Resolution on the Conservation of Sharks Caught in Association with Fisheries in the Eastern Pacific Ocean, Resolution C-05-03, adopted at the 73rd Meeting, 20–24 June 2005; ICCAT, Recommendation by ICCAT Concerning the Conservation of Sharks Caught in Association with Fisheries Managed by ICCAT, [Rec. 04-10], adopted in 2004, and the amendment, Recommendation [Rec. 05-5], adopted in 2005; IOTC, Resolution 05/05 Concerning the Conservation of Sharks Caught in Association with Fisheries Managed by

IOTC, adopted in 2005. At its 27th Annual Meeting in September 2005. NAFO adopted measures banning shark finning and the transshipment and landing of shark fins, cf. NAFO/FC Doc. 05/8.

[121] The International Plan of Action for Reducing Incidental Catch of Seabirds in Longline Fisheries (IPOA-Seabirds) was developed through the meeting of a Technical Working Group held in Tokyo, Japan, 25–27 March 1998, and the Consultation on the Management of Fishing Capacity, Shark Fisheries and Incidental Catch of Seabirds in Longline Fisheries held 26–30 October 1998 and its preparatory meeting held in Rome, Italy, 22–24 July 1998. See FAO, "Report of the FAO Technical Working Group Meeting on Reduction of Incidental Catch of Seabirds in Longline Fisheries", Tokyo, Japan, 25–27 March 1998, FAO Fisheries Report No. 585 (Rome: FAO, 1999); and FAO Fisheries Report No. 584.

[122] Technical Note on Some Optional Technical and Operational Measures for Reducing the Incidental Catch of Seabirds in Longline Fisheries, part of the IPOA-Seabirds, includes both technical and operational measures.

[123] IPOA-Capacity part III, ss. I and II, and at paras. 30, 35, and 42.

[124] IPOA-Capacity para. 18.

[125] For further discussion on the issue of overcapacity please see Chaps. 5 and 7. For a historical examination and overview of the issue of global capacity growth see S.M. Garcia and C. Newton, "Current Situation, Trends and Prospects in World Capture Fisheries" in E. Pikitch, D. Huppert, and M. Sissenwine, eds., *Global Trends: Fisheries Management* (AFS Symposium 20) (Bethesda, Md.: American Fisheries Society Publication, 1997) 3–27.

[126] Introduced at the 23rd Session of COFI in 1999, the IPOA-IUU was approved by COFI on 2 March 2001 and formally adopted by the FAO Council in June 2002; see FAO, "Report of the Twenty-fourth Session of the Committee on Fisheries", Rome, Italy, 26 February – 2 March 2001, FAO Fisheries Report No. 655. FAO Doc. FIPL/R655.

[127] Testimony of D. Balton, 11 September 2003.

[128] Testimony of D. Balton, 11 September 2003. See IPOA-IUU, Part III, sec. 9, which outlines principles and strategies for eliminating IUU fishing.

[129] The IPOA-IUU is discussed in more detail in Chaps. 4 and 9, secs. 4.5.3 and 9.4, respectively.

[130] *Reykjavik Conference on Responsible Fisheries in the Marine Ecosystem*, Reykjavik, Iceland, 4 October 2001, available at <http://www.fao.org/docrep/meeting/004/Y2211e.htm> (accessed 17 January 2006) Referred to as the Reykjavik Declaration.

[131] See Reykjavik Declaration, para. 10 of the Preamble.

[132] Reykjavik Declaration, para. 3.

[133] *Johannesburg Declaration on Sustainable Development*, Report of the World Summit on Sustainable Development, Johannesburg, South Africa, 4 September 2002, UN Doc. A/CONF.199/20. Referred to as the Johannesburg Declaration.

[134] *Johannesburg Plan of Implementation*, Report of the World Summit on Sustainable Development, Johannesburg, South Africa, 4 September 2002, UN Doc. A/CONF.199/20.

[135] S. Garcia and D. Doulman, at 174.

[136] Decision V/6 on the Ecosystem Approach, adopted by the Conference of Parties to the Convention on Biological Diversity at its fifth meeting in Nairobi in May 2000, is reproduced at <http://www.biodiv.org/decisions>.

[137] Plan of Implementation, at paras. 31(b), (c), (e), and (f).

[138] See LOSC Arts. 61(2) and 119(1a); UNFSA Arts. 5(b), 6(3)(a),6(3)(b), 6(7), and 16(1). A fundamental feature of UNFSA, as expressed in Annexes I and II.

[139] FAO, "Strategy for Improving Information on Status and Trends of Capture Fisheries" in *Report of the Twenty-Fifth Session of the Committee on Fisheries, Rome, 24–28 February 2003*, FAO Fisheries Report No. 702 (Rome: FAO, 2003) at 80. Para. 4 of the Strategy outlines the international instruments requiring information on the status and trends of fisheries, including UNFSA, the FAO Compliance Agreement (Arts. VI), the Code of Conduct (Arts. 6.4, 7.4, 12), IPOA-Capacity; Chap. 40 of Agenda 21, which requires States to report on sustainability factors, and international conventions, such as the CITES and the CBD, which require collection and exchange of information on the status of biota.

[140] FAO Fisheries Report No. 702, Appendix H. The Strategy was adopted by consensus at the 25th Session of COFI in February 2003 and endorsed by the FAO Council in June 2003.

[141] Strategy-STF para. 12.

[142] Strategy-STF para. 29.

[143] Strategy-STF para. 31.

[144] Strategy-STF paras. 23–45.

[145] Strategy-STF para. 1.

[146] United Nations Dag Hammarskjöld Library, "General Assembly: Resolutions/Decisions", United Nations Documentation: Research Guide. Online <http://www.un.org/Depts/dhl/resguide/gares.htm>.

[147] See UN General Assembly, *Oceans and the Law of the Sea*, UN Doc. A/60/L.22, 60th session (17 November 2005) (adopted as Resolution 60/30, 29 November 2005). The annual resolution on the oceans and law of the sea emphasizes the integrated management and sustainable development of the oceans and the seas. The resolution does not deal specifically with fisheries, although it does elaborate on measures related to the marine environment, marine resources, marine biodiversity, and the protection of vulnerable marine ecosystems (paras. 61–84) and calls on States to become a Party to UNFSA (para. 3).

[148] UN General Assembly, *Large-scale pelagic drift-net fishing and its impact on the living marine resources of the world's oceans and seas*, UN Doc. A/RES/46/215, 46th Session, at 147. Adopted 20 December 1991. Adopted in 1991 as a follow-up on UNGA Resolutions 44/225 and 45/197, adopted in 1989 and 1990, respectively, these resolutions called attention to large-scale pelagic driftnet fishing and its detrimental impact on the living marine resources; see UN General Assembly, *Large-scale pelagic drift-net fishing and its impact on the living marine resources of the world's oceans and seas*, UN Doc. A/RES/44/225, 85th Session, adopted 22 December 1989, and UN Doc. A/RES/45/197, 71st

Session, adopted 21 December 1990. UNGA Resolution 46/215 established a global moratorium on large-scale pelagic driftnet fishing on the high seas, including enclosed and semienclosed seas, effective from 31 December 1992. It encouraged all members of the international community to take measures individually and collectively to prevent large-scale pelagic driftnet fishing operations on the high seas.

[149] UN General Assembly, *Fisheries by-catch and discards and their impact on the sustainable use of the world's living marine resources*, UN Doc. A/RES/49/118, 49th Session (2 February 1995). Adopted by consensus on 19 December 1994. The resolution called on FAO to include provisions on bycatch and discards in the Code of Conduct, and the UN Fish Stocks Conference to elaborate bycatch and discard provisions. Likewise, RFMOs were asked to review the impact of bycatch and discards on the sustainable use of living marine resources within their respective areas of competence.

[150] For example, Resolution 58/14 of 24 November 2003, specifically mentioned UNFSA, the FAO Compliance Agreement, the FAO IPOAs and the Strategy for Improving Information on Strategy and Trends, the Reykjavik Declaration, Decisions V/6 and V/12 of the Conference of Parties to the Convention on Biological Diversity (discussed below at 2.4.14.4), and the WSSD Plan of Implementation.

[151] UN General Assembly, *Oceans and the law of the sea: sustainable fisheries, including through the 1995 Agreement for the Implementation of the Provisions of the United Nations Convention on the Law of the Sea of 10 December 1982 relating to the Conservation and Management of Straddling Stocks and Highly Migratory Fish Stocks, and related instruments*, UN Doc. A/60/L.23, 60th session (17 November 2005) (adopted as Resolution 60/31, 29 November 2005).

[152] See UN General Assembly, UN Doc. A/60/L.23, 60th session at paras. 33–46.

[153] See Report of the UN Secretary-General, *Report submitted in accordance with paragraph 17 of General Assembly resolution 59/25 of 17 November 2004, to assist the Review Conference to implement its mandate under paragraph 2, article 36 of the United Nations Fish Stocks Agreement*, UN Doc. A/CONF.210/2006/1 at para. 6.

[154] UN Doc. A/59/24 of 4 February 2005, para. 73.

[155] For the CSD recommendation, see the CSD Report of the Seventh Session. Supplement No. 9, 1999, Decision 7/1. Oceans and seas, at paras. 38–45.

[156] UNGA Resolution 54/33, 24 November 1999.

[157] UN Doc. A/54/33 of 18 January 2000, para. 2.

[158] UN General Assembly, *Report on the work of the United Nations Open-ended Informal Consultative Process on Oceans and the Law of the Sea at its sixth meeting*, UN Doc. A/60/99, 60th session (7 July 2005); see also "Summary of the Sixth Meeting of the Open-Ended Informal Consultative Process on Oceans and the Law of the Sea: 6–10 June 2005" (2005) 25/18 *Earth Negotiations Bulletin* 1–14 at 2.

[159] UN Doc. A/56/13, para. 8.

[160] The outcomes of the meeting include a timeline and program of work for the preparation for the conference, a draft agenda for the preparatory meeting, and a set of recommendations to the 60th session of the General Assembly related to the preparatory work and convening of the conference and its preparatory meeting, cf. the meeting report, UN Doc. ICSP4/UNFSA/REP/INF.1 of 18 July 2005.

[161] *The Rome Consensus on World Fisheries*, adopted by the FAO Ministerial Conference on Fisheries, Rome, Italy, 14–15 March 1995, FAO Doc. AC441/E (1995), para. 5.

[162] *The Rome Consensus on World Fisheries*, para 5.

[163] *The Rome Consensus on World Fisheries*, para. 7.

[164] *The Rome Consensus on World Fisheries*, para. 8.

[165] FAO, *The Rome Declaration on the Implementation of the Code of Conduct for Responsible Fisheries*, adopted by the FAO Ministerial Meeting on Fisheries, Rome, Italy, 10–11 March 1999, FAO Doc. X2220/E (1999).

[166] *The Rome Declaration on the Implementation of the Code of Conduct for Responsible Fisheries*, para. 12.

[167] *The Rome Declaration on Illegal, Unreported and Unregulated Fishing*, adopted by the FAO Ministerial Meeting on Fisheries, Rome, Italy, 12 March 2005. Online: <ftp://ftp.fao.org/fi/DOCUMENT/ministerial/2005/iuu/declaration.pdf> (accessed 19 December 2005).

[168] Ministerial Declaration of the *Conference on the Governance of High Seas Fisheries and the UN Fish Agreement*, St. John's, N.L., 5 May 2005. See online <http://www.dfo-mpo.gc.ca/fgc-cgp/declaration_e.htm> (accessed 7 January 2006). The conference was hosted by Canada in St. John's, Newfoundland and Labrador, from 1 to 5 May 2005. It was convened to discuss and agree on specific actions to improve the governance of high seas fisheries. To find possible, practical ways by which States and RFMOs/As might move forward to strengthen the implementation of measures for promoting sustainable fisheries, the conference addressed the following five core themes: ecosystem-based considerations in fisheries management, compliance and enforcement, decision making in RFMOs/As, balancing fishing capacity and fishing aspirations, and new areas and gaps in fisheries management. These five themes were discussed in workshops held during the conference. A summary report of the workshops is included in the final conference report from the St. John's Conference. See online: <http://www.dfo-mpo.gc.ca/fgc-cgp/index_e.htm> (accessed 7 January 2006). Also published in "Conference on the Governance of High Seas Fisheries and the UN Fish Agreement: Moving from Words to Action", (2005) 20 *International Journal of Marine and Coastal Law* 605–629.

[169] K. Gjerde, "Editor's Introduction to the Special Issue on High Seas Fisheries Governance: Moving from Words to Action" (2005) 20 *International Journal of Marine and Coastal Law* 323 at 324.

[170] K. Gjerde, at 325–326.

[171] See *2005 Annual Meeting Press Release*, 23 September 2005, available at <http://www.nafo.int/about/frames/about.html>.

[172] Cf. speaking notes for Jim Jones, *Technical Briefing [on] Canada's participation in the International Commission for the Conservation of Atlantic Tunas and the importance of ICCAT for Canada's tunas and swordfish fisheries*, Halifax, N.S., 8 November 2005, available at <http://www.dfo-mpo.gc.ca/overfishing-surpeche/media/sp-20051108_e.htm> (accessed 7 January 2006).

[173] *International Convention for the Regulation of Whaling*, done at Washington, D.C., 2 December 1946, entered into force 10 November 1948, 161 UNTS 72 (ICRW).

[174] ICRW Art. III.1. See also IWCOFFICE online: <http://www.iwcoffice.org/> (accessed 19 January 2006).

[175] *Schedule [to the] International Convention for the Regulation of Whaling, 1946*, as amended by the Commission at the 57th Annual Meeting, Ulsan, Republic of Korea, 20–24 June 2005; IWCOFFICE online: <http://www.iwcoffice.org/_documents/commission/schedule.pdf> (accessed 19 January 2006). The Schedule forms an integral part of the Convention, see ICRW Art. I.1.

[176] ICRW Art. V.2.b.

[177] See ICRW Arts. IV–VI.

[178] IWC, "Chairman's Report of the Thirty-Fourth Annual Meeting" (1983) 33 *Reports of the International Whaling Commission* 20–21. The full text of the moratorium, produced in resolution form, can be found in M. Leich, "Contemporary Practice of the United States Relating to International law: Environmental Affairs" (1985) 79 *American Journal of International Law* 431–441 at 435.

[179] See also "IWC Information — Conservation and Management", IWCOFFICE online <http://www.iwcoffice.org/commission/iwcmain.htm#conservation> (accessed 19 January 2006). A Revised Management Procedure (RMP) was developed subsequently, which the Commission accepted and endorsed in 1994 but has yet to implement.

[180] "Cetaceans" are listed in No. 17 of Annex I to LOSC.

[181] *Convention on International Trade in Endangered Species of Wild Fauna and Flora* (CITES), done at Washington, D.C., 3 March 1973, entered into force 1 July 1975, 993 UNTS 243, 244(E), 272(F).

[182] Recommendation by ICCAT Concerning the ICCAT Bluefin Tuna Statistical Document Program [Rec. 92-1], adopted in 1992, entered into force 25 July 1993.

[183] Recommendation by ICCAT Regarding Belize and Honduras Pursuant to the 1994 Bluefin Tuna Action Plan Resolution [Rec. 96-11], adopted in 1996, entered into force 4 August 1997; Recommendation by ICCAT Regarding Panama Pursuant to the 1994 ICCAT Bluefin Tuna Action Plan Resolution [Rec. 96-12], adopted in 1996, entered into force 4 August 1997. See D. Agnew, "The Illegal and Unregulated Fishery for Toothfish in the Southern Ocean, and the CCAMLR Catch Documentation Scheme" (2000) 24 *Marine Policy* 361–374 at 369.

[184] See D. Rothwell on "forum shopping and treaty parallelism" at 131.

[185] See CCAMLR, *Report of the Twenty-First Meeting of the Commission*, 21 October – 1 November 2002 (Hobart, Australia: CCAMLR, 2002) paras. 8.62–8.84 and 10.1–10.75; with regard to CITES, see 993 UNTS 243.

[186] Office of Sustainable Fisheries, National Marine Fisheries Service, *International Agreements Concerning Living Marine Resources of Interest to NOAA Fisheries* (Silver Spring, Md.: National Oceanic and Atmospheric Administration, 2004) at 100.

[187] *International Agreements Concerning Living Marine Resources of Interest to NOAA Fisheries*, at 100.

[188] FAO, "Report of the Twenty-Sixth Session of the Committee on Fisheries, Rome, Italy, 7–11 March 2005", FAO Fisheries Report No. 780 (Rome: FAO, 2005) at paras. 54–56.

[189] FAO Fisheries Report No. 780 at paras. 57–62.

[190] *Convention on the Conservation of Migratory Species of Wild Animals*, done at Bonn, Germany, 23 June 1979, entered into force 1 November 1983, 19 ILM 15. Referred to here as the Bonn Convention.

[191] Bonn Convention Art. I.

[192] Bonn Convention Art. I.

[193] Bonn Convention Art. VII. At the 8th Conference of the Parties, delegates discussed migratory sharks and marine turtles, and agreed to list basking sharks under Appendix II. See "Summary of the Eighth Conference of the Parties to the Convention on Migratory Species: 20–25 November 2005" (2005) 18/27 *Earth Negotiations Bulletin* at 8–9.

[194] Bonn Convention Art. VI.2.

[195] *Convention on Biological Diversity*, done at Rio de Janeiro, Brazil, 22 May 1992, 31 ILM 823 (1992), Preamble.

[196] CBD Art. 1.

[197] CBD Art. 3.

[198] CBD Art. 5.

[199] UNEP, Conference of the Parties to the Convention on Biological Diversity (COP), "Appendix 1, The Jakarta Ministerial Statement on the Implementation of the Convention on Biological Diversity", *Report of the Second Meeting of the Conference of the Parties to the Convention on Biological Diversity*, UNEP/CBD/COP/2/19, Jakarta, Indonesia, 6–17 November 1995, at 40.

[200] The Jakarta Ministerial Statement, at para. 14.

[201] Convention on Biological Diversity, COP 7, Decision VII/5, at paras. 57–62. See also Appendix 3, Elements of a Marine and Coastal Biodiversity Management Framework, at C(8). Online <http://www.biodiv.org/decisions/default.aspx?m=COP-07&id=7742&lg=0> (accessed 19 December 2005).

[202] See COP 7, Decision VII/5, at paras. 3 and 54–56.

[203] Agreement Establishing the World Trade Organization, in GATT Secretariat, *The Results of the Uruguay Round of Multilateral Trade Negotiations: The Legal Texts* (Geneva: GATT, 1994) at 14 and reprinted in (1994) 33 *International Legal Materials* 1144. WTO was created at the conclusion of the Uruguay Round of multilateral negotiations and, on 1 January 1995, replaced the General Agreement on Tariffs and Trade (GATT) (55 U.N.T.S. 194), which had been the framework for international trade law since 1947. The principal set of rules to be applied by the WTO are those originally set out in the 1947 GATT Agreement, although they have been recreated as part of the WTO Agreement and renamed GATT 1994 (Art. II(4)). Over 30 agreements and understandings relating to trade in goods, services, trade-related intellectual property, and dispute settlement that emerged as part of the Uruguay Round of Multilateral Trade Negotiations must be added to GATT 1994. Together, GATT 1994 and the Uruguay Round agreements and understandings administered by the WTO constitute the multilateral trade law regime.

[204] R. Tarasofsky, *Regional Fisheries Organizations and the World Trade Organization: Compatibility or Conflict?* (Cambridge: TRAFFIC International, 2003) at vi.

[205] UNFSA Art. 17(4).

[206] UNFSA Art. 23(3).

[207] These matters are discussed further in Chap. 5 and 9, secs. 5.4.2 and 9.4.3, respectively.

[208] *The Results of the Uruguay Round*, at 404.

[209] *United States — Restrictions on Imports of Tuna: Reports of the Panel*, (Tuna–Dolphin I and II), DS21/R and DS29/R, released on 3 September 1991 and 20 May 1994, respectively. Often regarded as the trigger of the trade–environment debate, the two GATT Tuna–Dolphin Disputes (Tuna–Dolphin I and II) were the first to test the legitimacy of using environmentally unfavourable foreign process and production methods as justification for trade restrictions.

[210] *Chile — Measures Affecting the Transit and Importation of Swordfish*, WT/DS193/3, 6 April 2001.

[211] Para. 31(i) of the Ministerial Declaration of the 4th WTO Ministerial Conference, held in Doha, Qatar, from 9 to 14 November 2001, WT/MIN(01)/DEC/1, adopted on 14 November 2001 (referred to as the Doha Declaration).

[212] Cf. paras. 28, 31, and 32(i) of the Doha Declaration.

[213] The Cancun Ministerial Conference ended after five days of negotiations on 14 September 2003 after Chairperson Luis Ernesto Derbez concluded that despite considerable movement in consultations, members remained entrenched. See WTO online <http://www.wto.org/english/thewto_e/minist_e/min03_e/min03_e.htm> (accessed 21 January 2006).

[214] "An ecolabel is a tag placed on a product that certifies that it was produced in a sustainable, environmentally-friendly way. Such tags let consumers make informed choices about what they are buying, so that those who wish to can support responsible food production. In essence, they create a market mechanism that promotes sustainable production methods. The new guidelines are aimed at providing guidance to governments and organizations that already maintain, or are considering establishing labelling schemes for certifying and promoting labels for fish and fishery products from well-managed marine capture fisheries". FAO Web site <http://www.fao.org/newsroom/en/news/2005/100302> (accessed 22 January 2006).

[215] For example in the case of ecolabelling, COFI in March 2005 noted "the special circumstances, conditions and concerns applying to developing countries and countries in transition that required time, financial and technical assistance to develop and maintain appropriate fisheries management arrangements in order to participate in, and benefit from, voluntary ecolabelling schemes. It was additionally agreed that direct support towards the often high cost of accreditation and certification would also be necessary". See FAO COFI Report at para. 66 available at <http://www.fao.org/documents/show_cdr.asp?url_file=/docrep/008/a0008e/a0008e09.htm> (accessed 22 January 2006).

[216] *Agreement on the Conservation of Albatrosses and Petrels*, Canberra, Australia, 19 June 2006, (2004) A.T.S. 5.

[217] ACAP Art. II(1).

[218] Of the world's 24 species of albatrosses, 23 are considered endangered or vulnerable according to the "red-list" criteria of the World Conservation Union, see the ACAP Web site at <http://www.acap.aq/index.php/acap> (accessed 6 January 2006).

[219] As of 1 July 2005; see ACAP online <http://www.acap.aq/acap/parties> (accessed 6 January 2006).

[220] ACAP Art. II.

[221] ACAP Art. V.

[222] ACAP Art. X.

[223] ACAP Art. IX(1).

[224] *Report of the First Session of the Meeting of the Parties*, Hobart, Australia, 10–12 November 2004, Item 9.1.1 and *Resolution on the Establishment of the Advisory Committee for the Agreement on the Conservation of Albatrosses and Petrels* (Resolution 1.5), adopted at the first session of the Meeting of the Parties, Hobart, Australia, 10–12 November 2004.

[225] ACAP Art. X.

[226] D. Rothwell, at 131.

[227] UNFSA Art. 17

[228] W. Edeson, "Tools to Address IUU Fishing: The Current Legal Situation" Background papers for the Expert Consultation on Illegal, Unreported and Unregulated Fishing Organized by the Government of Australia in Co-operation with FAO, Sydney, Australia, 15–19 May 2000, Document AUS:IUU/2000/8 (Rome: FAO, 2000), FAO online <http://www.fao.org/documents/show_cdr.asp?url_file=/DOCREP/005/Y3274E/y3274e0a.htm> (accessed 24 October 2005).

[229] B. Kuemlangen, "Legal implications and options for regional cooperative arrangements for the management of coastal and oceanic large pelagic resources for CARICOM States" in R. Mahon and P.A. McConney, eds., *Management of Large Pelagic Fisheries in CARICOM Countries*, FAO Fisheries Technical Paper No. 464 (Rome: FAO, 2004) 5–14 at 6.

Chapter 3

Juridical Status of Fish Stocks

3.1 Introduction

Political jurisdictions and delimitation of maritime boundaries often divide aquatic ecosystems and the resources they support. Under the United Nations Convention on the Law of the Sea (LOSC), fish stocks are ascribed legal classifications that are distinct from their scientific definitions. The Agreement for the Implementation of the UN Convention of December 1982 relating to the Conservation and Management of Straddling Fish Stocks and Highly Migratory Fish Stocks (UNFSA) pertains to two juridical categories of fish stocks: straddling fish stocks and highly migratory fish stocks. This chapter discusses working definitions for these fish stocks as well as for joint or shared stocks, now commonly referred to as transboundary stocks, and discrete high seas or purely high seas fish stocks. Some of the associated issues arising from these definitions are also discussed below.

The juridical status of a fish stock has significant legal and political implications for its conservation and management. While fish typically move throughout a given geographic range and are associated with a marine ecosystem, the migratory and distribution patterns of fish stocks are often not coincident with political or legal boundaries. Accordingly, with the extension of coastal-State jurisdiction pursuant to LOSC, the maritime boundaries delimiting the exclusive economic zones (EEZs) determine the area demarcating the high seas. This, in turn, determines whether fish occur in waters under the sovereignty and sovereign rights of coastal States or on the high seas, and therefore under whose jurisdiction the fish and the fishing vessels fall. Given that jurisdictional principles establish the "locus of decision-making authority",[1] it is "crucial to understand how fish stocks are distributed in the ocean relative to coastal State jurisdiction",[2] flag-State jurisdiction, and the competency of regional fisheries management organizations (RFMOs) under contemporary international fisheries law.

The juridical status of some stocks is unclear where maritime boundaries are disputed or not yet delimited. In other cases, where the boundaries are clear and accepted, States can still debate the juridical character of fish stocks. They can, for example, contest the extent to which a stock occurs in coastal-State waters, contend that a stock occurring in the high seas is part of a separate population, or purport that fish found beyond the EEZ are purely high seas stocks. With the advent of genetic testing and improved tagging surveys, scientists can now resolve many such disputes and identify the competent management authority.

No single State has authority or control over an entire stock when it occurs both within and beyond the EEZ or entirely on the high seas. Therefore, States must cooperate to ensure that these international fisheries are managed sustainably and responsibly.

3.2 Regional fishery bodies with competence for straddling fish stocks

There are now four RFMOS and two regional arrangements with competence over straddling fish stocks: the Commission for the Conservation of Antarctic Marine Living Resources (CCAMLR), the Northwest Atlantic Fisheries Organization (NAFO), the North East Atlantic Fisheries Commission (NEAFC), and the South East Atlantic Fisheries Organisation (SEAFO). SEAFO is the newest of these, and the convention establishing SEAFO came into force in April 2003. Arrangements exist for the Bering Sea Donut Hole (BSDH) and South Tasman Rise. The Agreement establishing the Southern Indian Ocean Fisheries Arrangement (SIOFA) has been concluded. It is expected that the Agreement will be adopted in July 2006. The Agreement will enter into force when the required four ratifications are deposited, including at least two from the adjacent coastal States. The high seas enclave in the Sea of Okhotsk known as the Peanut Hole is situated entirely within the Russian EEZ and is managed by the Russian Federation as an UNFSA Article 16 arrangement.[3] See maps in Part III.

As shown on the global overview map of straddling fish stocks (Figure S.1), there are several areas where commercially harvested straddling fish stocks are not regulated by an RFMO. While the Galapagos Agreement covering the Eastern Pacific is not expected to achieve the requisite number of ratifications or accessions to come into force,[4] a proposed RFMO for the high seas area in the South Pacific is under negotiation and is envisaged to embrace the South Tasman Rise, the Challenger Plateau, and the area beyond the EEZs of Chile, Peru, and possibly Ecuador.[5] This leaves the Southwest Atlantic and the North Pacific as the other principal areas where straddling fish stocks are known to occur but are not managed by RFMOs.[6]

3.3 Straddling fish stocks

With the creation of EEZs, it was clear that the natural distribution of certain fish stocks would straddle areas of national jurisdiction and the adjacent high seas (see Figure 3.1). As noted in the UN Secretary-General's Report to the UNFSA Review Conference,[7] LOSC does not use the term straddling stock, and while the term is used extensively in UNFSA, straddling fish stock is not defined. To clarify the term, the following discussion is offered as a working definition of the concept known in international fisheries law as "straddling stocks". Because of the characteristics of these stocks' distribution or migration pattern, the categorization of straddling fish and resulting regional management authority are determined by both species and geographic location (i.e., a stock of mackerel occurring within the regulatory area of NEAFC would be referred to as NEAFC mackerel as opposed to the mackerel stock(s) found further south in SEAFO).

Under UNFSA, the term "straddling fish stocks" or "straddling stocks" commonly applies to those fish referred to in LOSC Article 63(2) as belonging to "the same stock or stocks of associated species occur[ring] both within the exclusive economic zone and in an area beyond and adjacent to the zone…". Article 63(2) does not pertain to transboundary stocks that are also known as joint or shared stocks, as referred to in Article 63(1) of LOSC (see Figure 3.1, illustration 6). As many species of marine living resources occur within areas of national jurisdiction and on the high seas, it may be useful to define straddling fish stocks by what they do *not* include. Straddling fish stocks do not include anadromous, catadromous, or highly migratory species (as defined by LOSC Annex I), sedentary species (as defined by Article 77 of LOSC[8]), or discrete high seas stocks, i.e., those species whose migration patterns do not include areas of national jurisdiction.

In practice it seems to be broadly understood that straddling fish stocks occur under many scenarios (see Figure 3.1). Straddling fish stocks can be either demersal or pelagic,[9] and depending on the width of the continental shelf extending beyond 200 miles and on oceanographic structures, they can be neritic or oceanic.[10] Their migratory patterns can vary, with some stocks "straddling out" of the EEZ, while others "straddle into" the EEZ, or both.[11] Some straddling fish stocks have a migratory pattern that is highly mobile such as the Chilean horse mackerel, whose geographic range extends beyond the maritime zones of Chile and Peru across the South Pacific reaching close to New Zealand[12] (see Figure 3.1, illustration 4). Oceanic redfish, which are found in the Northeast and Northwest Atlantic, straddle the jurisdiction of NAFO and NEAFC, and therefore the management of this stock requires inter-RFMO cooperation.[13]

Fish stocks can also occur within the coastal-State EEZ and on high seas or RFMO regulatory area without individual fish migrating across the line.[14] According to accepted interpretations of LOSC Article 63(2), it is not required that straddling fish stocks migrate between areas under national jurisdiction and the high seas.[15] This is clearly stipulated in the FAO glossary that defines straddling stocks as stocks "which occur[s] both within the EEZ and in an area beyond and adjacent to EEZ" or "[f]ish stocks that migrate between EEZs and the high seas".[16] The distribution, diffusion, or migration pattern of straddling fish stocks during the adult life-cycle stage can vary, thereby compounding the difficulty in determining the degree of the transboundary effects for particular fish stocks, such as noted by Molenaar[17] and Hedley et al.[18]

A related and unresolved question is whether the term straddling stocks refers to the migratory behaviour of the "adult" population or whether the notion of straddling embraces the distribution, diffusion, and seasonal migrations of fish stocks "throughout their life cycle". Hey's definition of straddling stocks "is understood to be cyclical movement by more or less the entire *(adult)* population from one particular place to another".[19] A fish may occur predominantly on the adjacent high seas during its adult phase yet spend its nursery or juvenile periods within one or more EEZs, or vice versa. Arguably, responsible, coherent, ecosystemic management of these stocks is critical not only over their entire "adult geographical range" but also throughout their life cycle, to ensure their biological unity and sustainability.

Another difficulty related to ascertaining the juridical status of fish stocks is determining whether the fish stocks in question actually straddle the maritime boundary delimiting the EEZ. Given that the legal categorization of fish stocks has significant implications for fisheries management and access to a fishery, whether the stocks fished on the high seas also occur within the adjacent EEZ is often

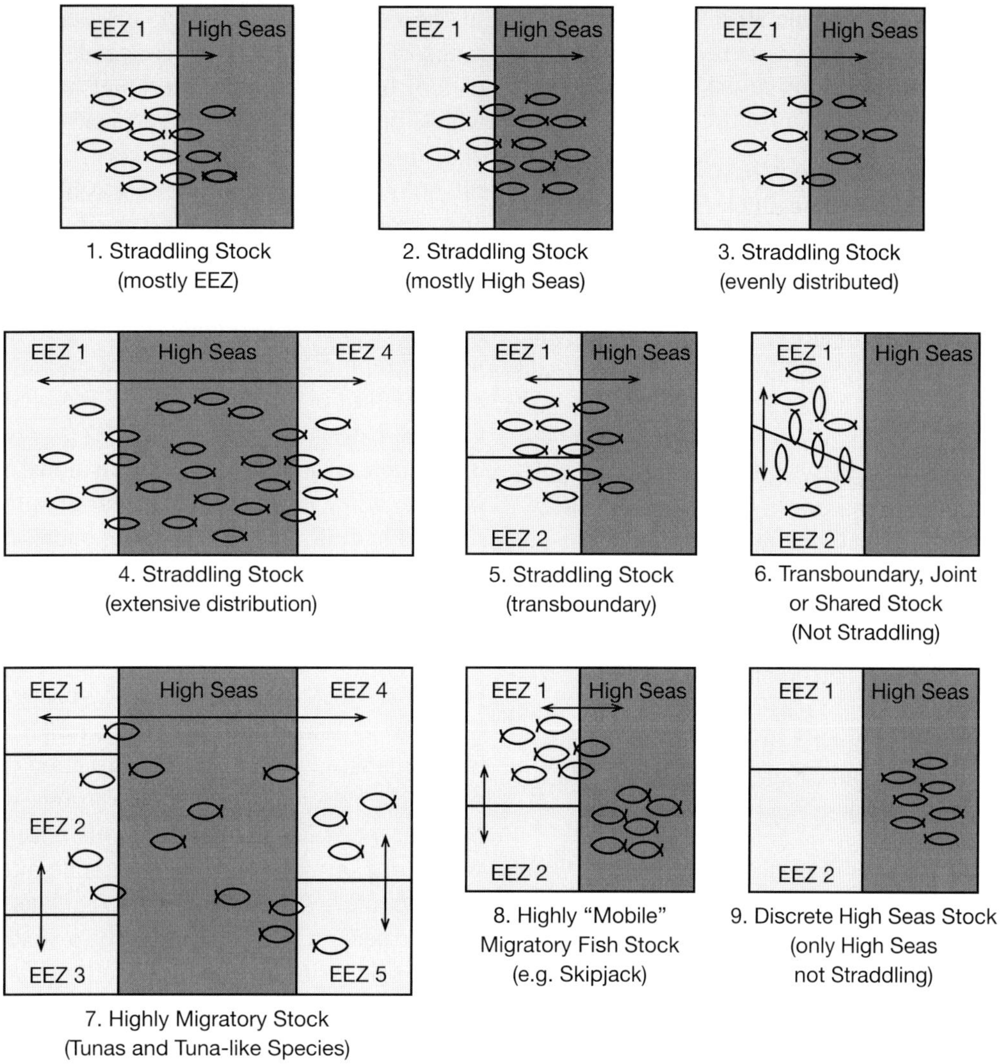

Figure 3.1. Straddling fish stocks occur both inside the EEZ and on the high seas. Some stocks may occur primarily in one EEZ (1) or primarily on the high seas (2), while others may be evenly distributed between an EEZ and the high seas (3). Straddling fish stocks of broadly distributed species and species with life cycle shifts in distribution may occur on the high seas and in one or more EEZs (4 and 5). Illustrations 6 and 9 are not straddling fish stocks or highly migratory fish stocks but are included here for clarity to define transboundary stocks and discrete high seas stocks. Illustrations 7 and 8 depict highly migratory (tuna and tuna-like) fish stocks.

disputed in high seas fisheries. This was a contentious issue in the Bering Sea Donut Hole in the early 1990s, when foreign fishing nations contended that a portion of the Alaska pollock found in the high seas enclave (created following extension of jurisdiction by the United States and Russia) was purely a high seas stock.[20] Similarly, in the late 1990s, it was not clear at the outset of the orange roughy fishery on the South Tasman Rise offshore Australia whether the orange roughy stock in question was a single straddling stock (i.e., one that straddled the Australian EEZ and the high seas) or whether there were two stocks as New Zealand contended,[21] one within the Australian EEZ and another on the high seas, i.e., a discrete high seas stock. At the time, only Australia and New Zealand were fishing for orange roughy on the South Tasman Rise and the categorization of the stock was contentious: the "point of definition was a key issue of dispute during the early stages of the negotiations".[22] The underlying issue was whether Australia would be entitled to coastal-State rights over straddling stocks, or whether it, like New Zealand,[23] would engage in fishing on the high seas as a flag State if the stocks were categorized as discrete high seas.[24] Subsequent research[25] determined that the fish within the Australian EEZ and the adjacent high seas South Tasman Rise were not genetically distinct. Therefore, the stock is not a discrete high seas stock, but one which occurs in the EEZ and on the high seas.

Efforts to define a straddling stock have not been limited to determining whether a single stock migrates into the high seas from an area under national jurisdiction or vice versa, but also the extent to which a stock occurs in the EEZ. It was proposed in the South Tasman Rise orange roughy fishery that the biomass distribution or percentage of orange roughy found within areas under national jurisdiction compared with that on the high seas should be a determinant in defining whether the fish stock was "straddling". Orange roughy was found within the EEZ of Australia and in the adjacent high seas area. However, the orange roughy stock occurred primarily outside the Australian EEZ. According to Molenaar,[26] New Zealand's position was that the stock in question was not a straddling stock, as the principal concentration of the stock was on the high seas and catches within the Australian zone were low.[27] In this case, it was also argued that a straddling stock was defined as one where the biomass of a single stock was approximately evenly distributed between the EEZ and the high seas.[28]

"The extent to which the stocks occur and are fished in areas under national jurisdiction", noted in Article 7(2)(d) of UNFSA, is generally not a consideration in determining the juridical classification of straddling fish stocks. Rather the percentage or abundance of the biomass occurring within the EEZ[29] is one factor in determining compatible measures between the high seas regime and that of the coastal State and may influence the prescriptive authority over the stock (for a discussion of "compatibility" see Chapter 8). This notion of "zonal attachment" is sometimes used by certain RFMOs as a criterion in allocating quota or ascertaining the share of a particular straddling stock that occurs in a regulatory area of an RFMO.[30]

In the case of the Barents Sea Loop Hole, stocks occur mostly within areas of national jurisdiction. In the early 1990s, when the migration and distribution pattern of cod in the Barents Sea moved into the Loop Hole, a high seas enclave, high seas fishing activity met with considerable protests from the coastal States of Norway and Russia.[31] As the cod stock principally occurred within the EEZ and was fully exploited by the coastal States within their zones, an unregulated high seas fishery would have undermined the conservation regime and exceeded the total allowable catch (TAC). This dispute over fishing in international waters led to the multilateral agreement over fishing in the Barents Sea ("Loop Hole Agreement").[32]

In the case of Chilean jack mackerel, the geographic range of the stock has not changed, but the distribution pattern has varied. Jack mackerel is a straddling stock whose distribution pattern is typically aggregated within 200 miles from the Chilean coast. In recent years, the offshore fishery for this species has been prosecuted several miles beyond Chile's EEZ, as there has been a noted shift in biomass distribution into the high seas area beyond and adjacent to the EEZ. Chile continues to manage this fishery for its nationals as a single stock throughout its geographic range, allocating a quota to be harvested either within or beyond the EEZ. Any fish harvested by Chilean nationals beyond the EEZ is deducted against their quota. As there is no international fisheries organization with competence over this area for the management of straddling stocks, fishing undertaken by foreign fishing nations remains unregulated.[33]

Given that a straddling stock derives its juridical definition from its geographic location, the legal status of a particular stock may be altered by changes in maritime boundary delimitation, geographic range, and (or) distribution pattern. A change in distribution pattern may be due to changes in the marine ecosystem, whether natural or anthropogenic. A stock that was once found entirely within one or more EEZs may migrate into the high seas and become a straddling fish stock. This is particularly true for pelagic stocks.[34] Similarly, a stock that once occurred solely on the high seas and was therefore classified as a discrete high seas stock may move into an EEZ and thus become a straddling stock, or a stock may move entirely within an area of national jurisdiction and thereby be subject solely to coastal-State management.

When any portion of stock (adult, juvenile, or other life stage) straddles areas of national and international jurisdiction, regardless of the abundance or biomass distribution between the two areas, coherent and compatible management is required to avoid conflict and overfishing and to ensure long-term sustainability of the resource. It is important to understand that it is the whole species assemblage (and ecosystem) that is straddling or likely to straddle. Therefore, related bycatch species need to be taken into consideration in any management plan.[35] UNFSA underscores the need for cooperatively managing the fish stock as one single unit by applying compatible management measures throughout the geographic range and across the entire distribution of the stock. The "compatibility principle", found in Article 7 of UNFSA, is a basic tenet of conservation for both straddling fish stocks and highly migratory fish stocks.

3.4 Highly migratory fish stocks

Highly migratory fish stocks are those species whose distribution and migration pattern cover great distances. While no definition is provided in LOSC,[36] Article 64 uses the term highly migratory species to refer to species listed in Annex I.[37] UNFSA, however, does not refer to the species listed in LOSC Article 64 and confines the Agreement to "fish stocks", specifically excluding cetaceans listed in LOSC Annex I. According to Garcia,[38] it was a concerted decision during the negotiation of UNFSA not to open the debate on cetaceans. Most regard the term "highly migratory fish stocks" used in UNFSA to be limited to tuna and tuna-like species.

The RFMOs established to manage highly migratory fish stocks[39] generally limit their competence[40] to tuna and tuna-like species, as well as to the incidental capture of associated and dependent species. As Donohue points out, "UNFSA does not ignore these other species (listed in Annex I) — it requires RFMOs to take into account interactions with other marine living resources, e.g., Articles 5 (d) (e) (f) (g) and 7(2)(f) without purporting to regulate their harvesting".[41] It is interesting to note that the recently established Western and Central Pacific Fisheries Commission (WCPFC) adopts LOSC Annex I as the list of living marine resources over which the organization has competence.

Six RFMOs have competence over highly migratory fish stocks: the Commission for the Conservation of Southern Bluefin Tuna (CCSBT), the Inter-American Tropical Tuna Commission (IATTC), the International Commission for the Conservation of Atlantic Tunas (ICCAT), the General Fisheries Commission for the Mediterranean (GFCM), the Indian Ocean Tuna Commission (IOTC), and WCPFC, which was established in June 2004 (see Figure S.2). The global institutional infrastructure for regulating fishing of tuna and tuna-like species through RFMOs is now complete.

Jurisdictional debates hinging on the characterization of a stock are not limited to straddling fish stocks. A current debate concerns the structure of Atlantic (formerly referred to as northern) bluefin tuna and the degree of mixing between the Western and Eastern Atlantic populations. Atlantic bluefin tuna has been managed as two separate populations, the Western Atlantic and the Eastern Atlantic, since 1981 with the establishment by ICCAT of the east–west management boundary.[42] The Western Atlantic population is currently under a rebuilding plan with associated conservation measures.[43] By contrast, there is a significantly larger fishery (estimated to be 15 times larger) in the east, including the Mediterranean.[44] The western tuna is reportedly mixing with the eastern tuna and is being harvested in the Eastern Atlantic fishery, thereby undermining the rebuilding strategy for the western population.[45]

Recent tagging data indicates[46] that there is mixing of Atlantic bluefin tuna in the Central Atlantic Ocean with considerable movement across the west–east boundary. This new scientific information calls into question previous understanding of stock structure and distribution. The challenge now is to incorporate this knowledge into ICCAT and GFCM management strategies and the management boundary.

3.5 Discrete high seas stocks

In the post-UNFSA period another concept has evolved concerning the issue of fish stocks that are found only on the high seas and are not thought to occur in areas under national jurisdiction, i.e., not within EEZs. There are now three terms used to describe these stocks: (1) "discrete high seas fish stocks"; (2) the term more recently introduced by FAO, "purely high seas stocks"; and (3) the term used in the UN Secretary-General Report, "other high seas stocks", to refer to stocks that are not highly migratory or straddling.[47] For the purposes of this work, the term discrete high seas fish stocks will be used.

Through genetic sampling, scientists are now capable of determining whether a stock of fish is straddling (i.e., it occurs both within areas of national jurisdiction and on the high seas) or whether the stock found on the high seas is genetically distinct and therefore a discrete high seas stock that does straddle into or out of the EEZ. The genetic characteristics of fish stocks found on the high seas at the present time are generally unknown.[48] As a result it is not always clear whether a stock is a discrete high seas fish stock, e.g., fish stocks like orange roughy that congregate around seamounts.

Discrete high seas stocks fall outside UNFSA's application unless they are species that are associated with or dependent on straddling or highly migratory fish stocks. All RFMOs having competence over straddling fish stocks also manage discrete high seas stocks occurring within their convention areas. These RFMOs derive their legal authority to manage discrete high seas fish stocks from the general high seas obligation under Article 117 of LOSC.[49] NAFO, for example, manages four discrete high seas stocks — 3M cod, 3M redfish, 3M shrimp, and 3M American plaice — all found within the NAFO Regulatory Area in division 3M on the Flemish Cap. The post-

UNFSA SEAFO Convention[50] pertains to both straddling and discrete high seas fishery resources within its area of competence, i.e., the designated waters beyond national jurisdiction. The SEAFO Convention explicitly excludes all highly migratory species listed in Annex I of LOSC and those sedentary species falling under coastal-State jurisdiction pursuant to Article 77(4) of LOSC.

The management of discrete high seas stocks has emerged as an important fisheries governance issue in the decade since UNFSA was adopted. With increasingly large, technologically advanced vessels, sophisticated fish-finding and harvesting gear, and declining fish stocks inshore, fishing nations have pursued new and exploratory fisheries on the high seas. Some of the targeted discrete high seas stocks are vulnerable deep-sea fisheries. Other than NAFO, SEAFO, and recent efforts by NEAFC, the fishery for discrete high seas stocks is unregulated. Without adoption of the precautionary approach or timely introduction of conservation and management measures, discrete high seas stocks have collapsed after only a short period of large-scale fishing activity (e.g., orange roughy occurring in the Southern Indian Ocean and on the Louisville Ridge, situated on the high seas east of New Zealand). Without monitoring and control, such unregulated species may become extirpated or possibly extinct.

Some governments and nongovernmental organizations have called for the possible extension of UNFSA's mandate to include discrete high seas stocks as a means of addressing this governance lacuna.[51] The UN General Assembly (UNGA) Fisheries Resolution A/60/31, paragraph 12, calls for UNFSA principles to be extended to discrete high seas stocks.[52] Others have proposed that unregulated discrete high seas species be encompassed within the competence of a management body, whether an existing RFMO or some other mechanism, to ensure that they are not overexploited as target or bycatch.[53] There have also been calls to suspend fishing for discrete high seas fish stocks in areas that are not under the competence of an RFMO, as well as proposals to ban bottom trawling to protect vulnerable marine ecosystems (typically associated with discrete high seas stocks or deep-sea fisheries).[54]

3.6 Deepwater species

A related and often overlapping issue to discrete high seas stocks over the past 10 years is the unregulated harvesting of deep-sea fishery resources. Demersal species found in the deep ocean beyond 200 miles may be straddling or discrete high seas fish stocks, depending on whether their geographic range extends within the EEZ. This demonstrates that determining the biomass distribution or migratory pattern of deepwater species is central to determining their juridical status.

Deepwater species represent an unknown but limited potential for marine capture fisheries and could pose a serious management problem.[55] For example, NEAFC in 2002 introduced a temporary measure to limit fishing effort for species inhabiting deepwaters. In 2004 and 2005 NEAFC introduced a 30% reduction in effort. Lack of data and information about the state of these stocks has hampered progress in establishing permanent measures for these fisheries, but new data in 2006 may make it possible to regulate individual fisheries. NEAFC has also closed five areas to demersal fishing gear for 2005–2007 to protect vulnerable marine habitat and deep-sea fish stocks. In 2005 GFCM also introduced a prohibition against the use of towed dredges and trawl fisheries below 1000 m to protect deepwater fisheries habitat.[56]

3.7 Transboundary, joint, and shared stocks

Straddling fish stocks are to be distinguished from other transboundary fish stocks by their distribution pattern. Fish stocks that migrate between two or more EEZs are generally known as "shared stocks"[57] or "joint stocks" or the now-preferred term "transboundary stocks".[58] Should these transboundary stocks also occur on the high seas, they would be considered straddling fish stocks for the purposes of UNFSA.[59] This latter case was not explicitly foreseen by LOSC[60] but is still captured within the UNFSA management regime for stocks that have a migratory distribution pattern between areas of national jurisdiction and the high seas.

Two regional fishery organizations with competence for transboundary fish stocks are the recently created South West Indian Ocean Fisheries Commission (SWIOFC), an FAO Article VI body, and the Permanent Commission for the South Pacific (CPPS), headquartered in Ecuador. These regional organizations may play an important role in ensuring the compatibility of conservation and management measures within the area of competence of the regional EEZs and the adjacent high seas regime. SWIOFC is expected to work closely with the not yet in force SIOFA, which will be given the competence to manage the high seas areas in the region. And in the case of CPPS, were the Galapagos Agreement to come into force, CPPS would play a major role. Perhaps CPPS will play a coordinating role in promoting coherent management between the proposed South Pacific

RFMO and the coastal States of the Southeast Pacific.

3.8 Conclusion

In the post-UNFSA era, the definitions of the juridical categorizations of straddling fish stocks and highly migratory fish stocks have been increasingly elaborated and clarified, although there are still areas of uncertainty. A distinction is now made between shared, joint, and transboundary stocks, which migrate between two or more areas of national jurisdiction, and straddling stocks, which migrate between areas of national jurisdiction and the high seas. While discrete high seas fish stock is not a juridical classification under LOSC, the recognition that this term applies to those fisheries confined to the high seas, which are distinct from straddling or other fish stocks, is an important development. With the emergence of genetic sampling, some disputes over the categorization of high seas stocks as straddling or discrete will be increasingly resolved scientifically. Other disputes over the cross-jurisdictional status of a stock and therefore the appropriate management response will persist, e.g., in those cases where it is unclear whether we are dealing with a "population" or a "stock", and in cases where the extent of distribution within the EEZ or on high seas is contentious. As well, it is not clear whether straddling fish stocks refers to the fragmented jurisdictional distribution during the adult phase or throughout the life cycle of the fish stock.

In distinguishing between highly migratory fish stocks and highly migratory species, UNFSA explicitly excludes cetaceans. In an effort to provide an overview of the state of implementation of UNFSA, this monograph will highlight in the following chapters the efforts of the six RFMOs with competence to manage highly migratory fish stocks and the four RFMOs, as well as the regional fisheries management arrangement where relevant, with competence to manage straddling fish stocks.

Endnotes

[1] W. Burke, "Compatibility and Precaution in the 1995 Straddling Stock Agreement" in H.N. Scheiber, ed., *The Law of the Sea: The Common Heritage and Emerging Challenges* (Leiden: Martinus Nijhoff Publishers, 2000) at 105.

[2] W. Burke, at 107.

[3] There are few examples of high seas enclaves or high seas areas, such as the Peanut Hole, that are surrounded entirely by the EEZ of a single State. Michael Lodge notes three other similar enclaves: two tiny enclaves within the EEZ of New Zealand and a high seas enclave within the EEZ of Japan. See M. Lodge, "The Fisheries Regimes of Enclosed and Semi-Enclosed Seas" in E. Hey, ed., *Developments in International Fisheries Law* (The Hague: Kluwer Law International, 1999) 193–216 at 208–209.

[4] *Framework Agreement for the Conservation of Living Marine Resources on the High Seas of the South Pacific*, "The Galapagos Agreement" done at Santiago, Chile, 14 August 2000, not yet in force. The Agreement shall enter into force on the 30th day after all four coastal States of the Southeast Pacific have deposited their instruments of ratification, see Art. 19(1).

[5] The convention area for the proposed RFMO could also have included the high seas area beyond other coastal and island States in the Pacific Ocean. International consultations on the establishment of the South Pacific Regional Fisheries Management Organization, online <http://www.southpacificrfmo.org/Home/> (accessed 18 January 2006). The first international meeting for discussing the establishment of the proposed RFMO took place in Wellington, New Zealand, 14–17 February 2006.

[6] See E. Molenaar, "Addressing Regulatory Gaps in High Seas Fisheries" (2005) 20(3) *International Journal of Marine and Coastal Law* 533–570 at 541. During the March 2006 Preparatory Meeting for the UNFSA Review Conference (ICSP-5), there were calls to establish a regional framework to regulate the non highly migratory fish stocks in the North Pacific.

[7] Report of the UN Secretary-General, *Report Submitted in Accordance with Paragraph 17 of General Assembly Resolution 59/25 of 17 November 2004, to Assist the Review Conference to Implement its Mandate Under Paragraph 2, Article 36 of the United Nations Fish Stocks Agreement*, UN Doc. A/CONF.210/2006/1 at 13.

[8] Some molluscs and crustaceans, while sedentary, are also considered to be straddling stocks under the competence of certain RFMOs.

[9] Demersal species: species that live in close relation with and depend on the sea bottom. For example, cods, groupers, and lobsters are demersal resources. The term "demersal fish" usually refers to the living mode of the adult. Pelagic fish: fish that spend most of their life swimming in the water column with little contact with or dependency on the bottom. Usually refers to the adult stage of a species. Neritic species: species with a life cycle and distribution that are largely confined to the continental shelf and upper slope. Oceanic species: species distributed and caught mainly beyond the shelf and (or) that migrate extensively across the open ocean. A key stage of their life cycle may take place inshore, and they may yield large catches in coastal waters. These resources are essentially pelagic for most of their life cycle but some demersal oceanic resources exist on seamounts. See FAO Fisheries Glossary, FAO online <http://www.fao.org/fi/glossary/default.asp> (accessed 23 January 2006).

[10] S. Garcia, "Straddling Stocks" (May 2001), UN Atlas of the Oceans, online <http://www.oceansatlas.com/world_fisheries_and_aquaculture/html/resources/capture/highseas/straddling_stocks.htm> (accessed 24 October 2005).

[11] G. Bryden, "UN Fish Stocks Agreement" (May 2001), UN Atlas of the Oceans, online <http://www.oceansatlas.com/world_fisheries_and_aquaculture/html/govern/instit/intlagr/unfsa.htm> (accessed 23 October 2005).

[12] E. Meltzer, "Global Overview of Straddling and Highly Migratory Fish Stocks: The Nonsustainable Nature of High Seas Fisheries" (1994) 25 *Ocean Development and International Law* 255–344 at 260.

[13] It should be noted that Rayfuse argues that the "emerging phenomenon of stocks which are said to 'straddle' between two RFO regulatory areas …[t]echnically, under the LOSC, these are not straddling fish stocks. They could, perhaps more accurately be referred to as 'internationally shared stocks'". R. Rayfuse, *Non-Flag State Enforcement in High Seas Fisheries* (Leiden: Martinus Nijhoff Publishers, 2004) at 206. The author (E. Meltzer) would suggest that these are still straddling stocks, as there is still geographic continuity between the occurrence of these species within EEZs and a common high seas area.

[14] C. Carr, "Recent Developments in Compliance and Enforcement for International Fisheries" (1997) 24 *Ecology Law Quarterly* 847–860 at 848. Carr does not differentiate between transboundary or shared stocks and those stocks straddling a high seas maritime boundary under his understanding of straddling stocks. This interpretation is generally no longer widely held.

[15] See R. Churchill and V. Lowe, *The Law of the Sea* 3rd ed. (Manchester: Manchester University Press, 1999) at 305 [emphasis added] where they define straddling stocks as "stocks of fish that *migrate* between, *or* occur in both, the EEZ of one or more States and the high seas…". E. Hey states that "migration is understood to be cyclical movement by more or less the entire (adult) population from one particular place to another". See E. Hey "The Regime for the Exploitation of Transboundary Marine Fisheries Resources" (Dordrecht: Martinus Nihoff Publishers, 1989) at 82-83. See also C. Hedley, E.J. Molenaar, and A.G. Oude Elferink, *The Implications of the UN Fish Stocks Agreement (New York, 1995) for Regional Fisheries Organisations and International Fisheries Management* (Brussels: European Parliament, 2003) at 5.

[16] FAO, Fisheries Glossary, FAO online <http://www.fao.org/fi/glossary/default.asp> (accessed 18 January 2006). See "straddling stock".

[17] E. Molenaar, "The South Tasman Rise Arrangement of 2000 and other Initiatives on Management and Conservation of Orange Roughy" plus Appendix 'Arrangement between the Government of Australia and the Government of New Zealand for the Conservation and Management of Orange Roughy on the South Tasman Rise' (2001) 16(1) *International Journal of Marine and Coastal Law* 77–124 at 86–87.

[18] C. Hedley, E.J. Molenaar, and A.G. Oude Elferink, at 6. These authors put forward the following issue related to defining straddling stocks: "In practice, the seriousness of these transboundary effects determines whether management authorities should manage the stock as a single (straddling) stock or as two discrete (non-straddling) stocks. The term 'stock' is first of all used by (fisheries) managers to distinguish groups of individuals of the same species. Stocks are distinguished from each other on account of, for example, their range

of distribution. From a biological or genetic perspective, however, two stocks may belong to the same 'population'. No agreement exists on the level at which transboundary effects are so significant that management should be based on one and not on two stocks (even though they belong to the same population). Added thereto are the many complexities related to measuring the seriousness of the consequences of different types of transboundary effects (e.g., loss of genetic diversity). Only a very general definition for straddling stocks may therefore be in reach but during the negotiations of the Fish Stocks Agreement even this could not be achieved".

[19] See E. Hey, at 82-83. Also see E. Molenaar, at 87 in footnote 46 quoting Churchill and Lowe, and Hey.

[20] E. Meltzer, at 284.

[21] E. Molenaar, at 86–87.

[22] J. Willing, "Arrangement between the Government of Australia and the Government of New Zealand for the Conservation and Management of Orange Roughy on the South Tasman Rise" in FAO, *Report of the Norway–FAO Expert Consultation on the Management of Shared Fish Stocks, Bergen, Norway, 7–10 October 2002*, FAO Fisheries Report No. 695, Supplement (Rome: FAO, 2003) at 200.

[23] J. Willing, at 201.

[24] This argument that stocks are discrete high seas and not straddling has been made in other contexts, e.g., by high seas fishing nations harvesting pollock in the Central Bering Sea Donut Hole. See E. Meltzer, at 284.

[25] Genetic testing as part of the certification process for the Marine Stewardship Council recently determined that a stock of Patagonian toothfish previously thought to be a straddling stock was, in fact, a distinct stock found within the EEZs of the South Georgia and South Sandwich Islands.

[26] E. Molenaar, personal communication, May 2005.

[27] This scenario of stocks straddling into the EEZ was not fully appreciated when LOSC was negotiated. S. Garcia, personal communication, January 2006.

[28] See J. Willing, generally. In that case, the scientific framework for assessing the biological data that define straddling stocks defined the stocks as "EEZ" or "high seas" or "straddling stocks", depending on the biomass distribution. A straddling stock was defined as one where the biomass of a single stock was approximately evenly distributed between the EEZ and the high seas.

[29] E. Molenaar, at 93, indicates that the concept of zonal attachment refers to the notion expressed in UNFSA Art. 7(2)(d) that "'the extent to which the stocks occur and are fished in areas under national jurisdiction' shall be taken into account in determining compatible conservation and management measures".

[30] E. Molenaar, at 93–94, notes that this notion of "zonal attachment" is sometimes used as a criteria to determine allocation of quota in RFMOs such as ICCAT, NAFO, and NEAFC and observes that the noninclusive list in Art. 11 of UNFSA does not identify zonal attachment as an allocation criterion for new entrants. Terje Løbach, personal communication, December 2005, notes that zonal attachment is not an allocation criteria used by RFMOs, rather it is a consideration used to determine the share of a particular straddling stock that occurs in the regulatory area of an RFMO. Zonal attachment calculations are mostly used in determining how much of a stock occurs in various zones, including areas managed by RFMOs. Based on factors such as spawning areas, where juveniles and fry occur, where the fishing takes place, feeding grounds etc., it is in some cases (based on biological information) possible to follow the stock over a year and calculate the distribution pattern (biomass multiplied with time).

[31] For discussion of Barents Sea Loop Hole fisheries issues see, e.g., R. Churchill, "The Barents Sea Loophole Agreement: A 'Coastal State' Solution to a Straddling Stock Problem" (1999) 14(4) *International Journal of Marine and Coastal Law* 467–490; O. Stokke, "Managing Fisheries in the Barents Sea Loophole: Interplay with the UN Fish Stocks Agreement" (2001) 32 *Ocean Development and International Law* 241–262; O. Stokke, "Sub-Regional Cooperation and Protection of the Arctic Marine Environment: The Barents Sea" in D. Vidas, ed., *Protecting the Polar Marine Environment: Law and Policy for Pollution Prevention* (Cambridge: Cambridge University Press, 2000) 124–148; O. Stokke, "The Loophole of the Barents Sea Fisheries Regime" in D. Vidas, ed., *Governing High Seas Fisheries: The Interplay of Global and Regional Regimes* (Oxford: Oxford University Press, 2001) 273–301; O. Stokke, L. Anderson, and N. Mirovitskaya, "The Barents Sea Fisheries" in O. Young, ed., *The Effectiveness of International Environmental Regimes: Causal Connections and Behavioral Mechanisms* (Cambridge, Mass.: MIT Press, 1999) 91–154.

[32] *Agreement Between the Government of Iceland, the Government of Norway and the Government of the Russian Federation Concerning Certain Aspects of Cooperation in the Area of Fisheries*, done at St. Petersburg, Russia, 15 May 1999, entered into force 15 July 1999; 2070 UNTS 204; Norwegian Parliamentary Paper St. Prp. No. 74 (1998–99) [*Loop Hole Agreement*].

[33] Chilean Ambassador Fernando Zegers Santa Cruz and Roberto Izquierdo Menéndez, Sonapesca, personal communication, May 2005. Chile, Australia, and New Zealand have initiated negotiations for the establishment of an RFMO in the South Pacific to regulate non highly migratory stocks.

[34] S. Garcia, personal communication, January 2006.

[35] S. Garcia, "Straddling Stocks" (May 2001), UN Atlas of the Oceans, online <http://www.oceansatlas.com/world_fisheries_and_aquaculture/html/resources/capture/highseas/straddling_stocks.htm> (accessed 23 January 2006).

[36] S. Kaye, *International Fisheries Management* (The Hague: Kluwer Law International, 2001) at 123–124.

[37] Annex I to LOSC, "Highly Migratory Species", lists 17 species: albacore tuna, bluefin tuna, bigeye tuna, skipjack tuna, yellowfin tuna, blackfin tuna, little tuna, southern bluefin tuna, frigate mackerel, pomfrets, marlins, sail-fishes, swordfishes, sauries, dolphins, oceanic sharks, and cetaceans.

[38] S. Garcia, personal communication, March 2005.

[39] Another term, "highly mobile species", is now used to differentiate between the geographically contained migration pattern of skipjack tuna and the migrations patterns of the other oceanic tunas and tuna-like species (Judith Swan, personal communication, March 2005).

[40] Apart from the cetaceans, these RFMOs tend not to include the pomfrets and sauries within their management strategy, i.e., set TACs and quotas. For the purposes of UNFSA are the species that are not now included within the scope of the RFMOs considered to be unregulated species? Should these unregulated highly migratory stocks be incorporated within the mandate of the RFMO management regime in the relevant geographic range or area?

[41] Amos Donohue, personal communication, 28 January 2005.

[42] See ICCAT Recommendation on Bluefin Management Measures [Rec. 81-1], adopted in 1981.

[43] The rebuilding plan was established in 1998, see ICCAT Recommendation to Establish a Rebuilding Program for Western Atlantic Bluefin Tuna [Rec. 98-7], entered into force 21 June 1999.

[44] E.g., M. Miyake, N. Miyabe, and H. Nakano, "Historical Trends of Tuna Catches in the World" FAO Fisheries Technical Paper No. 467 (Rome: FAO, 2004) at 15, para. 2.4.

[45] Cf. ICCAT, *Report of the Standing Committee on Research and Statistics (SCRS)*, Madrid, Spain, 4–8 October 2004, at 83.

[46] A. Boustany, S. Davis, P. Pyle, S. Anderson, B. Le Boeuf, and B. Block, "Satellite Tagging: Expanded Niche for White Sharks" (2002) 415(6867) *Nature* 35–36.

[47] S. Garcia, personal communication, January 2006. See also Report of the UN Secretary-General, *Report submitted in accordance with paragraph 17 of General Assembly resolution 59/25 of 17 November 2004, to assist the Review Conference to implement its mandate under paragraph 2, article 36 of the United Nations Fish Stocks Agreement*, UN Doc. A/CONF.210/2006/1 at para. 14.

[48] This technique will undoubtedly be used increasingly to determine the biological unity and ecosystem considerations required for coherent management as well as to resolve jurisdictional disputes.

[49] Art. 117 of LOSC requires States to cooperate on the conservation and management of living marine resources found in the area beyond national jurisdiction. The issues of cooperation and compatibility in the context of SEAFO are discussed by D. Miller and E. Molenaar, "The SEAFC Convention: A Comparative Analysis in a Developing Coastal State Perspective" in A. Chircop, S. Coffen-Smout, and M. McConnell, eds., *Ocean Yearbook 20* (Ardsley, N.Y.: Transnational Press, 2006) 305–375.

[50] Miller and Molenaar argue that the provisions in the SEAFO Convention on dispute settlement do not apply to discrete stocks. "With the exception of the dispute settlement procedures in [SEAFC's] *Article 24,* the SEAFC Convention's provisions apply equally to both straddling stocks and discrete high seas stocks". See D. Miller and E. Molenaar, at 328. However Art. 24 of the convention could be interpreted to apply to discrete stocks as well as straddling fish stocks. Where disputes are not resolved in a timely manner Art. 24(4) triggers recourse to either Part XV of LOSC *or* where the matter concerns one or more straddling stocks, the provisions of Part VIII of UNFSA shall apply. Hence an unresolved dispute concerning a discrete stock would invoke a Part XV LOSC procedure.

[51] Miller and Molenaar write that "it would therefore be reasonable to assume that at least the SEAFC Convention Parties have indicated their willingness to apply the Fish Stocks Agreement provisions to discrete high seas stocks". See D. Miller and E. Molenaar, at 328.

[52] UN General Assembly, *Oceans and the law of the sea: sustainable fisheries, including through the 1995 Agreement for the Implementation of the Provisions of the United Nations Convention on the Law of the Sea of 10 December 1982 relating to the Conservation and Management of Straddling Stocks and Highly Migratory Fish Stocks, and related instruments*, UN Doc. A/60/L.23, 60th session (17 November 2005) (adopted as Resolution 60/31, 29 November 2005, at para. 12.

[53] See the recent initiative of Australia, New Zealand, and Chile to commence international consultations on the establishment of a new RFMO to regulate currently unregulated fish stocks in the South Pacific — one of the last remaining areas of the high seas without a comprehensive international management regime for discrete and straddling stocks, online <http://www.southpacificrfmo.org/Home/> (accessed 23 January 2006).

[54] For example, at the Deep Sea Conference 2003 it was agreed to "prohibit destructive fishing methods/gear pending agreement on international regulation of discrete high seas fisheries". See FAO, "Report on DEEP SEA 2003, An International Conference on Governance and Management of Deep-Sea Fisheries", Queenstown, New Zealand, 1–5 December 2003, FAO Fisheries Report No. 772 (Rome: FAO, 2005) at 30. The Parties to the Convention on Biological Diversity (CBD) in 2004 suggested an interim prohibition of destructive practices that adversely impact the marine biological diversity associated with the vulnerable marine areas of the high seas. See Decision VII/5 of the 2004 Conference of the Parties to the CBD, *Marine and coastal biological diversity: Review of the programme of work on marine and coastal biodiversity*, CBD online <http://www.biodiv.org/convention/result.aspx?id=7742>, at para. 2.4(b). A moratorium on deep-sea bottom trawling was also discussed at the UN Open-Ended Consultative Process on Oceans and the Law of the Sea (UNICPOLOS) in June 2004, which, however, failed to reach agreement; see UN General Assembly, *Report on the work of the United Nations Open-ended Informal Consultative Process on Oceans and the Law of the Sea at its fifth meeting*, UN Doc. A/59/122, 59th session (1 July 2004) at paras. 77 *et seq*. The NGOs Deep Sea Conservation Coalition (DSCC) is calling on the UN General Assembly to secure a moratorium on high seas bottom trawling, see DSCC online <http://www.savethehighseas.org/> (accessed 23 January 2006).

[55] J. A. Moore, "Deep-Sea Finfish Fisheries: Lessons from History" (1999) 24 *Fisheries* 16–21.

[56] General Fisheries Commission for the Mediterranean (GFCM), *Report of the Twenty-Ninth Session*, Rome, Italy, 21–25 February 2005, Appendix G, "Recommendation GFCM/2005/1 on the Management of Certain Fisheries Exploiting Demersal and Deepwater Species", at para. 2.

[57] Rayfuse also refers to shared stocks as "joint stocks" or transboundary stocks in her book, see R. Rayfuse, *Non-Flag State Enforcement* at 205. Munro uses the term "shared fish stocks" as an umbrella term to include straddling, highly migratory, transboundary, and discrete stocks. G. Munro, "On the Management of Shared Fish Stocks" in FAO, *Report of the Norway–FAO Expert Consultation on the Management of Shared Fish Stocks, Bergen, Norway, 7–10 October 2002*, FAO Fisheries Report No. 695, Supplement (Rome: FAO, 2003) at 2. The terms joint stocks and shared stocks are not normally used by FAO, S. Garcia, personal communication, 13 March 2005.

[58] S. Garcia, personal communication, March 2005, noted that the terms "shared stocks" or "joint stocks" can be misconstrued when translated into other languages to infer an equal sharing of the fish occurring within the EEZs of two or more coastal States, and that transboundary is a preferred term to define the migration or distribution of fish occurring in two or more adjacent EEZs. Molenaar notes his preference is to use shared or joint for Art. 63(1) and transboundary as an overarching term to simply indicate that they are not discrete to either the maritime zones of a single coastal State or to the high seas. Erik Jaap Molenaar, personal communication, December 2005.

[59] B. Kuemlangen, "Legal Implications and Options for Regional Cooperative Arrangements for the Management of Coastal and Oceanic Large Pelagic Resources for CARICOM States" in R. Mahon and P.A. McConney, eds., *Management of Large Pelagic Fisheries in CARICOM Countries*, FAO Fisheries Technical Paper No. 464 (Rome: FAO, 2004) at 13, refers to "shared straddling" stocks. R. Churchill and V. Lowe, at 305, state that "straddling stocks are stocks of fish that migrate between, or occur in both, the EEZ of one or more States and the high seas". Hedley, Molenaar and Oude Elferink, in citing Churchill and Lowe, and Hey, note that "migration is understood to be cyclical movement by more or less the entire (adult) population from one particular place to another". See C. Hedley, E.J. Molenaar, and A.G. Oude Elferink, at 5.

[60] S. Garcia, "Straddling Stocks" (May 2001), UN Atlas of the Oceans, online <http://www.oceansatlas.com/world_fisheries_and_aquaculture/html/resources/capture/highseas/straddling_stocks.htm> (accessed 23 January 2006).

Chapter 4

International Fisheries Governance*

4.1 Introduction

International governance of straddling fish stocks and highly migratory fish stocks is based on the framework established by the Law of the Sea Convention (LOSC),[1] which is further elaborated in the United Nations Fish Stocks Agreement (UNFSA). These two treaties are supported by the Food and Agriculture Organization (FAO) Code of Conduct for Responsible Fisheries and the FAO Compliance Agreement, as well as by subsequent instruments that address the need for responsible, sustainable fishing, as outlined in Chapter 2.[2] A central element of the governance regime under UNFSA is coherent management of international fisheries through a network of regional fisheries management organizations (RFMOs). RFMOs are the primary mechanism through which States are expected to fulfill their duty to cooperate and are the expressed governance mechanism for achieving a sustainable, compatible conservation and management regime[3] to regulate these fisheries throughout the geographic range of the straddling or highly migratory fish stocks. UNFSA calls upon States to strengthen existing RFMOs and establish new ones where appropriate in accordance with the functions and characteristics stipulated in the Agreement. By strengthening the institutional construct of RFMOs with a view to creating a global network of regional management organizations and arrangements, UNFSA has established a governance system capable of overcoming the tragedy of the commons and securing the sustainable development of international fisheries.[4] Fundamental to the success of UNFSA is for RFMOs to be effective[5] "vehicles of good governance"[6]. For these reasons RFMOs are considered by some to constitute "the heart of the Agreement".[7]

This chapter provides a brief historical overview of the governance framework for regulating straddling fish stocks and highly migratory fish stocks. To this end, it examines the establishment of regional fisheries organizations, recognizing that while there are many bilateral arrangements and understandings in place, implementation under UNFSA is primarily through RFMOs. The evolution of these regional fishery bodies and other arrangements is examined as the mechanism for overcoming the tragedy of the commons in the pursuit of sustainable development, building upon the post-UNCED and post-UNFSA principles of good governance.

4.2 New model for global good governance

Twenty years after the Stockholm Conference of 1972, the need for developing a new governance model to address the reality of rapid industrialization, economic disparity, and resource despoliation was evident during the preparatory meetings and at the 1992 United Nations Conference on Environment and Development (UNCED). The international community sought to reconcile the values and objectives of a wide variety of stakeholders, recognizing that environmental protection, social welfare, and economic development are not separate but complex, interdependent problems, which have intergenerational as well as intragenerational consequences. UNCED and its associated instruments (the Rio Declaration, Agenda 21, the Convention on Biological Diversity, and the Convention on Climate Change) introduced the concept of sustainable development and the need for an integrated, holistic model to avoid and resolve resource-use conflicts. New principles were espoused for anticipatory, risk-averse, adaptive, and sustainable management, including precaution, ecosystem approach, intergenerational equity, transparency, openness, participation, and subsidiarity.[8] In responding to the challenges of conserving and protecting the environment, reducing poverty, and redistributing global wealth, this model endeavoured to reaffirm the centrality of human beings to sustainable development.

Following UNCED, there was another call for improved global governance and for the reinforcement of the principles of cooperation and collaboration to effect peaceful and sustainable development through the United Nations. This world view was expressed in the 1995 UN Commission on Global Governance:

* The author gratefully acknowledges the research assistance of Leonie Renwrantz, Jay L. Batongbacal, and David Dzidzornu in preparing early drafts of this chapter.

> Effective global governance calls for a new vision, challenging people as well as governments to realize there is no alternative to working together to create the kind of world they want for themselves and their children.[9]

The principles and approaches emanating from UNCED and the trends in global governance expressed by the Commission were incorporated in UNFSA and underpin contemporary international fisheries governance.

4.2.1 Beyond the tragedy of the commons: collaborative regulation of the commons

From the time of Grotius in the early 17th century until UNFSA, the concept of the freedom of the seas, *mare liberum* (open sea), dominated world fisheries. The concept is based on the natural law theory that the high seas are a global commons owned by all, to be used by all — *res communis*.[10] Most fisheries were considered inexhaustible, unbounded, and a common property resource, owned by no individual, no vessel, nor no one nation State until harvested.[11] Governing the commons and common property resources is one of the fundamental challenges underlying high seas fishing. This challenge is often referred to as the "tragedy of the commons". The notion that exploitation of any commons inevitably and eventually results in despoliation and nonsustainable development was popularized by Hardin. While Hardin's work has been influential, it is now considered oversimplified.[12] Important work on alternative governance approaches has been undertaken by Elinor Orstom and others to demonstrate that collaborative regulation of the commons is possible, indeed imperative:

> Systematic multidisciplinary research has, however, shown that a wide diversity of adaptive governance systems have been effective stewards of many resources.... As the human footprint on the Earth enlarges, humanity is challenged to develop and deploy understanding of large-scale commons governance quickly enough to avoid the large-scale tragedies that will otherwise ensue.[13]

UNFSA provided the opportunity to further strengthen and promote the regional approach to fisheries governance in high seas common areas of international interest. By reinforcing and expanding the institutional structure of RFMOs with the objective of establishing a global management network to responsibly regulate access to and allocation of fishing opportunities for straddling or highly migratory fish stocks, UNFSA has engineered a governance shift capable of overcoming the tragedy of the commons, inherent in high seas fishing. As a result, the inevitability of overexploitation and stock collapse due to the tragedy of the commons is no longer an incontrovertible notion.

4.2.2 Regional governance and regime building[14]

Regionalism is an important governance concept in relation to RFMO management of straddling fish stocks and highly migratory fish stocks. Though adopted and elaborated in UNFSA, the concept of regionalism has a long history in international relations,[15] is inherent in LOSC, and is embedded in international fisheries law. The notion of marine regions within the global system was first advanced through FAO in the 1950s and the subsequent establishment of regional arrangements or regimes for fisheries. LOSC adopted regionalism as an element of good oceans governance, declaring that responsible management of cross-boundary resources is founded on cooperation among participants within a region, which include not only coastal States with geographic interests, but also distant-water flag States and user States who may have other legitimate interests.[16] While the term "region" is typically associated with an identified geographic area, regions can also be defined by a specific function such as fishing or by the geographic range of a particular species. Functional regions such as regional fishery bodies (RFBs) focus on patterns of resources, activities, and associated policy problems.[17] These functional marine regions may be and often are coterminous with physical regions.[18]

A region may also be considered an institutional framework or political construct in international affairs.[19] In these instances, the region arises from a group of States endeavouring to address common objectives or a shared problem using a special arrangement defined by a political instrument (e.g., a treaty or international institution).[20] How the region is defined has biological, political, legal, and institutional dimensions. Hence, in addition to geographic attributes, cultural relations, political organizations, economic systems, and levels of development can determine a marine region and its mandate and participation. These are all important considerations in the establishment and effective operation of RFMOs.

4.3 The evolution of regional fishery organizations

The notion of cooperative management through international agreement to address overharvesting of marine living resources was introduced in the late 19th century. The precept that ocean resources were inexhaustible had been challenged in Europe, as conservation concerns and the increasing levels of fishing, particularly of seals, led to the North Sea Fisheries Convention in 1882.[21] A coalition of European States established the first international fisheries organization, the International Council for the Exploration of the Sea (ICES), in 1902 to promote scientific marine research, mainly for the Baltic and North seas. As Kaye notes, ICES also provided technical advice under several fisheries conventions and spawned similar scientific organizations for the Mediterranean Sea in 1914 and the Western Atlantic Ocean in 1920.[22] Other regional fishery organizations with a mandate to promote conservation and management of marine capture fisheries beyond the jurisdiction of the narrow pre-UNCLOS III (Third United Nations Conference on the Law of the Sea) coastal-State territorial sea were also established.[23]

In 1949 the first of the current generation of RFMOs, the General Fisheries Council (now Commission) for the Mediterranean (GFCM), was established. Following the Second Word War, several other regional organizations were created, including the Inter-American Tropical Tuna Commission (IATTC) in 1950; the International Commission for the Northwest Atlantic Fisheries (ICNAF) (the antecedent to the Northwest Atlantic Fisheries Organization (NAFO)) in 1950; the North East Atlantic Fisheries Commission (NEAFC), which came into force in 1963; and the International Commission for the Conservation of Atlantic Tunas (ICCAT), which came into force in 1969. These organizations were intended to establish a mechanism for avoiding overexploitation of a particular species occurring within an identifiable geographic location. These agreements and their associated commissions were based on flag-State enforcement and covered a vast high seas convention area. Generally, the agreements did not allow "for any independent authority by a management agency to regulate a fishery and none allowed for any enforcement authority except by flag States".[24]

A 1955 preparatory meeting[25] prior to the First United Nations Conference on the Law of the Sea (UNCLOS I)[26] and the 1958 Convention on Fishing and the Conservation of Living Resources of the High Seas (i.e., beyond three miles)[27] formally recognized that the nature and scope of fishery conservation problems would be best addressed through international cooperation at the regional level.[28] The pivotal role of regional fisheries agreements and their associated organizations was clearly central to the international fisheries regime envisaged prior to the extension of national jurisdiction.[29] As Kaye notes, conference delegates "stressed the need for international cooperation to ensure that the resources of the high seas were conserved".[30]

In examining the inadequacy of international measures for resolving international fishery conservation problems, the 1955 Rome Conference recommended several guiding principles that remain pertinent today in understanding the regime for regulating straddling fish stocks and highly migratory fish stocks under LOSC and UNFSA:[31]

(1) Regional agreements should cover one or more stocks of marine animals that can be distinguished from other stocks and regulated separately.

(2) All States fishing in the resource and the adjacent coastal States should have the opportunity of joining the convention and participating in the adoption of regulatory measures.

(3) Conservation measures should be based on scientific advice.

(4) Conventions should have clear rules regarding the rights and duties of member nations and clear operating procedures.

(5) Conventions should provide for effective enforcement.

Clearly, the delegates at the 1955 Rome Conference and the 1958 High Seas Convention intended to increase the number of existing regional agreements to regulate high seas fisheries worldwide.[32]

FAO assisted with recognizing this vision by establishing regional fisheries commissions open to both coastal and distant-water fishing nations under Articles VI and XIV of the FAO Constitution. This network of FAO RFBs promotes cooperation by facilitating and coordinating regional fisheries research efforts, including littoral and high seas interests. As the principal intergovernmental organization dedicated to fisheries, FAO has served as a repository for fisheries data, coordinated fisheries research, advanced technical guidelines, and promoted cooperation. Other pre-UNFSA RFBs established pursuant to a treaty have long-established informal or formal links with FAO.

The inadequacies and ineffectiveness of regional cooperation to regulate high seas fisheries following the 1958

Conference led, in part, to the enclosure movement, i.e., the extension of coastal-State jurisdiction over fisheries and the establishment of the exclusive economic zone (EEZ). By the beginning of UNCLOS III in 1973, there was broad acceptance of the EEZ concept.[33] Jurisdiction on the high seas area was reconfigured, and as a result distant-water fishing nations and coastal States were required to negotiate agreements for access to fish that were no longer in international waters but within national jurisdiction. The new maritime boundaries also required the modification or creation of new regional fishery organizations, as in the case of NAFO, and under LOSC, the call for new regional bodies to address straddling fish stocks and highly migratory fish stocks.

Articles 63 and 64 of LOSC call upon States to provide coherent and coordinated management for fisheries beyond national control through international agreement, particularly through regional fisheries organizations. This institutional arrangement is consistent with two fundamental principles of international law, the protection of State sovereignty and the duty to cooperate, developed in the pre-UNCLOS I period.[34]

4.4 The typology of RFBs

Regional fishery organizations are generally referred to collectively as regional fishery bodies (RFBs). There are now over 30 RFBs worldwide.[35] RFBs are diverse in scope and mandate; they can be species specific; FAO or non-FAO organizations; science advisory bodies; or multilateral management entities known as RFMOs[36] or RFMAs.

The term RFMO refers to those RFBs that have management[37] and enforcement responsibilities, a constitutive instrument legally in force, and a permanent Secretariat with regular organizational meetings. A regional fishery management arrangement (RFMA) is different in that it is a cooperative fisheries management mechanism that does not establish a formal organization or entity.

The typology of RFBs can be divided into three categories: (1) multilateral *management* bodies (RFMOs or RFMAs) through which members directly establish management measures; (2) *advisory* bodies, which provide members with scientific and management advice, including those that provide a regional coordination and development function; and (3) *scientific research* organizations, which provide only scientific and data advice.[38]

RFMOs are thus a geopolitical construct overlaid with operational or management functions in fisheries through agreement among interested States, regional economic integration organizations (REIOs), or fishing entities. Four basic elements characterize current RFMOs: (1) a geographic area of application, or range in the case of the Commission for the Conservation of Southern Bluefin Tuna (CCSBT), for a given fish stock or stocks, (2) an institutional arrangement that establishes a body or structure to carry out the management of the given area, (3) an array of functions necessary for carrying out its management objectives, and (4) a constitutive instrument that sets out the terms of cooperation among the participating States, REIO, and fishing entity. Organizations with only recommendatory or advisory mandates and no decision-making and enforcement powers are not considered RFMOs or RFMAs but rather RFBs under UNFSA.

Article 1(1)(d) of UNFSA defines an RFMA as a "cooperative mechanism established in accordance with the Convention and this Agreement by two or more States for the purpose, *inter alia*, of establishing conservation and management measures in a subregion or region for one or more straddling fish stocks or highly migratory fish stocks". This definition gives States considerable flexibility in devising a practical and effective management mechanism. Establishing a formal organization or fishery entity is not necessary if a less-structured mechanism is agreeable and effective. A practical cost-effective regional arrangement enables the participants to rotate annual meetings and to collect data, with FAO or some other appropriate international organization serving as the repository. Examples of RFMAs include the Central Bering Sea Donut Hole (BSDH) Convention;[39] the Southern Indian Ocean Fisheries Agreement used UNFSA as its reference and if adopted will establish the Southern Indian Ocean Fisheries Arrangement (SIOFA)[40]; the South Tasman Rise Agreement;[41] the 1999 Loop Hole Agreement among Russia, Norway, and Iceland;[42] the bilateral fisheries agreements between Russia and Japan, Korea, China, and Poland over the Peanut Hole in the Sea of Okhotsk;[43] and the swordfish agreement between the European Union and Chile.[44]

Before leaving the subject of typology, it is important to note that RFBs established within the framework of FAO under Article VI must be distinguished from those established under Article XIV of the FAO Constitution. Article VI bodies have only recommendatory or advisory mandates and no decision-making or enforcement powers. Further, their budgets are primarily funded by FAO, which also provides the secretariat. Bodies in this category are the Fishery Committee for the Eastern Central Atlantic (CECAF),the Western Central Atlantic Fishery Commission (WECAFC), and the South West Indian Ocean Fisheries Commission (SWIOFC, established in November 2004 by the FAO Council). These bodies would not be considered RFMOs or RFMAs but rather RFBs under UNFSA.[45]

Article XIV marine fisheries bodies, on the other hand, are bodies with an autonomous budget based on contributions from the members, and they have, subject to the specific provisions of the treaty in question, autonomous decision-making powers, including the power to enact binding international conservation and management measures. They are established under their own treaty regime adopted by the parties in accordance with Article XIV of the FAO Constitution and are registered with the UN. These include the General Fisheries Commission for the Mediterranean (GFCM), the Asia Pacific Fishery Commission (APFIC), the Indian Ocean Tuna Commission (IOTC), and the Regional Commission for Fisheries (RECOFI). With the possible exception of APFIC, which does not have the power to adopt binding measures, such bodies would be considered RFMOs. There have been repeated efforts to transform FAO Article VI organizations (in particular CECAF and WECAFC) into Article XIV bodies with regulatory functions, an autonomous budget, and revised mandates in keeping with UNFSA.[46] However, progress has been slow.

4.5 RFMOs as vehicles of good governance

The 10 RFMOs and one RFMA considered in this book correspond most often with the structure and systems of functional regions. The management scope of RFMOs is constructed from three functional elements: the geographic contiguity of the marine areas and user States; the participation of other interests such as foreign (often distant-water) fishing nations; and the perceived geographic distribution or biological range of the stocks. Thus, RFMOs and RFMAs need not be coterminous with geographic boundaries, even though they sometimes are.[47]

The continuing evolution of international fisheries governance in response to the challenge of sustainable development is based on the premise that through collaboration, shared responsibility, and collective problem solving, the global commons can be sustained. UNFSA incorporates these governance principles and provides a global framework or an international regime for the management of straddling fish stocks and highly migratory fish stocks through RFMOs. The goal of this governance system is to improve fisheries management, restore fish stocks, and ensure long-term sustainability of the resource. RFMOs and other regional or bilateral arrangements give effect to this global system at the regional level. To varying degrees, RFMOs have also adopted these principles. RFMOs may thus be seen as vehicles of good governance[48] in collaborative management and participatory, transparent decision making.

The operational structure of fisheries management regimes[49] is strengthened, but remains flexible, under UNFSA. Articles 8 to 16 of UNFSA provide for the mechanisms to be used for international cooperation, notably RFMOs and RFMAs.[50] Article 8.1 elaborates on the broad duty to cooperate "either directly or through appropriate subregional or regional fisheries management organizations or arrangements". UNFSA has expanded the scope and coverage of reciprocal State duties and obligations and opened the traditional fisheries management regime, formerly confined more or less to geographically contiguous States, to the participation of all States with a real interest in a fishery. Generally there are two key approaches for such cooperation: bilateral or multilateral.

As mentioned above, the initiative to transform the existing Article VI FAO fisheries bodies to Article XIV bodies with expanded management functions has made little progress. This may be illustrated by the arrangements put in place to implement the newly created SWIOFC. It was decided after many negotiations that the contiguous coastal States of the Southwest Indian Ocean having duties, rights, and interests in conserving joint or transboundary fisheries resources were best served by becoming an Article VI FAO advisory body.[51] The need for regulating straddling fish stocks and discrete high seas fish stocks in the international waters beyond the SWIOFC will be under the regulatory competence of the proposed SIOFA. The intention is for the two bodies to work together, to coordinate meetings, and to endeavour to have compatible management plans.[52]

The adoption of UNFSA and the widespread acceptance of the new global fisheries regime have been accomplished through the converging efforts of States within areas under national jurisdiction and within international organizations. These efforts have included deliberate leadership efforts of individual States to promote the adoption and implementation of UNFSA within RFMOs. Political leadership of States through the Ministerially-led High Seas Task Force and through the 2005 FAO Committee on Fisheries (COFI) ministerial meeting to address illegal, unreported, and unregulated (IUU) fishing and the 2005 St. John's Declaration have strengthened support for regional and international fisheries institutions. Coordination of RFMO activities through initiatives such as the global vessel register for tuna RFMOs and FAO efforts to establish international standards for fisheries management, research, and regulation also contribute to the emerging post-UNFSA global fisheries regime.

It is important to note that while UNFSA provides a global framework there is no formal juridical relationship between the Agreement and the RFMOs. RFMOs are separate entities governed by their respective agreements or constituent arrangements. There has been, in consequence, a discernible institutional disconnect between the global and regional levels. However, given the nature of international law, globalization, the highly mobile nature of international fisheries and global markets, and the transboundary and dynamic character of straddling fish stocks and highly migratory fish stocks, RFMOs are increasingly perceived not as stand-alone regimes, but as being influenced by, and functioning within, the broader global regime created by LOSC, UNFSA, and other post-UNCED instruments. The rules upheld in one regime can conflict with or reinforce those established in another fisheries regime.[53] Thus, coordination, cooperation, and networking are integral parts of the regime interplay envisaged under UNFSA and modern international fisheries law to strengthen the conservation and management regime for straddling fish stocks and highly migratory fish stocks.

There are limitations on the role that RFMOs can play in the international arena. RFMOs are international organizations, but are neither global in scope nor in application.[54] Regionalism has its limitations: RFMOs are not comprehensive institutional mechanisms that coordinate multi-State actions arising from multiple uses of the oceans. Rather, RFMOs operate as function-based and issue-oriented entities that revolve around specific problems and issues. Addressing issues and activities outside of their primary scope or beyond their identified region requires expanded cooperation with other ocean management bodies. This raises particular challenges with regard to implementation of the ecosystem approach (discussed in detail in Chapter 7), as RFMOs are designed only to manage a juridically classified stock and not to undertake multiple-use management.

These limitations must be kept in mind when considering the extent to which RFMOs are expected to implement global agreements and the broader principles of ocean governance. Even the preeminent principles such as precaution and the ecosystem approach must be applied within the context of the RFMOs' competence and cannot be used to extend their management scope inappropriately to other users and resources of the sea not directly associated with the fishery, even if they are located within the convention area. Other international agencies may have mandates better suited than RFMOs to address ocean governance.

Within the main field of their concern (straddling fish stocks and highly migratory fish stocks), however, the ecosystem approach and the other principles of sustainable development provide an imperative for inter-RFMO cooperation to account for the entire range of the species. A perusal of the 11 convention areas in this study shows considerable overlap in many of their areas of competence and species ranges[55] (see Figures S.1 and S.2 and maps in Part III).These distinctions in jurisdictional scope indicate the vestiges of the focused and sectoral approach that emerged under the previous governance regime. This fisheries governance regime has been superseded by contemporary fisheries management approaches, including UNFSA.

4.5.1 Strengthening and modernizing existing pre-UNFSA RFMOs

Article 10 of UNFSA lists a set of criteria that should be adhered to by States through RFMOs and RFMAs in their obligation to cooperate. Since most of the existing RFMOs and RFMAs predate UNFSA, many of them do not live up to its standards;[56] States are under the obligation to review and, where necessary, to amend the constitutive instruments of the RFMOs and RFMAs of which they are members.[57] As expressed by Jim Jones,

> The RFMOs of today need to be equipped with the proper tools to address the challenges they face. They need reforming. To make these changes, political will is imperative. There is already a discernable shift in international attitudes toward the management of high seas fish stocks. This was clearly evident in the Ministerial Declaration that resulted from the international conference on the *Governance of High Seas Fisheries and the UN Fish Agreement* that was held last May in St. John's in Newfoundland and Labrador. In the Declaration, fisheries ministers agreed to modernize RFMOs so that these organizations will be mandated to make decisions based on sound science; use the precautionary approach to ensure fish stocks conservation; ensure its rules are clear, understandable and consistent with international agreements; and enforce catches and fishing effort to ensure compliant fishing behaviours.[58]

UNFSA does not expressly require the updating of the conventions of existing RFMOs and RFMAs, but merely calls on States to "cooperate to *strengthen* existing subregional and regional fisheries management organizations and arrangements in order to improve their effectiveness in establishing and implementing conservation and management measures for straddling fish stocks and highly migratory fish stocks".[59] However, at its 60th session in November 2005, the UN General Assembly welcomed and urged "further efforts by regional fisheries management organizations and arrangements, as a matter of priority, to strengthen *and modernize* their mandates to include an ecosystem approach to fisheries management and biodiversity considerations, where those aspects are lacking, to ensure that they effectively contribute to long term conservation and management of marine living resources".[60] As Molenaar stresses, formal amendment of RFMO conventions might not be the most preferable way to implement UNFSA, as this is a very time-consuming process, and even after entering into force, the amendments might not be binding on all members.[61] As the case of IATTC and its Antigua Convention shows,[62] changing the constituent instrument of an RFMO or RFMA is not a straightforward task.[63] Further, States that are not parties to UNFSA may not consider themselves obliged to reform the RFMO or RFMA of which they are members and may not necessarily support initiatives taken to this end.[64] Implementation of the standards set by UNFSA requires guidance, political will, and resources.[65] There is a crucial deficiency in UNFSA by not providing a mechanism whereby RFMOs and RFMAs can be held to account; there is, in other words, no oversight of RFMOs or RFMAs at the global level that would promote and reinforce measures to effectively implement the Agreement.[66] Therefore some suggest that "amendment by practice", or the adoption of resolutions or decisions aiming at reinterpreting or elaborating upon existing mandates might be a more feasible way of bringing RFMOs and RFMAs in line with UNFSA.[67] Such an approach has the added advantage that it avoids the need to go through domestic treaty approval processes, which can be slow and, in some countries, difficult because of already overcrowded parliamentary agendas.

Although the adoption of UNFSA has not, as some may have hoped, led to a complete overhaul of the conventions of all pre-UNFSA RFMOs and RFMAs,[68] and despite the difficulties outlined above, in the last few years a number of these entities have been undergoing or commencing a reform process aimed at making them "compatible" with UNFSA and its objectives. This process up to the May 2006 Review Conference will now be outlined.

4.5.1.1 IATTC

IATTC, which was established in 1950, is the oldest of the RFMOs responsible for the conservation and management of highly migratory fish stocks. With the coming into force of several relevant international instruments including LOSC, the FAO Compliance Agreement, the FAO Code of Conduct, and UNFSA, the need for updating the IATTC constituent instrument was evident.[69] In 2003 after a lengthy review, IATTC adopted a new convention — the Convention for the Strengthening of the Inter-American Tropical Tuna Commission (the "Antigua Convention").[70] It has yet to enter into force. The Antigua Convention represents a complete overhaul of the constitution of the Commission and will replace the existing IATTC Convention when it enters into force. It reflects modern developments in fisheries management, including the application of the precautionary approach and the compatibility principle, as described in UNFSA and the Code of Conduct.[71]

4.5.1.2 ICCAT

A reform process was launched by ICCAT at its Annual Meeting in November 2005,[72] one that is consistent with the St. John's Declaration,[73] as well as with other recent political commitments, such as the FAO Rome Declaration on IUU Fishing and the Bali Plan of Action.[74] To this end, the ICCAT members decided that at the 2006 Annual Meeting, the Commission should review ICCAT's conservation and management program taking account of the provisions set out in relevant international fisheries instruments.[75] It was also decided that following the review, the Commission should develop a work plan to address the strengthening of the organization. Over the last decade ICCAT has strived to make improvements in line with UNFSA.[76] For example, to improve conservation and enforcement measures, ICCAT adopted rebuilding plans for North Atlantic swordfish, West Atlantic bluefin tuna, and marlin.[77] The organization has also adopted a comprehensive trade-sanctions scheme to deter incidents of noncompliance, which includes the establishment of negative and positive vessel lists.[78] To improve monitoring, control, and surveillance on the water, ICCAT has adopted a port-inspection scheme and rules for vessel chartering. Moreover, at its 2005 Annual Meeting, ICCAT agreed to introduce a regional observer program to control transshipments at sea.[79] ICCAT has also taken steps to improve the organization overall, by adopting criteria for the allocation of fishing opportunities[80] and for attaining the status of Cooperating Non-Contracting Party, entity, or fishing entity within the waters that are regulated by the organization.[81]

4.5.1.3 NAFO

The need for strengthening the conservation and management regime for the sustainable development of straddling fish stocks in the Northwest Atlantic Ocean was an underlying factor in the 1992 UNCED call for a conference on high seas fishing leading to the negotiations and ultimately the adoption of UNFSA in 1995. Since that time, certain member States, particularly Canada, have repeatedly urged for the adoption of measures that would render NAFO compatible with UNFSA. One of the first areas to be examined was the precautionary approach which was adopted by NAFO in 2003. At its Annual Meeting in September 2005, NAFO launched a review of the NAFO Convention, especially with a view to incorporating more integrated oceans management approaches, reforming decision-making processes, and examining the current structure of NAFO.[82] To this end, the organization has established an ad hoc Working Group on NAFO Reform (WG Reform), which began work in April 2006. Its mandate comprises the evaluation and recommendation of changes to the NAFO Convention to reform the decision-making process, as outlined in paragraphs 4 A and B of the St. John's Declaration of May 2005; the examination of the current structure of NAFO constituent bodies and their subsidiary bodies; and the recommendation of changes to streamline the structure and operation of NAFO to make it a more effective RFMO.[83] Moreover, NAFO's Standing Committee on International Control (STACTIC) will review, in accordance with paragraphs 4 C and D of the St. John's Declaration, the effectiveness of the existing NAFO monitoring, control, and surveillance regime to determine the changes required to strengthen such instruments and make them more effective, in terms of operation, results, and costs. Most notably, STACTIC will review and provide recommendations on the establishment of guidelines for sanctions, the role of observers, and follow-up on infringements.[84] Finally, the Fisheries Commission, NAFO's management body, will address the issues outlined in paragraph 5 of the St. John's Declaration.[85] Generally, NAFO's approach to modernization of the organization can be considered a piecemeal approach, as the initiatives to implement the provisions of UNFSA are being taken one by one and through different processes.

4.5.1.4 NEAFC

In 1996 NEAFC set up a Working Group on the Future of NEAFC to review the constitutional rules of the organization in light of the new developments such as UNFSA and the FAO Code of Conduct.[86] In 2004 the Working Group was given a mandate to evaluate the role of NEAFC in taking a broader ecosystem approach to fisheries management. In that regard, the Working Group examined how to strengthen the role of NEAFC in addressing overall ocean management in the convention area and looked into the possible restriction of the Convention in this respect and the consequent need of interpretation and (or) amendment.[87] On the basis of the recommendations of the Working Group, at its 24th Annual Meeting in 2005, the NEAFC Commission agreed to modernize the NEAFC Convention and bring it in line with developments since 1982, by giving a clearer mandate to pursue the ecosystem approach, to protect biodiversity, and to apply the precautionary approach. Final adoption will take place when two Contracting Parties have finalised internal procedures.[88] Moreover, NEAFC decided at the meeting to try to strengthen its role in overall ocean management in the convention area. There was consensus that strengthening of cooperation, through RFMOs, is the way to improve ocean governance in the high seas.[89]

4.5.2 Creation of new RFMOs

UNFSA Article 8(5) provides that where a particular straddling fish stock or highly migratory fish stock is not under the jurisdiction of an RFMO or RFMA, States shall cooperate to establish such an organization or enter into other appropriate arrangements to ensure conservation and management of such stock. This was reiterated on numerous occasions, e.g., by the UN General Assembly at the 59th session in January 2005[90] and at the 60th session in November 2005,[91] and by the participants of the 2005 St. John's Conference.[92] Since the adoption of UNFSA, new RFMOs and RFMAs have been negotiated (e.g., the Galapagos Agreement, which is not yet in force, SIOFA, which is also not yet in force, and the proposed South Pacific RFMO, which had its first negotiating meeting in February 2006). The RFMOs established after UNFSA include the South East Atlantic Fisheries Organisation (SEAFO) in the Southeast Atlantic Ocean and the Western and Central Pacific Fisheries Commission (WCPFC) in the Western and Central Pacific Ocean. For tuna and tuna-like species, RFMO global coverage is now complete with the establishment of WCPFC.[93]

The remaining principal gaps in RFMO coverage of the high seas for straddling stocks are the Southwest Atlantic Ocean and the high seas areas covered by the Fishery Committee for the Eastern Central Atlantic (CECAF) and the Western Central Atlantic Fishery Commission (WECAFC),[94] both of which are advisory bodies established under Article VI of the FAO Constitution. Efforts are underway by Japan, Republic of Korea, the Russian Federation, and the United States to examine the options for establishing a management framework for the high seas bottom fisheries operating in the Northwest Pacific Ocean. The immediate focus of concern is the Emperor Seamount Chain, which extends from the northwestern Hawaiian Islands to the tip of the Aleutian Island Chain, in particular, the seamounts located south of 45°N. Approximately 10 to 15 vessels from Japan, Korea, and Russia have fished in the area in recent years.

4.5.2.1 SEAFO

The 2001 Convention on the Conservation and Management of Fishery Resources in the South-East Atlantic Ocean[95] is the first international fisheries agreement to establish a regional organization for the management and conservation of straddling fish stocks since the 1995 adoption of UNFSA.[96] Following its entry into force in April 2003, the Convention established SEAFO, which is headquartered in Namibia. Commercially important species in the SEAFO Convention Area, which covers a sizeable part of the high seas of the Southeast Atlantic, include sedentary and straddling species such as alfonsino, orange roughy, oreo dories, armourhead, wreckfish, deepwater hake, and red crab. To give effect to the Convention, the Contracting Parties shall apply general principles such as science-based decision making, the precautionary approach, the ecosystem approach, and the protection of marine biodiversity.[97]

4.5.2.2. WCPFC

WCPFC was established following the entry into force of its Convention in June 2004.[98] The headquarters of the Commission is in Kolonia, Pohnpei State, Federated States of Micronesia. WCPFC is responsible for the management and conservation of all species of highly migratory fish stocks listed in Annex I of LOSC that occur in the convention area and such other species of fish as the organization may determine, except sauries.[99] The treaty itself draws heavily on UNFSA and is intended to be UNFSA compatible.

This post-UNFSA convention introduces new decision-making processes that seek to avoid the problems experienced in some other fisheries regimes with opting-out or objection procedures. Decisions adopted by the Commission are binding on all members 60 days after the date of their adoption. A member who has voted against a decision or who was absent during the meeting at which the decision was made may, within 30 days of the adoption of the decision, seek a review of the decision by a review panel established under Annex II on two grounds, namely (1) that the decision is inconsistent with the provisions of the WCPFC Convention, UNFSA, or LOSC; or (2) that the decision unjustifiably discriminates in form or in fact against the member concerned. The implementation of the decision in question is suspended for all members of the Commission pending the findings and recommendations of the review panel.[100] This process enables minority or dissenting perspectives to be addressed in a timely and effective manner (see Chapter 8).

4.5.2.3 SIOFA

In the post-UNFSA period, consultations have also been taken up on the establishment of a regional fisheries body for the Southern Indian Ocean, where exploitation of straddling fish stocks has been largely unchecked. In January 2004 participants in the intergovernmental consultation agreed that there should be separate coastal and high seas arrangements and to split negotiations accordingly.[101] On 25 November 2004 the FAO Council established SWIOFC as an advisory body under Article VI of the FAO Constitution,[102] responsible for the maritime zones of coastal States in the Southwest Indian Ocean. For the high seas, it was agreed that there should be a separate instrument, albeit linked to SWIOFC, outside the framework of FAO, with the power to take binding decisions on conservation and management measures.[103] The negotiation process for the establishment of the high seas arrangement SIOFA culminated satisfactorily at the Fifth Intergovernmental Consultation in Mombasa, Kenya, 20–22 April 2005. (The meeting was held immediately after the first session of SWIOFC.)[104] The technical edit exercise to finalize the text of the Southern Indian Ocean Fisheries Agreement was completed in late 2005. The Agreement has not yet been adopted formally and signed but a meeting is in the offing (probably July–August 2006 at FAO in Rome, given that FAO is the designated depositary).[105] Interim measures for collection and submission of catch data have been introduced with the expectation that the Agreement will be adopted and ratified in the near future. SIOFA is an RFMA similar to the Bering Sea Donut Hole Arrangement, which relies on Annual Meetings of Parties to carry out its objectives. The geographical competence of SIOFA, which consists exclusively of high seas areas, extends to the west coast of Australia[106] (see map for SIOFA in Part III) (see Figure S.1).

4.5.2.4 Galapagos Agreement

The Galapagos Agreement of 14 August 2000[107] was signed by Colombia, Chile, Ecuador, and Peru with the objective of the conservation of living marine resources in the high seas zones of the Southeast Pacific. Special reference is made to straddling fish stocks and highly migratory fish stocks; however, conservation standards will be "first adopted for particular species judged to be of high priority".[108] An organization is to be set up under the Agreement (Article 11). The Galapagos Agreement is widely regarded as being incompatible with UNFSA,[109] and in the *swordfish* dispute, compatibility of the Galapagos Agreement with Articles 64 and 116–119 of LOSC was contested by the European Community (EC).[110] This agreement is not expected to come into force, and other efforts to regulate these fisheries as part of a broader South Pacific initiative are underway as outlined next.

4.5.2.5 South Pacific RFMO

In March 2005 Australia, New Zealand, and Chile[111] proposed the establishment of a new RFMO in the high seas areas of the South Pacific Ocean. The proposed South Pacific RFMO is intended to close the existing gap in the international conservation and management of non highly migratory stocks (i.e., straddling fish stocks and discrete high seas stocks)[112] and in the protection of marine biodiversity in this area.[113] The geographic area of the proposed RFMO could extend from the most eastern part of the South Indian Ocean (abutting the eastern boundary of the proposed convention area to be managed by SIOFA) through the Pacific Ocean towards the EEZs of South America[114] (see Figure S.1).The first international meeting to discuss the establishment of the organization took place in Wellington, New Zealand, 14–17 February 2006. The process of developing the RFMO will seemingly take four to five meetings over the next three to four years. Participation is open to States, REIOs, and fishing entities that have an interest in South Pacific fisheries.[115] Another negotiating session is planned for November 2006 in Australia.

Areas without RFMOs to manage straddling fish stocks or highly migratory fish stocks are unregulated. In these geographic regions the vessels fishing may be regulated by the flag State but as there is no regime the competence to oversee the conservation and management of these stocks, these fisheries are vulnerable to overexploitation and the systemic consequences of the tragedy of the commons.

4.5.3 IUU fishing

The concept of illegal, unreported, and unregulated (IUU) fishing has emerged in parallel with the post-UNFSA development of tools for encouraging the establishment of RFMOs in geographic areas where no regional regime exists and for promoting membership in RFMOs. In the last few years, international action to combat IUU fishing has gathered momentum, as reflected in the establishment of RFBs and RFMOs with competence to prescribe conservation and management measures over areas of the high seas.[116] Throughout the world, there is increasing awareness of the extent to which IUU fishing contributes to undermining and nonattainment of management goals; fishery collapses; thwarting stock rebuilding efforts for depleted fisheries; and loss of social and economic opportunities related to fishing.[117] The governance regime established by international legal instruments, including LOSC, the FAO Compliance Agreement, UNFSA, the FAO Code of Conduct, and the IPOAs on seabirds, sharks, and capacity, address the international legal basis of IUU fishing. By the end of the 1990s, IUU fishing became a prominent subject of attention in the UN system and other forums.[118] The FAO International Plan of Action to Prevent, Deter, and Eliminate Illegal, Unreported and Unregulated Fishing (IPOA-IUU) was adopted in 2005.

Paragraphs 78 through 84 of the IPOA-IUU recognize the key role of RFMOs in developing and implementing national and regional plans of action and in monitoring and identifying IUU fishing vessels. In fact, RFMOs have taken a substantial number of measures and actions to address IUU fishing and related activities, as evidenced by the rapidly increasing number of resolutions, recommendations, and related decisions adopted by them.[119]

An emerging practice of many RFMOs is that fishing vessels fly in the flag of noncooperating nonmembers of an RFMO found within the RFMO regulatory area are generally presumed, especially for purposes of monitoring and enforcement, to be engaged in IUU fishing. This presumption is based on the explicit participation in RFMOs and authorization for only members and cooperating nonmembers to fish the regulatory area. Membership in the RFMO is the primary mode of cooperation under UNFSA, followed by cooperating nonmember status, and participation normally entails implementation by flag States and their vessels of elaborate monitoring and surveillance systems intended to ensure compliance with the RFMO conservation and management measures. Vessels that are not from member States or from cooperating nonmembers cannot be assumed to be conducting fishing in accordance with the RFMO's conservation and management measures if they have not previously and expressly committed to compliance. RFMOs would thus be justified in presuming that such "unidentified" nonmember vessels found within their regulatory areas are IUU fishing vessels.

The international framework for IUU fishing complements the governance regime ensuing from UNFSA to regulate fishing and control access to high seas fishing opportunities through participation in RFMOs. The IPOA-IUU also provides a focal point and coordinating mechanism for promoting cross-jurisdictional responses to address IUU fishing, whether within or beyond national jurisdiction. This allows States to attempt to address the problems of overexploitation of fisheries on both national and international levels.

4.6 Membership in RFMOs

The efforts to strengthen and modernize existing RFMOs, create new RFMOs, and combat IUU fishing outlined above have led to a heightened interest on the part of States to join RFMOs, regardless of whether these States have ratified or acceded to UNFSA. This has led to increased pressure upon the RFMOs to open their doors to greater participation and membership. Recent movements in the membership of RFMOs are one of the most obvious indicators that the regime established by UNFSA is gaining acceptance by the international community and that there is an attempt to abide by and implement the relevant obligations under international fisheries treaties (see Table 1.2). Increased membership in RFMOs will presumably lead to better governance as more and more fishing nations come under the coverage of the international fisheries management regime. However, membership in RFMOs is in itself a complicated issue with significant implications at both national and international levels, suggesting that improvement in governance will not necessarily come easily.

4.6.1 Regulating access to the high seas fisheries resources

Prior to UNFSA, the effectiveness of RFMO management was directly dependent on the compliance of member States. Membership could not be forced or imposed upon nonconsenting States, and fishing vessels flying nonmember flags were beyond the jurisdiction and competence of the RFMOs. The traditional exclusivity of flag-State jurisdiction did not permit RFMOs or their members to substantially and actively intervene in the operations of such vessels. Moreover, fishing by nonmembers usually had competitive advantages over fishing by members and increased the risk of overexploitation. It also provided an incentive for members to opt out of allocation decisions if they perceived themselves as being unfairly treated in the light of the uncontrolled harvest by competitors.[120]

UNFSA gives effect to the general duty to cooperate in the conservation and management of straddling fish stocks and highly migratory fish stock, originally expressed in LOSC,[121] by calling on States to become members of RFMOs, or to agree to apply the conservation and management measures of such RFMOs. Article 8(3) of UNFSA provides that "where a subregional or regional fisheries management organization or arrangement has the competence to establish conservation and management measures for particular straddling fish stocks or highly migratory fish stocks, States fishing for the stocks on the high seas and relevant coastal States shall give effect to their duty to cooperate by becoming members of such organization or participants in such arrangement, or by agreeing to apply the conservation and management measures established by such organization or arrangement". While UNFSA does not force States to join RFMOs, it does oblige States to apply all management and conservation measures adopted by an RFMO regardless of whether they are members of the RFMO in question.[122] This obligation is reinforced by providing that "only those States which are members of such an organization or participants in such an arrangement, or which agree to apply the conservation and management measures established by such organization or arrangement, *shall have access to the fishery resources* to which those measures apply".[123] Therefore, the right of access to the resources is directly linked to the obligation either to become a member of an RFMO or to apply its management measures; in other words, States that do not comply with management and conservation measures adopted by an RFMO are "effectively excluded from exercising the freedom of fishing on the high seas".[124] Cooperation with an RFMO should be rewarded with catch allocations.[125] However, this is only true in theory, as States that merely agree to comply with an RFMO management regime without being a member of the RFMO in question are faced with the problem of not being taken into consideration for the allocation of fishing opportunities.[126] Hence, in practice, membership in RFMOs has become a crucial prerequisite of access to the high seas fisheries resources.

4.6.2 Real interest

The question that first logically arises is which States or entities have the right to become a member of an RFMO. UNFSA does not set detailed criteria for membership; however, Article 8(3) of UNFSA establishes that States "having a real interest in the fisheries concerned" are eligible to become members of RFMOs or participants in RFMAs and cannot be precluded from, or discriminated against, becoming such. Membership in RFMOs is not intended to be open to all and should apply only to States with a "real interest".[127] A need was apparently felt not to repeat the key element of Article 118 of LOSC, which refers to "States whose nationals exploit … living resources" on the high seas.[128] However, UNFSA does not define what real interest means.[129] The concept of real interest must therefore be studied and applied on a case-by-case basis, taking into account the particular situation of each RFMO.[130] However, from the structure of Article 8(3), it is evident that "States fishing for the stocks on the high seas and relevant coastal States",[131] i.e., States whose maritime zones are included in or adjacent to the RFMO convention area are at any rate included, for their real interest is implicit in their duty to cooperate.[132] As a result, these States not only have the duty but also the right to become a member of the RFMO. It is further argued that "States having a real interest" not only include these two categories of States, for otherwise it would have been sufficient to make a reference to the first sentence of Article 8(3).[133] As Molenaar observes, post-UNFSA developments in RFMOs and negotiation processes aimed at establishing RFMOs indicate that real interest is considered broadly and is not widely used to limit membership in RFMOs per se.[134] Rather, it seems more relevant to the determination of the legitimacy of applications for membership in RFMOs and in discouraging the effects of reflagging and flags of convenience.[135] Further, as Plé emphasizes, RFMOs should generally be open to "all responsible States with a genuine connection to the fishery in question".[136] He argues that it is better to bring non-members under the umbrella of an RFMO regime than to allow their vessels to remain "free riders".[137]

By providing that all States "having a real interest in the fisheries concerned may become members" of the RFMO in question, and that RFMOs shall not preclude such States from membership or participation,[138] States wishing to become a member of an RFMO have, in fact, a right to do so. As Munro et al. note, Articles 8, 10, and 11 make it very clear that RFMOs must be prepared to accommodate new entrants.[139] These new entrants include flag States that do not have a catch history in a regulatory area and wish to fish in the future, and those that previously fished in the area and wish to resume their fishing activities.[140] Article 11 of UNFSA lists various considerations that RFMOs are to take into account upon receiving applications from prospective new members, including *inter alia* the status of the straddling fish stocks and highly migratory fish stocks and the existing level of fishing effort in the fishery; the needs of coastal fishing communities whose livelihoods are dependent mainly on fishing for the stocks; and the needs of coastal States whose economies are overwhelmingly dependent on the exploitation of living marine resources.[141] The issues associated with the acceptance of new entrants are discussed further below.

4.6.3 The "free-rider" problem

One of the main reasons for linking access to the high seas fisheries resources to membership in an RFMO was to reduce the problem of the so-called "free riders", which exercise their right to fish on the high seas without complying with their duty to cooperate.[142] This "phenomenon of non-member fishing" has undermined considerably the effectiveness of RFMOs in managing their respective regulatory areas.[143] Making membership or compliance with the management regime mandatory and limiting access to the resources only to those States that are members or that closely cooperate place considerable pressure on fishing States to come under the management umbrella of RFMOs. Balton takes the view that if all flag States complied with these rules, "non-member fishing would, almost by definition, largely disappear".[144] However, there are still a large number of States that are not parties to UNFSA, and it is unclear whether such States are under an obligation to join RFMOs or to comply with their conservation measures. This latter consideration raises an important but still unresolved issue concerning many aspects of UNFSA, namely to what extent have its provisions become binding on States as part of customary international law. It also raises the question whether the obligation to cooperate in Articles 116 to 119 of LOSC has become part of customary international law. These questions still lack an authoritative answer in the contemporary law of the sea.

Munro et al. suggest a new type of free-rider problem. They argue that a new member may also be seen as a free rider in the sense that a State that joins an RFMO enjoys, at no cost to itself, a pro-rata share of the return on the investment from costs and sacrifices incurred over time to conserve and responsibly manage the resource made by the existing members. This could lead both existing and prospective members of RFMOs to conclude that they would be better off by refusing to cooperate.[145]

4.6.4 Efforts to increase membership or participation

The emphasis of UNFSA on membership has encouraged States within and outside RFMOs to undertake various efforts to further increase membership in RFMOs and compliance with RFMO management regimes. The urgent need for States to join or agree to apply the conservation and management measures established by RFMOs has been reiterated by the UN General Assembly, the Secretary-General, and the 2005 St. John's Conference.[146]

To achieve broad cooperation and participation and therefore compliance with the RFMO conservation and management regime, RFMOs created new categories of participation, which, in some cases, are akin to membership but do not qualify as such. Likewise the terminology used to refer to such participation varies considerably from one RFMO to another. RFMOs are thus increasingly open to include cooperating nonmembers and fishing entities. Cooperating nonmember status normally allows States to participate in meetings and receive quota allocations, and exempts them from measures intended to deter unregulated fishing,[147] even though they may not be allowed to vote and are not, strictly speaking, bound by the provisions of the RFMO convention. Cooperating status generally requires that such States formally accept to be bound by the conservation and management measures of the RFMO in question, as well as agree to provide data and information on the measures taken to ensure compliance by its vessels. CCAMLR, CCSBT, IATTC, ICCAT, IOTC, NEAFC, and WCPFC have all introduced such categories of participation in the last decade.[148] CCAMLR has an additional special subcategory of cooperating nonmember States, those States participating in CCAMLR's catch documentation scheme (CDS), which may be port States through which fish catch is channelled, rather than fishing nations.[149] Further, RFMOs increasingly have expanded membership to include REIOs, such as the European Union (EU). REIO membership recognizes the major role of such inter-State arrangements in international fisheries management. The EU has concluded access arrangements with many countries and is a member of most RFMOs (CCAMLR, GFCM, ICCAT, IOTC, NAFO, NEAFC, SEAFO, and WCPFC). Only the CCSBT and the IATTC do not permit REIOs to join as members, although the Antigua Convention will allow IATTC to accept REIO membership once it is ratified and enters into force.[150]

4.6.4.1 Fishing entities

Membership or participation of fishing entities in RFBs has been problematic, in particular in respect of the Fishing Entity of Taiwan (also known as Chinese Taipei and, within the UN system, as Taiwan Province of China). The ongoing controversy over Taiwan's legal status prevents it from becoming a member of FAO bodies or RFMOs such as the GFCM and IOTC, as membership is defined by reference to being a member of FAO, the UN, its specialized agencies, or the International Atomic Energy Agency. Taiwan is not a member of the UN or any of these UN-affiliated organizations, primarily because of the application of the "one China" policy within the UN system. However, Taiwan is a major participant in high seas fishing and has a significant presence in all oceans.[151] The adoption of the new concept of "fishing entity" pursuant to Article 17(3) of UNFSA as a subject of duties and obligations has enabled Taiwan to participate in some RFMOs. Provision is made for Taiwan to participate in the IATTC - Antigua Convention (not yet in force), in the CCSBT as part of the Extended Commission, and as a cooperating fishing entity in ICCAT and IATTC.[152] Provision is also made in SEAFO for fishing entities to cooperate in the implementation of conservation and management measures "with a view to having such measures applied *de facto* as extensively as possible to fishing activities in the Convention Area. Such fishing entities shall enjoy benefits from participation in the fishery commensurate with their commitment to comply with conservation and management measures in respect of the stocks".[153]

The WCPFC Convention goes beyond UNFSA by creating a mechanism for such entities to formally accept the obligations under the instrument, a situation that is almost analogous to ratification. It is through this process that Taiwan is able to participate in the WCPFC in the Western and Central Pacific Ocean under the name "Chinese Taipei". Annex 1(1) to the WCPFC Convention, entitled "Fishing Entities", provides as follows:

> After the entry into force of this Convention, any fishing entity whose vessels fish for highly migratory fish stocks in the Convention Area, may, by a written instrument delivered to the depositary, agree to be bound by the regime established by this Convention. Such agreement shall become effective thirty days following the delivery of the instrument. Any such fishing entity may withdraw such agreement by written notification addressed to the depositary. The withdrawal shall take effect one year after the date of receipt of the notification, unless the notification specifies a later date.

The Convention draws a subtle distinction between the role of members of the Commission and the role of Contracting Parties. Contracting Parties are States or entities entitled to ratify or accede to the Convention. Members of the Commission include all Contracting Parties and others entitled to membership of the Commission, including fishing entities. With few exceptions, most of the provisions of the Convention apply to members of the Commission. According to Annex 1, "Such fishing entity shall participate in the work of the Commission, including decision-making, and shall comply with the obligations under this Convention".[154] The provisions of the Convention that specifically apply to Contracting Parties include election of the Chairman and Vice Chairman of the Commission (Article 9(4)); the determination of the location of the Headquarters of the Commission (Article 9(7)); invitation of other States and REIOs to accede to the Convention (Article 35); and the power to withdraw from the Convention (Article 42).

So far, no fishing entity has been permitted to participate in any of the fisheries bodies within the FAO framework, with the exception of the IOTC, where its participants are present under the heading of "invited experts". As of May 2006, the IOTC is the only tuna RFMO that has not found a practical solution to the participation of fishing entities in its work.[155]

4.6.4.2 Vessel registers: positive and negative lists

To encourage States to join RFMOs and to promote compliance with conservation and management measures, RFMOs have developed fishing vessel registers or records (so-called "positive lists" and "negative lists") as primary tools of RFMO management.[156] The positive list constitutes the official register or record of fishing vessels authorized to fish in the regulatory area; noninclusion in the register or record raises the rebuttable presumption that a vessel is engaged in IUU fishing. Vessels on the negative list will be subject to a range of measures to be taken against them by Contracting Parties, including denial of landing, denial of import and transshipment, refusal of flag, and denial of chartering arrangements.[157]

For practical purposes, these lists, when combined with the presumption that vessels not listed as authorized to fish are engaged in IUU fishing, could, if widely applied and effectively maintained by RFMOs, come to constitute an important fetter on the ability of vessels to exercise the remnants of the freedom of fishing on the high seas.

4.6.4.3 Catch and trade documentation schemes

Another management tool of RFMOs aimed at increasing membership and promoting compliance is the use of trade measures, such as catch documentation and trade documentation schemes. Vessels of States authorized to fish in the convention area must be licensed and registered to harvest specific species and can only land or export the authorized catch with valid documentation. CCAMLR has a catch documentation scheme in place, and, CCSBT, IATTC, ICCAT, IOTC, and WCPFC have developed trade documentation schemes (statistical trade documents). Additionally, port States that are members of RFMOs have adopted regulations empowering the relevant national authorities to prohibit landings and transshipments when the catch has been taken in a manner that undermines the effectiveness of the management regime.[158]

4.6.5 Implications of increased membership

UNFSA has been effective in promoting membership in RFMOs, as evidenced by the marked increase in the number of fishing nations participating in RFMOs, not only as Contracting Parties, but also under the special categories noted above. For example, as of May 2006, 41 States and one REIO from five continents are members of ICCAT, 23 of whom joined since 1995.[159] All major fishing nations are now members in RFMOs. Table 1.2 illustrates the membership in RFMOs as of May 2006.

The right to membership entails the right of new entrants to a share of the available fishing opportunities.[160] However, membership may become so numerous as to make allocation decisions difficult,[161] as in the case of ICCAT. Most high seas fisheries are already fully exploited or overexploited, and many RFMOs are therefore bound to set the total allowable catches (TAC) in their areas of competence at low levels. The fishing opportunities are often already fully used or will be so in the near future.[162] In these cases, new entrants in an RFMO may affect the allocation of TAC among RFMO members by diminishing the already comparatively small shares of existing members. This situation creates an underlying resentment on the part of current members against accommodating new entrants[163] and may discourage nonparticipating States from becoming RFMO members if catch quota allocations are viewed as being inequitable[164] or, in the worst case, not even worth fishing. Hence, while the increase in membership and participation in RFMOs is a positive development, paradoxically it also raises problems that may discourage membership or participation.

In a resolution to guide the expectations of future members, NAFO formally advises new entrants that fish stocks managed by NAFO are fully allocated and that fishing opportunities are limited to new fisheries and to the "others" category in the NAFO Quota Allocation Table, rather than the major stocks.[165] NEAFC has stated that new members will participate on the same basis as existing members only in future allocations of stocks that are unregulated at the time the application is made and that they can only fish existing stocks that fall under the "cooperative quota" category.[166] However, to prevent unregulated and illegal fishing, expectations need to be managed and fishing opportunities need to be allocated responsibly.[167]

Experience has shown that difficulties in reaching agreement can arise when membership is broad and interests so different, e.g., coastal States and States fishing on the high seas. Though crucial, ensuring the responsible, sustainable management of these fisheries is not always a strong basis for cooperation.[168] In practice this divergence of interests and the promotion of self-interest by participants usually translate into managerial and organizational problems for the fisheries bodies. RFMOs are typically hampered by disagreement over scientific advice or resource allocation; the inability to make binding decisions; or disagreement on effort controls, observer programs, and penalties.[169]

Large membership is also likely to create an increased administrative load for RFMO secretariats as well as additional budgetary requirements for the organization. Budgetary requirements may only be met by full members in cases where the financial obligations pertain only to those States under the terms of the relevant convention. Expanded membership may further burden RFMOs that have members who technically should not be voting and participating in decisions and yet are allowed to do so for practical considerations of ensuring compliance. The financial implications of fully participating in an RFMO can be significant, particularly for developing States. Some RFMOs also have members that are in financial arrears, which may create a lack of adequate financial resources and additional stress for the RFMO administration.[170]

4.7 Conclusion

RFMOs are an essential component of the international regime to promote the sustainable development of fisheries. States are to give effect to their duty to cooperate directly through participation in RFMOs. Under UNFSA, States are obliged to join RFMOs or must agree to comply with or not undermine conservation measures; arguably, such an obligation is now part of customary international law independent of the provisions of UNFSA. RFMOs are the principal cooperative arrangement fostering international cooperation, balancing the rights and interests of coastal States and high seas fishing States for enhanced fisheries management to ensure the long-term sustainability of fish stocks and related fisheries over their entire geographic range. The corollary of this objective is for RFMOs to avoid overfishing, rebuild overfished stocks, and protect nontarget species affected by fishing through precautionary, responsible, and regulated conservation programs. "Indeed, in FAO and elsewhere it has been recognized that regional fishery bodies are the only real mechanism for this task".[171]

RFMOs are distinct from their member States with a separate and independent international legal personality. This personality can be derived either explicitly from the constituent treaty establishing the RFMO or from the fact that they are intergovernmental bodies performing acts on the international plane, which include adopting international conservation and management measures, making resolutions on subjects as diverse as trade-related measures capable of impacting on both parties and non-parties, and the rights of Cooperating Non-Contracting Parties. While RFMOs are creatures of their respective conventions or agreements with independent powers, the powers of the commission are only those granted to them by member States. As such, RFMOs are organizations mandated to regulate the fisheries in their respective convention areas, governed by their member States, not their secretariats. RFMOs thus reflect the policies and commitment of the Contracting Parties and members.

RFMOs are not perfect constructs but are essential in the development of a coherent global management regime for responsible regulation of high seas fisheries. Long-term conservation and sustainable development require coordination, cooperation, and collaboration as well as compliance on the part of all participants: RFMOs, coastal States, flag States, port States, and the fishing industry. As noted by Miles,

> The management of the world ocean is characterized by a haphazard quilt of global, regional, bilateral, and national arrangements differentiated by activity and ocean, but without effective mechanisms for achieving coordination across either oceans or activities. This fragmentation creates occasions for significant conflict and interferes with productive use of the oceans to a noticeable degree. As new technologies increasingly permit a variety of uses of ocean space, policies designed to achieve coordinated management of the oceans are required if conflict is to be avoided and if human use of the marine environment is to be less wasteful. Such policies necessarily transcend national jurisdictional zones and are therefore beyond the capability of individual nation-states, irrespective of their level of development.[172]

UNFSA has achieved a certain measure of success in reforming and strengthening regional fisheries management, in introducing new principles and management functions, in prompting the reactivation of previously dormant organizations, in establishing new organizations, and in negotiating the creation of regional management bodies where none currently exist. The Agreement has become the point of reference or benchmark against which all RFMOs with competence for straddling fish stocks or highly migratory fish stocks are to be held accountable. To reverse the decline in straddling fish stocks and highly migratory stocks, rehabilitate marine habitats, conserve associated marine ecosystems, and protect marine biodiversity, the regime established by UNFSA and complimentary instruments must be implemented by all RFMOs (old and new). As Ambassador Nandan stated in his preparatory statement as Chair of the Honolulu Convention to negotiate the establishment of WCPFC:

> It is no longer acceptable to base new regimes for conservation and management on precedents from other fisheries management organizations, most of which pre-date UNCED, Agenda 21 and the UN Fish Stocks Agreement, let alone the 1982 [LOS] Convention. The world has moved on. Any new arrangement must reflect the reality of international fisheries... [UNFSA] has already become the reference point for the review of fisheries management organizations.[173]

The next chapter will discuss the concepts of overfishing and overcapitalization, which reflect the reality of many international fisheries. As well, the chapter will provide an overview of the status of key commercial straddling fish stocks and highly migratory fish stocks regulated by RFMOs.

Endnotes

[1] The UNFSA regime builds upon the duty of States to cooperate through regional, multilateral organizations under LOSC for the conservation and management of straddling fish stocks and highly migratory fish stocks. See, in particular, LOSC Arts. 63, 64, and 117 and UNFSA Art. 8. As noted in endnote 1 of Chap. 2, the language in many international legal instruments and technical reports refers to the "conservation and management of fish stocks". Indeed, the full title of UNFSA includes this phrase: The Agreement for the Implementation of The Provisions of the United Nations Convention on the Law of the Sea of 10 December 1982 relating to the *Conservation and Management of Straddling Fish Stocks and Highly Migratory Fish Stocks.* [emphasis added.] Some reviewers took exception to the notion that humans can manage fish stocks. While this book uses this terminology, it is understood that the contemporary management objective is to responsibly regulate fisheries (i.e., to primarily manage human activity, rather than the fish stocks per se) so that the exploitation of fish stocks is within sustainable levels of biomass taking into account the ecological interconnections and the possible adverse impacts on the marine ecosystem. However, it should also be emphasized that with the increasing adoption of ecosystem considerations and the ecosystem approach to fisheries described in Chap. 7, the factors taken into account go beyond human predation in the efforts to understand ecological integrity and the health of the oceans.

[2] K. Gjerde, "Editor's Introduction to the Special Issue on High Seas Fisheries Governance: Moving from Words to Action" (2005) 20 *International Journal of Marine and Coastal Law* 323–344 at 329.

[3] Regimes are understood to be governance systems that need not, though often do, have a formal organization apparatus. Sydnes has identified three stages in the process of regime formation or regime building: agenda formation, negotiation, and operationalization. In practice, these three stages may overlap. A. Sydnes, "Establishing a Regional Fisheries Management Organization for the Western and Central Pacific Tuna Fisheries" (2001) 44 *Ocean and Coastal Management* 787–811 at 788–789.

[4] A. Sydnes, "Regional Fishery Organizations: How and Why Organizational Diversity Matters" (2001) 32 *Ocean Development and International Law* 349–372. "RFOs are generally established by states as political solutions to collective action problems. States are seeking to achieve specific objectives by cooperating" at 353. He also observed that regional fisheries organizations are increasingly being viewed as 'vehicles of good governance' created to secure the sustainable development of international fisheries" at 349. Sydnes concludes in his article that "the ability of RFOs to be 'vehicles of good governance' to secure sustainable management has to be proven" at 364.

[5] M. Lodge and S. Nandan, "Some Suggestions Towards Better Implementation of the United Nations Agreement on Straddling Fish Stocks and Highly Migratory Fish Stocks of 1995" (2005) 20 *International Journal of Marine and Coastal Law* 345–379 at 354.

[6] See A. Sydnes, at 349. The term "vehicles of good fishery governance" was used during the first meeting of FAO and non-FAO RFBs 11–12 February 1999 and appears in the report of the meeting FAO, "Report of the Meeting of FAO and Non-FAO Regional Fishery Bodies and Arrangements", FAO Fisheries Report No. 597 (Rome: FAO, 1999) in abstract and at para 27.

[7] G. Munro, A. Van Houtte, and R. Willmann, "The Conservation and Management of Shared Fish Stocks: Legal and Economic Aspects", FAO Fisheries Technical Paper No. 465 (Rome: FAO, 2004) at 40.

[8] This model of governance, variously known as integrated ocean management, integrated coastal and oceans management, and integrated resource management, is characterized by multidisciplinary, cross-sectoral, intergovernmental, multistakeholder, participatory, and integrated approaches. For a discussion of these principles see, e.g., R. Costanza et al., "Principles for Sustainable Governance of the Oceans" (1998) 10 *Science* 198–199; B. Cicin-Sain and R. Knecht, eds., *Integrated Coastal and Ocean Management: Concepts and Practices* (Washington, D.C.: Island Press, 1998)

[9] Commission on Global Governance, *Our Global Neighbourhood: The Report of the Commission on Global Governance* (Oxford: Oxford University Press, 1995).

[10] According to Grotius in 1633, "[t]he sea can in no way become the private property of any one, because nature not only allows but enjoins its common use…Nature does not give a right to anybody to appropriate such things as may inoffensively be used by everybody and are inexhaustible, and therefore, sufficient for all". Quoted in J. Morell, *The Law of the Sea: An Historical Analysis of the 1982 Treaty and Its Rejection by the United States* (Jefferson, N.C.: McFarland, 1992) at 2. See Hugo Grotius, *Mare Liberum* (1633) Translation by R.V.D. Magoffin (Oxford: Oxford University Press, 1916). See also discussion in Chap. 2. For a historical perspective see L. Juda, *International Law and Ocean Use Management: The Evolution of Ocean Governance* (London: Routledge, 1996) at 8–17; and J. Talhelm, "Curbing International Overfishing and the Need for Widespread Ratification of the United Nations Convention of the Law of the Sea" (2000) 25 *North Carolina Journal of International Law and Commercial Regulation* 381–418 at 386; R.P. Anand, "Changing Concepts of Freedom of the Seas: A Historical Perspective" in J. Van Dyke, D. Zaelke, and G. Hewison, eds., *Freedom of the Seas in the 21st Century* (Washington, D.C.: Island Press, 1993) 72–86.

[11] See Chap. 5 for further discussion on overfishing and the tragedy of the commons. As noted by E. Larocque, "The Convention on the Conservation and Management of Highly Migratory Fish Stocks in the Western and Central Pacific Ocean: Can Tuna Promote Development of Pacific Island Nations", published online by the University of Hawaii's William S. Richardson School of Law Environmental Program at <http://www.hawaii.edu/elp> (accessed January 2006): "This theory, combined with the technological advances and changing fishing practices, led to stock over-exploitation. This in turn led to territorial dispute and demand for new legal arrangements to regulate activities on the high seas". As well there have been numerous governance initiatives to regulate access to the commons, control fishing practices, and regulate output through the establishment of national, community, corporate, or individual quotas. For a discussion of trading quotas and other initiatives to overcome the "tragedy of the commons", see A. Serdy, "Fishery Commission Quota Trading Under International Law" in A. Chircop, S. Coffen-Smout, and M. McConnell, eds., *Ocean Yearbook 21* (Leiden: Brill, 2007).

[12] See G. Hardin, "The Tragedy of the Commons" (1968) 162 *Science* 1243–1248; and T. Dietz, E. Ostrom, and P. Stern, "The Struggle to Govern the Commons" (2003) 302 *Science* 1907–1912 at 1907.

[13] T. Dietz, E. Ostrom, and P. Stern, at 1910.

[14] The term "governance" derives from Latin origins that suggest the notion of "steering". Governance is the "exercise of authority, control, management, power of government" (*Webster's New Universal Unabridged Dictionary* (London: Dorset & Baber, 1979)) and includes aspects such as the processes by which authority is exercised in the management of resources, and the capacity of governments to design, formulate, and implement policies. Cf. The World Bank, *Governance and Development* (Washington, D.C.: The International Bank for Reconstruction and Development/The World Bank, 1992) at 58, note 1. In other words, governance refers to planning, policy-making, and decision-making processes to manage human activities and also encompasses initiatives at the regional and international level. Governance systems are generally referred to as "regimes". More specifically, regimes are "[s]ocial institutions composed of agreed-upon principles, norms, rules and decision-making procedures that govern the interaction of actors in specific issue areas". O. Young and G. Oshrenko, eds., *Polar Politics: Creating International Environmental Regimes* (Ithaca and London: Cornell University Press, 1993). Regimes need not, though often do, have a formal organization apparatus.

[15] M. Gonçalves, "Concepts of Marine Region and the New Law of the Sea" (1979) 3 *Marine Policy* 255–263 at 255. Marine regionalism dates back at least to the first decade of the 20th century. In the years after World War II, with the establishment of the United Nations and its specialized agencies, regional approaches to problem solving in the ocean environment were taken by three agencies, the United Nations Environment Programme (UNEP) with its regional seas activities, the Food and Agriculture Organization (FAO), which inaugurated regional commissions and councils, and the Intergovernmental Oceanographic Commission (IOC), which initiated regional scientific organizations and programs. In addition, there were other international efforts in the fields of fisheries and environmental management. A. Chircop, "Participation in Marine Regionalism: An Appraisal in a Mediterranean Context" in E. Mann Borgese, N. Ginsburg, and J. Morgan, eds., *Ocean Yearbook 7* (Chicago: University of Chicago Press, 1989) 402–416 at 402. See also L. Alexander, "New Trends in Regionalism" in E. Mann Borgese, N. Ginsburg, and J. Morgan, eds., *Ocean Yearbook 11* (Chicago: University of Chicago Press, 1994) 1–8 at 2.

[16] Chircop notes that the term "region" and its derivatives appear no less than 53 times in 25 different articles of LOSC, spanning the provisions on the EEZ, high seas, enclosed and semienclosed seas, land-locked states, the Area, protection and preservation of the marine environment, and transfer of technology. A. Chircop, at 405.

[17] See L. Alexander, "New Trends in Regionalism" in E. Mann Borgese, N. Ginsburg, and J. Morgan, eds., *Ocean Yearbook 11* (Chicago: University of Chicago Press, 1994) 1–8.

[18] See L. Alexander, at 2.

[19] See L. Alexander, 1–8.

[20] M. Gonçalves, at 255.

[21] *North Sea Fisheries Convention*, 6 May 1882, entered into by Great Britain, Germany, France, the Netherlands, Denmark, and Belgium for a period of five years.

[22] S. Kaye, *International Fisheries Management* (The Hague: Kluwer Law International, 2001) at 47.

[23] International Pacific Halibut Commission in 1923; International Pacific Salmon Fisheries Commission in 1937; North Pacific Fur Seal Commission in 1958.

[24] W. Burke, "Compatibility and Precaution in the 1995 Straddling Stock Agreement" in H. Scheiber, ed., *Law of the Sea: The Common Heritage and Emerging Challenges* (The Hague: Nijhoff Publishers, 2000) 105–126 at 107. A notable exception of a RFB providing for an independent decision-making authority was the International Pacific Salmon Fisheries Commission created under the 1930 (Canada–US) Convention for the protection, preservation, and extension of the salmon fisheries of the Fraser River system. This authority was largely removed by the successor agreement negotiated in 1985.

[25] The Rome Technical Conference on the Conservation of the Living Resources of the Sea, 1955.

[26] *First United Nations Conference on the Law of the Sea: Final Act*, Geneva, Switzerland, 24 February – 27 April 1958, A/CONF.13/38 (58.V.4), Vol. II, pp. 143–147.

[27] *Convention on Fishing and the Conservation of the Living Resources of the High Seas*, done at Geneva, Switzerland, 29 April 1958, entered into force 20 March 1966 after deposit of the 30th instrument of ratification, and as of 2005 has been ratified by 37 State Parties.

[28] As Lodge and Nandan note, "The emphasis on regional agreements as the basic pattern for management of international fisheries was apparent as long ago as 1955, at the Rome Technical Conference on the Conservation of the Living Resources of the Sea. That Conference agreed for the first time that conservation and management of high seas fisheries resources could only be carried out through international cooperation in research and regulation and that the best way of achieving this was through the establishment of regional conventions, based on geographical and biological distribution of the marine populations concerned". M. Lodge and S. Nandan, at 347 note 7. The Conference was held in April and May 1955 and attended by some 45 State representatives. For an account of this Conference, see W. Herrington, "The Convention on Fisheries and Conservation of Living Resources: Accomplishments of the 1958 Geneva Conference" in L. Alexander, ed., *The Law of the Sea: Offshore Boundaries and Zones* (Columbus: Ohio State University Press, 1967) at 26.

[29] S. Kaye, at 255.

[30] S. Kaye, at 63.

[31] M. Lodge and S. Nandan, at 347. Also see *Foreign Relations of the United States*, Vol. X (Washington, D.C.: Department of State, 1955–1957); G. Lugten, *The Impact of Extra Legal Factors in the Historical Development of International Fisheries Law* (Ph.D. thesis, University of Tasmania, 1996) at 144; F. Garcia Amador, *The Exploitation and Conservation of the Resources of the Sea* (Leiden: Sijthoff, 1959) at 136–137; and S. Kaye, at 62–65.

[32] M. Schaefer, "Some Recent Developments Concerning Fishing and Conservation of the Living Resources of the High Seas" (1970) 7 *San Diego Law Review* 371–407 at 397. This world view is exemplified by the work of Myers Smith McDougal. See M.S. McDougal, "Perspectives for an International Law of Human Dignity" in M.S. McDougal et al., eds., *Studies in World Public Order* (New Haven, Conn.: Yale University Press, 1960) at 987, where McDougal said the ultimate goal of international law is "to create and preserve an universal order of human dignity…". He believed that the creation of a network of international organizations, with the UN as the focal point, was

the primary vehicle for overcoming the constraints of the political boundaries of nation States to achieve and maximize universal respect for human dignity. McDougal understood there were "inescapable interdependencies throughout the world in resource potentialities and institutional structures". See M.S. McDougal and D. Haber, *Property, Wealth, Land: Allocation, Planning, and Development — Selected Cases and other Materials on the Law of Real Property*, Vol. iii (Charlottesville,Va.: Michie Casebook Corp., 1948) at 1202. Quoted by M.P. Ryan, W.C. Renhardt, and K. Tamai, "International Governmental Organization Knowledge Management for Multilateral Trade Lawmaking" (2000) 15(6) *American University International Law Review* 1347–1378 at 1352.

[33] S. Kaye, at 87. Kaye also indicates at p. 86 that by the "mid-1960s there was a broad consensus that fisheries jurisdiction could extend beyond the territorial sea, although the dimensions of valid regulation were still open to some dispute".

[34] State-centred regimes rely on the duty to cooperate and the willingness of States to negotiate (*pactum de negotiando* under Art. 63), the duty to conclude an agreement (*pactum de contrahendo* under Art. 64), the degree of willingness to be bound, and the degree of commitment and capacity of flag States to exercise control over fishing vessels and enforce conservation measures on the high seas. A State's refusal to have its sovereign rights restricted by an international fisheries agreement is best exemplified by an objection procedure, which allows individual States to legitimately opt out of the organization's management strategy.

[35] See FAO, "Regional Fishery Bodies and Governance: Issues, Actions and Future Directions", FAO Fisheries Circular No. 959 (Rome: FAO, 2000); FAO, "Report of the Third Meeting of Regional Fishery Bodies", FAO Fisheries Report No. 703 (Rome: FAO, 2003); and FAO, "Report of the Fourth Meeting of Regional Fishery Bodies. Rome, 14-15 March 2005", FAO Fisheries Report No. 778 (Rome: FAO, 2005).

[36] The term RFMO is sometimes used interchangeably with RFB, but RFMOs are, in fact, a subset within the global network of RFBs.

[37] A.K. Sydnes, "Regional Fishery Organizations: How and Why Organizational Diversity Matters" (2001) at 361, notes that "fisheries management is seen as comprising three interrelated functions: gathering, analyzing, and disseminating scientific information of the fisheries; deciding upon regulatory measures; and establishing mechanisms to enforce the regulatory measures. Subsequently the main task of regional fisheries management organizations is to make decisions. Accordingly regional fisheries management organizations need to have in place procedural arrangements so that they can establish regulatory measures. Without making decisions, such organizations would not be fulfilling their roles". An important aspect of decision making is whether decisions are binding on member countries. While most of the RFMOs do have authority to make binding decisions, the RFMOs do not have supranational authority. All RFOs have institutionalized safeguards (escape mechanisms) for their member States, "either by making decisions by consensus, by making majority decisions subject to objection procedures or by making decisions advisory rather than binding on member countries" at 363.

[38] See A. Sydnes, "Regional Fishery Organizations: How and Why Organizational Diversity Matters" 349–372. See also C. de Fontaubert and I. Lutchman, *Achieving Sustainable Fisheries: Implementing the New International Legal Regime* (Gland: IUCN, 2003); R. Barston, "The Law of the Sea and Regional Fisheries Organisations" (1999) 14(3) *International Journal of Marine and Coastal Law* 333–352; and
C. Schmidt, *Why Fish Piracy Persists: The Economics of Illegal, Unreported and Unregulated Fishing* (Paris: OECD, 2005) Chap. 3 at 101.

[39] *Convention on the Conservation and Management of the Pollock Resources in the Central Bering Sea*, done at Washington, D.C., 16 June 1994, entered into force 8 December 1995 (BSDH Convention). The multilateral agreement on the pollock resources of the Central Bering Sea illustrates an "arrangement" as opposed to an "organization". The BSDH Convention does not call for a formal Secretariat. The parties meet annually to decide upon the annual harvest level and the individual national quota of pollock for each party. The hosting Contracting Party is responsible for logistical and administrative support for the Annual Conference. The only other organ of the group is the Scientific and Technical Committee, whose plan of work is determined by the Annual Conference. The Committee's main functions are to "compile, exchange, and analyze information on fisheries harvests, and pollock and other living marine resources covered by this Convention" (Art. IX) and to make recommendations to the Annual Conference regarding conservation and management of pollock, including the annual harvest level.

[40] The Southern Indian Ocean Fisheries Agreement when adopted will establish SIOFA. The drafting process of the text was completed in late 2005. See *Report of the Fifth Intergovernmental Consultation on the SIOFA (2005)*, at para. 5.1, cited in E. Molenaar, "Addressing Regulatory Gaps in High Seas Fisheries" 20(3) *International Journal of Marine and Coastal Law* 533–570 at 542 and notes 19 and 50. See Afterword for update.

[41] *Arrangement between the Government of Australia and the Government of New Zealand for the Conservation and Management of Orange Roughy on the South Tasman Rise*, signed for New Zealand on 17 February 2000 and for Australia on 25 February 2000, entered into force 1 March 2000; text at (2001) 16 *International Journal of Marine and Coastal Law* 119–124 (South Tasman Rise Agreement).

[42] *Agreement Between the Government of Iceland, the Government of Norway, and the Government of the Russian Federation Concerning Certain Aspects of Cooperation in the Area of Fisheries,* 15 May 1999.

[43] These agreements establish a delicately balanced set of reciprocal fishing rights with catch quotas and specify that scientific and technical consultations should be held. In some cases, joint regulatory zones are prescribed stipulating number and size of trawlers, types of gear, dates of operation, and catch. A number of these agreements are nongovernmental. P. Hayes and L. Zarsky, "Environmental Issues and Regimes in Northeast Asia" (1994) 6(4) *International Environmental Affairs* 283–319.

[44] On 24 January 2001 the European Union (EU) and Chile reached a provisional settlement to resolve the 10-year WTO–ITLOS "swordfish dispute", WTO case WT/DS193 (Chile — Measures Affecting the Transit and Importation of Swordfish), and ITLOS case No. 7. The agreement covers both access for EU fishing vessels to Chilean ports and bilateral and multilateral scientific and technical cooperation on conservation of swordfish stocks. See press release "EU and Chile settle WTO–ITLOS swordfish dispute" of 24 January 2001, Europa–EU–UN online <http://europa-eu-un.org/articles/en/article_2230_en.htm> (accessed 5 March 2006).

[45] It is interesting to note that CCAMLR was not considered an RFMO by some for a certain time but rather a "regional fishery body with the characteristics of an RFMO", as its mandate from the outset has been broader than fisheries.

[46] The FAO held a meeting of all RFBs in 1999 to examine how to strengthen all RFBs and promote inter-RFB cooperation. See FAO, "Report of the Meeting of FAO and Non-FAO Regional Fishery Bodies or Arrangements", Rome, Italy, 11–12 February 1999, FAO Fisheries Report No. 597 (Rome: FAO, 1999). This has now become a regular biennial occurrence and is timed to coincide with COFI. See report: FAO Fisheries Report No. 778. The Secretaries of the tuna fisheries bodies meet on an annual basis. A meeting of the tuna RFMOs is planned to take place in January 2007 hosted by the Government of Japan. See Afterword for update.

[47] For example, membership in GFCM, CCAMLR, ICCAT, and Sea of Okhotsk is determined geographically. Participation in IATTC, ICCAT, IOTC, NAFO, NEAFC, and SEAFO is not confined to contiguous or coterminous States, but also includes other interested parties in the fisheries of the region. In the case of CCSBT, the notion of region and the area of competence of the RFMO responsible for southern bluefin tuna are determined by the geographic range and distribution of the target species.

[48] See A. Sydnes, "Regional Fishery Organizations: How and Why Organizational Diversity Matters" 349–372 at 349.

[49] For a discussion on international regimes and regime formation see A. Sydnes, "Establishing a Regional Fisheries Management Organization for the Western and Central Pacific Tuna Fisheries" at 788–791.

[50] Molenaar introduces the term "regional fisheries management mechanism", along with the acronym RFMM, as an overarching term to include both organizations and arrangements under UNFSA, E. Molenaar, "The South Tasman Rise Arrangement of 2000 and other Initiatives on Management and Conservation of Orange Roughy" (2001) 16 *International Journal of Marine and Coastal Law* 77–124 at 90.

[51] See FAO, "Report of the Third Intergovernmental Consultation on the Establishment of a Southwest Indian Ocean Fisheries Commission", Nairobi, Kenya, 27–30 January 2004, FAO Fisheries Report No. R742 (Rome: FAO, 2004), paras. 22–23 and 28–29.

[52] FAO Fisheries Report No. R742 para. 26.

[53] For example, the impact of having no regime in the Southern Indian Ocean adjacent to CCAMLR created difficulties monitoring and enforcing the Patagonian toothfish strategies. This problem will be addressed with the creation of the SIOFA. NAFO and NEAFC cooperate on sighting vessels and the use of a vessel monitoring system (VMS) across the North Atlantic; Dr. Johanne Fischer, NAFO Executive Secretary, personal communication, December 2005.

[54] The only international organization that may also be similar to a RFB but with a global scope and application is the International Whaling Commission (IWC). See International Convention for the Regulation of Whaling, 1946.

[55] These RFMO convention areas also overlap with the those of the IWC.

[56] M. Lodge and S. Nandan, at 357.

[57] This is reiterated by the UN General Assembly, *Oceans and the law of the sea: sustainable fisheries, including through the 1995 Agreement for the Implementation of the Provisions of the United Nations Convention on the Law of the Sea of 10 December 1982 relating to the Conservation and Management of Straddling Stocks and Highly Migratory Fish Stocks, and related instruments*, UN Doc. A/60/L.23, 60th session (17 November 2005) (adopted as Resolution 60/31, 29 November 2005), at para. 10.

[58] Speaking notes for Jim Jones, Regional Director General, Gulf Region, Fisheries and Oceans Canada, Technical Briefing, *Canada's participation in the International Commission for the Conservation of Atlantic Tunas and the importance of ICCAT for Canada's tunas and swordfish fisheries*, Halifax, N.S., 8 November 2005. See online <http://www.dfo-mpo.gc.ca/overfishing-surpeche/media/sp-20051108_e.htm> (accessed 25 January 2006).

[59] UNFSA, at Art. 13 [emphasis added]. See also E. Molenaar, "Addressing Regulatory Gaps in High Seas Fisheries" at 547.

[60] UN General Assembly, *Oceans and the law of the sea: sustainable fisheries, including through the 1995 Agreement for the Implementation of the Provisions of the United Nations Convention on the Law of the Sea of 10 December 1982 relating to the Conservation and Management of Straddling Stocks and Highly Migratory Fish Stocks, and related instruments*, UN Doc. A/60/L.23, 60th session (17 November 2005) (adopted as Resolution 60/31, 29 November 2005), at para. 58 [emphasis added].

[61] E. Molenaar, "Addressing Regulatory Gaps in High Seas Fisheries" at 547.

[62] See further below at 4.5.1.1.

[63] See M. Lodge and S. Nandan, at 357–358.

[64] However, as Molenaar notes, "the fact that one or more parties to an RFMO are not parties to [UNFSA] is not always a valid excuse for not upgrading RFMOs. An obligation to upgrade may already exist on the basis of the [LOSC] or customary international law". See E. Molenaar "Addressing Regulatory Gaps in High Seas Fisheries" at 547.

[65] S. Garcia and I. de Leiva Moreno, "Global Overview of Marine Fisheries" in M. Sinclair and G. Valdimarsson, eds., *Responsible Fisheries in the Marine Ecosystem* (Rome: FAO, 2003) 1–24 at 15.

[66] M. Lodge and S. Nandan, at 357.

[67] For example, NAFO in 2004 adopted a framework for the precautionary approach by resolution, cf. *Report of the 26th Annual Meeting of the Fisheries Commission* (2004), at 8. See M. Lodge and S. Nandan; E. Molenaar "Addressing Regulatory Gaps in High Seas Fisheries" at 547.

[68] E. Molenaar "Addressing Regulatory Gaps in High Seas Fisheries" at 546.

[69] The Preamble of the Antigua Convention particularly identifies LOSC, the Rio Declaration, and Agenda 21, particularly Chap. 17, the Johannesburg Declaration and Plan of Implementation, the Code of Conduct, the FAO Compliance Agreement, the FAO International Plans of Action, and UNFSA.

[70] *Convention for the Strengthening of the Inter-American Tropical Tuna Commission established by the 1949 Convention between the United States of America and the Republic of Costa Rica*, "Antigua Convention", done at Washington, D.C., 14 November 2003, not yet in force. IATTC online <http://www.iattc.org/PDFFiles2/Antigua_Convention_Jun_2003.pdf> (accessed 26 January 2006).

[71] Antigua Convention, Arts. IV and V. See also the Report of the UN Secretary-General, *Report submitted in accordance with paragraph 17 of General Assembly resolution 59/25 of 17 November 2004, to assist the Review Conference to implement its mandate under paragraph 2, article 36 of the United Nations Fish Stocks Agreement*, UN Doc. A/CONF.210/2006/1, at para. 219.

[72] Resolution by ICCAT to Strengthen ICCAT [Res. 05-10], adopted in 2005.

[73] Ministerial Declaration of the *Conference on the Governance of High Seas Fisheries and the UN Fish Agreement*, St. John's, N.L., 5 May 2005, online <http://www.dfo-mpo.gc.ca/fgc-cgp/declaration_e.htm> (accessed 18 January 2006).

[74] FAO, *The Rome Declaration on Illegal, Unreported and Unregulated Fishing*, adopted by the FAO Ministerial Meeting on Fisheries, Rome, Italy, 12 March 2005, online <ftp://ftp.fao.org/fi/DOCUMENT/ministerial/2005/iuu/declaration.pdf> (accessed 19 December 2005); *Bali Plan of Action "Towards Healthy Oceans and Coasts for the Sustainable Growth and Prosperity of the Asia–Pacific Community"*, Joint Ministerial Statement of the 2nd APEC Ocean-Related Ministerial Meeting (AOMM2), Bali, 16–17 September 2005, APEC online <http://www.apec.org/apec/ministerial_statements/sectoral_ministerial/ocean-related/2005_ocean-related/bali_plan_of_action.html> (accessed 18 January 2006).

[75] Resolution by ICCAT to Strengthen ICCAT [Res. 05-10], adopted in 2005, at para. 1.

[76] Cf. speaking notes for Jim Jones, *Technical Briefing on Canada's participation in the International Commission for the Conservation of Atlantic Tunas and the importance of ICCAT for Canada's tunas and swordfish fisheries*, Halifax, N.S., 8 November 2005, online <http://www.dfo-mpo.gc.ca/overfishing-surpeche/media/sp-20051108_e.htm> (accessed 18 January 2006).

[77] 1999 Recommendation by ICCAT to Establish a Rebuilding Program for North Atlantic Swordfish [Rec. 99-2], entered into force 15 June 2000; Recommendation by ICCAT to Establish a Rebuilding Program for Western Atlantic Bluefin Tuna [Rec. 98-7], entered into force 21 June 1999; Recommendation by ICCAT to Establish a Plan to Rebuild Blue Marlin and White Marlin Populations [Rec. 00-13], entered into force 26 June 2001.

[78] Resolution by ICCAT Concerning Trade Measures [Res. 03-15], adopted in 2003; 2002 Recommendation by ICCAT to Establish a List of Vessels Presumed to Have Carried out Illegal, Unreported and Unregulated Fishing Activities in the ICCAT Convention Area [Rec. 02-23], entered into force 4 June 2003; 2002 Recommendation by ICCAT Concerning the Establishment of an ICCAT Record of Vessels over 24 m Authorized to Operate in the Convention Area [Rec. 02-22], entered into force 3 June 2003.

[79] Resolution by ICCAT on Port Inspection [Res. 79-2], adopted in 1979, and the Recommendation by ICCAT for a Revised ICCAT Port Inspection Scheme [Rec. 97-10], entered into force 13 June 1998; Recommendation by ICCAT on Vessel Chartering [Rec. 02-21], entered into force 3 June 2003.

[80] ICCAT Doc. ICCAT Criteria for the Allocation of Fishing Possibilities [01-25], adopted in 2001.

[81] See Resolution by ICCAT on Becoming a Cooperating Party, Entity or Fishing Entity [Res. 01-17], adopted in 2001, and Recommendation by ICCAT on Criteria for Attaining the Status of Cooperating non-Contracting Party, Entity or Fishing Entity in ICCAT [Rec. 03-20], adopted in 2003.

[82] See "NAFO Starts a Reform Process", *2005 Annual Meeting Press Release*, 23 September 2005, NAFO online <http://www.nafo.int/about/frames/about.html> (accessed 18 January 2006).

[83] NAFO/GC Doc. 05/02, at para. 1.

[84] NAFO/GC Doc. 05/02, at para. 2.

[85] NAFO/GC Doc. 05/02, at para. 3.

[86] E.g., the *Report of the Working Group on the Future of NEAFC*, London, UK, 24–26 April 2001, at para. 1. See NEAFC online <http://www.neafc.org/reports/future-neafc/neafc_wgfn/wgfn_2001.pdf> (accessed 25 January 2006).

[87] See "Terms of Reference for the Future Working Group", NEAFC Doc. AM 2004/56 of the 23rd Annual Meeting, London, UK, 8–12 November 2004.

[88] *Report of the 24th Annual Meeting of NEAFC*, London, UK, 14–18 November 2005, NEAFC Commission Report AM2005, at para. 12. NEAFC online <http://www.neafc.org/reports/annual-meeting/docs/full_reports/24neafc_annual_vol-1-main-report_2005.pdf> (accessed 25 January 2006).

[89] Press release, 21 November 2005, para. 7, NEAFC online <http://www.neafc.org/news/docs/2005-press-release_final.pdf> (accessed 18 January 2006). See also the *Report of the 24th Annual Meeting of NEAFC*, London, UK, 14–18 November 2005, NEAFC Commission Report AM2005. NEAFC online <http://www.neafc.org/reports/annual-meeting/docs/full_reports/24neafc_annual_vol-1-main-report_2005.pdf> (accessed 25 January 2006).

[90] UN General Assembly, *Sustainable fisheries, including through the 1995 Agreement for the Implementation of the Provisions of the United Nations Convention on the Law of the Sea of 10 December 1982 relating to the Conservation and Management of Straddling Fish Stocks and Highly Migratory Fish Stocks, and related instruments*, UN Doc. A/RES/59/25, 59th Session, adopted 17 January 2005, at para. 53.

[91] UN General Assembly, *Oceans and the law of the sea: sustainable fisheries, including through the 1995 Agreement for the Implementation of the Provisions of the United Nations Convention on the Law of the Sea of 10 December 1982 relating to the Conservation and Management of Straddling Stocks and Highly Migratory Fish Stocks, and related instruments*, UN Doc. A/60/L.23, 60th session (17 November 2005) (adopted as Resolution 60/31, 29 November 2005), at para. 57.

[92] Ministerial Declaration of the Conference on the Governance of High Seas Fisheries and the UN Fish Agreement, St. John's, N.L., 5 May 2005. See online <http://www.dfo-mpo.gc.ca/fgc-cgp/declaration_e.htm> (accessed 24 January 2006), at para. 8; see also the Workshop Summary of the Conference, at para. 5(c).

[93] For a discussion of gaps in coverage and some efforts to establish new RFMOs, see M. Lodge and S. Nandan, at 356–358. See also E. Molenaar "Addressing Regulatory Gaps in High Seas Fisheries".

[94] E. Molenaar "Addressing Regulatory Gaps in High Seas Fisheries" at 541.

[95] *Convention on the Conservation and Management of Fishery Resources in the South-East Atlantic Ocean*, done at Windhoek, Namibia, 20 April 2001, entered into force 30 April 2003, (2002) 41 ILM 257 (SEAFO Convention). The Convention was signed by the European Community, Iceland, Norway, Republic of Korea, and United States of America. As of July 2005, the Contracting Parties of SEAFO are the European Community, Namibia, and Norway.

[96] See SEAFO online <http://www.seafo.org/> (accessed 24 January 2006).

[97] SEAFO Convention Art. 3.

[98] *Convention on the Conservation and Management of Highly Migratory Fish Stocks in the Western and Central Pacific Ocean*, done at Honolulu, Hawaii, 5 September 2000, entered into force 19 June 2004, 40 ILM 277 (WCPFC Convention). As at November 2004, the following States had ratified or acceded to the Convention: Australia, China, Cook Islands, Federated States of Micronesia, Fiji Islands, Korea, Kiribati, Marshall Islands, Nauru, New Zealand, Niue, Papua New Guinea, Samoa, Solomon Islands, Tonga, and Tuvalu. See WCPFC online <http://www.wcpfc.org/> (accessed 24 January 2006) later changed to www.wcpfc.int.

[99] WCPFC Convention Arts. 1(f) and 3(3).

[100] WCPFC Convention Art. 20.

[101] FAO Fisheries Report No. R742. On the whole, see also E. Molenaar, "Addressing Regulatory Gaps in High Seas Fisheries" at 541–542.

[102] FAO, "Report of the Council of FAO, Hundred and Twenty-Seventh Session", Rome, Italy, 22–27 November 2004, FAO Doc. I CL 127/REP, at para. 99. The statutes of the SWIOFC are contained in Appendix E. This is an interesting case study to examine as the coastal States opted to establish an RFB under Art. VI not Art. XIV, primarily because of socioeconomic considerations.

[103] FAO Fisheries Report No. R742.

[104] South West Indian Ocean Fisheries Commission (SWIOFC) — 1st Session, held in Mombassa, Kenya, 18–20 April 2005. For more information see FAO online <ftp://ftp.fao.org/fi/DOCUMENT/safr/swiofc_1_2005/default.htm> (accessed 24 January 2006).

[105] Jean-François Pulvenis de Séligny, personal communication, March 2006.

[106] See the April 2005 Draft of the SIOFA Convention, cited in E. Molenaar, "Addressing Regulatory Gaps in High Seas Fisheries" at 542 and notes 19 and 50.

[107] *Framework Agreement for the Conservation of Living Marine Resources on the High Seas of the South Pacific*, "The Galapagos Agreement", done at Santiago, Chile, 14 August 2000, (2001) 45 *UN Law of the Sea Bulletin* 70 (not yet in force).

[108] Galapagos Agreement at Art. 4(1).

[109] See E. Molenaar, "Addressing Regulatory Gaps in High Seas Fisheries" at 547.

[110] See, e.g., M.A. Orellana, "The Swordfish Dispute Between the EU and Chile at the ITLOS and the WTO" (2002) 71 *Nordic Journal of International Law* 55–81.

[111] The Minister of Fisheries, Forestry and Conservation of Australia and the Fisheries Minister of New Zealand announced their intention to start a negotiation process for the establishment of this new RFMO in March 2005. See E. Molenaar, "Addressing Regulatory Gaps in High Seas Fisheries" at 542. See also "South Pacific fisheries to be protected", media release DAFF05/034MJ of 12 March 2005, online <http://www.mffc.gov.au/releases/2005/05034mj.html> (accessed 24 January 2006); and "Chile commits to South Pacific Fisheries Management", press release of 14 June 2005, online <http://www.southpacificrfmo.org/news/current-news/new-news-2/> (accessed 25 January 2006).

[112] These non highly migratory fish stocks include discrete high sea stocks as well as straddling fish stocks. Commercially important species include orange roughy, oreos, alfonsino, bluenose, jack mackerel, moonfish, and escolar. See the *First International Meeting on the Establishment of the South Pacific Regional Fisheries Management Organization*, Wellington, New Zealand, 14–17 February 2006, *Information Paper*, Meeting Doc. SP/01/Inf3, at paras. 3, 10, and 12. Online <http://www.southpacificrfmo.org/Event.first-meeting/?operation=documents> (accessed 25 January 2006).

[113] See the *First International Meeting on the Establishment of the South Pacific Regional Fisheries Management Organization*, Wellington, New Zealand, 14–17 February 2006, *Proposed Elements of a Legal Framework*, Meeting Doc. SP/01/WP1. Online <http://www.southpacificrfmo.org/Event.first-meeting/?operation=documents> (accessed 25 January 2006).

[114] Molenaar notes that "it is likely that the [South Pacific RFMO] leads to the discontinuation of the bilateral 2000 South Tasman Rise Agreement between Australia and New Zealand, in view of the geographical and species overlap. The pursuance of a multilateral approach by means of the [South Pacific RFMO] also appears to imply that Australia and New Zealand no longer envisage a bilateral solution to the conservation and management of Tasman Sea fisheries generally". See E. Molenaar, "Addressing Regulatory Gaps in High Seas Fisheries" at 543.

[115] See online <http://www.southpacificrfmo.org/Home/> (accessed 24 January 2006).

[116] FAO, "International Action and Responses by Regional Fishery Bodies or Arrangements to Prevent, Deter and Eliminate Illegal, Unreported and Unregulated Fishing", FAO Fisheries Circular No. 996 (Rome: FAO, 2004) at 1. Marine Resources Assessment Group Ltd. (MRAG), "Review of Impacts of Illegal, Unreported and Unregulated Fishing on Developing", Final Report prepared for UK Department of International Development (DFID) (London: MRAG, 2005).

[117] FAO Fisheries Circular No. 996, at 1. E. Molenaar states that the earliest formal use of the acronym IUU at the international level was in 1997 in CCAMLR. It was then used in 1998 in ICCAT and then the term IUU was used with increasing frequency in other international fora. See E. Molenaar, "Participation, Allocation and Unregulated Fishing: The Practice of Regional Fisheries Management Organisations" (2003) 18(4) *International Journal of Marine and Coastal Law* 457–480 at 459.

[118] FAO Fisheries Circular No. 996, at 1.

[119] FAO Fisheries Circular No. 996, at 12.

[120] E. Molenaar, "Participation, Allocation and Unregulated Fishing: The Practice of Regional Fisheries Management Organisations" (2003) 18(4) *International Journal of Marine and Coastal Law* 457–480 at 460.

[121] LOSC Arts. 63, 64, and 118.

[122] UNFSA Art. 7(1). See also M. Lodge and S. Nandan, at 355.

[123] UNFSA Art. 8(4) [emphasis added].

[124] E. Molenaar, "The Concept of 'Real Interest' and Other Aspects of Co-operation through Regional Fisheries Management Mechanisms" (2000) 15(4) *International Journal of Marine and Coastal Law* 475–531 at 491. Tahindro remarks that notwithstanding the benefit of

these provisions for the conservation of straddling fish stocks and highly migratory fish stocks, this limitation of the open-access regime of the high seas as provided in UNFSA marks a "significant departure" from the relevant provisions of LOSC. See A. Tahindro, "Conservation and Management of Transboundary Stocks: Comments in Light of the Adoption of the 1995 Agreement for the Conservation and Management of Straddling Fish Stocks and Highly Migratory Fish Stocks" (1997) 28 *Ocean Development and International Law* 1–58 at 50.

[125] M. Lodge and S. Nandan, at 356.

[126] On this issue, see further below at 4.6.2, 4.6.3, and 4.6.4.

[127] E. Molenaar "Participation, Allocation and Unregulated Fishing: The Practice of Regional Fisheries Management Organisations" at 462.

[128] E. Molenaar "The Concept of 'Real Interest' and Other Aspects of Co-operation through Regional Fisheries Management Mechanisms" at 493–494. Molenaar refers to Barston, who submits that the decision to include the wording "real interest", but not a cross-reference to Art. 118 LOSC, which refers to "States whose nationals exploit... living resources" on the high seas, was aimed at interpreting Art. 118 LOSC narrowly to "prevent the proliferation of regional organizations and limit distant water fishing states' entry". See R. Barston, at 350.

[129] UNFSA Art. 8(3); see also UNFSA Art. 9(2). In LOSC, the term "real interest" is not used.

[130] J.-P. Plé, "Responding to Non-Member Fishing in the Atlantic: The ICCAT and NAFO Experiences" in H. Scheiber, ed., *Law of the Sea: The Common Heritage and Emerging Challenges* (The Hague: Nijhoff Publishers, 2000) 197–207 at 206.

[131] UNFSA Art. 8(3), first sentence.

[132] E. Molenaar "The Concept of 'Real Interest' and Other Aspects of Co-operation through Regional Fisheries Management Mechanisms" at 494–495.

[133] E. Molenaar "The Concept of 'Real Interest' and Other Aspects of Co-operation through Regional Fisheries Management Mechanisms" at 495. Molenaar submits that there are three additional categories of States that would be eligible to become members of an RFMO: (1) flag States that want to resume fishing in the area managed by the RFMO; (2) flag States that want to fish in the future; (3) States with no intention to fish that nevertheless want to participate at 495-496.

[134] See E. Molenaar "The Concept of 'Real Interest' and Other Aspects of Co-operation through Regional Fisheries Management Mechanisms" at 475.

[135] E. Molenaar "The Concept of 'Real Interest' and Other Aspects of Co-operation through Regional Fisheries Management Mechanisms" at 531.

[136] J.-P. Plé, at 198–199 and 206.

[137] The insistence of UNFSA on the open character of RFMOs, the membership of which should be qualified only by the "real interest" of States in the fishery concerned, is also noted by Tahindro. He points out that the concern of UNFSA that RFMOs might preclude certain States from membership derives from the fact that at the time of drafting, the membership of some fisheries commissions and organizations did not include all States engaged in the fishery. In some cases, they included only the coastal States in the region and not the States fishing on the high seas. In other cases, some States had refrained from becoming parties to the regional arrangement. And in still further cases, the RFMOs have traditionally reflected the interests of those concerned with the exploitation of the resource and have inadequately reflected the need for conservation and management. See A. Tahindro, at 21.

[138] UNFSA Art. 8(3).

[139] G. Munro, A. Van Houtte, and R. Willmann, FAO Fisheries Technical Paper No. 465 at 42 referencing P. Orebech, K. Sigurjonsson, and T. McDorman, "The 1995 United Nations Straddling Fish Stocks and Highly Migratory Fish Stocks Agreement: Management, Enforcement and Dispute Settlement" (1998) 13 *International Journal of Marine and Coastal Law* 119–141.

[140] See E. Molenaar "The Concept of 'Real Interest' and Other Aspects of Co-operation through Regional Fisheries Management Mechanisms" at 495–496.

[141] G. Munro, A. Van Houtte, and R. Willmann, FAO Fisheries Technical Paper No. 465 at 42.

[142] J.-P. Plé, at 198–199 and 206.

[143] D. Balton, "Dealing with the 'Bad Actors' of Ocean Fisheries". Paper presented to the OECD Workshop on Illegal, Unreported and Unregulated (IUU) Fishing Activities, Paris, France, 19–20 April 2004, AGR/FI/IUU/RD(2004)3 (Paris: OECD, 2004).

[144] D. Balton, at 5–6.

[145] G. Munro, A. Van Houtte, and R. Willman, FAO Fisheries Technical Paper No. 465 at 42 and 48–49, see also at 24.

[146] UN General Assembly, *Oceans and the law of the sea: sustainable fisheries, including through the 1995 Agreement for the Implementation of the Provisions of the United Nations Convention on the Law of the Sea of 10 December 1982 relating to the Conservation and Management of Straddling Stocks and Highly Migratory Fish Stocks, and related instruments*, UN Doc. A/60/L.23, 60th session (17 November 2005) (adopted as Resolution 60/31, 29 November 2005), at para. 55; Report of the UN Secretary-General, *Report submitted in accordance with paragraph 17 of General Assembly resolution 59/25 of 17 November 2004, to assist the Review Conference to implement its mandate under paragraph 2, article 36 of the United Nations Fish Stocks Agreement*, UN Doc. A/CONF.210/2006/1), at para. 233; para. 9 of the Ministerial Declaration of the *Conference on the Governance of High Seas Fisheries and the UN Fish Agreement*, St. John's, N.L., 5 May 2005. See online <http://www.dfo-mpo.gc.ca/fgc-cgp/declaration_e.htm> (accessed 5 February 2006).

[147] E. Molenaar "Participation, Allocation and Unregulated Fishing: The Practice of Regional Fisheries Management Organisations" at 466.

[148] For example, in the WCPFC and some other RFMOs a distinction is made between Contracting Parties and members. Members in the WCPFC Convention include both States Parties to the Convention as well as cooperating nonparties, see Art. 32 WCPFC Convention. In the case of WCPFC, Indonesia and United States have recently joined as cooperating nonmembers.

[149] CCAMLR Resolution 24/XXIV on Non-Contracting Party Cooperation Enhancement Program.

[150] Antigua Convention Art. XXX.

[151] An estimated 90% of high seas catch is taken by nationals of six States or entities: Japan, Spain, Poland, the Republic of Korea, the Russian Federation, and Taiwan. See L. Juda, "Rio Plus Ten: The Evolution if International Marine Fisheries Governance" (2002) 33 *Ocean Development and International Law* 109–144 at 123.

[152] For a comprehensive discussion of the participation of Taiwan in RFBs, see A. Serdy, "Bringing Taiwan into the International Fisheries Fold: the Legal Personality of a Fishing Entity" (2005) LXXV *British Yearbook of International Law* 183–221 at 200.

[153] SEAFO Art. 22.4.

[154] WCPFC Annex 1(2).

[155] A meeting of IOTC is scheduled for Goa, India, in May 2006 to consider the possibility of placing IOTC outside the FAO framework.

[156] See further Chap. 9 at 9.4.1.2 and 9.4.1.3 where these lists are discussed in more detail.

[157] R. Rayfuse, *Non-Flag State Enforcement in High Seas Fisheries* (Leiden, Boston: Nijhoff Publishers, 2004).

[158] UNFSA Art. 23(3). Catch and trade documentation schemes are discussed further in 9.4.3

[159] ICCAT as of May 2006 has 42 Contracting Parties and membership is expected to continue to increase. ICCAT online <http://www.iccat.es/contracting.htm> (accessed 5 May 2006).

[160] E. Molenaar, "Participation, Allocation and Unregulated Fishing: The Practice of Regional Fisheries Management Organisations" at 459.

[161] See A. Tahindro, at 24–25.

[162] See also E. Molenaar, "Participation, Allocation and Unregulated Fishing: The Practice of Regional Fisheries Management Organisations" at 459.

[163] E. Molenaar, "Participation, Allocation and Unregulated Fishing: The Practice of Regional Fisheries Management Organisations" at 498.

[164] E. Molenaar "Participation, Allocation and Unregulated Fishing: The Practice of Regional Fisheries Management Organisations" at 459.

[165] NAFO Resolution to Guide the Expectations of Future New Members with regard to Fishing Opportunities in the NAFO Regulatory Area, GC Doc. 99/9, Annex 13.

[166] *Indications to Guide the Expectations of Interested Non-Contracting Parties of NEAFC*, discussion paper of a Norway–FAO expert consultation on the management of shared fish stocks in Bergen, Norway, 7–10 October 2002. See FAO, "Report of the Norway–FAO Expert Consultation on the Management of Shared Fish Stocks", FAO Fisheries Reports No. R695 (Rome: FAO, 2003) at para. 62.

[167] This can be done by *inter alia* developing new allocation criteria. For example, after ICCAT had experienced a significant increase in membership from distant-water fishing nations as well as interest in participating from adjacent coastal States, many of whom are new entrant developing coastal States without a record of substantial past performance, ICCAT reviewed its practice to allocate quota based on historical catch for those species subject to ICCAT TAC controls. In November 2001, ICCAT adopted Criteria for the Allocation of Fishing Possibilities containing an extensive list of some 15 factors and categories. See 2001 ICCAT Criteria for the Allocation of Fishing Possibilities [01–25].

[168] A. Tahindro, at 21.

[169] See A. Tahindro, at 27.

[170] A. Tahindro, at 27.

[171] S. Garcia and D. Doulman, "FAO's Fisheries Programme and the Plan of Implementation of the World Summit on Sustainable Development" in S. Ebbin, A. Håkon Hoel, and A. Sydnes, eds., *A Sea Change: The Exclusive Economic Zone and Governance Institutions for Living Marine Resources* (Dordrecht: Springer, 2005) 169–193 at 188.

[172] E. Miles, "The Management of Marine Regions: The North Pacific" (1979) 6 *Ocean Development and International Law* vii–xxx at vii.

[173] Report of the Multilateral High Level Conference on the Conservation and Management of Highly Migratory Fish Stocks in the Western and Central Pacific, Seventh Session (April 2000) Annex 8.

Chapter 5

State of the Fisheries*

By Evelyne Meltzer and Susanna D. Fuller

5.1 Introduction

Avoiding and reducing overfishing, together with the obligation to maintain or restore populations of harvested species at levels that can produce maximum sustainable yield (MSY), have been objectives of the international regime governing fisheries management since the adoption of the United Nations Convention on the Law of the Sea (LOSC) in 1982.[1] In theory, scientists would determine the maximum amount to be removed, and managers would limit the harvest to that level. But even before 1982, it had become increasingly clear that fish populations were being overexploited throughout the world, and that the misapplication of MSY as a target was a contributing factor. Several other factors were also at work. Although the 1982 LOSC[2] exhorted use of the "best scientific information available", excessive reduction of population levels was not uncommon because of inherent uncertainties in estimating population levels and a limited understanding of the factors affecting population abundance. Indeed, a major obstacle to achieving sustainable fisheries is the inability to accurately assess the stock, let alone determine how fishing affects other organisms in the environment. These problems were exacerbated by the inability or failure of many States (and regional fisheries management organizations (RFMOs)) to collect accurate, complete, and timely data. The lack of reliable data contributed to assessment difficulties. This management strategy was further undermined by the difficulty of ensuring compliance with conservation measures where a regulatory regime existed. Ineffective regulation or the absence of regulation, combined with the inability to control individual vessel and fleet capacity, resulted in overfishing and depletion of straddling fish stocks and highly migratory fish stocks.

> The crises in marine fisheries of the 1990s that led to the collapse of a number of valuable fish stocks, such as cod in the Northwest Atlantic and pollock in the Bering Sea, and the resulting tensions that this created between coastal states and high seas fishing states have been extensively chronicled...by the early 1990s it had become abundantly apparent that the regime for high seas fishing established by the 1982 Convention was inadequate to prevent the depletion of the world's fish stocks.[3]

Accordingly, international initiatives have focussed on the need to strengthen the regime for fisheries management by introducing new concepts and approaches, and by reexamining the interpretation and application of LOSC provisions concerning MSY, with the objective of avoiding or eliminating overfishing.[4]

Calls to eliminate and avoid overfishing can be found in the United Nations Fish Stock Agreement (UNFSA)[5], the 1995 Rome Consensus on World Fisheries, the 1995 Food and Agriculture Organization (FAO) Code of Conduct for Responsible Fisheries, the 2002 World Summit on Sustainable Development (WSSD) Johannesburg Declaration, and several recent UN General Assembly (UNGA) resolutions. Indeed, the international community has waited with increasing expectations since the London Conference on Overfishing highlighted the issue in 1946. Despite efforts by RFMOs and States to strengthen the global fisheries framework, overfishing continues and is now considered one of the greatest environmental scourges of the planet. In less than a century, the notion of overexploitation of fish stocks has gone from unimaginable, if not impossible, to being at the forefront of global environmental, socioeconomic, and political concern.

Unlike LOSC, which assumes that MSY is a target, UNFSA now uses MSY as a limit rather than a target reference point. UNFSA also incorporates additional concepts and approaches to marine fisheries management, including the introduction of reference points, precaution, the ecosystem approach, data collection and sharing rules, transparency, protection of biodiversity, and the control of excess capacity. These substantive principles have strengthened the international minimum standards required under LOSC to regulate overfishing, bycatch, and allowable catches of straddling fish stocks and highly migratory fish stocks. Not only are target species and

* Susanna D. Fuller is completing her doctorate in marine biology at Dalhousie University. The authors would like to acknowledge the research assistance of Dr. Cornelia den Heyer and are grateful for the comments and contribution of Mr. Michael Gardner in preparing earlier drafts of this chapter.

their associated and dependent species to be taken into account when adopting management measures, but UNFSA requires that nontarget species (both fish and nonfish species) be considered as well. But not until these principles become a matter of general practice will they contribute to the conservation of straddling fish stocks and highly migratory fish stocks and improved global fisheries management.

It bears emphasizing that overfishing is not always deliberate, and that the reduction of fish populations is not always due to fishing activity. Stock abundance can be adversely impacted and influenced by environmental factors, including climate change, marine pollution, habitat degradation, natural variations, and natural oscillations. However, in the aggregate, stocks that have collapsed and not recovered are less resilient, and the adverse effects of fishing activities become all the more significant. The recognition that stock status or population health can be affected by a variety of nonfisheries factors underscores the need for applying precautionary management and setting precautionary reference points provided for under UNFSA.

Likewise, overfishing is not only due to large-scale industrial fishing fleets but also can be the result of excess harvesting by recreational, small-scale, and artisanal fishing activities. The sustainable development issues associated with small-scale and artisanal fishing are particularly relevant in developing coastal States where access to the sea to fish for food, which is primarily consumed locally, is critical to the subsistence of coastal communities. However, inadequate fisheries management capacity, together with lack of other employment or income generating opportunities, or alternative sources of food, has resulted in coastal overfishing and declines in nearshore catches, thereby exacerbating the plight of coastal communities that depend on the sea for their livelihoods. This biological and socioeconomic problem is further aggravated when industrial fleets fish close to shore and compete with small-scale fleets.

The history of fishing straddling fish stocks and highly migratory fish stocks is characterized by increased exploitation as a result of technological changes, increased catch efficiency, expansion into new waters, and the targeting of new species. Overexploitation of stocks has occurred where these factors were allowed to play out against a backdrop of a poor understanding of allowable exploitation levels and an inability to control fishing effort. Determining stocks status is a critical obligation under UNFSA and key to understanding the relative state of fisheries and the corresponding need for effective conservation and management as provided in UNFSA and in other instruments. As stocks have declined, fisheries issues have been recast as part of the broad-scale environmental and biodiversity concerns about food security, ecological security, and, ultimately, human security.

As a preface to a discussion of overfishing, this chapter provides a summary of current stock status of straddling fish stocks and highly migratory fish stocks under the competence of RFMOs. This provides a context within which to further examine the phenomenon of overfishing, including a brief overview of terminology, the evolution of the principles pertaining to the concept of MSY, the interrelationship of overexploitation with overcapacity and subsidization, and the collapse of fish stocks to the point where they become species at risk. Examples of how RFMOs address these issues are outlined.

5.2 Status of straddling fish stocks and highly migratory fish stocks

5.2.1 Global overview

The oceanic ecosystem is by far the largest on Earth, covering more than 70% of the planet by area and an even larger percentage by volume. Individual effects of overfishing, habitat destruction, pollution, and climate change, as well as the combination of such environmental stresses, have given rise to serious conservation concerns. Most dramatic perhaps is the possibility of biological or functional extinction of marine fishes,[6] as well as the more widespread and well-published stock collapses. Some of these stocks are experiencing slower than expected recovery despite conservation efforts. Additionally, there are broader marine ecosystem concerns, including trophic regime shifts.[7] Reduction of aquatic diversity and impaired ecosystem functioning can trigger cascading ecosystem effects[8] and could therefore result in ecosystem collapse. If this were to occur in the open ocean, changes would likely be massive in scale and probably difficult to reverse. Responsible and effective fisheries management is then imperative to ensure the restoration and preservation of the world's largest source of naturally produced protein for future generations.

As noted in Chapters 2 and 4, the pre-UNFSA call for a new governance regime to conserve and manage high seas fisheries emerged from the 1992 Earth Summit recognition that there was an urgent and immediate need to address the world fisheries crises.[9] The 2005 FAO review of the state of world fishery resources provides an overview of the status of global fisheries.[10] In measured language, the FAO Review suggests that with an estimated marine catch of some 85 × 10^6 tonnes (t), "the maximum long-term potential of the world marine capture fisheries has been reached, with some stocks and areas being overfished and some stocks not producing their full expected long-term potential".[11]

The limits to growth are clearly evident from global catch trends since the 1950s. As seen in Figure 5.1, 1950 to 1966 reported catches increased threefold as a result of the rapid build-up of harvesting capacity (more and larger vessels with improved gear and electronics). Moderate catch increases followed in most areas up until 1975. Catches stabilized or declined after 1975 in 11 of the 16 FAO regions. These regions, as illustrated in Figure 5.1, account for about 35% of total global catch. The northwest and southeast regions of the Pacific Ocean are among the areas experiencing a generally rising trend in catches, though with some fluctuations and declines in recent years. Only in the Western Central Pacific and Indian oceans have catches (mainly large pelagics) continued to grow steadily.

In the 2005 FAO Review, catch data and (or) stock-assessment information are available for about 76% of stock or species groups, which produced approximately 80% of the total marine catches as of 2002. For the remaining 24% of stock or species groups, the information is not sufficiently reliable to allow an estimate of their global state of exploitation.

In its detailed breakdown of stock status, the 2005 FAO Review notes that 77% of the stocks (accounting for over 80% of the total marine catch) are fully exploited, overexploited, depleted, or recovering. These stocks have reached or are above their maximum sustainable production limit. Only 23% are regarded as under — or moderately — exploited, offering the prospect of some room for growth.

Since 1974 the proportion of overexploited, depleted, and recovering stocks has *more than doubled*, rising from 10% to 25%. Conversely, the proportion of underexploited and moderately exploited stocks has declined steadily, dropping from about 40% to 23%. The combined negative effect of these trends caused the proportion of stocks exploited at least at the level of MSY to decline initially from about 50% in 1974 to just over 40% by the late 1980s, as more stocks shifted to a state of overexploitation. That proportion then climbed back to the 50% range by 2004, as stocks shifted from under to full exploitation (the overexploited proportion stabilized).

The present global situation, as evidenced over the past five decades, can be characterized as a "chronic, degenerative trend, with occasional localized acute crises".[12] Academic studies, using much of the data compiled by FAO and States through logbook records and research surveys, have highlighted the severity of the global fish stock decline.[13]

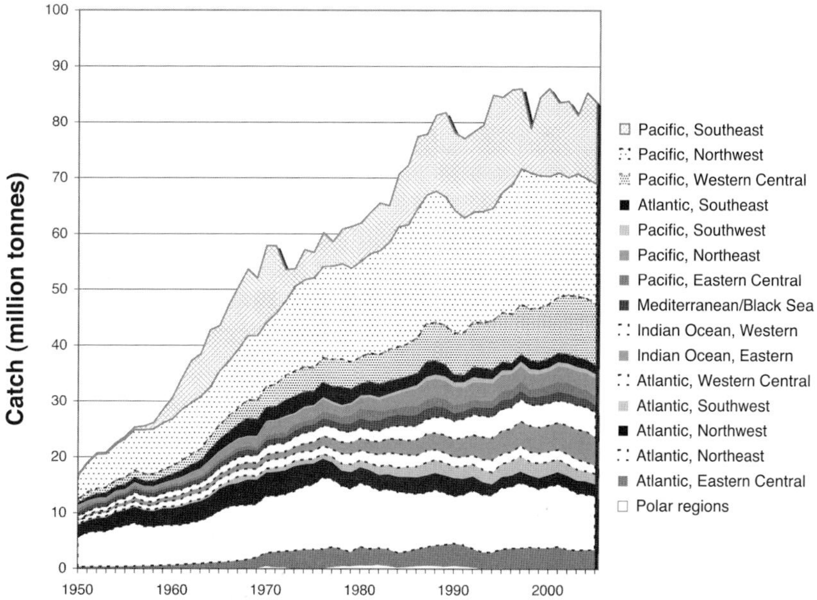

Figure 5.1. Global catch trends by FAO region (information derived from Fisheries Global Information System (FIGIS), 2005).

5.2.2 UNFSA provisions for stock assessment

The scientific exercise to predict stock abundance is known as stock assessment. Stock assessment is the process of collecting and analysing biological and statistical information to determine the changes in fish stock populations due to fishing, and where possible, to predict future trends of stock abundance.[14]

The objective of UNFSA is to ensure long-term conservation and sustainable use of straddling fish stocks and highly migratory fish stocks, and to assess these stocks in accordance with these principles. Under Article 10 of UNFSA, which elaborates the functions of RFMOs, States are to "obtain and evaluate scientific advice, review the status of the stocks and assess the impact of fishing on non-target and associated or dependent species" (Article 10(d)); "agree on standards for collection, reporting and verification and exchange of data on fisheries for the stocks" (Article 10(e)); "compile and disseminate accurate and complete statistical data, as described in Annex I, to ensure that the best scientific evidence is available" (Article 10(f)); and "promote and conduct scientific assessments of the stocks and relevant research and disseminate the results thereof" (Article 10(g)). Determining stock status is therefore a critical obligation under UNFSA and is key to understanding the relative state of fisheries and the corresponding need for effective implementation of management strategies and plans, as proposed under UNFSA.

5.2.3 Stock assessment in practice

Modern fishery science, a subset of marine biology that includes fish stock assessments, is a relatively new discipline, which started to become institutionalized in the 1950s.[15] "At its inception, fishery science was limited by a 19th century mechanistic worldview and by lack of computational technology".[16] Stock assessment has become increasingly sophisticated, relying on complex and data-rich computer models[17] to generate the probable status of the stock population. These assessments are based on resource surveys; knowledge of the habitat requirements, life history, and behaviour of the species; environmental indices that determine impacts on stocks; and catch statistics.

Unlike assessing the population of many land-based species or easily identifiable and long-lived marine mammals, counting fish that are actually in the sea is difficult and remains one of the major challenges faced by fisheries scientists and mathematicians. Fish stock assessment is determined by calculating abundance based on what was caught in the past and not what is currently in the ocean to determine an appropriate harvest limit for next year, i.e., stock assessments use catch data from past years to estimate biomass. Biomass for the total stock is often calculated by estimating how many fish from each year class were not caught as well as how many juveniles survived to sexual maturity.[18] As stock assessment is typically based upon historical information, it is often difficult, and sometimes impossible, to accurately predict or track the impact of current and future ecological regime changes on marine fish populations and marine ecosystem productivity. Including environmental uncertainty and direct and indirect impacts on fish stocks can be done to some extent through precautionary decision rules and adaptive management.[19]

In summary, although stock assessments are required to provide estimates of stock biomass, they are inherently uncertain, as they are dependent on data quality; are not robust to environmental or biological variation; and are only as accurate as the parameters used to determine their output. In addition, the use of past data to predict future fish stock abundance is problematic in that it cannot include future environmental variation. Since the early 1990s fisheries scientists and fisheries managers have used a process known as management strategy evaluation (MSE)[20] to address the deficiencies of stock assessment. This technique examines the robustness of management through simulations in which the uncertainties and data insufficiencies are explicitly stated. The robustness of the management strategy is tested through model simulations. The intent of MSE is not to improve stock assessment, necessarily, but to improve the overall management strategy. MSE is still an emerging field, and international fisheries management, which is often fraught with data uncertainties, will benefit from its use.

FAO remains the primary repository for all information on fish stocks and catch records and hence the reference source for global information and supplier of knowledge. FAO has introduced various initiatives, including the FAO Global Information System (FIGIS), to improve the collection of data as well as its quality, timeliness, and transparency. Other initiatives are aimed at assisting individual States in revising data collection systems and at establishing a global cooperative system for information available in regional fishery bodies and national centres of excellence. The FAO partners with a number of RFMOs in developing the Fisheries Resources Monitoring System (FIRMS), which is designed to provide a comprehensive and easily accessible repository of information on the status of stocks.

The FAO Strategy for Improving Information on Status and Trends of Capture Fisheries (Strategy-STF) is a voluntary instrument, aiming to provide a framework, strategy, and plan for the improvement of knowledge and understanding of fishery status and trends. Strategy-STF serves as a basis for fisheries policy making and management for the conservation and sustainable use of fishery resources within ecosystems. The Strategy applies to all FAO members, nonmembers, and fishing entities, and requires actions such as

> data collection systems in small-scale fisheries and multispecies fisheries; better information on status and trends of fisheries, including integration of ecosystem considerations into fisheries management; global inventory of fish stocks and fisheries; participation in the FIGIS; structuring and capacity-building; development of criteria and methods for ensuring information quality and security; development of arrangements for the provision and exchange of information; establishment of working groups to assess the status and trends of fisheries; sustainable data collection and information on the status and trends of fisheries; and capacity-building.[21]

Increasingly, RFMOs, with the support of FAO, are collecting data for use in stock assessments, are basing management decisions on science, and are participating in FIRMS.

5.2.4 RFMO variability on stock assessments

How stock assessment is conducted varies from one RFMO to another, and the status of the fish stocks under the competence of RFMOs is expressed in different ways. For example, the North East Atlantic Fisheries Commission (NEAFC) receives stock assessment and status advice as well as management recommendations from the International Council for the Exploration of the Sea (ICES). In accordance with ICES policies, NEAFC now describes the status of stocks under its competence as a measure of their reproductive potential rather than in relation to MSY. Fisheries are described in terms of biomass and reproduction as well as whether or not the stock is harvested sustainably and how biomass and fishing mortality are related to preagreed precautionary limits. While RFMOs have different biological reference points, most refer to MSY as either a limit or a target, and hence the status of the stock is generally referred to as below or above MSY, whether the variable is fishing mortality, biomass, or landings. This allows an assessment of stocks that are overfished and stocks that are experiencing overfishing. However, what is meant by "overfished" varies not only from RFMO to RFMO but also within RFMOs.[22]

5.3 What overfishing means

5.3.1 Concepts

The freedom of the seas doctrine espoused by Grotius was based on the notions that living marine resources were infinitely renewable and available to anyone fishing on the high seas. With the advent of increasingly efficient fish extraction methods, together with sophisticated fish detection and processing technology, the capacity of high seas fishing fleets to overexploit a particular stock reached critical dimensions by the early 1990s.[23] Despite initiatives such as the UN Fish Stocks Conference resulting in the legally binding UNFSA, the High Seas Task Force on Illegal, Unreported and Unregulated Fishing (IUU), WSSD, and the International Plan of Action - Illegal, Unreported and Unregulated Fishing (IPOA-IUU), international fisheries are still largely characterized by overfishing and successive depletion of commercial fish stocks. The regulation of high seas fisheries to avoid overfishing and rebuild depleted stocks remains a central preoccupation of contemporary international fisheries law and regional management regimes.

As a first step, it is useful to examine the meaning of the terms associated with overfishing. In the fisheries management lexicon, *overfishing* is most commonly expressed as fishing mortality (F) or rate of fishing that is beyond the fishing mortality rate of MSY (F_{MSY}) or beyond any agreed reference point up to, and including, MSY (defined in greater detail below). Overfishing is a general term that can be further defined, in terms of management reference points, by fishing mortality rates that can have consequences for the stock in question. There are several types of overfishing:

> **Growth overfishing** occurs when too many small fish are being harvested, usually because of excessive effort and poor gear selectivity. The fish are not given the time to grow to the size at which the maximum yield per recruit would be obtained from the stock. A reduction of fishing mortality on juveniles or their outright protection would lead to an increase in yield from the fishery.[24]

Recruitment overfishing occurs when the rate of fishing is (or has been) such that annual recruitment to the exploitable stock has become significantly reduced. This situation is characterized by a greatly reduced spawning stock and generally very low recruitment year after year. If prolonged, recruitment overfishing can lead to stock collapse, particularly under unfavourable environmental conditions.[25]

Biological overfishing is described as catching such a high proportion of one or all age-classes in a fishery as to reduce yields and drive stock biomass and spawning potential below safe levels. Biological overfishing occurs when fishing levels are higher that those required for extracting the MSY of a resource and when recruitment starts to decrease statistically.[26] This is a generic category in which both growth and recruitment overfishing occur.

Ecosystem overfishing occurs when species composition and dominance are significantly modified by fishing (e.g., with reductions of large, long-lived demersal predators and increases of small, short-lived species at lower trophic levels).[27] Ecosystem overfishing encompasses issues such as biodiversity loss, serial depletion, habitat degradation, and changes in food web structure caused by fishing.[28]

Economic overfishing exists in circumstances in which a fishery expands beyond the point where maximum economic yield (MEY) is generated or where there is less than maximum (or no) economic rent[29] (see Figure 5.15). MEY occurs before MSY is reached. In open-access or competitive fisheries, there is an incentive to expand fishing effort and overcapitalize. Even where there is a TAC, participants in the fishery still have an incentive to maximize their share and will therefore continue to invest in vessels and gear. This process will continue until net revenues are reduced to zero. Economic overfishing does not always lead to biological overfishing but increases the likelihood that it will occur.

Though sometimes used interchangeably in the literature and in discussions, there is a distinction between the terms *overfishing* and *overfished*. Overfishing is described as ongoing excess fishing "pressure" or "effort", while overfished refers to the state of the stock when the biomass is below an agreed reference biomass (B) or below the biomass of MSY (B_{MSY}). Overfishing is the act of exploiting a stock at a rate that if continued in the medium term would be unsustainable and lead to an equilibrium biomass or reference point at which the stock is considered overfished. At this level, a stock is overfished even if fishing is reduced or stopped. While not specific to MSY, the concepts of overfishing and overfished can be explained or illustrated using MSY and the surplus production model (or any other agreed level) (see Figure 5.2).

The term *collapse* is often used to refer to a stock that has been overfished. Collapse describes a marked decline in stock abundance that may or may not be attributed to overexploitation. Collapse is often used now to describe declines that result in abundances that are lower than those required to support a fishery. FAO describes collapse as a reduction of stock abundance by fishing and (or) other causes to levels at which the production is negligible compared with historical levels. The term is normally used when the (reduction) process is sudden compared with the likely time scale of recovery; however, it is sometimes misused.[30] Decline in fish stock abundance can also occur as a result of natural oscillations, global climate change, degradation of the marine environment, and destruction of critical habitat. These factors are not always known nor their impact always forecasted, thereby creating a precarious situation when accompanied by overfishing.

Stocks at these low levels may also be described as biologically *depleted*, as well as commercially or economically depleted (see Figures 5.2, 5.15, and 5.16). With overcapitalization and subsidies (including IUU fishing), many depleted fish populations that would otherwise be considered uneconomic can continue to support commercial fishing. Depletion is particularly worrisome in the case of some tuna species (e.g., southern bluefin tuna, Western and Eastern Atlantic bluefin, and Eastern Pacific bigeye) and other highly valued commercial pelagic stocks. Given the perceived high concentrations of these individual schools (all pelagic stocks school or aggregate), the total biomass can sometimes be deceptive. The density of one school or several aggregations cannot be projected to the entire stock.

5.3.2 MSY

5.3.2.1 Overview

Determining how much fish can be responsibly harvested from the sea is a preoccupation and challenge for many involved in fisheries, particularly fisheries scientists and managers. Securing MSY as a target has been the traditional management objective for fisheries, both within national jurisdiction and on the high seas.[31] Despite its well-known shortcomings,[32] MSY and related concepts such as maximum sustained catch are widely used[33] and included in the constitutive agreements of RFMOs such as the Inter-American Tropical Tuna Commission (IATTC), the International Commission for the Conservation of Atlantic Tunas (ICCAT), the International Commission for the Northwest Atlantic Fisheries (ICNAF)[34], and now the Northwest Atlantic Fisheries Organization (NAFO).[35] In fact, the IATTC Convention and possibly others have a target of keeping stocks at levels of abundance that will permit the maximum sustained catch. That management goal is quite separate from setting MSY itself , or F_{MSY}. Most comments on the shortcomings of MSY apply to its use as a catch target on a continuing basis.

The development of quantitative methods for determining fishing mortality or fishing effort corresponding to MSY, and support from the FAO, led to the wide adoption of MSY as a management goal (either explicitly or implicitly) in the 1950s, 1960s, and even 1970s. Several international fisheries commissions (e.g., IATTC, ICNAF, and the North Pacific Fur Seal Commission) had adopted MSY as a management goal by the mid-1960s.[36] MSY was the only reference point explicitly referred to in LOSC.[37]

LOSC incorporated MSY, as qualified by environmental and economic factors, as a management strategy to ensure that living resources would not be threatened by overexploitation. Article 61(2) enjoins coastal States to ensure through "proper conservation and management measures", taking into account "the best scientific evidence available", that living resources will not be endangered by overexploitation. Article 61(3) declares that

> Such measures shall also be designed to maintain or restore populations of harvested species at levels which can produce the maximum sustainable yield, as qualified by relevant environmental and economic factors, including the economic needs of coastal fishing communities and the special requirements of developing States, and taking into account fishing patterns, the interdependence of stocks and any generally recommended international minimum standards, whether subregional, regional or global.

While Article 61 relates to coastal States, Articles 116–119 provide that stocks be maintained at levels that can produce MSY on the high seas. Since this is the primary language regarding fisheries management, the interpretation and application of setting MSY are of some importance from socioeconomic, political, and biological perspectives. From an international legal perspective, MSY serves as the benchmark for determining overfishing.[38] MSY thus became the internationally agreed target reference for commercial species and the basis for calculating total allowable catch (TAC) and quotas.

UNFSA refers to MSY with the same qualifications as LOSC. However, UNFSA defines MSY as a "limit", which is to be applied in combination with the precautionary approach, the minimization of waste and bycatch, the protection of biodiversity, and the prevention of excess capacity.[39] UNFSA guidelines for the application of the precautionary approach change the way that MSY has been used, by referring to F_{MSY} as a limit. By doing so, the focus is shifted from maintaining stocks at adequate levels to establishing a target for fishing mortality and managing fishing effort, rather than managing the stock abundance. From a harvest perspective MSY as a target is no longer considered sustainable; however, it is still considered useful as a minimum target in stock recovery plans.[40]

While other, more conservative reference points have been developed and extensively used in RFMOs and at the national level (such as $F_{0.1}$ or $F_{2/3MSY}$) MSY remains an important concept in fisheries management. For example, the 2002 WSSD Plan of Implementation calls for fish stocks to be maintained or restored where possible to levels that can produce MSY by 2015.[41] MSY has been progressively complemented or replaced by more sustainable and precautionary objectives, including other reference points for development and management targets.[42] It is therefore important to understand this term and the nuances associated with its use. The following sections define MSY and examine its evolution from a target to a limit with examples drawn from several RFMOs.

5.3.2.2 Defining MSY

A fundamental characteristic of fish is that they are renewable living resources. The principle that underlies managing the exploitation of renewable resources is that they must be harvested responsibly and sustainably, i.e., that the amount extracted or removed from the population should not adversely affect the ability of the population to replace itself.

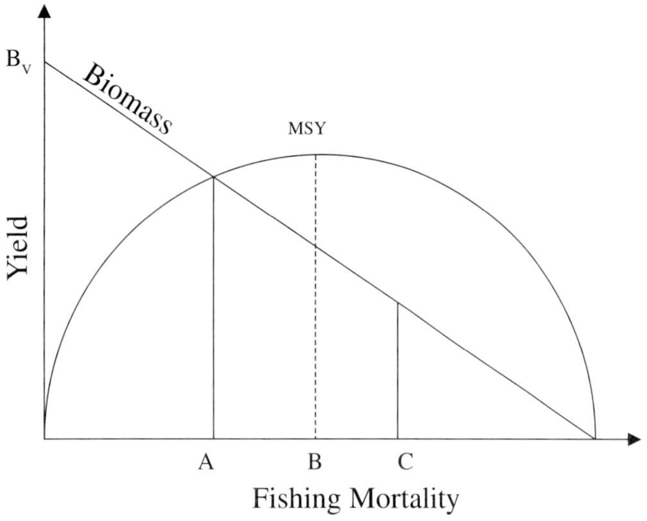

Figure 5.2. Simplified illustration of the mathematical framework of indicators and reference points used in fisheries management, based on the surplus production model. This model does not include cycles of productivity, supply–demand reactions, changes in fishing efficiency, or density dependence, all of which affect fisheries worldwide and are increasingly being considered by fisheries scientists and managers. B_v represents the virgin biomass of the stock, before fisheries exploitation. Line A represents $F_{MEY,}$ a conservative level of fishing mortality producing maximum economic yield. Line B represents F_{MSY} or fishing effort that can achieve MSY, the maximum sustainable catch. In this model F_{MSY} coincides with F_{lim}, the level of fishing mortality (F) above which fishing should not occur. Line C represents a level of fishing mortality at which the stock is at risk of collapse.

Since World War II, fisheries scientists and managers have developed various models and associated reference points to quantify TAC and fisheries quotas that can be sustainably extracted. MSY is the maximum catch that can be harvested on a sustainable basis. This biological concept is illustrated in Figure 5.2 as the highest point (or peak) on the curve of the surplus production model.[43] Theoretically, MSY is the point that can sustain the maximum regenerative ability of the resource.[44] The virgin prefisheries biomass is controlled primarily by population-dependent biological processes. As fishing mortality increases, the spawning biomass initially declines, which, in turn, reduces density-dependent pressures and allows recruitment and growth, and therefore sustainable yield, to increase. At even higher fishing mortality rates, removals increase faster than the stock can compensate with growth and productivity, and the sustainable yield declines. When this occurs, the stock is considered overfished (See Figure 5.16).

5.3.3 Shortcomings of MSY

Even during negotiations at the Third United Nations Conference on the Law of the Sea (UNCLOS III), the use of F_{MSY} as a target reference point was subject to much criticism.[45] Risks associated with using F_{MSY}[46] as a target reference point include errors in estimation; uncertainty surrounding its exact value; its year-to-year variability; its negative properties in terms of stock stability and resilience; its lack of reference to the state of associated or dependent species or the resource habitat; its lack of relation to the cost of fishing; and the inability to effectively implement harvesting strategies based on F_{MSY}.[47] Further, data required to calculate F_{MSY} and determine appropriate TACs are most often obtained through catch reporting and, where available, research surveys. The quality of the data is therefore a critical determinant in the accuracy of stock assessments and hence the estimation of sustainable TACs and quotas. A slight error in MSY estimation or achievement of F_{MSY} as a target reference point can result in declines in stock abundance and overfishing. "In the case of MSY, undetected error will result in 'overfishing' 50% of the time and this has rendered this particular reference point a dangerous one for fisheries management".[48] These risks raise concerns about whether or not MSY is an appropriate management goal. During UNFSA negotiations, MSY was perceived as "virtually to invite problems because the margin of error in projecting abundance is so great".[49]

F_{MSY} as a target reference point continues to be criticized for being a high-risk management strategy, often leading to overfishing. Together with a suite of issues ranging from competing objectives and competitive fishing to inappropriate management institutions, the use of F_{MSY} is considered one of the primary reasons for fisheries management failure in the last two decades.[50] Given that fully exploited stocks are those fished close to MSY, and overfished stocks are beyond MSY, "the risk of collapse (with its related socio-economic dislocations) increases as the exploitation rates (and stress) approach the MSY level and increase beyond it".[51]

5.3.4 MSY: from target to limit

The juridical expression of the evolution of MSY from a target reference point to a limit reference point is found within the precautionary framework set out in UNFSA Annex II[52] and subsequent FAO technical guidelines.[53]

As a result of overfishing, declining fish stocks, serial depletions of commercial fish, and the impacts of fishing on ecosystems, MSY has been progressively complemented or replaced by more sustainable and precautionary objectives, including other reference points for development and management targets.[54] During the FAO Technical Consultation on High Seas Fishing in September 1992, Garcia stressed the "non-precautionary nature of MSY and the need for more and different reference points to be used as a basis for more precautionary management strategies".[55] "The supplementary concept of a limit reference point is proposed that measures when a dangerous zone is being approached or entered".[56] MSY is no longer universally accepted as a valid development target but rather as the "upper limits to the rate of fishing or lower limits to the spawning biomass".[57]

With UNFSA, MSY shifted from a target reference point, as applied under LOSC, to a limit reference point.[58] "Indeed the UN Fish Stock Agreement has established that, for precautionary purposes, MSY should be considered as a limit"[59] to be avoided and not a target to be reached. Accordingly, under UNFSA, MSY is regarded as an absolute minimum target for stock rebuilding. "The related objective is to allow catch levels (or fleets sizes) compatible with the maintenance of a stock at or above the MSY level".[60]As Garcia notes, the shift from a target to a "conceptually precautionary limit" not only redefines the legal convention but also includes the associated concepts that resources are limited, effort needs to be controlled, and catch capacity needs to be aligned with resources.[61]

5.3.5 Use of MSY as a limit by RFMOs

Many RFMOs have moved towards managing for MSY as a limit and developing additional precautionary reference points and thresholds to reduce inherent uncertainty and risks. The desire to avoid overfishing and ensure sustainable fisheries have led to more complex stock-assessment models and to the introduction of harvest control rules. Harvest-level advice continues to be given in terms of F_{MSY} in the interests of maintaining or obtaining B_{MSY}, but these are seen as levels that must not be surpassed.[62]

The 1949 IATTC Convention provides that populations of fish will be maintained at the levels of abundance that will permit the maximum sustained catch.[63] However, IATTC now interprets (the stock size that will produce) MSY as a limit reference point, and if catches reach or go beyond levels that will sustain MSY, management measures are imposed. The Antigua Convention reflects the UNFSA text and requires that stocks be maintained and restored to a biomass level producing MSY.[64] ICCAT, which contains similar wording in its constitutive agreement, continues to treat MSY as a target[65] for species under its competence in the Atlantic Ocean. ICCAT and the General Fisheries Commission for the Mediterranean (GFCM) cooperate on the management of highly migratory fish stocks in the Mediterranean, where MSY is considered a target as well.

The Commission for the Conservation of Antarctic Marine Living Resources (CCAMLR) recognized the shortcomings[66] of MSY and abandoned it as a reference point in the late 1980s. After experimenting with $F_{0.1}$, CCAMLR developed a probability based approach to estimating sustainable yield for harvested stocks and has used that method since that time. In addition, CCAMLR recognized early on, the need for a broader approach to stock assessment, including precaution and the ecosystem approach. Principles and methods instituted by CCAMLR have informed UNFSA and the progressive changes in fisheries management on the high seas. NAFO interprets MSY as a limit and uses precautionary reference points for stocks that are not currently under moratoria.[67] The NEAFC Convention does not make specific reference to yield, but requires the Commission to act in the interests of conservation and optimum utilization of the fisheries resources.[68] NEAFC receives scientific advice from ICES. ICES advice is provided using spawning stock biomass and fishing mortality as the main biological traits on which management reference points are based. These criteria are not directly linked to surplus production models — or MSY.[69] NEAFC has begun to manage certain species using precautionary reference points. RFMOs also use age-based stock-assessment models[70] that produce non-MSY reference points.

Post-UNFSA RFMOs, including the Indian Ocean Tuna Commission (IOTC), the South East Atlantic Fisheries Organisation (SEAFO), the Western and Central Pacific Fisheries Commission (WCPFC), and the Southern Indian Ocean Fisheries Arrangement (SIOFA), all interpret MSY as a limit. SEAFO's Convention contains many provisions in accordance with UNFSA, and while it does not refer specifically to MSY or sustained fisheries yield, it does call for the long-term conservation and sustainable use of the fishery resources in its convention area. WCPFC Convention calls for maintaining or restoring stocks to levels that produce MSY and states that any international minimum standards related to maintaining these levels may be considered in management measures.[71] IOTC manages according to MSY; however, data quality and reliable estimates of fishing mortality make maintaining or achieving B_{MSY} as a limit or a target a difficult objective. SIOFA's Agreement states that stocks should be maintained at levels capable of producing MSY and that depleted stocks should be rebuilt to such levels.[72]

5.4 Status of major stocks

5.4.1 Straddling fish stocks

The poor state of many straddling fish stocks remains a significant concern. Several stocks have not recovered despite moratoria on directed harvesting for over a decade. Currently, NAFO has nine stocks under moratoria, six of which have been closed to directed fisheries for over 10 years; a tenth, Greenland halibut, is now under a rebuilding plan, and an eleventh stock, yellowtail flounder, is recovering. Alaskan pollock in the Central Bering Sea Donut Hole (BSDH) has been under a moratorium since 1993, with 2005 pollock biomass in the convention area estimated at 422 000 t, which is substantially below the 1.67×10^6 t required to trigger the reopening of the fishery. There is no directed fishery for pollock in the Sea of Okhotsk Peanut Hole. CCAMLR has several stocks that are depleted (the directed fishery for some species including Antarctic rock cod is prohibited, and there is no longer a fishery for crab, squid, or lightfish). Krill have been assessed in some areas. The status of Patagonian toothfish (commonly known as Chilean seabass) has been the subject of international concern because of large-scale IUU fishing.[73] In NEAFC, blue whiting is considered to have full reproductive capacity, though catches in recent years have exceeded recommended TACs. Mackerel, which has been classified as having reduced reproductive potential, is considered at risk of being overfished. Atlanto-Scandian (Norwegian) spring-spawning herring is considered to have full reproductive capacity. Stock status is uncertain for redfish, Rockall haddock, and deep-sea species in the NEAFC area, and therefore these stocks are considered to be vulnerable to overfishing.[74] SEAFO has yet to assess stocks under its competence.

Some RFMOs responsible for straddling fish stocks also face the challenge of assessing and managing deep-sea straddling fish stocks such as orange roughy, which are typically long lived, slow growing, and vulnerable to overexploitation. Figures 5.3 to 5.7 illustrate stock status for key commercial species for which reliable abundance data are available. For other stocks, catch data are used as a proxy to illustrate trends in abundance (catch per unit effort (CPUE) used as an index).

5.4.1.1 Cod in Northwest Atlantic Ocean

NAFO's Convention Area covers both areas under national jurisdiction and adjacent high seas areas, whereas the NAFO Regulatory Area (NRA) encompasses only the international waters. Catch data are reported throughout the convention area. NAFO manages northern cod

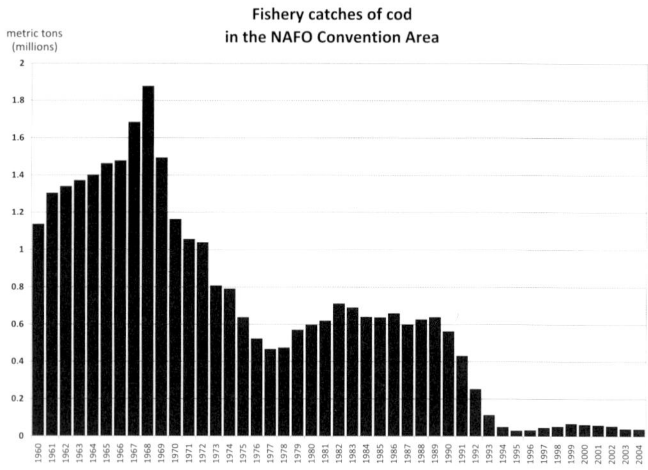

Figure 5.3. Cod Catches in the NAFO Convention Area. (Source: NAFO Fisheries, http://www.nafo.int/fisheries/frames/fishery.html (accessed 30 October 2006), reproduced with permission of NAFO.)

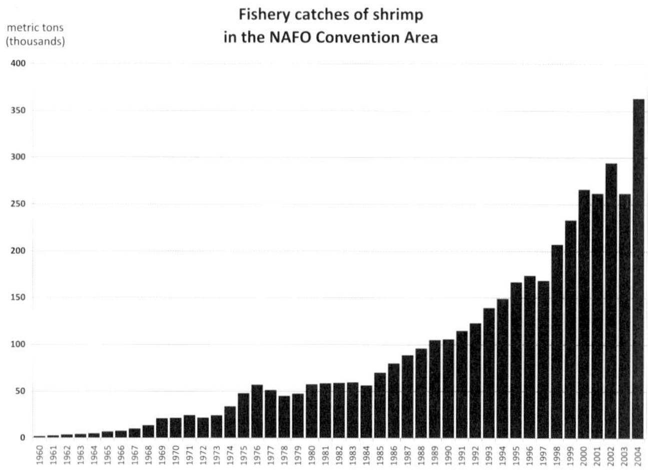

Figure 5.4. Shrimp catches in the NAFO Convention Area. Trends in invertebrate catches are directly opposite of those of most groundfish species. (Source: NAFO Fisheries, http://www.nafo.int/fisheries/frames/fishery.html (accessed 30 October 2006), reproduced with permission of NAFO.)

stocks in Divisions 3LMNO outside the 200-mile limit (see NAFO area map in Part III of book).[75]

Northern cod stocks in Divisions 3NO, which straddle the 200-mile limit on the Tail of the Grand Banks, reached peak catches of 2×10^6 t in the late 1960s (Figure 5.3). Catch limits in the form of TACs were first implemented in 1974 by ICNAF, and after 1979, ICNAF's successor organization, NAFO, enforced catch limits. Despite these measures, by 1992, some 12 years after the establishment of NAFO and the extension of Canadian fisheries jurisdiction out to 200 nautical miles, the stock had seriously declined. In 1993, NAFO introduced an initial

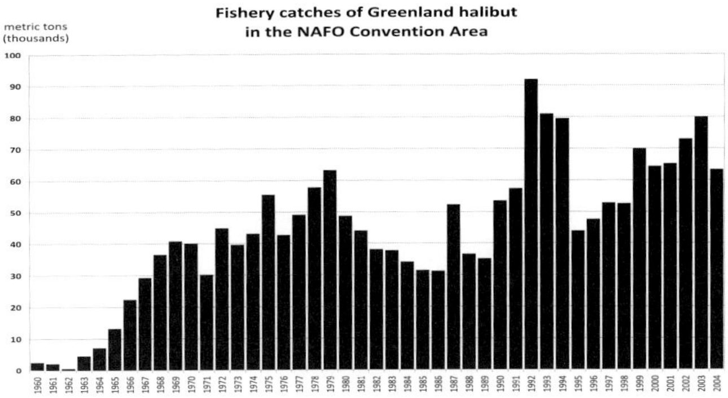

Figure 5.5. Greenland halibut catches in the NAFO Convention Area. (Source: NAFO Fisheries, http://www.nafo.int/fisheries/frames/fishery.html (accessed 30 October 2006), reproduced with permission of NAFO.)

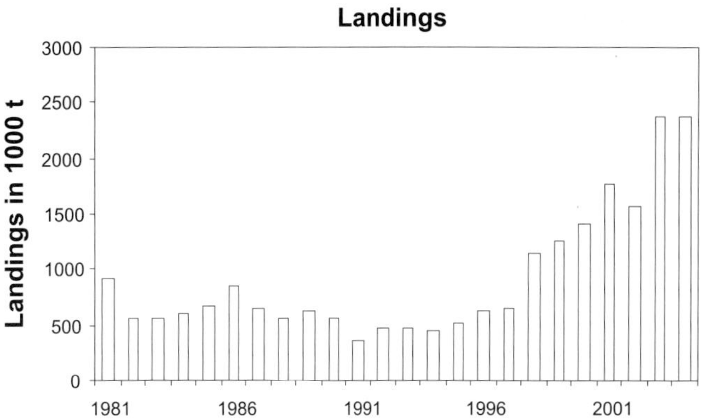

Figure 5.6. Blue whiting landings in the NEAFC Convention Area. (From ICES AFCM Stock Advice 2005: Blue whiting combined stock (Subareas I, IX, XII, and XIV), reproduced with permission of ICES.)

moratorium on the directed cod fishery in Division 3L, and additional stocks in 3NO and 3M (discrete high seas stocks) were closed in 1995.

Currently, there is concern that bycatch of cod in other ongoing fisheries has increased since 1995, significantly affecting the fishing mortality of the collapsed stock,[76] with a maximum catch of 5400 t recorded in 2003.[77] The 2005 assessment for 3LMNO cod indicates that all stocks remain at very low levels, and moratoria will remain in place until at least 2007–2008. The collapse of this iconic species and failure of the cod to rebound, despite the moratoria and other stock restoration management measures, remain major environmental and fisheries management issues in the Northwest Atlantic Ocean.

5.4.1.2 Shrimp in Northwest Atlantic Ocean

Following the collapse of the groundfish stocks, there was a marked increase in the abundance of shrimp and a redirection of fishing effort to the shrimp fishery (Figure 5.4). 3NO shrimp have been under moratoria since 1995,[78] with active fisheries occurring in 3L and in 3M. Shrimp in 3M are considered discrete high seas stocks and are managed through an effort allocation scheme, as stock assessment for setting TACs is difficult. Effort is allocated through the number of vessels and days at sea.

5.4.1.3 Greenland halibut in Northwest Atlantic Ocean

The Greenland halibut fishery has increased in recent years, primarily because fishing efforts shifted from stocks under moratoria to this species (Figure 5.5). In 2004, following a decline in catches, NAFO implemented a 15-year rebuilding plan for stocks in Subarea 2 and 3KLMNO.[79] Under this plan, the allocated TAC will gradually decline until 2007, at which point the rebuilding plan will be reassessed; however, TACs may not be set at greater or less than 15% of the TAC in the previous year. As of 2005, the harvestable biomass of 3LMNO stocks was at the lowest known levels. Recent assessment of this stock included concerns that young fish, which had not yet spawned, were being caught, thus jeopardizing future year classes.[80]

5.4.1.4. Blue whiting in Northeast Atlantic Ocean

Blue whiting is a member of the Gadoid family (similar to cod and haddock) and is an important part of the ecosystem in the Northeast Atlantic Ocean, where it preys on smaller fish and zooplankton and acts as a forage species for a range of predators.[81] It is typically found at depths between 200 and 600 m along the continental shelf. Blue whiting reaches sexual maturity between two and four years and spawns between February and April, with key spawning grounds in the European Community's waters and international waters, and spawning migrations into Faroese waters twice a year. The main fishery is on spawning aggregations,[82] with landings directed mainly to the production of fish meal, although additional markets for the fish are being sought.[83]

The directed fishery for blue whiting began in the 1970s, partially as a result of the decline in the Atlanto-Scandian (Norwegian) spring-spawning herring fishery. Prior to 1997 catches remained relatively stable at ~600 000 t/year, which is consistent with the TAC of 650 000 t set by NEAFC since 1994. Catches began to increase in the late 1990s. The blue whiting fishery soon became the largest fishery in the Atlantic Ocean, with catches reaching a high of 2.3×10^6 t in 2003 (Figure 5.6) despite

ICES advice to maintain a combined TAC of 600 000 t. In 2004 the international catch reached 2.4 × 10^6 t, with catches divided among Norway (40%), Iceland (17%), Russia (15%), the European Union (15%), and the Faroes (13%).[84] Recent large landings are based on the current high recruitments, which are considerably above those of previous years.[85]

Stock assessment of this species is difficult because of insufficient fishery-independent data and uncertainties in estimating stock biomass and interpreting acoustic survey results[86]. However, the blue whiting stock, as a result of high recruitment to the fishery and consecutive years of good year classes, appears to be resilient to the fishing pressure[87] and is described by ICES as having full reproductive capacity.

5.4.1.5 Patagonian toothfish in Southern Ocean

The Patagonian toothfish is the most valuable finfish species in Antarctic and sub-Antarctic waters. Toothfish is a large predatory fish that occurs in the Southern Ocean, to depths of 3000 m, and can grow up to 2 m. As toothfish are long-lived species, they are vulnerable to overexploitation. Since the initial interest in the fishery in the late 1980s, the species has experienced high market demand.[88] By 1997 the IUU fishery was taking much of the catch, in violation of CCAMLR's Conservation Measures[89] (Figure 5.7). Another serious problem is that there is almost no data reporting for IUU catches, making decisions about the status and future management of this fishery very difficult. Since the late 1990s, significant efforts have been made to deter IUU fishing, and one fishery for Patagonian toothfish is now Marine Stewardship Council certified. This is considered by some a testament to the improved management of this species in South Georgia.[90]

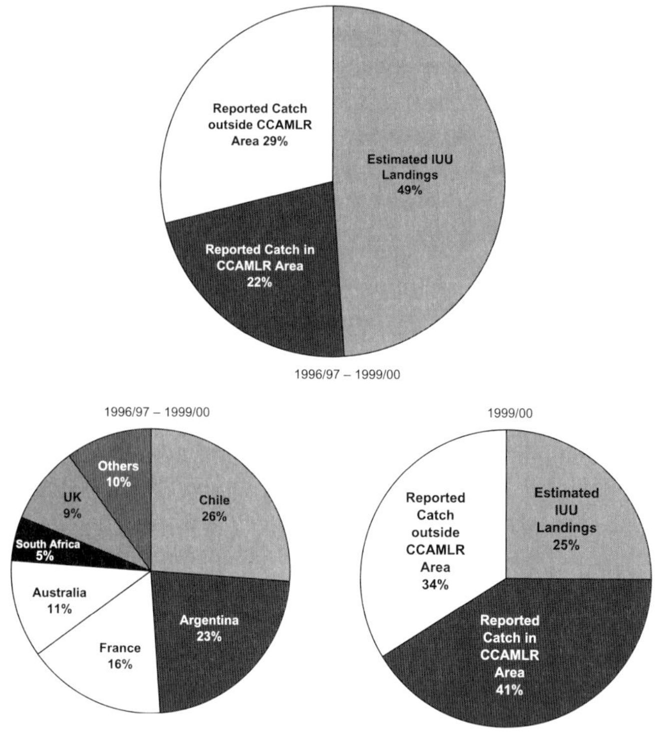

Figure 5.7. Catches of Patagonian toothfish inside and outside the CCAMLR area, as of 2000. (Source: Traffic 2001. Data from CCAMLR, 1998b, 1999d, 2000d and 2000e.)

Table 5.1 presents a summary of the status of straddling fish stocks managed by RFMOs as of March 2006.

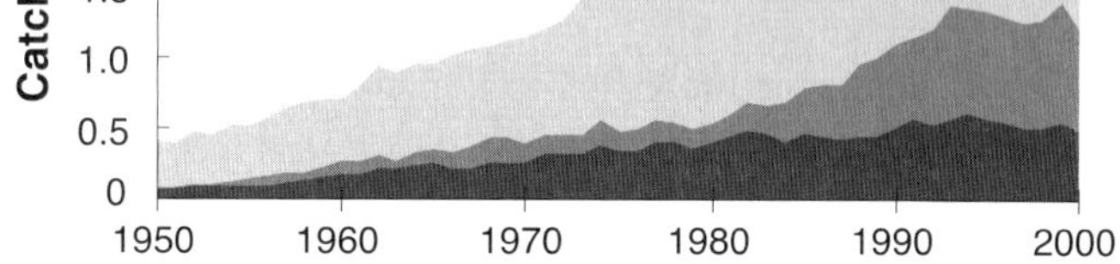

Figure 5.8. World cumulative tuna catches, by ocean. (From FAO, "Historical Trends of Tuna Catches in the World", FAO Fisheries Technical Report No. 467 (Rome: FAO, 2004), reproduced with permission of FAO.)

5.4.2 Highly migratory fish stocks

5.4.2.1 Overview

In the last century, fisheries for highly migratory fish stocks have greatly expanded. Figure 5.8 shows the increasing tuna catches in all of the world's oceans over the last 50 years, with greatest expansion in the Pacific Ocean. The history of the tuna fishery is punctuated by technological changes, expansion into new waters, and the targeting of new species. At the inception of commercial harvesting of tuna and tuna-like species, fisheries were concentrated in coastal areas and were seasonal in nature, taking advantage of the migratory nature of these stocks. Prior to the 1950s, predominantly artisanal fisheries occurred in all oceans, prosecuted with small boats and limited technology.

Following the pattern for many fisheries, there was expansion after World War II, as vessel capacity and the demand for fish products increased. By the 1960s Japanese

Table 5.1. Status of straddling fish stocks managed by RFMOs (as of March 2006*).

RFMO	Key species	Stock status	Bycatch
CCAMLR	Antarctic krill, sub-Antarctic lantern fish, mackerel icefish, Patagonian toothfish, sevenstar flying squid, Antarctic rock cod, crabs	Antarctic krill: status assessed in some areas Patagonian toothfish: depleted Mackerel icefish (Area 48): highly variable recruitment, restricted fishery, TAC imposed around South Georgia (Subarea 48.3) Antarctic rock cod: depleted in Areas 48 and 58, no directed fishery Antarctic toothfish: status unknown, regulated by precautionary TACs Sub-Antarctic lantern fish: no fishery since 1991–1992 Sevenstar flying squid: precautionary TACs imposed Crabs: no fishery since 2002	Seabirds, Antarctic rock cods, ice fishes, skates, rays, noncommercial benthic species
BSDH	Walleye or Alaskan pollock	Stock under moratoria	
NAFO	Cod, Greenland halibut, redfish, skates, American plaice, yellowtail flounder, white hake, witch flounder, capelin, squid, shrimp Discrete high seas stocks: 3M cod, 3M American plaice, 3M redfish, 3M shrimp	Collapsed and under moratoria: since 1993: 3NO capelin, 3L cod; since 1995: 3NO cod, 3M cod, 3LNO American plaice, 3M American plaice, 3NO witch flounder, 3L witch flounder; since 1998: 3LN redfish Status uncertain: 3LNO thorny skates, 3O redfish, 3NO white hake (current TAC unsustainable), Greenland halibut (under 15-year rebuilding plan, 2004 catches exceeded TAC by 27%) Recovered (moratorium lifted in 1998): 3LNO yellowtail flounder, in 2004 B_{lim} set at 30% of B_{MSY} Other: 3M shrimp (under effort regulation, 2005 biomass > B_{lim}); 3M redfish (spawning biomass increasing as of 2005)	Species under moratoria (all nontarget species), sharks, rays
NEAFC	Redfish (pelagic and oceanic), blue whiting, mackerel, Atlanto-Scandian (Norwegian) spring-spawning herring, Rockall haddock, deep-sea species (blue ling, black scabbardfish, orange roughy)	Uncertain: redfish, Rockall haddock Full reproductive capacity: blue whiting (may be harvested unsustainably), Atlanto-Scandian (Norwegian) spring-spawning herring At risk of reduced reproductive capacity: mackerel Generally unknown: deep-sea species (according to ICES, many are beyond safe biological limits)	Deep-sea species, corals and sponges, juvenile haddock
SEAFO	Alfonsino, orange roughy, armourhead, wreckfish, deepwater hake, mackerel, sharks, octopus, squids, cardinal fish, horse mackerel, oreodories Discrete high seas stocks: red crab	Stock status is generally unknown for high seas areas; Scientific Committee in process of gathering existing data	Data on specific bycatch species has yet to be collected

* See discussion in Chapter 3 on defining straddling fish stocks. Straddling fish stocks are not highly migratory fish stocks as listed in the LOSC Annex I, nor do they include anadromous or catadromous species.

long-liners were fishing in all oceans, targeting albacore tuna and yellowfin tuna for canning. The development of vessel-based cold-storage capabilities at the end of the decade meant that bluefin tuna and bigeye tuna could be harvested and sold as fresh frozen. Korea and Taiwan, learning from the Japanese, also began long-lining for tuna. European vessels moved south to fish off Africa. In the Pacific Ocean the industrial fishery was predominantly US purse-seiners targeting yellowfin tuna.[91] The move from bait boats to purse-seine gear that had occurred in the Pacific soon spread to the Atlantic.

Continued expansion occurred in the 1970s, with increasingly efficient cold storage. Concern about dolphin by-catch in the Pacific Ocean brought about increased management measures. During the 1980s fisheries expanded into the Indian Ocean and the Western Pacific Ocean, and several new countries entered the fishery. As a result, RFMOs introduced management measures to address flags of convenience and to mitigate overfishing. These new management measures, along with an increased demand for products, an expansion of fisheries both in area and species targeted, and an increase in the number of States wanting to participate in the fisheries, have met with increased IUU fishing, particularly in the last two decades.[92] The use of fish aggregating devices (FADs) began in the 1990s, and "fishing efficiency, the size of the fish taken, the species compositions of the catch, and the by-catch of species caught incidentally have all changed drastically with the adoption of this fishing method".[93] RFMOs responsible for highly migratory fish stocks began to introduce measures to limit fleet capacity in the late 1990s. Globally, tuna and tuna-like species are generally fully exploited or overexploited.

Of the RFMOs responsible for highly migratory fish stocks, the Commission for the Conservation of Southern Bluefin Tuna (CCSBT) faces the greatest challenges with stock status, as southern bluefin tuna is estimated to have declined by 90% since 1960. Some Atlantic Ocean stocks under the competence of ICCAT are overfished or below target levels ($B < B_{MSY}$), including northern albacore tuna, Western Atlantic bluefin tuna (depleted), blue marlin, and white marlin. The latter three are currently under recovery plans. The status of Eastern Atlantic bluefin tuna, with reference to B_{MSY}, is uncertain, and there is concern that this stock is being overexploited. There is also concern about the magnitude of stock mixing between the Western and Eastern Atlantic bluefin tuna populations, and that potential overfishing in the Eastern Atlantic is compromising the rebuilding efforts in the Western Atlantic.

ICCAT's rebuilding plan for North Atlantic swordfish appears to have met with some success. In 2002, after four years into a 10-year recovery program, the stock had increased from 65% to 94% of its target stock size (B_{MSY}). Areas were closed to fishing, and spawning grounds were protected. The recovery plan is in place until 2008. South Atlantic albacore and blue sharks are above levels required to produce MSY. In 2005, GFCM and ICCAT developed data collection procedures for tuna farming to improve the estimation of amounts fished and size composition of the catches (i.e., to ensure that fattening operations are properly controlled to avoid overfishing, consistent with conservation and management measures). Swordfish in the Mediterranean are stable or declining.

In the Eastern Pacific Ocean (EPO), several stocks under the competence of IATTC are above the biomass needed to produce MSY, including skipjack tuna, southern albacore tuna, striped marlin, and swordfish, with yellowfin tuna biomass at MSY levels. Bigeye tuna is at its lowest abundance since 1975, and therefore this stock presents a significant rebuilding challenge. The status of other pelagics, such as black marlin and sailfish, remains uncertain. Blue marlin is assessed to be near full exploitation albeit with some uncertainty. North Pacific albacore, for which management is being coordinated between IATTC and WCPFC, is considered to be fully exploited, although there is some uncertainty about the status of the stock.

In the Western Pacific, in the area regulated by the recently formed WCPFC, skipjack and South Pacific albacore are above B_{MSY}, while bigeye and yellowfin are considered to be fully exploited. Fisheries for bigeye and yellowfin face expansion, and ensuring harvest levels commensurate with the resource is a challenge for WCPFC and its members. Two tuna species of concern in the Indian Ocean under the competence of IOTC are yellowfin and albacore. A further increase in effort beyond 2000 levels is not recommended for yellowfin, and there is currently no reliable assessment for albacore. Skipjack stocks are seen as stable, as are those of swordfish, although overfishing ($F > F_{MSY}$) of swordfish is suspected.

In summary, highly migratory fish stocks are difficult to assess, and hence TACs are set on relatively few species. Regulation of the fisheries focuses much more on capacity limitations and vessel registries to monitor and control catch levels (Table 5.2). For many species, there are no stock assessments, and data collection on stock abundance, demographics, and biological parameters is highly fishery dependent. Assessments available from RFMOs indicates that most stocks of tuna and tuna-like species are fully exploited, with some overfished, and a

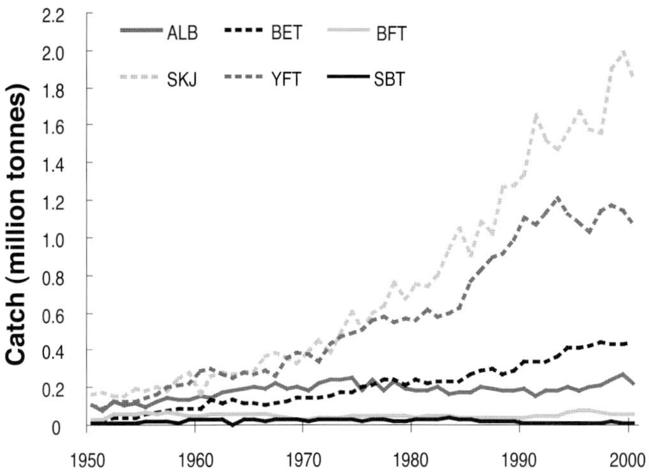

Figure 5.9. World catches of major tuna species. (From FAO, "Historical Trends of Tuna Catches in the World", FAO Fisheries Technical Report No. 467 (Rome: FAO, 2004), reproduced with permission of FAO.)

Abbreviations: ALB, Albacore tuna; BET, Bigeye tuna; BFT, Bluefin tuna; SKL, Skipjack tuna; YFT, Yellowfin tuna; SBT, Southern bluefin tuna.

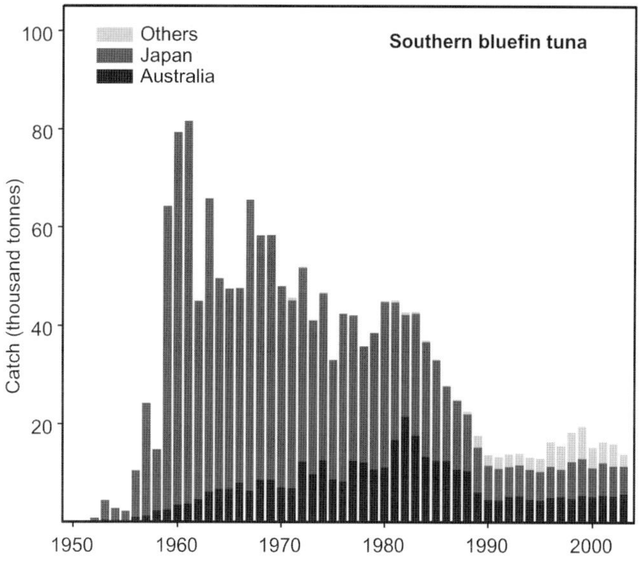

Figure 5.10. Catches of southern bluefin tuna. (From Fishery Status Reports 2004, reproduced with permission of the Bureau of Rural Science.[94])

few depleted (Table 5.2). Few opportunities to increase exploitation exist, except in some areas of the Pacific, and possibly in the Indian Ocean, where increases in catches of skipjack tuna might be sustainable. However, if current fishing techniques are used, this can only be done at the expense of detrimental bycatches of other species, particularly yellowfin and bigeye tuna.

Figures 5.8 to 5.14 illustrate stock status for key commercial species where reliable abundance data were available. For other stocks, catch data are used as a proxy to illustrate trends in abundance (CPUE as an index). Tuna and tuna-like species have been increasingly exploited as fleets move to new waters, target new species, and increase catch efficiency. Faster-growing species, such as skipjack and yellowfin, have experienced much of the fishing pressure since 1970 (Figures 5.8 and 5.9). Table 5.2 provides an overview of key species, stock status, and bycatch for stocks managed under CCSBT, GFCM (joint management with ICCAT), IATTC, ICCAT, IOTC, and WCPFC.

Figure 5.9 illustrates reported catches by major fishing nations of tunas since 1950. Catches of bluefin have remained at low levels compared with catches of smaller, faster-growing tunas such as skipjack and yellowfin.

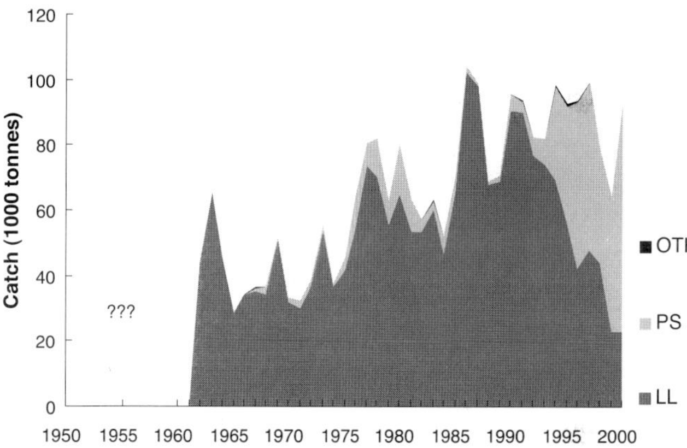

Figure 5.11. Catches of bigeye in the Eastern Pacific Ocean, by fishing gear. PS, purse seine; LL, longline; OTH, other. (From FAO, "Historical Trends of Tuna Catches in the World", FAO Fisheries Technical Report No. 467 (Rome: FAO, 2004), reproduced with permission of FAO.)

5.4.2.2 Southern bluefin tuna

Southern bluefin tuna are mainly found between 30°S and 50°S, with spawning grounds within and adjacent to the exclusive economic zone (EEZ) of Indonesia. The species does not reach maturity until at least eight years of age and has a life-span of approximately 40 years. Catches for southern bluefin tuna peaked in the early 1960s and have declined ever since (Figure 5.10). The southern bluefin tuna stock under the jurisdiction of the CCSBT provides perhaps the most dramatic example of stock depletion, with current estimates of biomass at 3%–10% of pre-exploitation levels. The stock remains at extremely low levels, and prospects of stock recovery given current catches are improbable. Additionally, the high level of unreported catch over the last two decades significantly

Table 5.2. Status of highly migratory fish stocks managed by RFMOs (as of March 2006).

RFMO	Key Species	Stock Status	Bycatch
CCSBT	Southern bluefin tuna	May have declined by up to 90% since 1960; currently estimated to be at 3%–10% of preexploitation biomass	Sharks, rays, seabirds, turtles, juvenile tuna
GFCM	Eastern Atlantic bluefin tuna, swordfish	Above B_{MSY}: Eastern Atlantic bluefin tuna Declining: swordfish	Sharks, juveniles of target species.
IATTC	Yellowfin tuna, bigeye tuna, albacore tuna, skipjack tuna, bonito, Pacific bluefin tuna, sailfish, billfishes including marlin and swordfish	Above B_{MSY}: skipjack ($F < F_{MSY}$ level stock abundant, southern stock catches < MSY); striped marlin (at AMSY as of 2002); swordfish At AMSY: yellowfin tuna Below B_{MSY}: bigeye tuna (2003 biomass at lowest levels since 1975) Fully exploited: albacore tuna (northern stock: $F > F_{MSY}$) Status uncertain: blue marlin (stock appears to be fully exploited); black marlin or sailfish (no recent assessment)	Sharks, rays, juvenile tunas, dorado and other large fishes, turtles
ICCAT	30 key highly migratory stocks, including yellowfin tuna, bigeye tuna, albacore tuna, skipjack tuna, bonito, Atlantic bluefin tuna, swordfish, blue marlin, white marlin, shortfin mako, blue shark, porbeagle	Above F_{MSY} (overfishing): North Atlantic (NA) albacore tuna, Western Atlantic (WA) bluefin tuna, blue marlin, white marlin Near F_{MSY}: bigeye tuna, NA swordfish, yellowfin tuna Below F_{MSY}: South Atlantic (SA) albacore tuna Above B_{MSY}: SA albacore tuna, NA and SA blue shark Near B_{MSY}: bigeye tuna, NA swordfish, yellowfin tuna Below B_{MSY} (overfished): NA albacore tuna, Western Atlantic (WA) bluefin tuna, blue marlin, white marlin Under recovery plan: WA bluefin tuna (by 2018), NA swordfish (by 2009), blue marlin and white marlin (by 2006) Uncertain: skipjack tuna, Eastern Atlantic bluefin tuna, sailfish, SA swordfish, Mediterranean swordfish, Mediterranean albacore tuna (never assessed), NA shortfin mako, small tunas (never assessed)	Seabirds, turtles, sharks, juveniles of target species
IOTC	Yellowfin tuna, skipjack tuna, bigeye tuna, albacore tuna, longtail tuna, kawakawa, frigate tuna, bullet tuna, narrow barred Spanish mackerel, Indo-Pacific king mackerel, Indo-Pacific blue marlin, black marlin, striped marlin, Indo-Pacific sailfish, swordfish (also southern bluefin tuna but CCSBT has primary responsibility)	Above MSY: yellowfin tuna, bigeye tuna Stable: skipjack tuna, swordfish (current catch level unsustainable, possible overfishing in SW Indian Ocean) Uncertain: albacore tuna (no reliable assessment, catch increases to consider PA if adopted) Assessments not available for other species	Sharks, rays, juvenile target species, other fishes, seabirds, turtles, marine mammals
WCPFC	Skipjack tuna, yellowfin tuna, albacore tuna, bigeye tuna	Above B_{MSY}: skipjack tuna, bigeye tuna (although $F_{current} > F_{MSY}$), SP albacore tuna Near fully exploited: bigeye tuna and to some extent yellowfin tuna	Sharks, rays, juvenile target species, bill fishes, other large fishes, seabirds, turtles

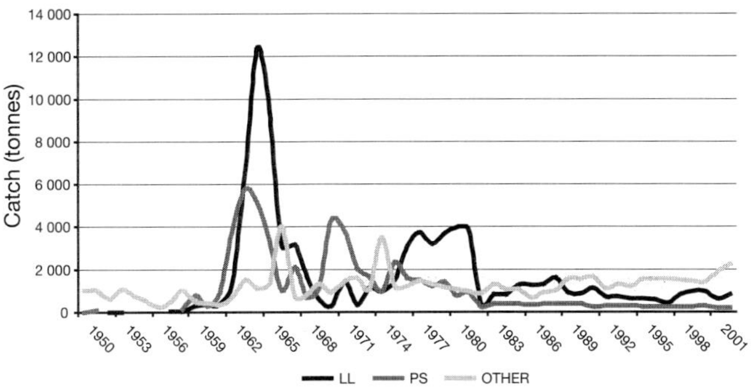

Figure 5.12. Bluefin tuna catches in the Western Atlantic Ocean. (From FAO, "Historical Trends of Tuna Catches in the World", FAO Fisheries Technical Report No. 467 (Rome: FAO, 2004), reproduced with permission of FAO.)

Abbreviations: LL, Longline; PS, Purse Seine

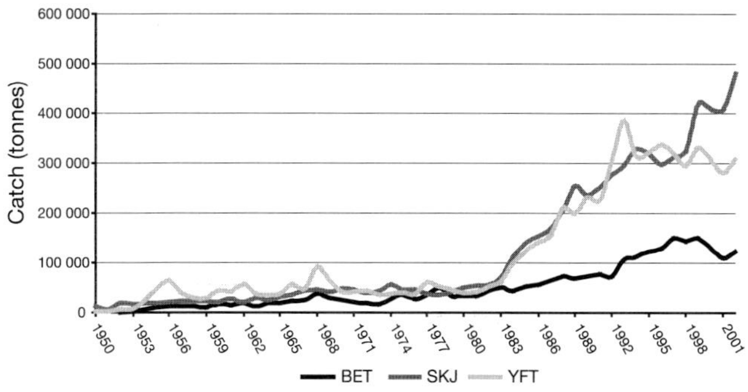

Figure 5.13. Catches of tuna in the Indian Ocean. (From FAO, "Historical Trends of Tuna Catches in the World", FAO Fisheries Technical Report No. 467 (Rome: FAO, 2004), reproduced with permission of FAO.)

Abbreviations: BET, Bigeye tuna; SKJ, Skipjack tuna; YFT, Yellowfin tuna.

increases the need for meaningful control of fishing and the use of the precautionary approach. Southern bluefin was listed as critically endangered on the International Union for the Conservation of Nature (IUCN) Red List in 1996.

5.4.2.3 Bigeye tuna in Eastern Pacific Ocean

Under the jurisdiction of IATTC, bigeye tuna in the Eastern Pacific Ocean is at its lowest biomass levels since 1975. Bigeye is the major stock conservation concern for IATTC. The last assessment, prior to the May 2006 UNFSA Review Conference, was in 2004. Since 1998 there has been poor recruitment and high fishing mortality. Under current fishing effort, the spawning stock biomass is predicted to decline further. The population is unlikely to rebuild unless fishing mortality is greatly reduced or recruitment is above average for a number of consecutive years.[95] Several management measures have been implemented, including the reduction of fleet capacity, seasonal closures, and a statistical documentation scheme (SDS). One of the challenges for bigeye conservation is that fish destined for the cannery (generally smaller fish) are not included in the SDS and therefore are not adequately monitored through trade and processing.

5.4.2.4 Western Atlantic bluefin tuna

Bluefin tuna fisheries in the Western Atlantic have not been exploited to the level of those in the Eastern Atlantic and Mediterranean. In the 1950s the only fisheries were Canadian trap and harpoon fisheries and US trap and recreational rod-and-reel fisheries. When Japanese long-liners started operating in the Atlantic in 1957, they found a high concentration of bluefin tuna in waters off Brazil, where the fishery peaked in 1964 at well over 10 000 t, and then ended by the late 1960s. Following the stock depletion off South America, Japanese long-liners found new grounds off New England and in the Gulf of Mexico, with catches increasing again in the late 1970s until 1981.

The US purse-seine fishery started in the late 1950s and peaked in 1962 and again in 1970. The catch fluctuated widely during this period. The catches declined, due partly to domestic regulations, towards 1980. In the meantime, Canadian trap catches were reduced to almost nil. The catches of the American recreational rod-and-reel fishery started to increase substantially, though there were very wide fluctuations, particularly in the 1970s.

ICCAT adopted strict catch-quota regulations for Western Atlantic bluefin tuna in 1981, and these came into effect in 1982. As only three countries (Canada, Japan, and the United States) were fishing bluefin in this area until recent years, and the regulations were strictly implemented, the catches since 1982 are believed to correspond to TACs set by ICCAT.[96] ICCAT has implemented recovery plans for the Western Atlantic bluefin tuna (by 2018), North Atlantic swordfish (by 2009), and blue and white marlin (by 2006). The status of at least eight other stocks is uncertain, and some stocks have never been assessed.

5.4.2.5 Yellowfin tuna, skipjack tuna, and bigeye tuna in Indian Ocean

Tuna catches in the Indian Ocean began to increase in the early 1980s (Figure 5.13). Yellowfin tuna biomass in the Indian Ocean is above that required to produce MSY, but any increase in fishing effort beyond 2000 levels is to be avoided. Bigeye biomass is also above MSY, but overfishing may be occurring, as fishing effort has increased in recent years. Skipjack and swordfish populations are stable, although there is possible overfishing of swordfish in the Southwest Indian Ocean. As of 2005, there is no reliable assessment of albacore tuna.

5.4.3 Stock recovery

5.4.3.1 Overview within RFMOs

The restoration of fish stocks is one of the objectives included in Article 5 of UNFSA. The commitments of the WSSD to restore fisheries to levels capable of producing MSY by 2015 puts pressure on both national and international management bodies to develop strategies that will result in stock restoration. While a desirable outcome of fisheries management, restoring stocks has proven to be challenging, even when there is political will to reduce fishing effort. Most recovery and rebuilding plans implemented in the last decade have not yet met the rebuilding objectives. Fishing effort on collapsed species often continues through bycatch, and factors other than fishing affect stock structure and biomass. Environmental effects (e.g., North Atlantic Oscillation (NAO), El Niño, and ocean acidification) can have a dramatic impact on recruitment to the fishery. Habitat destruction and pollution are additional factors that impede stock recovery. Reducing a stock below a certain biomass effectively removes much of its resilience to withstand natural fluctuations in the environment, reinforcing the need for applying risk-averse conservation measures. Of course such stock reductions may be indicative of broader and far-ranging ecosystem changes, and recovery might not be possible without subsequent ecosystem adjustment.[97]

While many fish species are considered to be recovering from being overfished, there are several specific examples of both straddling fish stocks and highly migratory fish stocks where recovery plans have been implemented to varying degrees of success. In the NAFO Convention Area, the implementation of national and international moratoria designed to reduce fishing effort on groundfish stocks and allow them to recover have met with little success. More than a decade later, fish stocks remain at an historical low, with only yellowtail flounder considered to be recovering.[98] Additionally, NAFO established a rebuilding plan for Greenland halibut in 2003 to achieve stock stability.[99] Since the implementation of the plan, catches have exceeded allocated TACs.[100] In 1999 ICCAT's North Atlantic swordfish assessment determined that the population had decreased by two-thirds as a result of increasing fishing pressure. As previously discussed, ICCAT's North Atlantic swordfish recovery plan has resulted in an increase of the spawning stock, and the plan will be in place until 2008. The 2006 stock assessment indicated that the North Atlantic swordfish stock had increased to the level estimated to produce B_{MSY}.[101] The Western Atlantic bluefin tuna is under a recovery plan to 2018, and blue marlin and white marlin are under a recovery plan and are to be reassessed in 2006.

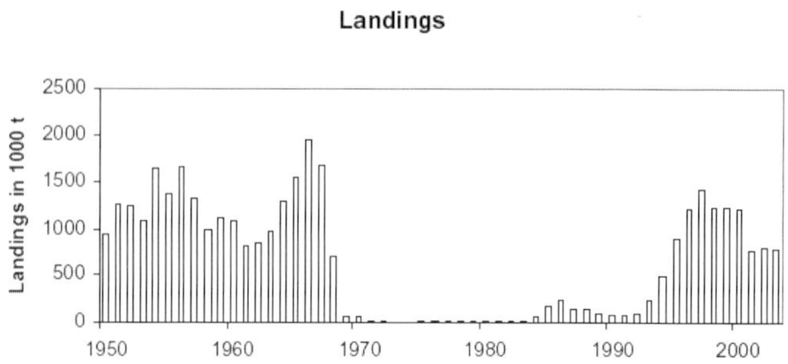

Figure 5.14. Catches of Atlanto-Scandian (Norwegian) spring-spawning herring. The stock collapsed in 1969, but has recovered to precollapse spawning stock biomass. (From ICES 2004 Advice to NEAFC: Norwegian spring-spawning herring, reproduced with permission of ICES.)

5.4.3.2 Atlanto-Scandian (Norwegian) spring-spawning herring

Perhaps one of the most notable examples of stock recovery is that of the Atlanto-Scandian (Norwegian) spring-spawning herring. Prior to industrial exploitation, this stock was one of the most abundant in the world, with total landings of 15×10^6 to 20×10^6 t (Figure 5.14). When the stock is healthy, it is both a transboundary and a straddling fish stock, spawning in Norwegian seas, migrating from within the EEZ, through the EEZs of European Union nations, the Faroë Islands, and Iceland as well as through the high seas area known as the Loop Hole. Herring is a schooling fish; hence it is vulnerable to overfishing, as intense effort on a school can drastically reduce the population. By the late 1960s the stock had collapsed as a result of overfishing and remained at low levels, primarily within Norwegian waters. A moratorium on fishing was declared in 1969. The stock remained at low levels until the recruitment of a strong year class in 1992, which prompted recovery of the spawning stock biomass. As of 2004, ICES describes the stock as having full reproductive capacity and as being harvested sustainably.[102]

Given that most recovery and rebuilding plans have been put in place within the last decade, the success of many has yet to be seen. In addition to specific recovery plans, trade documentation schemes and Convention in International Trade of Endangered Species of Wild Fauna and Flora (CITES) listings of marine species are also expected to reduce the amount of illegal fishing and to reduce effort on listed species. These initiatives, along with vessel registries, IUU vessel lists, and port State control measures have been introduced by some RFMOs to address non-compliance with fisheries management measures (see Chapter 9). With the WSSD commitment of sustainable fisheries by 2015, UNGA resolutions[103] calling for sustainable fisheries, and continual pressure from NGOs[104] and certain governments,[105] stock recovery will likely receive increased focus over the coming decade.

5.4.4 Unregulated species

While the term "unregulated" is now often used in reference to IUU fisheries, there are unregulated legal fisheries. Not all straddling fish stocks and highly migratory fish stocks are subject to regulatory measures. There are three general scenarios:

(1) Regulated vessel fishing unregulated species in regulated waters, e.g., NAFO fisheries for grenadier; ICCAT fisheries for which there is no TAC or other controls.

(2) Regulated vessel fishing unregulated species in an unregulated area, e.g., *Illex* spp. (squids) in waters beyond and adjacent to Argentina in the Southwest Atlantic; the New Zealand fishery for orange roughy in the Southern Indian Ocean prior to the establishment of SIOFA; and fishing for straddling fish stocks in the South Pacific and Eastern Pacific high seas' areas where there is currently no RFMO.

(3) Unregulated vessels fishing unregulated species in an unregulated area, e.g., IUU fishing of straddling fish stocks in the Southwest Atlantic, South Pacific, Eastern Pacific, Northeast Pacific, and Southern Indian Ocean prior to establishment of management regimes for high seas areas.

Many stocks on the high seas continue to have no harvest limits, though there may be regulatory measures in place. In some instances, the stock may be a relatively new target fishery for which quotas have yet to be set. Such fisheries can continue for some time before regulation is imposed. This is evident in the skate fishery in NAFO, which was regulated as of 2005, despite being targeted for several years prior. Some RFMOs may consider stocks with no quotas as regulated. For example, NEAFC considers its deep-sea fisheries regulated, since it imposed a 30% reduction of fishing effort on these stocks. No TACs, quotas, or bycatch limits are set, but the regulation is an effort reduction.

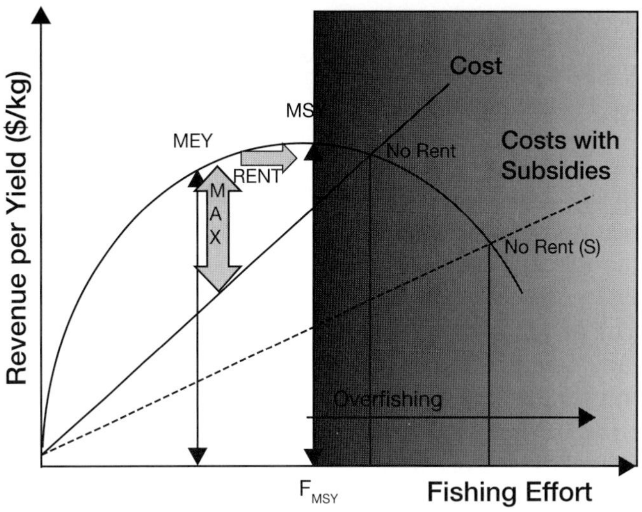

Figure 5.15. Cost-benefit model of a stock assuming constant value per unit yield and the Schaefer surplus production model. Effort is the combination of vessels, labour, and time catching fish. MSY is based solely on the biological capacity of the stock. MEY is where the benefit for unit cost is the highest. With this revenue-yield model, MEY is always lower than MSY. With subsidization and overcapitalization of the fishery, MEY approaches MSY. Commercial depletion occurs when the value of the yield is less than the cost of fishing.

5.4.5 New and exploratory fisheries

Under UNFSA, new and exploratory fisheries are to be regulated as part of the precautionary approach (PA),[106] as data are often unavailable or inadequate, and these fishing activities may result in serious or irreversible harm. The PA is an essential component of contemporary fisheries management, because as traditional stocks are becoming overexploited and depleted, fishing effort is shifting to species further down the food chain (including larger predator species) and moving into deeper waters, where previously untargeted species are being caught and seafloor (benthic) ecosystems are being affected. Typically, new fisheries are prosecuted without management measures being imposed, and such measures are only imposed once the stock begins to decline or if catches increase and allocation of the stock to Contracting Parties becomes necessary. One of the difficulties in regulating new and exploratory fisheries is that there is often little data upon which to base a stock assessment, estimate biomass, or set harvest limits. It is therefore essential to invoke the PA in such instances. CCAMLR has endeavoured to ensure that new fisheries do not begin until a management plan is in place and has adopted the PA in making decisions pertaining to new and exploratory fisheries.

5.5 Beyond optimum economic yield

5.5.1 Overcapacity and overfishing

MSY as a target reference point is not only undesirable for conservation purposes, but is also inefficient at producing maximum economic return (Figure 5.15). For centuries, the overexploitation of fish stocks was inconceivable, not only because the supply of fish was considered inexhaustible, but also because overfishing would render fishing uneconomical. As Huxley stated in 1883, "any tendency to over-fish will meet with its natural check in the diminution of the supply,... this will always come into operation long before anything like permanent exhaustion has occurred". Huxley did not foresee the advances in technology, the failure of market control, the introduction of subsidies, or the demand for highly valued species, all of which contributed to making fishing for depleted species economically feasible.

The traditional fisheries concept of MSY was adopted by economists to generate a bio-economic model and to articulate the concept of maximum economic yield (MEY) (Figure 5.15).[107] This bio-economic model is based on harvesting one species at a constant price without subsidies. An abstraction of the real world situation, it is nevertheless useful in understanding conventional fisheries management. MEY is defined as "the optimal difference between fishing costs and income. It relates to overall yield from a fishery that provides the maximum economic return as defined by the difference between the monetary cost of fishing and the economic value of the yield".[108] MEY is accordingly the harvest level that provides the highest economic return, or rent.[109] This definition indicates that effective fisheries management or an economically efficient system exists at a level below F_{MSY}, whereby the biomass harvested for each stock is set at an economically optimal target level.[110]

Figure 5.15 shows the total sustainable revenue curve, indicating the revenues that can be obtained from the fishery at every level of fishing effort. Bisecting the total revenue curve is the line for total costs of harvesting at each effort level. Effort is the combination of vessels, labour, and time catching fish. As levels of effort increase along the x axis (fishing effort), the catch and associated revenues increase to a maximum return shown on the curve at MEY. As can be seen, the harvest level that provides the highest economic returns, or rent, is found considerably before catch reaches the biological reference of MSY.[111] Catches continue to increase with additional effort, until MSY, and then begin to decline as the stock is overfished. At the intersection of the total cost and the total revenue, return is dissipated, and theoretically, there are no longer any incentives to increase effort.

Fisheries managers understand that they must cap extraction of fisheries resources considerably below MSY to maximize both economic (MEY) and conservation benefits. Curiously, MEY, which is more "related to social and economic performance of fisheries, has not achieved international recognition nor been included in international law, as is the case with MSY".[112] The reasons for this paradox, where there is theoretical understanding of MEY yet no practical application, is in large part due to incentives, market structure, and governance regimes, including subsidies, open access, overcapitalization, fuel prices,[113] and inadequate regulation.[114] The conceptual illustration of MEY depicts how overexploitation can occur despite economic inefficiencies, particularly within an open access system, resulting in the tragedy of the commons, popularized by Hardin,[115] and the well-worn adage "too many boats chasing too few fish".

5.5.2 Subsidies and overcapacity

Bedevilling fisheries management within an open-access or limited-access regime is the incentive to maximize catch (share maximization) in the short run. This incentive structure inevitably leads to excessive investment in harvesting capacity (i.e., number of vessels, size of vessels, gear, days at sea), which leads to excessive fishing pressure, catches above sustainable yields, stock depletion, and sometimes, stock collapse (e.g., orange roughy in the South Tasman Rise). As a commercial stock is depleted, vessels move on to the next fishery. This fishing pattern, characteristic of high seas fishing, is known as "serial depletion" (and is illustrated by the depletion of pollock stocks in the Bering Sea Donut Hole and subsequently in the Sea of Okhotsk Peanut Hole).[116] It is estimated that the maximum harvest levels the ocean is capable of sustaining can be caught by 50% of the existing world fishing fleet, underscoring the relationship among overcapitalization, overcapacity, and overfishing.[117] Subsidization, as illustrated in Figure 5.15, can contribute to overcapitalization, thereby providing an alternative cost structure and incentive to further increase fishing effort beyond overfishing. This level of effort can result in population reductions to the point where fish populations are considered to be species at risk (Figure 5.16).

Recognizing the correlation among overcapitalization, subsidization, and overexploitation, UNFSA, the IPOA-Capacity, and the FAO Code of Conduct address the structural problems that exacerbate the institutional incentive to overfish. Article 5 of UNFSA requires States

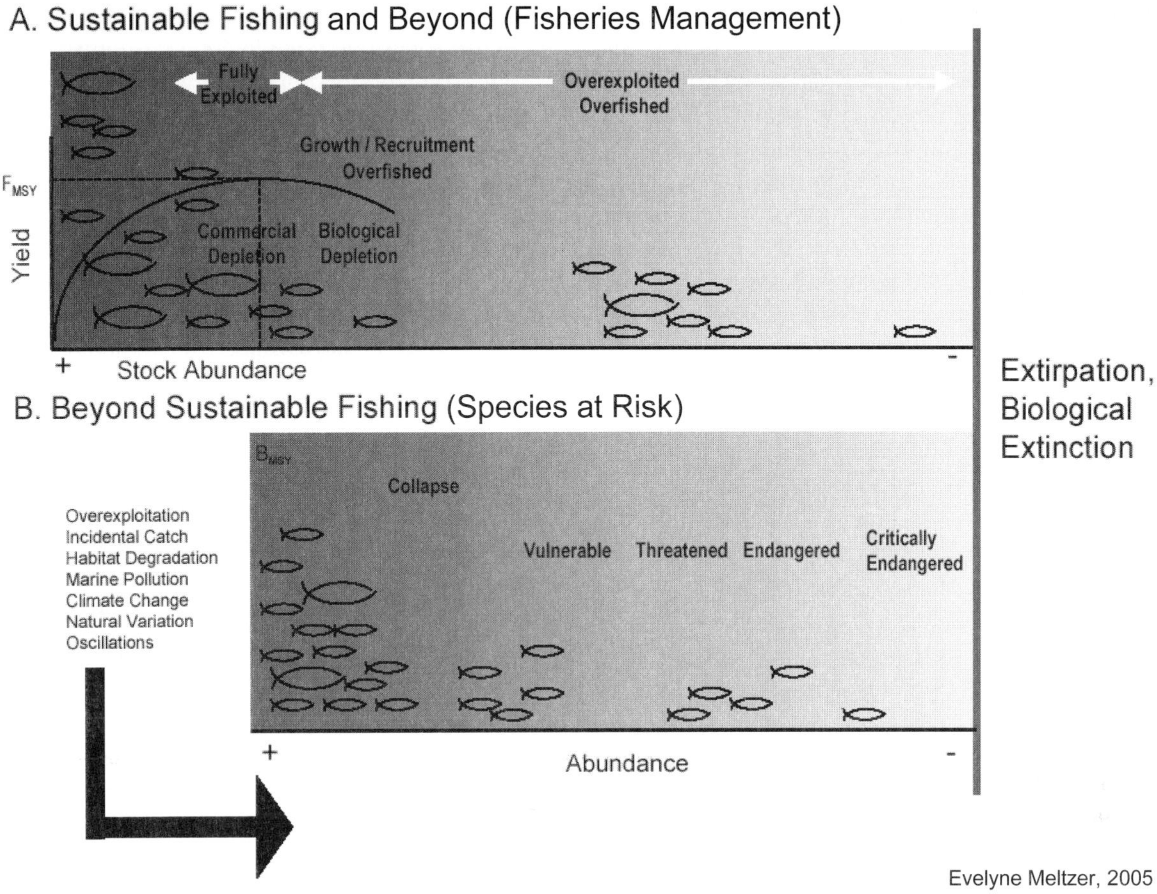

Figure 5.16. Figure depicts different lexicon to describe status of fish stocks when population is overexploited. (A) Fisheries science and management based on density-dependent population growth. (B) Beyond overexploitation, extinction risk criteria are based on rates of population decline from virgin or reference population abundances (e.g., IUU fishing). (See J. Hutchings, "Collapse and Recovery of Marine Fishes" (2000) 406 *Nature* 882–885; J. Hutchings, "Conservation Biology of Marine Fishes: Perceptions and Caveats Regarding Assignment of Extinction Risk" (2001) 58 *Canadian Journal of Fisheries and Aquatic Sciences* 108–121; N. Dulvy, Y. Sadovy, and J. Reynolds, "Extinction Vulnerability in Marine Populations"; R. Myers and B. Worm, "Rapid Worldwide Depletion of Predatory Fish Communities".)

to take measures to reduce fishing capacity.[118] RFMOs have also undertaken initiatives to reduce capacity in accordance with the IPOA-Capacity. This point was underscored in the Chairman's closing statement at the UNFSA Conference, "I need not remind the Conference that the Food and Agriculture Organization of the United Nations (FAO) has warned that disastrous social and economic consequences await the entire fishing industry unless fishing fleets are reduced in size, subsidies are eliminated and the activities of fishing fleets are more effectively regulated".[119]

The challenge facing RFMOs, member States, and the fishing industry is to reduce capacity and alter the incentive structure. RFMOs with competence for straddling fish stocks have reduced fishing effort through TACs, closed areas, and associated measures but not by reducing fleet capacity. In contrast, RFMOs responsible for highly migratory fish stocks (i.e., IATTC, ICCAT, IOTC, and CCSBT) have used fleet-capacity assessment and management plans, and placed limits on capacity for member States. The initiatives of these four organizations are examined briefly below.

5.5.2.1 IATTC

IATTC has used fleet capacity reduction as a key element in its program for the conservation of tuna stocks in the EPO.[120] IATTC established a Working Group on Fleet Capacity in June 1998[121] to collect information on the fleet capacity of member States, and to develop recommendations for management to ensure that capacity did not increase at a rate that would jeopardize the tuna stocks and drive them below biomass levels needed to produce MSY. In October 1998 the Working Group tabled the first Resolution on capacity[122] and set purse-seine vessel capacity limits for all Contracting Parties and cooperating nonmembers authorized to fish in the EPO. In 1999 an additional Resolution was passed concerning the capacity of long-line vessels,[123] in response to Japan's initiative to reduce the number of large-scale tuna long-line fishing vessels by 20% (132 vessels) by scrapping them. The Resolution called upon other Contracting Parties to reduce their fleets as well to protect juvenile tunas and to reduce fishing mortality on tuna stocks. The 2000 Resolution on fleet capacity[124] further clarified the measurement of capacity. It called for maintenance of capacity limits established in 1999 and for the development of a draft plan for regional management of fishing capacity, as called for in the IPOA-Capacity. This Resolution also established the basis for a vessel registry. In a second Resolution in 2000,[125] Contracting Parties agreed to develop and implement a plan to achieve a target level of 135 000 t of carrying capacity (or such other limit as the Commission may decide) as of 1 January 2005. Fleet-capacity measures will be examined again at the IATTC meeting in June 2006.

While these Resolutions laid the groundwork for a comprehensive fleet-capacity management plan, the onus was, and continues to be, on the Contracting Parties to reduce their fleets or to cease fleet expansion. The 2002 Resolution prohibited new vessels from fishing in the EPO or an increase in the capacity of existing vessels, but provided special allowances for developing countries that did not have a significant fleet. While the Resolution limited increases in the fleet size, the current fleet is considerably too large. In 2005, the IATTC adopted the Regional Plan of Action, which included a target capacity for the purse-seine fleet. However, as of 2006, no progress has been made in reducing the fleet size toward the target.[126]

5.5.2.2 ICCAT

Following the binding Resolution on capacity passed by IATTC at its 62nd meeting in 1998, ICCAT passed a binding Recommendation in 1999 to limit capacity fishing on northern albacore tuna.[127] This Recommendation required Contracting Parties to limit their number of vessels to the average number fishing between 1993 and 1995 and to submit a list of vessels to the Commission. The Standing Committee on Research and Statistics was also asked to evaluate fishing capacity according to the gear type of vessels fishing in waters under the jurisdiction of ICCAT.

5.5.2.3 IOTC

The 1998 IOTC Resolution on Cooperation with Non-Contracting Parties[128] invokes the IPOA-Capacity in an attempt to address fishing by nonmembers. This binding Resolution was reiterated in 2003,[129] requiring that Cooperating Non-Contracting Parties with more than 50 vessels on the 2003 Vessel List not increase the number of vessels greater than 24 m beyond the 2003 level. In addition, corresponding tonnage could not exceed the overall tonnage listed in 2003. IOTC adopted a Resolution on capacity in 1999[130] to manage the potential for fisheries to increase in its area of competence. Based on recommendations from the Scientific Committee on the catch of juvenile bigeye tuna in the purse-seine fishery, and on the increase in IUU fishing in the area, the Commission resolved to take action on capacity reduction, including seasonal and area closures and the use of FADs. Facing continued decline of tuna stocks in 2002, the Commission adopted a further Resolution, requesting advice from the Scientific Committee on measures for addressing overcapacity in the purse-seine fishery.[131] In determining whether States may attain Contracting Party status, a 2003 Resolution on new members requires that the Compliance Committee review requests for cooperating status and use caution in accepting States as cooperating nonmembers, so as not to "introduce excessive fishing capacity of other regions".[132] To manage for the conservation of tuna stocks and also provide for the participation of developing States in the fisheries managed by IOTC, the Commission adopted a Resolution in 2005[133] requiring Cooperating Non-Contracting Parties and developing coastal States with catches under 1000 t who would like to increase their bigeye catches to submit "fleet development plans". In this way, increased fishing capacity could be measured and accommodated through the implementation of additional conservation strategies.

5.5.2.4 CCSBT

Unlike the other tuna RFMOs, CCSBT is responsible for only one stock. CCSBT has an authorized vessel list and sets quota for Contracting Parties and Cooperating Non-Contracting Parties. In this context and despite the status of the southern bluefin tuna, the CCSBT has not regarded fleet capacity restrictions at the international level to be necessary. However, some Contracting Parties have imposed domestic fleet capacity restrictions as a measure to manage to their national quotas. Other post-UNFSA RFMOs, such as SEAFO, SIOFA, and WCPFC,[134] are in the process of adopting a suite of capacity management measures, through catch limits and closed areas.

5.5.3 Dealing with subsidies

Subsidies, as illustrated in Figure 5.15, play a major role in supporting and expanding fishing effort, either by reducing capital or operating costs for vessel owners, or by increasing income.[135] Subsidies cause adverse impacts in two ways: by making the subsidy recipients better off than their nonsubsidized counterparts, and by allowing a type or level of fishing effort that would not otherwise be economic.[136] The impact on fishing effort is widely considered the more serious outcome because of the potentially adverse effects on sustainability, international trade, and socioeconomic circumstances. These impacts are transmitted through either or both reduced fish stocks and lower prices in the short term, and potentially depleted stocks in the long term. In recent years the link between subsidies and IUU fishing has also emerged as a matter of considerable concern.

Although the subsidy issue has been on the international fisheries management agenda for some time,[137] the issue of overcapacity and fisheries management was only explicitly addressed in the 1999 IPOA-Capacity.[138] However, the IPOA-Capacity is nonbinding, and nations have shown limited enthusiasm for adopting the guidelines. One observer notes that IPOA documents "… essentially reproduce the broad principles outlined in the Code of Conduct … instead of providing innovative and definite measures".[139]

Parallel discussion on subsidies opened up within the World Trade Organization (WTO) negotiation process in 1999[140] in response to frustration at the ineffectiveness of other mechanisms in addressing overfishing and the availability of apparently more effective procedures under WTO. The WTO Committee on Trade and the Environment initially addressed subsidies in 1997, publishing a 1998 paper describing the relevant WTO rules and an overview of fishing industry subsidies notified under Article 25 of the Agreement on Subsidies and Countervailing Measures (SCM). The 2001 Doha Declaration included a commitment to "…clarify and improve WTO disciplines on fisheries subsidies, taking into account the importance of this sector to developing countries".[141]

Building on this commitment, in 2002 a group of eight countries submitted a paper aimed at clarifying how subsides impinge on trade, identifying gaps in the SCM, and determining the role of the WTO discipline regarding fisheries subsidies.[142] They argued that in the case of straddling fish stocks and highly migratory fish stocks, subsidized fleets take a disproportionate share of a stock and, at a minimum, put nonsubsidized fleets at a disadvantage.[143] This subsidized competition may also make it uneconomic for coastal States to fully develop their own fisheries. In the long run, subsidies that result in overcapacity put excessive pressure on stocks and therefore cause their depletion, which carries adverse trade and economic consequences for States sharing the resource. Because of the clear trade implications, it is argued that WTO rules ought to apply. This debate continues.

WTO arguably has a role to play in addressing the subsidy issue as it pertains to overcapacity and the consequences for overfishing and stock depletion, but there are limits to its effectiveness.[144] Also, and perhaps more importantly, while subsidies pose an undeniable threat, they are just one contributing factor to the larger issue of unsustainable fishing practices.[145] The key to addressing unsustainable fishing practices lies in implementing effective management regimes. The United Nations Environment Program (UNEP) states the challenge clearly and succinctly:

> The tools of fisheries management are increasingly well understood, but are yet to be fully and consistently applied. A series of international instruments — including the FAO *Code of Conduct for Responsible Fisheries* and its associated FAO International Plans of Action, and various regional fisheries management agreements — provide an increasingly coherent framework for effective fisheries management.[146]

5.6 Beyond MSY and excess capacity

Stocks at risk of being overfished have little or no margin of error at the MSY target level given the uncertainties and risks associated with this approach as described above. Immediate corrective action to reduce fishing effort on a stock subject to overfishing is required and, in theory, should produce a corresponding increase in abundance. Fish, however, have variable recovery periods and variable responses to a reduction in fishing mortality.[147] In the cases of Central Bering Sea Alaskan pollock, northern cod, and southern bluefin tuna, the overfished stocks require a long recovery period. An example of a successful relatively short recovery plan is North Atlantic swordfish within the ICCAT Convention Area.

Overfishing of straddling fish stocks and highly migratory fish stocks has led to concerns that certain species are at risk and should be listed under CITES. Given the market structure, aggregation pattern of pelagic species, and vessel technology, it is feasible that overfished stocks could remain economic until the last school is harvested. The harvesting capacity of the new generation of fishing vessels raises the spectre of the "single-vessel phenomenon". Both of these issues are examined in more detail below.

5.6.1 The spectre of the single vessel

There are differing views on the extent to which the world's fisheries have declined in the last 50 years. There is general agreement that the current fisheries management regime has failed to ensure sustainable fisheries, and that without changes to catch levels, exploitation rates, and management regimes, fish stocks will continue to decline beyond the level where recovery may be possible. The capacity and capability for distant-water fishing have increased dramatically as a result of the introduction of increasingly sophisticated vessels, navigation aides and gear for capturing fish, as well as satellite imagery and other devices for fish finding. And even if the number of vessels decreases, with technological advances the overall fishing pressure most likely will not decline and indeed might increase.[148] Furthermore, in the last decade it has also been generally agreed that fishing activities do impact the marine ecosystem, and that the extent of this impact has significantly increased in the last 50 years.[149]

Given the fragile state of much of world's marine fish stocks, and the technological capability for fish finding and fish harvesting, it is feasible that one vessel, in contravention of conservation measures, could single handedly cause serious or irreversible harm to a particular stock or to the regional marine ecosystem. This dystopia can be referred to as the "phenomenon of the single vessel". As a single oil spill from a highly regulated or wayward tanker can cause dramatic adverse consequences not only for the marine environment but also for coastal communities, so can a single fishing vessel. Though, unlike an oil spill, the devastation caused by IUU fishing is not so graphic, or patent. However, it is equally or more damaging to marine ecosystems and the livelihood of all fishing communities, whether distant-water fishing nations or coastal States, which are dependent on the sustainability and recovery of fish stocks. This spectre of the serious or irreversible harm caused by a single vessel is exacerbated by the systemic overcapacity in the world fishing fleet. In 1995 it was estimated that the maximum harvest levels that the ocean is capable of sustaining could be caught by 50% of the existing world fishing fleet.[150] In 1999 the world fishing fleet was thought to be two and a half times greater than it should be if fish stocks were to be sustained at a safe level.[151] According to FAO the overall number of fishing vessels worldwide has not changed significantly in recent years.[152] This excessive fishing pressure underscores the importance of effective implementation of UNFSA and reinforces the need for examining the efforts of coastal States and RFMOs to reduce fishing capacity, rebuild fish stocks, protect marine biodiversity, and rehabilitate living marine resources.[153]

5.6.2 Species at risk

In recent years, as living resources have been overexploited and habitats degraded or destroyed, there has been an increasing recognition of the impacts of human activity on biodiversity and ecosystems as well as the need for conserving and protecting "species at risk". While not well understood, the removal of any single species, whether prey or top predator, has an ecological impact in the marine ecosystem. Widespread depletion of top predators, together with the recent depletions of deepwater species, presents serious implications for the integrity of the marine ecosystem. The biomass of many key species under the competence of RFMOs is at a critical level. Many straddling fish stocks in the NAFO area, Bering Sea Convention Area, the high seas enclave of the Sea of Okhotsk, the Challenger Plateau, and the South Tasman Rise are under moratoria. Despite the cessation of targeted fishing of these stocks, in some cases for more than a decade (e.g., Central Bering Sea Alaska pollock, NAFO northern cod), the species were so overexploited that recovery is slow, much slower than estimated. Despite RFMO oversight, tuna and tuna-like fish stocks are being harvested beyond a sustainable level. With the fully exploited status of most stocks of tuna and tuna-like species and the overexploited status for certain stocks (e.g., bluefin, albacore, bigeye, South Atlantic swordfish, and other billfishes), it is feared that some of these fisheries are beyond overfished and have moved onto the trajectory of species at risk (Figure 5.16). Northern Atlantic and Southern Atlantic bluefin stocks are of particular concern.[154] Overfishing in this century has moved beyond an expression of overexploitation to one signifying widespread collapse[155] and now poses serious risks to the long-term viability of many fish stocks and associated ecosystems.

With fishing (both permitted and illegal) at nonsustainable levels coupled with bycatch of moratoria species, some commercial fish stocks face extirpation and possible extinction[156]. Such fishing activity may lead to irreversible perturbations to the marine ecosystem.[157] Given the aggregation pattern of pelagic species, it is theoretically possible, especially with modern technology, to extract the last school of a particular population of tuna or orange roughy or indeed an entire species. The biodiversity implications of overfishing are further complicated by the significant bycatch of associated and dependent species in many commercial fisheries. Thus, the objective of UNFSA to ensure the long-term sustainability of stocks goes beyond the objective to avoid overfishing of target species and the associated and dependent species, but also addresses the contemporary problem of ecosystem overfishing in an effort to protect biodiversity and the health of the marine ecosystem. Figure 5.16 illustrates where the area beyond overfished overlaps with the lexicon associated with species at risk to describe the status of fish stocks, i.e., vulnerable, threatened, endangered, extirpated, and extinct.[158]

Reduction of biodiversity *vis-à-vis* extirpations caused by intensive fishing, habitat destruction, and rapid environmental change has been well publicized.[159] There is the growing realization that biodiversity is important to ecosystems resilience.[160] There is also scientific consensus that the high fecundity of marine fishes is not sufficient to protect them against biological extinction. Certain life histories such as long life-span and late maturation, common in deepwater species, may result in a greater extinction risk and slower recovery. Biodiversity conservation and protection of fish species are now central preoccupations of many governments, scientists, and NGOs. There have been efforts to list certain fish species on CITES,[161] the IUCN Red List,[162] and national species at risk lists.[163] In 2000 FAO reviewed the life history characteristics of exploited aquatic organisms in relation to the risk that they may become extinct. The report provides an evaluation of the CITES criteria and its applicability to exploited marine species.[164]

Overfishing is no longer an issue solely within the domain of international fisheries organizations but is now a preoccupation of those involved with CITES as well as the Convention on Biological Diversity (CBD). This merging of fisheries issues and fisheries governance with biodiversity conservation and integrated ocean governance has forced a dialogue between formerly separate endeavours of examination. Traditional international law for fisheries was related to, but quite separate from, the international law pertaining to CITES[165] and protection of biodiversity. As a result of the developments in world fisheries, UNFSA incorporated provisions for biodiversity and habitat protection, bridging these two legal areas. Similarly, the more narrowly defined disciplines of fisheries biology and fisheries science have converged with a wide range of biological sciences, including conservation biology, restoration ecology, fish genetics, and the science underpinning species at risk criteria and

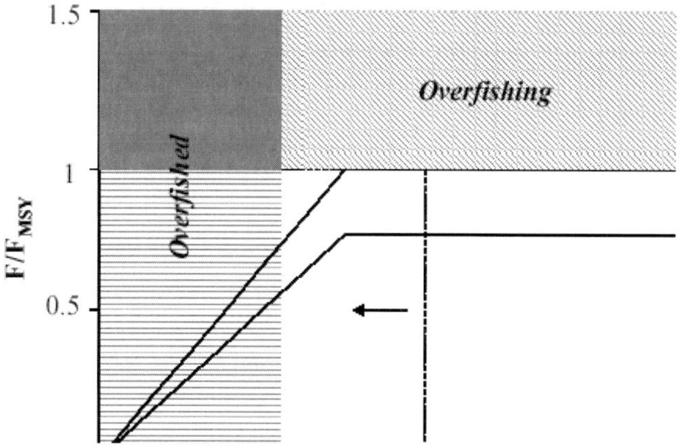

Figure 5.17. This figure provides an example of harvest control rules and illustrates the differences between overfished and overfishing and shows where the two overlap. Overfishing refers to the *action* and overfished refers to the *state* of the stock. The target is set below the fishing limit as a buffer to avoid overfishing. (From P. Mace, "A New Role for MSY in Single-Species and Ecosystem Approaches to Fisheries Stock Assessment and Management" (2001) 2 *Fish and Fisheries* 2–32, reproduced with permission of Blackwell Publishing Ltd.)

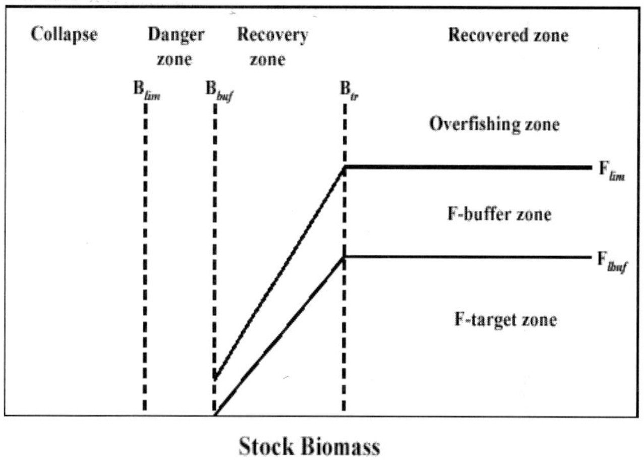

Figure 5.18. Harvest control rule schema as developed by NAFO, indicating biomass targets for recovery (B_{tr}), thresholds (B_{buf}), and limits (B_{lim}) and fishing mortality thresholds (F_{buf}) and limits (F_{lim}). (From P. Mace, "A New Role for MSY in Single-Stock and Ecosystem Approaches to Fisheries Stock Assessment and Management" (2001) 2 *Fish and Fisheries* 2–32, reproduced with permission of Blackwell Publishing Ltd.)

recovery plans. A broader application of fisheries science is required to better understand and examine overfishing, ecosystem impacts, and effects on fisheries. There is also a need for national and international agencies responsible for these different issue areas to collaborate to eliminate overfishing, restore fish stocks where possible, and protect biodiversity. FAO has advised CITES on revised criteria suitable to list fishery species, as the original criteria were considered ineffective if applied to fisheries.[166] UN agencies, in addition to FAO, are now liaising with RFMOs concerning management of selected target fish species.

5.6.3 RFMOs' response to setting nonsustainable TACs: harvest control or decision rules

RFMOs have begun to address overfishing that has resulted, in part, from setting nonsustainable TACs, through the introduction of harvest control rules or decision rules. These rules specify preagreed course of management actions, depending on the fishing mortality and stock biomass levels (e.g., Central Bering Sea Donut Hole Convention). NAFO has developed such rules during meetings of the Ad Hoc Working Group on the Precautionary Approach, which was formed in 1997, and the Joint Scientific Council and Fisheries Commission Working Group on Precautionary Approach (see Figures 5.17 and 5.18).[167] CCAMLR also uses decision rules in the management of its krill fishery. The development of limits, buffers, and targets for stock biomass and fishing mortality will help to avoid overfishing, or at least help RFMOs understand where on the trajectory of stock status the population of a particular species lies.

The challenge both for managers and stakeholders is how to operationalize harvest control rules. Within these management regimes, it must be clear what management actions will be taken once these targets or references points are reached.[168] For example, in some organizations a fishery is closed once a target is reached. In others, the fishing season may be limited, or where the target has been exceeded, quota reductions are implemented in the following years. For harvest control rules to be successful, particularly within a management evaluation framework, both managers and stakeholders must be committed to transparency, accountability, and precaution.

5.7 Conclusion

The regulation of high seas fisheries to avoid overfishing and rebuild depleted stocks remains the central preoccupation of contemporary international fisheries law and regional management regimes. The problem of overfishing is thus complex and multifaceted. Overcapacity; nonprecautionary management decisions; opting out of regional conservation schemes; noncompliance with agreed management measures; IUU fishing, both within and beyond EEZs; destructive and other nonsustainable fishing practices; sophisticated fish-finding, fish-attracting, and fish-harvesting technology; and modern vessels have all contributed to serial depletion of commercial species and many associated and dependent species. Overexploitation of fish stocks has moved beyond commercial depletion and "dwindling fish stocks"[169] to the point of some fish stocks being classified as species at risk (endangered or extirpated). This alarming and previously unthinkable development has resulted in focussing the concerted attention of international institutions in addition to FAO to address overfishing. The CBD, the International Labour Organization, the International Maritime Organization, UNEP, UNGA, and WTO all are examining modalities to further strengthen the conservation and management of fisheries since the adoption of UNFSA in 1995. Another high-profile initiative to combat overfishing is the High Seas Task Force. This Task Force is endeavouring to coordinate a global restructuring of systemic problems to render IUU fishing uneconomic and unacceptable.

High seas fisheries have expanded, particularly in the last three decades, as many coastal stocks collapsed and as demand for fish products, both for food and for fish meal, increased. There continues to be some uncertainty regarding the rate of fleet expansion, both on the high seas and within national jurisdictions, since the establishment of the EEZ regime and the reconfiguration of the high seas' area.[169] Nonetheless, it is clear that current levels of harvesting capacity are substantially higher than that needed to harvest the world's resources at sustainable levels. The state of world fisheries provides the biological context within which much of the existing international legal framework has developed. UNFSA clearly requires the avoidance of overfishing and the restoration of fish stocks to levels that are considered sustainable as well as the reduction of capacity commensurate with the available resource.

In addition to reformulating MSY from a target to a limit, UNFSA introduced two major procedural innovations to prevent overfishing: the precautionary approach to fisheries as a risk-averse strategy given the inherent uncertainty and risk associated with overfishing; and the ecosystem approach to fisheries, invoking the need for taking into account the marine ecosystem beyond associated and dependent species, in making conservation and management decisions. These two fundamental principles and their application by RFMOs will be discussed in the next chapters.

Endnotes

[1] S. Garcia and D. Doulman, "FAO's Fisheries Programme and the Plan of Implementation of the World Summit on Sustainable Development" in S. Ebbin, A. Håkon Hoel, and A. Sydnes, eds., *A Sea Change: The Exclusive Economic Zone and Governance Institutions for Living Marine Resources* (Dordrecht: Springer, 2005) 169–193 at 186.

[2] *United Nations Convention on the Law of the Sea*, done at Montego Bay, Jamaica, 10 December 1982, entered into force 16 November 1994, 33 ILM 1309. Referred to as LOSC.

[3] M. Lodge and S. Nandan, "Some Suggestions Towards Better Implementation of the United Nations Agreement on Straddling Fish Stocks and Highly Migratory Fish Stocks of 1995" (2005) 20 *International Journal of Marine and Coastal Law* 345–379 at 346–347.

[4] As J. Van Dyke states, the goal of UNFSA "was to stop the dramatic overfishing that has decimated the fish stocks in many parts of the world", "The Evolution and International Acceptance of the Precautionary Principle" in D. Caron and H. Scheiber, eds., *Bringing New Law to Ocean Waters* (Leiden: Brill, 2004) 357–379 at 366. He cites that among the stocks that are seriously depleted are New Zealand orange roughy, bluefin tuna, rockfish, and herring.

[5] *Agreement for the Implementation of the Provisions of the United Nations Convention on the Law of the Sea of 10 December 1982 Relating to the Conservation and Management of Straddling Fish Stocks and Highly Migratory Fish Stocks*, done at New York, 4 December 1995, entered into force 11 December 2001, 2167 UNTS 88; (1995) 34 ILM 1542. Referred to as UNFSA.

[6] Extinction in marine systems has only just begun to be investigated on a broad scale by the scientific community. For discussion and examples of marine extinctions, see N. Dulvy, Y. Sadovy, and J. Reynolds, "Extinction Vulnerability in Marine Populations" (2003) 4 *Fish and Fisheries* 25–64; J. Reynolds, N. Dulvy, N. Goodwin, and J. Hutchings, "Biology of Extinction Risk in Marine Fishes" (2005) 272 *Philosophical Transactions of the Royal Society B* 2337–2344. Known marine extinctions to date can be viewed at <www.uea.ac.uk/bio/people/faculty/reynoldslab/extinction_tableB.htm> (accessed May 2005).

[7] K. Frank, B. Petrie, J. Choi, and W. Leggett, "Trophic Cascades in a Formerly Cod-Dominated Ecosystem" (2005) 308 *Science* 1621–1623. See also Committee on Ecosystem Effects of Fishing: Phase II — Assessments of the Extent of Change and the Implications for Policy, National Research Council, *Dynamic Changes in Marine Ecosystems: Fishing, Food Webs, and Future Options* (Washington, D.C.: National Academies Press, 2006).

[8] J. Jackson, M. Kirby, W. Berger, K. Bjorndal, L. Botsford, B. Bourque, R. Bradbury, R. Cooke, J. Erlandson, J. Estes, T. Hughes, S. Kidwell, C. Lange, H. Lenihan, J. Pandolfi, C. Peterson, R. Steneck, M. Tegner, and R. Warner "Historical Overfishing and the Recent Collapse of Coastal Ecosystems" (2001) 293 *Science* 629–637.

[9] See overview by E. Meltzer, "Global Overview of Straddling and Highly Migratory Fish Stocks: The Nonsustainable Nature of High Seas Fisheries" (1994) 25 *Ocean Development and International Law* 255–344, and FAO, "World Review of Highly Migratory Species and Straddling Fish Stocks", Fisheries Technical Paper No. 337 (Rome: FAO, 1994).

[10] FAO, "Review of the State of World Marine Fishery Resources", FAO Fisheries Technical Paper No. 457 (Rome: FAO, 2005).

[11] FAO Fisheries Technical Paper No. 457 at 3, note 4.

[12] S. Garcia and J. Grainger, "Gloom and Doom? The Future of Marine Capture Fisheries" (2005) 360 *Philosophical Transactions of the Royal Society B* 21–46 at 33.

[13] R. Myers and B. Worm, "Rapid Worldwide Depletion of Predatory Fish Communities" (2003) 423 *Nature* 280–283. Myers and Worm estimate that predatory fish have declined by 90% of preexploitation levels. D. Pauly and J. Maclean, *In a Perfect Ocean* (Washington, D.C.: Island Press, 2002).

[14] FAO Fisheries Glossary. FAO online <http://www.fao.org/fi/glossary/default.asp> (accessed 18 October 2005).

[15] M. Mangel and P. Levin, "Regime, Phase and Paradigm Shifts: Making Community Ecology the Basic Science for Fisheries" (2005) 360 *Philosophical Transactions of the Royal Society B* 95–105 at 95, claim stock assessment began with J. Beverton and S. Holt in 1957. Recorded interest in quantitative fishery population dynamics is thought to have begun with the 1916 publication of William Francis Thompson's work on the population dynamics of Pacific halibut for the province of British Columbia (see W.F. Thompson, "Statistics of the Halibut Fishery in the Pacific". Report of the British Columbia Commissioner of Fisheries for 1915 (Victoria, B.C.: Government of British Columbia, 1916) 65–126. The Russian scientist F.I. Baronov was independently working on estimating fishing mortality, and his 1918 landmark paper developed the principles of quantitative fisheries science and began the process of fishery science becoming recognized as a separate part of aquatic sciences. (See F.I. Baranov, "On the Question of the Biological Basis of Fisheries (1918) 1 *Nauchnoe Issledodovatelski Ikhtiologicheskii Institut Izvestiya* 81–128 [in Russian]). Thompson and Bell are considered to have published the first North American application of population dynamics in 1934. (see W.F. Thompson and F.H. Bell, "Biological Statistics of the Pacific Halibut Fishery (2) Effect of changes in Intensity upon Total Yield and Yield Per Unit of Gear". Report of International Fisheries Commission, 8). See discussion by J.L. McHugh, "Trends in Fishery Research" in N.G. Benson, ed., *A Century of Fisheries in North America*, American Fisheries Society Special Publication No. 7 (Washington, D.C.: American Fisheries Society 1970) 25–56.

[16] M. Mangel and P. Levin, at 95. Estimating fish population was done by examining fish biology and reproduction to determine an acceptable level of fish harvesting for a single stock or species.

[17] The bases of all stock assessments are models that use both available data and collective knowledge about fish, fisheries, and environment to assess populations. Models are merely simplifications of reality, and the more complex a model, the greater demand there is for data. Inevitably, assumptions must be made about the data to use the models, and it is therefore part of the stock assessment process to evaluate how robust a particular model is to misspecifications of its basic assumptions. A number of methods and applications are used to assess stocks: general fishery indicators, indices of relative abundance, production models, sequential population analyses, and yield-per-recruit and projections. Scientists conduct quantitative assessments to help RFMO decision makers make management decisions, using a variety of tools (models) and the best scientific knowledge (research) and data sets (statistics and surveys) available.

[18] "Fishing mortality estimates and SRRs [spawner recruit relationship] come from retrospective analysis of catches and vessel survey using best estimates of age and size and composition, analysed retrospectively, using constant parameter values of natural mortality rate". J. Caddy and J. Seijo, "This Is More Difficult Than We Thought! The Responsibility of Scientists, Managers and Stakeholders To Mitigate the Unsustainability of Marine Fisheries" (2005) 360 *Philosophical Transactions of the Royal Society B* 59–75.

[19] "What is at issue is our understanding of the biological processes at the population and ecosystem level and how these affect stocks. Ecosystem processes about which we have very limited understanding are affecting many stocks at historically low levels. This is exemplified by the lack of recovery of many of the Northwest Atlantic fish stocks…they are likely in predator traps from which they may not be able to recover without broader ecosystem change", Robert O'Boyle, Canadian Department of Fisheries and Oceans, Maritimes Region, personal communication, 6 July 2007.

[20] For an outline of the theory behind management strategy evaluation (MSE) see A.D.M. Smith, *Management Strategy Evaluation — The Light on the Hill* (North Beach, W.A.: Australian Society for Fish Biology, 1994) at 249–253. More recently, MSE has been used in Australian fisheries management, see *Stock Assessment and Management Strategy Evaluation for Sub-Antarctic Fisheries 2002–2004*. Available at <http://www.afma.gov.au/research/reports/2005/r2001_1291.pdf> (accessed 25 July 2007). The Australian approach is also detailed by A.D.M. Smith, K.J. Sainsbury, and R.A. Stevens, "Implementing Effective Fisheries Management Systems — Management Strategy Evaluation and the Australian Partnership Approach" (1999) 56 *ICES Journal of Marine Science* 967–979. In the United States, the National Marine Fisheries Service (NMFS) incorporated MSE into its evaluation of stock assessments in 2001. NMFS 2001, *Stock Assessment Improvement Plan*, available at <http://www.st.nmfs.gov/StockAssessment/tm_spo56/saipafrontmatter.pdf> (accessed 25 July 2007).

[21] United Nations 2005, *Sustainable Fisheries, including through the 1995 Agreement for the Implementation of the Provisions of the United Nations Convention on the Law of the Sea of 10 December 1982 relating to the Conservation and Management of Straddling Fish Stocks and Highly Migratory Fish Stocks, and related instruments*, available at <http://daccessdds.un.org/doc/UNDOC/GEN/N03/453/75/PDF/N0345375.pdf?OpenElement> (accessed 1 October 2005).

[22] Victor Restrepo, personal communication, 30 March 2006.

[23] See E. Meltzer, at 261–262.

[24] FAO Fisheries Glossary. FAO online <http://www.fao.org/fi/glossary/default.asp> (accessed 18 October 2005).

[25] FAO Fisheries Glossary. FAO online <http://www.fao.org/fi/glossary/default.asp> (accessed 18 October 2005).

[26] FAO Fisheries Glossary. FAO online <http://www.fao.org/fi/glossary/default.asp> (accessed 18 October 2005).

[27] FAO Fisheries Glossary. FAO online <http://www.fao.org/fi/glossary/default.asp> (accessed 18 October 2005).

[28] UNEP, *Ecosystem-based Management of Fisheries — Opportunities and Challenges for Coordination between Marine Regional Fishery Bodies and Regional Seas Conventions*, Regional Seas Reports and Studies No. 175 (Nairobi: UNEP, 2001).

[29] Rent in a fisheries context is defined by FAO as the difference between the total revenues obtained from the fishery resource and the total costs of production, i.e., capital and labour valued at their opportunity costs. The total costs of production include a reasonable profit, and the rent is often considered a "surplus profit", over and above what would be considered a "normal" rate of return. For this reason, the decision as to who receives the rent (e.g., the society, the management authority, or the fishermen) remains a key policy issue.

[30] FAO Fisheries Glossary. FAO online <http://www.fao.org/fi/glossary/default.asp> (accessed 18 October 2005).

[31] The basis for the concept of MSY was developed in the 1930s: see review by T. Smith, *Scaling Fisheries* (Cambridge: Cambridge University Press, 1994). See also M. Shaeffer, "Some Aspects of the Dynamics of Populations Important to the Management of the Commercial Marine Fisheries" (1954) 1 *Bulletin of Inter-American Tropical Tuna Commission* 25–56.

[32] See also P. Larkin, "An Epitaph for the Concept of MSY" (1977) 106(1) *Transactions of the American Fisheries Society* 1–11; W. Gabriel and P. Mace, "A Review of Biological Reference Points in the Context of the Precautionary Approach" Fisheries" in V.R. Restrepo, ed., *Proceedings of the 5th National NMFS Stock Assessment Workshop: Providing Scientific Advice to Implement the Precautionary Approach Under the Magnuson–Stevens Fishery Conservation and Management Act*, U.S. Dep. Commer. NOAA Technical Memorandum NMFS-F/SPO-40 (Silver Springs, Md.: US Department of Commerce, NOAA, 1999) 34–45; P. Mace, "A New Role for MSY in Single-Stock and Ecosystem Approaches to Fisheries Stock Assessment and Management" (2001) 2 *Fish and Fisheries* 2–32.

[33] Most international stock assessments for straddling fish stocks and highly migratory fish stocks are linked to MSY. Many RFMOs are moving towards including additional precautionary reference points and harvest control rules.

[34] ICNAF was established in 1950 as an intergovernmental organization charged with coordinating the management of key fisheries for straddling fish stocks in the Northwest Atlantic. The organization was superceded by the Northwest Atlantic Fisheries Organisation (NAFO) in 1979 following the extension of national fisheries jurisdiction out to 200 nautical miles, pursuant to LOSC.

[35] A. Punt and A. Smith, "The Gospel of Maximum Sustainable Yield in Fisheries Management: Birth, Crucifixion and Reincarnation", Chap. 3, in J. Reynolds, G. Mace, K. Redfnod, and J. Robinson, eds., *Conservation of Exploited Species* (Cambridge: Cambridge University Press, 2001) 41–66 at 45.

[36] FAO, "International Fisheries Bodies", FAO Fisheries Technical Paper No. 64 (Rome: FAO, 1966).

[37] A. Punt and A. Smith, at 45.

[38] The definition of MSY and question of what is a permissible harvest level pursuant to Article 61 of LOSC (beyond which would be deemed overfishing) have been the subject of much dispute and in some cases litigation (as in *New Zealand Fishing Industry Association et al.* v. *Minister of Fisheries et al.*) New Zealand Federation of Commercial Fishermen (Inc.), New Zealand Fishing Industry Association (Inc.), Simunovich Fisheries Limited, North Harbour Nominees Limited and Moana Pacific Fisheries, Area 1 Maori Fishing Consortium and Ngapuhi Fisheries Limited, Hauraki Maori Trust Board, and Paepae/Taumata 2 v. Minister of Fisheries, Chief Executive of the Ministry of Fisheries, New Zealand Recreational Fishing Council Inc. Consolidated CP237/95. The opinions in this case have not been reported, but are available through the *Maori Law Review* on the Internet. For a discussion of this case, see W. Burke, "Evolution in the Fisheries Provisions of UNCLOS" in N. Ando, E. McWhinney, R. Wolfrum, and B. Röben, eds., *Liber Amicorum Judge Shigeru Oda* (The Hague: Kluwer Law International, 2002) 1355–1362.

[39] A. Punt and A. Smith, at 45. For a discussion on MSY as a limit, see J. Caddy and R. McGarvery, "Targets or Limits for the Management of Fisheries?" (1996) 16 *North American Journal of Fisheries Management* 479–487.

[40] S. Garcia and D. Staples, "Sustainability Reference Systems and Indicators for Responsible Marine Capture Fisheries: A Review of Concepts and Elements for a Set of Guidelines" (2000) 51(5) *Marine and Freshwater Research* 385–426 at 410. Garcia further indicates (personal communication, January 2006) that MSY as a biological concept could be deployed as a F_{MSY} target reference point or a F_{lim} limit reference point; however, UNFSA provides that MSY is clearly not acceptable as a management target for sustainable fisheries. UNFSA makes it clear that MSY is heretofore to be considered a limit reference point.

[41] World Summit on Sustainable Development, *Johannesburg Plan of Implementation* (2002). United Nations Division for Sustainable Development online <http://www.un.org/esa/sustdev/documents/WSSD_POI_PD/English/WSSD_PlanImpl.pdf> (accessed 18 October 2005).

[42] See S. Garcia, "The Precautionary Approach to Fisheries: Progress Review and Main Issues (1995–2000)" in M. Nordquist and J. Norton Moore, eds., *Current Fisheries Issues and the Food and Agriculture Organization of the United Nations* (The Hague: Kluwer Law International, 2000) 470–560 at 480.

[43] FAO describes MSY as "the highest theoretical equilibrium that can be continuously taken (on average) from a stock under existing (average) environmental conditions without significantly affecting reproductive success". MSY as a target reference point for commercial species can be calculated using various models including Shaefer's surplus production model (which can include density dependence) and stock-recruitment and yield-per-recruit models.

[44] MSY identifies the population size at which population production is highest, such that production in excess of natural mortality can be harvested in a sustainable manner.

[45] MSY is sometimes characterized as having failed; however, many fisheries scientists argue that the failure of MSY-based management is not a failure of the MSY concept but is the result of an inappropriate application of surplus production model, or unsuccessful implementation. See also P. Larkin 1977.

[46] W. Gabriel and P. Mace, at 34–45, summarize the different conceptual models for fish stock assessment developed to accommodate various life histories and availability of catch data; relative abundance data; stock-recruitment data; and age-specific mortality, growth, and maturity data. A fundamental issue in fish stock models and the reference points generated is the inherent uncertainty and associated risk. Harwood and Stokes identify four sources of uncertainty: (1) process stochasticity, (2) observation error, (3) model error, and (4) implementation error. They advocate the use of computer simulations and scenario planning to assess the most robust model. J. Harwood and K. Stokes, "Coping with Uncertainty in Ecological Advice: Lessons from Fisheries" (2003) 18 *Trends in Ecology and Evolution* 617–622. But as S. Garcia, A. Zerbi, C. Aliaume, T. Do Chi, and G. Lasserre, "The Ecosystem Approach to Fisheries. Issues, Terminology, Principles, Institutional Foundations, Implementation and Outlook", FAO Fisheries Technical Paper No. 443 (Rome: FAO, 2003), highlight, risk assessment can only assess the known sources of uncertainty and therefore produces conservative estimates of uncertainty. Further, as succinctly summarized by S. Garcia, "Fishery Science and Decision-Making: Dire Straights to Sustainability" (2005) 76 *Bulletin of Marine Science* 171–196, "scientific certainty is not sufficient condition for sustainability, as the best scientific advice is often badly implemented or simply disregarded".

[47] S. Garcia and D. Staples, "Sustainability Reference Systems and Indicators for Responsible Marine Capture Fisheries: A Review of Concepts and Elements for a Set of Guidelines" at 410. Some fisheries stock assessment scientists observe that the failure of MSY-based management throughout the world is due, in part, to the inappropriate application or unsuccessful implementation of the surplus production model (i.e., inadequate, incomplete, and inaccurate catch data reporting and collection and (or) failure of managers to apply scientific advice) (Bob O'Boyle, Department of Fisheries and Oceans, personal communication, March 2005). Arguably, in an ideal world where reliable, complete, and timely data were available, scientists could fairly estimate MSY (Dr. Julie Porter, Department of Fisheries and Oceans, personal communication, August 2005).

[48] FAO, "Reference Points for Fisheries Management: Their Potential Application to Straddling Stocks and Highly Migratory Resources", FAO Fisheries Circular No. 864 (Rome: FAO, 1993) at para. 104.

[49] W. Burke, "Compatibility and Precaution in the 1995 Straddling Stock Agreement" in H. Scheiber, ed., *The Law of the Sea: The Common Heritage and Emerging Challenges* (Leiden: Martinus Nijhoff Publishers, 2000) 105–126 at 112.

[50] For an overview of the problems associated with using F_{MSY}, see A.M. Parma, R. Hilborn, and J.M. Orensanz, "The Good, the Bad, and the Ugly: Learning from Experience To Achieve Sustainable Fisheries" (2006) 78(3) *Bulletin of Marine Science* 411–427.

[51] S. Garcia and D. Staples, "Sustainability Reference Systems and Indicators for Responsible Marine Capture Fisheries: A Review of Concepts and Elements for a Set of Guidelines" at 409.

[52] UNFSA, Annex II, paras. 2 and 5.

[53] FAO, "Fisheries Management", FAO Technical Guidelines for Responsible Fisheries No. 4 (Rome: FAO, 1997) at 13 states "This reflects current thinking that providing a margin for safety that takes into account normal variability and uncertainty requires using MSY as a limit for fisheries management, rather than as a target".

[54] See S. Garcia, "The Precautionary Approach to Fisheries: Progress Review and Main Issues (1995–2000)" at 480.

[55] S. Garcia, "The Precautionary Approach to Fisheries: Progress Review and Main Issues (1995–2000)" at 482.

[56] FAO Fisheries Circular No. 864 (Rome: FAO, 1993) para. 104. See also discussion by W. Burke, "Compatibility and Precaution in the 1995 Straddling Stock Agreement".

[57] See S. Garcia, "The Precautionary Approach to Fisheries: Progress Review and Main Issues (1995–2000)" at 410.

[58] S. Garcia and D. Staples, "Sustainability Reference Systems and Indicators for Responsible Marine Capture Fisheries: A Review of Concepts and Elements for a Set of Guidelines" at 410, recognize that the notion of MSY has shifted from that of a target reference point to limit reference points according to UNFSA.

[59] S. Garcia, A. Zerbi, C. Aliaume, T. Do Chi, and G. Lasserre, "The Ecosystem Approach to Fisheries. Issues, Terminology, Principles, Institutional Foundations, Implementation and Outlook" at 23.

[60] S. Garcia, A. Zerbi, C. Aliaume, T. Do Chi, and G. Lasserre, "The Ecosystem Approach to Fisheries. Issues, Terminology, Principles, Institutional Foundations, Implementation and Outlook" at 23.

[61] S. Garcia, personal communication, March 2005.

[62] For example, the 2005 recommendation for management of yellowtail flounder stocks in the NAFO Regulatory Area states "Scientific Council recommended that total catches should not exceed 15 000 tons in 2005 and 2006. This corresponds to catch projections based on $F = 2/3\ F_{msy}$ and an assumed catch of 14 500 tons (= TAC) in the year 2004. Scientific Council noted that catches exceeded TACs in 1998–2001, but were lower than the TACs in 2002 and 2003. Scientific Council again notes that the advice applies to all removals (directed plus by-catch)". NAFO online <www.nafo.ca> (accessed 18 October 2005).

[63] IATTC Convention, at Art. II(5).

[64] Art. VII(c) of the Antigua Convention states that the Commission "adopt measures that are based on the best scientific evidence available to ensure the long-term conservation and sustainable use of the fish stocks covered by this Convention and to maintain or restore the populations of harvested species at levels of abundance which can produce the maximum sustainable yield, *inter alia*, through the setting of the total allowable catch of such fish stocks as the Commission may decide and (or) the total allowable level of fishing capacity and (or) level of fishing effort for the Convention Area as a whole".

[65] The ICCAT Convention, in Art. VIII(a), states: "The Commission may, on the basis of scientific evidence, make recommendations designed to maintain the populations of tuna and tuna-like fishes that may be taken in the Convention area at levels which will permit the maximum sustainable catch …".

[66] In the early years of CCAMLR, the derivative methods from the MSY approach were applied. Although MSY is no longer the specific goal of stock assessments in CCAMLR, the models being developed in CCAMLR rely on similar kinds of information, including knowledge of the dynamics of populations (birth (recruitment) and death (mortality) rates); productivity of individuals, including growth and reproduction; and the interaction between the fishery and the population, i.e., the ages or sizes of fish being exploited. See A. Constable, "The Status of Antarctic Fisheries Research", Institute of Antarctic and Southern Ocean Studies VIAS online <http://www.vias.studies.aq/publications/40th/Constable.pdf> (accessed 14 August 2005).

[67] See June 2004 report on Limit Reference Points Workshop, also Scientific Council advice and Fisheries Council adoption of management measures and recommendations for 2005. Available at <www.nafo.int> (accessed January 2006).

[68] Art. 4(1) states, "The Commission shall perform its functions in the interests of the conservation and optimum utilisation of the fishery resources of the Convention Area and shall take into account the best scientific evidence available to it". *Convention on Future Multilateral Cooperation in the Northeast Atlantic Fisheries*, done at London, UK, 18 November 1980, 1285 UNTS 129.

[69] See NEAFC, *Report of the 23rd Annual Meeting of the North-East Atlantic Fisheries Commission, 8–12 November 2004* (London: NEAFC, 2005) at 4. NEAFC online <http://www.neafc.org/reports/annual-meeting/docs/23neafc_annualmeeting_report_2004.pdf> (accessed 15 July 2005).

[70] For example, the scientific advice provided to NAFO by the Scientific Council is generated through a joint effort by NAFO members and makes use of different data sampling programs carried out by members States as well as available statistics on the resources and their environment. For more detail see <http://nafo.int/science/frames/science.html> (accessed 25 July 2007).

[71] Art. 5(b) states that the Commission is to "ensure that such measures are based on the best scientific evidence available and are designed to maintain or restore stocks at levels capable of producing maximum sustainable yield, as qualified by relevant environmental and economic factors, including the special requirements of developing States in the Convention Area, particularly small island developing States, and taking into account fishing patterns, the interdependence of stocks and any generally recommended international minimum standards, whether subregional, regional or global". *Convention on the Conservation and Management of Highly Migratory Fish Stocks in the Western and Central Pacific Ocean,* done at Honolulu, Hawaii, 5 September 2000, 2004 ATS 15.

[72] Southern Indian Ocean Fisheries Agreement, Art. 4(d).

[73] According to an article by Paul Greenberg, "The Catch", *New York Times Magazine*, 23 October 2005, in some fisheries for Chilean seabass "the pirate haul may be four times the legal catch". Online <http://www.nytimes.com/2005/10/23/magazine/23bass.html> (accessed 30 May 2006).

[74] ICES Working Group on the Biology and Assessment of Deep-Sea Fisheries Resources (WGDEEP) has the responsibility of conducting stock assessments for deep-sea fisheries. Online <http://www.ices.dk/marineworld/deepseafish.asp> (accessed January 2006). According to ICES 2004, the population levels of many deep-sea species are beyond safe biological limits.

[75] The Canadian Department of Fisheries and Oceans manages cod in the other NAFO areas, outside of the NAFO Regulatory Area, which include 2J3KL, 3Ps, 3Pn, 4RS, 4T, 4Vsw, 4X, and 5Zjm.

[76] NAFO 2005 cod stock assessment. Available at <www.nafo.int/fisheries/frames/fishery.html> (accessed 3 October 2005).

[77] NAFO STACFIS 2003 catch estimate for cod. Available at <www.nafo.int/fisheries/frames/fishery.html> (accessed 3 October 2005).

[78] NAFO 2005 shrimp stock assessment. Available at <http://www.nafo.ca/fisheries/fishery/species/shrimp.html> (accessed 3 October 2005).

[79] NAFO 2005, "NAFO Conservation and Enforcement Measures. Chap. 1 — Conservation and Enforcement Measures. Article 7 — Greenland Halibut in Subareas 2 and Divisions 3KLMNO", NAFO FC Doc. 07/1, Serial No. N5335. Available at <http://www.nafo.int/fisheries/frames/regulations.html> (accessed 27 May 2007).

[80] NAFO 2005 Greenland halibut stock assessment. Available at <http://www.nafo.int/fisheries/frames/fishery.html> (accessed 3 October 2005).

[81] ICES background information on blue whiting. Available at <www.ices.dk/marineworld/bw2.asp> (accessed 8 October 2005).

[82] According to FAO 2001 "Shoals large enough to give a catching rate of 5–10 tonnes an hour or higher are found at the southern end of the Faroe–Shetland channel from mid February and extend over a wide area to the west of Scotland and Ireland until the fish move northward and disperse again by the end of May. Blue whiting are caught mainly by large midwater trawl, towed at a depth of 200–500 m, either

in very deep water or close to the sea bed when the fish layer is near to the edge of the shelf. Trawlers over 40 m long and at least 1200 hp are required to give the necessary maneuverability and power…". Torry Research Station, "Handling and Processing Blue Whiting", Torry Advisory Note No. 81 (Edinburgh: HMSO Press, 1980). Available at <http://www.fao.org/wairdocs/tan/x5952e/x5952e00.htm> (accessed April 2006).

[83] Torry Advisory Note No. 81 (Edinburgh: HMSO Press, 1980). Available at <http://www.fao.org/wairdocs/tan/x5952e/x5952e00.htm> (accessed April 2006).

[84] "Blue Whiting — The Fish Stock and Its Utilization". Available at <http://www.srmjol.is/srmjol/content/view/30/51/> (accessed 8 October 2005).

[85] ICES 2004. Advice to NEAFC on blue whiting. Available at <http://www.neafc.org/fisheries/docs/acfm_reports-2004/blue-whiting_2004.pdf> (accessed 3 October 2005).

[86] See "Report of the Nordic Blue Whiting Network Meeting Bergen, 24.-26.3.2003" for overview of acoustic surveys by NEAFC member States.

[87] ICES 2004. Advice to NEAFC on blue whiting. Available at <http://ww.neafc.org/fisheries/docs/acfm_reports-2004/blue-whiting_2004.pdf> at 345 (accessed 3 October 2005).

[88] Patagonian Toothfish has been referred to as "white gold" by the fishing industry. ISOFISH, *The Chilean Fishing Industry: Its Involvement in and Connections to the Illegal Unreported and Unregulated Exploitation of Patagonian Toothfish in the Southern Ocean.* ISOFISH Occasional Report No. 2 (Hobart: International Southern Oceans Longline Fisheries Information Clearing House (ISOFISH), 1999). As noted by P. Greenberg, "The Catch", *New York Times Magazine*, 23 October 2005, in 2005 Chilean seabass sold for as much as US$20 per pound in the United States — "nearly five times what it cost when it first appeared in US markets…A whole animal may go for more than $1000". The fish was declared the "dish of the year" by Bon Appetit magazine in 2001 much to the dismay of some environmental groups. See <http://72.14.205.104/search?q=cache:e3muqVYdCaIJ:www.ecofish.com/pressroom/news_DenverPost2002.htm+Bon+Appetit+%22dish+of+the+year%22+2001&hl=en&ct=clnk&cd=10&gl=ca> (accessed 30 May 2006). See also National Environmental Trust <http://www.net.org/proactive/newsroom/release.vtml?id=27749> (accessed 30 May 2006).

[89] CCAMLR XVI. See, in particular, Items 4 and 5. The IUU fishery affects not only target species, but also has an effect on bycatch of seabirds, which is currently dealt with through a series of Conservation Measures under CCAMLR.

[90] See discussion in P. Greenberg, "The Catch", *New York Times Magazine*, 23 October 2005. See also J. Jacquet, "The Toothfish that Bit Al Gore" *The Tyee*, 20 July 2007, at <http://72.14.205.104/search?q=cache:YnABkrANclYJ:thetyee.ca/Views/2007/07/20/Toothfish/+Bon+Appetit+%22dish+of+the+year%22+2001&hl=en&ct=clnk&cd=6&gl=ca> (accessed July 2007).

[91] It was then that dolphin mortality became a bycatch issue.

[92] For a discussion on the rise of IUU fishing and measures for addressing this problem, see High Seas Task Force 2006, "Closing the Net: Stopping Illegal Fishing on the High Seas", Governments of Australia, Canada, Chile, Namibia, New Zealand, and the United Kingdom, WWF, IUCN, and the Earth Institute at Columbia University.

[93] M.P. Miyake, N. Miyabe, and H. Nakano, "Historical Trends of Tuna Catches in the World", FAO Fisheries Technical Paper No. 467 (Rome: FAO, 2004). This report offers a detailed history of the tuna fishery and expansion to new areas as well as new species.

[94] See A. Caton and K. McLoughlin (eds.), *Fishery Status Reports 2004* (Canberra: Bureau of Rural Sciences, 2004).

[95] S. Harley and M. Maunder 2003, "Status of Bigeye Tuna in the Eastern Pacific Ocean in 2003 and Outlook for 2004". Available at <http://www.iattc.org/PDFFiles2/SAR5%20_BET_%20ENG.pdf> (accessed 10 October 2005).

[96] FAO Fisheries Technical Paper No. 457.

[97] See K.T. Frank, B. Petrie, and N.L. Schackell, "The Ups and Downs of Trophic Control in Continental Shelf Ecosystems" (2007) 22 *Trends in Ecology and Evolution* 236–242.

[98] See 2004 NAFO Scientific Council Report, available at <http://www.nafo.int/publications/frames/general.html> (accessed 27 May 2007), and related press release from NAFO 2004 General Council Meeting, available at <http://www.nafo.int/about/media/press/press04.html> (accessed 27 May 2007).

[99] See 2003 NAFO Annual Meeting of the General Council and Fisheries Commission, Report of the Fisheries Commission, Annex 15. Available at <http://www.nafo.int/publications/frames/general.html> (accessed 27 May 2007).

[100] NAFO 2005, "NAFO Conservation and Enforcement Measures. Chap. 1 — Conservation and Enforcement Measures. Article 7 — Greenland Halibut in Subareas 2 and Divisions 3KLMNO", NAFO FC Doc. 07/1, Serial No. N5335. Available at <http://www.nafo.int/fisheries/frames/regulations.html> (accessed 27 May 2007).

[101] ICCAT, *Report of the 2006 Atlantic Swordfish Stock Assessment Session*, Madrid, Spain, 4–6 September 2006. Available at <http://www.iccat.int/Documents/SCRS/DetRep/DET_swo.pdf> (accessed 27 May 2007).

[102] ICES 2004 advice to NEAFC on Atlanto-Scandian (Norwegian) spring-spawning herring. Available at <http://www.neafc.org/fisheries/docs/acfm_reports-2004/norwegian_ss-herring.pdf> (accessed 4 October 2005).

[103] See, for example, UNGA Sustainable Fisheries Resolution 51/25.

[104] See, for example, Deep Sea Conservation Coalition 2004, "A Net with Holes: The Regional Fisheries Management System" Deep Sea Conservation Coalition Policy Paper. Greenpeace Australia Pacific, "Legally binding UN fisheries agreement could be used to protect world's oceans", Media Release, 11 December 2001, available at <http://www.greenpeace.org/australia/news-and-events/media/releases/overfishing/legally-binding-un-fisheries-a> (accessed 11 December 2001); Greenpeace, "Saving Our Seas From Piracy: A New Era in Fisheries Management", available at <http://weblog.greenpeace.org/pacific/background/new_era.html> (accessed 15 October 2006); Greenpeace, "Too Many Ships" (accessed 8 March 2005); L.A. Kimball, *International Ocean Governance: Using International Law and Organizations to Manage Marine Resources Sustainably* (Gland: IUCN–NOAA, 2001); M. Lack, K. Short, and A. Willock, *Managing Risk And Uncertainty In Deep-Sea Fisheries: Lessons From Orange Roughy* (NSW, Australia: TRAFFIC Oceania and WWF Australia, 2003); C.J. Small, *Regional Fisheries Management Organizations: Their Duties and Performance in Reducing Bycatch of Albatrosses and Other Species* (Cambridge:

Birdlife International, 2005); WWF, *The Footprint of Distant Water Fleets on World Fisheries* (Godalming: WWF, 1998).

105 For example, see initiatives of the Ministerially-led Task Force on Illegal, Unreported and Unregulated Fishing on the High Seas (www.high-seas.org) and the Ministerial Declaration from the St. John's Conference in 2005.

106 The precautionary approach is discussed in detail in Chap. 6.

107 R. Beverton and S. Holt, *On the Dynamics of Exploited Fish Populations* (London: Chapman Hall, 1993).

108 M. Onestini, "Resource Impacts of Fisheries Subsidies: Is Effective Management Likely in the Presence of Subsidies?". Paper presented at the UNEP Workshop on Fisheries Subsidies and Sustainable Fisheries Management, Geneva, Switzerland, 26–27 April 2004. See also the FAO definition of MEY: "when relating total revenues from fishing to total fishing effort in a surplus production model, the value of the largest positive difference between total revenues and total coasts of fishing (including the cost of labour and capital) with all inputs valued at their opportunity costs". FAO Glossary, note 5.

109 For further discussion on MEY, see S. Schonberger and J. Agar, *Argentina: Towards Rights Based Fisheries Management*, World Bank Departmental Working Paper (Washington, D.C.: World Bank, 2001); and S. Kaye, *International Fisheries Management* (The Hague: Kluwer Law International, 2001) at 53–59.

110 R. Hannesson, "Effects of Liberalizing Trade in Fish, Fishing Services and Investment in Fishing Vessels", OECD Papers Offprint No. 8 (Paris: OECD, 2001).

111 C. Stone, D. Downes, and C. de Fontaubert, "Biodiversity, Trade and the Fishing Sector Case Study: West Africa", Comment Draft 1-05-99 (1999) (copy on file with author).

112 S. Garcia and D. Staples, "Sustainability Reference Systems and Indicators for Responsible Marine Capture Fisheries: A Review of Concepts and Elements for a Set of Guidelines", at 409.

113 As noted by the FAO, the recent rapid rise of fuel prices could "change the economics of the fishing industry. Distant water fishing will be particularly affected. Most likely, fish carriers will be used in an attempt to cut overall fuel costs by reducing the time fishing vessels spend steaming to and from the fishing grounds". See FAO <http://www.fao.org/fi/website/FIRetrieveAction.do?dom=topic&fid=3461> (accessed 18 October 2007).

114 As expressed by S. Schonberger and J. Agar, "economic theory predicts that fishermen will continue to fish until total revenues equal total costs. At this effort level, where there is open access, rents are completely dissipated and there are no incentives to increase effort. Fishing over this level, occurs because fishermen consider only the private costs of harvesting and not the social costs. This is clearly a market failure, as private cost is less than the social costs".

115 G. Hardin, "The Tragedy of the Commons" (1968) 162 *Science* 1243–1248, see also S. Kaye, at 57.

116 See overview on Donut Hole and Sea of Okhotsk Peanut Hole fisheries by E. Meltzer, at 283–293.

117 S. Garcia and C. Newton, "Current Situation, Trends and Prospects in World Capture Fisheries" in E. Pikitch, D. Huppert, and M. Sissenwine, eds., *Global Trends: Fisheries Management* (AFS Symposium 20) (Bethesda, Md.: American Fisheries Society Publication, 1997).

118 Art. 5 calls on States to take measures to prevent or eliminate overfishing and excess fishing capacity and to ensure that levels of fishing effort do not exceed those commensurate with the sustainable use of fishery resources.

119 S. Nandan, "Statement of the Chairman Upon the Adoption of the Agreement on the Implementation of the Provisions of the United Nations Convention on the Law of the Sea of 10 December 1982 Relating to the Conservation and Management of Straddling Fish Stocks and Highly Migratory Fish Stocks" in J.-P. Levy and G. Schram, eds., *United Nations Conference on Straddling Fish Stocks and Highly Migratory Fish Stocks: Selected Documents* (The Hague, Boston, London: Martinus Nijhoff Publishers, 1995) 749–754 at 750.

120 R. Rayfuse, *Non-Flag State Enforcement in High Seas Fisheries* (Leiden: Martinus Nijhoff Publishers, 2004) at 142–143.

121 IATTC Resolution on the Establishment of a Working Group on Fleet Capacity in the Eastern Pacific Ocean. C-98-06. IATTC online <http://www.iattc.org/PDFFiles/C-98-06%20Capacity%20WG%20resolution%20Jun%2098.pdf> (accessed 15 August 2005).

122 IATTC Resolution on Fleet Capacity. C-98-11. IATTC online <http://www.iattc.org/PDFFiles/C-98-11%20Capacity%20resolution%20Oct%2098.pdf> (accessed 15 August 2005).

123 IATTC Resolution on the Management of Fishing Capacity of Large Scale Tuna Long line Fishery. C-99-04. IATTC online <http://www.iattc.org/PDFFiles/C-99-04%20Longline%20capacity%20resolution%20Jun%2099.pdf> (accessed 15 August 2005).

124 IATTC Resolution on Fleet Capacity C-00-01. IATTC online <http://www.iattc.org/PDFFiles/C-00-01%20Fleet%20capacity%20resolution%20Feb%2000.pdf> (accessed 15 August 2005).

125 IATTC Resolution on the Capacity of the Tuna Fleet Operating in the EPO. C-00-10. IATTC online <http://www.iattc.org/PDFFiles/C-00-10%20Fleet%20capacity%20resolution%20Aug%2000.pdf> (accessed 15 August 2005).

126 Dr. Robin Allen, Director, IATTC, personal communication, May 2006.)

127 ICCAT Recommendation Concerning the Limitation of Fishing Capacity on Northern Albacore (in force 21 June 1999). ICCAT online <http://www.iccat.es/Documents%5CRecs%5Ccompendiopdf-e%5C1998-08-e.pdf> (accessed 15 August 2005).

128 IOTC Resolution 98/05 On Cooperation With Non-Contracting Parties. IOTC online <http://www.iotc.org/English/resolutions/reso_detail.php?reso=5> (accessed 15 August 2005).

129 IOTC Resolution 03/01 On the Limitation of Fishing Capacity of Contracting Parties and Cooperating non-Contracting Parties. IOTC online <http://www.iotc.org/English/resolutions/reso_detail.php?reso=28> (accessed 15 August 2005).

130 IOTC Resolution 99/01 On the Management of Fishing Capacity and on the Reduction of the Catch of Juvenile Bigeye Tuna by Vessels, Including Flag of Convenience Vessels, Fishing for Tropical Tunas in the IOTC Area of Competence. IOTC online <http://www.iotc.org/English/resolutions/reso_detail.php?reso=6> (accessed 15 August 2005).

131 IOTC Resolution 02/08 On the Conservation of Bigeye and Yellowfin Tuna in the Indian Ocean. IOTC online <http://www.iotc.org/English/resolutions/reso_detail.php?reso=26> (accessed 15 August 2005).

[132] IOTC Resolution 03/02 On Criteria for Attaining the Status of Co-operating Non-Contracting Party. IOTC online <http://www.iotc.org/English/resolutions/reso_detail.php?reso=29> (accessed 15 August 2005).

[133] IOTC Resolution 05/01 On Conservation and Management Measures for Bigeye Tuna. IOTC online <http://www.iotc.org/English/resolutions/reso_detail.php?reso=35> (accessed 15 August 2005).

[134] The WCPFC Convention requires that it take measures to limit capacity, in wording that reflects Art. 5 of UNFSA.

[135] In his comprehensive study for the World Bank (1998), Milazzo places the global subsidy estimate at $14–$20 billion annually, against a world harvest valued at $60–$70 billion (20%–25%). The capacity of the world fleet is estimated to be 150%–250% greater than that needed to catch what the oceans can produce sustainably. M. Milazzo, "Subsidies in World Fisheries: A Re-examination", World Bank Technical Paper No. 406 (Washington, D.C.: World Bank, 1998).

[136] Several sources provide excellent discussions of conceptual and practical issues in specifying subsidies and measuring their resource- and trade-distorting impacts: M. Milazzo; APEC, *Study into the Nature and Extent of Subsidies in the Fisheries Sector of APEC Members Economies* (Singapore: APEC Secretariat, 2000); OECD, *Transition to Responsible Fisheries — Economic and Policy Implications: Part 3, Government Financial Transfers and Resource Sustainability* (Paris: OECD, 2000); Cox, A., and C.-C. Schmidt, "Subsidies in the OECD Fisheries Sector: A Review of Recent Analysis and Future Directions". Paper presented to the FAO Expert Consultation on Identifying, Assessing and Reporting on Subsidies in the Fishing Industry, Rome, Italy, 3–6 December 2002 (Paris: OECD, 2002) .; W. Shrank, "Introducing Fisheries Subsidies", FAO Fisheries Technical Paper No. 437 (Rome: FAO, 2003); L. Westlund, "Guide for Identifying, Assessing and Reporting on Subsidies in the Fisheries Sector", FAO Fisheries Technical Paper No. 438 (Rome: FAO, 2004); UNEP, *Analyzing the Resource Impact of Fisheries Subsidies: A Matrix Approach* (Geneva: UNEP, 2004).

[137] FAO, *Marine Fisheries and the Law of the Sea: A Decade of Change* (Rome: FAO, 1992). The path taken by the subsidy issue initially as a fisheries management matter within the UN-FAO process and more recently as a trade question within the WTO framework is well documented in W. Shrank. He notes that subsidies are not mentioned explicitly in the Cancun Declaration (1992), UNCED Agenda 21 (1992), the UN Conference on Straddling Fish Stocks and Highly Migratory Fish Stocks (1995), nor the FAO Code of Conduct for Responsible Fisheries (1995). At best, subsidies are referenced implicitly in clauses calling for prevention of overfishing and overcapacity.

[138] The Plan (Part III) lists several urgent actions aimed at assessing and monitoring fishing capacity, and urges adoption of national plans to manage, and if necessary, reduce fishing capacity.

[139] M. Tsamenyi, L. Manarangi-Trott, and S. Rajkumar, "The International Legal Regime for Fisheries Management", paper presented at UNEP Workshop on Fisheries Subsidies and Sustainable Fisheries Management, Geneva, Switzerland, 26–27 April 2004.

[140] Five nations, Australia, Iceland, New Zealand, the Philippines, and the United States, submitted a joint statement calling for the elimination of fisheries subsidies, as these nations considered them to be harmful to trade and the environment. WTO, Committee on Trade and Environment, *Benefits of Eliminating Trade Distorting and Environmentally Damaging Fisheries Subsidies*, WTC/CTE/W/121 (28 June 1999).

[141] WTO Ministerial Conference — Fourth Session, Doha, Qatar, 2001: Ministerial Declaration, WT/MIN (01)/DEC/1, para. 28.

[142] These are the five countries submitting the original statement on subsidies to the WTO, in addition to Chile, Ecuador, and Peru. WTO Negotiating Group on Rules, *The Doha Mandate to Address Fisheries Subsidies: Issues*, TN/RL/W/3 (24 April 2002). Available at <http://www.fao.org/documents/show_cdr.asp?url_file=/DOCREP/006/Y4647E/y4647e07.htm> (accessed 15 March 2006).

[143] WTO Negotiating Group on Rules, *The Doha Mandate to Address Fisheries Subsidies: Issues*, TN/RL/W/3 (24 April 2002). Available at <http://www.fao.org/documents/show_cdr.asp?url_file=/DOCREP/006/Y4647E/y4647e07.htm> (accessed 15 March 2006).

[144] In its 2002 report, *Turning the Tide on Fishing Subsidies*, WWF argues that the WTO rules fall short because WTO defines subsidy too narrowly, does not cover situations where the harm takes the form of reduced production levels (catches), does not address overcapacity and the impact on overfishing and sustainability, and is not strong enough to force governments to provide clear and comprehensive subsidy data. WWF, *Turning the Tide on Fishing Subsidies: Can the World Trade Organization Play a Positive Role?* (Geneva: WWF, 2002). See also, W. Shrank, who makes the point that any action under WTO would face the difficult test of having to prove there was excessive fishing effort, that it resulted from subsidization, and that it was the effort of the subsidized fleet that was responsible for the reduction of fish stocks.

[145] In response to the submission by the group of eight, Japan, drawing on recent conceptual studies, notes that in circumstances where management regimes are effective (e.g., with precautionary catch limits and full compliance), subsidies would have little or no impact on resource sustainability. So, the emphasis should be on securing effective management. WTO, WT/CTE/W/226 (24 April 2003). For a conceptual discussion on the environmental effects of subsidies, see FAO, "Report of the Technical Consultation on the Use of Subsidies in the Fisheries Sector, Rome, 30 June–2 July 2004", FAO Fisheries Report No. 752 (Rome: FAO, 2004).

[146] WTO, WT/CTE/W/236 (17 June 2004), UNEP Workshop on Fisheries Subsidies and Sustainable Fisheries Management, Geneva, Switzerland, 26–27 April 2004.

[147] See J. Hutchings, "Collapse and Recovery of Marine Fishes" (2000) 406 *Nature* 882–885 and J. Hutchings, "Conservation Biology of Marine Fishes: Perceptions and Caveats Regarding Assignment of Extinction Risk" (2001) 58 *Canadian Journal of Fisheries and Aquatic Sciences* 108–121.

[148] "The overall fishing pressure continues to increase even though the number of boats might decrease" quoting Daniel Pauly in P. Greenberg "The Catch", *New York Times Magazine*, 23 October 2005. The challenge for RFMOs such as IATTC is to implement comprehensive fleet-capacity limits and fleet management plans so that fishing pressure is, in fact, reduced commensurate with the available resource.

[149] See, for example, J. Jackson, M. Kirby, W. Berger, K. Bjorndal, L. Botsford, B. Bourque, R. Bradbury, R. Cooke, J. Erlandson, J. Estes, T. Hughes, S. Kidwell, C. Lange, H. Lenihan, J. Pandolfi, C. Peterson, R. Steneck, M. Tegner, and R. Warner, "Historical Overfishing and the Recent Collapse of Coastal Ecosystems" (2001) 293 *Science* 629–637, and FAO Fisheries Technical Paper No. 457.

[150] C. Safina, "The World's Imperiled Fish", *Scientific American*, November 1995, at 50.

[151] WWF, "Underwriting Overfishing", Issue Summary No. 1 9/99, available at <http://assets.panda.org/downloads/Issuesum.pdf> (accessed 18 October 2007). See also Garcia and Newton, who convincingly argue that overall global fleet capacity doubled as a consequence of the extended jurisdiction in 1977. S. Garcia and C. Newton, "Current Situation, Trends and Prospects in World Capture Fisheries".

[152] FAO FIGIS bulletin on "Fleets". Available at <http://www.fao.org/fi/website/FIRetrieveAction.do?dom=topic&fid=3461> (accessed 18 October 2007).

[153] The 2004 United Nations Report of the Secretary General on Fisheries Issues highlights the expectation that RFMOs will take the lead role in the implementation of measures for conserving biodiversity. National and international fisheries law is entwined with national and international biodiversity legislation. In fact, widespread and continuous overfishing has triggered the application of biodiversity legislation to commercially fished species, both nationally and internationally. UN online <http://daccessdds.un.org/doc/UNDOC/GEN/N04/471/20/PDF/N0447120.pdf?OpenElement> (accessed 19 October 2005).

[154] Overfishing to the point of risking biological extinction has been largely avoided for key commercial tuna and tuna-like species (other than bluefin tuna) because of "the high productivity of tuna species and the decreases in tuna prices to uneconomical as markets become saturated". See paper by J. de Leiva and J. Majkowski, "Status of the Tuna Stocks of the World" (on file with author).

[155] Some scientists have warned that commercial fisheries could collapse by 2048 if the current levels of habitat destruction and overfishing around the world are maintained. See B. Worm, E.B. Barbier, N. Beaumont, J.E. Duffy, C. Folke, B.S. Halpern, J.B.C. Jackson, H.K. Lotze, F. Micheli, S.R. Palumbi, E. Sala, K. Selkoe, J.J. Stachowicz, and R. Watson, "Impacts of Biodiversity Loss on Ocean Ecosystem Services", (2006) 314 *Science* 787–790.

[156] N. Dulvy, Y. Sadovy, and J. Reynolds, "Extinction Vulnerability in Marine Populations"; J. Reynolds, N. Dulvy, N. Goodwin, and J. Hutchings, "Biology of Extinction Risk in Marine Fishes"; R.A. Myers and B. Worm, "Extinction, Survival, or Recovery of Large Predatory Fishes" (2005) 360 *Philosophical Transactions of the Royal Society B* 13–20. Known marine extinctions to date can be viewed at <www.uea.ac.uk/bio/people/faculty/reynoldslab/extinction_tableB.htm> (accessed 15 May 2005).

[157] B. Worm et al., "Impacts of Biodiversity Loss on Ocean Ecosystem Services"; K.T. Frank, B. Petrie, J. Choi, and W. Leggett, "Trophic Cascades in a Formerly Cod-Dominated Ecosystem"; K.T. Frank, B. Petrie, and N.L. Schackell, "The Ups and Downs of Trophic Control in Continental Shelf Ecosystems".

[158] The April 2005 COSEWIC meeting in Halifax, N.S., brought together international fisheries scientists to discuss the implications of biodiversity species status measures for marine fish and particularly how accepted diversity status measures are related to fisheries management reference points. Further, several NGOs have arisen in the last decade with the express mandate to advocate and facilitate conservation of marine diversity. Oceana (www.oceana.org) and the Marine Conservation Biology Institute (www.mcbi.org), both based in the United States, work with scientists and fishermen on marine conservation issues. A coalition of NGOs, the Deep Sea Conservation Coalition, has developed in response to the threat posed by high seas bottom trawling (www.savethehighseas.org).

[159] For a discussion on the importance of resilience to ecosystem function and a review of recent science see J.E. Duffy and J.J. Stachowicz, "Why Biodiversity Is Important to Oceanography: Potential Roles of Genetic, Species, and Trophic Diversity in Pelagic Ecosystem Processes" (2006) 311 *Marine Ecology Progress Series* 179–189.

[160] Several marine fish and invertebrate species are listed on Appendix II and the listing of marine species is likely to increase. See K. Cochrane and D. Doulman, "The Rising Tide of Fisheries Instruments and the Struggle to Keep Afloat" (2005) 360 *Philosophical Transactions of the Royal Society B* 77–94.

[161] The World Conservation Union (IUCN) develops a Red List of species each year. Available at <http://www.iucn.org/themes/ssc/redlists/RLcats2001booklet.html> (accessed 15 October 2005) .

[162] The Committee in the Status of Endangered Wildlife in Canada (COSEWIC) lists northern wolfish and northern cod, with many more species undergoing status assessments. See <www.cosewic.gc.ca> for species status reports and current listings.

[163] FAO, "An Appraisal of the Suitability of the CITES Criteria for Listing Commercially-exploited Aquatic Species", FAO Fisheries Circular No. 954 (Rome: FAO, 2000). The relationship of CITES with fisheries management is further discussed in Chap. 2 at 2.4.14.2.

[164] The use of CITES to combat IUU fishing is not without opposition, as evidenced by discussion at COFI in March 2005. This stems from concerns that CITES (1) is not an appropriate instrument for fisheries management (because of its listing criteria and how they are applied, concerns about the validity of scientific evaluation process, and the adequacy of decision-making procedures); and (2) is too blunt an instrument, creating too many indirect effects (e.g., stock-specific listings would create trade barriers for all fish from a listed stock, not just for IUU fish; the difficulty of distinguishing look-alike species; and a CITES listing would undermine confidence in the competency of the RFMO).

[165] See COFI 24th Session, Rome, Italy, 26 February – 2 March 2001. Review of the CITES Listing Criteria for Species Exploited by Fisheries in Marine and Large Freshwater Bodies. Available at <http://www.fao.org/DOCREP/MEETING/003/X8772E.HTM> (accessed March 2006).

[166] See F. Serchuk, D. Rivard, J. Casey, and R. Mayo, *Report of the Ad Hoc Working Group of the NAFO Scientific Council on the Precautionary Approach*, NAFO SCS Doc. 97/12 (Dartmouth, N.S.: NAFO, 1997); NAFO, *Report of the Joint Scientific Council and Fisheries Commission Working Group on Precautionary Approach*, NAFO/FC Doc 99/2 Serial No. N4051 (Dartmouth, N.S.: NAFO, 1999); NAFO, *Report of the Joint Scientific Council and Fisheries Commission Working Group on Precautionary Approach*, NAFO/ FC Doc 00/2 Serial No. N4207 (Dartmouth, N.S.: NAFO, 2000).

[167] Fisheries scientist Carl Walters has called for more formal "gaming" exercises to allow stakeholders to explore the consequences of various management options. See C. Walters, "Use of Gaming Procedures in Evaluation of Management Experiments" (1994) 51 *Canadian Journal of Fisheries and Aquatic Sciences* 2705–2714.

[168] S. Nandan, "Statement of the Chairman Upon the Adoption of the Agreement on the Implementation of the Provisions of the United Nations Convention on the Law of the Sea of 10 December 1982 Relating to the Conservation and Management of Straddling Fish Stocks and Highly Migratory Fish Stocks" in J.-P. Levy and G. Schram, eds., *United Nations Conference on Straddling Fish Stocks and Highly Migratory Fish Stocks: Selected Documents* (The Hague, Boston, London: Martinus Nijhoff, 1995) 749–754.

[169] Garcia and Newton convincingly argue that overall global fleet capacity doubled as a consequence of the extended jurisdiction in 1977. S. Garcia and C. Newton, "Current Situation, Trends and Prospects in World Capture Fisheries".

Part II: Selected Themes Pertaining To RFMO Regulation Of Straddling Fish Stocks And Highly Migratory Fish Stocks

Chapter 6

The Precautionary Approach to Fisheries*

6.1 Introduction

A major achievement of the UNFSA regime is the inclusion of the precautionary approach (PA) in international fisheries management. The PA, as set out in UNFSA and the Food and Agriculture Organization (FAO) Code of Conduct for Responsible Fisheries, is considered a major post-UNCED structural innovation towards realizing sustainable development objectives in fisheries. The PA in fisheries management shares the same conceptual basis as the precautionary principle articulated in UNCED but has distinct juridical and scientific applications. The PA is thus a new paradigm applied to fisheries under UNFSA. It requires the application of risk-averse science and a risk-averse management regime throughout the geographic range of the fish stocks, whereby decision makers err on the side of caution and marine conservation when information is uncertain, unreliable, or inadequate.

The PA takes into account, and allows for, the inherent uncertainty, error, and associated biological risk in predicting stock abundance, as well as the challenges associated with accurately assessing fishing mortality and effectively controlling fishing activity. Additionally, the PA recognizes the difficulty in determining the effects of the marine environment on the fish stocks and vice versa. To decrease uncertainty and error in stock assessment and improve the quota-allocation process, precautionary reference points and preagreed decision rules were established under UNFSA along with guidelines for prosecuting new and exploratory fisheries. The wide application of the PA calls for rigorous regulatory-appraisal and decision-making processes, including a shift in the "burden of proof", whereby the proposed activity must be shown not to pose a risk of serious or irreversible harm to the target stocks, associated or dependent species, or their environment (i.e., fishing activities must be within an acceptable level of harm in keeping with international minimum standards).

Contemporary fisheries management recognizes the need for taking into account ecological interactions and the impact of fishing on the marine ecosystem. This has lead to another major innovation in fisheries governance known as the ecosystem approach to fisheries (EAF) (further discussed in Chapter 7). The ecosystem approach greatly increases the data requirements and related uncertainty beyond the conventional, albeit often unrealized, demand for accurate, adequate, and timely information. UNFSA accordingly calls for enhanced data collection and submission together with improved monitoring, control, and surveillance (MCS) and enforcement (see Chapter 9). Improved MCS reduces the uncertainty and error in assessments and the risks of overfishing and of ecosystem shifts due to fishing activities. Given the inherent uncertainty, environmental variability, and dynamic nature of the marine ecosystem, adopting a holistic ecosystem approach further reinforces the need for the wide application of the PA.

The PA is an important doctrinal shift in international conservation and fisheries governance that is being increasingly adopted by regional fisheries management organizations (RFMOs) responsible for the management of straddling fish stocks and highly migratory fish stocks. The Convention for the Conservation of Antarctic Marine Living Resources, which established the Commission for the Conservation of Antarctic Marine Living Resources (CCAMLR), was the first international fisheries instrument to outline precaution as a guiding conservation principle. Since 1995, the International Council for the Exploration of the Sea (ICES) and FAO have elaborated the PA to give practical expression to the concept and to assist RFMOs in implementing the PA. While it is clear that the PA is an accepted normative concept and management principle, RFMOs, through their member States, have, with some notable exceptions, been hesitant to adopt and apply the PA as widely as called for under UNFSA and the FAO Code of Conduct. If RFMOs are to give full effect to the new regime established under UNFSA and related instruments, greater efforts to adopt and implement the PA must be made by RFMOs and by coastal States within their areas of national jurisdiction.

This chapter briefly outlines the history and definition of PA and its incorporation into international fisheries law. The PA is discussed in terms of its specific applications in determining reference points and decision rules as well as its wider application in fisheries management decision making. This chapter also provides a discussion on

* The author gratefully acknowledges the research assistance of Susanna D. Fuller in preparing this chapter.

efforts to adopt and apply precaution by RFMOs as well as on the degree to which these RFMOs have established measures that reflect the PA. Challenges in implementing the PA as well as opportunities to integrate the concept into management of straddling fish stocks and highly migratory fish stocks are also reviewed.

6.2 International legal underpinnings

The precautionary principle is a relatively recent[1] norm in international environmental law that crystallized during the UNCED process. In its simplest form, the precautionary principle requires that in the case of scientific uncertainty regarding an activity that could result in serious or irreversible harm, the decision should be in favour of the environment. Article 7 of the Bergen Ministerial Declaration was an early statement of the principle:

> In order to achieve sustainable development, policies must be based on the precautionary principle. Environmental measures must anticipate, prevent and attack the causes of environmental degradation. Where there are threats of serious or irreversible damage, lack of scientific certainty should not be used as a reason for postponing measures to prevent environmental degradation.[2]

In addition, where consequences of an action are unknown but are judged to have some potential for major or irreversible harm, then the action should be avoided. Several international declarations and agreements subsequently adopted the principle: Principle 15 of the Rio Declaration on Environment and Development; Article 3(3) of the Framework Convention on Climate Change; the preamble of the Convention on Biological Diversity; and Article 103(R)(2) of the Treaty of European Union. The pre-UNCED 1982 CCAMLR Convention refers to the concept of precaution in Article II(3)(c), thereby making it the first legally binding international fisheries instrument to include precaution. The first practical application of the precautionary principle in international fisheries management was addressing the large-scale pelagic driftnet fishery, a fishing practice regarded by many as causing serious environmental damage.[3] The precautionary principle was invoked, and the burden of proof shifted to driftnet fishing States to prove that the fishery did not cause serious or irreversible harm.[4] On 22 December 1989 the United Nations General Assembly (UNGA) adopted Resolution 44/225 calling for a ban on large-scale pelagic driftnet fishing on the high seas. High seas driftnet fishing States were unable to discharge this evidentiary burden or reverse public opprobrium.

Given the inherent uncertainty and risk associated with fisheries stock assessment and management, application of the precautionary principle to fisheries was rejected by many States during UNCED and was therefore not incorporated in the 1992 Agenda 21 with respect to fisheries. Nonetheless, it was readily apparent that a new paradigm for fisheries management was required, as past failures of fisheries managers to take conservation action in the face of scientific uncertainty had led to nonsustainable fishing practices and overexploitation of key stocks. Over the course of the UNFSA negotiations, an acceptable formulation was developed, which enabled the application of the PA in the fisheries context. As Lodge and Nandan state, "[o]ne of the most important contributions of UNFSA is the way in which it seeks to operationalise the precautionary approach in the context of fisheries management. In so doing, it builds on the precautionary principle adopted at UNCED and applies this principle directly to fisheries management".[5] Article 6 of UNFSA, together with Article 7 and Annex II,[6] explicitly includes the general principle for the PA as well as operational indicia for invoking and applying the approach. Article 6(1) of UNFSA requires that States "apply the precautionary approach widely to conservation, management and exploitation of straddling fish stocks and highly migratory fish stocks in order to protect the living marine resources and preserve the marine environment". In implementing the PA, States are to improve decision making regarding fishing operations by "obtaining and sharing the best scientific information available" and by employing "improved techniques for dealing with risk and uncertainty" (Article 6(3)(a)).

The PA was also incorporated in the non legally binding FAO Code of Conduct for Responsible Fisheries in Article 7.5. The Code and supplementary soft law instruments have been useful in elaborating the approach as well as effecting widespread support. In an effort to promote the practical expression of the PA, FAO has published several reports on biological reference points[7] and the PA[8] and has developed technical guidelines for the application of the PA to facilitate implementation of the Code. Given that precautionary measures must cover the entire range of a stock, both Article 6 of UNFSA and the Code are applicable within coastal waters and on the high seas.

6.3 Defining the precautionary approach

The cornerstone of the definition of the PA is that caution is required when information is uncertain, unreliable, or inadequate.[9] UNFSA Article 6(2) directs that States "be more cautious when information is uncertain, unreliable or inadequate". At the same time, the PA emphasizes collection and analysis of complete and reliable data to reduce scientific uncertainty and risks to the fisheries and the marine environment. Arguably, all fisheries, even relatively data-rich regimes, are operating within an uncertain, unreliable, or inadequate setting.[10] For data-poor regimes, or for regimes where only industry-dependent data are used, the necessary information is often incomplete or unreported, and the inherent uncertainty is exacerbated. It is when data are uncertain, coupled with the risk of serious *or* irreversible harm, that the PA is invoked.[11]

The second aspect of the PA is the injunction that the absence of adequate scientific information shall not be used as a reason for postponing or failing to take conservation and management measures (Article 6(2)). This formula allows action to be taken based on the best scientific information available, consistent with the United Nations Convention on the Law of the Sea (LOSC), despite incomplete or missing information. Application of precaution in fisheries management calls for scientists and managers to consider uncertainty or error in the regulatory appraisal of possible options and the associated consequences. In so doing, they are to err on the side of caution.

A third critical feature of the PA is the reversal in the burden of proof. The onus of scientific proof shifts to the State or party that intends to derive benefits from the resource to demonstrate that a proposed management strategy is within an acceptable level of harm taking into consideration the associated risks (i.e., no serious *or* irreversible harm will ensue).[12] This replaces the traditional approach of the laissez-faire "wait and see" doctrine of the freedom of the seas, where harmful activities were tolerated unless, and until, serious or irreversible harm could be demonstrated. This is a fundamental shift in the governance paradigm.[13]

The 1995 FAO Technical Consultation on the Precautionary Approach to Capture Fisheries examined the issue of burden of proof. It concluded that

- All fishing activities have environmental impacts and it is not appropriate to assume that these are negligible until proved otherwise.
- Although the PA to fisheries may require cessation of fishing activities that have potentially serious adverse impacts, it does not imply that no fishing can take place until all potential impacts have been assessed and found to be negligible.
- The PA to fisheries requires that all fishing activities be subject to prior review and authorization; that a management plan be in place that clearly specifies management objectives and how impacts of fishing are to be assessed, monitored, and addressed; and that specified interim management measures should apply to all fishing activities until such time as a management plan is in place.
- The standard of proof to be used in decisions regarding authorization of fishing activities should be commensurate with the potential risk to the resource, while also taking into account the expected benefit of the activities.[14]

The standard of proof required as well as the determination of an acceptable level of harm will vary in a given circumstance. The range of interpretation of what is an acceptable impact or what constitutes serious or potentially irreversible harm may lead to irreducible conflict among different interests. Further, adopting risk-averse strategies implies that there is an exercise to assess risk and prevent unacceptable harm. Additionally, the variation in the standard of proof required (from "reasonable grounds for concern" of the European Commission's Communication on the Precautionary Principle,[15] to the "balance of probabilities" of common law civil actions, to the criminal law requirement of proof "beyond a reasonable doubt") or what is deemed appropriate has not been determined and is subject to interpretation. The standard of proof that is appropriate may vary with the risk of harm and (or) the degree of harm.[16]

In practice, the PA is increasingly applied through the development of biological reference points that complement the existing fisheries management objectives, based on sustaining and maintaining maximum sustainable yield (MSY). These reference points and corresponding decision rules, coupled with specific management strategies, are established and implemented, to varying degrees of success, as part of attempts to restore and maintain fish stocks.

6.3.1 Reference points and decision rules

As UNFSA recognized the potential for conflicting perspectives and aimed to prevent overexploitation, the Agreement adopted a framework for States to establish precautionary reference points[17] as well as a priori decision rules and harvest controls, in instances where limit reference points are approached or exceeded. Limit reference points (commonly referred to in fisheries science

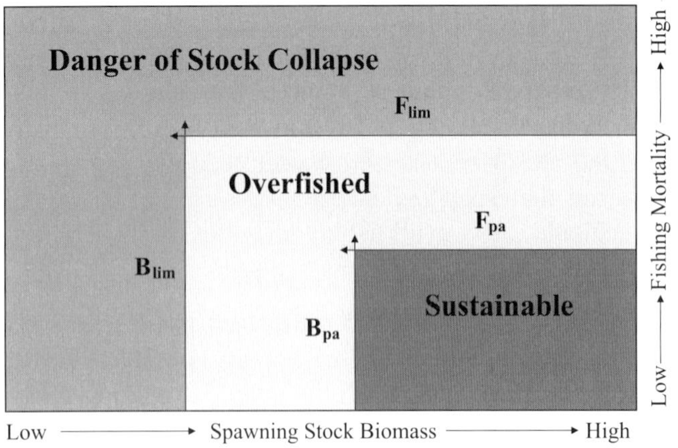

Figure 6.1. Precautionary approach reference points, as defined by ICES.[21] Redrawn from CEFAS, "Overfishing and the ICES Precautionary Approach to Management".[22] This figure illustrates the ICES model of PA reference points in relation to stock status. This model incorporates uncertainty by leaving considerable space around reference points to accommodate data inadequacy, environmental factors, and other uncertainties related to fisheries assessment and management. Limit references points used by ICES are not specifically related to MSY or the surplus production model. A fishery can be considered sustainable when spawning stock biomass is high and corresponding fishing mortality low.

as LRPs) are indicators that "mark the onset of dangerous conditions. Appropriate pre-negotiated responses are called for when approaching such threshold conditions. The parameters of such conditions, together with guidance as to the appropriate level of response, are set out in Annex II of UNFSA and include the need to establish provisional reference points when information is poor or absent".[18] FAO describes LRPs as the limit "beyond which the state of a fishery and (or) a resource is not considered desirable". As expressed by Lodge and Nandan, "the use of limit reference points represents a progressive development from the identification of 'target' reference points based on maximum sustainable yield (MSY) towards a 'traffic-light' system in order to identify the point when a resource system leaves a safe (green) condition and before it reaches the 'red' condition associated with overfishing, recruitment collapse and ecosystem change"[19] (see Figure 6.1, also Figures 5.17, 5.18). According to ICES, limit reference points should not be used when there is a greater than 95% probability that the limit will be exceeded. "ICES stresses that precautionary reference points should not be treated as management targets, but as lower bounds on spawning biomass and upper bounds on fishing mortality."[20]

RFMOs are to determine stock-specific reference points on the basis of the best scientific information available, taking into account uncertainties in stock size and productivity (Article 6(3) (c) of UNFSA), reference points, stock condition in relation to such reference points, levels of distributions of fishing mortality, and the impact of fishing on nontarget species and their environment. As well, RFMOs are to adopt plans to conserve such species and protect habitats of special concern (Article 6(3) (d) of UNFSA). Precautionary management calls for different and more cautious measures for fisheries depending on the level of exploitation and availability of data (e.g., new and exploratory fisheries). Read together with Article 5, which calls for taking into account broader ecosystem considerations,[23] the PA is a fundamental component of the ecosystem approach, given frequent data gaps and inherent uncertainties in understanding the impacts of fishing operations on the ecosystem.

Moritaka Hayashi distinguishes between limit reference points and target reference points as set out in the technical guidelines for biological reference points in Annex II of UNFSA. "Limit reference points set boundaries which constrain utilization within safe biological limits and beyond which resource-building programs are required. They are designed for conservation and warn against the risk of over-exploitation. Target reference points guide policy makers in resource utilization and are designed to indicate when an objective is being approached".[24] Since the MSY concept as understood under LOSC was no longer adequate,[25] it was reformulated to serve as a limit reference point rather than a target (see Chapter 5.3.4). The reformulation of MSY and the adoption of precautionary reference points under UNFSA form part of a continuum in fisheries management that aims to avoid serious or irreversible harm. Accordingly, for broad application of the PA to stock conservation, F_{MSY}, the fishing mortality corresponding to maintenance of MSY, is to be adopted by RFMOs and States as a limit reference point rather than a target. For already depleted stocks, the biomass that can produce MSY (B_{MSY}) may serve as an initial rebuilding target. This represents a significant paradigm shift in conventional fisheries management and an important post-LOSC evolution of the concept.

Annex II requires that "previously agreed precautionary reference points … shall be used to trigger pre-agreed conservation and management action". Under this provision, UNFSA calls on RFMOs through their constituent States to identify specific management measures and a set of decision rules that outline in advance what decisions should be made at different stock status levels. Preagreed decision rules are an important component of precautionary management. Such rules preempt controversy, prolonged debate, and stonewalling, and are intended to protect stocks that are approaching or exceeding limit reference points from decisions that are based on short-term political horizons rather than science. "More generally, they specify the data to collect, how to analyze those data, and the management actions to take given the results from the analyses. The performances of decision rules are evaluated by [model] simulation".[26]

In addition to target and limit points, fisheries scientists and managers have identified threshold reference points. Such threshold points[27] are intended to serve as a warning that the limit reference point is being approached and can trigger preagreed measures to ensure that the limit will not be exceeded. With the "technical and institutional revision of the use of MSY"[28] from a target to a limit and the development of alternative and complementary limit and threshold reference points, precaution has become inextricably linked to reference points. Reference points as well as decision or control rules are increasingly institutionalized within RFMO regimes to ensure precautionary fisheries management strategies.

6.3.2 Precautionary management strategies and plans

Under UNFSA, RFMOs are to develop management strategies based on stock-specific precautionary reference points (Annex II, paragraphs 3, 4, and 5) and to adopt *plans* to conserve nontarget, associated, and dependent species and their environment (Article 6(3)(d)). The management strategy for specific stocks must ensure that as reference points are approached, they will not be exceeded and that the risk of exceeding limit reference points remains very low. Should such reference points be exceeded, corrective and restorative action must be taken not only for the target population but, where necessary, also for associated and dependent species. Accordingly, the PA requires RFMOs to assess the impact of fishing on other species in the same ecosystem and to protect habitats of special concern.

While Article 6(d) explicitly calls for the generation of "plans",[29] conceptually and practically implementing an ecosystem approach, in accordance with UNFSA Articles 6(3)(c), 5(d)(e)(g), and 7(2)(f), requires that all fishing activity be subject to prior review and authorization as part of a management plan or other management frameworks as provided for in the FAO Code of Conduct (Article 7.3.3). It is this interplay between the ecosystem approach and the PA found in UNFSA Article 6 that calls upon RFMOs and States to generate plans that will enable them to proceed in an ecologically responsible way and promulgate compatible measures for the stock over its entire range. The adoption of management plans is therefore a necessary governance mechanism for ensuring that the PA is applied widely and that the overarching objectives of the ecosystem approach to fisheries (EAF)[30] are achieved (i.e., to avoid overfishing by maintaining populations of stocks within the agreed precautionary reference points, to restore stocks by taking corrective action to facilitate stock recovery, and to ensure that fishing activity does not result in harmful impact on the living marine resources as a whole by adopting and implementing PA and EAF).

The FAO Code of Conduct and the FAO technical guidelines on PA underscore that to effect precautionary, compatible, and coherent management of the whole stock over its entire area of distribution, reference points and decision rules should be used together in a clear management plan. A model management plan should include:

(1) identification of broad management objectives, with specific objectives to protect key target species, limit the ecosystem effects of the fishery, and promote recovery of depleted species;

(2) identification of reference points that provide triggers for action at predetermined states of the stock or system;

(3) identification of specific management measures and a set of decision rules that outline in advance what decisions should be made at different stock status levels;

(4) plans for monitoring and evaluating the effectiveness of the management plan in achieving the broad and specific management objectives.[31]

Article 6(3)(c) of UNFSA clarifies that RFMOs and States can take socioeconomic conditions into account in applying the PA to the specific stocks involved. While socioeconomic considerations can be factored into the evaluation of various scenarios in risk analysis and regulatory appraisal, e.g., the timing or specific choice of reference point, no State can dispense with or disregard the need to err on the side of caution in the designation of, and adherence to, specific precautionary reference points. The degree to which socioeconomic considerations and political pressures can influence decision making is stopped by the safeguards afforded under the PA by preagreed thresholds and decision rules, together with the international minimum standards against which management measures can be assessed. In so doing, the

PA endeavours to balance the needs of the present generation with those of future generations. Accordingly, decision makers have an obligation to consider the intergenerational imperative of sustainable development when evaluating the merits and consequences of regulatory options and strategies.

Under this definition and formulation of the PA, the task for RFMOs and States, according to Burke, is "to evaluate the overall context of uncertainty including developing the potential magnitude of the harm, the probability of its occurrence, the availability of alternative measures, the costs of precautionary measures and alternative actions, and, if risk averse action is to be taken, whether it can be part of an adaptive approach leading to better understanding of the problem and to more appropriate and less costly measures".[32] The objecting State or party has the burden of demonstrating that the proposed precautionary regulation is not necessary, as the projected risk and potential impact of the fishing activity are within an acceptable range.

The PA is thus not a "simple add-on to conventional management" as Garcia observes; instead, it should reshape fisheries management "*de novo* into a significantly different system of governance".[33] The PA is accordingly a revolutionary structural innovation in fisheries management that demands a reorientation in the institutional mindset to effect a systemic change. Management objectives need to be clear and transparent and the regulatory-appraisal process must be risk averse and rigorous, aware of the need for balancing complex policy issues and decisions based on a spectrum of scenarios and a multitude of socioeconomic factors. The PA applied widely over the entire range of fish stocks (which is not always known) embraces all facets of the management strategy. Excellent science can be defeated by inadequate or unreliable data, lack of consensus, poor control rules, non-risk-averse management decisions, overcapacity, irresponsible fishing and noncompliance, ineffective enforcement, and (or) perverse market incentives.

The incorporation of broader ecosystem considerations in fisheries management, as prescribed under UNFSA and included in the FAO Code of Conduct, requires the wide application of the PA to implement the new governance paradigm. The PA, as part of the EAF, imposes an obligation on all those concerned to recognize the imperative to err on the side of caution and marine conservation, given inherent uncertainties, intensification of fishing, and severe socioeconomic consequences of overexploitation for present and future generations.

6.4 Efforts to adopt and implement the PA to fisheries

To date much attention has focused on defining, elaborating, and developing frameworks within which to operationalize the new paradigm. Efforts to give practical expression and meaningful implementation to the PA at both international and national levels have proven a greater challenge.

ICES and the FAO have contributed significantly to defining and establishing science-based implementation of the PA. Through its advisory role to the North East Atlantic Fisheries Commission (NEAFC) and the Norwegian-Russian Fisheries Commission for the Loop Hole, ICES has made significant contributions to PA implementation. ICES develops statistical models and harvest-control rules, actively considers ecosystem impacts of fishing, and also examines methods for estimating limit and precautionary reference points. In efforts to assist States and RFMOs in implementing the PA, FAO has convened international meetings and produced technical guidelines on PA implementation as part of its efforts to implement the FAO Code of Conduct. FAO also annually reports on the progress of States and RFMOs in the implementation of the Code of Conduct.[34]

In 2000 Garcia summarized key requirements when defining and establishing science-based implementation of the PA:[35]

(1) A level of precaution commensurate to risk is to be applied at all times to all fisheries.

(2) PA is to be applied systematically, i.e., in research, management, and fishing operations.

(3) Potentially irreversible changes are to be avoided (to maintain options for future generations).

(4) Undesirable outcomes are to be anticipated and measures are to be taken to reduce their likelihood.

(5) Corrective measures are to be applied immediately and are to be effective within an acceptable time frame.

(6) Priority is to be given to conserving the productive capacity of the resource.

(7) Precautionary limits are to be put on fishing capacity when resource productivity is highly uncertain.

(8) The burden of proof is to be appropriately (realistically) placed.

(9) Standards of proof commensurate with the potential risk to the resource are to be established.

(10) A comprehensive legal and institutional management framework formalizes the approach.

Thus, the PA can be operationalized and applied widely through specific management measures, objectives, and risk-averse decision making. Given that the PA under UNFSA applies throughout the geographic range of the stocks, both within areas under national jurisdiction and on the high seas, RFMOs and coastal-State governments are required to develop institutional and legal criteria, standards, and frameworks to effect a systemic change in fisheries management. To date, nonscientific legal and management aspects of delineating the PA have not received the same level of effort that has been made to develop, for example, precautionary reference points.

Coastal States[36] and RFMOs with competence for straddling fish stocks or highly migratory fish stocks, with the assistance of FAO or ICES, have developed limit reference points and precautionary reference points, revisited traditional target reference points such as MSY,[37] set decision rules, and undertaken risk assessment to achieve fisheries management that is consistent with the PA. In the case of new and exploratory fisheries, where data are unavailable or inadequate, the PA is increasingly invoked. In other data-poor fisheries, the presumption of uncertainty and the need for applying the precautionary approach are increasingly included in management strategies. Where regulated species are fully exploited or overfished, there is a presumption that at this level of fishing, given the inherent uncertainties, these species are subject to serious or irreversible harm if precautionary measures are not adopted and implemented.

The PA has been adopted by the scientific bodies[38] of five RFMOs (CCAMLR, General Fisheries Commission for the Mediterranean (GFCM), Indian Ocean Tuna Commission (IOTC), Northwest Atlantic Fisheries Organization (NAFO), and NEAFC).[39] The PA is expressly mandated in the treaties of three post-UNFSA RFMOs (Inter-American Tropical Tuna Commission (IATTC) - Antigua Convention, South East Atlantic Fisheries Organisation (SEAFO), and Western and Central Pacific Fisheries Commission (WCPFC)).[40] The ecosystem approach is also manifested in the treaty provisions of five RFMOs (CCAMLR, Commission for the Conservation of Southern Bluefin Tuna (CCSBT), IATTC-Antigua, SEAFO, WCPFC),[41] often complementing the PA. The Central Bering Sea Donut Hole Arrangement does not contain an express reference to either approach, but as a framework document, its terms are sufficiently flexible to permit its scientific and management bodies to likewise adopt these approaches operationally.

From a review of RFMO practice, it would appear that the actual adoption and implementation of the PA consistent with UNFSA are slow and protracted processes. Most RFMOs and States have been preoccupied with the definition and methodological development of reference points, the (re)definition of MSY as a limit, and the development of decision rules. RFMOs have generally neither elaborated nor implemented the wider application of the PA, for a variety of reasons, including the lack of political will, the lack of consensus, inadequate financial resources, and the need for capacity building. While the burden of proof has rarely shifted in practice, the application of the PA to date has translated into greater demands on science. The actual practice of individual RFMOs is detailed below.

6.4.1 Pre-UNFSA RFMOs

6.4.1.1 CCAMLR

While the 1980 CCAMLR Convention, which entered into force in 1982, does not expressly refer to the PA, the Convention was the first international fisheries instrument to outline precaution as a guiding conservation principle.[42] Since that time, CCAMLR has pioneered efforts in operationalizing and implementing the PA in fisheries management. From the principles outlined in the CCAMLR Convention, two central concepts regarding the application and implementation of the PA have developed and can be stated as

(1) Management strives to follow a PA. CCAMLR collects the data it can, then weighs the extent and effect of the uncertainties and gaps in such data before making a management decision.

(2) The approach aims to minimize the risk of long-term adverse effects rather than delaying decisions until all necessary data are available.[43]

CCAMLR was created to restore and protect the living marine resources in the Southern Ocean. Central to CCAMLR's establishment was the need for avoiding over-exploitation of the major prey species (krill), given the associated perturbations on the food chain and adverse impacts on predator species, including whales and seals. As such, "concerns about uncertainty, risk and irreversibility of fisheries impacts on resources and the ecosystem" were included in the CCAMLR constitutive agreement and shaped its direction. Precautionary catch limits for the krill fishery were first considered in 1990[44] and adopted in 1991 with a total catch of krill in Statistical Area 48 set at 1.5×10^6 tonnes (t).[45] Lack of data and uncertainty in estimating total krill biomass have precluded CCAMLR from adopting precautionary total allowable catches (TACs) prior to this time, as a result of disagreement among Contracting Parties.[46] Precautionary TACs were set for krill in additional statistical areas in subsequent years.[47] Current Conservation Measures have built on earlier efforts. In setting the TAC for this forage species, a share of the TAC is allocated to top predators.

Despite CCAMLR's commitment to precautionary management[48] and the establishment of precautionary catch limits, estimated krill biomass is below the long-term objective of precautionary yield.[49] Illegal, unreported, and unregulated (IUU) fishing by member and nonmember nations with a variety of gear types has resulted in a greater uncertainty in long-term predictions of krill catches.[50]

CCAMLR incorporates uncertainty in its management strategy in respect to the possible effects on both fisheries and the ecosystem by adopting a risk-averse management paradigm (i.e., uncertainty is dealt with in the interests of minimizing risk of irreversible changes in the system as well as unsustainable fishing practices).[51] As noted by Constable et al., "while the precautionary approach is now entrenched in its work, essential work is still required to develop management procedures that are robust against the unknown and uncertain behaviors of the Antarctic ecosystem".[52] CCAMLR's experience with the essential principles of conservation and efforts to operationalize these principles through practical measures and achievable goals can serve as a lesson to newer RFMOs and those struggling to achieve sustainable fisheries. "First, conservation objectives can only be achieved by implementing management measures even when very little is known. Secondly, methods have been found for achieving scientific consensus despite the uncertainties surrounding estimates of parameters and the behavior of the system".[53] Scientific consensus has been achieved through analyses that provide a series of outcomes that can be tested through simulated management measures, whereby scientists are not required to agree that any given effect may or may not happen, only that it cannot be ruled out.[54]

CCAMLR determines PA catch limits using a generalized yield model (GYM) for all assessed fisheries, taking into account the biology of the species and the known stock size.[55] While UNFSA does not specify a preferred scientific model, the GYM model is more robust and sophisticated than most single-species models, as it considers factors other than the spawner-recruit relationship. Long-term annual yields are assessed against "objectives which are expressed in terms of probabilistic decision rules, including limit reference points adopted by the Commission, based on advice of the Scientific Committee".[56] The mackerel icefish fishery is limited by precautionary catch limits, area restrictions, and gear restrictions in the area surrounding South Georgia.[57] The crab fishery in Statistical Subarea 48.3 is limited by a precautionary catch limit of 1600 t; each member of the Commission may only have one vessel targeting crab, and gear is limited to pots.[58] The fishery is also subject to an experimental harvest regime, which includes limits on time spent fishing (pot time limits) and restrictions on areas where pots may be set.[59] In the past several years, there has not been any directed fishing for crab in the CCAMLR Convention Area. Because of IUU fishing, particularly on Patagonian and Antarctic toothfish stocks, a continued challenge for the Commission is setting precautionary TACs that reflect the unknown catch levels.

In the CCAMLR Convention Area, the PA is applied on an ecosystem-wide basis and not only just for target-species management measures. The scientific bodies of CCAMLR undertake annual assessments of population status and trends of both harvested and dependent species. As Parkes observes, "CCAMLR's application of the Precautionary Approach extends not only to managing fisheries in the single species sense, but also to making explicit allowance for the requirements of dependent species, and the uncertainty in ecological relationships within the Antarctic ecosystem".[60] Given the difficulty in identifying and understanding possible causes of perturbations to the marine ecosystem and the subsequent impact on nontarget species from fishing krill, precaution and the ecosystem approach have been the twin principles guiding CCAMLR's management strategy since its inception. Causes of ecosystem change "can result from the krill fishery, or from environmental changes, or natural biological variability or cycles of the target species. That is why it is important to apply appropriate levels of precaution when taking decisions regarding the impacts of the krill fishery on the ecosystem".[61]

In keeping with the precautionary objective of Article II of its Convention — to ensure that management decisions have a low risk of long-term adverse effects — CCAMLR continues to develop biological reference points upon which to base its decision rules. According to Denzil Miller, "these rules have benefited enormously from recent initiatives to refine plausible models for testing approaches to krill management".[62] In the interests of precaution, CCAMLR interprets the dependent predator population's greatest net annual increment (GNAI) as set out in Article II of its Convention, as being that which would occur if its prey were not subject to fishing. However, since this relationship is not known for most marine species, CCAMLR applies an interim decision-based management approach. Rather than trying to monitor abundance of dependent species in relation to a specified GNAI level, CCAMLR applies precautionary catch limits. These aim to ensure that fishing for prey species is limited to a level that is unlikely to impact persistently on predators; not complying with these limits would violate the principles of Article II.[63]

In addition to developing PA TACs for target species, since 1991 CCAMLR has developed mechanisms for the precautionary management of new fisheries[64] and since 1993 has included exploratory fisheries and appropriate guidelines. In CCAMLR, a "new" fishery refers to a species and (or) fishing on a ground that has not been previously fished (so the designation encompasses proposals to expand the fishing area). For a new fishery, information on the target species and dependent species must be collected, and catch or effort may be limited. A fishery is considered a new fishery in its first year,[65] and in the second year, the fishery becomes an "exploratory" fishery.[66] Since 2001 all new fisheries in the CCAMLR Convention Area must adhere to conservation rules before fisheries commence, in accordance with the PA[67], as stipulated in Article 6(6) of UNFSA, which calls for States to implement conservation measures for new and exploratory fisheries.[68] These "measures are intended to control the development and prevent expansion of the fishery"[69] when the actual size of the exploited stock is uncertain or data on a new stock being targeted are poor or unavailable.

When data are insufficient to estimate PA catch limits, calculations of limits for new and exploratory fisheries are adjusted for fishable area using biological data from existing fisheries in the GYM, especially if the new fishery is to expand beyond existing fishing grounds. The catch limit calculations should also allow for recent catch history and unreported catches. Where the proposed new fishery is for a species that has not previously been targeted, management needs to be even more cautious. Worldwide, new fisheries are often exploited — even overexploited — prior to regulatory appraisal and well before the necessary information is available or even collected for scientifically defensible management. CCAMLR's PA attempts to acknowledge this reality by striving to minimize risks of irreversible changes in the status of the target stock and by recognizing that fisheries need to be managed from the outset.[70]

Evidence that this management approach is working to some extent is seen through the 2004 proposal by the United Kingdom to expand bottom trawling for icefish in Statistical Subarea 48.3. The proposal was withdrawn after concerns were expressed by delegations about potential impacts on fragile coral and sponge communities.[71] This decision is consistent with the PA and with the requirement for risk assessment and reversal of the burden of proof, before approving a new fishery.[72] As of June 2005, directed fishing for all finfish is prohibited in Subareas 48.1 and 48.2 and for specific finfish in other areas. These measures are to remain in place until further information is gathered and analyzed by the Scientific Committee and the Working Group on Fish Stock Assessment (WG-FSA).[73]

In efforts to combat the IUU fishery for toothfish, CCAMLR has instituted a requirement that exploratory fisheries follow clearly defined experimental fishing plans. This approach strives to maximize the data collection potential of fishing vessels while ensuring that unacceptable damage is not inflicted on stocks for which key management data are missing. Fishing vessels are required to undertake research on stock distribution and abundance as part of their development of either new or exploratory fisheries. Once a new or exploratory fishery commences, all relevant Conservation Measures, data collection, and submission procedures, as well as bycatch mitigation measures, apply. Fishery research plans are implemented, allowing further refinement of precautionary yields. Although concern has been voiced regarding further expansion in the number of exploratory fisheries,[74] CCAMLR has had difficulty restricting the expansion of exploratory fisheries in the face of commercial pressure to increase harvest levels.[75]

Despite CCAMLR's adoption of precautionary and ecosystem approaches and implementation of management measures for new and exploratory fisheries, the Commission continues to battle with enforcement issues and IUU fishing of toothfish.[76] CCAMLR established PA limits for the Patagonian toothfish fishery in 1996 and 1997 following the expansion of this fishery and an increase in IUU fishing. However, ongoing illicit activities have undermined science-based approaches to management, often rendering ineffective the precautionary and sustainable management of the toothfish fishery. It is patent that effective control and enforcement are required to achieve a viable regime. CCAMLR has made a concerted effort to reduce IUU fishing, improve compliance, and reduce fishing capacity through a suite of mechanisms discussed in Chapter 9 on MCS and enforcement.[77]

6.4.1.2 IATTC

Since the 1980s, IATTC has included precaution as part of management decisions in the absence of adequate scientific information.[78] IATTC has also included an adaptive management approach when assessing impacts of expanded fisheries on stocks, particularly with the expansion of the purse-seine fishery for bigeye tuna. In keeping with the PA and implementation of decision rules, if catches exceed MSY, now defined by IATTC as a limit reference point, management measures are imposed. For example, as a result of a tenfold increase (from 5000 to 50 000 t) in the purse-seine catch of bigeye tuna, particularly juvenile and midsize fish, the IATTC 1998 binding Resolution on bigeye tuna invoked the PA consistent with the FAO Code of Conduct and UNFSA. The Resolution called for a reduction in catch, specifically by limiting the use of fish aggregating devices (FADs) and transshipment, and by prohibiting the use of FADs once the catch limit is reached.

IATTC permanent scientific staff continually improve the basis for stock assessments through more rigorous statistical and modeling approaches. While the incorporation of Bayesian methods in IATTC stock assessment takes uncertainty and risk into account,[79] the Commission still uses B_{MSY} and F_{MSY} as limit reference points and has not developed B_{pa} or F_{pa} (see Figure 6.1). In October 2003, a workshop on reference points was held to review the reference points for assessments and advice for the fisheries of the Eastern Pacific Ocean. In particular, the workshop focused on a number of areas of uncertainty in assessments and how these would be best addressed to provide a basis for taking precautionary measures. Several general and IATTC-specific changes were recommended that have been since incorporated in the assessment techniques and presentation of advice to the Commission.[80] IATTC held a workshop on stock assessment methods in November 2005 to compare current methods used by the Commission with methods used by other RFMOs and coastal States.[81]

While the Antigua Convention includes the general principle of maintaining populations of targeted species that can produce MSY, Article VI of this Convention explicitly requires the application of the PA, reflecting Article 6 of UNFSA and the FAO Code of Conduct. Accordingly, members of the Commission are required to be more cautious when information is uncertain, unreliable, or inadequate.

6.4.1.3 ICCAT

Shortly after the adoption of UNFSA, the question of distinguishing between the precautionary principle and the precautionary approach, as outlined in the Agreement, was raised during the 14th and 15th meetings of the International Commission for the Conservation of Atlantic Tunas (ICCAT). As Barston reports, the discussion centered on Canada's request for "elucidation of whether ICCAT used in its statistical calculations the precautionary *principle*, or precautionary *approaches*. The committee chairman endorsed the latter; Japan concurred and felt that the effect of the Straddling Stocks Agreement on ICCAT needed further discussion. Canada, in cooperation with the United States, proposed at the fifteenth meeting of the Commission in Madrid, 14–21 November 1997 that a new working group on the precautionary approach should be established by ICCAT".[82] In keeping with the trend in international fisheries management to examine adoption of the PA, the ICCAT Standing Committee on Research and Statistics (SCRS) created the Ad Hoc Working Group on the Precautionary Approach in 1997.

The ICCAT Ad Hoc Working Group on PA "reviewed the progress made in other international fora. It examined the issues of by-catch, migrations, selectivity, abundance estimates, habitat, environmental and ecological considerations, spatial patterns, data collection, measures of stock status, candidates for reference points (mainly MSY related, including proxies), and debated on the opportunity to use MSY as a limit instead of as a target (as required by [UNFSA]). It also examined the issues of harvest control rules, simulation of management strategies, sources of uncertainty, communication between scientists and managers, and made a number of recommendations".[83] The Working Group recommended that scientific evaluation of proposed management options be conducted prior to implementation to ensure consistency with the PA.[84]

Following recommendations of the Working Group and recommendations from the 1996 ICCAT Tuna Symposium in Portugal, an FAO-organized Expert Consultation on Implications of the Precautionary Approach for Tuna Biological and Technological Research was held in March 2000. Key outcomes of this consultation included a need for the reduction of uncertainty in knowledge of target species and bycatch of tuna and related species, as well as the development of new methods to both quantify and reduce such uncertainties.[85] The Working Group on PA met in conjunction with the 2001 SCRS Plenary. At this time, it was noted that the Working Group had fulfilled its immediate mandate and that it could continue to serve as an umbrella group for reporting specifically on management strategy evaluations. The Working Group was given no recommendations for new activities at that time and was effectively suspended by the Commission.[86]

ICCAT continues to use MSY as a target harvest level rather than a limit reference point, as called for in UNFSA.[87] The SCRS presents stock status for species for which adequate data exist and calculates relevant biological reference points. The Commission uses the SCRS advice to construct management options to be implemented by the Contracting Parties. There are several target species that have never been assessed, including the small tunas and the Mediterranean albacore, as a result of data inadequacies and the fact that these fisheries are not the key fisheries under ICCAT competence. Inadequate data preclude conducting stock assessments for skipjack tuna, Eastern Atlantic bluefin, sailfish, South Atlantic swordfish, Mediterranean swordfish, and the North Atlantic shortfin mako.[88]

Part of PA implementation is instituting management measures to address failed management or enforcement strategies that have led to overfishing. This can be done by management strategies to rebuild fish stocks to a predetermined level, which often, but not always, corresponds to a biomass level that can at least produce MSY. ICCAT has developed recovery plans for the Western Atlantic bluefin tuna that are in effect until 2018. Populations of blue marlin and white marlin are managed by recovery plans as of 2006.

ICCAT's precautionary management of North Atlantic swordfish through a recovery plan may be seen as a success story in efforts to restore a population to a sustainable level. The North Atlantic swordfish is under a recovery plan until 2009. Four years into the recovery plan, the 2002 assessment showed strong signs of recovery. This recovery is attributed to a strong year class in 1997 (one year before the implementation of the recovery plan) and to precautionary management of catch levels and discards.[89] The 2004–2005 ICCAT report states "the biomass at the beginning of 2002 was estimated to be 94% (range: 75 to 124%) of the biomass needed to produce MSY. The 2001 fishing mortality rate was estimated to be 0.75 times the fishing mortality rate at MSY (range: 0.54 to 1.06). The replacement yield for the year 2003 was estimated to be the MSY level. As the TAC for North Atlantic swordfish for 2002 was 10 400 t, it was considered likely that biomass would increase further under those catch levels. The TAC set for 2003–2005 is 14 000 t".[90] The stock will be assessed again in late 2006.

In 2004 Canada introduced to the Commission a draft recommendation concerning the PA. Some delegations were of the opinion that ICCAT is unlike any other fisheries organization in that it has clear management objectives and is historically rich in data. This view is not shared by all Contracting Parties, independent research scientists, or NGOs. To date, the Commission has no formalized position on adopting the PA as espoused in UNFSA, but has implemented aspects of precautionary management despite the lack of an overall framework.

6.4.1.4 GFCM

Amendments to the GFCM Agreement in 1997 included the incorporation of the PA in Article III(2). At the 23rd GFCM meeting in 1998 "delegations urged that the precautionary approach be implemented in order to protect biodiversity especially for marine vertebrates (including sharks)".[91] At the Second Session of the Sub-committee on Stock Assessment (SCSA), the need for establishing biological reference points to improve fisheries management within a PA was emphasized. As a follow-up, the 2004 SCSA workshop on biological reference points identified a set of related indicators and favoured the "traffic light approach" (similar to the approach of NAFO) on reference points. This approach was further endorsed by the Scientific Advisory Committee (SAC) and the Commission. Concomitantly, the Sub-committee on Economics and Social Sciences (SCESS) agreed upon a set of social and economic indicators that can be used, for each of SAC fleet segments, in various GFCM management units. The process of defining reference points based on the PA for specific fisheries is a standing activity of SAC, as reiterated in 2005 by the Commission. The 2005 closure of trawl fisheries below 1000 m, because of the uncertainty about the effects of fisheries in those areas and the concern over shrimp spawning grounds, and the 2006 protection of sensitive habitats through prohibition of trawling in three areas of the Mediterranean are also significant precautionary management measures adopted by GFCM.

6.4.1.5 NEAFC

ICES has been a leader in defining, elaborating, and recommending precautionary management strategies. Through its Advisory Committee on Fishery Management (ACFM), ICES offers scientific advice and management recommendations on request to NEAFC. Since the mid-1990s, NEAFC has requested that ICES give PA reference points for all commercial stocks in its regulatory area. Annual ICES advice includes management recommendations for stock rebuilding plans. Efforts have been made to develop precautionary reference points that adhere specifically to UNFSA. Annual advice from ICES on the five main stocks (redfish, blue whiting, Atlanto-Scandian (Norwegian) spring-spawning herring, mackerel, and Rockall haddock) is based on biological reference points, particularly B_{pa} and B_{lim}. ICES advice includes management scenarios with a high probability that the spawning stock biomass does not fall below a certain level (i.e., B_{lim}) where recruitment is thought to be impaired or the dynamics of the stock are unknown. To ensure this, a precautionary spawning stock biomass, B_{pa}, is calculated. If the spawning stock biomass stays above B_{pa}, the probability of falling below B_{lim} should be low. In these management scenarios there are implicit assumptions on levels of probability and risk in relation to the biological condition of the stock.

Between 2002 and 2005 NEAFC adopted several management measures for Rockall haddock. In 2002 NEAFC closed areas adjacent to Rockall Bank in efforts to protect juvenile haddock. In 2003 the European Community (EC) proposed a recovery plan to ICES and requested that ICES comment on the plan, with particular reference to the PA, impacts of fishing on sensitive habitat, and levels of fishing mortality that would lead to the sustainability of the stock. Stock status of Rockall haddock is uncertain because of data deficiencies. At the 2005 Annual Meeting, NEAFC members agreed to consider a recovery plan for Rockall haddock.[92] The fishery remains closed as of 2006. In 2005 members also agreed on long-term management plans and harvest-control rules for Atlanto-Scandian (Norwegian) spring-spawning herring, mackerel, and blue whiting.[93] Despite there being management plans for herring and blue whiting since 2002, there has been no agreement on coastal-State allocations for these species. However, in 2006 an agreement was reached among coastal States on allocations for blue whiting. Herring allocations remain unresolved. The ICES ACFM also issues statements as to whether management by NEAFC is consistent with the PA.

In 2002 NEAFC established a list of previously unregulated deep-seas species found within the NEAFC Regulatory Area. The issue of deepwater fisheries is a major concern for many scientists and NGOs, and one that NEAFC is endeavouring to address. In 2004 ICES noted that most deepwater fisheries[94] "have developed rapidly and the resources which they exploit are especially vulnerable to overfishing…and are considered to be harvested unsustainably".[95] One of the major difficulties in advising on deepwater species is lack of reliable, available data, and therefore there is a need for precautionary management.[96] In response to ICES advice, NEAFC agreed to a 30% reduction in the deep-sea trawl-fishery effort for 2005 onwards.[97] However, the NEAFC management measure states that "the effort shall not exceed 70% of the highest level put into deep-sea fishing in previous years for the relevant species".[98] The NEAFC approach to managing deep-sea fisheries has provoked debate and concern among international NGOs that this RFMO's regulatory response is not precautionary and therefore inadequate to prevent potential and arguably foreseeable harm to these deepwater species.

6.4.1.6 NAFO

In line with UNFSA, NAFO adopted MSY as a statutory limit reference point in 1997. To develop the scientific basis for implementing the PA, NAFO's Scientific Council established a dedicated Working Group in June 1997.[99] Workshops were held in March and May 1998, and a joint session was held with the NAFO Fisheries Commission in May 1999. According to Garcia,[100] the May 1999 meeting recognized that other management tools were needed in addition to reference points and control rules to apply the PA widely. Harmonization of terminology between ICES and NAFO was also deemed important, not only to manage the Northwest Atlantic Ocean in a coherent manner, but also because scientists and managers participate in both fora. In an effort to develop a simpler and transparent formulation, the Working Group proposed a "traffic light approach"[101] for some stocks with limited data. In September 1999, the NAFO General Council endorsed the PA as described in UNFSA and agreed to use the FAO guidelines to develop a PA framework.[102]

Prior to developing a PA framework, NAFO agreed on fisheries closures specifically to address stock depletion. Since 1992 and throughout the 1990s, NAFO has imposed moratoria on cod in Divisions 3M, 3L, and 3NO; redfish in Division 3LN; American plaice in Divisions 3M and 3LNO; witch flounder in Divisions 3NO and 3L (that portion within the regulatory area); and capelin in 3NO.[103] To date, these moratoria are still in place and only the moratorium on yellowtail flounder in 3LNO was lifted in 1998.[104] Whether the closure of a depleted fishery can be characterized as a precautionary measure is arguable. As noted by Shelton et al., "where fisheries have been closed, this is generally because harvesting the remaining fish is no longer commercially profitable and can no longer be justified in any circumstances. Although *post facto* such closures may be described as precautionary, the real test will be what happens if and when the stock recovers".[105]

In 2000 the Fisheries Commission developed a PA implementation plan for 3LNO yellowtail flounder and some other stocks.[106] In 2003 the Scientific Council proposed a PA framework,[107] which was adopted at the September 2004 meeting of the Fisheries Commission. In 2005 the framework was tested on two stocks — 3LNO yellowtail flounder (considered to be recovering as of 1998) and 3M shrimp. At the request of the Fisheries Commission, the Scientific Council provided advice in 2005 for these two selected stocks consistent with the NAFO PA framework. This initial application was a pilot that would serve as a guide for the Commission in applying the PA to the remainder of NAFO-regulated stocks.[108] NAFO has also developed a 15-year rebuilding plan for Greenland halibut.

In 2005 the Fisheries Commission requested that the Scientific Council provide PA information for 2006 for 3LNO yellowtail flounder and 3M shrimp,[109] including limit and PA reference points (as described in Annex II of UNFSA); stock biomass and fishing mortality trajectory over time overlaid on the NAFO PA framework; information regarding the current "zone" the stock is within; proposals for harvest strategies to move the stock to the "safe zone"; and finally, a comparison of PA advice with traditional advice.[110] The Fisheries Council also elaborates on specific elements to be taken into account when considering the PA.[111] Given that many stocks in the NAFO area are well below the levels required to produce MSY, the Scientific Council has also been asked to consider rebuilding plans within a series of time frames, to determine possible scenarios for recovery.[112]

6.4.1.7 Bering Sea Donut Hole

A moratorium on fishing pollock in the Bering Sea Donut Hole has been in effect since 1994, as a result of stock collapse. The Convention on the Conservation and Management of the Pollock Resources in the Central Bering Sea entered into force in 1995. This multilateral agreement, concluded prior to the adoption of UNFSA, does not expressly incorporate the PA but includes a decision-rule mechanism for determining when the fishery can be reopened. The Convention provides that the pollock fishery will not be authorized until a threshold biomass of 1.67×10^6 t of Aleutian Basin pollock is reached. This provision is to ensure the recovery of the pollock stock. Efforts to reopen the fishery and revisit the decision rule prior to reaching the preagreed threshold have been discouraged. Though not specifically identified as "precautionary", the threshold limit represents a precautionary reference point.[113] In 2005 the estimated pollock biomass level in the convention area was 422 000 t, which is substantially below the biomass level that would trigger reopening of the fishery.[114]

6.4.1.8 CCSBT

CCSBT was established in 1994 to manage the fishery for southern bluefin tuna. The CCSBT, unlike other RFMOs, has competence for one stock only and the Commission's competence is effectively defined by the geographic range and distribution of southern bluefin tuna. This species is primarily found between 30°S and 50°S, with one known spawning ground located within, and adjacent to, the Indonesian exclusive economic zone (EEZ), south of Java. Southern bluefin tuna is considered the most commercially valuable fish in the world with approximately 90% of the catch purchased by a monopsonist buyer and consumed in Japan. The stock is "currently at some risk" and "the current status of the stock is now under intense scrutiny".[115]

The CCSBT Convention does not contain any reference to the PA. The Commission's goal is to "ensure, through appropriate management, the conservation and the optimum utilisation of southern bluefin tuna".[116] The Commission's strategy has evolved to recognize the need for incorporating broader ecosystem considerations given existing uncertainties, such as including discarded and undeclared catch into stock assessments for southern bluefin tuna.[117]

CCBST's short history has been shaped by the fact that neither scientists nor the parties were able agree on the TAC for several years following creation of the organization. The dispute between Australia – New Zealand and Japan over a proposed increase in TAC led to the disputed Japanese experimental fishing program (EFP) and ultimately resulted in the southern bluefin tuna cases. The ruling of the International Tribunal for the Law of the Sea (ITLOS) on this case is significant to this discussion on PA in that many commentators contend that the "tribunal applied the precautionary approach to fisheries, although it did not expressly say so".[118]

In 1995 Japan proposed a 50% increase in the TAC as part of an EFP, with the goal of increasing knowledge of stock recovery and population growth projections. It was also proposed that this additional part of the TAC could be used to entice nonparties to join CCSBT's Convention and bide by its conservation and management measures.[119] (Allocations in recent years have been granted to Korea and the Fishing Entity of Taiwan. Indonesia continues to fish within its EEZ. Allocations have also been made to the Philippines as a cooperating nonmember.)[120]

Australia and New Zealand were not in agreement with Japan on the proposed EFP. At the heart of this impasse was the lack of scientific consensus on stock status, stock abundance (biomass), projected recovery times, or level of fishing effort that would cause irreversible harm.[121] Further, there was no previously agreed procedure for the population projections and no effective dispute settlement process. The inability of the parties to reach a consensus on a TAC from 1995 to 1998 resulted in no TAC being set. While the goals of the EFP were to "identify and narrow the range of uncertainty in the stock assessment",[122] New Zealand and Australia maintained the position that to increase the TAC would be to increase the risks of irreversible harm and compound uncertainty.[123]

In 1998 Japan decided to unilaterally begin a pilot EFP. As dispute resolution within the Commission was not successful, Australia and New Zealand submitted the matter to judicial resolution through the LOSC and requested provisional measures through ITLOS.[124] The complaint was based on the general uncertainty about the resource, biological concern of irreversible harm caused by the EFP catches, and the urgency of the situation given the declining stock[125] and the fact that no TAC had been agreed upon since 1995.On the other hand, Japan argued that the goal of the EFP was to reduce uncertainty and that there was no urgency or risk of irreparable harm to the stock.[126]

In its Order of 27 August 1999, ITLOS held that the parties should "act with prudence and caution to ensure that effective conservation measures are taken to prevent serious harm to the stock of southern bluefin tuna"[127] and, as a provisional measure, decided that the parties should refrain from conducting an EFP involving the "catch of southern bluefin tuna, except with the agreement of the other parties or unless the experimental catch is counted against its annual allocation."[128]

In its Award of 4 August 2000, the Arbitral Tribunal decided it lacked jurisdiction to handle the case.[129] The award therefore did not settle the dispute but it did break the decision-making impasse at CCSBT. Following an independent scientific determination that the EFP could not be considered a valid conservation or protection measure, the parties reached consensus in 2001 to end the EFP.

While ITLOS avoided an explicit reference to the PA, in his separate opinion, Judge Laing opined, "it becomes evident that the Tribunal has adopted the precautionary approach for the purposes of provisional measures in such a case as the present".[130] He notes that the Tribunal had not "*per se*, engaged in an explicit reversal of the burden of proof", as that was a matter "best reserved for the stage of the merits, i.e., for the arbitral tribunal".[131] In his opinion, Judge Laing explored the concept of precaution in the face of scientific uncertainty referring to LOSC Articles 116–119 and 290[132] as well as Articles 6 and 7 of the FAO Code of Conduct and Articles 5(c) and 6 of UNFSA.

The ITLOS Order and its application of precaution in a fisheries management dispute were the subject of several subsequent commentaries in international fisheries law.[133] As noted by Schiffman, "the fact that the order of provisional measures was ultimately dissolved by the later constituted arbitral tribunal for lack of jurisdiction need not detract from the ITLOS's important legal reasoning on the issue of Precaution".[134] The case is viewed by many as an historical application of the PA, referencing the LOSC fisheries provisions and Article 290, which allows the prescription of provisional measures to "prevent serious harm to the marine environment". As noted earlier, the precautionary approach to fisheries, under UNFSA and the FAO Code of Conduct, was a specific formulation designed to distinguish it from other applications of precaution with respect to the marine environment, where lack of scientific uncertainty in the face of serious or irreversible harm would result in an injunction or moratoria.

The inability of CCSBT to agree on a TAC for several years resulted in continued nonprecautionary fishing pressure on the southern bluefin tuna stock, high rates of high-grading and discarding, and targeted fishing on spawning grounds, all of which have left the stock at historically low levels. Estimates by an independent scientific panel state that the spawning biomass is 3%–14% of the original spawning biomass[135]. At this level, the stock is extremely vulnerable to environmental variations or variations in recruitment and spawning success. Given its depleted state, it has been suggested that the species be listed under CITES.[136]

While CCSBT did set targets for rebuilding the stock, prospects for stock recovery at current catches are unlikely — it is more likely that the stock will decline than increase. To reduce this trend, CCSBT is considering independent scientific advice to reset targets to lower levels with reduced catch limits and improved enforcement. In 2005 CCSBT adopted a management procedure (CMP-2) and quota reductions to decrease the probability that spawning stock biomass will further decline (projections are to ensure a 2004 spawning stock biomass level by 2014). In 2005 market reviews indicate that 10 000 t of southern bluefin tuna were illegally caught,[137] and this situation was addressed at the October 2006 annual meeting of the CCSBT.[138] CCSBT must urgently adopt and widely apply the PA in keeping with UNFSA and international minimum standards, given the high risk of serious or irreversible harm to the stock from fishing operations.

6.4.2 Post-UNFSA RFMOs

6.4.2.1 IOTC

The IOTC Convention does not make specific reference to the PA; however, the Commission has adopted the PA to some degree by incorporating uncertainty in stock assessments. IOTC does not have specific procedures for establishing precautionary reference points, and no such reference points have been established for specific stocks. In 1999 IOTC did invoke the PA in its Resolution to reduce catches of bigeye tuna, in particular catches of juveniles. This binding Resolution called for the reduction of capacity and area closures to address unsustainable levels of fishing for bigeye tuna.[139] IOTC also noted that the PA was needed in management actions given the difficulty in assessing the status of the bigeye tuna stocks.[140]

In 1999 the Scientific Committee reported that the Working Party on Tropical Tunas had been unable to conclude whether the bigeye tuna stock was currently fully exploited or overexploited. Taking all the indicators into consideration, the Committee agreed that if the catches remain high, the stock is likely to become overexploited, and that taking into account the PA, there is a need for immediate management action[141]. Adoption and implementation of the PA by the IOTC remain a challenge, which has been further compounded by the impacts of the December 2004 tsunami.

6.4.2.2 WCPFC

The WCPFC Convention contains the PA in Article 5(c), and Article 6 further outlines the PA with reference to Annex II of UNFSA. The adoption of the PA is a major strength of the Convention.[142] The PA is required for designing and implementing management measures. The adoption of the PA may mean that target reference points will be set such that there is a reasonably small probability of the limit reference point being exceeded. At the first meeting of the Scientific Committee in 2005, emphasis was placed on the ecosystem approach, of which the PA is an inherent component. With regard to stock assessments, the Commission restated its commitment to a PA and urged the Scientific Committee to take precaution into account in formulating future conservation and management measures.[143] A practical issue is that the successful adoption and implementation of the PA in the WCPFC (as well as other RFMOs) will be determined by the level of consensus within the Scientific Committee and the degree to which scientific advice is taken into account by the Commission in its decision-making process.

Another practical issue is the need for a reduction in bigeye tuna catches, which will subsequently affect the skipjack tuna fishery, as this species is caught in association with the fishery targeting bigeye tuna. While the status of the bigeye tuna stock is of concern, skipjack tuna, as of 2006, is fished sustainably (stock above B_{MSY})[144] and is a significant fishery for Forum Fisheries Agency (FFA) countries. The challenge for the Commission is to adopt precautionary catch limits for bigeye tuna and to determine whether a viable skipjack tuna fishery can continue. The question becomes one of regulatory appraisal, taking into account the social and economic impact of various science-based management scenarios. Under a precautionary management strategy, the associated catches of bigeye tuna would be within an acceptable level of harm (i.e., that the proposed management strategy would not result in serious or irreversible harm).[145] This example illustrates the practical dilemmas of sustainable development and fisheries management throughout the world, where socioeconomic imperatives influence conservation choices and decisions.

6.4.2.3 SEAFO

The Southeast Atlantic Ocean was formerly covered by the International Commission for Southeast Atlantic Fisheries (ICSEAF), with a new convention first proposed in 1997.[146] SEAFO's Convention includes the PA as a guiding principle in Article 7. In implementing the PA, the Commission must consider international best practices regarding its application, including the guidelines in UNFSA, Annex II of UNFSA, and the FAO Code of Conduct. Article 7 of the Convention was agreed upon after a series of modifications based on the concept as expressed in UNFSA.[147]

The first meeting of the Scientific Committee was held in September 2005, and the second meeting of the Commission was held in October 2005. While the PA was not formally adopted, the Scientific Committee recommended that there be no increase in catch for all species fished in the SEAFO Convention Area in 2006.[148]

6.4.3 RFMOs–RFMAs not yet in force

6.4.3.1 SIOFA

The Agreement establishing the Southern Indian Ocean Fisheries Arrangement (SIOFA) contains the PA in Article 4(c), which requires its application in accordance with UNFSA and the FAO Code of Conduct. The Meeting of the Parties has yet to develop reference points or apply the PA to fisheries under its jurisdiction. This Agreement was not adopted prior to May 2006.

6.5 Challenges and opportunities for the implementation of the PA

The development and implementation of management strategies based on the PA and the best available scientific information is a central principle of UNFSA. Indeed, this conception is the first essential pillar of the Agreement articulated by Chairman Nandan in his concluding statement of August 1995.[149] Given that the lack of human understanding of the natural world is immense, and the risks of human overexploitation and irrevocable harm are high, there is urgent need for wide application of the PA. Since the entry into force of UNFSA, the UN General Assembly (UNGA) has repeatedly urged the international community (see, *inter alia*, UNGA Resolutions 59/25 and 60/31) to apply the PA together with the EAF, for the conservation and management of straddling fish stocks and highly migratory fish stocks as a matter of priority.

Despite the admonishments of UNGA resolutions and the juridical expression of the PA in UNFSA and the FAO Code of Conduct, few States and RFMOs have fully implemented the PA. The PA is not merely a laudable goal. It is a fundamental paradigm shift and management imperative to err on the side of conservation and caution. The framework as negotiated under UNFSA and further elaborated by the FAO Guidelines for Responsible Fisheries is within reach of all RFMOs. Full implementation of the PA provides the world community with the potential for an immediate, practical, cost-effective, and rigorous set of tools for responsibly managing fish stocks, avoiding overfishing, and restoring depleted stocks. The PA is a sound and wise concept that enables decision makers to balance conservation objectives with shorter term socioeconomic interests. The principal challenges in implementing the PA within the sustainable development framework are political will and leadership.

Listed below are challenges and opportunities for coastal States and RFMOs in their efforts to adopt and implement the PA.

(1) Developing precautionary reference points, decision rules, and harvest-control rules

MSY as a limit — The post-LOSC redefinition of MSY as part of the elaboration of the PA was a major achievement of UNFSA and represents a significant paradigm shift in international fisheries governance.[150] Although the international community has recognized widely that MSY is an inappropriate target, some RMFOs have been reluctant to reduce fishing mortality and adopt F_{MSY} as a limit reference point, i.e., the maximum yield to be allowed on an exploited stock in accordance with UNFSA.[151] Broad acceptance of MSY as a limit is part of implementing the PA. Arguably, the failure of some RFMOs (who continue to manage using MSY) to reconceptualize MSY is not only a missed opportunity, but also calls into question the Contracting Parties' duty to conserve and *bona fides*.

Control and decision rules — To implement the PA, harvest controls and other preagreed decision rules need to be specified for all commercial stocks under the competence of RFMOs[152] so that limits are not exceeded and target reference points are achieved. Where fisheries are closed, precautionary thresholds need to be developed by RFMOs to predetermine the level at which fisheries will reopen.

Science-based decision making — It is essential to appreciate that the formulaic expression of the PA with reference points, preagreed thresholds, and other decision rules are still products of negotiations. As highlighted in the southern bluefin tuna cases, resolving potential problems related to lack of scientific consensus requires transparent and fair governance with provisional modalities and dispute settlement procedures. This is particularly relevant as there are multiple possible reference points, each with different biological and socioeconomic implications. Given that scientific analysis can result in competing alternatives, yet scientifically objective interpretations, there is a need for a robust governance regime to ensure that scientific consensus can be reached and that preagreed decision rules and protocols are established wherever possible to avoid disputes. This is particularly relevant given that different interpretations can lead to different socioeconomic consequences. Lack of agreement can still occur when advice is nonbiased and apolitical. Lack of agreement can also occur when diplomatic and political pressures influence scientific advice and objectivity.

(2) Data challenges

Knowledge gaps — UNFSA calls for science-based management using the best available scientific information. The PA acknowledges the limitations of science, problems with data, and weaknesses in all scientific and management regimes, underscoring the inherent uncertainty and lack of knowledge involved in predicting the population of specific stocks and impacts on and from marine ecosystems. At the same time, the PA calls for increased data and knowledge to reduce uncertainty. Given the complexity of fisheries management, scientists and managers must endeavour to identify sources of uncertainty, error, and risk within the scientific and management framework. Thus, there is a need for RFMOs and their constituent States to acknowledge what they do and do not know; to address this information gap through improved data collection, exchange, management, and analysis; and to provide scientific advice and decisions based on risk assessment, risk-averse decisions, and precautionary management strategies. As Garcia observes, "[t]he recognition of and focus on uncertainty and risk should lead to a better accounting of (and transparency about) what is poorly known as well as greater efforts to at least identify the major gaps in knowledge as well as the potential consequences, for the resources and, ultimately, for the people".[153]

Data uncertainties — Provision of timely, accurate, and complete data is a basic obligation of all States fishing within the regulatory areas of RFMOs and a key element of UNFSA, the FAO Code of Conduct, and the FAO Compliance Agreement. Whereas LOSC sets a relatively low standard for data collection, UNFSA underlines the need for providing high-quality data that are comprehensive and available in a timely manner. Indeed, Annex I describes the timely collection, dissemination, and analysis of data as "fundamental". As Chairman Nandan emphasized, the "Agreement provides the necessary framework for the collection, provision and exchange of data to ensure proper conservation and management of straddling fish stocks and highly migratory fish stocks".[154] Focussing on what we do know requires a reliable data collection and management system, as well as an effective MCS and enforcement regime throughout the range of the fish stocks. However, the data for many species under the competence of RFMOs are inadequate and there is also a lack of confidence in the available data, thereby precluding the setting or estimation of PA reference points. Under these circumstances, a precautionary management strategy would estimate undeclared catch and take into account increased uncertainty resulting from poor or nonexistent data.

Data reporting — Collecting reliable data to generate stock assessment models and manage in accordance with precautionary reference points is problematic for all regimes because of the lack of reporting and misreporting by nonmembers as well as members of RFMOs. Enhanced monitoring and enforcement by RFMOs and flag States and improved compliance, particularly by member States fishing in RFMO convention areas, with this basic reporting obligation would go a long way toward reducing uncertainty and strengthening the management regime.

Adaptive management — Another important component of implementing the PA is adaptive management. Together with identifying data gaps and areas of uncertainty, Rosenberg[155] stresses the need for establishing a mechanism for incorporating new information as it is acquired and the importance of developing the ability to make rapid systemic adjustments based on the additional knowledge. This demands intersessional (i.e., between Annual General Meetings) adjustments and oversight of the fishery requiring ongoing scientific and management engagement and an enhanced monitoring and coordination role for States and RFMO secretariats.

(3) Wide application of PA

PA regulatory appraisal — Article 6 of UNFSA recognizes the need for rigorous regulatory appraisal where socioeconomic and environmental considerations will influence decisions. The PA decision-making framework calls for precise criteria and defined burden of proof (depending on the data available, the degree of risk, and the consequences of error). In so doing, the current focus on identifying reference points will need to expand to encompass all aspects of decision making. The wide application of the PA has major implications for RFMO science-based decision making, where management objectives are identified and transparent, and where scientists and managers work together to develop management scenarios that are consistent with PA in recognizing that different scenarios have varying biological and socioeconomic consequences.

Socioeconomic considerations — In implementing the PA, the extent to which socioeconomic considerations attenuate scientific findings as well as the determination of what is an acceptable level of harm need to be addressed within an effective decision-making framework. In developing a management strategy, RFMOs must balance short-term and long-term socioeconomic concerns, the status of the stock, and the projected time frame over which stocks are estimated to recover to meet target objectives such as those agreed at the World Summit on Sustainable Development (WSSD). In applying the PA, RFMOs and coastal States must avoid exceeding precautionary limit reference points as well as strive to meet target reference points.

Addressing overcapacity — A major impediment to the implementation of the PA in many RFMOs is overcapacity. Despite the International Plan of Action (IPOA) on Capacity and the explicit obligation to reduce excess capacity under UNFSA, few RFMOs or coastal States have introduced measures to reduce capacity commensurate with a precautionary TAC to avoid overfishing and restore stocks (i.e., limiting fleet capacity at a precautionary level).

Shifting the burden of proof — While the FAO technical guidelines on the PA recognize that the shift in the burden of proof, together with appropriate standards of proof, is a fundamental feature of PA, these important policy and legal questions require delineation and elaboration. Internal challenges of management plans or external legal challenges may precipitate the need to address this evidentiary and paradigmic shift. Further, as RFMOs and coastal States seek to ensure compatible measures and the application of the PA over the entire geographic range of the stocks, compatible standards of proof may also be required, thereby raising the issue of how does one determine compatibility.

Ensuring transparency — With the shift in the burden of proof, science-based management plans within a PA decision framework must be accountable and transparent. The reasoning behind decisions must be provided, outlining the biological assessment and the socioeconomic considerations for the specific stocks and marine ecosystem. Managers may be called upon to explain or prove that the proposed or adopted management plan or strategy is consistent with the PA framework.

Applying spatial management tools — Marine reserves, closed areas (including seasonal closures), and other spatial management techniques could be used to implement the PA. Reserves are increasingly viewed as an important fisheries management and biodiversity conservation tool.[156] Both UNFSA and the CBD's Jakarta Mandate call on States to protect marine biodiversity. Given the history of irresponsible fishing and management errors, exacerbated by natural variation, climate change, and natural oscillations, marine conservation–protection areas are a critical precautionary measure. This is especially true for straddling fish stocks and highly migratory fish stocks in areas where stocks that were once overfished have been typically very slow to recover or appear to be irreparably depleted. In some cases, spatial management measures may provide better resource protection and conservation, ecosystem rebuilding strategies, and controls for fish stocks and critical habitats than conventional fisheries management measures.

(4) Improving financial and technical capacity

Increased costs — Increased science and good data are indispensable elements of precautionary management under the UNFSA regime. However, there are significant costs associated with increased data collection, management and analysis, and monitoring and enforcement. The use of the best available science requires a financial commitment to fund research that will facilitate procurement of high-quality fishery-independent data, develop methodologies to reduce uncertainty, error and risk, and carry out independent research and scientific assessment. All States, even economically developed coastal States, are having difficulty allocating funds to basic marine scientific research within their areas of national jurisdiction.

Financial trade-offs — Managers must weigh the financial burden associated with these programs against the potential socioeconomic and biological costs of overfishing. Some RFMOs have opted not to commit significant budgetary funds to enhancing MCS and (or) data collection and research programs. The lack of a reliable MCS program compromises data quality and the understanding of the fishery. In these circumstances, RFMOs, in accordance with the PA, must acknowledge the increased uncertainty, error, and risk when developing management strategies.

Cooperation and development assistance — While the lack of adequate funds presents a real constraint and underscores the need for precaution, there are increasing opportunities for multilateral research assistance as well as collaborative research programs (e.g., partnerships with academic institutions). A coordinated and collaborative effort amongst RFMOs and between RFMOs and coastal States will be increasingly required to effectively apply an ecosystemic precautionary approach over the entire geographic range of a stock to ensure compatible management.

Development assistance — The lack of financial and technical capacity to undertake data collection and management as well as MCS responsibilities is a major impediment for developing coastal States. This shortfall can be addressed through technical assistance and other initiatives to ensure that all coastal States are able to meet their conservation and management obligations. Some RFMOs have introduced development assistance programs to address this challenge, thereby improving the integrity of the data set.

6.6 Conclusion

While ignorance escapes self-recognition by definition, it is widely acknowledged that there is much we do not know about the marine ecosystem or the impact of fishing operations. The PA recognizes this ignorance and replaces the hubris of previous attempts to manage the resource under the freedom of the seas doctrine with the cautionary, conservation ethos based on humankind's ability to irrevocably damage ocean ecosystems. This paradigmic shift and evolution in fisheries management are in keeping with Socrates' observation that "ignorance is a source of wisdom". Accordingly, the PA calls on managers to cautiously undertake fisheries based on the best available science while being mindful of the inherent uncertainties; the knowledge gaps; the perennial problems with collecting adequate, reliable, and timely data; and the ubiquitous problems of IUU and noncompliance with the conservation regime. RFMO and coastal States' management strategies therefore must always be precautionary, armed with the knowledge that the lack of human understanding of the natural order is immense and that the risks of overexploitation and irrevocable harm are high.

As stated by Caddy and Seijo, "The management of marine fisheries needs to undergo dramatic change in the new millennium, in response to the well-documented evidence of global overfishing and the general depletion of commercial fish stocks".[157] Given the diminished stock status for most straddling fish stocks and highly migratory fish stocks, collapse of many coastal fisheries, degradation of marine ecosystems, and failure of most stocks to rebound despite recovery plans and efforts to reduce catches, the application of the PA is a global imperative. Applying the PA throughout the range of commercial stocks is perhaps the single greatest hope for resource recovery and sustainable development.

The post-UNFSA era is a period of transition and shifting paradigms that holds many challenges for RFMOs and coastal States. UNFSA, through the institutionalization of the PA, provides the international community with an enormous opportunity to make great strides in international fisheries management. RFMOs have a pivotal role in the new regime to ensure that the international law of fisheries, together with its guiding principles, concepts, and minimum standards, is implemented. Since the adoption of UNFSA in 1995, the scientific components of many RFMOs have made considerable progress in giving practical expression to the PA provisions through the definition and methodological development of reference points and harvest controls. This process has largely been a scientific exercise and most RFMOs have not widely applied the PA to encompass all aspects of fishery management. However, in tandem with the move towards adopting and implementing EAF, there is an ever-increasing recognition by many coastal States and RFMOs of the need for implementing the PA, not only through the establishment of stock-specific reference points and harvest-control rules but also through the broader development of precautionary management plans and risk-averse decision frameworks. As Shelton et al. observe, "The scale may only tip towards the implementation of sustainable fisheries practices when the number of disasters (collapses and non-recoveries) outweighs the optimism and denial of many resource users and political systems, and when the general public, concerned with the environmental and resource degradation that has occurred, decides to lean in favour of practices such as those embodied under the precautionary approach".[158] Similarly, Freestone notes that the precautionary approach is an essential component of the new paradigm for international fisheries management, and "it remains to be seen whether the regional bodies charged with its implementation will be able to retain the strength of its original vision...what is most important is that the basic tenets of the approach are mainstreamed into the practice of fishery management regimes throughout the world".[159]

Endnotes

[1] It is generally agreed that the precautionary principle originated in Germany, where it is known as the "vorsorgeprinzip" or principle of precaution. The concept includes risk prevention, cost effectiveness, ethical responsibilities towards maintaining the integrity of natural systems, and the fallibility of human understanding. The substance of the principle is hardly new in so far as it underlies the cautionary aphorism "better safe than sorry" and the Roman maxim *non laedus* — do no harm. For a discussion of the evolution of the precautionary principle and the development of the precautionary approach in fisheries conservation and management, see D. Freestone, "Implementing Precaution Cautiously: The Precautionary Approach in the Straddling and Highly Migratory Fish Stocks Agreement" in E. Hey, ed., *Developments in International Fisheries Law* (The Hague: Kluwer Law International, 1999) 287–325.

[2] *Bergen Ministerial Declaration on Sustainable Development in the ECE Region*. Adopted at the Regional Conference on Action for a Common Future for the Economic Commission for Europe held in Bergen, Norway, 14–16 May 1990. Available at <http://www.dep.no/md/nsc/declaration/022001-990330/dok-bn.html>.

[3] W. Burke, "Compatibility and Precaution in the 1995 Straddling Stocks Agreement" in H. Scheiber, ed. *Law of the Sea: The Common Heritage and Emerging Challenges* (The Hague: Martinus Nijhoff Publishers, 2000) 105–126 at 123, opines that "One sorry lesson from the driftnet experience in the United Nations General Assembly is the demonstration, perhaps not needed, that politics continues to trump science when emotions are high. In this instance, as in others, the use of partial and inadequate information had two consequences: it created pressure on those who make decisions and it was cited to justify the intervention that resulted".

[4] J. Ellis, "The Straddling Stocks Agreement and the Precautionary Principle as Interpretive Device and Rule of Law" (2001) 32 *Ocean Development and International Law* 289–311.

[5] M. Lodge and S. Nandan, "Some Suggestions Towards Better Implementation of the United Nations Agreement on Straddling Fish Stocks and Highly Migratory Fish Stocks of 1995" (2005) 20(3–4) *International Journal of Marine and Coastal Law* 345–379 at 359 (footnote excluded quoting Principle 15 of the Rio Declaration).

[6] UNFSA Annex II: Guidelines for the application of precautionary reference points in conservation and management of straddling fish stocks and highly migratory fish stocks.

[7] J. Caddy and R. Mahon, "Reference Points for Fisheries Management", FAO Fisheries Technical Paper No. 347 (Rome: FAO, 1995); and J. Caddy, "A Short Review of Precautionary Reference Points and Some Proposals for their Use in Data Poor Situations", FAO Fisheries Technical Paper No. 379 (Rome: FAO, 1998).

[8] FAO, "Research Implications of Adopting the Precautionary Approach to Management of Tuna Fisheries", FAO Fisheries Circular No. 963 (Rome: FAO, 2001).

[9] The FAO Code of Conduct (1995) describes the precautionary approach as, *inter alia*, (1) being more cautious when information is less certain; (2) not using the absence of information as a reason to postpone or fail to implement conservation and management measures; (3) when developing management plans, taking account of uncertainties relating to the size and productivity of stocks, reference points, stock condition relative to those reference points, levels and distribution of fishing effort, and impacts on nontarget and associated or dependent species; (4) taking a very cautious approach to the management of newly developing fisheries until sufficient data are available to assess the impact of the fishery on the long-term sustainability of the resource; (5) defining and implementing limit and target reference points; and (6) defining, in advance, decision rules for stock assessment. FAO Code of Conduct for Responsible Fisheries, adopted by the Twenty-eighth Session of the FAO Conference, 31 October 1995. FAO online <http://www.fao.org/documents/show_cdr.asp?url_file=/DOCREP/005/v9878e/v9878e00.htm> (accessed 21 October 2005).

[10] See discussion of 7th North Pacific Rim Fisheries Conference, Busan, Korea, Wrap Up Panel Discussion, 20 May 2004, <http://pr-fisheries.alaskapacific.edu/uploads/Wrap%20Up%20Panel%20Discussion-May%2020.pdf> (accessed 21 April 2006).

[11] S. Garcia, "The Precautionary Approach to Fisheries: Progress Review and Main Issues (1995–2000)" in M. Nordquist and J.N. Moore, eds., *Current Fisheries Issues and the Food and Agriculture Organization of the United Nations* (The Hague: Kluwer Law International, 2000) 470–560.

[12] The shift in the burden of proof is clearly part of the FAO Technical Consultation on the Precautionary Approach to Marine Capture Fisheries : FAO, "Precautionary Approach to Fisheries. Part 1: Guidelines on the Precautionary Approach to Capture Fisheries and Species Introductions". Elaborated by the Technical Consultation on the Precautionary Approach to Capture Fisheries (Including Species Introductions), Lysekil, Sweden, 6–13 June 1995 (A scientific meeting organized by the Government of Sweden in cooperation with FAO), FAO Fisheries Technical Paper No. 350, Part 1 (Rome: FAO, 1995) at sec. 1, and as discussed in S. Garcia, at 485–487. This aspect of the PA remains to be fully understood and requires further practical elaboration. Those who put forward an alternative management strategy that does not err on the side of conservation or is considered less precautionary must prove that their proposal will not cause serious or irreversible harm.

[13] As J. Ellis observes, "The influence of the precautionary approach can be seen in four major categories of obligations in the Agreement: first, obligations related to information collection and exchange; second, obligations related to the manner in which decision making is to proceed; third, obligations related to the nature of measures to be implemented; and fourth, obligations relating to measures to be taken when certain triggering events occur. Beyond these specific categories, the approach may be employed as a basis for interpreting the text of the Agreement and other instruments within the regime in a manner favorable to conservation goals", at 299.

[14] FAO Fisheries Technical Paper No. 350, Part 1, at sec. 1(7).

[15] European Commission Communication on the Precautionary Principle was delivered 2 February 2000. Full text of the communication is available at <http://europa.eu.int/comm/environment/docum/20001_en.html> (accessed 15 October 2005).

[16] Although the term burden of proof is used here in relation to the precautionary approach, the term is not without its difficulties. The principal problem is that its meaning can be confused with that used in court proceedings, especially in countries that have the common law system. Thus used, the analogy is not entirely helpful. For example, strictly speaking, Art. 61.2 of the LOSC sets a low standard (of proof)

for conservation measures, namely reliance on the best scientific evidence available. On the other hand, the PA, which is to be applied widely, urges caution when information is uncertain, unreliable, or inadequate. Further, specific obligations are placed on States in Art. 6.3 of UNFSA, which have the effect of both increasing the standard of proof with respect to conservation and management measures and placing a burden on the State wishing to undertake fishing to demonstrate that fishing can reasonably be undertaken. This is the sense given to the term "shift in the burden of proof' in the PA.

[17] The precautionary reference points are target levels of fishing effort designed to ensure that the abundance of fish stocks is maintained at a level above that which can produce MSY. In setting precautionary reference points, the resilience and reproductive capacity of stocks, as well as major sources of uncertainty related to the knowledge of the stocks and the fisheries exploiting them, are to be taken into account. "Precautionary management reference points are those intended to prevent recruitment overfishing, even if a direct link between spawning stock size and subsequent recruitment cannot be established statistically. In order to be precautionary, the appropriate null hypothesis is that the stock does not compensate for reduced abundance by increasing productivity. There are many limit reference points that can be used to prevent recruitment overfishing. However, it is unlikely that a single limit reference point can provide protection for the resource under all foreseeable circumstances and, thus, it is useful to define harvest control laws which combine several thresholds. A precautionary management strategy should contain, in addition to limit reference points, *a priori* decision rules on the acceptable probability (risk) that recruitment overfishing will take place, given the target harvest and the estimated stock status. The choice for a particular definition of risk should be tied to uncertainty and precautionary management dictates that more uncertain situations require more conservative measures". A. Rosenburg and V. Restrepo, "Precautionary Management Reference Points and Management Strategies" in FAO, *Precautionary Approach to Fisheries. Part 2. Scientific Papers*, FAO Fisheries Technical Paper No. 350, Part 2 (Rome: FAO, 1996) 129–140 at 29.

[18] M. Lodge and S. Nandan, at 359 (footnotes excluded).

[19] M. Lodge and S. Nandan, footnote 54 at 359 referring to report by J. Caddy and R. Mahon, FAO Fisheries Technical Paper No. 347. See also ICES Advice on Reference Points, available at <http://www.marine.ie/NR/rdonlyres/510E2331-F1E6-456C-95C1-FB-36CA23302F/0/ICEStheFormofAdvice.pdf> (accessed 23 May 2007).

[20] See ICES advice on reference points, available at <http://www.marine.ie/NR/rdonlyres/510E2331-F1E6-456C-95C1-FB-36CA23302F/0/ICEStheFormofAdvice.pdf> (accessed 23 May 2007).

[21] A fishery is sustainable when the spawning stock biomass is within the precautionary biomass levels (B_{pa}) and when fishing mortality is below the precautionary level (F_{pa}). The range leaves room for year-to-year variation in recruitment and year-class success. A fishery is overfished when the spawning stock biomass is less than B_{pa}, or when fishing mortality is higher than F_{pa}. A stock is in danger of collapse when the spawning stock biomass is less than B_{lim} or when fishing mortality is over F_{lim}. A model with such thresholds allows for decisions to be made within the precautionary approach paradigm.

[22] See CEFAS document for general description of ICES precautionary approach, including growth overfishing and recruitment overfishing. Centre for Environment, Fisheries and Aquaculture Science (CEFAS), "Overfishing and the ICES Precautionary Approach to Management" (Lowestoft: CEFAS, n.d.). CEFAS online <http://www.cefas.co.uk/data/fisheries-information/management-and-advice/overfishing-and-the-ices-precautionary-approach-to-management.aspx> (accessed 21 October 2005).

[23] Because general knowledge of the complex marine ecosystem is limited and the possible effects of fisheries are poorly understood, it is important to incorporate uncertainty into fisheries science and decision making. Chap. 7 elaborates on the ecosystem approach to fisheries.

[24] M. Hayashi, "The Role of the United Nations in Managing the World's Fisheries" in G. Blake et al., eds., *The Peaceful Management of Transboundary Resources* (Leiden: Martinus Nijhoff Publishers, 1995) 373–393 at 382.

[25] As MacDonald notes, using MSY as a management tool led to many variables being ignored in allocation decisions, particularly biological variables such as minimum reproductive biomass, safe biological limits, optimum recruitment levels, and maximum statistical probability of ecological and economic collapse. J. MacDonald, "Appreciating the Precautionary Principle as an Ethical Evolution in Ocean Management" (1995) 26 *Ocean Development and International Law* 255–286 at 255.

[26] A. Punt and A. Smith, "The Gospel of Maximum Sustainable Yield in Fisheries Management: Birth, Crucifixion and Reincarnation" Chap. 3 in J. Reynolds, G. Mace, K. Redfnod, and J. Robinson, eds., *Conservation of Exploited Species* (Cambridge: Cambridge University Press, 2001).

[27] See S. Garcia, at 510.

[28] See S. Garcia, at 483.

[29] This is supported by the Code of Conduct, which states in Art. 7.3.3 "Long-term management objectives should be translated into management actions, formulated as a fishery management plan or other management framework".

[30] Following the adoption of UNFSA and the FAO Code of Conduct for Responsible Fisheries, the concept of the ecosystem approach to fisheries was discussed during the Reykjavik Conference and elaborated through the FAO Guidelines to Ecosystem Based Management. See further the discussion in Chap. 7.

[31] Requirements for PA management are summarized by I. Lutchman in C. de Fontaubert and I. Lutchman with D. Downes and C. Deere, *Achieving Sustainable Fisheries: Implementing the New International Legal Regime* (Gland: IUCN, 2003) 31–44, and are specified under the FAO guidelines, see FAO, "Fisheries Management 2. The Ecosystem Approach to Fisheries", FAO Technical Guidelines for Responsible Fisheries No. 4, Suppl. 2 (Rome: FAO, 2003).

[32] W. Burke, at 122.

[33] See S. Garcia, at 531.

[34] See FAO Technical Paper No. 350. Part 2 (1996); J. Caddy, FAO Technical Paper No. 379 (1998); and FAO, "Indicators for Sustainable Development of Marine Capture Fisheries", FAO Technical Guidelines for Responsible Fisheries No. 8 (Rome: FAO, 1999); and FAO, "Precautionary Approach to Capture Fisheries and Species Introductions", FAO Technical Guidelines for Responsible Fisheries No. 2 (Rome: FAO, 1996).

[35] Largely quoted from S. Garcia, at 485–487.

[36] ICES, upon the request of the European Union (EU), has developed a procedure for implementing a PA in its advice to the Commission on fish stocks and future catch levels. The precautionary reference figures produced by ICES are used by member States to negotiate catch quotas. Unfortunately, these negotiations often result in quotas exceeding the ICES recommendations. Many fish stocks are now at levels below the precautionary reference point and some are below the limit reference point and therefore these stocks require drastic recovery plans. See also, ICES, *Report of the Precautionary Approach to Fisheries Management*, Copenhagen, Denmark, 5–11 February 1997, ICES CM 1997/Assess: 7, ICES online <http://www.ices.dk> (accessed 21 October 2005) and ICES, *Report of the Precautionary Approach to Fisheries Management*, Copenhagen, Denmark, 3–6 February 1998, ICES CM 1998/ACFM: 10. Canada has incorporated the PA into policy documents on fisheries management, but continues to develop practical implementation measures. See discussion in D. VanderZwaag, S. Fuller, and R. Myers, "Canada and the Precautionary Principle/Approach in Ocean and Coastal Management: Wading and Wandering in Tricky Currents" (2002–2003) 34 *Ottawa Law Review* 117. For efforts in the United States, see V. Restrepo, ed., *Proceedings of the Fifth National NMFS Stock Assessment Workshop: Providing Scientific Advice to Implement the Precautionary Approach Under the Magnuson–Stevens Fishery Conservation and Management Act*, U.S. Dep. Commer. NOAA Technical Memorandum NMFS-F/SPO-40 (Silver Springs, Md.: US Department of Commerce, NOAA, 1999). For efforts in Canada see DFO, 2006 "A Harvest Strategy Compliant with the Precautionary Approach", DFO Can. Sci. Advis. Sec. Sci. Advis. Rep. 2006/023.

[37] J. Caddy and R. Mahon, FAO Fisheries Technical Paper No. 347, and J. Caddy, FAO Fisheries Technical Paper No. 379, compare and contrast efforts by NAFO and ICES to develop precautionary reference points and the relation of such points to MSY. In summary, at the time NAFO's PA can be summarized by the "control law" (as in Serchuk et al. 1997) where managers are to ensure that spawning stock biomass (SSB) >> buffer biomass (B_{buf}) > biomass limit (B_{lim}) and maintain target stock (F_{target}) ≤ buffer fishing mortality (F_{buf}) < limit stock (F_{lim}). ICES, which serves as an advisory group to NEAFC, developed control laws as follows: $F = F_{0.1}$ for $B >$ precautionary approach biomass (B_{pa}); $F = F_{0.1}[(B - B_{lim})/(B_{pa} - B_{lim})]$ for $B_{lim} < B \leq B_{pa}$; and $F > 0$ for $B < B_{lim.}$ These control laws reflect the fishing mortality based on the biomass of the stock and its abundance in relation to the biomass at maximum sustainable yield (B_{MSY}).

[38] The decision to pursue a PA is a political or management decision. The management body will then instruct the scientific body to operationalize the PA. Based on this operationalization, the scientific body will offer scientific advice to the management body, which then again may take a decision based on a PA.

[39] See GFCM online <http://www.fao.org/fi/body/rfb/GFCM/gfcm_home.htm> (accessed 26 August 2005); ICCAT online <http://www.iccat.es/> (accessed 26 August 2005); NAFO online <http://www.nafo.ca/About/FRAMES/AbFrMand.html> (accessed 26 August 2005); NEAFC online <http://www.neafc.org/> (accessed 26 August 2005); CCAMLR online <http://www.ccamlr.org/pu/e/gen-intro.htm> (accessed 26 August 2005); IOTC online <http://www.iotc.org/English/index.php> (accessed 26 August 2005).

[40] IATTC under the Antigua Convention, at Art. IV, *Convention for the Strengthening of the Inter-American Tropical Tuna Commission established by the 1949 Convention between the United States of America and the Republic of Costa Rica*, "Antigua Convention", done at Washington, D.C., 14 November 2003, IATTC online <http://www.iattc.org/PDFFiles2/Antigua_Convention_Jun_2003.pdf> (accessed 21 October 2005) not yet in force; *Convention on the Conservation and Management of Fishery Resources in the South-East Atlantic Ocean*, done at Windhoek, Namibia, 20 April 2001, entered into force 30 April 2003, SEAFO online <http://www.seafo.org/> (accessed 21 October 2005), Arts. 3(b) and 7; *Convention on the Conservation and Management of Highly Migratory Fish Stocks in the Western and Central Pacific Ocean*, done at Honolulu, Hawaii, 5 September 2000, entered into force 19 June 2004, 2004 ATS 15 Art. 6.

[41] IATTC- Antigua Convention at Art. VII.1(f); *Convention on the Conservation of Antarctic Marine Living Resources*, done at Canberra, Australia, 20 May 1980, entered into force 7 April 1982, ATS 1982 No. 9; 19 ILM 841, at Art. II(3)(b); *Convention for the Conservation of Southern Bluefin Tuna*, done at Canberra, Australia, 10 May 1993, entered into force 20 May 1994, 1994 ATS 16, at Art 9.2; SEAFO, at Arts. 3(c), 3(d), and 3(e); WCPFC at Arts. 5(b), 5(d), 5(e), 5(f), 8.2(a), and 10.1(c)

[42] CCAMLR Convention, at Art. II(3)(c), embodies the risk-averse nature of the PA, stating that any harvesting should consider the "prevention of changes or minimization of the risk of changes in the marine ecosystem which are not potentially reversible over two or three decades, taking into account the state of available knowledge of the direct and indirect impact of harvesting, the effect of the introduction of alien species, the effects of associated activities on the marine ecosystem and of the effects of environmental changes, with the aim of making possible the sustained conservation of Antarctic marine living resources".

[43] C. Hedley, E. Molenaar, and A. Oude Elferink, *The Implications of the UN Fish Stocks Agreement (New York, 1995) for Regional Fisheries Organisations and International Fisheries Management* (Brussels: European Parliament, 2003).

[44] CCAMLR IX, Item 8. Consideration of the implications of possible limits on krill catches in Subarea 48.3.

[45] The 1991 advice of CCAMLR's Scientific Committee, as quoted in C. de Fontaubert and I. Lutchman, *Achieving Sustainable Fisheries: Implementing the New International Legal Regime* (Gland: IUCN, 2003) and as specified under the FAO (FAO Technical Guidelines for Responsible Fisheries No. 4, Suppl. 2, at 27) was "the reactive management, the practice of taking management action when the need for it has become apparent is not a viable long-term strategy for the krill fishery. The adoption of a long-term strategy that entailed adjustments to management measures with increasing information was endorsed. In the interim, a precautionary approach is desirable and in particular, a precautionary limit on annual catches should be considered". Conservation Measure 32/X, adopted in 1991, was the first binding regulation in CCAMLR. For a detailed map of the CCAMLR Convention Area with subsections see <http://www.ccamlr.org/pu/e/conv/map.htm> (accessed 6 July 2007). See also map in Part III of this book.

[46] CCAMLR IX, Item 8.3.

[47] CCAMLR XI, Item 9.6; CCAMLR XIV, Item 8.10; CCAMLR XV, Item 8.5.

[48] CCAMLR Conservation Measure 51-03 (2005) "Precautionary catch limitation on *Euphausia superba* in Statistical Division 58.4.2". Conservation Measures 51-01 and 51-02 contain catch limitations to other areas. In addition, krill stock assessments include additional components in the generalized yield models, to account for seasonal growth; fishing seasons; and uncertainties in biomass estimates, recruitment, and natural mortality.

[49] A. Constable, W. de la Mare, D. Agnew, I. Everson, and D. Miller, "Managing Fisheries to Conserve the Antarctic Marine Ecosystem: Practical Implementation of the Convention on the Conservation of Antarctic Marine Living Resources (CCAMLR)" (2000) 57 *ICES Journal of Marine Science* 778–791 at 789. Parties have requested expansion of the fishery at annual meetings of the Commission since 2000.

[50] E. Fanta, *CCAMLR Report 2004: CCAMLR XXIII: Commission for the Conservation of Antarctic Marine Living Resources*, Hobart, Australia, 25 October – 5 November 2004 (Curitiba, Brazil: Universidade Federal do Paraná, 2004). Scientific Committee on Antarctic Research (SCAR), available at <http://www.scar.org/researchgroups/lifescience/ccamlr/hobart04/fantaCCAMLR2004.pdf> (accessed 21 October 2005), at 5.

[51] For detailed discussions of CCAMLR's management of specific stocks, see A. Constable, "CCAMLR Ecosystem Monitoring and Management: Future Work" (2002) 9 *CCAMLR Science* 233–256; A. Constable, "A Possible Framework in which to Consider Plausible Models of the Antarctic Marine Ecosystem for Evaluating Krill Management Procedures" (2005) 12 *CCAMLR Science* 99–117; I. Everson, "Consideration of Major Issues in Ecosystem Monitoring and Management" (2002) 9 *CCAMLR Science* 213–232; and D. Miller, "Antarctic Krill and Ecosystem Management — From Seattle to Siena" (2002) 9 *CCAMLR Science* 175–212.

[52] A. Constable, W. de la Mare, D. Agnew, I. Everson, and D. Miller, at 789.

[53] A. Constable, W. de la Mare, D. Agnew, I. Everson, and D. Miller, at 789. The authors consider this principle to be an essential element in international fisheries management, and it is increasingly being implemented in the fisheries management of many coastal States.

[54] W. de la Mare, "Some Recent Developments in the Management of Marine Living Resources" in R. Floyd, A. Sheppard, and P. Barro, eds., *Frontiers of Population Ecology* (Melbourne, Australia: CSIRO Publishing, 1996) 599–616.

[55] See A. Constable and W. de la Mare, "A Generalized Yield Model for Evaluating Yield and the Long Term Status of Fish Stocks Under Conditions of Uncertainty" (1996) 3 *CCAMLR Science* 31–54. The GYM was originally developed from the krill yield model, which involves the proportion of unexploited biomass that can be caught each year, see D. Miller and D. Agnew, "Krill Fisheries in the Southern Ocean" in I. Everson, ed., *Krill: Biology, Ecology and Fisheries. Fish and Aquatic Resources Series* 6 (London: Blackwell Science, 2000) 300–337. The approach was later adopted by CCAMLR for other species (A. Constable and W. de la Mare, "Bias in the Estimation of Krill Yield from Using a Deterministic Formulation of the Median Unexploited Spawning Biomass" (1998) 5 *CCAMLR Science* 1–8. G. Kirkwood and A. Constable, "Integration of CPUE Data into Assessments Using the Generalized Yield Model" (2001) 8 *CCAMLR Science* 65–74.

[56] G. Parkes, "CCAMLR's Application of the Precautionary Approach" (1999) Proceedings, 5th NMFS NSAW. NOAA Technical Memo. NMFS-F/SPO-40, 87–95 at 87. See a similar version of this paper in G. Parkes, "Precautionary Fisheries Management: The CCAMLR Approach" (2000) 24 *Marine Policy* 83.

[57] CCAMLR Conservation Measure 42–01 (2004) "Limits on the fishery for *Champsocephalus gunnari* in Statistical Subarea 48.3 in the 2004/05 season".

[58] CCAMLR Conservation Measure 52-01 (2005) "Limits on the fishery for crab in Statistical Subarea 48.3 in the 2004/05 season".

[59] CCAMLR Conservation Measure 52-02 (2005) "Experimental harvest regime for the crab fishery Statistical Subarea 48.3 in the 2004/05 season".

[60] G. Parkes, at 89.

[61] E. Fanta, at 2.

[62] Compilation of Reports, Fourth Meeting of Regional Fishery Bodies, Rome, Italy, 14–15 March 2005, RFB/IV/2005/5 at 3. Denzil Miller is the Executive Secretary of CCAMLR, as of 2006.

[63] Compilation of Reports, Fourth Meeting of Regional Fishery Bodies, Rome, Italy, 14–15 March 2005, RFB/IV/2005/5 at 3

[64] See CCAMLR, *Report of the Ninth Meeting of the Commission*, CCAMLR IX, Hobart, Australia, 22 October – 2 November 1990.

[65] CCAMLR Conservation Measure 21-01 (2002) "Notification that Members are considering initiating a new fishery".

[66] CCAMLR Conservation Measure 21-02 5 (2002) "Exploratory fisheries".

[67] See CCAMLR Conservation Measure 21-01 (2002) "Notification that Members are considering initiating a new fishery".

[68] Art. 6(6) reads "For new or exploratory fisheries, States shall adopt as soon as possible cautious conservation and management measures, including, *inter alia*, catch limits and effort limits. Such measures shall remain in force until there are sufficient data to allow assessment of the impact of the fisheries on the long-term sustainability of the stocks, whereupon conservation and management measures based on that assessment shall be implemented. The latter measures shall, if appropriate, allow for the gradual development of the fisheries". For a discussion of new and exploratory fisheries, in relation to the application of the PA, see D. Miller, E. Sabourenkov, and D. Ramm, "Managing Antarctic Marine Living Resources: The CCAMLR Approach" (2004) 19(3) *International Journal of Marine and Coastal Law* 317–363.

[69] See S. Garcia, at 489.

[70] CCAMLR Conservation Measure 21-01 (2002) "Notification that Members are considering initiating a new fishery".

[71] CCAMLR, *CCAMLR XXIII — Meeting Report — Antarctic Southern Ocean Coalition*, 25 October – 5 November 2004, Hobart, Australia, Sec. 8, at 15.

[72] As noted by E. Molenaar, UNGA Resolution 59/25 calls for (1) a case-by-case approach, (2) an approach that is based on science, and (3) the precautionary approach: E.J. Molenaar, "Addressing Regulatory Gaps in High Seas Fisheries" (2005) 20 *International Journal of Marine and Coastal Law* 533–570 at 536. This proposal was contentious in that it would open an area to bottom trawling that had not been previously fished with this gear. Certain sectors of the fishing industry are concerned because bottom trawling is presumed to be destructive in the absence of area-specific scientific information. Under the PA, the onus shifts to the industry to prove that such a proposal is within an acceptable level of harm.

[73] See NOAA 2005. Draft Programmatic Environmental Impact Statement on Codified Regulations at 50 CFR Part 300 Subparts A and G. Implementing Conservation and Management Measures Adopted by CCAMLR. Map of closed areas as of 2005 at 46.

[74] See CCAMLR XXIII Item 9 at 9.4 and 9.5.

[75] See CCAMLR XXIII Item 9 at 9.2, "The Commission noted that the Secretariat had encountered significant difficulties in monitoring fisheries in 2003/04. This had resulted in eight instances where catches exceeded their catch limits, and most of these instances occurred in exploratory fisheries". Item 14.7, which addresses cooperation with other organizations, includes the submission from the Antarctic and Southern Ocean Coalition (ASOC), which states that "ASOC remains concerned about the highly unsustainable levels of IUU fishing in the Convention Area. Current levels of IUU fishing seem to be an underestimate. Fewer sightings of illegal vessels does not necessarily mean that there is less illegal fishing, unless there is a corresponding decline in trade flows and market availability of toothfish, which we have not observed".

[76] See E. Sabourenkov and D. Miller, "The Management of Transboundary Stocks of Toothfish *Dissostichus* spp. under the Convention on the Conservation of Antarctic Marine Living Resources" in A.I.L. Payne, C.M. O'Brien, and S.I. Rogers, eds., *Management of Shared Fish Stocks* (Oxford: Blackwell Publishing, 2004) 68–94. The authors illustrate the development and expansion of IUU fishing in the Southern Ocean and the history of the Patagonian toothfish fishery off Chile and Argentina, then in areas both inside and outside CCAMLR boundaries. As industrial longline fleets depleted fishing areas in the early 1990s, they continued into new areas, just as CCAMLR was formulating its measures for new and exploratory fisheries.

[77] See FAO, "Implementation of the International Plan of Action to Deter, Prevent and Eliminate Illegal, Unreported and Unregulated Fishing", FAO Technical Guidelines for Responsible Fisheries No. 9 (Rome: FAO, 2002) at sec. 8.

[78] *Convention for the Establishment of an Inter-American Tropical Tuna Commission*, as amended, done at Washington, D.C., 31 May 1949, entered into force 3 March 1950, TIAS 2044 establishing IATTC. IATTC has considered the effects of fishery expansion on the target species' population, particularly for bigeye tuna. As fisheries were expanded into new areas, an adaptive management approach was adopted so that effects on the stock were considered (Robin Allen, IATTC Director, personal communication, March 2005).

[79] IATTC permanent scientific staff publish regularly in peer-reviewed journals and have contributed to the progression of fisheries science in estimating stock abundance. For example, see M. Maunder, "Paradigm Shifts in Fisheries Stock Assessment: From Integrated Analysis to Bayesian Analysis and Back Again" (2003) 16(4) *Natural Resource Modeling* 465–475.

[80] Robin Allen, IATTC Director, personal communication, 10 December 2005.

[81] IATTC, November 2005. Workshop on Stock Assessment Methods, La Jolla, California, Meeting Agenda. Available at <http://www.iattc.org/PDFFiles2/Stock-assessment-methods-meeting-noticeNov05ENG.pdf> (accessed 20 December 2007).

[82] R. Barston, "The Law of the Sea and Regional Fisheries Organisations" (1999) 14(3) *International Journal of Marine and Coastal Law* 333–352 at 347, footnote 57.

[83] See S. Garcia, at 496, and S. Garcia, *Report of the Meeting of the ICCAT Ad Hoc Working Group on the Precautionary Approach*, Miami, Fla., 13–14 May 1998 (Madrid: ICCAT, 1998).

[84] See <http://www.iccat.es/Documents/SCRS/DetRep/PA_dublin.pdf> at 16 (accessed November 2005).

[85] See FAO Fisheries Circular No. 963.

[86] See ICAAT, "Report of the Standing Committee on Research & Statistics (SCRS) (Madrid, Spain - October 8 to 12, 2001)" in *Report for Biennial Period, 2000–2001, Part II (2001) - Vol. 2* at 123. Available at <http://www.iccat.int/Documents/BienRep/REP_EN_00-01_II_2.pdf>

[87] P. Mace and W. Gabriel, "Evolution, Scope and Current Applications of the Precautionary Approach in Fisheries"in V.R. Restrepo, ed., *Proceedings of the 5th National NMFS Stock Assessment Workshop*: *Providing Scientific Advice to Implement the Precautionary Approach Under the Magnuson–Stevens Fishery Conservation and Management Act*, U.S. Dep. Commer. NOAA Technical Memorandum NMFS-F/SPO-40 (Silver Springs, Md.: U.S. Department of Commerce, NOAA, 1999). 65–73. In P. Mace and W. Gabriel, "Evolution and Implications of the Precautionary Approach in Fisheries", (1999) SCRS/98/124, Collated Volume of Scientific Papers ICCAT (49)4 266–272 at 270, these authors stress that ICCAT does not recognize formally that fishing beyond MSY represents overfishing or overexploitation, however, ICCAT nevertheless "tends to evaluate the status of stocks relative to MSY-based reference points. Typically, ICCAT classifies stocks as 'overexploited' when the exploitable biomass falls and stays below B_{MSY}". S. Garcia, "The Precautionary Approach to Fisheries: Progress Review and Main Issues (1995–2000)" at 496, notes that ICCAT evaluates stock status relative to "MSY-based reference points and considers them over-exploited when their biomass falls below the biomass corresponding to MSY". On the other hand, the Ad Hoc Working Group established by SCRS has stated that "there are very few examples where fishing mortality has been limited to F_{MSY} over a significant period of time, even where MSY has been the stated management objective, and the Committee was not aware of any examples where stocks have collapsed despite fishing mortality being maintained near F_{MSY} over a substantial period" (report of the 1999 meeting, Dublin, Ireland).

[88] See E. Meltzer, "Global Overview of Straddling and Highly Migratory Fish Stocks: Maps and Charts, Detailing RFMO Coverage and Implementation", (2005) 20(3–4) *International Journal of Marine and Coastal Law* (Special Issue) 571–604 at 600. See also updated to May 2006 ICCAT matrix in Part III of this book.

[89] North Atlantic swordfish stock assessment in ICCAT, *Report for Biennial Period 2002–2003 Part 1 (2002) - Vol. 2* at 115–124. Available at <http://www.iccat.es/Documents/BienRep/REP_EN_02-03_I_2.pdf> (accessed 15 October 2005).

[90] Swordfish stock assessment in ICCAT, *Report for Biennial Period 2004–2005, Part I (2004) - Vol. 2* at 122–131. Available at <http://www.iccat.es/Documents/BienRep/REP_EN_04-05_I_2.pdf> (accessed 20 December 2007).

[91] See General Fisheries Commission for the Mediterranean (GFCM), *Selected Global Issues in Fisheries of Relevance to GFCM*, GFCM 25th Session, Sliema, Malta, 12–15 September 2000. GFCM online <http://www.fao.org/DOCREP/MEETING/X7663E.HTM> (accessed 20 October 2005).

[92] NEAFC, *Report of the 24th Annual Meeting of the North-East Atlantic Fisheries Commission*, 14–18 November 2005 (London: NEAFC, 2005). Available at <http://www.neafc.org/reports/annual-meeting/docs/24neafc_annual_vol-1-main-report_2005.pdf> (accessed 20 December 2005).

[93] NEAFC 2005. Information on the North-East Atlantic Fisheries Commission. Draft September 2005, at 5.

[94] The trend towards fishing in deeper and deeper water as coastal stocks become overexploited and no longer commercially viable presents a challenge for the EAF and the PA. Scientific information about deep-sea species and habitats is limited, and fisheries tend to expand into new areas long before scientific research on the impacts of these fisheries takes place. The 2003 FAO Meeting on Deep Sea Fisheries concluded that there was a general need for improving governance of deepwater resources. See FAO, "Report on DEEP SEA 2003, an International Conference on Governance and Management of Deep-Sea Fisheries", Queenstown, New Zealand, 1–5 December 2003, FAO Fisheries Report No. 772 (Rome: FAO, 2005). Further consideration of the issue at the March 2005 COFI meeting summarized that "Deepwater demersal fisheries resources continue to be vulnerable to overexploitation, because of their low productivity and weaknesses in existing governance mechanisms. These fisheries are generally to be found on the high seas and the information needed for their effective management is often deficient and legal means to ensure that data are provided and that participants manage their operations responsibly are often lacking". Full implementation of UNFSA and the Code of Conduct as well as World Summit on Sustainable Development (WSSD) commitments will address some conservation issues in the deep sea, but there is an immediate and pressing need for increased data collection that can lead to science-based decision making and management recommendations for deep-sea fisheries. Several actions were proposed to better document existing management challenges, which include paucity of data, environmental impacts, difficulty in conducting stock assessments, and management inaction. There have also been recommendations to strengthen regional fisheries bodies through the incorporation of legal framework and an overview of deep-sea fisheries operations. NGOs have expressed concern about the sustainability of deep-sea fisheries as well as the ability of RFMOs to manage deep-sea fisheries in a manner that is consistent with UNFSA, the FAO Code of Conduct, and international biodiversity agreements. See COFI 2005, Background document on Deep Sea Fisheries. Available at <http://www.uicn.org/themes/marine/pdf/FAO-DeepSeaFisheries-3862e.pdf> (accessed 15 October 2006).

[95] See ICES 2004 advice to NEAFC on deep sea fisheries. Available online <http://www.neafc.org/measures/docs/deepsea-2005.pdf> (accessed 21 October 2005)

[96] See ICES 2004 advice to NEAFC on deep sea fisheries. Available online <http://www.neafc/measures/docs/deepsea-2005.pdf>. (accessed 21 October 2005

[97] The ICES advice to NEAFC on deepwater fisheries is more conservative: "Fisheries on deepwater species have developed rapidly and the resources which they exploit are generally especially vulnerable to overfishing". Within the ICES area, species–stocks have been depleted before appropriate management measures have been implemented. It is also of concern that the landings statistics that are available may not reflect the true scale of the recent fishing activity, especially in waters outside the national EEZs. The ICES 2004 recommendation for deepwater fisheries was "Most exploited deepwater species are considered to be harvested unsustainably; however, it is currently not possible to provide advice for specific fisheries for deep-sea species. Consistent with a precautionary approach, ICES recommends immediate reduction in established deep-sea fisheries unless they can be shown to be sustainable. Measures should also be implemented to reduce exploitation of deep-sea species by fisheries primarily targeting shelf species (hake, anglerfish, and megrim). New deep-sea fisheries or expansion of existing fisheries into new fishing areas should not be permitted unless the expansion is very cautious, and is accompanied by programmes to collect data which allow evaluation of stock status as the basis for determining sustainable exploitation levels". NEAFC online <http://www.neafc.org/fisheries/docs/acfm_reports-2004/deep-water-general-2004.pdf> (accessed 21 October 2005).

[98] See Recommendation III from the 23rd Annual Meeting, "NEAFC Recommendation for Conservation and Management Measures for Deep-Sea Species in the NEAFC Regulatory Area in 2005" (London: NEAFC, 2005). NEAFC online <http://www.neafc.org/measures/docs/deepsea-2005.pdf> (accessed 21 October 2005).

[99] F. Serchuk, D. Rivard, J. Casey, and R. Mayo, *Report of the Ad Hoc Working Group on the NAFO Scientific Council on the Precautionary Approach*, NAFO SCS Doc. 97/ 12 (Dartmouth, N.S.: NAFO, 1997).

[100] S. Garcia, "The Precautionary Approach to Fisheries: Progress Review and Main Issues (1995–2000)" at 493–494.

[101] J. Caddy and R. Mahon, FAO Fisheries Technical Paper No. 347.

[102] Contracting Parties agreed in a NAFO Resolution 2/99 to (1) determine precautionary reference points for stocks where sufficient information exists; (2) for all other stocks, determine provisional precautionary reference points, whenever possible, and a precautionary approach otherwise; (3) provide mechanisms for filling data gaps; (4) implement precautionary management strategies (harvest control) that are consistent with the first two items; (5) consider additional management strategies that are complementary to the precautionary approach; (6) define and adopt precautionary strategies for reopening closed fisheries and for developing fisheries; and (7) harmonize terminology with other relevant fisheries organizations.

[103] See NAFO, *Annual Report 1999* (Dartmouth, N.S.: NAFO, 1999) at 57.

[104] See NAFO, *Annual Report 1999* (Dartmouth, N.S.: NAFO, 1999) at 43.

[105] See P. Shelton, P. Mace, W. Brodie, and J. Mahé, *A Proposal for a More Flexible Framework for Implementing the Precautionary Approach on NAFO Stocks*, NAFO SCR Doc. 03/58 (Revised.) (Dartmouth, N.S.: NAFO, 2003) at 3.

[106] This implementation plan included the following principles: maintain harvest levels that will continue to rebuild and maintain the stock biomass above the rebuilt biomass level; continue with a comprehensive suite of management measures; ensure that the fishery will not jeopardize recovery of other stocks in the area under moratorium, specifically Atlantic cod and American plaice; for managers, stock performance measures could be expressed in terms of biomass, including how the biomass levels relate to reference points and catch levels; production models do not permit determination of all reference points; scientists should move toward using age-structured models; production modeling could be used to evaluate real *F* limits and to provide insight in what will happen if there are lower or higher *F*'s; and the need for developing "target" biomass levels that are higher than the biological limits so that fisheries management objectives such as economics can be taken into account. See NAFO Annual Report 2000 at 74–76 <http://archive.nafo.int/open/ar/ar00.pdf> (accessed October 2005).

[107] NAFO, Scientific Council, *Proposed NAFO Precautionary Approach Framework*, SCS Doc. 03/23 (Dartmouth, N.S.: NAFO, 2003).

[108] See NAFO, *Report of the General Council Meeting, 26th Annual Meeting, 13–17 September 2004, Dartmouth, Nova Scotia, Canada*, GC Doc. 04/5 (Dartmouth, N.S.: NAFO, 2004).

[109] See NAFO, "Fisheries Commission Request for Scientific Advice on Management in 2006 of Certain Stocks in Subareas 2, 3 and 4", NAFO 26th Annual Meeting, Dartmouth, N.S., 13–17 September 2004. Serial No. N5031, NAFO/FC Doc. 04/7. NAFO online <http://www.nafo.int/science/advice/2006/FCrequest.html> (accessed 21 October 2005).

[110] See NAFO, Serial No. N5031, NAFO/FC Doc. 04/7. NAFO online <http://www.nafo.int/science/advice/2006/FCrequest.html> (accessed 21 October 2005).

[111] Elements to be considered include: (1) References to "risk" and to "risk analyses" should refer to estimated probabilities of stock population parameters falling outside biological reference points. (2) Reference points that are proposed by the Scientific Council as indicators of biological risk should be accompanied by a description of the nature of the risk associated with crossing the reference point, such as recruitment overfishing or impaired recruitment. (3) When a buffer reference point is proposed in the absence of a risk evaluation to maintain a low probability that a stock, measured to be at the buffer reference point, may actually be at or beyond the limit reference point, the Scientific Council should explain the assumptions made about the uncertainty with which the stock is measured. (4) Wherever possible, short- and medium-term consequences should be identified for various exploitation rates (including no fishing) in terms of yield, stability in yield from year to year, and the risk or probability of maintaining the stock within or moving it to the "safe zone". Whenever possible, this information should be cast in terms of (*i*) risk assessments relating fishing mortality rates to the trends in biomass (or spawning biomass), (*ii*) the risks of stock collapse and recruitment overfishing, (*iii*) the risks of growth overfishing, and (*iv*) the consequences in terms of both short- and long-term yields. (5) When providing risk estimates, the time horizon should be clearly spelled out. By way of consequence, risks should be expressed in time frames of 5, 10, and 15 years (or more), or in terms of other appropriate time intervals, depending on stock-specific dynamics. Furthermore, to provide the Fisheries Commission with the information necessary for considering the balance between risks and yield levels, each harvesting strategy or risk scenario should include, for the selected time intervals, the risks and yields associated with various harvesting options in relation to B_{lim}, and F_{lim} and target F reference points selected by managers. NAFO, Scientific Council, *Proposed NAFO Precautionary Approach Framework*, SCS Doc. 03/23 at 3–4.

[112] J. Caddy and R. Mahon, FAO Fisheries Technical Paper No. 347; J. Caddy, FAO Fisheries Technical Paper No. 379.

[113] Dr. Loh Lee Low, National Marine Fisheries Service, Seattle, Wash., personal communication, April 2005 and March 2006.

[114] National Oceanic and Atmospheric Administration (NOAA), 10th Annual Conference of the Parties to the Convention on the Conservation and Management of Pollock Resources in the Central Bering Sea. Available at <http://www.afsc.noaa.gov/refm/cbs/Docs/10th%20Annual%20Conference/10thS&T_FINAL.pdf> (accessed 15 October 2005). Paul Niemier, National Oceanic and Atmospheric Administration, personal communication, March 2006 at 3.

[115] Compilation of Reports, Fourth Meeting of Regional Fishery Bodies, Rome, Italy, 14–15 March 2005. RFB/IV/2005/5 at 4. It should be noted that prior to the establishment of CCSBT, a voluntary management agreement existed among Australia, New Zealand, and Japan — the principal fishing nations of the stock. Following an increase in the fishery in the 1950s, catches peaked at 80 000 t in the 1960s and by the 1980s, the stock was in significant decline, and fishing nations set an annual TAC of 38 650 t in 1985. The TAC was further reduced in 1989 by 70% of 1985 levels. The voluntary management agreement evolved into a legally binding convention that entered into force in 1994, primarily to formalize existing cooperative management schemes and to enforce voluntary catch levels.

[116] *Convention for the Conservation of Southern Bluefin Tuna*. Done at Canberra, Australia, 10 May 1993, entered into force 20 May 1994. Art. 3.

[117] Brian MacDonald, Executive Secretary, CCSBT, personal communication, March 2005.

[118] See C. Romano, "The Southern Bluefin Tuna Dispute: Hints of a World to Come…Like It or Not" (2001) 32 *Ocean Development and International Law* 313–348 at 315.

[119] See C. Romano.

[120] CCSBT catch limits for 2004–2005 are allocated as follows: Japan, 6065 t; Australia, 5625 t; Republic of Korea, 1140 t; Fishing Entity of Taiwan 1140 t; New Zealand, 420 t. For cooperating nonmembers the following catch limits have be set: 50 t for the Philippines; 30 t for South Africa, and 800 t for Indonesia. Catch limits for South Africa and Indonesia would apply if the two countries become cooperating nonmembers.

[121] CCSBT, *Report of the 1997 Scientific Committee Meeting*, Canberra, Australia, 28 July – 8 August 1997.

[122] CCSBT, *Report of the Resumed Third Annual Meeting*, Canberra, Australia, 18–22 February 1997, at 10.

[123] CCSBT, *Report of the Resumed Third Annual Meeting*, Canberra, Australia, 18–22 February 1997, at 10.

[124] C. Romano, at 317–320, explains the legal basis and accompanying strategy for commencing dispute settlement under the LOSC and requesting a tribunal under UNCLOS rather than through other dispute settlement mechanisms.

[125] See S. Garcia, "The Precautionary Approach to Fisheries: Progress Review and Main Issues (1995–2000)" at 500.

[126] See S. Garcia, "The Precautionary Approach to Fisheries: Progress Review and Main Issues (1995–2000)" at 501.

[127] Order, para. 77 (text at <www.itlos.org>).

[128] Order, para. 90(1)(d) (text at <www.itlos.org>)

[129] See Southern Bluefin Tuna Cases Award Text available at <http://www.worldbank.org/icsid/> (accessed 15 October 2005) and the discussion of the Award in C. Romano, at 329. The Tribunal found it had no jurisdiction to deal with the case. Whether or not the Arbitral Tribunal Award was correct in its interpretation of international law and specifically dispute resolution mechanisms has been the subject of some debate, as is evidenced by the contrasting views of D. Colson and P. Hoyle, "Satisfying the Procedural Prerequisites to the Compulsory Dispute Settlement Mechanisms of the 1982 Law of the Sea Convention: Did the *Southern Bluefin Tuna* Tribunal Get It Right?" (2003) 34 *Ocean Development and International Law* 59–92 and B. Kwiatkowska, "The *Southern Bluefin Tuna* Arbitral Tribunal Did Get It Right: A Commentary and Reply to the Article by David Colson and Dr. Peggy Hoyle" (2003) 34 *Ocean Development and International Law* 369–395.

[130] Southern Bluefin Tuna Cases (New Zealand v. Japan; Australia v. Japan) Provisional Measures, Separate Opinion by Judge Laing at para 19. See <http://www.worldlii.org/int/cases/ITLOS/1999/2.html> (accessed 20 October 2005).

[131] Southern Bluefin Tuna Cases (New Zealand v. Japan; Australia v. Japan) Provisional Measures, Separate Opinion by Judge Laing at para 21. See <http://www.worldlii.org/int/cases/ITLOS/1999/2.html> (accessed 20 October 2005).

[132] Additionally Judge Tullio Treves noted that a PA seems inherent in the very notion of provisional measures contained in Art. 290 (of LOSC).

[133] See S. Garcia, "The Precautionary Approach to Fisheries: Progress Review and Main Issues (1995–2000)" at 500–502, and S. Marr, "The Southern Bluefin Tuna Cases: The Precautionary Approach and Conservation and Management of Fish Resources" (2000) 11(4) *Environmental Journal of International Law* 815–831.

[134] H. Schiffman, "The Precautionary Approach at the International Tribunal for the Law of the Sea: The Southern Bluefin Tuna Cases" (2005) 5(1–2) *International Journal of Global Environmental Issues* 78–95.

[135] M. Basson, D. Kolody, T. Polacheck, A. Preece, and J. Hartog, 2004. "Assessment and Projections of SBT Stock Assessment and Summary of Agreed Stock Status Reference Points", CCSBT-ESC/0904/23.

[136] A. Cox, M. Stubbs, and L. Davies, "Southern Bluefin and CITES: An Economic Perspective", Report for Fisheries Research Fund and Environment Australia, ABARE Research Report 99-2 (Canberra: Australian Bureau of Agricultural and Resource Economics, 1999)

[137] A. Darby. Japan Forced to Halve Bluefin Catch. The Age Company Ltd. 1–2. 16 October 2006. Available at <www.theage.com.au/news/national/japan-forced-to-halve-bluefin-catch.html> (accessed October 2006).

[138] Report of the 13th Annual Meeting of the Commission, at 8. Available at <http://www.ccsbt.org/docs/pdf/meeting_reports/ccsbt_13/report_of_CCSBT13.pdf> (accessed 10 July 2007).

[139] IOTC Resolution 99/01, IOTC online <http://www.iotc.org/English/resolutions/reso_detail.php?reso=6> (accessed 20 October 2005).

[140] See IOTC, *Report of the Fourth Session of the IOTC*, Kyoto, Japan, 13–16 December 1999, IOTC/S/04/99/R[E] (Victoria, Seychelles: IOTC, 2000). The summary of the meeting held during the Second Session of the IOTC Scientific Committee is available at IOTC online <http://www.iotc.org/English/meetings/sc/history/sc2.php> (accessed 31 August 2005).

[141] Stock status and management recommendations for bigeye tuna were outlined in the *Report of the First Session of the Working Party on Tropical Tunas*, Victoria, Seychelles, 1–4 September 1999, at 21–23. Available at <http://www.iotc.org/files/proceedings/1999/wptt/IOTC-1999-WPTT-R%5BEN%5D.pdf> (accessed 10 July 2007).

[142] D. Balton, 29 April 2004 presentation to Subcommittee on Fisheries <http://canada.usembassy.gov/content/textonly.asp?section=issues&subsection1=&subsection2=&document=headlines_archive > (accessed September 2005). L. Cordonnery, "A Note on the 2000 Convention for the Conservation and Management of Tuna in the Western and Central Pacific Ocean" (2002) 33 *Ocean Development and International Law* 1–15 at 6.

[143] WCPFC Summary Record of Second Session, Pohnpei, Federated States of Micronesia, 12–16 December 2005, at para. 23. Available at <http://www.wcpfc.org/wcpfc2/pdf/WCPFC2_Records_Summary.pdf> (accessed 11 February 2006).

[144] WCPFC Summary Record of Second Session, Pohnpei, Federated States of Micronesia, 12–16 December 2005, at para. 23. Available at <http://www.wcpfc.org/wcpfc2/pdf/WCPFC2_Records_Summary.pdf> (accessed 11 February 2006).

[145] Transform Aqorau, FFA Legal Advisor, personal communication, 5 March 2006.

[146] See S. Garcia, "The Precautionary Approach to Fisheries: Progress Review and Main Issues (1995–2000)" at 498 for a brief history of the negotiation of SEAFO Convention and inclusion of PA provisions.

[147] "Article 7 of the SEAFC Convention, entitled 'Application of the Precautionary Approach', bears little resemblance to the way in which the initial coastal State draft addressed the precautionary approach. The latter drew heavily on Article 6 and Annex II to [UNFSA] and the FAO Code of Conduct, even to the point of producing two annexes detailing the approach's general implications and guidelines for the application of precautionary reference points. Article 7 consists of three paragraphs. While the first two are essentially similar to Article 6(1) and (2) of [UNFSA], the third requires the SEAFO to 'take cognisance of best international practices', including the aforementioned Annex II and the FAO Code of Conduct. The significant modification of the original coastal State proposal was caused by the reluctance of certain participants in the SEAFO process (particularly Japan and the EC) to draw too heavily on the Fish Stocks Agreement, in absence of their formal adherence thereof". D. Miller and E. Molenaar, "The SEAFC Convention: A Comparative Analysis in a Developing Coastal State Perspective" in A. Chircop, S. Coffen-Smout, and M. McConnell, eds., *Ocean Yearbook 20* (Ardsley, N.Y.: Transnational Press, 2006) 305–375 at 311–312 (footnotes excluded). This view is not shared by all involved in the negotiation process nor by the current administration of the RFMO.

[148] Report of the SEAFO Scientific Committee, 2005 at para 8.8. Full report available at <www.seafo.org/Scientific%20Committee/reports/SC%20Report%202005%20Eng_final.pdf> (accessed 5 March 2006).

[149] "[M]anagement must be based on the precautionary approach and the best available scientific information". S. Nandan, "Statement of the Chairman Upon the Adoption of the Agreement on the Implementation of the Provisions of the United Nations Convention on the Law of the Sea of 10 December 1982 Relating to the Conservation and Management of Straddling Fish Stocks and Highly Migratory Fish Stocks" in J.-P. Levy and G.G. Schram, eds., *United Nations Conference on Straddling Fish Stocks and Highly Migratory Fish Stocks: Selected Documents* (The Hague, Boston, London: Martinus Nijhoff Publishers, 1995) 749–754 at 749.

[150] P. Shelton, P. Mace, W. Brodie, and J. Mahé. These authors observe on the negotiations of the NAFO PA framework: "The change from Fmsy as a target to Fmsy as a limit requires a change in the mindset of fisheries managers and resource users — not just a subtle change, but a fundamental shift towards long-term sustainable fishing practices, including accurate reporting of removals, elimination of illegal effort, minimization of bycatch, adherence to mesh size regulations and respecting closed area and season regulations".

[151] See further, the discussion in Chap. 5.3.2. As Cordonnery explains, "The precautionary approach purports to be a paradigm shift from the maximum sustainable yield (MSY) approach previously used as a method for determining the potential productivity of a stock and the setting of a catch limit". L. Cordonnery, at 7–8.

[152] As J. Caddy suggested in 1998, "harvest control laws could be elaborated in more detail, negotiated with industry, and set to function largely independently of annual assessments in the form of fishery management procedures". FAO Fisheries Technical Paper No. 379.

[153] See S. Garcia, "The Precautionary Approach to Fisheries: Progress Review and Main Issues (1995–2000)" at 531.

[154] S. Nandan, "Statement of the Chairman Upon the Adoption of the Agreement on the Implementation of the Provisions of the United Nations Convention on the Law of the Sea of 10 December 1982 Relating to the Conservation and Management of Straddling Fish Stocks and Highly Migratory Fish Stocks" at 751.

[155] A. Rosenberg, "How Can Managers Use Precautionary Management Advice?" in V. Restrepo, ed., *Proceedings of the Fifth National NMFS Stock Assessment Workshop*, U.S. Dep. Commer. NOAA Technical Memorandum NMFS-F/SPO-40 (Silver Springs, Md.: US Department of Commerce, NOAA, 1999) at 102.

[156] High seas marine protected areas (MPAs) have been discussed; however, none have been established. For a discussion of strategies and potential pilot areas, see K. Gjerde and C. Breide, *Towards a Strategy for High Seas Marine Protected Areas Proceedings of the IUCN, WCPA and WWF Experts Workshop on High Seas Marine Protected Areas*, 15–17 January 2003, Malaga, Spain (Gland: IUCN, 2003). Roberts et al. outline the use of MPA as a management tool. C. Roberts, J. Hawkins, and F. Gell, "The Role of Marine Reserves in Achieving Sustainable Fisheries" (2005) 360 *Philosophical Transactions of the Royal Society B* 123–132.

[157] J. Caddy and J. Seijo, "This Is More Difficult Than We Thought! The Responsibility of Scientists, Managers and Stakeholders to Mitigate the Unsustainability of Marine Fisheries" (2005) 360 *Philosophical Transactions of the Royal Society of London Series B* 59–75 at 59.

[158] See P. Shelton, P. Mace, W. Brodie, and J. Mahé.

[159] D. Freestone, "Implementing Precaution Cautiously: The Precautionary Approach in the Straddling and Highly Migratory Fish Stocks Agreement" in E. Hey, ed., *Developments in International Fisheries Law* (The Hague, London, Boston: Kluwer Law International, 1999) 287–325 at 325.

Chapter 7

The Ecosystem Approach to Fisheries*

7.1 Introduction

The United Nations Fish Stocks Agreement[1] (UNFSA) was the first legal instrument to include ecosystem considerations in fisheries management. The word ecosystem is a 20th century construct recognizing the dynamic, complex, and systemic interrelationships among living organisms and between these organisms and their physical environment, as well as the ecological influences on the variety and variability of living organisms (biological diversity).[2] The ecosystem approach to fisheries (EAF) therefore recognizes the need for a transition from traditional single-species management to a broader and more comprehensive management regime that takes into account both target and nontarget species (fish and nonfish) throughout a large marine ecosystem. The EAF accepts that fishing operations can adversely affect ecosystem function and productivity. The EAF also acknowledges that fish stocks are affected by changes in the marine ecosystem due to natural fluctuations as well as anthropogenic-driven changes from other sectors. This holistic, ecological approach requires additional knowledge and data about the ocean and living marine resources. Given the complexities of biological relationships, unknown impacts of a changing marine environment, and the need for quantifying the impacts of fishing activities on noncommercial species and marine habitat, there is inherent uncertainty and risk involved in managing fisheries. Hence, the precautionary approach, which was the subject of Chapter 6, is an integral part of the ecosystem approach.

Since 1995, the international community, through a variety of efforts, has elaborated a theoretical framework for management of fish stocks both within and beyond areas of national jurisdiction. The inclusion of ecosystem and biodiversity concerns in international law obligated States, regional fisheries management organizations (RFMOs), and ultimately the scientific community to define particular objectives of the EAF and envisage its application to fisheries management. Despite the challenges to achieving full implementation of the ecosystem approach, there exists sufficient knowledge to manage the fisheries for straddling fish stocks and highly migratory fish stocks in accordance with this new paradigm. The EAF builds upon conventional fisheries management, with the establishment of a broad spectrum of practical management measures and programs from bycatch mitigation to area closures, habitat protection, and fleet-capacity reduction.

This chapter briefly discusses the legal bases for the use of the ecosystem approach in fisheries management, as well as the development of associated terminology (including ecosystem-based management, ecosystem-based fisheries management, and ecosystem approach to fisheries). Both pre- and post-UNFSA RFMOs are incrementally moving towards adopting and implementing EAF. Specific efforts to this end are detailed in this chapter, particularly those related to controlling bycatch of target and nontarget species, mitigating impacts of fishing gear, collecting and using information on the physical environment, improving data collection and information management, and cooperating with other organizations to ensure compatible and transparent management objectives. The chapter concludes with an overview of opportunities and challenges to implementing EAF as envisaged in the 2003 FAO Technical Guidelines for Responsible Fisheries: *Fisheries Management 2. The Ecosystem Approach to Fisheries*, which includes the entire spectrum of fisheries management from understanding the ecosystem to ensuring precautionary fisheries allocations and fishing capacity commensurate with the resource.

7.2 International legal underpinnings[3]

Straddling fish stocks and highly migratory fish stocks are components of the marine ecosystem that contains a large biomass of many species and stocks, including nonfish species, found within areas under national jurisdiction and on the high seas. These species are interdependent and are part of an intricate, dynamic web of biological, physical, and chemical interactions. The removal of one species or a group of species by fishing will impact not only the particular targeted population but other species and functions within the ecosystem as well. Conversely, the status of commercial fish stocks will be influenced by perturbations in the marine ecosystem caused by factors other than directed fishing operations.

* The author gratefully acknowledges the research assistance and contribution of Susanna D. Fuller in preparation of this chapter.

Article 61(3) of the 1982 UN Convention on the Law of the Sea[4] (LOSC) recognizes the interdependence of fish stocks and calls on States in Articles 61(4) and 119(1)(b) to take into account associated and dependent species that may be impacted when adopting management measures for targeted fish stocks. While the Convention does not specifically mandate ecosystem management for fisheries beyond associated and dependent species, it does allude to a wider perspective in the preamble, recognizing the need for an integrated approach to ocean governance.[5] Further, under Part XII regarding protection and preservation of the marine environment, Article 194(5) provides that measures be taken to "protect and preserve rare or fragile ecosystems as well as habitat of depleted, threatened or endangered species and other forms of marine life".

The shortcomings of regulating fisheries on a stock-by-stock basis and the impact of fishing operations on nontarget species (fish, seabirds, mammals, turtles) resulted in the recognition of the need for a new imperative, consisting of a more holistic and integrated approach. In the years following LOSC's adoption and the UN Conference on the Environment and Development (UNCED), it was also increasingly recognized that fisheries management issues were no longer exclusively limited to the sustainability of fish stocks, but needed to take into account the health of the ocean ecosystem, its contribution to human well-being, and its natural variability and vulnerability to habitat degradation. Harmful fishing practices such as overfishing, excessive bycatch, and discarding called for urgent corrective action and formal inclusion of the ecosystem approach in the new international law of fisheries (see Box 7.1).

While the ecosystem approach in relation to fisheries was not an explicit obligation in international law prior to UNFSA, the broad consideration of biodiversity and the importance of the natural environment and its related functions and services were implicit in the 1971 Convention on Wetlands of International Importance especially as Waterfowl Habitat (RAMSAR),[6] the 1973 Convention on International Trade in Endangered Species of Wild Fauna and Flora (CITES),[7] and the 1979 Bonn Convention on the Conservation of Migratory Species of Wild Animals (CMS).[8] It was not until UNCED in 1992 that the terminology used today for the ecosystem approach was introduced in Agenda 21 and subsequently expressed in the legally binding instrument of the 1992 Convention on Biological Diversity (CBD)[9] and its 1995 Jakarta Mandate on Marine and Coastal Biodiversity.[10]

The 1980 convention establishing the Commission on the Conservation of Antarctic Marine Living Resources (CCAMLR)[11] incorporated, among others, the principle of interdependence of species and the related objective to minimize discards and bycatch.[12] CCAMLR's constitutive agreement provides that ecological relationships between harvested species and dependent and related species must be maintained.[13] Figure 7.1 illustrates these relationships among the main predator and prey species in the Antarctic marine environment. CCAMLR monitors noncommercial predator species as part of its ecosystem-based management of the krill fishery. CCAMLR pioneered the inclusion of a holistic approach in its efforts to manage the living resources under its competence in the Southern Ocean.

Reflecting the post-LOSC developments to adopt a more holistic, multispecies approach to sustainable use and biodiversity conservation, UNFSA provides juridical direction, expressly referring to ecosystems considerations in the management and conservation of straddling fish stocks and highly migratory fish stocks. UNFSA Articles 5 and 6 outline the principles of EAF and expand and reinforce the LOSC provisions both within and beyond areas under national jurisdiction. Article 5 (b), (d), (e), (f), and (g) requires that conservation and management measures be adopted for nontarget species, both fish and nonfish species. Article 6 (3) and (5) calls for States to assess the status of nontarget, dependant, and associated species sharing the same ecosystem as target species to determine the impact of fishing on these other species, and to ensure their conservation. Parties to UNFSA are required to adopt improved fishing practices so that bycatch[14] and the overexploitation of nontarget species are reduced; they are also required to assess and reduce the impacts of fishing on marine habitat. Annex II further specifies guidelines for application of precautionary reference points to account for the characteristics of the fishery as well as other sources of mortality, and calls for the implementation of management strategies to restore and maintain stocks.[15]

Since the adoption of UNFSA, the EAF has been substantively expressed in several soft law international instruments, including the 1995 FAO Code of Conduct,[16] the 1995 Jakarta Mandate on Marine and Coastal Diversity,[17] the 1999 FAO International Plan of Action for Reducing Incidental Catch of Seabirds in Longline Fisheries (IPOA-Seabirds),[18] the 1999 FAO International Plan of Action for the Conservation and Management of Sharks (IPOA-Sharks),[19] the 2001 Reykjavik Declaration on Responsible Fisheries in the Marine Ecosystem,[20] the CBD Decision V/6 on the Ecosystem Approach, and the 2002 World Summit on Sustainable Development (WSSD) Plan of Implementation calling for development and full

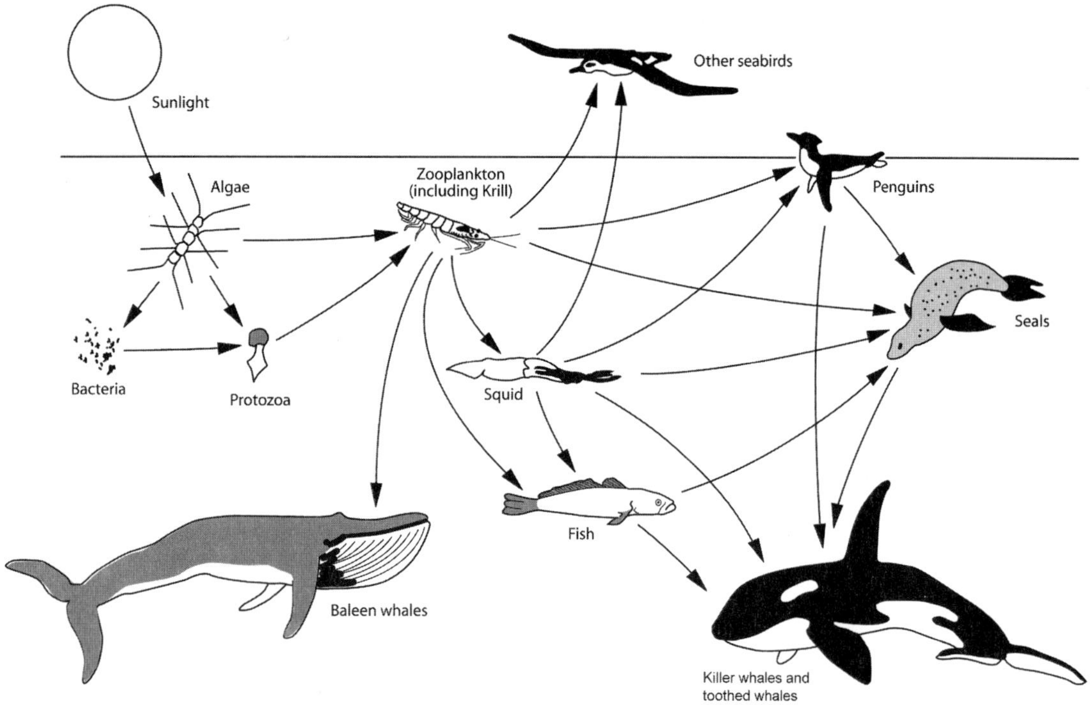

Figure 7.1. Simplified trophic food web in the Southern Ocean. (From CCAMLR, "Understanding CCAMLR's Approach to Management", reproduced with permission of CCAMLR.)

implementation of the ecosystem approach by 2010,[21] the UN General Assembly (UNGA) Resolution 59/25 reiterating the WSSD call upon States to implement the ecosystem approach,[22] and the FAO Technical Guidelines for Responsible Fisheries on the ecosystem approach to fisheries, completed in 2003.[23]

Box 7.1. Consequences of overfishing.[24]

"The world's ocean ecosystems are experiencing biological change at an unprecedented rate. Potentially permanent influences on species interactions, food web structure, and trophic dynamics are most dramatically reflected by staggering declines in the abundance of marine fishes. Large pelagic sharks… have declined more than 75% in the Northwest Atlantic since 1986… Similar reductions over the past 50 years have been reported for other predatory marine fishes …Ultimately, failure to arrest these declines will result in the loss of species from parts of their current geographical ranges, a fate already experienced by more than 55 marine fish species and one that has resulted in the probable global extinction of at least three of these species over the last two centuries…"[25]

7.3 Terminology associated with the ecosystem approach

There are different terms associated with incorporating ecosystem considerations in fisheries management, including EAF, ecosystem approach to fisheries management (EAFM), and ecosystem-based fisheries management (EBFM).[26] The Reykjavik Conference adopted the term EAF rather than EBFM. This term was selected to emphasize that taking ecosystem considerations into account in fisheries was an evolution in fisheries management, not a call for fisheries managers to manage the ecosystem.[27] Some conference participants did not perceive any conceptual difference between the two expressions and were unaware of the issue; however, other participants recall that EBFM had a much broader and unintended connotation when translated into other languages, and therefore the consensus was to use EAF to express the concept.[28]

Nonetheless, because the application of the ecosystem approach to fisheries has evolved, both terms are now being used, and distinctions between EAF and EBFM have been made (see Box 7.2). Both are ecosystem-conscious fisheries management approaches, but they differ in their primary focus. EBFM's primary concern is sustaining healthy marine ecosystems and controlling fisheries for that purpose. The primary concern of EAF is the sustainability of the fisheries with due consideration to the need for maintaining the ecosystem supporting it.[29] Benedict Satia notes that both EAF and EBFM reflect a "merging of two related but different paradigms — ecosystem management that focuses on ecosystem well-being, and fisheries management that focuses more on human well-being".[30] This hybrid paradigm is an expression of sustainable development, recognizing that the sustainability of human social systems depends on the maintenance and restoration of ecological resilience. The ecological consequences of irresponsible, nonsustainable fishing are environmental degradation and socioeconomic dislocation. Therefore, responsible and sustainable regulation of human social and economic systems associated with fisheries is embedded in EAF.[31]

Despite these distinctions, the EAF, EAFM, and EBFM are sectoral approaches to adopting integrated management of commercial fisheries, within ecologically meaningful boundaries. These terms are nested within the broader framework of integrated oceans management (IOM) (See Box 7.2) and represent a growing number of strategies for achieving a more comprehensive approach to the management of human use of the marine environment, including fisheries resources.[32] The key difference among these strategies is the scope of the issues, human activities, and the geographic area to be managed.

The EAF according to the FAO formulation expressly encompasses all aspects of fisheries conservation and management, including research, operations, processing, and trade.[33] While these specific terms are not yet common in international instruments or national legislation, the underlying principles and conceptual objectives appear in many instruments, as described above. The FAO formulation is used as a basis in this chapter for discussing RFMO implementation of EAF.

Box 7.2. Defining EAF, EBFM, and IOM.

Ecosystem approach to fisheries (EAF)

The term EAF was adopted by the FAO Technical Consultation on Ecosystem-Based Fisheries Management (EBFM) held in Reykjavik, Iceland, in October 2001. EAF is defined by Ward et al.[34] as "an extension of conventional fisheries management recognizing more explicitly the interdependence between human well-being and ecosystem health and the need to maintain ecosystems productivity for present and future generations, e.g. conserving critical habitats, reducing pollution and degradation, minimizing waste, protecting endangered species". According to the FAO, EAF "strives to balance diverse societal objectives, by taking into account the knowledge and uncertainties about biotic, abiotic and human components of ecosystems and their interactions and applying an integrated approach to fisheries within ecologically meaningful boundaries. The purpose of EAF is to plan, develop and manage fisheries in a manner that addresses the multiple needs and desires of societies, without jeopardizing the options for future generations to benefit from the full range of goods and services provided by marine ecosystems."[35] The EAF thus is an expression of sustainable development that incorporates the principles of inter- and intra-generational equity in the pursuit of ecosystem well-being.

Ecosystem-based fisheries management (EBFM)

A distinction is made by some between ecosystem management, on the one hand, and EAF and EBFM, on the other hand, as the latter two deal only with fisheries, and EBFM is limited to the "management" of fisheries.[36] EBFM represents a sectoral approach that is not exclusively science based, but involves balancing varied fisheries interests and determining the appropriate level of risk in accordance with the precautionary approach. The National Research Council of the United States defines EBFM as "an approach that seriously takes all major ecosystem components and services — both structural and functional — into account in managing fisheries and one that is committed to understanding larger ecosystem processes for the goal of achieving sustainability in fishery management".[37] The need to manage human activities, rather than the ecosystem is echoed by Gavaris et al. who state, "the word 'based' in ecosystem-based management explicitly acknowledges that [it is not the ecosystem but the people that are] managed".[38] Other scientists have suggested that EBFM requires an alteration in focus, whereby management priorities start with the wider perspective of the ecosystem as opposed to the narrower purview of the target species.[39]

Box 7.2. Defining EAF, EBFM, and IOM. *continued*

Integrated ocean management (IOM)

The ecosystem concepts of EAF and EBFM are related to, but are clearly distinct from, the concept of integrated ocean management as well as other cross-sectoral approaches such as integrated coastal management (ICM), integrated coastal area management (ICAM), and watershed management. IOM is a post-LOSC oceans governance concept that recognizes the interdependence and interrelationship of the multiple uses, activities, institutions, socioeconomic systems, stakeholders, and resources (both living and nonliving) in coastal and ocean space and the corresponding need for adopting a comprehensive and integrated approach to planning, regulation, and management to avoid, reduce, and resolve the conflicting and competing interests and objectives of governments and user groups. This concept recognizes that the traditional fragmentation of policies, programs, and services along sectoral lines and political boundaries is inconsistent with the need for managing human use of the oceans in a coordinated, integrated manner. EAF is one component of IOM. The conservation and management of straddling fish stocks and highly migratory fish stocks taking ecosystem considerations into account are part of the broader global EAF initiative. Under UNFSA, RFMOs are institutional vehicles for delivering EAF for fish stocks under their competence; these narrowly and well-defined species-specific organizations are neither mandated nor capable of undertaking the multiple-ocean-use function envisioned under IOM for their respective geographic marine areas.

7.4 Efforts to adopt and elaborate EAF

In contrast to the elaboration of the precautionary approach, there is no attempt in UNFSA to expand upon the nature of the ecosystem approach or to provide guidance as to its practical implementation.[40] Concerted efforts have been made since UNFSA by FAO, the International Council for the Exploration of the Sea (ICES),[41] RFMOs, coastal States,[42] and research scientists[43] to give practical expression to the somewhat daunting task of incorporating an "all of the marine ecosystem" approach in developing fisheries management strategies. EAF includes the estimation of a permitted level of harvest for target species and incidental catches together with other measures for achieving conservation, biodiversity protection, and management objectives. The 2001 Reykjavik Conference on Responsible Fisheries in the Marine Ecosystem was an international meeting to examine the practical implementation of the ecosystem approach. The resulting Reykjavik Declaration[44] recognized the "complex inter-relationship between fisheries and other components of the marine ecosystems" but stressed that the inclusion of ecosystem considerations such as predator–prey relationships in fisheries management "would enhance management performance". The conference concluded that despite inherent uncertainties and lack of adequate knowledge about marine ecosystems, enough information was available to begin to incrementally implement EAF. The Declaration was a reaffirming call to move forward, recognizing the challenges of implementing EAF while acknowledging that the task was not insurmountable. Following this conference, the FAO developed guidelines for practical implementation of the EAF and endorsed the inclusion of EAF in the workings of regional and international fisheries management organizations.

The 2003 FAO Technical Guidelines for Responsible Fisheries — *Fisheries Management 2. The Ecosystem Approach to Fisheries* — built upon the conceptual basis and institutional requirements for improved fisheries management found in UNFSA, the FAO Code of Conduct, and the IPOAs. The FAO guidelines represent the current international consensus on the meaning of EAF and how it might be achieved.[45] Key components of the guidelines include avoiding overfishing; ensuring reversibility and rebuilding; minimizing fisheries impacts; considering species interactions; ensuring compatibility; protecting ecosystems and habitats; conserving biodiversity; protecting endangered species; applying the precautionary approach; allocating user rights; promoting sectoral integration; reducing pollution; eliminating ghost fishing; broadening stakeholder participation; maintaining ecosystem integrity; and protecting selected marine areas.[46] The proposed EAF methodology is to "introduce a series of modifications to conventional fisheries governance", incorporating the interrelated guiding principles of sustainable development, "with the view to improving poor performance".[47] Various countries, RFMOS, and ICES have endorsed this practical, multipronged, incremental approach to adopting EAF.[48]

7.4.1 Increased data needs for understanding marine ecosystem impacts

EAF represents an important shift or extension in the conventional fisheries paradigm from single-species fisheries management to a multispecies approach throughout a large marine ecosystem or marine geographic area. Traditionally, fisheries science assessed each species in isolation. However, it was long suspected that interactions among species and between the physical environment and exploited stocks could be important factors behind the dramatic changes in fish stocks and "should ideally be considered in the analysis of population dynamics and fishery management".[49] Under EAF, mitigating the adverse ecosystem impacts from fishing requires an awareness of the synergistic and cascading effects resulting from the removal of important forage species, e.g., krill, or from the removal and loss of top predators, such as tunas, sharks, and cod.[50] Additionally, because of the potential impacts of fishing on nontarget and nonfish species within the target species' ecosystem, there is an exponential increase in the area of inquiry and in the information required to undertake the EAF. Moving from conventional fisheries management to EAF therefore increases both the quantity and type of information needed about fish stocks and associated ecosystems.

Article 5(j) of UNFSA requires RFMOs to "collect and share, in a timely manner, complete and accurate data concerning fishing activities on, *inter alia*, vessel position, catch of target and non-target species and fishing effort, as set out in Annex I, as well as information from national and international research programmes". Increasingly, RFMO scientific staff and commission members are meeting together to interpret the data and consider the consequences of different management options consistent with precautionary and ecosystem approaches. Efforts to share information and standardize catch and research information are being made through the Fisheries Global Information System (FIGIS)[51] and the Fishery Resources Monitoring System (FIRMS),[52] although not all RFMOs are currently members. This is an important step in improving the quality, timeliness, and transparency of information available in regional fisheries commissions and national centres of excellence.[53] Data sharing, particularly among RFMOs with overlapping geographic area (i.e., the International Commission for the Conservation of Atlantic Tunas (ICCAT) and the Northwest Atlantic Fisheries Organization (NAFO); the Commission for the Conservation of Southern Bluefin Tuna (CCSBT), the Inter-American Tropical Tuna Commission (IATTC), and the Western and Central Pacific Fisheries Commission (WCPFC)), will be a significant asset in determining EAF objectives, indicators, and monitoring. The requirement to manage fisheries with an ecosystem approach has major implications for RFMOs, as they must broaden their data collection and analysis capacity and foster integration among their working groups and committees, and among other RFMOs. As data collection is an integral and necessary facet of developing and implementing EAF, national and regional observer programs as well as independent scientific observer programs are collecting much of the data needed for calculating bycatch mortality and effects of fishing on related and dependent species in some RFMOs (CCAMLR, IATTC, and NAFO).[54]

7.4.2 Increased uncertainty and the precautionary approach

To successfully implement the EAF will require information not only on the target species but also associated and dependent species and nontarget species as well as the marine environment. This immediate increase in required information by fisheries scientists presents a challenge in a discipline that already acknowledges the paucity of reliable data, lack of understanding, and inherent uncertainty. Some studies have been conducted in parts of the world on trophic relationships in the marine environment and the impacts of fisheries on target and nontarget species. Even in these instances, the cascading effects and implications for fisheries management are still frequently unknown or only partly understood.[55] Compounding the difficulty to effect the EAF is that marine environmental changes are much more difficult to observe, document, understand, predict, and correct than those in terrestrial ecosystems.[56] As a result, with the introduction of the EAF, there is an increase in scientific uncertainty. As Garcia et al observe, "the EAF broadens significantly the fields of considerations to be made, increasing the amount of scientific uncertainty and related risk for the resources and the people depending on them".[57]

Given the uncertainties associated with the lack of knowledge, data, and understanding about the ocean and living marine resources, the precautionary approach is a fundamental and inextricable feature of implementing EAF. As Kaye notes, "decision-makers will rarely, if ever, have meaningful data on all the elements, or relationships between those elements, of an ecosystem."[58] Even where it is possible to control the catch of individual species by limiting incidental mortality through gear restrictions and season controls, incorporating and assessing ecosystem impacts and ecological interdependencies among marine species (including marine mammals) greatly complicates the analysis of appropriate management measures.[59] As a result, "the precautionary approach, already adopted as a necessary complement to conventional fisheries management, is also systematically and a fortiori seen as an integral part of EAF".[60]

7.4.3 Cooperation, compatibility, and transparency

The 1982 LOSC division of ocean space and maritime boundaries is a political construct that does not correspond to specific ecosystems. The recognized implication is that management measures need to be compatible and coherent throughout the geographic range of the stocks. Given that straddling fish stocks and highly migratory fish stocks by definition move across national and international boundaries within an ecosystem, the need for cooperative arrangements and compatible measures as provided for in UNFSA is another *sine qua non* of managing these fisheries. As management expands its focus from a single target stock to the ecosystem implications of fishing operations, there is an ever-increased need for cooperation, coordination, and compatibility of conservation and management measures within the fisheries sector as well with other users of the marine environment throughout the geographic range of the directed fishery.

No single RFMO, intergovernmental organization (IGO), or international organization or entity has competence over all ecosystem components, resources, and human uses, including human activities. Even CCAMLR, with a relatively circumscribed ecosystem and few human uses in the Southern Ocean, defers competence over the largest predators, whales and seals, south of 60° in the Antarctic marine ecosystem to the International Whaling Commission (IWC) and the Convention for the Conservation of Antarctic Seals (CCAS Convention).[61] Cooperation among RFMOs and between RFMOs and coastal States is therefore imperative in an ecosystem approach, as is compatibility among management measures for species that cross RFMO boundaries and migrate between national and international jurisdictions. RFMOs are an important component of international oceans governance efforts to ensure sustainable development and achieve integrated ocean management. Given the multiple and often conflicting uses of the ocean, in taking account of ecosystem impacts both on and from fisheries, RFMOs need to coordinate activities and cooperate with national agencies and international organizations with mandates for other human uses.

Since 1995 there have been increased efforts for inter-RFMO cooperation, as evidenced by the annual meeting of tuna RFMOs and the biennial meeting of regional fishery bodies on the margin of the FAO Committee on Fisheries (COFI) meeting. With overlapping areas of competence, many RFMOs (e.g., CCSBT, IATTC, the Indian Ocean Tuna Commission (IOTC), and WCPFC) have met jointly to promote coherent, ecosystem-based management. RFMOs have begun to meet with other UN agencies such as the United Nations Environment Programme (UNEP) and the International Maritime Organization (IMO) to discuss issues of common concern. In addition to cooperation among RFMOs, IGOs, and other bodies, there is a further need for coordination within RFMOs, particularly between the scientific staff or advisory committees and the management bodies.

Another key element of EAF is enhanced transparency to improve institutional development, research, and decision-making functions in conventional fisheries management. This includes increased data sharing and availability, participation in meetings and advisory processes, and understanding the rationale for management decisions. As noted by Garcia, "it is clear that the monitoring of the state of the fisheries and their resources (and environment) needs to be strengthened substantially...for better informed and improved governance as well as for more transparency and better public information".[62] Cooperation, coordination, participation, and transparency through the implementation of EAF present opportunities for positive change. The challenge for States and RFMOs is to realize these opportunities.

7.4.4 Precautionary, rule-based decision making and efficient dispute settlement

To ensure that competing stakeholder objectives and management strategies are compatible, the EAF governance regime requires precautionary decision-making processes, based on the best scientific evidence available, to avoid serious or irreversible harm. Implementation of EAF therefore relies heavily on science-based decision making and requires closer cooperation and communication between scientists and managers.[63] Failure to cooperate, to reach agreement, and to comply with management measures has resulted in unsustainable and irresponsible harvest of fish stocks and in incidental mortality of other related and dependent species, with an impact on the broader marine ecosystem.

In making decisions regarding EAF, environmental impact assessments are an important tool for gathering the information needed to ascertain whether a proposed activity is within international minimum standards and will avoid serious or irreversible harm. Arguably, to demonstrate that this requirement has been met, such assessments should be part of a responsible fisheries management decision process.

The governance regime to support EAF dictates the establishment of timely and effective dispute settlement, together with preagreed decision rules and harvest controls, to reduce and avoid conflict and to quickly resolve disputes whenever possible. Accordingly, the establishment of dispute settlement mechanisms and precautionary, transparent decision making underlying EAF is a central pillar of UNFSA.

7.4.5 Reducing fishing pressure through capacity reduction

Another fundamental feature of EAF, related to realizing the UNFSA objectives of avoiding overfishing and ensuring reversibility, is the need for reducing fishing pressure by controlling fleet capacity.[64] (See also discussion of overfishing and overcapacity in Chapter 5.) The control of fishing capacity includes ensuring fleet size, individual vessel size, gear, days at sea, and fishing practices are commensurate with an agreed harvest level for an ecologically viable stock. Accordingly, the ability of RFMOs and States to avoid, reduce, and eliminate excess capacity, through precautionary, anticipatory, adaptive management and associated regulatory mechanisms, is a key aspect of the implementation of EAF.

EAF manages human activity given that it is impossible to manage the marine ecosystem. However, in managing fishing activities, it is essential to take into account, wherever possible and to the extent possible, ecosystem considerations. The EAF endeavors to predict and understand the impact of fishing activities on marine ecosystems and vice versa, to improve the responsible regulation of extracting living resources from the sea.

7.4.6. Ecosystem indicators and objectives

The principles embodied in EAF recognize the wide range of societal and ecosystem objectives, within the context of sustainable development. Garcia has observed "[t]he operational framework for [EAF] EBFM is practically identical to the sustainability indicators' framework, both from a scientific and decision-making point of view".[65] As specified in the FAO Guidelines for Responsible Fisheries, "in order to implement EAF, the broad objectives must be translated into operational objectives with direct and practical meaning in the context of the fishery, and against which the performance of the fishery and its management can be evaluated".[66] Given that there are often competing conservation and management objectives operating within an ecosystem or defined ocean space, clear objectives and conservation priorities are necessary within an integrated management framework.[67]

The task confronting scientists, managers, and others participating in the fisheries management process is to translate concepts, principles, and various conservation measures into an EAF management plan with specific operational objectives. This task is not trivial and poses many challenges. To develop indicators and objectives related to a particular area or fishery, it is necessary to identify and delimit the ecosystem, given jurisdictional constraints and the need for cooperation among management bodies. Agreeing on clear baselines for what is considered the natural structure or function of an ecosystem is the key to determining management objectives and goals. There remain considerable gaps in knowledge of large-scale dynamics of marine food webs and ecological patterns and processes that occur at the scale of fisheries.[68]

A broad approach involves converting conservation objectives for productivity, biodiversity, and habitat into explicit EAF strategies, whose performance can be measured by indicators,[69] to achieve management objectives and successfully regulate fishing activity. There have been collaborative efforts to develop these indicators[70] and a framework through which selection of appropriate indicators can be done.[71] Evaluating and assessing the achievement of ecosystem objectives, following consensus on indicators and management strategies to achieve ecosystem objectives, are challenges facing RFMOs.

7.5 Implementing EAF

While there is now a general understanding of the meaning and purpose of EAF and acceptance of the principle as intrinsically valuable and a laudable goal, the practical application is still being elaborated and is not without difficulties. The transition from a single-species approach to a broader ecosystem approach includes regulating, restricting, and monitoring bycatch of target and nontarget species (fish and nonfish species); reformulating maximum sustainable yield (MSY) from a target to a limit;

using precautionary reference points and adopting the precautionary approach; establishing closed areas or marine protected areas (MPAs); protecting sensitive marine habitats; limiting access and reducing capacity; adopting an adaptive management approach[72] to avoid overfishing; and rebuilding stocks. These are all incremental efforts and together represent a major shift in the high seas fisheries management paradigm.[73] Figure 7.2(*a*) illustrates the conventional view of fisheries management, where catch limits and other controls on fishing activity are determined and implemented. This view does not recognize the potential effects of biomass removal on dependent species, either directly through mortality or indirectly through food-web relationships. Nor does it consider impacts of fishing on habitat or the effects of habitat and environmental fluctuations on fish productivity.[74] Figure 7.2(*b*) outlines full implementation of the EAF, where an integrated assessment process includes the environment, the target stock, and related predators and prey. In reality, there is a continuum of incremental modifications and adjustments that can move current fisheries management strategies (Figure 7.2(*a*)) towards the more holistic goal of fisheries management that fully considers ecological relationships and ecosystem considerations (Figure 7.2(*b*)).

This progressive shift is accompanied by the need for additional data, analytical infrastructure, data sharing, new scientific research programs, additional budgetary commitments, human and institutional capacity building, and finding cooperative solutions to problems engendered by jurisdictional complications. Furthermore, there is a need for identifying "precise measures and indicators to assess performance of EAF including impacts on wealth generation and equity".[75] Recognizing that post-LOSC UNFSA efforts to improve fisheries management now include EAF and marine biodiversity protection and conservation, States and RFMOs with the assistance of FAO are confronting the challenge of meeting these obligations.

FAO recognizes that the challenge of implementing EAF is more acute in cases where there is a disparity in scientific capability, including research institutions and personnel. The special needs of developing coastal States are a significant concern in implementing the provisions of UNFSA. It is also important to encourage global efforts to maintain accurate, transparent, and complete data inventories and to strengthen the scientific regimes of RFMOs to ensure the best scientific information is available to all members. Additionally, management capacity and institutional strengthening are required at both the national and regional levels to improve performance and to meet international standards as outlined in international agreements and obligations. This is true for all RFMOs and all coastal States. In the case of developing States, FAO will, to the extent possible, apply the EAF directly through existing field projects in West Africa and the Caribbean, as well as in other areas or countries, on request and where funding is made available.[76] EAF is an integral component of sustainable development. States seeking to enhance food security and ecosecurity within their exclusive economic zones (EEZs) and through RFMOs "need to develop management strategies that balance sound ecological practices with economic needs and adopt an ecosystem approach to ensure long term sustainable use of the ocean and its resources".[77]

As outlined in the next section, RFMOs and some States have incorporated a broader ecosystem approach to managing fisheries, primarily through managing bycatch and area closures. Fisheries scientists and managers at all levels are developing tools, strategies, and institutions to implement EAF. The efforts of each RFMO are reviewed, including specific goals to adopt EAF and particular management measures that can be placed under the umbrella of EAF.

7.6 RFMO implementation of EAF

In part as a result of the adoption of UNFSA, and partially as a response to declining populations of commercial fish stocks and adverse impacts on nontarget species, many of the general principles of EAF are being applied incrementally to fisheries management through RFMOs. This should not be construed to imply that RFMOs are the governing bodies of any particular marine ecosystem, but rather that there is a requirement for fisheries scientific advice and fisheries managers to take ecosystem considerations into account when assessing stock status and developing management strategies. While EAF is new for most RFMOs, the CCAMLR Convention included a far-sighted provision, pre-dating the wider adoption of EAF,[78] that reflects a cognizance of the impact of harvesting krill, the key prey species, on the Antarctic ecosystem. IATTC has taken ecosystem considerations into account by addressing the dolphin bycatch associated with the yellowfin tuna purse-seine fishery, and more recently by addressing bycatch of juvenile target species, seabirds, and sharks. Both CCAMLR and IATTC have taken specific steps to institutionalize EAF as a viable fisheries management strategy and to implement EAF through several working groups or programs on bycatch and the physical environment. In negotiating the constitutive agreement to strengthen and modernize IATTC, States have expressly included the ecosystem approach in the Antigua Convention (not yet in force). The North East Atlantic Fisheries

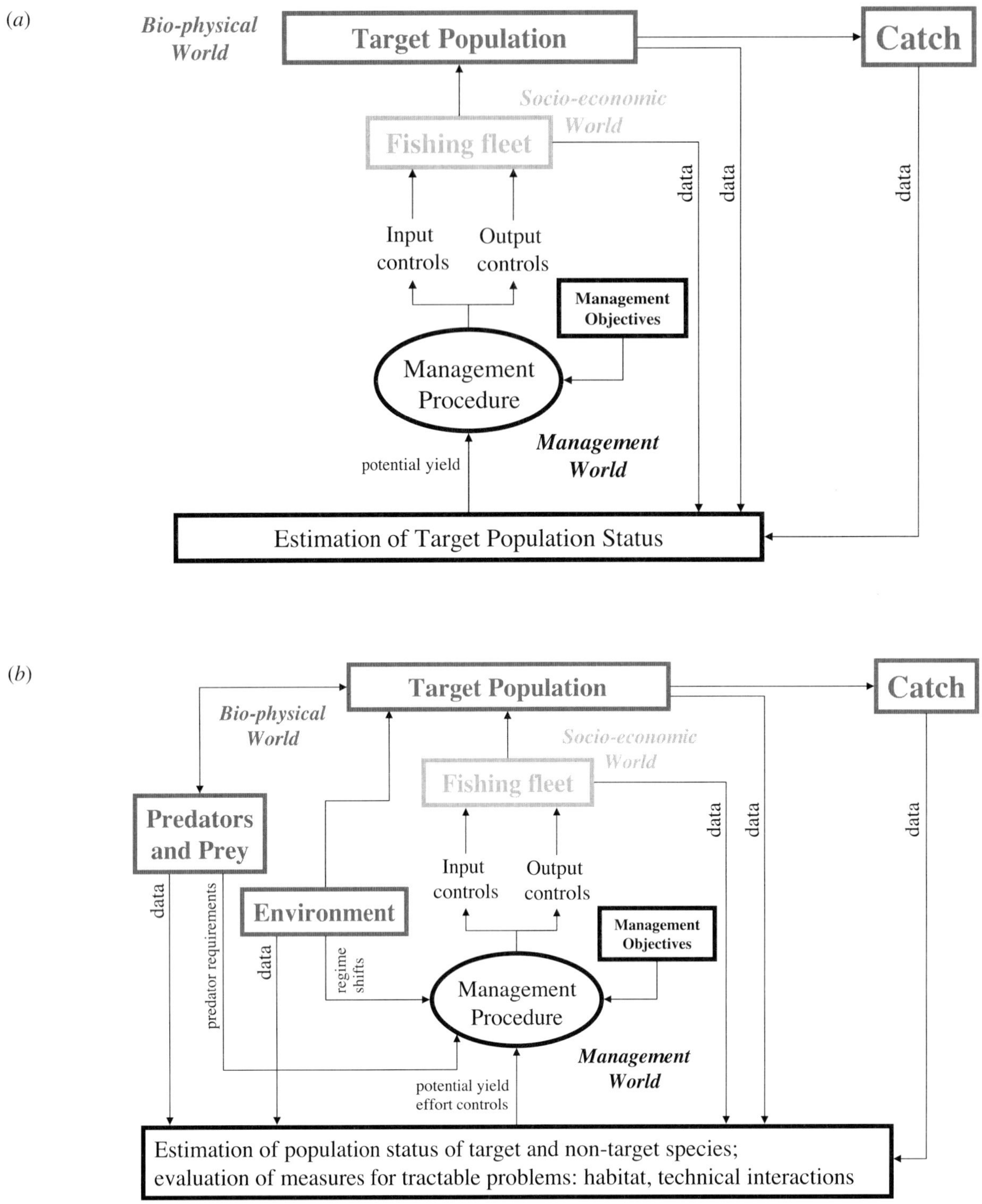

Figure 7.2. (*a*) Conventional fisheries management recognizes the biophysical world in which the stock exists, the socioeconomic world of the fishing community that harvests the stock, and the management world in which catch limits are determined. (*b*) The EAF requires that the environment, the target stock, and the associated predators and prey are integrated in the assessment before the management procedure is used to determine catch limits. (From MRAG Americas Inc., "Review of Ecosystem-Bycatch Issues for the Western Central Pacific Region", prepared for the Preparatory Conference for the Western and Central Pacific Fisheries Commission (WCPFC), Manila, Philippines, 12–18 November 2002, WCPFC Prep Con/WP.9, reproduced with permission of MRAG Americas Inc.)

Commission (NEAFC) has proposed new wording to be used in the revision of its convention, which was agreed in substance at the 2005 Annual Meeting.[79] Post-UNFSA RFMOs, including WCPFC, the South East Atlantic Fisheries Organisation (SEAFO), and the not yet established Southern Indian Ocean Fisheries Arrangement (SIOFA), refer to the marine ecosystem and the need for its protection in their conventions.

For the most part, pre-UNFSA RFMOs have acknowledged the general concept but have yet to establish the required scientific programs and institutional changes to implement EAF as a means of fisheries management. CCAMLR is a notable exception, having included ecosystem provisions from its inception. For many years, several RFMOs have been collecting and examining information on the physical environment (NAFO, CCSBT, the General Fisheries Commission for the Mediterranean (GFCM), ICCAT, IATTC, NEAFC through ICES) to better understand the implications of changes in the marine ecosystem on target commercial stocks. In the post-UNFSA era, many RFMOs have introduced, in an incremental and piecemeal fashion, measures that together reflect a more holistic and integrated management approach that is consistent with EAF. Most RFMOs have established working groups, through scientific councils and committees, to provide technical and management advice on incorporation and implementation of ecosystem considerations. NEAFC is examining EAF through the Working Group on the Future of NEAFC. In 2005, NAFO tasked its Scientific Council with collecting information on ecologically sensitive areas and further addressing bycatch concerns. GFCM is examining EAF through its Sub-committee on Marine Environment and Ecosystem (SCMEE). CCSBT deals specifically with bycatch through the Ecologically Related Species Working Group (ERSWG) and has begun a Scientific Research Program (SRP) to improve data collection on southern bluefin tuna. The Central Bering Sea Donut Hole Convention deals specifically with pollock; however, regime shifts in the Northern Pacific have influenced discussion on the management of the fishery and the pollock fishery moratorium. IOTC has a Bycatch Working Party. WCPFC has established an Ecosystem and Bycatch Specialist Working Group (EBSWG), which will collaborate with other overlapping working groups. While SEAFO has yet to establish similar working groups, the intention to implement EAF is included in its agreement. The soon to be adopted Southern Indian Ocean Fisheries Agreement also provides for the implementation of EAF.

Additionally, RFMOs have adopted since 1995, to varying degrees, management measures that consider the impacts of fishing on dependent and associated species (also a LOSC obligation). Such measures include bycatch regulation and mitigation measures; closed areas; gear restrictions; season limits to protect spawning populations; precautionary management of forage species; establishment of rebuilding plans for depleted stocks; cooperation with other IGOs and RFMOs for management of both highly migratory fish stocks and straddling fish stocks that migrate across RFMO boundaries; application of the precautionary approach in fisheries management; and the implementation of IPOAs. Ecosystem modeling, which may act as an alternative to single-species stock assessment models, is also being developed as part of EAF implementation. Included in these models are variables to incorporate uncertainty with respect to fisheries and potential ecosystem linkages.

Reducing or eliminating bycatch is a critical issue in managing the ecosystem impacts of fishing. It is also a problem that can have various technical solutions depending on the species, the geographic location, the technology, and the fishery. RFMOs have begun to address the bycatch problem in response to IPOAs, biodiversity conservation measures, nongovernmental organization (NGO), and consumer campaigns, and concerns of member States. RFMOs have adopted and implemented bycatch mitigation measures to reduce the effects of fishing on other fish species and particular habitats through gear modifications, including mesh-size restrictions, and bycatch limitations. In NAFO, gear selectivity of trawl-based fisheries has improved with the introduction of the Nordmore grate in shrimp fisheries; however, bycatch of stocks under moratoria and juvenile target species is a critical ecosystemic concern and has triggered additional management measures, such as vessel relocation. NEAFC has addressed bycatch of deep-sea species, first through data collection to provide information on the quantity and species composition of bycatch, and second through an overall effort reduction in the deep sea. Area closures have been instituted by CCAMLR to avoid fishing on seriously depleted stocks. CCAMLR also sets conditions for closures and openings of fisheries in areas that have not been fished and has seasonal closures of directed fisheries to reduce seabird bycatch. NEAFC has closed vulnerable areas to bottom trawling and longline fishing; IATTC, ICCAT, and NAFO have introduced spawning ground seasonal closures; and GFCM has imposed depth restrictions.

RFMOs have begun to address bycatch of nontarget species such as turtles, sharks, dolphins, and seabirds, all of which have been the subject of global biodiversity concerns. CCAMLR was the first RFMO to introduce measures (streamer or tori lines) to address accidental injury or mortality of seabirds caught as bycatch on surface longlines in the Southern Ocean. CCAMLR has subsequently implemented an extensive suite of conservation and bycatch mitigation measures. IATTC and CCSBT have also established specific management measures to mitigate seabird bycatch, while ICCAT, IOTC, and WCPFC do not have mandatory measures in place as yet.[80] IATTC, ICCAT, and IOTC have committed to collecting data on shark catches and reducing shark catches to less than 5% of landed catch. At its 2005 meeting, NAFO agreed to prohibit shark finning. RFMO efforts have also been made to address the declining marine turtle populations[81], which is primarily due to a number of human-related causes, including incidental capture in commercial fishing by longline fisheries (ingestion of hooks, external hooking, and (or) entanglement in gear). IATTC and IOTC have adopted resolutions on turtles and are addressing the bycatch problem primarily through gear modifications. IATTC, together with the International Dolphin Conservation Program pursuant to the Agreement on the International Dolphin Conservation Plan (AIDCP), has successfully reduced the incidental catch of dolphin associated with the yellowfin tuna purse-seine fishery.

7.6.1 CCAMLR

The CCAMLR Convention was the first regional fisheries agreement to stipulate that both precautionary and ecosystem considerations be incorporated in management decisions. CCAMLR continues to be a pioneer and leader in the development of what is now understood to be the EAF,[82] informing UNFSA and the constitutive agreements of post-UNFSA RFMOs. CCAMLR was established following ratification in 1982, in response to concerns raised by the Antarctic Treaty consultative parties[83] during the Ninth Antarctic Treaty[84] Consultative Meeting in 1977 about the sustainability of the exploitation of Antarctic marine living resources. Large-scale exploitation of fish stocks started in the late 1960s and resulted in the severe decline of targeted fish stocks including marbled rockcod and blackfin icefish.[85] Of particular concern were the expansion of the krill fishery and the potential adverse impact of these fishing operations on krill-dependent species such as whales, seals, penguins, and other seabirds such as albatrosses and petrels. While the predator–prey dynamics and interactions were not fully understood, krill, a small shrimp-like prey species suitable for human consumption and animal feed, was known to be a central component of the Southern Ocean food web upon which land-based breeding colonies of birds as well as seals depended and was also considered vital to the recovery of depleted baleen whale populations.[86]

CCAMLR was accordingly established with the objective of protecting and conserving the living marine resources in the waters surrounding Antarctica, an area of approximately 32.9×10^6 km^2 of the Southern Ocean, constituting about 15% of the world's total ocean surface.[87] These marine resources include krill, Patagonian and Antarctic toothfish, icefish, lanternfish, other finfish, mollusks, crustacean, birds, and all other commercial and noncommercial living organisms.[88] Article I of the CCAMLR Convention defines the marine ecosystem as the complex relationships of Antarctic marine living resources with each other and with their physical environment. This notion is reflected in the delimitation of the convention area, which is bounded by the Antarctic Convergence (commonly referred to as the Polar Front) to the south and by 60°S latitude to the north, thereby encompassing the Antarctic marine ecosystem[89] (Article I). The Convention incorporates standards designed to ensure conservation of individual populations and species as well as the Antarctic marine ecosystems as a whole. Its "ecosystem approach considers the entire Southern Ocean as a suite of interlinked and ecologically distinct systems".[90]

Article II of the Convention articulates key ecosystem principles,[91] including the maintenance of ecological relationships among harvested, dependent, and related populations and the restoration of depleted populations to levels that can produce the greatest net annual recruitment. CCAMLR is also required to prevent changes or minimize the risks of changes in the marine ecosystem that are not potentially reversible over two or three decades, taking into account the direct and indirect impact of harvesting, the introduction of alien species, the effects of associated activities on the marine ecosystem, and the effects of environmental changes. Recognizing the uncertainty and risk associated with harvesting resources that would impact the ecosystem status and species linkages, the Convention also pioneered the implicit adherence to the precautionary approach in its constitutive agreement. In recent years, the management of the toothfish fisheries and the need for controlling illegal, unreported, and unregulated (IUU) fishing, as well as the seabird bycatch and the entanglement of animals in marine debris, have preoccupied much of CCAMLR's efforts to reduce impacts on the Southern Ocean ecosystem and achieve conservation of Antarctic marine living resources.

In addition to CCAMLR's efforts to implement the EAF in accordance with its constitutive agreement, the

Commission also participates in the broader regime known as the Antarctic Treaty System. The Antarctic Treaty includes "preservation and conservation of living resources" (Article IX.1. (f)), but makes no particular reference to the environment. Since its entry into force, additional protocols and treaties relevant to the Antarctic environment were developed, including the 1991 Madrid Protocol on Environmental Protection, which *inter alia* includes basic obligations with regard to environmental impact assessments (Annex I), conservation of Antarctic fauna and flora (Annex II), waste disposal and management (Annex III), prevention of marine pollution (Annex IV), and area protection and management (Annex V);[92] and the Convention for the Conservation of Antarctic Seals (CCAS).[93] The objectives of CCAS are the "protection, scientific study and rational use of Antarctic seals and to maintain a satisfactory balance within the ecological system". Commercial sealing has not taken place in the convention area for decades and is not expected to recommence.[94] In addition, the IWC, established in 1948, pursuant to the International Convention for the Regulation of Whaling,[95] has competence over whale management globally, including in the Southern Ocean. This illustrates and underscores the need for RFMO cooperation and collaboration with organizations responsible for management in areas adjacent to the convention area or for species not included in the competence of CCAMLR. CCAMLR makes a concerted effort to establish links with adjacent Commissions on species and issues of common interest (e.g., regarding seabird incidental mortality with the various tuna commissions (ICCAT, IOTC, CCSBT, and WCPFC); and straddling stocks such as toothfish with SEAFO).

The CCAMLR institutional structure reflects the collaborative approach required to implement EAF. A number of CCAMLR committees report and make recommendations to the Commission, including a Scientific Committee, which has two working groups, and an ad hoc working group: the Working Group on Fish Stock Assessment (WG-FSA), the Working Group on Ecosystem Monitoring and Management (WG-EMM), and the Ad-hoc Working Group on Incidental Mortality Associated with Fishing (WG-IMAF).[96] Advice from the working goups is submitted to the Scientific Committee, which may take into account any additional information. The Scientific Committee then refers management advice to the Commission for consideration. All the working groups and the Scientific Committee, together with the Commission, develop research plans and management measures. Under Article IX, management measures agreed to by the Commission become legally binding Conservation Measures. A lack of sufficient knowledge and adequate tools makes regulating large and complex marine ecosystems currently unfeasible. "CCAMLR's [ecosystem] approach regulates human activities such as fishing, so that deleterious changes in the Antarctic ecosystems are avoided or, at worst, minimized and are essentially reversible".[97] As part of its management strategies, CCAMLR has adopted precautionary and risk-averse management, consistent with EAF.[98] EAF has been implemented through a variety of measures, from the development of the CCAMLR Ecosystem Monitoring Program (CEMP) in 1985, which assesses the impact of krill fishing in the Antarctic marine ecosystem on dependent species (seabirds and seals primarily),[99] to direct EAF management of target fish stocks (krill), bycatch mitigation measures (i.e., for seabirds), area closures (i.e., 1985 closure of the *Notothenia* fishery around South Georgia), gear restrictions and modifications, and risk-averse management procedures for other species to account for uncertainty (e.g., toothfish). Unlike the majority of other RFMO constitutive agreements, the CCAMLR Convention makes no reference to maximum sustainable yield (MSY); rather the target fishing mortality used in the early years as a management policy for finfish fisheries was $F_{0.1}$.[100] This decision rule was subsequently amended, as fishing at this level could have depleted the spawning stock biomass to a level where stock recruitment would have been at risk. CCAMLR now uses a decision rule based on limiting to 10% the probability of the spawning stock biomass falling below 20% of the preexploitation median level over a 20-year harvesting period or limiting the median stock escapement over a 20-year period to 75% or the lower of these two scenarios as the criterion to determine the allowable level of fishing mortality.[101]

From the outset, "CCAMLR's Scientific Committee realized … that monitoring the entire ecosystem would be highly impractical. Frequent surveys of the population status of krill would be beyond the financial capacity of the Antarctic research vessels".[102] To monitor the state of the marine ecosystem, given the geographic scope and variability of the Southern Ocean, the Scientific Committee selected a range of species in different areas (including fished and nonfished areas) "to reflect changes in the availability of harvested species, especially krill… Environmental parameters, such as hydrography and ice cover, were also considered to be important indicators of environmental change and thus, were also included in the [CEMP]".[103] Additionally, information is collected to enhance management decisions through improved understanding of natural variability and potential human-induced effects on stock dynamics.

CEMP's major functions continue to be monitoring and evaluating significant changes in critical components of the ecosystem (indicator species).[104] This work serves as the basis for Conservation Measures and other management decisions, "detecting and recording significant changes in critical components of the ecosystem"[105] as well as improving the ability to differentiate between changes due to fishing operations and those due to environmental variability.[106] The information base and data acquisition undertaken by CCAMLR members form part of the annual ecosystem assessments. An important feature of CEMP is that all data from the participating country surveys are collected in a consistent manner to assist in appropriate decision-making rules. CEMP works closely with WG-EMM, which is mandated to synthesize the data. WG-EMM has adopted a strategic modeling approach to address the difficulties inherent in implementing EAF. Computer simulated models are key tools in determining scientific priorities and recommending management options.[107] Rather than developing a comprehensive ecosystem model, CCAMLR develops simpler models for strategic purposes that capture important features of the ecosystem, while recognizing the multiple linkages that exist among components.[108] CCAMLR is also addressing the question of combining CEMP indices with environmental indices to get an impression of "ecosystem health".[109] It is now understood that indirect effects of harvesting species other than krill can also cascade through trophic levels. As noted by Everson,

> In the early years of CCAMLR, krill was viewed as the central component of the food web and was therefore the focus of CEMP. It is now clear that similar approaches to those developed for krill need to be developed for other important species of the food web. Exploitation of lanternfish in the second half of the 1980s and the recent interest in harvesting squid have highlighted the need to look at some other food chains. Lanternfish are the staple food of king penguins, and also of fur seals in the Indian Ocean sector. Squid feed on zooplankton, including krill, and lanternfish. They are preyed on by toothfish, albatrosses, larger penguins, seals and toothed whales. The life histories of commercial squid species are quite unlike those of finfish and krill. Consequently, although the general principles of ecosystem assessment currently applied to krill might also be used for lanternfish and squid, specific procedures will need to be developed for such assessments.[110]

Currently, the primary target species in the CCAMLR area are krill, Patagonian toothfish, and mackerel icefish. Krill fisheries in the CCAMLR area are managed according to EAF,[111] with precautionary catch limits (TACs),[112] decision rules, and close monitoring of dependent species through CEMP. WG-EMM has focused on the krill fishery and associated modeling for the past several years.[113] In 2004, WG-EMM reported that populations of gentoo, eastern and western rockhopper penguins, and Crozet shags continued to decline near Marion Island. The reason for these declines was cited as decreased prey abundance for species foraging in the area.[114] The focus on prey and predator abundance is an approach to EAF that has yet to be replicated by other RFMOs. Consistent with EAF, the toothfish fisheries are managed with precautionary TACs, cognizant of the IUU fishing[115] that has undermined the conservation regime. The toothfish management scenario also adheres to CCAMLR's decision-rule approach to management.

7.6.1.1 Seabird bycatch

The viability of seabird populations that are adversely impacted by fishing operations is a major global concern. The life history and status of seabirds known to occur in waters managed by CCAMLR are well documented. WG-IMAF has identified 20 species of seabirds (including albatrosses, petrels and shearwaters, penguins, and Antarctic skua) that are most at risk from longline fishing. According to Birdlife International, the CCAMLR Convention Area is ranked fifth among all RFMOs in albatross abundance, with grey-headed and wandering albatrosses being the most common species in the area.[116] The Amsterdam albatross, discovered on Amsterdam Island in the Indian Ocean (just beyond the CCAMLR Convention Area) and described in 1983 as a "new species", is now listed as endangered under US legislation.[117]

Since 1984, CCAMLR has requested that its members maintain records and report number and, where appropriate, additional biological data on any seabirds and marine mammals that are caught as bycatch in target fisheries. In the ensuing years, CCAMLR has introduced several measures aimed at reducing bycatch of bird species. The introduction of longline fishing in the CCAMLR area in 1989 was noted as a possible threat to seabird populations, particularly albatrosses.[118] Following increased seabird bycatch, CCAMLR adopted a non legally binding resolution to protect seabirds from this threat in 1990.[119] The 1991 binding Conservation Measure 29-10 required the minimization of seabird mortality in longline fishing or longline fishing research and was followed by the establishment of the Woking Group on Incidental Mortality

Associated with Longline Fishing (WG-IMALF) in 1992 (now WG-IMAF). These steps, together with the development of tori (streamer) lines on long-liners in the Southern Ocean, were the first efforts to develop mitigation measures to reduce the incidental mortality of seabirds in longline fisheries in the world. The Conservation Measures specifying the mandatory elements to be used for minimizing incidental seabird mortality have been subsequently refined and improved, particularly for risk assessment of seabird bycatch in each of the 18 fishery subdivisions of the convention area.[120] All fisheries must abide by seabird Conservation Measures.

The need for accurate data on seabird mortality and bycatch rates spurred development of the Scheme of International Scientific Observation, established by CCAMLR in 1992, which has specific objectives aimed at the overall reduction in seabird mortality. The increase in IUU fishing of Patagonian toothfish in the 1990s, with unreported catches of five to six times the TAC established by CCAMLR, decimated toothfish stocks and also greatly impacted seabirds. At the height of the toothfish fishery in 1996–1997, bycatches of black-browed albatross and white-chinned petrels by IUU fishing were estimated to be 20 times greater than reported bycatches.[121] While the compliance of Contracting Parties with Conservation Measures has been high, effectively reducing seabird mortality, the IUU fishery and associated incidental catch remain a conservation challenge. In 2004, as a result of increased enforcement and corresponding reduction in IUU fishing, seabird mortality decreased by an estimated 30%.[122]

In addition to Conservation Measures, CCAMLR has developed educational material for the industry and observers in efforts to increase data quality on incidental catches. CCAMLR published *Fish the Sea, Not the Sky* in 1992 and in 1999 published an identification guide to seabirds (in cooperation with the Museum of New Zealand) for use by this program, in an effort to improve data collection and hence bycatch mitigation measures.

7.6.1.2 Bycatch and impact on other species

Bycatch is taken into considerations in WG-IMAF and WG-FSA. To reduce the impact of fishing gear on non-target species, CCAMLR has linked TACs for target species to bycatch TAC for particular species. Reaching a bycatch TAC can result in a fishery closure regardless of whether the TAC for the target species has been reached. CCAMLR first adopted resolutions on finfish bycatch in 1985 and 1986. CCAMLR Conservation Measures 33-01 and 33-02 specify bycatch limits for several finfish species in the CCAMLR area.[123] Bycatch limits are in place for skates, rays, and *Macrourus* spp.[124] As of June 2005, directed fishing for all finfish is prohibited in several statistical subareas (48.1, 48.2) and for specific finfish in other areas. These measures are to remain in place until further information is gathered and analyzed by the Scientific Committee and WG-FSA.[125]

Trawling, other than by using midwater trawls, is prohibited by CCAMLR in the South Georgia region (Statistical Subarea 48.3). Because the midwater trawls used in icefish fisheries in this area were found to kill albatrosses and petrels, in 2004 the United Kingdom suggested that it should be allowed to use bottom trawls, fished off the bottom in this fishery. Concern about potential adverse impacts on fragile coral and sponge communities caused the United Kingdom to withdraw this request.[126]

In addition to bycatch mitigation measures for associated and dependent species, CCAMLR has also protected areas of importance to specific species. In 2004, Conservation Measure 91-03[127] established protection measures on Seal Islands, which are a CEMP monitoring site. The measure prohibits access to the islands during breeding seasons and restricts the building of any new structures unless approved by CEMP. Strict waste-disposal rules are also in place.

CCAMLR is also making efforts to develop a marine protected area (MPA) strategy and plan of action.[128] The Commission has tasked the Scientific Committee with this initiative. An intersessional workshop was held in 2005 to consider the establishment of MPAs as a management tool within the CCAMLR area. At this meeting it was agreed that "the primary aim is to establish a harmonized regime for protection of the Antarctic marine environment across the ATS [Antarctic Treaty System]".[129] At the 24th Meeting of the Commission in November 2005, it was noted that there was a need to develop a strategic approach to MPA development in the Southern Ocean, and that collaboration at the technical and policy levels was needed.[130] In seeking accelerated progress on this issue, the Commission urged the Scientific Committee to hold a second workshop in 2007.[131]

In recognition of the decreasing toothfish stocks, particularly from IUU fishing, some environmental NGOs have called for a moratorium on all catches of toothfish and also consideration of a CITES Appendix II Listing.[132] In 2002 a CITES resolution was adopted (Res. Conf. 12.4)[133] on cooperation between CCAMLR and CITES, with particular regard for the catch documentation scheme (CDS) on toothfish; however, as of April 2006, a formal resolution on cooperation has not yet been adopted.[134]

7.6.1.3 Pollution and marine debris

At its 1990 meeting (CCAMLR-1X) and following the 1991 adoption of the Madrid Protocol, which directly addresses marine pollution and debris,[135] CCAMLR began to address marine debris and potential ecological consequences of waste disposal during fishing operations at sea. At recent CCAMLR meetings, reports from States on observations of marine debris on beaches and entanglement have been requested.[136] Guidelines for conducting marine-debris surveys were first developed in 1993. In 1996 a Conservation Measure was passed[137] that made reference to the International Convention for the Prevention of Pollution from Ships (MARPOL Convention)[138] and addressed the entanglement of fur seals in fishing gear and plastic. This Conservation Measure requires that nonplastic implements be used for securing bait boxes, that all looped plastic be cut, and that all plastic be retained on board the vessel until it offloads in port.

7.6.1.4 Data needs

Practical implementation of EAF, in a holistic manner, requires the identification of a wide range of data to be collected and then used in a manner that assists in making ecosystem-based management recommendations. Although CEMP and scientific observer program have successfully increased data-collection capacity, there are still tremendous data management needs for long-term observations of the Southern Ocean ecosystem, particularly in respect of acquiring the ability to discriminate between natural variability and human-induced effects.

Despite its commitment to the EAF, compliance with conservation and management measures presents an ongoing challenge,[139] as does IUU fishing for Patagonian toothfish. As a result, CCAMLR's progressive management has been undermined, and the stocks are suffering. CCAMLR is vigorously endeavouring to enforce its binding Conservation Measures, *inter alia,* through its *Dissotichus* spp. CDS, port-State control, and domestic control. In 2005, CCAMLR adopted a nonbinding

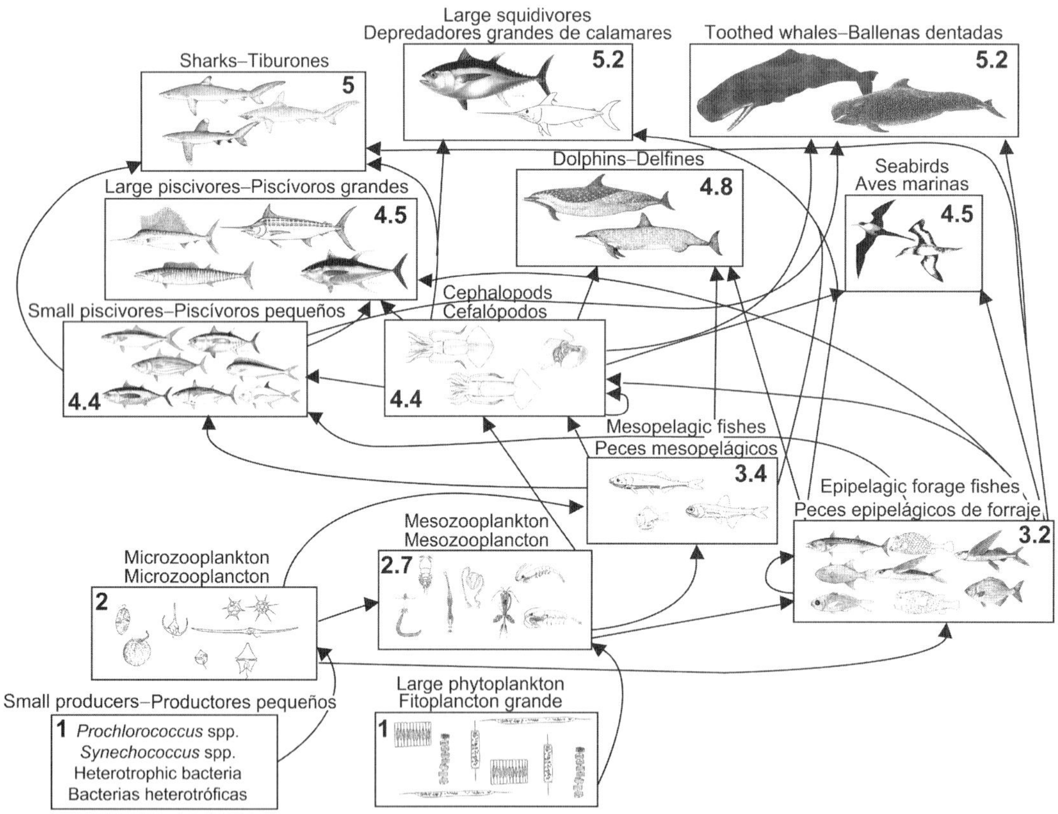

Figure 7.3. Simplified food-web diagram of the pelagic ecosystem in the tropical Eastern Pacific Ocean. The numbers inside the boxes indicate the approximate trophic levels of each group. (From IATTC, "Tunas and Billfishes in the Eastern Pacific Ocean in 2003", IATTC Fishery Status Report No. 2 (La Jolla, Calif.: IATTC, 2004), reproduced with permission of IATTC.)

resolution (Resolution 24/XXIV) on enhancing cooperation with Non-Contracting Parties for combating IUU and implementing CDS, to improve compliance with its CDS program.

7.6.2 IATTC

Although it is the oldest regional fisheries organization dedicated to managing tuna, IATTC's 1949 Convention "clearly envisages EAF",[140] thereby providing the Commission with competence to consider target species in the context of the wider marine ecosystem. IATTC is responsible for all fish taken by tuna fishing vessels, including target species, fish bycatch, and nonfish bycatch (dolphins, turtles, and seabirds) (Figure 7.3 illustrates the food web in the Eastern Pacific Ocean). The IATTC Convention also expresses the need for conducting research on natural and anthropogenic effects on target stocks.[141]

7.6.2.1 Tuna–dolphin interactions

A long-standing concern for IATTC has been the ecosystem impacts of the yellowfin tuna fishery. In the late 1950s, fishermen discovered that yellowfin tuna in the Eastern Pacific Ocean (EPO) aggregated beneath schools of dolphins. With this discovery, fishermen began using purse seines as the preferred method for harvesting these tuna, tracking and encircling their nets around dolphin herds to capture the tuna found in concentrations below. The incidental catch and mortality of dolphins associated with the yellowfin tuna purse-seine fishery was particularly high during the early years of this fishery.[142]

IATTC began regulating the yellowfin tuna purse-seine fishery in 1961, and in the late 1970s, the Commission recognized the need for examining the impacts on dolphin populations (in particular, spotted dolphins, spinner dolphins, and common dolphins). In 1976, IATTC established the Tuna–Dolphin Program to monitor, study, and analyse the causes and effects of dolphin mortality, with a view to maintaining EPO dolphin stocks at or above levels that would assure their long-term survival. "However, due, at least in the early stages, to a lack of funding, progress under the program was slow".[143] Nonetheless, since 1978, IATTC has assumed responsibility to undertake research on both the effects of the fishery on dolphin stocks and the benefits of measures to mitigate dolphin bycatch.

The establishment of IATTC's program was, in part, in response to domestic legislation in the United States in 1972 enacted to protect marine mammals and control the activities[144] of the large US tuna purse-seine fishery operating in the EPO.[145] Although an American domestic law, the 1972 Marine Mammal Protection Act (MMPA), together with its 1984 and 1988 amendments, had "a definitive impact on the development of measures at the international level".[146] The law prohibited "taking" of marine mammals and later imposed strict measures for dolphin-associated fishing by US vessels. It also introduced a trade embargo on tuna and tuna products that were not deemed "dolphin safe".[147] Many tuna fishing nations and processors (Colombia, Mexico, Vanuatu, and Venezuela) were effectively locked out of the US market by this embargo, as they did not meet the US standard for incidental catch of dolphins, which required that no dolphins be encircled during the fishing trip.[148]

In response to the limited success of reducing dolphin bycatch[149] under the IATTC Tuna–Dolphin Program, the US trade embargo, the General Agreement on Trade and Tariffs (GATT) challenges by Mexico and the European Union (EU),[150] and US efforts to introduce a global moratorium on the use of purse seines to harvest tuna,[151] IATTC convened meetings in 1992 in La Jolla, California, to address US concerns about dolphin-associated fishing as well as the concerns of other fishing countries seeking access to the US market to have a viable regime. The 1992 La Jolla Agreement, a voluntary agreement, established the International Dolphin Conservation Program (IDCP). Under the IDCP, voluntary limits were set on the maximum number of dolphins that could be incidentally killed annually in the fishery, with the yearly maximum being progressively lowered over seven years,[152] with a goal of eliminating dolphin mortality in the yellowfin tuna fishery. This program required 100% observer coverage of vessels greater than 400 tonnes (t), with at least half the observers from the IATTC observer program. The program also created a multistakeholder review panel to monitor and report on compliance, and established a scientific advisory board to assist IATTC with research on reducing dolphin mortalities and gear selectivity modifications. The data gathered through the IDCP and a list of all vessels involved in the purse-seine fishery were to be provided to the Director of the IATTC. Based on reported figures, dolphin mortality was significantly reduced within the first year of the voluntary agreement to 3601 individuals (below the agreed 1993 target of 19 500 individuals).[153]

Despite the reduction in dolphin mortality under the IDCP, the US trade embargoes continued. The interested States (Belize, Columbia, Costa Rica, Ecuador, France, Honduras, Mexico, Panama, Spain, and the United States) met and negotiated the 1995 Panama Declaration. The signatory States agreed to negotiate and enter into a binding regional multilateral agreement reaffirming their commitment to reducing dolphin bycatch consistent with sound ecosystem management (as expressed in UNFSA and the FAO Code of Conduct). The United States, in turn, undertook to modify its legislation so that participation in the IDCP by signatory States would satisfy US standards under the MMPA, and accordingly these Sates would not be subject to trade restrictions on yellowfin tuna and related products.

The United States amended its legislation in 1997 and adopted the International Dolphin Conservation Program Act. Under this Act, the capture of tuna in sets in which there is no mortality or serious injury of dolphins satisfies the dolphin-safe standard.[154] Shortly after its adoption, the IATTC convened a special session of interested States and hosted other negotiating sessions in 1998. The Panama Declaration process culminated in May 1998 with the adoption of the legally binding Agreement on the International Dolphin Conservation Program (AIDCP),[155] which entered into force in February 1999. This agreement essentially formalized the La Jolla Agreement and the IDCP, creating a stable and sustainable regime for dolphin conservation and ecosystem management in the EPO. The AIDCP has been subsequently amended to strengthen the dolphin conservation agreement. Costa Rica, Ecuador, El Salvador, EU,[156] Guatemala, Honduras, Mexico, Nicaragua, Panama, Peru, the United States, Vanuatu, and Venezuela have ratified the 1998 agreement. Additionally, Bolivia and Columbia are applying the agreement provisionally.

The AIDCP also provides that tuna harvest is permitted for fishing operations where dolphins are not mortally wounded or seriously injured.[157] Objectives of the Agreement include reduction of incidental dolphin mortalities in the tuna purse-seine fishery through the setting of annual limits for each stock in the EPO to levels approaching zero; seeking alternative means of capturing large yellowfin tunas not in association with dolphins; ensuring the long-term sustainability of tuna stocks and marine resources in the EPO; and avoiding, reducing, and minimizing bycatch of juvenile yellowfin tuna and other nontarget species such as sea turtles. It is noteworthy that the 1998 Agreement also requires Contracting Parties to take measures to protect other species taking into consideration interrelationships among species in the same ecosystem. To the maximum extent practicable, vessels are required to release incidentally caught live sea turtles and other threatened or endangered species.

The Parties to the AIDCP are required to implement the Agreement within the framework of IATTC. Accordingly, IATTC serves as the secretariat, operates the observer program, and conducts research and analysis based on data collected through the program. IATTC undertakes general studies of the association of tuna with dolphins,[158] an important component of understanding key interactions of the ecosystem, as well as particular studies to determine whether the purse-seine fishery for tuna is having a significant adverse impact on any depleted dolphin stock in the EPO. In 2002, the Director of IATTC reported that "the best scientific evidence supports the conclusion that the purse seine fishery is not having a significant adverse impact on any depleted dolphin stock in the EPO...specifically that the intentional deployment on or encirclement of dolphins with purse seine nets is not having a significant adverse impact on any depleted dolphin stock in the eastern tropical Pacific Ocean".[159] This conclusion, particularly in light of the low mortality levels and no evidence indicating problems from stress, is based on IATTC independent analyses of the stocks status and observations of their interactions with the tuna fishery over the last 30 years. Based on these findings, the United States has further amended its procedures regarding what constitutes dolphin-safe tuna.[160]

> This Agreement and its predecessor, the 1992 La Jolla Agreement, have been spectacularly successful in meeting t[heir] objective, as shown by the reduction in mortality of dolphins incidental to fishing. During 2003, 94% of all sets made on tuna associated with dolphins were accomplished with no mortality or serious injury to the dolphins. Furthermore, the total mortality of dolphins in the fishery has been reduced from about 132,000 in 1986 to less than 1,500 in 2004 — about 0.015% of the population.[161]

During 2004, the trend of low incidental dolphin mortalities to levels approaching zero continued (see Figure 7.4).

7.6.2.2 IATTC bycatch initiatives

The IATTC Working Group on Bycatch (WG-Bycatch) first met in July 1998 to address bycatch issues not covered by the AIDCP. Scientific advice to the Commission since then includes information from the WG-Bycatch on the ecosystem effects of fishing and specific efforts to reduce bycatch in the bigeye tuna and yellowfin tuna fisheries, specifically in the purse-seine fishery. The WG-Bycatch includes two subgroups: Ecological Studies and Modeling; and Technology and Fishing Techniques.

The first Resolution on Bycatch was passed in 1999[162] in recognition of the FAO IPOA-Sharks. The Resolution called for IATTC to develop methods for estimating

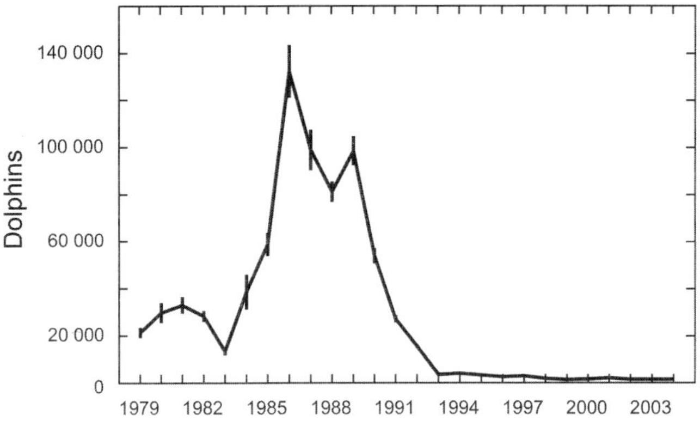

Figure 7.4. Reduction in dolphin mortality. (From IATTC, "Tunas and Billfishes in the Eastern Pacific Ocean in 2003", IATTC Fishery Status Report No. 2 (La Jolla, Calif.: IATTC, 2004, reproduced with permission of IATTC.)

discards of nontarget and target species and, in particular, for estimating catch of sharks and rays. In accordance with the AIDCP, States were required to release all sea turtles and report on the number and condition of those released. IATTC scientific staff was asked to investigate gear modifications and other measures to reduce incidental catches. Appropriate measures, as outlined in the IPOA, were to be included in an IATTC bycatch management scheme. The 2000 recommendation[163] from the WG-Bycatch brought to the 2000 IATTC meeting called for a reduction in bycatch of juvenile target species and turtles in the purse-seine fishery, as well as for further research on bycatch mitigation measures. The recommendation was followed by the 2000 Resolution on Bycatch, which further elaborated on past bycatch Resolutions. Specifically, the Resolution called for the establishment of a one-year pilot program (effective 1 January 2001), which required that all target species be retained, with the exception of those caught in the final set of the trip. The Resolution also called for the release of live sea turtles, rays, sharks, billfish, mahi mahi, and other nontarget species and specified measures to be taken in the release of encircled or entangled sea turtles. An evaluation of sorting grids in the purse-seine fishery to reduce catch of juvenile tuna was requested, as well as an evaluation of bycatch reduction measures including time–area closures, effort limits, bycatch limits, reduction of net depth in the purse-seine fishery, and limits to the numbers of fish aggregating devices (FADs).[164] Increased data collection on bycatch, particularly for vessels not covered by the observer program, were called for. Further development of management measures based on research and data collection was also recommended.

Binding bycatch Resolutions were subsequently passed in 2001[165] and 2002,[166] with each successive Resolution building on that of the previous year in response to research. The 2002 Resolution included provisions for reducing marine pollution, prohibiting the disposal of plastic, and requiring the recovery of FADs, as both plastic and FADs are implicated in sea turtle mortality. The 2002 Resolution also called for FAO to develop a comprehensive approach to addressing turtle bycatch. FAO published guidelines for reducing sea turtle bycatch in 2004.[167]

In 2003, IATTC passed a bycatch Resolution consolidating all measures taken to that time[168]on the reduction of juvenile tuna incidental mortality and nontarget species bycatch; on gear modifications; on data-collection requirements for billfish, sharks, and rays bycatch; and on prohibitions on the discarding of tuna catch.[169] The 2004 consolidated Resolution on Bycatch built upon that passed in 2003, further elaborating on reducing turtle bycatch and training of crew and fishers in release of live nontarget species. The tuna retention program, first begun as a pilot project in 2001, was extended to 2006, and in 2005, was again extended to 2007.[170]

Turtle incidental mortality associated with the fishery in the EPO is a long-standing concern that was addressed in the AIDCP for the yellowfin tuna purse-seine fishery. Turtles have been included in bycatch Resolutions since 1998 and in consolidated bycatch Resolutions since 2003. In 2003 a separate a nonbinding recommendation on turtles[171] was passed to develop a sea turtle conservation program. This recommendation was elaborated and adopted as a Resolution in 2004,[172] establishing a three-year conservation program.[173] While these measures are generally consistent with the FAO Guidelines to Reduce Sea Turtle Mortality,[174] IATTC staff recommended additional measures for sea turtle conservation, including a requirement to release turtles entangled in FADs whenever possible and the introduction of a system for marking FADs.[175] In 2003 IATTC, with support from the World Wildlife Fund, the Ocean Conservancy, and the US Western Pacific Fisheries Management Council, started a program that included the exchange of traditional hooks for circle hooks, the introduction of turtle release devices, and observer coverage for the longline fleets of member countries in Central and South America.[176] The 2004 Resolution requires gear modification in longline fleets to reduce turtle bycatch.[177]

Sharks are not a directed fishery but are often caught incidentally while targeting tuna and tuna-like species. Given the market for shark fins, these species typically have their fins removed and then are discarded. IATTC first addressed elasmobranch (shark and ray) bycatch in 1999. Successive bycatch Resolutions have included data collection and release of live elasmobranchs. At the 2004 IATTC meeting, Spain and Japan proposed resolutions to address the practice of finning, through which the recommendations of IPOA-Sharks could be implemented.[178] The 2005 Resolution[179] calls for data collection by Contracting Parties, cooperating nonparties, cooperating fishing entity, or regional economic integration organization (collectively CPCs) to be used in stock assessments for sharks in 2006. The Resolution also calls for the reduction of shark finning and requires that landings of shark fins represent less than 5% of the catch on board.

In the past, seabird bycatch has not been a significant concern for the IATTC. The extension of the IATTC boundaries by 10° toward the south, as established by in the Antigua Convention, expands the IATTC's jurisdiction into areas where seabird bycatch is likely to increase. During its Annual Meeting in June 2005, IATTC Resolutions on bycatch species were further strengthened to incorporate seabirds.[180] A specific Resolution on seabird bycatch was passed[181] that calls upon CPCs to report interactions with, and incidental catch of, seabirds to the Commission and to inform the Commission on the status of their National Plans of Action (NPOAs) for reducing incidental catches of seabirds in longline fisheries.

In addition to addressing incidental mortality of juvenile tuna through requirements for full retention, IATTC has also undertaken initiatives to address bycatch through the reduction of fishing effort by way of area closures and control of fleet capacity. IATTC has established seasonal area closures for the purse-seine fishery for bigeye, yellowfin, and skipjack in the EPO[182] and has been reducing capacity.[183] In addition to management measures, the IATTC offers a training program for captains on how to identify species and reduce bycatch of turtles and the species covered by the IPOAs in an effort to implement EAF.

7.6.2.3 IATTC's current ecosystem activities

The management practices adopted by IATTC in the last two decades (precautionary and adaptive management approaches with regard to the expansion of fisheries and associated impacts) and its efforts to reduce and eliminate, where possible, the dolphin bycatch associated with the yellowfin tuna purse-seine fishing operations took into account ecosystem considerations and were based on some of the key concepts and principles of the EAF. However, IATTC has only recently "focused its attention on the entire ecosystem in which the target species, the tunas and billfishes, reside".[184] This shift, formalized in June 2005, is reflected in the Work Program[185] and advice provided to the Commission in the last few years. A proposed 2005 Resolution on the ecosystem approach to management[186] requires that all IATTC decisions consider the effect of management measures on the marine ecosystem, including changes in trophic structure, bycatch of untargeted species, and physical ecosystem changes; however, this Resolution was not adopted at the Commission's meeting in June 2005.

The permanent scientific staff's annual advice to the Commission has included an extensive chapter on ecosystem effects of the fishery since 2002. Their chapter summarizes what is known about the direct impact of the fisheries upon various species and species groups of the EPO and reviews what is known about the environment and about other species not directly impacted by the fishery. The purpose of the chapter "is to provide the Commission the opportunity to consider the ecosystem as a whole as part of its consideration into fisheries management" and "to offer the Commission the opportunity to ensure that ecosystem considerations are clearly part of its agenda".[187] With the subsequent adoption of the 2005 Resolution, the Director of IATTC is now required to report on the implementation of EAF.

EAF is also reflected in the Antigua Convention. The Convention (not yet in force) modernizes IATTC, as it addresses the need to take into account the ecosystem considerations called for in UNFSA, includes the IPOAs, and embodies the relevant EAF provisions of the FAO Code of Conduct, particularly in the Convention's subparagraphs (a), (c), (f), (g), (h), (k), and (n) of Article VII(1).[188]

7.6.3 ICCAT

While ICCAT has no specific provisions for EAF in its Convention, the Commission has interpreted Article IV[189] of its constitutive agreement to expand its mandate to include the collection of data on sharks and other fish caught in its fisheries.

The Standing Committee on Research and Statistics (SCRS), the primary body for scientific data collection and advice to the ICCAT Commission, has several subcommittees, working groups, panels, and special research programs that are tasked with collecting specific information on target species, stock interactions, bycatch, and general environmental influences on regulated species.[190] For example, the Bluefin Year Program (BYP) collects information on Atlantic bluefin tuna biology, including larval dispersal, location of spawning grounds, stock structure through genetic testing, and the effects of the oceanic environment on distribution of juvenile tunas.[191]

Under the SCRS, ICCAT established the Sub-committee on Bycatch[192] in 1996, following concerns raised through CITES about ICCAT managed species. The Sub-committee primarily focuses on shark and turtle bycatch in fisheries targeting tunas in response to the FAO IP-OAs.[193] Shark bycatch reduction measures are outlined in Recommendation 04-10, including limiting shark catch to less than 5% of the total catch, identifying shark spawning grounds and nursery areas, and continuing data collection on shark bycatch.[194] The 2004 ban on shark finning instituted by ICCAT[195] if respected, will significantly help curb unsustainable shark mortality and keep the ocean ecosystem in balance. Further to the 2004 Recommendation, and out of concern for the declining status of the shortfin mako shark population, ICCAT strengthened the 2004 Recommendation in 2005 by requiring that all Contracting Parties, Cooperating Non-Contracting Parties, entities or fishing entities (hereafter referred to as CPCs) report to the Commission on the status of its implementation.[196] The 2003 nonbinding resolution on sea turtles encourages CPCs to reduce the bycatch of sea turtles and to support FAO efforts to address conservation concerns through a "holistic approach". Subsequent to the 2003 measures, though voluntary in nature, ICCAT increased measures to reduce sea turtle bycatch, citing the UNFSA call to "take ecosystem considerations into account". The nonbinding Resolution 05-08[197] encourages CPCs to undertake research on the use of circle hooks and other fishing methods and gear modifications that may reduce sea turtle mortality, as these hooks increase postrelease survivorship. This resolution also notes the effect of circle hooks on survivorship of blue and white marlin, both of which are currently under rebuilding plans.

Collection of bycatch data on nontarget species by Contracting Parties is encouraged. To date, the Commission has not implemented specific management measures to mitigate, reduce, or conduct research on seabird or turtle bycatch. The Sub-committee also provides data on the diversity of species caught in target fisheries, but does not estimate bycatch abundance.[198]

ICCAT introduced measures in the early 1980s to reduce bycatch of target species, particularly juvenile tunas, through the Working Group on Juvenile Tropical Tunas. This Working Group last met in 1984. Seasonal moratoria on FADs have been implemented since 1999.[199] Binding Recommendations have been passed on size limits on Atlantic bluefin tuna, prohibiting landing of bluefin less 10 kg,[200] and in the past size limits have been imposed for bigeye, yellowfin, and swordfish. In 2005, ICCAT members agreed to continue discussions on integrated management measures between Eastern and Western Atlantic bluefin tuna and eastern and western swordfish stocks. This is primarily out of concern that increasing uncontrolled harvest practices of bluefin in the Eastern Atlantic can adversely affect the western population, which is under a strict 20-year rebuilding plan. In 2006 there will be a fouth intersessional meeting of the Working Group to Develop Integrated and Coordinated Atlantic Bluefin Tuna Management Strategies on eastern and western stock mixing, with a report expected to be tabled at the 2006 ICCAT Annual Meeting. ICCAT continues to deal with a lack of information on catches, but is endeavoring to increase funding for data collection and for the development of data sharing systems.[201]

In recognition of the varied scientific and technical capacity of Atlantic coastal States, ICCAT is involved in numerous initiatives to increase understanding of stock and population structure of species under its competence. ICCAT works in cooperation with national fisheries management departments and academic research centres in tagging programs, specifically to determine stock structure of tuna species migrating across the Atlantic and to develop further understanding of tuna biology.[202]

ICCAT has been concerned about the high catches of juvenile tunas by purse-seine and bait-boat vessels fishing in the Gulf of Guinea using floating objects or FADs, because of the tendency of juvenile tunas to congregate around FADs. Since 1996, ICCAT has been taking steps to gather data on, and to enhance the protection of, juvenile tunas in the Gulf of Guinea.[203] In 1998 ICCAT adopted a binding measure that closed the Gulf of Guinea to purse-seine fishing using floating objects from 1 November 1999 through 31 January 2000.[204] To monitor this fishery, in 1996, observers were first recommended by ICCAT for bigeye and yellowfin fisheries, including the Gulf of Guinea fishery. In 1999 ICCAT extended its Gulf of Guinea time–area closure, and the measure was expanded to encompass all surface fleets in the Gulf flying the flag of Contracting Parties, Non-Contracting Parties, entities, and fishing entities.[205] The current recommendation prohibits fishing over floating objects and with auxiliary vessels from 1 November of one year to 31 January of the following year. The measure also requires vessels to carry an observer at all times for both compliance and the collection of biological data, and it requires the establishment of internal procedures by each party to penalize their vessels for noncompliance. All those fishing in the

Gulf are to present an annual report on implementation of the measures to the Secretariat for submission to the Commission.

ICCAT and GFCM have a Joint Working Group on Stocks of Large Pelagic Fishes in the Mediterranean Sea. In recent years, farming of Atlantic bluefin has increased in the Mediterranean, where tuna are seined, transferred live, and fattened in captivity. Recommendation 05-04, which was adopted at the 2005 ICCAT meeting, requires CPCs to collect data on where the farmed fish are harvested and requires that all farmed fish sold on the market be labeled as such.[206] In addition, the Commission must maintain a record of authorized farming facilities with name, location, and farming capacity operating in the convention area.

The SCRS established the Sub-committee on the Environment in 1991 to study the effect of the environment on tunas with the specific objectives of better understanding the impact of environmental changes on population dynamics and catchability of tunas and billfishes, better understanding the mechanisms of the food chain of the open ocean, and managing tuna resources more efficiently.[207] The Sub-committee's immediate concern is that of oceanographic factors influencing tuna biology and hence fisheries, rather than technical aspects or multispecies interactions. In a 2001 Sub-committee workshop on the effect of the marine environment on tuna recruitment, participants agreed,

> that attempts to define and quantify suitable recruitment habitat for the various species of direct concern to ICCAT could be useful. Suitable habitat should be defined on the basis of readily available environmental variables (e.g. salinity, temperature, dissolved oxygen) on the basis of inferences from field observations and (or) by direct experimentation. On the other hand, other oceanographic/meteorological variables (such as those mentioned above) should be taken into account in the study of environmental impacts on recruitment.[208]

At the 2004–2005 ICCAT meeting, the Sub-committee on the Environment recommended further expansion and examination of EAF by the SCRS and through the Sub-committee itself, with the caveat that implementing EAF will require collection of data not currently considered by ICCAT as well as comprehensive research programs to address ecosystem questions.[209] The Sub-committee report reflects an understanding of a pragmatic approach towards adopting EAF, which incorporates, in an incremental fashion, an increasing number of ecosystem considerations, as follows:

> The multi-species and ecosystem approaches imply a great amount of data and information (e.g., stomach contents, isotope ratios, prey abundance, etc.) which are not routinely collected by the statistical programs of the Contracting Parties. Consequently, this approach requires very important research efforts. Besides, the models currently available, such as multi-species VPAs [Virtual Population Analysis], the indicators (such as size spectrums), the statistical models (based on the time series) or the trophic models of the ECOPATH type, are not yet fully operative in terms of assessment and, above all, helpful for decisions on the management of the stocks exploited. In conclusion, it appears necessary to form a "think tank" on this issue within the SCRS, particularly to develop a work plan adapted to the general mandate of the SCRS.[210]

In this case, the SCRS agreed to the proposed work plan. At its October 2005 meeting, the Sub-committee recommended expanding its research further into ecosystem matters and recommended that the Sub-committee on Environment merge with the Sub-committee on Bycatch and evolve towards an ecosystem subcommittee.[211] This initiative has yet to be adopted by the SCRS or the Commission; however, the consideration of ecosystem factors is clear in the 2005 nonbinding resolution on pelagic *Sargassum*, which calls for data collection on impacts of pelagic *Sargassum* with particular regard to data on the status of the species under the competence of ICCAT and on the ecological importance of *Sargassum* to these species.

7.6.4 GFCM

The GFCM addresses ecosystem approach considerations through its Scientific Advisory Committee (SAC) in the framework of transversal activities coordinated by the Sub-committee on the Marine Environment and Ecosystem (SCMEE). The 1997 amended agreement refers to the protection of living marine resources under functions of the Commission (Article III(1) (e)). Accordingly, the Commission has begun to introduce measures consistent with EAF.[212] At the Sixth Meeting of GFCM's SAC,[213] SCMEE agreed to collect data on bycatch of protected species and on large migratory sharks, given the lack of data for these species in the Mediterranean. The Committee also collects data on algal blooms and analyses the data in relation to its effect on fisheries in the

GFCM area. GFCM works closely with COPEMED[214] and MedSudMed[215] to collect data on the implementation of the EAF. As part of collecting data and acquiring an understanding of the marine ecosystem, SCMEE has undertaken an exercise to map essential fish habitats, which is scheduled for completion in 2005. Another significant ecosystem initiative, proposed in 2004 and adopted in February 2005 by the GFCM, is the closure of areas below 1000 m to trawl fisheries in the Mediterranean and Black Sea. This measure, effective as of June 2005, is intended to conserve the spawning grounds of shrimp and to protect deep-sea habitat.[216]

The complexities of applying EAF in the Mediterranean were discussed at the 2004 meeting of the SAC. It was suggested that a multidisciplinary group be formed to expand on the EAF and address its application in the convention area.[217] The first Workshop on the EAF, held in 2005,[218] focused on building a common understanding of the key principles and tools for the approach and identifying elements for a work plan, including scientific research needs. "Presentations and discussions covered the geographical partition, time scales and classification of ecosystems as well as habitats; modelling; ecosystem indicators for spatial and temporal monitoring of fishing effort, selectivity and gears; and by-catches of vulnerable and protected species".[219]

At its eighth session in 2005, the SAC endorsed some principles for implementing the EAF, favoured the establishment of a network of fishing gear technologists to address gear selectivity issues, and reiterated the importance of further standardizing stock assessment methodologies.[220] Furthermore, at its 30th session, in January 2006, the GFCM adopted a Recommendation on the establishment of fisheries restricted areas to protect deep-sea sensitive habitats by prohibiting the use of towed dredges and bottom-trawl nets in three areas of the Mediterranean.

7.6.5 NAFO

The specter of irreversible change to the marine ecosystem and bycatch of moratoria species are major concerns for NAFO, especially as stocks long under moratoria have yet to recover. The NAFO Convention does not make reference to the ecosystem approach; however, ecosystem studies have long been part of the work of the Scientific Council. The Scientific Council established a Standing Committee on Fisheries and Environment (STACFEN) to include environmental considerations in the provision of scientific advice. For many years, STACFEN has investigated the impact of the physical environment on fisheries within the NAFO area. In 2004 NAFO committed itself to addressing broader ecosystem concerns of the impact of fishing on the marine environment and introduced EAF in its subsequent work plans.[221] Further developments pertaining to EAF were agreed at the 2005 NAFO Annual Meeting. The Commission tasked scientists with examining areas of biological and ecological significance.[222] Additionally, NAFO will hold a symposium in 2006 to further efforts to incorporate EAF and increase knowledge of the Northwest Atlantic marine ecosystems.[223]

Recognition of the impact of the marine environment on the health of fish stocks has been part of scientific studies in the NAFO area for the last 40 years. Starting in 1964, NAFO's antecedent organization, ICNAF, began decadal reviews of environmental conditions in the Northwest Atlantic and the corresponding influence on fish stocks.[224] The latest symposium was held in 2002 on the hydrographic variability of NAFO waters, comparing the period from 1991 to 2000 to conditions in past decades. NAFO collects data on physical environment conditions through various means, including the Marine Environmental Data Services (MEDS); vessels equipped with monitors of salinity and temperature; drift buoys that measure currents; moored buoys that measure wave data in the NAFO Regulatory Area; the Canadian Hydrographic Service for tide and water level data; and several other data collections by Contracting Parties. These data are used to further understanding of juvenile dispersal, stock status, and recovery options.

Recognizing that the design and implementation of an ecosystem approach are emerging requirements of international fisheries management,[225] NAFO committed itself to addressing EAF in its 2005 work plan,[226] and EAF was a key item on the agenda at both the June 2005 meeting of the Scientific Council[227] and the September 2005 Annual Meeting. At the 2005 Annual Meeting, FAO introduced EAF during a joint Fisheries Council – General Council – Scientific Council session to discuss the EAF, in keeping with the need for enhanced coordination and communication between science and management under EAF.[228] In addition, Contracting Party fishing vessels were asked to collect data on seamounts on a voluntary basis.[229]

Many measures have been taken over the years to manage ecosystem impacts of fishing. These include restrictions on bycatch, area closures, and gear modifications. Bycatch of nontarget species and juveniles of target species is particularly worrisome in the NAFO area given the adverse impact of incidental catches on stocks still under moratoria. Directed fisheries are not permitted for species for which bycatch limits are in place. Bycatch of species for which no quota is allocated to a particular Contracting Party is limited to no more than 10% of the total catch or 2500 kg, whichever is greater. Bycatch of species under moratoria, or for which a quota has been fully allocated, is limited to 5% or 1250 kg, whichever is greater. If bycatch limits are exceeded in any one net haul, the vessel must relocate to at least five nautical miles away from the last set and not return to the area for 48 hours. Bycatch limitations are also in place in the shrimp fishery and vary in allowed percentage, depending on the NAFO division in which the target fishing occurs.[230] At its 2005 Annual Meeting, NAFO adopted measures banning shark finning and the transshipment and landing of shark fins to improve the protection of sharks caught as bycatch in the NAFO area.[231] These measures are in keeping with the EAF and consistent with those adopted by regional tuna commissions (IATTC, ICCAT, and IOTC).

Articles 12 and 13 of the 2005 conservation and management measures outline gear requirements and minimum fish size restrictions, and are designed to minimize catch of juveniles of target species and juveniles of stocks under moratoria. Lining of nets at the cod end is prohibited.[232] The shrimp fishery has time–area restrictions in place for 2005 and is limited to below 200 m depth in certain areas.[233] Gear modifications have been made to reduce bycatch, particularly of species under moratoria. Nordmore grates are required on shrimp fishing nets, enabling non-target species to pass through the grates. In the NAFO area, bycatch could be significantly reduced as a result of these and other management measures and through changes in target species and reduced trawl effort.[234]

An important development introduced by NAFO in 2004 is the regulation of skate. NAFO was the first RFMO to establish a management plan for an elasmobranch species. This measure is considered significant in responding to the impact of fishing operations on nontarget species, in keeping with EAF.

7.6.6 NEAFC

The NEAFC Convention does not explicitly refer to EAF, but in recent years, as part of strengthening NEAFC and updating its constitutive agreement, efforts have been made to incorporate ecosystem concerns in NEAFC management practices. Since 2001, EAF has been a permanent item on the agenda of the Annual Meeting. At the June 2005 meeting of the Working Group on the Future of NEAFC, reference was made to the wording on EAF in both SEAFO's Convention and SIOFA's,[235] and textual changes to the NEAFC Convention preamble and Articles 1, 2, 4, 5, and 6 were proposed to take ecosystem considerations into account. The Commission, at its November 2005 Annual Meeting, agreed in substance to amend the NEAFC Convention, giving a clearer mandate to pursue the ecosystem approach, to protect biodiversity, and to apply the precautionary approach.[236] The recommended amendments will be formally adopted when two Contracting Parties finalize internal procedures. Until that time, the application of the amendments is voluntary.

NEAFC receives scientific advice from ICES; hence much of the move towards the EAF is generated through requests to and recommendations from ICES. In accordance with the general trend to manage fisheries within the broader context of the ecosystem, ICES has broadened its research scope in the last several years to incorporate ecosystem considerations.[237] At its 2004 Annual Meeting, NEAFC also committed itself to taking into consideration the environmental effects on the fisheries in recognition that environmental factors may affect the reproductive capacity of the stocks.[238]

In the last few years NEAFC has introduced conservation measures that manifest the Commission's increased awareness of broader ecosystem interactions. Of particular note is the closure of areas adjacent to the Rockall Bank in 2001 to protect the juvenile haddock population.[239] At its 2004 Annual Meeting, NEAFC also agreed to close five seamounts and a section of the Reykjanes Ridge in the NEAFC Regulatory Area to bottom fisheries, to be reviewed in 2007, so as to protect vulnerable deepwater habitats.[240]

The issue of deepwater fisheries is a major concern for many scientists and NGOs and one that NEAFC is endeavouring to address. ICES noted in 2004 that most deepwater fisheries[241] have developed rapidly, and the resources that they exploit are especially vulnerable to overfishing and are considered to be harvested unsustainably.[242] One of the major difficulties in advising on deepwater species is lack of reliable data, and therefore there is a need for precautionary management. ICES recommended that measures should be taken to reduce bycatch or secondary exploitation of deep-sea species by fisheries that primarily target shelf species and that there be a 30% reduction below 1998 levels for tusk and ling.[243] NEAFC agreed to a 30% reduction in deep-sea trawl-fishery effort for 2005 onwards.[244] However, the NEAFC management measure states that "the effort shall not exceed 70% of the highest level put into deep-sea fishing in previous years for the relevant species".[245] ICES advice for other deepwater species was for no targeted fishery and greater than 30% reductions. At the 2005 Annual Meeting, NEAFC agreed to maintain a 30% reduction in effort on deep-sea species. In addition to deepwater species, shark populations in the NEAFC area have also been of concern. In 2005 NEAFC agreed to no directed fishery for basking shark in 2006 and called for compilation of data to better assist stock assessments and advice on this and other pelagic shark species.

7.6.7 IOTC

The IOTC Agreement does not mention EAF; however, the Commission has incorporated aspects of an ecosystem approach, primarily through bycatch considerations. The Bycatch Working Party was established under the Scientific Committee in 2004, with the first meeting held in 2005. The working party is tasked with identifying major bycatch species in Indian Ocean tuna fisheries; investigating means to monitor and assess bycatch in general with initial emphasis on sharks; exchanging information on bycatch and identifying methods for carrying out assessments of bycatch; liaising with groups investigating bycatch issues for other regional bodies (e.g., CCSBT, IATTC, ICCAT) involved with the management of tunas; proposing measures to reduce unsustainable bycatch, as appropriate; and encouraging the conduct of research on ecosystems. The working party will address issues dealing with nontarget species, bycatch mitigation measures, and data collection on bycatch species.

In 2005 IOTC adopted non legally binding management recommendations[246] to reduce bycatch of seabirds and turtles. The recommendation on the incidental mortality of seabirds[247] requires parties to inform the Commission on the status of NPOAs, pursuant to IPOA-Seabirds; to collect data on incidental mortality of seabirds; and to provide assessments of seabird bycatch when feasible. The recommendation on sea turtles[248] specifically requires parties to collect data on sea turtle bycatch and to follow the FAO guidelines[249] for reducing bycatch in both the longline and purse-seine fisheries and is similar to the IATTC Resolutions on bycatch and its specific Resolutions to reduce sea turtle incidental mortality. There are requirements to release live sea turtles, to avoid encircling turtles with purse-seine nets, and to modify gear so that sea turtle bycatch is minimized. FADs must be monitored and removed when not in use. The Commission encourages Contracting Parties to submit data on sea turtle bycatch and other impacts, such as deterioration of nesting sites and swallowing of marine debris, to the Scientific Committee.

In keeping with the FAO IPOA-Sharks, IOTC adopted the legally binding Resolution Concerning the Conservation of Sharks Caught in Association with Fisheries by IOTC[250] in 2005. The Resolution requires Contracting Parties to report all data on shark catches. In addition, by 2006, the Scientific Committee, via the Bycatch Working Party, is to provide preliminary advice on stock status for key shark species and to develop a research plan to conduct full stock assessments. Sharks catches are limited to 5% of the total catch, and in nontarget fisheries, sharks are to be released alive when possible. The Resolution also calls for the identification of shark spawning grounds by Contracting Parties. Although IOTC has not adopted a comprehensive approach to EAF, it has begun to address bycatch of nontarget species, especially those organisms addressed by FAO IPOAs.

7.6.8 CCSBT

Although EAF is not explicitly referred to in the constitutive agreement of CCSBT[251], the Commission does take into account ecosystem considerations. Shortly after the Convention came into force, CCSBT established the Ecologically Related Species Working Group (ERSWG) in 1995, which was charged with reducing bycatch and evaluating fishery effects on associated species. ERSWG is also to advise on interaction between the southern bluefin tuna and predator and prey species that may affect the state of the stock. The current status of the southern bluefin stocks, together with the acknowledgement by the Commission that the stock will likely decline below 2001 levels, is a major preoccupation of the Commission, as is developing achievable management goals. The low stock abundance of southern bluefin tuna, fishing by nonmembers, and current market incentives to exceed current levels are factors considered by CCSBT in determining the TAC.[252]

At the 2004 meeting of ERSWG, Contracting Parties and Cooperating Non-Contracting Parties gave presentations on the status of national responses to the FAO IPOAs for seabirds and sharks. They concluded that the NPOAs are "a useful framework to define problems of incidental capture of seabirds and sharks, identify and implement workable and effective solutions, and allow for communication of the members' intent and activities".[253] CCSBT has introduced measures requiring the use of tori lines south of 30°S by members and nonmembers of the Commission to mitigate seabird bycatch. Research by members on the effectiveness of bait dyeing is ongoing.[254] Cooperation with CCAMLR, specifically through advice from WG-IMAF to the CCSBT Scientific Committee, has resulted in changes to the CCSBT seabird mitigation measures. Member compliance with this measure is unknown.[255] ERSWG produced species identification pamphlets in 2004 in an effort to improve data collection and educate fishers and observers on incidental mortality of seabirds, sharks, and rays. The next meeting of ERSWG, scheduled for early 2006, will be hosted by the Fishing Entity of Taiwan.

To address a lack of scientific information on the southern bluefin tuna and the fishery, CCSBT established a Scientific Research Program (SRP) in 2001. The SRP has the "overall objective of improving the quality of the data and information used as input to the stock assessment for southern bluefin tuna, contributing to the development of reliable indices to monitor future trends in stock size and identifying directions for further scientific research".[256] The major objectives of the SRP are to improve data collection on catch characterization (biological characteristics), to develop a scientific observer program, to improve measurements of catch per unit effort, and to implement a tagging program.

As of 2005, the Commission has created standards for a scientific observer program that is delivered through Contracting Parties. These standards have been followed by Contracting Parties since 2004.[257] Coverage is low (<10%), and data collected are maintained by members and not in a central database through CCSBT.

The CCSBT tagging program began in November 2001, elements of which are executed by members. Current tagging initiatives include Japanese tagging in the Western Indian Ocean, Australian tagging off the east coast of Australia, a five-year surface fishery tagging program off the southern coast of Australia (in its fourth year as of 2005) that is coordinated by the CCSBT Secretariat, a tagging program run by New Zealand within its EEZ, and a spatial dynamics archival tagging program conducted by Australia.[258] CCSBT also cooperates with other tuna bodies on research initiatives.

7.6.9 Central Bering Sea Donut Hole Arrangement

The Central Bering Sea Donut Hole Convention is concerned with the management of Alaskan pollock and does not refer to the wider ecosystem. The Record of Discussion, however, does note that the Central Bering Sea is a large marine ecosystem and that measures adopted should consider species that are ecologically related to Alaskan pollock within the convention area. Under the management of the Bering Sea Donut Hole Arrangement (BSDH), there has been no directed fishery in the Donut Hole since the inception of the management agreement in 1993, as the pollock population must reach a threshold level before the fishery can reopen. Interestingly, the North Pacific has been the subject of several ecosystem-level studies following the rapid decline in Stellar sea lions in the late 1980s and early 1990s. Some authors have suggested that overfishing of the pollock in the Bering Sea, which is an important forage species for sea lions, contributed to the marine mammals' collapse.[259] When the pollock population reaches agreed levels of abundance to sustain a fishery, management measures will likely examine ecosystem considerations, particularly regarding the pollock as forage species.

7.6.10 SEAFO

The SEAFO Convention acknowledges the need for a broad approach to fisheries management and the term "living marine resources", which is used throughout the convention, is defined in Article 1(n) as "all living components of marine ecosystems, including seabirds".[260] Article 3 requires that account be taken of various ecosystem considerations, as advocated by UNFSA Article 5.[261]

The first meeting of the Commission was held in March 2004, with the meeting report stating "SEAFO also presents an opportunity to further develop regional expertise in the research, management and compliance aspects of managing ecosystems. In doing this, SEAFO is encouraged to work together with other national and regional initiatives already established in the region".[262]

The first meeting of the Scientific Committee was held in September 2005. In accordance with Articles 2, 3, and 10(4) (d) of the SEAFO Convention, which concern the sustainability of fisheries and ecosystem conservation, it was noted that the Committee should address and make recommendations on the wider ecosystem impacts of fisheries activities, including overexploitation, bycatch depletion, and dumping; gear effects such as habitat destruction; impacts on seabirds, turtles, and marine mammals; establishment of closed areas; and linkages and impacts on adjacent EEZ areas.[263] In addition, pollution caused by fishing activities (e.g., bunker oil spillages, dumping of garbage, fishing gear) may also be addressed by the Scientific Committee. At the first Scientific Committee meeting, the Provisional Working Group (PWG) was established, and the Commission adopted the PWG at its second Annual Meeting in October 2005. The Committee is to assess baseline information that would pave the way for scientific advice based on the principles of EAF. The PWG also includes in its Terms of Reference, the evaluation of information on fisheries, oceanography, productivity, and biodiversity around seamounts in the SEAFO area, and where information exists, evaluation of the impact of fishing gear on seabed and benthic ecosystems. The Committee emphasized the need for cooperation with other organizations and programs working in the region, both within and beyond areas under national jurisdiction, underscoring the need for pooling and exchanging information on fisheries and the environment.

7.6.11 WCPFC

The WCPFC Convention clearly stipulates that ecosystem effects should play a large role in shaping fisheries management measures in the Western Central Pacific Ocean (WCPO).

The preamble of the WCPFC Convention[264] includes reference to the greater marine ecosystem, stating "the need to avoid adverse impacts on the marine environment, preserve biodiversity, maintain the integrity of marine ecosystems and minimize the risk of long-term or irreversible effects of fishing operations". Article 5 outlines principles and measures for conservation and management and specifically provides for ecosystem considerations, consistent with UNFSA. Article 6(1) (b), with respect to the application of the precautionary approach, requires that "the impact of fishing activities on non-target and associated or dependent species, as well as existing and predicted oceanic, environmental and socioeconomic conditions". Article 6(1) (c) subsequently addresses the need for data collection to address these impacts. Article 6(6) refers to the effects of a natural phenomenon on highly migratory fish stocks and the need for the adoption of emergency measures to ensure that fishing activity does not exacerbate adverse impacts.

The WCPO currently supports the largest industrial tuna fishery in the world, with estimated annual catches averaging about 1.5×10^6 t, between 1992 and 2002. All fisheries in the area discard a portion of the catch as a result of gear failures or because it may be of little or no economic value, damaged, or too small for processing. Depending on gear type and location of the fishery, marine mammals, seabirds, and sea turtles are also caught as bycatch.[265]

Following the inaugural meeting of the Commission in December 2004, the Ecosystem and Bycatch Specialist Working Group (EB-SWG) was established. EB-SWG has specific tasks of assessing bycatch, developing ecosystem models, and working with other specialist working groups in overlapping functions.[266] This Working Group met on the margins of the first regular session of the Scientific Committee in August 2005. The report of the EB-SWG, appended to the report of the Scientific Committee, addressed requests received from the Commission on seabird, turtle, and shark bycatch and included ecological and biological information on target species and bycatch mitigation measures.[267]

Several presentations during the Scientific Committee meeting,[268] many by independent scientists, provided information on a wide range of ecosystem considerations, including current estimates of bycatch of nontarget species (particularly albatrosses and other seabirds, turtles, and sharks), effects of fishing methods on target species, and distribution of target species. Data collection and data interpretation are the first steps towards understanding the WCPO ecosystem. WCPFC has begun to gather information on relevant species and ecosystem parameters. Through presentation of available data and consideration of the WCPO ecosystem, WCPFC has made progress towards implementing the ecosystem approach. Initiatives to develop and implement EAF have also been taken through the Forum Fisheries Agency (FFA).[269] In particular, the FFA has developed a draft guide to introduce the EAF[270] to the management of fisheries, particularly tuna, in the WCPO. This guide provides the tools to put into practice the ecosystem provisions outlined in Article 5 of the Convention and is intended to elaborate on a more sophisticated approach to fisheries management and the decision-making process, particularly for the sustainable development of the tuna resources in the WCPO.[271]

At the second regular session of the Commission, held in December 2005, a nonbinding resolution on bycatch[272] was adopted in addition to resolutions on seabirds[273] and sea turtles,[274] both in accordance with the FAO IPOAs. In consultation with the Technical and Compliance Committee, the Commission will consider measures to reduce seabird bycatch and mortality, including such measures applied and tested by CCAMLR (i.e., tori lines). Sea turtle bycatch measures, in accordance with the FAO Guidelines on the Reduction of Sea Turtle Mortality, include avoiding the encirclement of turtles by purse seines, ensuring safe release from net rolls, and monitoring FADs for sea turtle entanglement. Members are also encouraged to conduct research on the use of circle hooks. As of 2006, all members and cooperating members must report to the Commission the extent of implementation of turtle bycatch and mortality mitigation.

Consistent with the obligation for compatible and coherent management over the range of the stocks, there is a need to introduce EAF not only on the high seas, but also within the EEZs of Pacific coastal States. As such, in 2006, the FFA will pilot the EAF in accordance with the proposed draft guidelines[275] in one of the FFA member countries (yet to be determined), including the development of a series of subregional training programs. In addition, the FFA is developing tuna management plans for all countries based on the EAF guidelines and is drafting model fisheries legislation for the implementation of EAF. As Transform Aqorau observes, "[implementing EAF] is a challenge but it is common sense, that to manage fish stocks, one also has to [consider] the habitat in which the fish occur".[276] Despite the challenge confronting developing States, clear efforts are being made to take ecosystem considerations explicitly into account.

7.6.12 SIOFA

Though not yet in force, the agreement establishing SIOFA states that its objective is "to ensure the long term conservation and sustainable use of the fisheries resources in the Area through cooperation among the Contracting Parties and to promote the sustainable development of fisheries in the Area".[277] SIOFA's constitutive agreement expressly refers to the ecosystem approach to fisheries management in Article 4(a) as follows: "measures shall be adopted on the basis of the best scientific evidence available to ensure the long-term conservation of fishery resources, taking into account the sustainable use of such resources and implementing an ecosystem approach to their management".

Direct reference to the EAF in this provision was a strong request by delegations who wanted to go beyond the "minimization of fishing impact" concept.[278] This inclusion of the ecosystem approach comes closer to CCAMLR's objectives to integrate the notion of preserving ecosystem dynamics as a basis of sustainable management of fishery resources. Article 4(e) of the SIOFA Agreement includes as a general principle that "fishing practices and management measures shall take due account of the need to minimize the harmful impact that fishing activities may have on the marine environment" and paragraph (f) includes "the need to protect marine biodiversity". Article 6 outlines the functions of the Commission and includes, in Article 6(c), provisions for assessing the impact of fishing and other human activities on fish resources and the marine environment. Article 6(d) requires the Commission to take into account the need for protecting marine diversity in formulating conservation and management measures. The precautionary approach, which is inherent in EAF, is included in Article 4.

7.7 Practical challenges to EAF implementation

From a theoretical standpoint, EAF is understood and accepted. In reality, for many, the marine ecosystem is as complex and baffling as the challenge of implementing EAF is daunting. With regard to RFMO implementation of EAF, there exist numerous interrelated challenges:

> *Developing a system of ecological indicators and clear conservation objectives* — A system of indicators is needed that can be used to assess ecosystem health across regions. Clear and preagreed conservation objectives for achieving the EAF are essential.
>
> *Collecting data on a broad range of ecosystem components* — First, States and RFMOs committed to the EAF need to develop systematic data collection of fishery-dependent and fishery-independent data. Second, these data need to be recorded and stored in a manner that is accessible to fisheries scientists. Third, scientific and research capabilities must be strengthened for improved data analysis and to provide scientific advice taking into account ecosystem considerations. Fourth, the increased costs of data collection, analysis, and model development, together with the costs of administration and data management under EAF, are a significant challenge. Fifth, to ensure cost efficiency and strategic collection and use of ecosystem data, clear frameworks are needed. Sixth, the need for a reliable monitoring, control, and surveillance system must be recognized as part of an effective data-collection program. Finally, the data and its implications need to be communicated clearly, transparently, and in a timely fashion to contribute to regulatory appraisal and decision making.
>
> *Capacity building* — The entire infrastructure associated with implementing EAF is capital intensive. The increased costs of adopting EAF in developing countries, for which the costs associated with infrastructure required for single-species management already pose a hardship, will be difficult and may be impossible in some cases. As Burke observes, "the capability of applying an ecosystem approach is not widely distributed. It is a daunting task even for developed States with substantial

resources for management undertakings. Expecting adoption and implementation of this approach on a wide scale among developing States that even now are unable to determine an allowable catch of their EEZ living resources does not seem entirely realistic...".[279] It is therefore incumbent on developed States, RFMOs and other organizations to assist developing coastal States in building capacity so that EAF can be adopted and implemented through the transfer of technology, technical assistance, and institutional strengthening.

Expanding scientific and institutional infrastructure — Most RFMOs have not yet established the scientific or institutional infrastructure to compatibly regulate the fisheries for straddling fish stocks or highly migratory fish stocks over the geographic range of the stocks in a way that is consistent with EAF.

Modifying gear and reducing bycatch — Fishing gear technology and the impact of fishing on the marine ecosystem are concerns for biodiversity conservation and protection. Important developments have been made in improving the selectivity of gear to avoid bycatch, as in the case of the Nordmore grate, J-hooks, and streamer lines. Mitigating or eliminating incidental catches in trawl fisheries, longlines, and purse seines, however, is an ongoing challenge for RFMOs and the fishing industry. To achieve the required technological and scientific innovations and solutions, industry and States must invest in research and development. These investments in the future of the fishery require front-end financial commitments. Shifts to EAF technologies could necessitate significant capital outlay and potential socioeconomic dislocations. Deprivation due to structural and technological adjustments must be avoided, modified, or minimized wherever possible.

Protecting habitat — The impact of fishing operations on habitat is worrisome, and addressing this impact poses a well-recognized challenge. There is increasing international pressure to reduce the impacts of bottom trawling, as illustrated by the 2004 call to action from over 1000 scientists to the UN General Assembly for protection of deep-sea corals[280] and the call from a coalition of NGOs[281] for a moratorium on bottom trawling on the high seas, particularly in areas that are not currently under the jurisdiction of an RFMO. As a result, issues of habitat protection that are included within UNFSA have now become international biodiversity concerns and are being discussed in nonfisheries fora such as the CBD and the UN General Assembly.

Adopting new management techniques — The emerging institutional mindset of managing holistically that is encapsulated in EAF requires innovation, flexibility, and commitment. It is important to recognize that fisheries are now in a period of transition. RFMOs and participating States must ensure that the piecemeal and incremental approach to shifting from species-specific to multispecies management and then incorporating broader ecosystem concerns is done transparently with the active involvement of all stakeholders. Additionally, science-based decision making, environmental impact assessments, pre-agreed decision rules, and wide application of the precautionary approach are components of improved management and governance under the EAF.

Engaging the fishing industry — Moving towards adopting EAF within the context of the political economy of fisheries management will undoubtedly have impacts on the fishing industry (both positive and negative). If we are to achieve a viable regime employing the concepts and principles envisaged under EAF, the fishing industry must be actively engaged and committed to change. As Alastair MacFarlane observed, "States don't fish, companies fish".[282]

Encouraging good fishing practices — RFMOs need to ensure that the governance regime serves to regulate and protect those who are operating in good faith, and effectively controls "bad actors" who jeopardize the resource and undermine the conservation strategy. Similarly, it is crucial that fishing pressure and fleet capacity are reduced so as to be commensurate with ecologically sustainable fishing opportunities.

Deterring IUU fishing — Despite the efforts of CCAMLR to employ a precautionary and

> ecosystem approach since its inception, key commercial stocks in the convention area have been subject to overexploitation primarily because of IUU fishing. This experience underscores the global imperative for the improved regulation (registration, licensing, insuring), monitoring, and control of all vessels associated with fishing operations both within and beyond areas of national jurisdiction.

7.8 Conclusion

The post-LOSC development of EAF embedded in UNFSA and elaborated through the FAO guidelines, ICES, Reykjavik Declaration, individual coastal State,[283] and scientific efforts[284] is an important phase in the "process of continuous evolution of fisheries-related institutions".[285] The ecosystem approach is widely accepted in the context of fisheries management. The degree to which measures have been adopted and implemented to give practical expression to the new paradigm varies among RFMOs and coastal States. Implementing the new approach has challenges, but these are not insurmountable. EAF is incrementally being introduced into conventional fisheries management regimes, both within areas under national jurisdiction and on the high seas.

Nonetheless, the magnitude of the challenge should not be underestimated. In an important article by J. Caddy and K. Cochrane, the difficult task ahead for fisheries managers in embracing the ecosystem approach is put into the following context:

> Even while fisheries management struggles to get to grips with single-species issues, it is increasingly being called on to take a multispecies and ecosystem perspective. However, there are still few case studies with more than a few years duration which illustrate how these concepts have to be applied, and the difficulties are already apparent to all.[286]

Under LOSC Article 61(3), it is incumbent upon all States to adopt international minimum standards for fisheries. These standards have been continually elaborated and reinforced since 1982. The explicit inclusion of ecosystem considerations in UNFSA is a binding expression of the contemporary international minimum standards applicable to those fisheries within and beyond areas of national jurisdiction. Arguably these contemporary international minimum standards are also determined by what is considered responsible practice for all fisheries and for all States, as agreed under the FAO Code of Conduct, the IPOAs, and UNGA resolutions, and elaborated upon in FAO technical guidelines.[287] As such, the case may be made that the adoption and implementation of EAF does not only apply to straddling fish stocks and highly migratory fish stocks, but to all fisheries.

While EAF is viewed as a progression in the continuum of fisheries management, it requires fisheries managers to adopt a new governance and institutional model, and to work collaboratively with scientists, industry, and other stakeholders to implement a comprehensive and integrated management strategy. Additionally, implementation of EAF will require new and enhanced linkages among RFMOs dealing with specific species or groups of species as well as those related to specific geographic areas. The post-LOSC standards for responsible, precautionary, sustainable, and regulated fishing practices were developed not only to avoid and reduce the risk of serious or irreversible harm to the ecosystem, but also to ensure the recovery and long-term sustainability of fish stocks, associated and dependent species, and nontarget species in the same ecosystem. Effectively implementing EAF both within areas under national jurisdiction and on the high seas will greatly advance global efforts to secure the sustainable development of international fisheries.

Endnotes

[1] *Agreement for the Implementation of the Provisions of the United Nations Convention on the Law of the Sea of 10 December 1982 Relating to the Conservation and Management of Straddling Fish Stocks and Highly Migratory Fish Stocks*, done at New York, 4 December 1995, entered into force 11 December 2001, 2167 UNTS 88; (1995) 34 ILM 1542. Referred to as UNFSA.

[2] H. Wang, "Ecosystem Management and its Application to Large Marine Systems: Science, Law and Politics" (2005) *Ocean Development and International Law* 41–74, defines ecosystems as follows: "an ecosystem exists in a space with boundaries that may not be explicitly delineated. Ecosystems are distinguishable from each other based on their biophysical attributes and their location; an ecosystem includes both living organisms and their abiotic environment, including pools of organic and inorganic materials; organisms interact with each other and interact with the physical environment through fluxes of energy, organic and inorganic materials among the pools. These fluxes are mediated and functionally controlled by species behavior and environmental forces, an ecosystem is dynamic. This structure and functions change with time. An ecosystem exhibits emergent properties that are characteristic of its type and are invariant within the domain of existence". S. Garcia, A. Zerbi, C. Aliaume, T. Do Chi, and G. Lasserre, "The Ecosystem Approach to Fisheries. Issues, Terminology, Principles, Institutional Foundations, Implementation and Outlook", FAO Fisheries Technical Paper No. 443 (Rome: FAO, 2003) at 7, state that an ecosystem "can be defined as 'a system of complex interactions of populations between themselves and with their environment' or as 'the joint functioning and interaction of these two compartments (populations and environment) in a functional unit of variable size'...In this review, and in EAF, we will consider 'populations' as including people, and especially people involved in fisheries, with their technology and institutions".

[3] For a more detailed analysis of the international legal regime, see Chaps. 2 and 3.

[4] *United Nations Convention on the Law of the Sea*, done at Montego Bay, Jamaica, 10 December 1982, entered into force 16 November 1994, 33 ILM 1309. Referred to as LOSC.

[5] LOSC Convention Preamble, para. 3.

[6] The preamble of the Ramsar Convention recognizes the relationship between humans and the environment and the fundamental ecological functions of wetlands. *Convention on Wetlands of International Importance especially as Waterfowl Habitat*, Ramsar, Iran, 2 February 1971. Amended 1982 and 1987. Full convention available at <http://www.ramsar.org/key_conv_e.htm> (accessed 15 January 2006).

[7] The preamble of the CITES Convention recognizes "that wild fauna and flora in their many beautiful and varied forms are an irreplaceable part of the natural systems of the earth which must be protected for this and the generations to come". *Convention on International Trade in Endangered Species of Wild Fauna and Flora*, Washington, D.C., 3 March 1973. Amended 22 June 1979 at Bonn, Germany. Full convention online at <http://www.cites.org/eng/disc/text.shtml#II> (accessed 15 January 2006).

[8] The preamble of the CMS recognizes "that wild animals in their innumerable forms are an irreplaceable part of the earth's natural system which must be conserved for the good of mankind". *Convention on the Conservation of Migratory Species of Wild Animals* Bonn, Germany, 23 June 1979. Full convention available at <http://www.cms.int/> (accessed 15 January 2006).

[9] Art. 2 of the CBD defines ecosystem as "a dynamic complex of plant, animal and micro-organism communities and their non-living environment interacting as a functional unit". *Convention on Biological Diversity*, Rio de Janeiro, Brazil, 22 May 1992, 31 ILM 823 (1992).

[10] See review in S. Garcia and K. Cochrane, "Ecosystem Approach to Fisheries: A Review of Implementation Guidelines" (2005) 62(3) *ICES Journal of Marine Science* 311–318, and further discussion of the background to the EAF in S. Garcia, A. Zerbi, C. Aliaume, T. Do Chi, and G. Lasserre, FAO Fisheries Technical Paper No. 443, at 7.

[11] *Convention on the Conservation of Antarctic Marine Living Resources*, done at Canberra, Australia, 20 May 1980, entered into force 7 April 1982, ATS 1982 No. 9; 19 ILM 841 (CCAMLR).

[12] See Art. II of the CCAMLR Convention.

[13] Art. II, part 3(b) of the CCAMLR Convention.

[14] See Art. 5(f) UNFSA.

[15] See Annex II, paras. 3 and 4 of UNFSA for specific guidelines to establishing precautionary reference points and management strategies to maintain and restore stocks.

[16] See Art. 6.1 of the FAO Code of Conduct for Responsible Fisheries.

[17] UNEP, Conference of the Parties to the Convention on Biological Diversity (COP), "Appendix 1, The Jakarta Ministerial Statement on the Implementation of the Convention on Biological Diversity" in *Report of the Second Meeting of the Conference of the Parties to the Convention on Biological Diversity*, UNEP/CBD/COP/2/19 (second meeting, Jakarta, Indonesia, 6–17 November 1995) at 40. Available at <http://www.biodiv.org/programmes/areas/marine/default.asp> (accessed 6 February 2006).

[18] Rome: FAO, 1999.

[19] Rome: FAO, 1999.

[20] The 2001 Reykjavik Declaration on Responsible Fisheries in the Marine Ecosystem was a result of the Reykjavik Conference on Responsible Fisheries in the Marine Ecosystem, which was held 1–4 October 2001 and jointly organized by FAO and Iceland with the cosponsorship of Norway. Delegates from 59 countries attended.

[21] Para. 29(d) requires, *inter alia*, the development and implementation of an ecosystem approach by 2010, together with (1) the elimination of destructive fishing practices; (2) the establishment of marine protected areas (MPAs) and other time–area closures for the protection of nursery grounds; (3) the adoption of coastal land-use and watershed planning; and (4) the integration of economic sectors into marine and coastal-area management. World Summit on Sustainable Development, *Johannesburg Plan of Implementation* (2002). United Nations Division for Sustainable Development online <http://www.un.org/esa/sustdev/documents/WSSD_POI_PD/English/WSSD_PlanImpl.pdf> (accessed 18 October 2005) (WSSD POI).

[22] Para. 66 of the UN Resolution 59/25.

[23] FAO, "Fisheries Management 2: The Ecosystem Approach to Fisheries", FAO Technical Guidelines for Responsible Fisheries No. 4. Suppl. 2 (Rome: FAO, 2003).

[24] While not questioning the fact that some species are strongly depleted, some scientists have refuted the magnitude of these declines and questioned the analytical methods as well as data used to arrive at such conclusions; this was particularly true in the early days of the fisheries, i.e., 1960s. See, for example, J. Hampton, J. Sibert, P. Kleiber, M. Maunder, and S. Harley, "Fisheries: Decline of Pacific Tuna Populations Exaggerated?" (2005) 434 *Nature* E1–E2.

[25] J. Hutchings and J. Reynolds, "Marine Fish Population Collapses: Consequences for Recovery and Extinction Risk" (2004) 54 *Bioscience* 297–309 at 297.

[26] The terms ecosystem-based fisheries management, ecosystem-based management, ecosystem approach to fisheries management, and ecosystem approach to fisheries have gathered momentum in recent years and have entered the vernacular of fisheries management. The acronyms EBFM, EBM, EAFM, and EAF represent the same essential concept. All of these terms recognize that management must deal with the full set of ecological consequences of an activity and also understand the social and economic implications that the activity entails. The combination of the words may be new, but the definition and interpretation have been around for a long time. J. Scandol, M. Holloway, P. Gibbs, and K. Astles, "Ecosystem-Based Fisheries Management: An Australian Perspective" (2005) 18 *Aquatic Living Resources* 261–273; W. Fletcher, "Frameworks for Managing Fisheries Resources in Australia Through Ecosystem Approaches: Do They Fit Together and Are They Useful?" (2006) 78 *Bulletin of Marine Science* 691–704.

[27] S. Garcia, FAO, personal communication, January 2006: "The Group who drafted the Guidelines selected EAF (FAO Secretariat's proposal) to cover all aspects of fisheries and not only management. EAF is a broad framework that applies to fishing operations, fish processing, management, research, and trade. The first Guidelines produced were confined to management of capture fisheries, as a matter of priority and work is ongoing to extend the concept to fishing operations (e.g. bycatch issues) and aquaculture. EAF is nested within the more general concept of EA (yet to be formulated in any operational way (by the CBD or others)".

[28] S. Garcia (FAO) and J. Rice (Canadian Department of Fisheries and Oceans), personal communications, 2005–2006.

[29] The keynote address at the September 2005 meeting of ICES was delivered by K. Sainsbury. "The Ecosystem Approach to Fisheries", ICES online <http://www.ices.dk/iceswork/asc/2005/KeithSainsbury_talk.pdf> (accessed 20 October 2005).

[30] B. Satia, "The World Summit on Sustainable Development and Fisheries" (2003) *International Fisheries Law and Policy Review* (Special Issue) 11–23 at 22.

[31] J. Scandol et al. argue that "for all intents and purposes, ecosystem-based management is simply a re-expression of the processes and operational mechanisms already developed for the paradigm of (ecologically) sustainable development", J. Scandol, M. Holloway, P. Gibbs, and K. Astles, at 261.

[32] There is an evolving international consensus that the conservation and sustainable use of biodiversity beyond national jurisdiction require an integrated approach to management. Currently, there is no multilateral management mechanism for IOM. This governance gap has lead to calls for the establishment of regional organizations with a broader mandate and function to manage for a range of marine ecosystem components. These bodies, sometimes referred to as marine ecosystem resource management organizations (MERMOs) or regional oceans management organizations (ROMOs), would take into account all sectoral interests in an integrated fashion. See summary of the Third Global Conference on Oceans, Coasts and Islands, "Moving the Global Oceans Agenda Forward", Paris, France, 24–27 January 2006, available at <http://www.iisd.ca/ymb/globaloceans3/jan24.html> (accessed 28 January 2006).

[33] FAO Technical Guidelines for Responsible Fisheries No. 4, Suppl. 2.

[34] T. Ward, D. Tarte, E. Hegerl, and K. Short, *Ecosystem-Based Management of Marine Capture Fisheries* (Sydney: WWF Australia, 2002) at 5.

[35] FAO Fisheries Glossary. FAO online <http://www.fao.org/fi/glossary/default.asp> (accessed 18 October 2005).

[36] E. Molenaar, "Ecosystem Based Fisheries Management, Commercial Fisheries, Marine Mammals and the 2001 Reykjavik Declaration in the Context of International Law" (2002) *International Journal of Marine and Coastal Law* 561–595.

[37] National Marine Fisheries Service, *Ecosystem-Based Fishery Management. A Report to Congress by the Ecosystems Principles Advisory Panel* (Silver Spring, Md: NMFS - US Department of Commerce, 1999).

[38] S. Gavaris, J. Porter, R. Stephenson, G. Robert, and D. Pezzack, *Review of Management Plan Conservation Strategies for Canadian Fisheries on Georges Bank: A Test of a Practical Ecosystem-Based Framework*, ICES CM 2005/BB:05 (Copenhagen: ICES, 2005).

[39] See discussion in E. Pikitch, C. Santora, E. Babcock, A. Bakun, R. Bonfil, D. Conover, P. Dayton, P. Doukakis, D. Fluharty, B. Heneman, E. Houde, J. Link, P. Livingston, M. Mangel, M. Mcallister, J. Pope, and K. Sainsbury, "Ecosystem-Based Fishery Management" (2004) 305(5682) *Science* 346–347. The overall objective of EBFM, according to preeminent marine biologists, is to maintain, sustain, and restore health to marine ecosystems, and specific objectives include (1) avoiding degradation of ecosystems, as measured by indicators of environmental quality and system status; (2) minimizing the risk of irreversible change to natural assemblages of species and ecosystem processes; (3) obtaining and maintaining long-term socioeconomic benefits without compromising the ecosystem; and (4) generating knowledge of ecosystem processes sufficient to understand the likely consequences of human actions. In addition, where knowledge is insufficient, robust and precautionary fishery management measures that favor the ecosystem should be adopted.

[40] For discussion, see S. Kaye, *International Fisheries Management* (The Hague: Kluwer Law International, 2001) at 282.

[41] J. Rice, S. Trujillo, S. Jennings, K. Hvlland, O. Hagstrom, A. Astudillo, and J. Jenson, "Guidance on the Application of the Ecosystem Approach to Management of Human Activities in the European Marine Environment", ICES Cooperative Research Report No. 273 (Copenhagen: ICES, 2005). ICES, in cooperation with the EU, has made significant progress in establishing guidelines for EAF in terms of all human uses of the marine environment and with specific reference to the management of fisheries; the ICES 2004 Annual Report outlines clear objectives in implementing EAF. Again, rather than systematically adopting all aspects of EAF, States are implementing specific management measures and then including them under the EAF umbrella. ICES states, "all of the ICES member countries, and all of the Commissions we advise, are advancing an ecosystem approach. This is reflected by: measures to reduce bycatch and discarding in fisheries;

the identification, mapping, and protection of habitat; monitoring of ecosystem health indicators; integrated coastal management planning; measures to reduce land-based sources of nutrients and contaminants; use of climate information in stock assessments and in setting biological reference points that reflect regime shifts; setting aside some of the production of prey species to supply the needs of predators; protocols for applying the precautionary approach; increasing recognition of the importance of involving stakeholders in science and in decisions".

[42] For a review of Canadian, Australian, and EU efforts to develop and implement the ecosystem approach in fisheries management, see S. Parsons, "Ecosystem Considerations in Fisheries Management: Theory and Practice" (2005) 20 *International Journal of Marine and Coastal Law* 381–422 at 388–394. For an example of Canadian efforts, see R. O'Boyle, V. Kostylev, H. Breeze, T. Hall, G. Herbert, T. Worcester, D. Ricard, and M. Sinclair, *Developing an Ecosystem-Based Management Framework for Benthic Communities: A Case Study of the Scotian Shelf*, ICES CM 2005/BB:18 (Copenhagen: ICES, 2005). United States National Marine Fisheries Service efforts are outlined in W.-D. Busch, B. Brown, and G. Mayer, eds., *Strategic Guidance for Implementing an Ecosystem-based Approach to Fisheries Management* (Silver Spring, Md: United States Department of Commerce, National Oceanic and Atmospheric Administration, NMFS, 2003). Specific efforts of Australia are detailed in J. Scandol, M. Holloway, P. Gibbs, and K. Astles, at 261–273.

[43] See P. Mace, "A New Role for MSY in Single-Stock and Ecosystem Approaches to Fisheries Stock Assessment and Management" (2001) 2 *Fish and Fisheries* 515–525; E. Pikitch, C. Santora, E. Babcock, A. Bakun, R. Bonfil, D. Conover, P. Dayton, P. Doukakis, D. Fluharty, B. Heneman, E. Houde, J. Link, P. Livingston, M. Mangel, M. Mcallister, J. Pope, and K. Sainsbury, at 346–347; J. Link "Ecological Considerations in Fisheries Management: When Does It Matter?" (2002) 27 *Fisheries* 10–17.

[44] Reykjavik Declaration on Responsible Fisheries in the Marine Ecosystem. Available at <http://www.fao.org/docrep/meeting/004/Y2211e.htm> (accessed 15 January 2006).

[45] J. Scandol, M. Holloway, P. Gibbs, and K. Astles, at 261–273.

[46] FAO Technical Guidelines for Responsible Fisheries No. 4, Suppl. 2.

[47] Sutinen and Soboil 2003, quoted in S. Garcia, A. Aliaume, T. Do Chi, and G. Lasserre, FAO Fisheries Technical Paper No. 443, at 21.

[48] See S. Gavaris, J. Porter, R. Stephenson, G. Robert, and D. Pezzack, ICES CM 2005/BB:05 (Copenhagen: ICES, 2005), for discussion of an incremental approach to EAF in fisheries management of Georges Bank, in the Northwest Atlantic. As noted by S. Parsons, "the best way forward may be to build on existing successful single species approaches and expand them into an ecosystem approach to fisheries management. Realistically, we can only move to an ecosystem approach incrementally, starting with more rigorous (usually more cautious) application and extension of single species methods. This is recognized in the FAO Technical Guidelines for EAF which stress that EAF is neither inconsistent with, nor a replacement for, current fisheries management approaches and is likely to be adopted as an incremental extension of current fisheries management approaches. Others have also argued that expansion of single species reference points to take account of non-target species is a reasonable approach to pursue". See S. Parsons, "Ecosystem Considerations in Fisheries Management: Theory and Practice", prepared for Conference on the Governance of High Seas Fisheries and the United Nations Fish Agreement — Moving from Words to Action, St. John's, N.L., 1–5 May 2005; this paper was subsequently published as "Ecosystem Considerations in Fisheries Management: Theory and Practice" (2005) 20 *International Journal of Marine and Coastal Law* 381–422.

[49] OECD, "Synthesis Report for the Study on Economic Aspects of Management of Marine Living Resources by Ad Hoc Expert Group on Fisheries", AGR/FI/EG(96)1 (Paris: OECD Fisheries Committee, 1996) at 8.

[50] For documentation of the removal of pelagic predators, see R. Myers and B. Worm, "Rapid Worldwide Depletion of Predatory Fish Communities" (2003) 423 *Nature* 280–283. For discussion on the ecosystem effects of the removal of top predators in the Northwest Atlantic, see K. Frank, B. Petrie, J. Choi, and W. Leggett, "Trophic Cascades in a Formerly Cod Dominated Ecosystem" (2005) 308(5728) *Science* 1621–1623. Noettestad and Olsen chronicle the ecosystem role of marine mammals in "Whales and Seals: Top Predators in the Ecosystem" in H. Skjoldal, ed., *The Norwegian Sea Ecosystem* (Trondheim: Tapir Academic Press, 2004) at 395–434. The effects of predator depletion in coral reef communities are described by J. Bascompte, C. Melián, and E. Sala, "Interaction Strength Combinations and the Overfishing of a Marine Food Web" (2005) 102(15) *Proceedings of the National Academy of Sciences* 5443–5447.

[51] The Fisheries Global Information System (FIGIS) is a web-based information management tool that integrates fisheries information and interconnects groups of institutional partnerships to build up a network of subsystems for information and data sharing.

[52] The Fishery Resources Monitoring System (FIRMS) is a partnership of international organizations, regional fishery bodies, and national scientific institutes, collaborating within a formal agreement, that are willing to report and share information on status and trends of fishery resources.

[53] S. Garcia and I. de Leiva Moreno, "Global Overview of Marine Fisheries" in M. Sinclair and G. Valdimarsson, eds., *Responsible Fisheries in the Marine Ecosystem* (Wallingford: CABI International Publishing, 2003) 1–24 at 22.

[54] C. Small underlines the importance of an independent scientific observer program for RFMOs to effectively address seabird bycatch issues. Her report also includes recommendations for IATTC, ICCAT, IOTC, WCPFC, and CCSBT to mitigate seabird bycatch in target fisheries. See C. Small, *Regional Fisheries Management Organizations: Their Duties and Performance in Reducing Bycatch of Albatrosses and Other Species* (Cambridge: Birdlife International, 2005) at 24.

[55] For discussion of ecological relationships in the marine environment, see J. Link, at 10–17.

[56] A telling terrestrial example of unpredicted interrelationships between species and ecosystem functioning is illustrated by wolf depletions in Alberta. "Top predators keep life in order. When wolves were driven away from the region around Banff in Alberta, Canada, other things went with them, such as willow trees, willow warblers and beavers. That is because the elk multiplied, and browsed the young willows. Beaver lodges began to disappear as the willow supplies wilted, and sparrows chirped where once willow warblers and redstarts warbled. …The lesson is that big predators don't just take the lion's share; they orchestrate a share for everyone else too", *Guardian Weekly*, August 2005. A marine example of ecosystem effects of species removal is seen by the decline of the stellar sea lion as a result of decreased prey availability. See presentation by Gordon Krause for the North Pacific Fisheries Management Council for an overview. Available at <http:// www.fakr.noaa.gov/npfmc/current_issues/ssl/SSL603presentation.pdf> (accessed 14 January 2006). The decline of the sea

otter, a predator on sea urchins in the Pacific Northwest, radically altered kelp-forest dynamics in coastal areas. For further discussion see J. Estes and D. Duggings, "Sea Otters and Kelp Forests in Alaska: Generality and Variation in a Community Ecological Paradigm" (1995) 65(1) *Ecological Monographs* 75–100.

[57] S. Garcia, A. Zerbi, C. Aliaume, T. Do Chi, and G. Lasserre, FAO Fisheries Technical Paper No. 443, at 36.

[58] S. Kaye, at 273.

[59] OECD, "Synthesis Report for the Study on Economic Aspects of Management of Marine Living Resources by Ad Hoc Expert Group on Fisheries", AGR/FI/EG(96)1 (Paris: OECD Fisheries Committee, 1996) at 8.

[60] S. Garcia, A. Zerbi, C. Aliaume, T. Do Chi, and G. Lasserre, FAO Fisheries Technical Paper No. 443, at 36.

[61] *Convention for the Conservation of Antarctic Seals*, done at London, UK, 1 June 1972, entered into force 10 November 1978, 11 ILM 251.

[62] S. Garcia and I. de Leiva Moreno, at 40.

[63] S. Garcia, "Fishery Science and Decision-Making: Dire Straights to Sustainability" (2005) 76 *Bulletin of Marine Science* 171–196.

[64] For further discussion on capacity, see S. Garcia and C. Newton, "Current Situation and Prospects in World Capture Fisheries" in E. Pikitch, D. Huppert, and M. Sissenwine, eds., *Global Trends: Fisheries Management* (AFS Symposium 20) (Bethesda, Md.: American Fisheries Society Publication, 1997); R. Grainger and S. Garcia, "Chronicles of Marine Fishery Landings (1950–1994): Trends Analysis and Fisheries Potential", FAO Fisheries Technical Paper No. 359 (Rome: FAO, 1996); and M. Gianni, *High Seas Bottom Trawl Fisheries and Their Impacts on the Biodiversity of Vulnerable Deep-Sea Ecosystems: Options for International Action* (Gland: IUCN, 2004).

[65] S. Garcia, at 189.

[66] FAO Technical Guidelines for Responsible Fisheries No. 4, Suppl. 2, at 49.

[67] The Joint Nature Conservation Committee of the United Kingdom outlines steps to implementing EAF, including "1) the identification of the relevant ecosystems and their boundaries and characteristics; 2) the agreement of management objectives for each ecosystem (which should encompass wider ecosystem factors and not just the target stock);and all stake holder groups should be involved in their development; 3) long-term management objectives should be developed as well as short to medium-term objectives; 4) the establishment of sustainability indicators (including reference points, targets and limits) and the accompanying monitoring…" See online <http:// www.jncc.gov.uk/page-2518> (accessed 15 January 2006). The JNCC provides advice on wildlife and undertakes national and international conservation work.

[68] J. Scandol, M. Holloway, P. Gibbs, and K. Astles, at 269–270.

[69] The symposium "Quantitative Ecosystem Indicators for Fisheries Management", held 31 March – 3 April 2004 in Paris, France, was hosted by the Intergovernmental Oceanographic Commission at UNESCO headquarters. Papers presented at this meeting are published in one volume, see P. Cury and V. Christensen, "Quantitative Ecosystem Indicators for Fisheries Management" (2005) 62(3) *ICES Journal of Marine Science* 307–614.

[70] See P. Cury and V. Christensen, 307–614.

[71] For discussion of this framework, see J. Rice and M.-J. Rochet, "A Framework for Selecting a Suite of Indicators for Ecosystem Fisheries" (2005) 62(3) *ICES Journal of Marine Science* 516–527; and M.-J. Rochet and J. Rice, "Do Explicit Criteria Help in Selecting Indicators for Ecosystem-Based Fisheries Management?" (2005) 62(3) *ICES Journal of Marine Science* 528–539.

[72] C. Walters, *Adaptive Management of Renewable Resources* (New York: MacMillan, 1986), offers a solid introduction to adaptive management practices, particularly with reference to fisheries.

[73] See FAO Technical Guidelines for Responsible Fisheries No. 4, Suppl. 2; and S. Garcia, A. Zerbi, C. Aliaume, T. Do Chi, and G. Lasserre, FAO Fisheries Technical Paper No. 443.

[74] Figures from MRAG Americas Inc., "Review of Ecosystem-Bycatch Issues for the Western Central Pacific Region", prepared for the Preparatory Conference for the Western and Central Pacific Fisheries Commission (WCPFC), Manila, Philippines, 12–18 November 2002, WCPFC Prep Con/WP.9 at 10.

[75] B. Satia, at 22.

[76] B. Satia, at 22.

[77] Statement of S. Nandan, Agenda Item 49: Ocean and Law of the Sea, UNGA 59th Session, 17 November 2004.

[78] A.J. Constable, W. de la Mare, D. Agnew, I. Everson, and D. Miller, "Managing Fisheries to Conserve the Antarctic Marine Ecosystem: Practical Implementation of the Convention on the Conservation of Antarctic Marine Living Resources (CCAMLR)" (2000) 57 *ICES Journal of Marine Science* 778–791.

[79] NEAFC press release, 21 November 2005. See online <http://www.neafc.org/news/docs/2005-press-release_final.pdf> (accessed December 2005).

[80] As reported at the first meeting of the Advisory Committee for the Agreement on the Conservation of Albatrosses and Petrels (ACAP), which was held 20–22 July 2005 in Hobart, Australia.

[81] *Inter-American Convention for the Protection and Conservation of Sea Turtles*, concluded in Caracas, Venezuela, on 5 September 1996, entered into force 2 May 2001. This is the first international agreement devoted solely to protection of marine turtles. Each party agrees to implement broad measures for the conservation of sea turtles. This Convention arguably has implications for RFMOs under the UNFSA compatibility provisions if the adjacent coastal State has measures in place to conserve turtles that also occur in the convention area of the RFMO. See IAC online <http://www.seaturtle.org/iac/> (accessed 20 October 2005).

[82] See E. Molenaar, "CCAMLR and Southern Ocean Fisheries" (2001) 16(3) *International Journal of Marine and Coastal Law* 465–499; and J. Scandol, M. Holloway, P. Gibbs, and K. Astles, at 261–273. See also J. Croxall and P. Tranthan, "The Southern Ocean: A Model System for Conserving Resources" in L. Glover and S. Earle, eds., *Defying Oceans End* (Washington, D.C.: Island Press, 2004) 71–87, for an account of CCAMLR's accomplishments and challenges.

[83] The CCAMLR Convention forms part of the so-called "Antarctic Treaty System", a collection of international instruments arising from the Antarctic Treaty, which was negotiated to overcome the sovereignty issues and provide a basis for strong scientific cooperation.

[84] *Antarctic Treaty*, done at Washington, D.C., 1 December 1959, entered into force 23 June 1961, 402 UNTS 71.

[85] Many of these fish stocks were already overexploited in 1982 when CCAMLR came into force. See "Understanding CCAMLR's Approach to Management", at 3.2. CCAMLR online <http://www.ccamlr.org/pu/e/gen-intro.htm> (accessed 20 October 2005).

[86] S. Kaye, at 364; J. Heap, "Has CCAMLR Worked? Management Policies and Ecological Needs" in A. Jorgensen-Dahl and W. Ostreng, eds., *The Antarctic Treaty System in World Politics* (London: St. Martins Press, 1991) 43–53; G. Parkes, "CCAMLR's Application of the Precautionary Approach" in V.R. Restrepo, ed., *Proceedings of the 5th National NMFS Stock Assessment Workshop: Providing Scientific Advice to Implement the Precautionary Approach Under the Magnuson–Stevens Fishery Conservation and Management Act*, U.S. Dep. Commer. NOAA Technical Memorandum NMFS-F/SPO-40 (Silver Springs, Md.: U.S. Department of Commerce, NOAA, 1999) 87–95.

[87] Draft Programmatic Environmental Impact Statement on Codified Regulations at 50 CFR Part 3000 Subparts A and G. Implementing Conservation and Management Measures. Adopted by the Commission for the Conservation of Antarctic Living Resources. June 2005. United States Department of Commerce, National Oceanographic and Atmospheric Administration, National Marine Fisheries Service Office of Sustainable Fisheries, at 29.

[88] As noted earlier, Art. VI of the CCAMLR Convention recognizes the primacy of the IWC and the CCAS for whales and seals south of 60°S. CCAMLR Scientific Committee XXIII. Cooperation with Other Organizations at secs. 9.31 and 9.32. In 2005, CCAMLR, through the WG-EMM, will work with IWC to examine the relationship between blue whales and sea ice.

[89] See S. Kaye, at 369–370; and C. Joyner, *Governing the Frozen Commons* (Columbia, S.C.: University of South Carolina Press, 1998) at 120–122. According to Denzil Miller (personal communication, February 2006), the Convergence is an "ecological" boundary not just a geographical one. The reference to 60°S applies to the Antarctic Treaty Area and is important in respect to the relationship between the Treaty and CCAMLR (CCAMLR Arts. III–V).

[90] FAO, "Report of the Fourth Meeting of Regional Fishery Bodies. Rome, 14–15 March 2005", FAO Fisheries Report No. 778 (Rome: FAO, 2005) at 1. See also D. Miller, E. Sabourenkov, and D. Ramm, "Managing Antarctic Marine Living Resources: The CCAMLR Approach" (2004) 19(3) *International Journal of Marine and Coastal Law* 317–363.

[91] Art. II(3) of the CCAMLR Convention states that "any harvesting and associated activities in the area to which this Convention applies shall be conducted in accordance with the provisions of this Convention and with the following principles of conservation: (a) prevention of decrease in the size of any harvested population to levels below those which ensure its stable recruitment. For this purpose its size should not be allowed to fall below a level close to that which ensures the greatest net annual increment; (b) maintenance of the ecological relationships between harvested, dependent and related populations of Antarctic marine living resources and the restoration of depleted populations to the levels defined in sub-paragraph (a) above; and (c) prevention of changes or minimisation of the risk of changes in the marine ecosystem which are not potentially reversible over two or three decades, taking into account the state of available knowledge of the direct and indirect impact of harvesting, the effect of the introduction of alien species, the effects of associated activities on the marine ecosystem and of the effects of environmental changes, with the aim of making possible the sustained conservation of Antarctic marine living resources". *Convention on the Conservation of Antarctic Marine Living Resources*, done at Canberra, Australia, 20 May 1980, entered into force 7 April 1982, ATS 1982 No. 9; 19 ILM 841.

[92] CCAMLR cooperates with the Committee on Environmental Protection (CEP) of the Madrid Protocol particularly on scientific data collection and analysis.

[93] *Convention for the Conservation of Antarctic Seals*, done at London, UK, 1 June 1972, entered into force 10 November 1978, 11 ILM 251.

[94] See S. Kaye, at 361. As Denzil Miller notes (personal communication, February 2006), CCAS has, in practice, been administered to prohibit sealing of selected species and to manage any sealing that has occurred since its entry into force.

[95] *International Convention on the Regulation of Whaling*, done at Washington, D.C., 2 October 1946, entered into force 10 November 1948, 161 UNTS 72. This Convention does not form part of the Antarctic Treaty System.

[96] WG-IMALF was modified in 2001 to account for bycatch associated with trawl fishing. The group's name was changed in 2001 to the Ad Hoc Working Group on Incidental Mortality Associated with Fishing (WG-IMAF).

[97] FAO Fisheries Report No. 778, at 2.

[98] CCAMLR's implementation of the precautionary approach is discussed in Chap. 6. Uncertainty is dealt with in the interests of minimizing risk of irreversible changes in the system as well as unsustainable fishing practices. Such a philosophical and profound departure point is exactly why CCAMLR is considered advanced in its EAF implementation and is also the motivation for the collection of designated information under CEMP. A.J. Constable, "CCAMLR Ecosystem Monitoring and Management: Future Work" (2002) 9 *CCAMLR Science* 233–256; A.J. Constable, "A Possible Framework in which to Consider Plausible Models of the Antarctic Marine Ecosystem for Evaluating Krill Management Procedures" (2005) 12 *CCAMLR Science* 99–117; I. Everson, "Consideration of Major Issues in Ecosystem Monitoring and Management" (2002) 9 *CCAMLR Science* 213–232; and D. Miller, "Antarctic Krill and Ecosystem Management — From Seattle to Siena" (2002) 9 *CCAMLR Science* 175–212.

[99] CCAMLR completed a comprehensive review of CEMP in 2003. See Appendix D, Annex 4 of 22nd Meeting of the CCAMLR Scientific Committee.

[100] See G.E. Fogg, *A History of Antarctic Science* (Cambridge: Cambridge University Press, 1992) at 239, who notes that the concept of MSY was seen as inconsistent with an ecosystem approach. Coincidentally, $F_{0.1}$ was used as precautionary reference point by the Government of Canada in the 1980s and early 1990s.

[101] CCAMLR, "Understanding CCAMLR's Approach to Management". CCAMLR online <http://www.ccamlr.org/pu/e/gen-intro.htm> (accessed 20 October 2005).

[102] C. de Fontaubert and I. Lutchman with D. Downes and C. Deere, *Achieving Sustainable Fisheries Implementing the New International Legal Regime* (Gland: IUCN, 2003) at 40.

[103] C. de Fontaubert and I. Lutchman with D. Downes and C. Deere, at 40.

[104] CEMP monitors populations of crab eater and Antarctic fur seals, four species of penguins, the black-browed albatross, and two species of petrels. See also D. Agnew, "The CCAMLR Ecosystem Monitoring Programme" (1997) 9(3) *Antarctic Science*: 235–242; and G. Parkes, at 87.

[105] C. de Fontaubert and I. Lutchman with D. Downes and C. Deere, at 40.

[106] Prey species, environmental factors, and the links between these and predators are monitored. Through CEMP, 80 sites, species, and parameters are measured and data are collected in a consistent manner. See CCAMLR Web site for CEMP <http://www.ccamlr.org/pu/e/gen-intro.htm> (accessed 20 October 2005); and also G. Parkes, at 87.

[107] See A.J. Constable, "A Possible Framework in which to Consider Plausible Models of the Antarctic Marine Ecosystem for Evaluating Krill Management Procedures".

[108] See G. Parkes, at 6.

[109] A.J. Constable, "A Possible Framework in which to Consider Plausible Models of the Antarctic Marine Ecosystem for Evaluating Krill Management Procedures" (2005) 12 *CCAMLR Science* 99–117. See also A.J. Constable, W.K. de la Mare, D.J. Agnew, I. Everson, and D. Miller, "Managing Fisheries to Conserve the Antarctic Marine Ecosystem: Practical Implementation of the Convention on the Conservation of Antarctic Marine Living Resources (CCAMLR)" (2000) 57 *ICES Journal of Marine Science* 778–791, and CCAMLR Web site explaining CEMP <http://209.85.165.104/search?q=cache:25MO-3yI9T4J:www.ccamlr.org/pu/E/sc/cemp/intro.htm+ecosystem+health+and+Constable&hl=en&ct=clnk&cd=7&gl=ca> (accessed January 2006).

[110] See CCAMLR, "Understanding CCAMLR's Approach to Management" at sec.3.2(d). CCAMLR online <http://www.ccamlr.org/pu/e/gen-intro.htm> (accessed 20 October 2005).

[111] A. Brierley and K. Reid, "Kingdom of Krill" (1999) 2182 *New Scientist* 37–41.

[112] Krill TACs apply at area–subarea scales (i.e., much larger than those at which predator–prey interactions actually occur). Hence, the current CCAMLR focus on trying to establish ways of managing krill (including setting TACs) at small scale management unit (SSMU) levels. In practice, there are no SSMU measures as yet; however, the various krill precautionary measures (CCAMLR CMs 51-01–51-03) recognize the need for subdividing statistical area limits further when the fishery expands as well as the importance of avoiding an overconcentration of the fishery. In these terms there is a long, and ongoing, debate within CCAMLR on developing the SSMU approach. See A.J. Constable and S. Nicol, "Defining Smaller-Scale Management Units to Further Develop the Ecosystem Approach in Managing Large-scale Pelagic Krill Fisheries in Antarctica" (2002) 9 *CCAMLR Science* 117–131; A.J. Constable, "A Possible Framework in Which to Consider Plausible Models of the Antarctic Marine Ecosystem for Evaluating Krill Management Procedures"; and WG-EMM reports from 1993 onwards.

[113] CCAMLR Scientific Committee Meeting XXII 2004. Ecosystem Monitoring and Management at 5.

[114] CCAMLR, "Understanding CCAMLR's Approach to Management", at sec. 3.6. CCAMLR online <http://www.ccamlr.org/pu/e/gen-intro.htm> (accessed 20 October 2005).

[115] The occurrence of Patagonian toothfish beyond the convention area has, in recent years, caused serious difficulties in combating illegal, unreported, and unregulated (IUU) fishing for this species. A considerable percentage of the catch reported as having been taken outside the convention area is suspected to have actually been taken within the convention area. In accordance with the precautionary approach to fisheries, past IUU fishing is taken into account in setting TACs for toothfish (both Patagonian and Antarctic).

[116] C. Small, at 24.

[117] Draft Programmatic Environmental Impact Statement on Codified Regulations at 50 CFR Part 3000 Subparts A and G. Implementing Conservation and Management Measures Adopted by the Commission for the Conservation of Antarctic Living Resources. June 2005. United States Department of Commerce, National Oceanographic and Atmospheric Administration, National Marine Fisheries Service Office of Sustainable Fisheries. pp. 133–136 at 29.

[118] Draft Programmatic Environmental Impact Statement on Codified Regulations at 50 CFR Part 3000 Subparts A and G. Implementing Conservation and Management Measures Adopted by the Commission for the Conservation of Antarctic Living Resources. June 2005. United States Department of Commerce, National Oceanographic and Atmospheric Administration, National Marine Fisheries Service Office of Sustainable Fisheries. pp. 133–136 at 29.

[119] CCAMLR Resolution 5-08. Protection of Seabirds from Incidental Mortality Arising from Longline Fishing.

[120] John Croxall, personal communication, 16 December 2005.

[121] CCAMLR, *Report of the Sixteenth Meeting of the Commission*, Hobart, Australia, 27 October – 7 November 1997, at 20–23. Also see D. Agnew, "The Illegal and Unregulated Fishery for Toothfish in the Southern Ocean and the CCAMLR Catch Documentation Scheme" (2000) 24 *Marine Policy* 361–374.

[122] CCAMLR Scientific Committee Report XXIII. Incidental Mortality at sec. 5.18.

[123] CCAMLR Conservation Measure 33-01 (1995) "Limitation of the by-catch of *Gobionotothen gibberifrons*, *Chaenocephalus aceratus*, *Pseudochaenichthys georgianus*, *Notothenia rossii* and *Lepidonotothen squamifrons* in Statistical Subarea 48.3".

[124] CCAMLR Conservation Measure 22-01 (1986) "Regulation on mesh size measurement"; Conservation Measure 22-02 (1984) "Mesh size"; Conservation Measure 22-03 (1990) "Mesh size for *Champsocephalus gunnari*".

[125] See NOAA 2005. Draft Programmatic Environmental Impact Statement on Codified Regulations at 50 CFR Part 300 Subparts A and G. Implementing Conservation and Management Measures Adopted by CCAMLR. Map of closed areas as of 2005 at 46.

[126] CCAMLR, *CCAMLR XXIII — Meeting Report — Antarctic Southern Ocean Coalition*, 25 October – 5 November 2004, Hobart, Australia, sec. 8 at 15, with clarification from John Croxall, personal communication, 16 December 2005.

[127] CCAMLR Conservation Measure 91-03 (2004) "Protection of the Seal Islands CEMP site".

[128] See Preliminary Report of the XXIV Meeting of the CCAMLR Commission, paras. 4.12–4.18. See also Report of the 23rd Meeting of the CCAMLR Scientific Committee, paras. 3.28–3.51; Report of the 24th Meeting of the CCAMLR Scientific Committee, paras. 3.51–3.65; and CCAMLR, *CCAMLR XXIII — Meeting Report — Antarctic Southern Ocean Coalition*, 25 October - 5 November 2004, Hobart, Australia.

[129] Preliminary Report of the XXIV Meeting of the CCAMLR Commission, para. 4.12.

[130] Preliminary Report of the XXIV Meeting of the CCAMLR Commission, para. 4.13.

[131] Preliminary Report of the XXIV Meeting of the CCAMLR Commission, para. 4.18.

[132] Also see D. Agnew, "The Illegal and Unregulated Fishery for Toothfish in the Southern Ocean and the CCAMLR Catch Documentation Scheme".

[133] CITES Resolution 12.4. Available at <http://www.cites.org/eng/res/12/12-04.shtml> (accessed 14 January 2006). See also CCAMLR, *Report of the Twenty First Meeting of the Commission*, 21 October – 1 November 2002, at paras. 8.62–8.84 and 10.1–10.71, regarding CITES and Australia's bid to have toothfish listed under CITES.

[134] See CITES COP 13 report.

[135] CCAMLR, *Report of the XII Meeting of the Commission*, 25 October – 5 November 1993, Hobart, Australia, at para. 5.9.

[136] The CCAMLR Secretariat has made efforts to obtain a conservation measure that would consolidate all the measures on environmental protection. See Preliminary Report of the XXIV Meeting of the CCAMLR Commission, paras. 11.98–11.101.

[137] CCAMLR Conservation Measures 25-01(1996) "Regulation of the use and disposal of plastic packaging bands on fishing vessels".

[138] *International Convention for the Prevention of Pollution from Ships*, 1973, as modified by the Protocol of 1978 relating thereto (MARPOL 73/78).

[139] CCAMLR Scientific Committee Report XXIII. Incidental Catch at secs. 5.13 and 5.14. The slight increases in seabird bycatch in the last two years has raised concerns about compliance with conservation measures, particularly those regarding the discharge of offal.

[140] Robin Allen, IATTC Executive Secretary, personal communication, March 2005.

[141] Art. II(1) of the IATTC Convention, which specifies the functions and duties of the Commission, requires the Commission to conduct investigations on the abundance, biology, biometry, and ecology of yellowfin tuna and skipjack tuna in the waters of the Eastern Pacific Ocean fished by the nationals of the High Contracting Parties; on the fishes species commonly used as bait in the tuna fisheries, especially the anchovetta, and on other kinds of fish taken by tuna fishing vessels; and the effects of natural factors and human activities on the abundance of the populations of fishes supporting all these fisheries. *Convention for the Establishment of an Inter-American Tropical Tuna Commission*, done at Washington, D.C., 31 May 1949, entered into force 3 March 1950, TIAS 2044.

[142] Dolphin mortality associated with the tuna fishery is estimated to have been 200 000 to 500 000 individuals per year. See discussion in C. Hedley, "The 1998 Agreement on the International Dolphin Conservation Program: Recent Developments in the Tuna–Dolphin Controversy in the Eastern Pacific Ocean" (2001) 32 *Ocean Development and International Law* 71–92.

[143] C. Hedley, at 87. For further information on the history and components of the Dolphin Conservation Program, see J. Joseph, "The Tuna–Dolphin Controversy in the Eastern Pacific Ocean: Biological, Economic and Political Impacts" (1994) 25 *Ocean Development and International Law* 1–30.

[144] As Hedley notes, "The MMPA deals with various matters relating to all species of marine mammals, although the mortality of dolphins during tuna purse seining operations in the EPO was a major impetus for the passage of the act". See C. Hedley, at 72.

[145] According to R. Rayfuse, *Non-Flag State Enforcement in High Seas Fisheries* (Leiden: Martinus Nijhoff Publishers, 2004) at 146, at the time, 77% of the purse-seine capacity in the EPO was registered in the United States. Given the inability to harvest yellowfin tuna in the EPO without contravening US legislative standards, these vessels either left the area to fish in the Western and Central Pacific or reflagged to evade US controls. Hedley also cites the sale of US tuna vessels to private interests in Latin America and suggests that the transfer of US effort to the Western and Central Pacific may have resulted from difficulty in achieving access to EEZs of Latin American coastal States and from a reduction in the abundance of yellowfin tuna as a result of overfishing. See C. Hedley, at 73. See also G. Sakagawa, "Are US Regulations On Tuna–Dolphin Fishing Driving US Seiners to Foreign Flag Registry?" (1991) 11 *North American Journal of Fisheries Management* 241–252.

[146] C. Hedley, at 72.

[147] These embargoes were accompanied by changes in US law for labeling, which provided for the use of a "dolphin-safe" label on cans of tuna not caught in association with dolphins. This law was passed after the three largest US tuna processors announced they would not purchase tuna that was caught in association with dolphins. According to the Dolphin Protection Consumer Information Act [16 U.S.C. 1385], tuna can be labeled dolphin safe if "no dolphin were caught in the trip in which such tuna were harvested using a purse seine net intentionally deployed on or to encircle dolphins, and no dolphins were killed or seriously injured during the sets in which the tuna were caught". Sec. 1385 (h) (2) effectively banned the import of yellowfin tuna harvested in the northern EPO, as dolphins are associated with this target species in this area. There were additional stipulations concerning the requirements for dolphin-safe tuna: the fishing vessel must have been effectively incapable of deploying its purse-seine nets on dolphin; the IATTC must have confirmed that there was an approved observer on board the vessel during the entire trip in question; and the net must not have been intentionally deployed on or around dolphins. The US MMPA, adopted in 1972, required US fishers to use selective gear and prohibited dolphin bycatch, thereby restricting tuna catches.

[148] By contrast, the skipjack fishery in the southern EPO was not associated with dolphins and deployed fish aggregating devices, which are considered dolphin safe.

[149] Dolphin mortality was estimated to be 136 000 individuals in 1986 and 100 000 individuals in 1987.

[150] Mexico and later the EU challenged the United States before GATT dispute settlement panels, contending that the US legislation adopted for domestic marine conservation purposes constituted disguised restrictions on international free trade and contravened fundamental GATT principles. Although both challenges (Tuna–Dolphin I and Tuna–Dolphin II) resulted in decisions against the United States in the GATT dispute settlement panels, neither Mexico nor the EU pursued the matter further to GATT Council. For a discussion on these cases

see C. Joyner and Z. Tyler, "Marine Conservation Versus International Free Trade: Reconciling Dolphins with Tuna and Sea Turtles with Shrimp" (200) 31 *Ocean Development and International Law* 127–150.

[151] In response to the GATT panel decisions, in 1992 the US Congress passed the International Dolphin Conservation Program Act, which sought negotiations with Mexico and Venezuela to enter into a global moratorium on using purse-seine nets to harvest tuna. The US Senate passed the Act in 1997.

[152] The schedule of dolphin mortality progressively reduced the limit on the allowable annual dolphin bycatch each year until 1999, when the number was to be less than 5000 dolphins.

[153] See C. Hedley, at 75, and IATTC, *Annual Report of the Inter-American Tropical Tuna Commission, 2004* (La Jolla, Calif.: IATTC, 2006) at 41.

[154] The International Dolphin Conservation Protection Act (IDCPA) [16 U.S.C. 1413] was signed 15 August 1997 and became effective 3 March 1999. This act amended the MMPA, the Dolphin Protection Consumer Information Act [16 U.S.C. 1385], Tuna Conventions Act, and redefined the standard for "dolphin-safe" tuna. The Agreement on the International Dolphin Conservation Program (AIDCP) came into force two weeks earlier on 15 February 1999.

[155] All IATTC members, with the exception of France and Japan, are members of the International Dolphin Conservation Program (IDCP).

[156] Though the AIDCP provides for the participation of regional economic integration organizations (REIOs), such as the EU, the EU had stated it would not accede to the Agreement until it joined the IATTC (the 1949 Convention does not provide for REIOs) or its successor the Antigua Convention (Article XXX allows for REIOs; however, the Convention is not yet in force). The EU did accede to the AIDCP in 2005.

[157] AIDCP (amended October 2005), Annex VIII part 3(d).

[158] IATTC, "Tunas and Billfishes in the Eastern Pacific Ocean in 2003", IATTC Fishery Status Report No. 2 (La Jolla, Calif.: IATTC, 2004). The extent to which yellowfin tuna and dolphins compete for resources and whether either or both of them benefit from the interaction remain critical pieces of information given the large biomasses of both groups and their high rates of prey consumption.

[159] Letter written by Robin Allen, Director IATTC to Donald Evans, Secretary of Commerce, 30 October 2002, as a preface to the Scientific Report on the Status of Dolphin Stocks in the Eastern Pacific Ocean. IATTC online <www.iattc.org/PDFFiles/0653-800_D.Evans%20Final%20Report%20October%2030.pdf> (accessed 18 September 2005).

[160] Under the MMPA, nations harvesting tuna in the EPO with purse-seine vessels greater than 400 t carrying capacity are required to receive an "affirmative finding" based on documentary evidence provided by IATTC to export yellowfin tuna and yellowfin tuna products to the United States. In 2005 the United States renewed the affirmative finding for Mexico, Spain, and Ecuador.

[161] See AIDCP summary at <http://www.iattc.org/DolphinSafeENG.htm> (accessed 14 January 2006). In November 2005, the FAO awarded the Margarita Lizárraga medal to the AIDCP. The medal is awarded to a person or organization that has served with distinction in the application of the FAO Code of Conduct for Responsible Fisheries. According to J. Diouf, the Director-General of the FAO, the AIDCP was selected for the award "in recognition of its comprehensive, sustainable and catalytic initiatives in support of the Code of Conduct for Responsible Fisheries, in particular those aspects relating to the precautionary approach and to the utilization of fishing gear and techniques which minimize the catch of non-target species". Press release, November 2005. Available at <http://www.iattc.org/PDFFiles2/Press-release-re-Lizarraga-medal-ENG.pdf> (accessed 4 February 2006).

[162] IATTC Resolution on Bycatch. C-99-11.

[163] IATTC Working Group on Bycatch Recommendation on Measures to Reduce Bycatch in the Tuna Purse Seine Fishery in the Eastern Pacific Ocean. 2000.

[164] FADs are artificial or natural floating objects placed on the ocean surface, often anchored to the bottom, to attract several schooling fish species underneath, thus increasing their "catchability". Definition according to the FAO Glossary, FAO online <http://www.fao.org/fi/glossary/default.asp> (accessed 10 January 2006).

[165] IATTC Resolution on Bycatch. C-00-08.

[166] IATTC Resolution on Bycatch. C-01-04.

[167] FAO Guidelines to Reduce Sea Turtle Mortality in Fishing Operations, IGIFL online <http://www.intfish.net/docs/2004/faoturtles.pdf> (accessed 20 October 2005).

[168] IATTC Consolidated Resolution on Bycatch. C-03-08.

[169] Compliance with this resolution is thought to be low. See sec. 9, Report of the Working Group on Bycatch in the minutes of the 72nd IATTC Meeting. June 2004.

[170] IATTC Resolution on Full Retention. C-05-05.

[171] IATTC Recommendation on Sea Turtles. C-03-10.

[172] IATTC Resolution on a Three Year Program to Mitigate the Impact of Tuna Fishing on Sea Turtles. C-04-07.

[173] IATTC Document IATTC-73-08c. June 2005. This document outlines past Resolutions. The 1998 Resolution on FADs prohibited use of tender vessels and at-sea transshipment of tuna caught by purse-seine vessels.

[174] Other aspects of the FAO Guidelines to Reduce Sea Turtle Mortality concerning sea turtle conservation that have yet to be met by IATTC include research on survival rates of animals that are successfully untangled, research on socioeconomic impacts of the sea turtle conservation measures, and funding of educational materials on sea turtle conservation. Research on spatial and temporal closures for areas of high sea turtle bycatch have yet to be addressed by IATTC. As noted in Appendix A, p. 3, of Document IATTC-73-08c, "No research has been done on this, and no measures have been introduced. The spatial and temporal restrictions imposed on the purse-seine fishery by IATTC resolutions do not take sea turtle distributions into account. Longline fisheries are subject only to total catch restrictions, with no limits on when or where they may fish". The AIDCP observer program on purse-seine vessels collects information on sea turtle bycatch, but there is no such program on distant-water longline fleets, at 4.

[175] Marking of FADs was discussed at the 2004 IATTC Meeting. Also, see sec. 14 of the minutes of the 72nd Meeting.

[176] See IATTC Document BYC04005b, "Interactions of Sea turtles with Tuna Fisheries and Other Impacts on Turtle Populations"; and see World Wildlife Fund, "Circle hooks help save sea turtles in Ecuador", press release, 5 June 2005.

[177] See Report of the 4th Meeting of Working Group on Bycatch. January 2004.

[178] Document IATTC-73-08b. Conservation of Sharks. June 2005. This document outlines past Resolutions. The 1998 Resolution on FADs prohibited use of tender vessels and at-sea transshipment of tuna caught by purse-seine vessels.

[179] IATTC Resolution on the Conservation of Sharks Caught in Association with Fisheries of the Eastern Pacific Ocean. C-05-03.

[180] IATTC Resolution on the Incidental Mortality of Seabirds. C-05-01. In response to CCAMLR's invitation for IATTC to cooperate on the conservation of seabirds, IATTC responded that "The IATTC has not established any measures to reduce the catch of seabirds, and the staff is not aware of voluntary measures that may be in force in the EPO. There has been no systematic collection of mortality data by the IATTC". Document IATTC-73-08a, "Conservation of Seabirds", June 2005.

[181] See IATTC Resolution on the Incidental Mortality of Seabirds. C5-05-01.

[182] See IATTC Resolution for a Multi-Annual Program on the Conservation of Tuna in the Eastern Pacific Ocean for 2004, 2005 and 2006. C-04-09.

[183] IATTC Resolutions on Capacity began in 1998, with the call for the establishment of the Permanent Working Group on Capacity. The measurement of fleet capacity and the imposition of limits on nationals for carrying capacity of purse-seine fleets, as outline in the 1998 Resolution on Fleet Capacity, have been major undertakings. In 1999 the Resolution on the Management of Longline Fishing Capacity (C99-04) was passed.

[184] IATTC Fishery Status Report No. 2, at 90.

[185] Since 2001 scientific staff has used complex models to investigate how fisheries management measures might affect the ecosystem impacts of fishing. The Commission staff has also developed a multispecies mass-balance model for the pelagic species in the convention area to learn more about how the ecosystem functions and to investigate the relative ecological implications of alternative fishing strategies. The ecosystem model was developed using Ecopath with Ecosim (EwE). The mass-balance model was generated from estimates of abundance of the resources (biomasses), productivity and mortality rates of the resources, how they interact (diet compositions and food consumption rates), and how efficiently the resources are used in the ecosystem (IATTC Annual Report 2001, at 21).

[186] Document IATTC-73-12, "Proposed Resolution on an Ecosystem Approach to Management", June 2005.

[187] IATTC Fishery Status Report No. 2, sec. J at 90–97, Ecosystem Considerations.

[188] Art. VII of the Antigua Convention outlines the functions of the Commission and states that the Commission shall "(a) promote, carry out and coordinate scientific research concerning the abundance, biology and biometry in this Convention Area of fish stocks covered by the Convention and, as necessary, of associated or dependent species, and the effects of natural factors and human activities on the populations of these stocks and species"; "(c) adopt measures that are based on the best scientific evidence available to ensure the long-term conservation and sustainable use of the fish stocks covered by this Convention and to maintain or restore the populations of harvested species at levels of abundance which can produce the maximum sustainable yield, *inter alia*, through the setting of the total allowable catch of such fish stocks as the Commission may decide and (or) the total allowable level of fishing capacity and (or) level of fishing effort for the Convention Area as a whole"; "(f) adopt, as necessary, conservation and management measures and recommendations for species belonging to the same ecosystem and that are affected by fishing for, or dependent on or associated with, the fish stocks covered by this Convention, with a view to maintaining or restoring populations of such species above levels at which their reproduction may become seriously threatened"; "(g) adopt appropriate measures to avoid, reduce and minimize waste, discards, catch by lost or discarded gear, catch of non-target species (both fish and non-fish species) and impacts on associated or dependent species, in particular endangered species"; "(h) adopt appropriate measures to prevent or eliminate overfishing and excess fishing capacity and to ensure that levels of fishing effort do not exceed those commensurate with the sustainable use of the fish stocks covered by this Convention"; "(k) promote, to the extent practicable, the development and use of selective, environmentally safe and cost-effective fishing gear and techniques and such other related activities, including activities connected with, *inter alia*, transfer of technology and training"; "(n) promote the application of any relevant provision of the Code of Conduct and of other relevant international instruments including, *inter alia*, the International Plans of Action adopted by FAO in the framework of the Code of Conduct".

[189] Art. IV(1) of the ICCAT Convention states: "In order to carry out the objectives of this Convention the Commission shall be responsible for the study of the populations of tuna and tuna-like fishes (the Scombriformes with the exception of the families Trichiuridae and Gempylidae and the genus Scomber) and such other species of fishes exploited in tuna fishing in the Convention area as are not under investigation by another international fishery organization. Such study shall include research on the abundance, biometry and ecology of the fishes; the oceanography of their environment; and the effects of natural and human factors upon their abundance. The Commission, in carrying out these responsibilities shall, insofar as feasible, utilize the technical and scientific services of, and information from, official agencies of the Contracting Parties and their political subdivisions and may, when desirable, utilize the available services and information of any public or private institution, organization or individual, and may undertake within the limits of its budget independent research to supplement the research work being done by governments, national institutions or other international organizations". *International Convention for the Conservation of Atlantic Tunas*, as amended, Rio de Janeiro, Brazil, 4 May 1966, entered into force 21 March 1969, 673 UNTS 63; TIAS 6767; 6 ILM 293.

[190] These groups direct research and data collection and make recommendations to the SCRS on management measures.

[191] See 2005 report of BYP at <http://www.iccat.es/Documents/SCRS/ExecSum/BYP%20EN.pdf> (accessed 10 January 2006).

[192] Further to its mandate, the Sub-committee "guides research and analytical activities related to by-catch, especially of sharks with focus on blue, mako and porbeagle; it recommends methodological adaptations to the national statistical data collection systems in order to better quantify by-catch (logbook and observer programs designed to characterize total catch composition and disposition); and it coordinates data gathering and cooperation with other fishery or wildlife organizations on by-catch issues (e.g. FAO, CITES, ICES, *etc.*)". As

stated in the ICCAT, *Report for Biennial Period 2004–2005 Part I (2004) — Vol. 2*, Appendix 8 at 206. Available at <http://www.iccat.int/Documents/BienRep/REP_EN_04-05_I_2.pdf> (accessed January 2006).

[193] For details on the mandate and functions of the Sub-Committee on Bycatch, see ICCAT, *Report for Biennial Period 2004–2005 Part I (2004) — Vol. 2*, Appendix 8. Available at <http://www.iccat.int/Documents/BienRep/REP_EN_04-05_I_2.pdf> (accessed January 2006).

[194] ICCAT Recommendation 04/10 Concerning the Conservation of Sharks Caught in Association with Fisheries Managed by ICCAT.

[195] The ban was instituted following the shark 2004 stock assessment. See <www.iccat.es/Documents/SCRS/ExecSum/SHK%20EN.pdf> (accessed January 2006).

[196] ICCAT Recommendation by ICCAT to Amend Recommendation (Rec. 04-10) Concerning the Conservation of Sharks Caught in Association with Fisheries Managed by ICCAT. 05-05.

[197] ICCAT Resolution by ICCAT on Circle Hooks. 05-08.

[198] ICCAT, *Report for Biennial Period 2004–2005 Part I (2004) — Vol. 2*, Appendix 8 at 206–207. Available at <http://www.iccat.int/Documents/BienRep/REP_EN_04-05_I_2.pdf> (accessed January 2006).

[199] ICCAT Resolution 99-1.

[200] ICCAT Recommendation 04-07.

[201] *Report of the Fourth Informal Consultations of States Parties to the Agreement for the Implementation of the Provisions of the United Nations Convention on the Law of the Sea of 10 December 1982 relating to the Conservation and Management of Straddling Fish Stocks and Highly Migratory Fish Stocks*, New York, 31 May – 3 June 2005, at 7.

[202] An inventory of ICCAT tagging programs can be found at <http://www.iccat.es/tagging.htm> (accessed January 2006).

[203] FAD fishing is intensely practiced in the Gulf of Guinea, which is a main spawning ground of tropical tunas (yellowfin, skipjack, and bigeye) in the Atlantic and produces a high catch of juveniles, particularly after the peak spawning season. An important increase in the longline effort and the effect of the juvenile bigeye bycatch of the purse-seine fleet were identified as possible causes of suspected depletion of the bigeye population. J. Moron, *Report on Management Measures for the European Tuna Purse Seine Fleet*. Available at <http://www.spc.int/oceanfish/html/sctb/sctb14/FT-9%20Moron-Management.pdf> (accessed 16 January 2006).

[204] ICCAT. Recommendation by ICCAT Concerning the Establishment of a Closed Area/Season for the Use of Fish Aggregation Devices (FADs) 98-01 YFT & BET (no longer in force).

[205] ICCAT. Recommendation by ICCAT Concerning the Establishment of a Closed Area/Season for the Use of Fish Aggregation Devices (FADs) 99-1 YFT & BET (in force 15 June 2000).

[206] ICCAT Recommendation to Amend the Recommendation on Bluefin Tuna Farming. 05-04.

[207] See ICCAT, *Report for Biennial Period 2004–2005 Part I (2004) — Vol. 2*, Appendix 9, for details on the mandate and functions of the Sub-committee on the Environment. Available at <http://www.iccat.int/Documents/BienRep/REP_EN_04-05_I_2.pdf> (accessed January 2006).

[208] ICCAT, "ICCAT Workshop on Environment and Tuna Recruitment (Madrid, Spain, 7–12 May 2001)" (2002) 54 *Collective Volume of Scientific Papers* 895–952.

[209] ICCAT, *Report for Biennial Period 2004–2005 Part I (2004) — Vol. 2*, Appendix 8 at 207. Available at <http://www.iccat.int/Documents/BienRep/REP_EN_04-05_I_2.pdf> (accessed January 2006).

[210] ICCAT, *Report for Biennial Period 2004–2005 Part I (2004) — Vol. 2*, Appendix 8 at 213. Available at <http://www.iccat.int/Documents/BienRep/REP_EN_04-05_I_2.pdf> (accessed January 2006).

[211] International Commission for the ICCAT, "Report of the Sub-Committee on Environment" in *Report for Biennial Period 2004-2005 Part II (2005) — Vol. 2*. Available at <http://www.iccat.int/Documents/BienRep/REP_EN_04-05_II_2.pdf> (accessed 20 January 2006).

[212] The EU has formally adopted the EAF as part of the Common Fisheries Policy (CFP), which could influence conservation measures adopted in GFCM for coherent management, in keeping with compatibility obligations of EU members. The EU CFP was reviewed in 2002, and the new framework regulations for the CFP came in to force in January 2003. One of the major changes to the policy was the incorporation of a more ecosystem-based approach to fisheries management (which was inserted in the objectives in Art. 2(1) of Council Regulation (EC) No. 2371/2002 of 20 December 2002 on the conservation and sustainable exploitation of fisheries resources under the Common Fisheries Policy. See online <http://europa.eu.int/eur-lex> (accessed 15 January 2006)) *inter alia* through the introduction of recovery plans for threatened stocks and management plans for nonthreatened stocks.

[213] GFCM, *Report of the Sixth Session of the Scientific Advisory Committee of the GFCM*, Thessaloniki, Greece, 30 June – 3 July 2003. FAO online <http://www.fao.org/documents/show_cdr.asp?url_file=/DOCREP/006/Y5105B/y5105b01.htm> (accessed January 2006).

[214] COPEMED is an FAO project that provides advice and technical support, and establishes cooperation networks to facilitate coordination and support to fisheries research and management in the Western and Central Mediterranean, in support to the GFCM.

[215] MedSudMed is a GFCM-related FAO project that evolved from the FAO Code of Conduct for Responsible Fisheries and in particular from the requirement for an ecosystem approach. The aims of MedSudMed are to support research activities and training to increase and use knowledge on fisheries ecology and ecosystems, and to create a regional network of expertise. Particular attention is given to the technical coordination of research activities among countries, which should contribute to the implementation of EAF. Consideration is also given to the development of an appropriate tool for the management and processing of data related to fisheries and their ecosystems, under the form of a regional database and information system.

[216] GFCM has released no official figure on the size of the pending closure. However, biologist Sergi Tudela of WWF-Mediterranean estimates the area at 55% of the entire basin surface (including the Black Sea), or 1.63×10^6 km^2 in total. This would be larger than the 950 000 km^2 bottom-trawl closure recommended in February 2005 for the Aleutian Islands in Alaska. In comparison, the Great Barrier Reef Marine Park, widely considered the world's largest MPA, is roughly 350 000 km^2. As reported in (April 2005) 6(9) *MPA News* 4.

[217] See GFCM, *Report of the Seventh Session of the Scientific Advisory Committee of the GFCM*, Rome, Italy, 19–22 October 2004. FAO online <http://www.fao.org/documents/show_cdr.asp?url_file=/docrep/007/y5850b/y5850b01.htm> (accessed 1 August 2005).

[218] GFCM, *Report of the Workshop on Ecosystem Approach to Fisheries*, Salammbô, Tunisia, 7–9 September 2005. GFCM: SAC8/2005/Dma.6 (Rome: GFC, 2005).

[219] FAO, "Report of the Eighth Session of the Scientific Advisory Committee, Tirana, Albania, 25–28 October 2005", FAO Fisheries Report No. 789 (Rome: FAO, 2006) at para. 35 and see paras. 34–45 for additional elements of the EAF, including implementation of NPOA, interactions between fisheries and cetaceans, and establishment of protected areas.

[220] FAO Fisheries Report No. 789.

[221] NAFO, *Annual Report 2004* (Dartmouth, N.S.: NAFO, 2004) at 105. Available at <http://www.nafo.int/publications/frames/fisheries.html> (accessed 15 March 2006).

[222] See NAFO press release for 2005 meeting in Tallin, Estonia. See online <http://www.nafo.int/about/frames/about.html> (accessed 10 January 2006).

[223] The Fisheries Commission formally requested the Scientific Council to conduct these investigations. See "NAFO starts a reform process", NAFO press release, 23 September 2005. NAFO online <http://www.nafo.int/> (accessed 20 October 2005).

[224] 2002 Report of Scientific Council Meeting. Appendix I. Report of Standing Committee on Fisheries Environment (STACFEN), at 55. Available at <http://www.nafo.int/publications/frames/general.html> (accessed xxxx).

[225] See Note from Secretary in NAFO, *Annual Report 2004* (Dartmouth, N.S.: NAFO, 2004), where Johanne Fisher states "One emerging requirement of international fisheries management is the design and implementation of an ecosystem approach that NAFO will discuss over the next years". Available at <http://www.nafo.int/publications/frames/general.html> (accessed 15 March 2006).

[226] NAFO, *Annual Report 2004* (Dartmouth, N.S.: NAFO, 2004) at 105. Available at <http://www.nafo.int/publications/frames/fisheries.html> (accessed 15 March 2006).

[227] 2005 Report of Scientific Council Meeting, at 75. Available at <http://www.nafo.int/publications/frames/general.html> (accessed 15 March 2006).

[228] NAFO Executive Secretary, Johanne Fisher, personal communication, July 2005.

[229] While this request may result in an increased understanding of commercial and noncommercial species living on seamounts within the NAFO area, vessels must trawl on the seamounts to collect the data, which may be counter to both the precautionary and ecosystem approaches.

[230] NAFO, "NAFO Conservation and Enforcement Measures. Chap. 1 — Conservation and Management Measures. Article 9 — 3NO Cod Conservation Plan and Rebuilding Strategy", NAFO FC Doc. 08/1, Serial No. N5480. Available at <http://www.nafo.int/fisheries/frames/regulations.html> (accessed 27 May 2007).

[231] See NAFO press release 2005. See online <http://www.nafo.int/about/frames/about.html> (accessed 16 January 2006).

[232] NAFO, "NAFO Conservation and Enforcement Measures. Chap. 1 — Conservation and Management Measures. Article 10 — Quota Adjustments", NAFO FC Doc. 08/1, Serial No. N5480. Available at <http://www.nafo.int/fisheries/frames/regulations.html> (accessed 27 May 2007).

[233] NAFO, "NAFO Conservation and Enforcement Measures. Chap. 1 — Conservation and Management Measures. Article 12 — Gear Requirements", NAFO FC Doc. 08/1, Serial No. N5480. Available at <http://www.nafo.int/fisheries/frames/regulations.html> (accessed 27 May 2007).

[234] See K. Kelleher, "Discards in the World's Marine Fisheries. An Update", FAO Fisheries Technical Paper No. 470 (Rome: FAO, 2005) at 22.

[235] Meeting Minutes from the Working Group on the Future of the North East Atlantic Fisheries Commission, 28 June 2005, Brussels, Belgium. See online <http://www.neafc.org/reports/future-neafc/index.htm> (accessed 1 August 2005).

[236] NEAFC press release, 21 November 2005. See online <http://www.neafc.org/news/docs/2005-press-release_final.pdf> (accessed 16 January 2006).

[237] J.-J. Maguire "Fisheries Science and Management in the North Atlantic" in T. Pitcher, U.R. Sumaila, and D. Pauly, eds., *Fisheries Impacts on North Atlantic Ecosystems: Evaluations and Policy Exploration*, Fisheries Centre Research Reports FCRR 9(5) (Vancouver, B.C.: Fisheries Centre, 2001) 36–48 at 37. See ICES, "Report of the ICES Advisory Committee on Fishery Management, 1999", ICES Cooperative Research Report No. 236 (Copenhagen: ICES, 2000). See also J. Swan, "Summary Information on the Role of International Fishery Organizations or Arrangements and other Bodies Concerned with the Conservation and Management of Living Aquatic Resources", FAO Fisheries Circular No. 985 (Rome: FAO, 2003).

[238] NEAFC, *Report of the 23rd Annual Meeting of the North-East Atlantic Fisheries Commission, 8–12 November 2004*, Vol. 1 (London: NEAFC, 2004) at 6.

[239] Also found in these areas is the cold-water coral, *Lophelia pertusa*, known to provide complex and diverse habitat for fish and invertebrate species. For a complete review of the biology and ecology of *Lophelia pertusa* in the Northeast Atlantic, see A. Roger, "The Biology of *Lophelia pertusa* (Linneaus, 1758) and Other Deep-water Reef-forming Corals and Impacts from Human Activities" (1999) 84 *International Review of Hydrobiology* 315–406.

[240] NEAFC Briefing and Update on Commission Activities. September 2005.

[241] The trend towards fishing in deeper and deeper water, as coastal stocks become overexploited and no longer commercially viable, presents a challenge for EAF and the precautionary approach. Scientific information on deep-sea species and habitats is limited, and fisheries tend to expand into new areas long before scientific research on the impacts of these fisheries takes place. The 2003 FAO conference on deep-sea fisheries — Deep Sea 2003: Conference on the Governance and Management of Deep-sea Fisheries — concluded that there was a general need for improving governance of deepwater resources. Further consideration of the issue at the March 2005 COFI meeting summarized that "Deepwater demersal fisheries resources continue to be vulnerable to overexploitation, because of their low productivity and weaknesses in existing governance mechanisms. These fisheries are generally to be found on the high seas and the information needed for their effective management is often deficient and legal means to ensure that data are provided and that participants manage their operations

responsibly are often lacking". See COFI/2005/6 Deep Sea Fisheries. See online <http://www.iucn.org/themes/marine/pdf/FAO-DeepSeaFisheries-3862e.pdf> (accessed 16 January 2006). Full implementation of UNFSA and the Code of Conduct as well as WSSD commitments will address some conservation issues in the deep sea, but there is an immediate and pressing need for increased data collection that can lead to science-based decision making and management recommendations for deep-sea fisheries. Several actions were proposed to better document existing management challenges, including paucity of data, environmental impacts, difficulty in conducting stock assessments, and management inaction There have also been recommendations to strengthen regional fishery bodies, through legal framework and overview of deep-sea fisheries operations. NGOs have expressed concern about the sustainability of deep-sea fisheries as well as the ability of RFMOs to manage deep-sea fisheries in a manner that is consistent with UNFSA, the Code of Conduct, and international biodiversity agreements.

[242] ICES 2004 advice on deep sea fisheries. See online <http://www.neafc.org/fisheries/docs/acfm_reports-2004/deep-water-general-2004.pdf> (accessed 16 January 2006).

[243] Effort reductions for ling are advised for Subareas II, IV, VI, VII, and VIII and Division Va.

[244] The ICES advice to NEAFC on deepwater fisheries is more conservative. "Fisheries on deepwater species have developed rapidly and the resources which they exploit are generally especially vulnerable to overfishing. Within the ICES area, species/stocks have been depleted before appropriate management measures have been implemented. It is also of concern that the landings statistics that are available may not reflect the true scale of the recent fishing activity, especially in waters outside the national EEZs". The ICES 2004 advice for deepwater fisheries was "Most exploited deepwater species are considered to be harvested unsustainably; however, it is currently not possible to provide advice for specific fisheries for deep-sea species. Consistent with a precautionary approach, ICES recommends immediate reduction in established deep-sea fisheries unless they can be shown to be sustainable. Measures should also be implemented to reduce exploitation of deep-sea species by fisheries primarily targeting shelf species (hake, anglerfish, and megrim). New deep-sea fisheries or expansion of existing fisheries into new fishing areas should not be permitted unless the expansion is very cautious, and is accompanied by programmes to collect data which allow evaluation of stock status as the basis for determining sustainable exploitation levels", at 82. NEAFC online <http://www.neafc.org/fisheries/docs/acfm_reports-2004/deep-water-general-2004.pdf> (accessed 16 January 2006).

[245] See "NEAFC Management Measures for Deep Sea Species in 2005", NEAFC online <http://www.neafc.org/measures/docs/deep-sea-2005.pdf> (accessed 20 October 2005).

[246] IOTC recommendations enacted under Art. IX.8 are not binding and rely on voluntary compliance by Contracting Parties, while resolutions, if enacted under Art. IX.1, are binding and enforceable.

[247] IOTC Recommendation on the Incidental Mortality of Seabirds. 05/09.

[248] IOTC Recommendation on Sea Turtles. 05/08.

[249] FAO Guidelines to Reduce Sea Turtle Mortality in Fishing Operations, IGIFL online <http://www.intfish.net/docs/2004/faoturtles.pdf> (accessed 20 October 2005).

[250] IOTC Resolution Concerning the Conservation of Sharks Caught in Association with Fisheries Managed by the IOTC. 05/05.

[251] *Convention for the Conservation of Southern Bluefin Tuna*, done at Canberra, Australia,10 May 1993, entered into force 20 May 1994, 1994 ATS 16 (CCSBT).

[252] FAO Fisheries Report No. 778. Compilation of Reports by Regional Fishery Bodies Secretariats on the initiatives taken by their bodies to incorporate ecosystem consideration into fisheries management, at 3–4.

[253] CCSBT, "Report of the Fifth Meeting of the Ecologically Related Species Working Group", Wellington, New Zealand, 2–5 February 2004, para. 14 at 2. Available at <http://www.ccsbt.org/docs/pdf/meeting_reports/ccsbt_10/report_of_erswg5.pdf> (accessed January 2006).

[254] CCSBT, "Report of the Fifth Meeting of the Ecologically Related Species Working Group", Wellington, New Zealand, 2–5 February 2004, at 5, para. 33. Available at <http://www.ccsbt.org/docs/pdf/meeting_reports/ccsbt_10/report_of_erswg5.pdf> (accessed January 2006). "The results of research so far on the use of blue-dyed bait as a mitigation measure to reduce incidental capture of seabirds are encouraging". The objective of dyeing bait blue is to reduce the visibility to seabirds of baited hooks at the waters' surface.

[255] C. Small, at 24.

[256] See CCSBT online <http://www.ccsbt.org> (accessed 20 October 2005), and the mandate of the SRP.

[257] See CCSBT, "CCSBT Scientific Observer Program Standards". See online <http://www.ccsbt.org/docs/pdf/about_the_commission/observer_program_standards.pdf> (accessed 20 October 2005).

[258] See CCSBT Scientific Program. Available at <http://www.ccsbt.org/docs/research.html> (accessed 16 January 2006).

[259] See National Oceanographic and Atmospheric Administration, "Feds Provide Guidance to North Pacific Council to Better Protect Steller Sea Lions in Nation's Largest Fishery", press release, NOAA online <http://www.noaanews.noaa.gov/stories/s92.htm> (accessed 20 October 2005).

[260] Some authors state that SEAFO has not included EAF in the Convention per se. The SEAFO Convention, according to Molenaar, does not incorporate a holistic form of EAF such as that found in the CCAMLR Convention. Eric Molenaar, personal communication, January 2006. See also D.G.M. Miller and E. Molenaar, "The SEAFC Convention: A Comparative Analysis in a Developing Coastal State Perspective" in A. Chircop, S. Coffen-Smout, and M. McConnell, eds., *Ocean Yearbook 20* (Ardsley, N.Y.: Transnational Press, 2006) 305–375. See D. Miller, "Management and Governance Conventions and Protocols — SEAFC, WCPFC and SADC" in R. Shotton, ed., *Deep Sea 2003: Conference on the Governance and Management of Deep-sea Fisheries. Part 1: Conference Reports*. Queenstown, New Zealand, 1–5 December 2003. FAO Fisheries Proceedings No. 3/1 (Rome: FAO, 2005) 596–637.

[261] *Convention on the Conservation and Management of Fishery Resources in the South-East Atlantic Ocean*, Windhoek, Namibia, 20 April 2001, entered into force 30 April 2003. SEAFO online <http://www.seafo.org> (accessed 20 October 2005). The substance of what are now paras. (a), (e), and (f) was also contained in the original coastal-State draft. SEAFC Art. 3, specifies guiding principles for the Contracting Parties (CPs) and makes note of several principles implicating EA. CPs shall "(a) adopt measures, based on the best scientific evidence available, to ensure the long term conservation and sustainable use of the fishery resources to which this Convention applies; (b) apply the precautionary approach in accordance with Article 7; (c) apply the provisions of this Convention relating to fishery resources,

taking due account of the impact of fishing operations on ecologically related species such as seabirds, cetaceans, seals and marine turtles; (d) adopt, where necessary, conservation and management measures for species belonging to the same ecosystem as, or associated with or dependent upon, the harvested fishery resources; (e) ensure that fishery practices and management measures take due account of the need to minimize harmful impacts on living marine resources as a whole; and (f) protect biodiversity in the marine environment".

[262] See Report of Meeting, First Session, Swakopmund, Namibia, 9–13 March 2004, SEAFO online <http://www.seafo.org/reports/Meeting_Report_final_1.pdf> (accessed 20 October 2005). Ecosystem considerations are discussed at 14, among others.

[263] SEAFO, "Report of the SEAFO Scientific Committee", Windhoek, Namibia, 28–30 September 2005, DOC/SC/MEETING/06/2005.

[264] *Convention on the Conservation and Management of Highly Migratory Fish Stocks in the Western and Central Pacific Ocean,* signed 5 September 2000 in Honolulu, Hawaii, in force 19 June 2004.

[265] Ecosystem effects were considered at the preparatory conferences, leading to the adoption of the Convention. See report by MRAG Americas Inc., "Review of Ecosystem-Bycatch Issues for the Western Central Pacific Region", prepared for the Preparatory Conference for the Western and Central Pacific Fisheries Commission (WCPFC), Manila, Philippines, 12–18 November 2002, WCPFC Prep Con/WP.9 at 10.

[266] See the TOR for WCPFC Ecosystem and Bycatch Specialist Working Group as stated by the Chairman at the First Meeting of the Scientific Committee, 12–18 August 2005. WCPFC online <http://www.wcpfc.org/sc1/pdf/EB_SWG_Provisional_TOR.pdf> (accessed 20 October 2005).

[267] See Annex X of the WCPFC Report of the Scientific Committee. WCPFC online <http://www.wcpfc.org/sc1/pdf/EB_SWG_Provisional_TOR.pdf> (accessed 20 October 2005).

[268] All papers and presentations from the Ecosystem and Bycatch Specialist Working Group can be accessed at <http://www.spc.int/oceanfish/Html/WCPFC/SC1/scientific_committee.htm#ecoby> (accessed December 2005).

[269] The FFA, which includes many of the coastal island States participating in WCPFC, held a workshop on ecosystem-based fisheries management (EBFM) in July 2005, one of a series of workshops designed to develop and implement EBFM concept. One of the outcomes will be the production of a practical guide to EBFM for tuna fisheries in the Western and Central Pacific Ocean.

[270] The FFA uses the term ecosystem approach to fisheries management (EAFM).

[271] W. Fletcher, (Draft) *A Guide to Implementing an Ecosystem Approach to Fisheries Management (EAFM) within the Western and Central Pacific Region* (Honiara, Solomon Islands: Forum Fisheries Agency, Draft 2005).

[272] WCPFC Resolution on Non-Target Fish Species (adopted December 2005).

[273] WCPFC Resolution on the Incidental Catch of Seabirds (adopted December 2005).

[274] WCPFC Resolution to Mitigate the Impact of Fishing for Highly Migratory Fish Species on Sea Turtles (adopted December 2005).

[275] The FFA uses the term ecosystem approach to fisheries management (EAFM).

[276] Transform Aqorau, FFA Legal Counsel, Honiara, Solomon Islands, personal communication, January 2006.

[277] Draft of the Southern Indian Ocean Fisheries Agreement, Art. 2, 2004 (copy on file with author).

[278] Fuensanta Maria Candela Castillo, EU, Brussels, personal communication, January 2006.

[279] W. Burke, "Evolution in the Fisheries Provisions of UNCLOS" in N. Ando, E. McWhinney, R. Wolfrum, and B. Röben, eds., *Liber Amicorum Judge Shigeru Oda* (The Hague: Kluwer Law International, 2002) at 1361–1362.

[280] In February 2004, 1136 scientists signed a petition calling for the protection of deep-sea coral ecosystems. The text of the petition as well as the signatories can be seen online <http://www.mcbi.org/DSC_statement/sign.htm> (accessed 20 October 2005).

[281] The Deep Sea Conservation Coalition, its mission, and members as well as member publications are available at <http://www.savethehighseas.org> (accessed 20 October 2005).

[282] Alastair MacFarlane, International Coalition of Fisheries Associations, personal communication, June 2005.

[283] See S. Parsons, "Ecosystem Considerations in Fisheries Management: Theory and Practice", prepared for *Conference on the Governance of High Seas Fisheries and the United Nations Fish Agreement — Moving from Words to Action*, St. John's, N.L., 1–5 May 2005; this paper was subsequently published as "Ecosystem Considerations in Fisheries Management: Theory and Practice" (2005) 20 *International Journal of Marine and Coastal Law* 381–422.

[284] An international network of scientists has joined forces to develop and test properties of ecosystem indicators for fisheries management through the SCOR/IOC Working Group 119 on Quantitative Ecosystem Indicators for Fisheries Management. The general objective of the Working Group is to develop theory to evaluate changes in marine ecosystems (both states and processes) from environmental, ecological, social, and fisheries perspectives. During a period of four years (2001–2004), this Working Group defined generic indicators to be used for marine ecosystems, tested them, and applied them to a variety of ecosystems to evaluate their usefulness. This is expected to lead to new and refreshing insights, which will be of use for defining and evaluating new ecosystem management goals. The results of these efforts have been published in the *ICES Journal of Marine Science*, see P. Cury and V. Christensen, 307–614.

[285] S. Garcia, A. Zerbi, C. Aliaume, T. Do Chi, and G. Lasserre, FAO Fisheries Technical Paper No. 443, at 15.

[286] J. Caddy and K. Cochrane, "A Review of Fisheries Management Past and Present, and some Future Perspectives for the Third Millennium" (2001) 44 *Ocean and Coastal Management* 653–682 at 666.

[287] Professor Wolfrum suggests the failure of UNCLOS to provide for ecosystem management can be remedied by using the reference to "international minimum standards" in Art. 61(3) to import a code of conduct, albeit one that is voluntary, for fisheries management in general. "One cannot but summarize that the commitment of the Convention on the Law of the Sea concerning an ecosystem approach is not strong and does not meet the standards set by subsequent international agreements. The standards can be improved, though, through international minimum standards as referred to in Article 61, para. 3 of the Convention. This does not necessarily require the adoption of another international agreement. The establishment of codes of conduct implemented voluntarily may be quite sufficient. This is where FAO is playing an important role. If such codes of conduct may be considered to reflect 'international minimum standards' they may equally be applied by the International Tribunal for the Law of the Sea". Quoted in W. Burke, "Evolution in the Fisheries Provisions of UNCLOS" in N. Ando, E. McWhinney, R. Wolfrum, and B. Röben, eds., *Liber Amicorum Judge Shigeru Oda* (The Hague: Kluwer Law International, 2002) at 1361.

Chapter 8

Decision-Making*

8.1 Introduction

This chapter examines the decision-making processes associated with the management of straddling fish stocks and highly migratory fish stocks by regional fisheries management organizations (RFMOs) and focuses on the following aspects: the nature of RFMO prescriptive authority, State sovereignty, and the responsibilities of States to cooperate as members of international organizations; how decisions are made by RFMOs (voting, unanimity, consensus); and how States can opt out of RFMO decisions, including efforts made to qualify RFMO objection procedures to ensure that the conservation regime for these stocks is not undermined by unilateral action. It also highlights some of the new principles introduced under the United Nations Fish Stocks Agreement (UNFSA) that affect how decisions are made within RFMOs to ensure that the stocks are managed sustainably and coherently throughout their geographic range, such as the precautionary approach, the ecosystem approach to fisheries management, transparency, and compatibility.[1]

To increase the effectiveness of RFMOs, efforts have been made to strengthen their decision-making processes not only to enhance compliance and implementation, but also to avoid the possibility of disputes. The chapter also discusses RFMO efforts to adopt science-based decisions, preagreed decision rules including harvest controls, and allocation criteria, which are intended to provide for a more predictable, robust, and just system acceptable to all participants.[2]

Despite these improvements in the decision-making process of many RFMOs, disputes are unavoidable. Indeed Chairman Nandan in his concluding statement of August 1995 identified dispute resolution and the provisions introduced under UNFSA as the third pillar of the Agreement. The importance of dispute resolution for improving international fisheries governance is undeniable, and although this thematic area is not examined in detail in this book, it is briefly considered in the present chapter. Article 28 of UNFSA recognizes the link between "efficient and expeditious decision-making procedures" and the peaceful settlement of disputes.[3] Where disputes arise, a fair and timely process is required to reach agreement through peaceful means and to prevent impasses from compromising the conservation and management of the stocks in question and of the associated and dependent species. The lack of a resilient and efficient decision-making process, together with a fair and timely dispute settlement procedure, has often compromised the effectiveness of RFMOs and contributed to overexploitation of target and nontarget stocks.

8.2 State sovereignty and RFMO prescriptive authority

The ability to make management decisions is a defining feature and central task of RFMOs under the contemporary regime for regulating straddling fish stocks and highly migratory fish stocks. Under UNFSA Articles 6, 10(j), and 28, States are obliged to establish collaborative and cooperative arrangements and prevent disputes or impasses by agreeing on timely and effective decision-making procedures. UNFSA does not prescribe any particular approach to decision making, and therefore it rests with States to determine the procedures that will be applied within the context of the different RFMOs and arrangements. Each RFMO has an institutional structure and procedural arrangements to make decisions pursuant to its constitutive instrument. Existing RFMO treaties contain widely varying arrangements for making decisions. These variations in the RFMOs' respective treaties are based on the nature of the decisions to be taken and cover a whole range of voting options or requirements for their members on the one hand, to consensus building on the other.

While RFMOs have the authority to make legally binding decisions, States are careful to circumscribe the conditions and processes through which such decisions become binding on them. This tension between State sovereignty and RFMOs' ability to prescribe norms by adopting decisions has fundamentally shaped the legal character of RFMOs.[4] "All [RFMOs] have institutionalized safeguards for their member countries, either by making decisions by consensus, by making majority decisions subject to objection procedures, or by making

* The author gratefully acknowledges the research assistance of Jay L. Batongbacal, JSD doctoral candidate, Dalhousie Law School, in preparing early drafts of this chapter.

decisions advisory rather than binding on the member countries".[5]

Notwithstanding the formal vote-based procedures set forth in many RFMO constitutive agreements, the practice today in most RFMOs is for States to rely on consensus-based decision making. In reaching consensus, States avoid a vote whereby a minority or dissenting faction would be required to subsume its sovereignty to comply with the majority decision. While consensus can be an effective decision-making procedure and is increasingly used in international fora, consensus-based decision making has often been criticized for leading to protracted decision making, paralysis, or diluted conservation measures.

Another increasingly common practice for facilitating decision making without compromising State sovereignty in many RFMOs is the adoption of advisory or non legally binding decisions to effect change and achieve consensus (such as resolutions of the International Commission for the Conservation of Atlantic Tunas (ICCAT) rather than binding Recommendations). As in the case of soft law instruments discussed in Chapter 2, States are not legally required to comply with these nonbinding decisions of RFMOs. Nonetheless, non legally binding instruments can have a normative influence and affect State practice. For example, trade measures have proven to be an effective tool for improving compliance within the ICCAT management regime, though they were initially adopted as a nonbinding resolution. On the other hand, while non legally binding measures are useful, they must rely on moral suasion and evolving practice rather than on legal obligation to achieve compliance.

The third mechanism that safeguards State sovereignty provided for in many RFMO constitutive agreements is the objection procedure or opting-out provision, which provides dissenting members the option of not being bound by the particular measures. As originally envisaged, the objection procedure draws upon the concept of reservations in treaty law, which permits States, in entering into or acceding to treaties, to exclude the application of certain provisions or to modify their legal effect (subject, of course, to the specific provisions on reservations of the treaty in question).[6] An objection is similar in that it manifests the intention of a State not to be bound by a decision or measure adopted by the RFMO without affecting the original terms of its consent to the constitutive instrument. Both are based on the concept of State consent that is a basic norm, or Grundnorm, of international law. It bears emphasizing that these concepts, reservation and objection, are not exactly the same. Reservations must necessarily result in the severance and exclusion of particular parts of treaties from application to a State. Objections can have effects other than exclusion. Objections can delay the effect of measures for other RFMO members and can have other consequences as well, e.g., delaying the application of measures for other RFMO members may give rise to procedural remedies within the RFMO. This is highlighted by recent developments in RFMOs to ensure that minority interests do not undermine the overall conservation and management regime. The concept of objections in some RFMOs (e.g., the North East Atlantic Fisheries Commission (NEAFC), the South East Atlantic Fisheries Organisation (SEAFO), and the Western and Central Pacific Fisheries Commission (WCPFC)), described below, is evolving into a procedural step in the formulation of management measures, rather than just an instrument for exemption. This is a major procedural innovation that includes limiting the conditions under which an objection can be used; requiring the objecting State to provide justification for the objection; and requiring an alternative unilateral program that does not undermine the conservation regime. By qualifying its use, the objection safeguards State sovereignty while ensuring State responsibility for restoring and sustaining fish stocks and associated ecosystems.

These three procedural elements are founded in sovereignty and the consensual nature of international treaties. At times, they have thwarted the ability of RFMOs to prescribe the necessary norms to effect cooperation in the sustainable and responsible exploitation of the living marine resources under their competence. Similarly, the limited jurisdiction of RFMOs to enforce their decisions against State parties–members also limits the effectiveness of regional organizations. The measure of an RFMO's prescriptive authority or effectiveness of its decisions is in large part determined by the extent of compliance by States with the conservation and management measures. The obligation of member States to comply with an RFMO's conservation and management measures is based on the duty to cooperate and comply with treaty obligations or *pacta sunt servanda.*[7] Further, pursuant to UNFSA, RFMO decisions are also applicable to nonmember States that are Parties to UNFSA. Parties to UNFSA that are not members of an RFMO do not have an obligation to "comply" with RFMO measures, but rather an obligation to "apply" those measures to their own vessels engaged in the fishery in question. This legal distinction respects and underscores flag States' authority over their vessels and their duty to conserve living marine resources.

Noncompliance by member States of an RFMO and by unauthorized participants in a fishery (discussed in Chapters 4 and 9) is a pervasive and pernicious problem. Moreover, the consequences of noncompliance are often lacking, ineffective, or easy to evade. The need for implementing the improved monitoring, control, and surveillance and enforcement regime for flag States, nonflag States, and RFMOs introduced under UNFSA is discussed in Chapter 9.

Another aspect of State sovereignty in international law related to the concept of State consent and willingness to be bound is the well-established principle of *pacta tertiis nec nocent nec prosunt.* States are only bound by treaties to which they are party. Thus UNFSA, strictly speaking, does not apply to non-UNFSA States, and RFMO treaties do not apply to nonmembers. However, as discussed in Chapter 2, all States have a positive legal obligation under the 1982 Law of the Sea Convention (LOSC) to cooperate and to conserve living marine resources. Further, under Article 18 of the Vienna Convention on the Law of Treaties, States that have signed UNFSA but have not yet ratified or acceded to it are restricted from taking actions that undermine the object and purpose of the treaty. Arguably also, several of the basic principles of UNFSA might be considered to have become part of customary international law and are therefore binding even on nonparties to UNFSA. In addition, all States that are party to the Food and Agriculture Organization (FAO) Compliance Agreement who are not party to UNFSA or members of RFMOs are nonetheless legally obligated to ensure that their vessels do not engage in any activity that undermines the effectiveness of RFMO conservation measures.

8.3 Efforts to overcome misuse of State sovereignty

UNFSA addressed the difficulties arising from the possible misuse of these "sovereignty" safeguards (often "temporizing tactics designed to obfuscate opponents"[8]) by emphasizing State responsibility and the need for international minimum standards for fishing. UNFSA considerably strengthened the governance regime for straddling fish stocks and highly migratory fish stocks by incorporating the post-LOSC international minimum standards also found in the form of guidelines in the FAO Code of Conduct for Responsible Fisheries.[9] In so doing, it is arguable that the criteria for RFMO decisions either by consensus or majority vote are therefore subject to these international minimum standards. Similarly, States that invoke objection procedures are not exempt from observing international minimum standards or from complying with RFMO conservation and management measures. Accordingly, member States wishing to safeguard their sovereignty cannot legitimately undermine the overall management regime by taking unilateral action that falls below international minimum standards, or by only agreeing to be bound to decisions reached by consensus that do not meet international minimum conservation standards.

UNFSA establishes a global framework for the adoption of precautionary decisions, minimum conservation standards, compatibility, and dispute resolution. Under the Agreement, "the coastal State continues to have competence to prescribe within the EEZ, while competence to prescribe for high seas fishing outside the EEZ is by international agreement, including that of the flag State.... Both coastal States and high seas fishing States have obligations that burden their authority to act".[10] Thus, States are to implement their duty to cooperate within the "constraints of compatibility and precaution"[11] through the decision-making function of an RFMO. UNFSA establishes a direct link between the obligation to rely on the best scientific information available and the need to take into account the inherent uncertainties and risks associated with fisheries management by adopting the precautionary approach, which is considered "a major procedural innovation".[12]

The ability of RFMOs to address resource sustainability and to manage competing interests is considerably strengthened under UNFSA. The challenge and opportunity remain for States to adopt and implement these provisions within areas under national jurisdiction, as flag States on the high seas and as participants in RFMOs. While an effective conservation regime is in the best interest of all, there may be differing views on how this can be achieved. Reconciling dissenting views and competing interests still confounds and undermines RFMO efforts to achieve responsible fisheries management. As described by Michael Lodge, "One of the key factors which has paralyzed decision-making in most regional fishery bodies to date is insistence on decision-making mechanisms which either require full consensus of the parties or allow some kind of opting-out mechanism or objection procedure. If conservation and management measures are not to be undermined, it is important that effective decision-making procedures are adopted".[13] This invites the question — what is effective decision making both in terms of improving the process of reaching timely, fair, and transparent decisions and arriving at results that meet the conservation objectives required under UNFSA and other post-UNCED international instruments?

8.4 Effective decision making

A central challenge of international fisheries management through RFMOs is institutionalizing a decision-making process that ensures effective decisions are made guided by principles and standards found in UNFSA and related instruments (discussed in Chapter 2), while respecting State sovereignty.[14] Germane to effective[15] decision making is the extent to which decisions are considered to be binding, either because they are legally binding, as in the case of treaty obligations, or are nonetheless complied with even though they are derived from a nonbinding (soft law) instrument. Ted McDorman identifies three other criteria for determining what is meant by "effective" decision making on a formal level: acceptability to RFMO members, timeliness, and a conservation orientation.[16] To these may be added adoption and enforceability of international minimum standards as benchmarks for all fishing activities, transparency, and accountability. In an effort to improve the effectiveness of RFMO decisions and strengthen the decision-making process, there have been calls for RFMO accountability and performance reviews to examine whether States are cooperating through RFMOs in accordance with international minimum standards for conservation and management in keeping with contemporary international fisheries law.[17]

8.4.1 Institutional structure

RFMOs have different institutional structures and the locus of decision making varies among them. All RFMOs with jurisdiction for straddling fish stocks or highly migratory fish stocks (with the exception of the regional arrangements in the Central Bering Sea and proposed in the Southern Indian Ocean) employ a three-tiered structure. At the top is a primary organ, either a Council or a Commission composed of representatives of the parties. These Councils or Commissions are typically supported by a secondary level composed of bodies of a permanent or continuing nature that provide technical and administrative support. The exception is the Bering Sea Donut Hole Arrangement (BSDH), which does not have a formal Commission or Council, nor a Secretariat. Decisions for this arrangement are made at an Annual Conference.[18] The not yet in force Southern Indian Ocean Fisheries Agreement, which will establish the Southern Indian Ocean Fisheries Arrangement (SIOFA), has adopted this unstructured system as well. Finally, at a tertiary level are subcommittees, ad hoc panels, or temporary bodies that are assigned specific tasks or concerns by the RFMO.

It is possible for the locus of decision making, i.e., the burden of the substantive work for making decisions for purposes of conservation and management, to be effectively located at levels other than the Council or Commission. In the case of ICCAT and the Indian Ocean Tuna Commission (IOTC), decisions concerning particular stocks may be adopted by the respective Commissions based on the recommendations of subcommittees.[19] In the Commission for the Conservation of Southern Bluefin Tuna (CCSBT), the Commission and the extended Commission are mandated to establish total allowable catches (TACs) and national allocations on the basis of the report and recommendations of its Scientific Committee[20] (and extended Scientific Committee). The Northwest Atlantic Fisheries Organization (NAFO) is somewhat unique in that the decision-making power of its General Council, the highest-ranking body within the organization, is limited largely to policy issues and administrative decisions.[21] It is the NAFO Fisheries Commission that is tasked with formulating proposals for joint action, such as TACs, allocation of quota, and adoption of other conservation measures, as well as with adopting by consensus the STACTIC Report (the Standing Committee on International Control).[22]

Other secondary bodies are usually those responsible for the finance and administrative affairs of the RFMO, but notably, eight (the Commission for the Conservation of Antarctic Marine Living Resources (CCAMLR), the Inter-American Tropical Tuna Commission (IATTC), ICCAT, IOTC, NAFO, NEAFC, SEAFO, and WCPFC) have established continuing committees or permanent working groups for purposes of overseeing more specialized and continuing issues regarding the implementation, compliance, and enforcement of the measures adopted by the specific RFMOs.[23]

8.4.2 Quorum

One issue common in all decision-making processes is the need for a quorum and what constitutes a quorum. The post-UNFSA emphasis on State participation in RFMOs has resulted in an increase in their members and cooperating nonmembers. The ability of many States to attend meetings and actively participate in working groups and other committees, however, is often constrained by limited financial and human resources. Nonattendance can jeopardize or delay the decision-making process if a quorum is not achieved or a member is not present. Some means have been adopted to overcome this, such as the provision for contacting absent members or parties in the case of IATTC, WCPFC, and NEAFC, and the option of postal votes provided in NEAFC.

8.4.3 The vote-based decision-making process

Vote-based decision making is the traditional process prescribed in most RFMO constitutive agreements. A wide range of voting procedures and requirements exist in the various agreements, tailored specifically to each RFMO's needs. In practice today most States reach decisions through consensus.

Simple majority votes (50% + 1) can be used for decisions that are not considered to be of substance (BSDH, CCAMLR, the General Fisheries Commission for the Mediterranean (GFCM), ICCAT, NEAFC, SEAFO),[24] calling of special meetings (IATTC, SEAFO),[25] and procedural matters (WCPFC).[26] NAFO provides for a simple majority vote for all decisions and recommendations, whether substantive or not, although unanimity is required for an amendment to the NAFO Convention.[27] Other RFMO treaties require a two-thirds majority for matters of substance and for the adoption of amendments.[28] A three-quarters majority is used in three treaties for the adoption of amendments.[29] In the exceptional case of the WCPFC, a qualified and weighted vote, requiring a three-quarters majority vote simultaneously present in two "chambers" based on the geographic location and affiliation (i.e., the South Pacific Forum Fisheries Agency) of the members,[30] is used to decide in cases where consensus cannot be reached. The most stringent standard of a unanimous vote is required for all decisions in IATTC and CCSBT[31] and to invite new entrants into the fishery in BSDH.[32]

The use of vote-based decision making can be advantageous, as the issue is resolved at a specific point in time. By putting a question to a vote, positions of the participants are clarified and polled, and a decision is finally taken. The consensus approach, on the other hand, may involve protracted discussions or intractable parties, which can lead to stalemates and paralysis. However, where the gap between the majority and the minority is relatively small, the decisions may remain a matter of debate. Thus, the vote may not resolve the issue and instead may create contending factions within the organization, or trigger, where available, the use of the objection procedure. For this reason, qualified majorities, such as two-thirds and three-quarters, have been established to ensure that there is broader support for decisions. Still, vote-based decision processes are perceived as politically controversial for certain issues in some RFMOs (particularly as regards allocation decisions), hence the impetus to achieve consensus. Greater support for a proposed decision or measure within an organization is expected to result in better compliance and implementation.

8.4.3.1 Unanimous voting

IATTC's constitutive agreement provides for decisions to be made as the result of a unanimous vote. In recent years, the trend in IATTC has been to rely on consensus, and this is the formulation States adopted in the IATTC Antigua Convention. Although they are similar in many respects, unanimous voting and consensus decision making are distinct decision-making processes. In unanimous voting, each State has a veto over a decision. In consensus-based decision making, no vote is taken and the emphasis is on discussions and the absence of formal objections. There is also discussion prior to a vote, but there is typically extensive discussion using consensus to achieve the increased level of agreement required. Consensus may entail less than unanimity, as long as participants agree that the end result is preferable and everyone can live with the result. Ultimately, though the dynamics may be different, in practice the end result may be the same — in both the consensus and unanimous-voting processes, one member can block an agreement or obstruct the will of the majority. Some systems such as that of the WCPFC, which will be discussed further below, provide for fallback voting procedures where consensus cannot be achieved.

8.4.4 Consensus-based decision making

As noted earlier, with the exception of NEAFC, RFMOs today (including NAFO and IATTC) typically rely on consensus decision making despite the vote-based procedures found in their constitutive agreements. Of the 11 regional management treaties examined concerned with straddling fish stocks or highly migratory fish stocks, seven make express use of consensus as the criterion for a number of substantive decisions, particularly those relating to the setting of TACs and allocations (BSDH, CCAMLR, SEAFO, WCPFC),[33] financial and budgetary matters (CCAMLR, GFCM, IOTC, WCPFC),[34] and fisheries management recommendations (CCSBT).[35]

8.4.4.1 What is consensus?

Consensus decision making tends to involve significant deliberation and discussion that take into account and incorporate various perspectives to reach a collective decision. Consensus indicates the acceptance by the participating States of a decision and their intent to be bound by that decision. The Antigua Convention, a post-UNFSA treaty not yet in force negotiated to modernize IATTC, defines consensus as "the adoption of a decision without the expression of any stated objection".[36] WCPFC defines consensus as "the absence of formal objection made at the time the decision was taken".[37] More weight appears to be given in these definitions to the absence of formal objections rather than to the means of actual deliberation and determination of agreement. In practice, the successful resolution of a problem or a decision reached through consensus typically depends on how many actually agree with, and how many do not agree with but are at least not actively opposed to, the proposed decision or action. Silence may be considered an acquiescence and agreement to a decision.

8.4.4.2 Advantages of consensus-based decision making

Consensus-based decision making is a governance approach ideally suited to multilateral arrangements; it safeguards State sovereignty and is in keeping with the principle of cooperation underlying international law and State responsibility. To resolve contentious viewpoints so that agreement can be reached, consensus building requires the participants to acknowledge, understand, and reconcile differing perspectives and preferences. Consensus therefore can be an effective decision-making procedure that addresses contending concerns and avoids the creation of a marginalized and perhaps noncompliant minority. In theory, by adopting a consensus-based regime, RFMOs increase the likelihood that decisions and actions will be complied with and enforced by the participating parties.

8.4.4.3. Shortcomings of consensus-based decision making

As a procedural device, consensus-based decision making has proven to have certain disadvantages within RFMOs. Reaching consensus takes time, skill, effort, and a willingness to cooperate. Discussions and deliberations are often protracted, and the dissent of one member may result in a prolonged stalemate, preventing a decision from being reached in a timely manner.[38] Given the dynamic temporal and spatial characteristics of fish stocks, management decisions must be made in a timely fashion and not repeatedly deferred to the next Annual Meeting. As observed by Michael Lodge, it is "apparent that many … regional fishery bodies have failed to achieve real progress in addressing issues of sustainability. In particular, the functions of many of them do not adequately reflect the requirements of article 10 of UNFSA. One of the most important of these is article 10(j), which requires regional fishery bodies to agree on decision-making procedures which facilitate the adoption of conservation and management measures in a timely and effective manner".[39] As a result, the failure to reach consensus in a timely manner has compromised the decision-making process as well as the desired outcomes.

The use of consensus to reach decisions within RFMOs has been criticized for being an ineffective approach when the interests of the participants are essentially opposed and therefore not conducive to general agreement.[40] In such instances, without a default voting procedure or timely dispute resolution mechanism, the impasse typically leads to institutional paralysis, conflict, and the tragedy of the commons. Arguably all decision-making processes are vulnerable to such undesirable outcomes if States, in expressing their sovereignty, are unwilling or unable to cooperate, to agree, and to comply with the conservation regime.

Another serious shortcoming of the consensus approach is that the final outcome acceptable to all States may be considerably below the conservation standard required to avoid overfishing, rebuild fish stocks, and rehabilitate critical habitat. As a result, RFMO management decisions are often based on the "lowest common denominator"[41] and do not necessarily meet the international minimum standard for effective management. Due, in part, to this inability to take effective decisions (together with illegal, unreported, and unregulated (IUU) fishing), the exploitation of straddling fish stocks and highly migratory fish stocks has resulted in the overfishing and serious decline of many of these stocks.

An additional disadvantage of the consensus approach is that silence or lack of attendance is usually perceived as an indicator of agreement or acquiescence. This may pose difficulties, especially for developing States, since the reasons for silence may range from language barriers or lack of technical capacity to inability to participate fully within an RFMO. UNFSA prominently recognizes that developing States may have special requirements to be able to participate fully and effectively within the international system for straddling fish stocks and highly migratory fish stocks.[42]

8.4.4.4 Efforts to address the shortcomings of consensus-based decision making

The shortcomings associated with delay, stonewalling, defaulting to the lowest common denominator, and impasse need not be insurmountable obstacles that prevent consensus-based decision making from being an effective governance mechanism for RFMOs. The need for RFMOs and their members to address these shortcomings to ensure effective, responsible decision making for the sustainability of fish stocks has long been recognized. To this end, UNFSA introduced international minimum standards, harvest controls, precautionary reference points, and other science-based decision rules (discussed below) that can serve as the lowest common denominator to ensure the sustainability of the stocks. These standards can act as a default position even for a provisional period until agreement has been reached whether by consensus or an alternative mechanism to break a deadlock or dispute resolution. Additionally, several RFMO treaties (GFCM, IOTC, WCPFC) provide for the use of various levels of majority voting as an alternative should consensus not be reached.[43] Resorting to voting, however, may raise all the issues of noncompliance by those in the minority discussed earlier. Another mechanism provided for in UNFSA to help achieve timely decision making as well as avoid and reduce conflict (and promote agreement) is compulsory and binding procedures for dispute settlement. Effective, fair dispute settlement is crucial if RFMOs are to realize their respective management objectives when consensus cannot be reached within a reasonable period.This is further discussed later in this chapter.

As for the issue of agreement being reached in the absence of all parties, the revisions to IATTC under the Antigua Convention take into account the possibility of non-attendance by a party. The Antigua Convention includes a provision whereby a member who cannot join the consensus on a decision that was taken in its absence can notify the Commission of its position, which suspends the effect of the decision and requires the Commission to again call a meeting to seek consensus.[44] Similar consideration for the rights of absent member States is given by WCPFC.

8.4.5. Objection and opt-out procedures

Objection procedures are part of the decision-making procedures in some RFMOs; they are founded in sovereignty and the consensual nature of international agreements. Objection procedures typically arise in situations where States wish to avoid being outvoted and bound against their will by a measure with which they do not agree. They can be used responsibly and in good faith within the international cooperative arrangement, without undermining the conservation regime.[45] However, objection procedures are now often negatively perceived because of repeated use and abuse in a number of RFMOs over the years.[46] Typically, it is a member's unilateral actions following an objection rather the objection itself that have undermined the conservation efforts and have frustrated management decisions. This is most notably the case in relation to catch-allocation decisions,[47] particularly when States implement unilateral quotas exceeding recommended levels following an objection. UNFSA does not explicitly mention objection procedures but the implication of the obligation of UNFSA Parties to not undermine management measures may be seen as a constraint to the use of such procedures.[48]

The four regional treaties establishing BSDH, CCSBT, IATTC, and WCPFC[49] do not provide for an objection procedure. In three, the number of participating States is quite small and either unanimity (CCSBT[50] and IATTC) or consensus (BSDH)[51] is required for substantive decisions. Requiring consensus or unanimity obviates the need for an objection procedure insofar as all RFMO members have by definition agreed (or at least not objected) to the measure in question.[52] In the case of the WCPFC, this post-UNFSA Convention establishes comprehensive procedures to avoid an opt-out scenario, including a review mechanism, two-chamber voting, and dispute settlement in place of an objection procedure. The other RFMO treaties (creating CCAMLR, GFCM, ICCAT, IOTC, NAFO, NEAFC, and SEAFO) provide for an objection or opting-out procedure that permits any member to withhold consent to be bound by any measures that have been proposed, or that have already been established, for the management of the fishery.[53] It is noteworthy that while an objection procedure exists under the CCAMLR Convention, the full procedure has never been activated and only one technical objection to a Conservation Measure has been made since 1982.[54]

Objections must usually be made within a specified period from receipt of notification of adoption of a measure. Often, any initial objection is also used as a starting point for calculating additional periods within which other State Parties may file their objections. In many cases (GFCM, ICCAT, IOTC, NAFO, NEAFC), should a certain proportion of objections be received, the measure is deemed to have been rejected and will no longer apply.[55] While this is in keeping with the notion of cooperation and the need for consensus or majority approval of an initiative or management strategy, it is also a practical recognition that several objections would most likely render an adopted measure ineffectual.

8.4.5.1 Addressing abuse of objection procedures

To overcome abuse or misuse of the objection procedure, with a view to ensuring that management decisions conform with contemporary international minimum standards, recent RFMO treaties have introduced some procedural innovations. These new procedures are an important development in the effort to strengthen and ensure the integrity of RFMO decision-making processes.

In the WCPFC Convention, which does not have an explicit objection procedure, two anticipatory remedies have been provided that seek to address dissenting views. First, if it appears that there would be opposition to a proposal to be decided upon by consensus, the Commission may appoint a conciliator for the purpose of reconciling the differences among the members.[56] In this instance, disagreements are addressed early in the decision-making process. Second, any member who was absent when a decision was taken, or who voted against the proposal if consensus was not reached, may also seek a review of the decision by a review panel on the grounds either of unjustifiable discrimination against the member concerned or of inconsistency with WCPFC, UNFSA, or LOSC.[57] If the review panel finds in favour of the requesting member, the Commission is bound to modify, amend, or revoke the decision.[58] This provides timely means for resolving differences concerning decisions and for addressing minority concerns.

Innovations have also been proposed to improve State accountability to other RFMO members in making objections to conservation and management measures. In SEAFO, when making an objection, the objecting State is obliged to provide an explanation of its reasons together with proposals for alternative measures it intends to implement. Objections must also be based on specific grounds enumerated in the Convention, including: the Contracting Party considers the decision inconsistent with the SEAFO Convention; the member "cannot practicably comply with the measure"; the measure unjustifiably discriminates against the State; or "other specific circumstances apply". Any SEAFO member may request the Commission to review the measure and may request an ad hoc panel to examine the Commission's decisions and recommend interim measures.[59]

A similar procedure was agreed upon by the Parties to the NEAFC Convention at their 22nd Annual Meeting in November 2004. The NEAFC Parties adopted amendments to their decision-making provisions and agreed to apply such measures provisionally pending ratification by the member States. This amendment requires a party that objects to a binding Recommendation or gives notice of withdrawal of its acceptance to (1) state the reasons for its action and (2) declare its intentions following such action, including alternative conservation and management measures it intends to take or has already taken.[60] This procedural device adds to the transparency of the objection process and seeks to mitigate the impact of objection or withdrawal of acceptance by providing a basis for determining whether the objecting or withdrawing member may thereafter act in a manner that undermines the conservation and management measures that are still in place.[61]

It is important to highlight that associated with these mechanisms introduced by NEAFC, SEAFO, and WCPFC is the reversal of the onus to justify the objection. Opposition or dissent can only be expressed in certain circumstances, and the onus is put on the objecting or opposing State to prove that circumstances for review exist.[62] This process of requiring justification for the dissenting position and presentation of an alternative plan has the further benefit of enabling the member States of an RFMO at an early stage to examine the use of the objection and proposed measure to ensure the conservation regime is not compromised.

It appears, then, that there is a significant trend, particularly in post-UNFSA RFMO treaties, to place limits upon the use of objection procedures and to ensure that the objecting State demonstrates that any proposed postobjection behaviour is consistent with other general obligations of State Parties to UNFSA and to RFMO treaties. Principal among these is the duty to cooperate, which obliges all RFMO members (*a fortiori* those that are also members of UNFSA) to comply with conservation and management measures in their respective convention areas[63] and obliges flag States to ensure that their vessels do not engage in any activity that undermines the effectiveness of subregional and regional conservation measures.[64] This trend to circumscribe the use of the objection procedure is also consistent with the basic principle of treaty law that requires States to perform treaty obligations in good faith. These obligations should serve as additional safeguards against postobjection behaviour[65] that compromises RFMO management of the fisheries. Where the postobjection behaviour presents a risk of serious or irreversible harm thereby invoking the precautionary approach (PA) (see Chapter 6), the obligation rests with the objecting State to demonstrate that postobjection behaviour is consistent with international minimum standards, will not undermine the RFMO conservation regime, and as stipulated in UNFSA Article 7(2) (f) will "not result in harmful impact on the living marine resources as a whole". Formalizing the circumstances and conditions under which a State may invoke the objection procedure, together with the parameters under which subsequent fishing can proceed, is consistent with international efforts to promote rule-based decision making.[66]

8.5. Factors guiding RFMO decisions

8.5.1 Science-based and rule-based decision making

The role of science in decision making is enshrined in LOSC and UNFSA, where decisions are required to take account of the best scientific information available. The UNFSA regime builds upon LOSC and requires science-based and rule-based decision making to avoid overexploitation, rebuild fish stocks, and rehabilitate critical habitats. In establishing international minimum standards for conservation and management, UNFSA highlights the need for RFMOs to make decisions "based on" the best scientific evidence available, apply the precautionary approach, and take into account ecosystem considerations, including protection of marine biodiversity and critical habitats.[67] These three approaches (science-based, precautionary, and ecosystem) are also expressed in the FAO Code of Conduct, the International Plans of Action (IPOAs), the FAO Technical Guidelines for Responsible Fisheries, and the World Summit on Sustainable Development (WSSD).

The enhanced role of science in decision making is a cornerstone of the new regime, especially with the introduction of the PA and the ecosystem approach to fisheries (EAF) over the last decade. As Garcia observes, the scientific community is moving "from supporting conventional fisheries management towards advising for a precautionary approach to ecosystem management for sustainability".[68] Scientists under the new regime are not only to supply information to support fisheries management and resource rehabilitation but are also to inform policy and management decisions by providing a better understanding of the likely consequences of different policy options and to substantiate decisions given uncertainty, incomplete understanding, error, and risk.[69]

8.5.2. Science, precaution, and the ecosystem approach

McDorman describes the relationship between RFMO decision making and scientific information, evidence, advice, or recommendations as one of the central challenges for current RFMO reform and developments in decision making. He points out that the trend in treaty terminology is to formulate that management decisions are to be based on science, often including wording to the effect that the RFMO shall "take into account" or "take full account of" science. These stipulations indicate that science is intended to "inform but not predetermine" the decisions of RFMOs on management measures, and they require management decisions that are defensible or legitimized by existing scientific information. McDorman emphasizes that the challenge is to ensure congruity between management decisions and scientific advice.[70]

Given that "fishery management-oriented science is explicitly linked to decision-making",[71] what is required, as asserted by Levin, "are mechanisms for performing science that will guide society in making its decisions, and for building bridges between science and decision-making".[72] In addition to detailed international minimum standards and the use of the PA, Garcia notes that UNFSA and the FAO Code of Conduct provide for improved science-based decision-making systems in various ways, including "1) adoption of pre-agreed decision rules to apply without delay in case of emergency; 2) adoption of explicit target reference points, with limitation of reference points in relation to which such pre-agreed decision rules would apply; 3) a clear set of objectives; 4) a more active participation of the sector in decision-making; 5) improved transparency; and 6) improved accountability".[73]

The recognition of management failure and of uncertainty in decision making, as well as the need for a more precautionary approach to fisheries, has become evident in the post-UNCLOS III era. The explicit introduction of the legal obligation to apply the PA to the management of straddling fish stocks and highly migratory fish stocks is widely considered a major achievement of UNFSA. This is expressed as a fundamental guiding principle in Article 6 and is also expressed as the development of precautionary reference points in Annex II, guiding decision makers who under Article 7(2) (f) must take measures that would "not result in harmful impact on the living marine resources as a whole". With the introduction of precautionary reference points, maximum sustainable yield (MSY) has shifted from a target to a limit (See Chapter 5).

The PA significantly alters the decision-making and regulatory framework. Paradoxically, now more than ever, there is increased emphasis on science to satisfy the evidentiary requirements that a proposed management option will not result in serious or irreversible harm.

Anticipatory, precautionary science makes recommendations based on the information available. A key element in the PA is that "acknowledging the inevitable limits of knowledge leads to greater humility about the status of the available science, requiring greater care and deliberation in making the ensuing decisions. It also leads to a broadening of appraisals to include more scientific disciplines, more types of information and knowledge, and more constituencies".[74] Scientists and managers must continually adapt to changing conditions and information. "The adoption of robust, transparent and accountable approaches towards various aspects of risk, uncertainty and ignorance can be identified as one crucial means of regaining public confidence in regulatory decision-making".[75]

These PA concepts and principles underpin the EAF (as discussed in Chapter 7). Scientists and managers must take into account ecosystem considerations when examining options and making decisions. The EAF requires additional information, a broader field of inquiry, and a greater number of variables to acquire a better understanding of the biological context and ultimately make more complex management decisions. Given the quantity of information required, much of which has yet to be collected, and the inherent uncertainty and accompanying risks, the PA is an intrinsic component of the EAF.

It is not clear whether the risk-averse analyses and shift in the burden of proof intrinsic to PA have been adopted by States and institutional bodies charged with RFMO decision making. Rather, since UNFSA's adoption, most RFMOs, with the assistance of FAO or the International Council for the Exploration of the Sea (ICES), have examined the definition for PA with a view to implementing precautionary reference points associated with UNFSA Annex II (CCAMLR, GFCM, IATTC, ICCAT, IOTC, NAFO, NEAFC).[76] The PA has been an integral part of the CCAMLR regime since inception, and CCAMLR systematically uses a structured precautionary assessment procedure to determine catch limits.[77] The PA is expressly mandated in three post-UNFSA treaties (IATTC-Antigua (not yet in force), SEAFO's Convention, and WCPFC's Convention).[78] Five treaties (IATTC-Antigua, and the conventions for CCAMLR, CCSBT, SEAFO, and WCPFC)[79] also require application of the EAF, which often complements the PA. Only the Convention establishing BSDH does not contain an express or implied reference to either, but as a framework document, its terms do not preclude scientific and management bodies from adopting these approaches operationally, if and when the moratorium for fishing pollock is lifted. It needs also to be added that a conservation measure adopted by an RFMO can be precautionary in effect and be based on sound ecosystem considerations even though the measure makes no reference to such considerations and even if the constituent treaty of the RFMO makes no reference to these considerations.

8.5.3 Institutional structure to support science-based decision making

Decision making based on the best available scientific information is a fundamental principle of LOSC (Article 61 (2). It is also considered an essential pillar of UNFSA,[80] as reinforced under Annexes I and II. Almost all RFMOs include a form of scientific committee or commission to provide scientific information in support of decision making. The most common model is for the scientific body to be composed of scientists from member States who work autonomously within the organization, e.g., BSDH, CCAMLR, GFCM, ICCAT, IOTC, NAFO, and SEAFO. Other models[81] employed by RFMOs are those of NEAFC, which relies on a separate external organization, ICES, to provide scientific advice[82] (further to the NEAFC–ICES Memorandum of Understanding, ICES provides NEAFC with a full range of advice on all stocks in the convention area, including the stocks managed under the Joint Norwegian–Russian Fisheries Commission in the "Loop Hole"); IATTC, which has a large permanent scientific staff with offices in major fishing ports (interestingly, the Antigua Convention provides for the establishment of a Scientific Advisory Committee to review research programs, stock assessments, research, and recommendations of the permanent scientific staff); and, CCSBT and the WCPFC, which have adopted a multipronged approach, engaging independent (i.e., external) scientific expertise outside of the Commission[83] in addition to their own scientific committees.

CCSBT identified the need for separating the technical evaluation and advisory roles of its Scientific Committee. The Scientific Commitee, which serves as an advisory body to the Commission, has an independent Chair and an independent Advisory Panel. Stock assessment is conducted by the technical subcommittee or working group known as the Stock Assessment Group (SAG). The SAG also has an independent external Chair. The Independent Advisory Panel was created to assist national scientists with stock assessment, to provide support to the process, and to comment on members' papers. If members cannot agree on the science, the Advisory Panel will prepare independent advice. The Scientific Committee considers the advice of the SAG and then advises the Commission. Under the CCSBT model there is no permanent scientific staff. CCSBT has also established an Extended Scientific Committee to allow Taiwan (a fishing entity) to participate in scientific deliberations.

In WCPFC, in addition to member State representatives participating in a Scientific Committee, a small permanent scientific staff and external experts conduct stock assessments and provide scientific advice. However, during the transition period, the South Pacific Commission will be the interim provider of science advice to the Scientific Committee.

A shortcoming with the scientific council–committee model is that only those countries that can afford to send competent staff can participate. Sometimes, the best scientific evidence may not be available. The independence of the scientific advice may also be an issue in this model. In many cases, individual scientists as representatives of their country are constrained from deviating from their delegation's position.[84] The establishment of a permanent and independent scientific capacity within an RFMO can strengthen the scientific institutional capacity as well as ensure equal access to and transparency of scientific information. For example, IATTC's permanent scientific staff has significant research capacity and carries out intensive research activities in its area of competence. The scientific role in CCSBT, ICCAT, IOTC, NAFO, SEAFO, and WCPFC is mostly limited to facilitating and (or) coordinating key activities for member States such as data collection, collation, processing, and dissemination, and stock assessment. However, with the introduction of the PA and EAF, as well as decision rules (discussed below at 8.5.4), many RFMOs are reviewing the structure and composition of their respective scientific bodies to implement these approaches. It has also been suggested that an independent global scientific council should review and examine all RFMO scientific advice.[85]

The variety in the locus of decision making may be an indication of the extent to which scientific information, concern for environmental management, or other concerns dominate the decision-making philosophy within RFMOs. This is also relevant to the issue of transparency, as a clear and publicly perceivable locus of decision making is needed to respond to increasing calls for accountability, including performance reviews as discussed in Chapter 4. Clearly explained and recorded decisions are important given that there are legitimate disagreements over science (e.g., the reliability of the scientific information available, the assumptions used in different models).

As RFMOs begin to increasingly adopt and implement the PA and the EAF, there will be increased efforts to integrate science and management, thereby bridging the gap between those with the data and those with decision-making power. As a result, decisions should be increasingly science based.[86] As Mace states, "The move towards [ecosystem-based management] EBM provides an opportunity for examination of alternative, more workable institutional structures. One of the major issues is the dichotomy between greater separation of science and management vs. greater integration of the two".[87]

8.5.4 Decision rules

RFMOs rely on scientific advice in the making of decisions, but the extent to which the scientific input is adopted by the decision-making organs may be qualified by other "nonscientific" considerations that may or may not be expressed in the constitutive instruments. To strengthen science-based decision making and the relative effectiveness of RFMOs, decision rules have been added to the suite of fisheries management tools. Decision rules can be either quantitative (e.g., setting catch limits for a target species as preagreed fractions of abundance obtained from surveys) or qualitative (e.g., a certain value of an indicator triggers a decision to bring forward a review of management).[88] These decision rules specify what management action should be taken under different conditions, which is often determined by the value of an indicator in relation to a target or reference point. Such rules may range from very general criteria to specific formulae, including reference points and other harvest controls such as those used to determine TAC.

The use of preagreed decision rules is explicitly provided for in UNFSA. Annex II paragraph 4 states:

> Management strategies shall seek to maintain or restore populations of harvested stocks, and where necessary associated or dependent species, at levels consistent with previously agreed precautionary reference points. Such reference points shall be used to trigger pre-agreed conservation and management action. Management strategies shall include *measures* which can be implemented when precautionary reference points are approached. (Emphasis added.)

Accordingly, when these points are approached or exceeded, the corresponding preagreed action or conservation and management measures are triggered. Decision rules are therefore important to ensure that minimum standards and management measures are agreed upon in advance by member States, that the prior consent of member States to be bound is obtained, and that impasse or adoption of nonsustainable fishing practices is avoided.

Specific decision rules are possible and can be as precise as leading to a "default" or alternative decision that may be applied if consensus cannot be reached. One example is found in the Bering Sea Donut Hole Convention with respect to annual harvest levels for pollock. An annex to the agreement provides that where the annual Aleutian Basin pollock biomass is unavailable or is below 1.67×10^6 tonnes (t), the annual harvest level "shall be zero and, therefore, there shall be no directed fishing" on the pollock stock. If the biomass is equal to or above 1.67×10^6 t, the annual harvest level shall be determined in accordance with certain preagreed formulae.[89] This effectively sets a threshold level, expressed in terms of biomass, at which commercial fishing in the Central Bering Sea can recommence. To date this stock has not recovered to this level, and despite pressure to reopen the fishery, the agreed threshold in the Convention remains in force.

Other RFMOs have made use of decision rules in their management strategies. CCAMLR, for example, following the development of the krill yield model in 1990 designed to address uncertainties inherent in krill stock assessment, expanded the model in 1994, adding a three-part decision rule to determine precautionary catch limits used to establish the TAC.[90] More recently CCAMLR has adopted decision rules in managing the Patagonian toothfish fishery.

NAFO first examined the use of decision rules in the form of harvest controls during its initial meetings in 1997 to develop a framework for implementation of the PA.[91] These harvest-control rules are intended to create biomass targets for recovery, thresholds, and limits as well as fishing mortality thresholds and limits. NAFO adopted the framework in 2004 and is currently developing PA reference points for two stocks, on a pilot basis. The decision rules for harvest controls have yet to be used by the Scientific Council to provide advice to the Fisheries Commission. In the wake of the Southern Bluefin Tuna cases, CCSBT has been developing a management procedure to examine how rules about scientific data will be used to determine TACs and quotas.

In general, the use of decision rules can help reduce uncertainty and facilitate the making of decisions, for example, in setting the TAC and in those situations where the normal information requirements for making decisions have not been met. They may link preassigned reference points to certain management actions and can integrate the PA as described in Annex II of UNFSA, which urges the identification of precautionary reference points as aids to management. The development of decision rules may, at the very minimum, serve to provide fallback mechanisms in the case of delays in decision making. While the term decision rules in the fisheries context is generally associated with predetermining biological options, decision rules are also broadly recognized as an effective and transparent mechanism that results in a predetermined conservation-based outcome to which all States in an RFMO agreed.

8.5.5 Allocation criteria and fishing possibilities

The perennial problem confronting many fisheries managers nationally and within RFMOs is setting TACs and allocations. Several RFMO treaties and regulations have rule-based decision-making procedures, most notably for the important function of allocating catch quotas and (or) fishing opportunities.[92] However, the inability of RFMOs to successfully reach agreement on TACs and quotas remains a difficult issue that often results in a resource crisis. The use of decision rules, particularly determining the TACs and perhaps the allocation formulae, is a governance tool that will greatly facilitate this process. Preagreed fast-track dispute resolution or acceptable and substantiated postobjection behaviour would also contribute to ensure that the fishery proceeds in a sustainable manner.

While "UNFSA does not deal specifically with allocation",[93] and there is "*no* specific formulae for the allocation process" offered,[94] there are several related provisions, including Articles 10(b), 10(i), 11(c), and 17(3). Though allocation of participatory rights or fishing opportunities is central to RFMO decision making, this prescriptive process is often not rule based and as a result is often highly complex, contentious, and intractable.[95]

Where jurisdiction to prescribe falls within the authority of a coastal State within its exclusive economic zone (EEZ), determining the TAC and sharing arrangement for quotas is often controversial and can result in short-term political or economic decisions that lead to nonsustainable fishing.[96] Allocation is particularly challenging when there are several parties interested in fishing — it can be difficult to determine a sharing arrangement that will ensure sustainability even when there are only two parties. For example, in the case of the bilateral arrangement between Australia and New Zealand concerning the distribution of fishing opportunities for orange roughy on the South Tasman Rise, allocation was the key issue, and an inability to resolve the matter in a timely, effective manner led to the overexploitation of the stock and eventually to the closure of the fishery. Thus, while the right to fish is subordinate to the obligation to conserve, as Lodge and Nandan observe,

> The problem is that neither UNFSA, nor the decision rules of many existing RFMOs,

> provide mechanisms for allocation that balance conservation interests with the economic and social interests of states. In fact, within many RFMOs, negotiated criteria for catch allocation are often based on the notion of historical catch, which is a powerful incentive to indulge in a race to fish … The problem is exacerbated where there is also a problem of excess capacity, as is the case, for example, in most of the tuna RFMOs at the present time. In such cases, there can easily be pressure to agree [to] allocations in excess of safe biological limits.[97]

Currently, most RFMO treaties are silent or vague on what factors are to be considered in deciding fishing opportunities. Some include, as noted above, historical fishing activities (NAFO, SEAFO, WCPFC),[98] dependency of some national sectors (NAFO, WCPFC),[99] contribution to conservation and management efforts (NAFO, SEAFO, WCPFC),[100] adjacency to the convention area (NAFO, SEAFO),[101] the state of the fishery (SEAFO),[102] and a number of special interests and concerns including those of developing States (IOTC, SEAFO,WCPFC).[103] The Bering Sea Donut Hole Convention, which entered into force only after the once abundant pollock stock had collapsed and a fishing moratorium had been established in the Donut Hole, requires consensus for quota decisions but does not provide a formula for determining allocations. Again, this is an issue that would have to be addressed if the fishery was to resume.

Without clear guidelines or criteria, the determination of allocations is often a bitter behind-the-scenes "horse-trading" exercise (based on political leverage and negotiated exchanges) that can be a less than effective means of deciding how the fish pie is split. This process can be further charged by the decision of a dissenting party to opt out of, or object to, the TAC or allotted quota and then proceed to fish a unilaterally set quota, which may result in sustainable levels being exceeded. This scenario has often resulted in "fish wars" and failure of the co-operative regime to sustain the resource. Member States have rarely sought recourse to resolve an allocation dispute through third-party dispute settlement procedures. A notable example where this was attempted was in the Southern Bluefin Tuna Case, which is considered later in this chapter at 8.6.

To avoid or to resolve allocation issues, some RFMOs (CCAMLR, ICCAT,[104] and IATTC) have used what is sometimes referred to as "Olympic" or "derby" fishery, whereby the TAC is established for a specific species in a particular geographic area, and participating States have authorized access to the area until the TAC has been reached based on the catch reports submitted. Once the predetermined TAC is reached, the fishery is closed for the area.[105] This approach has the advantage of avoiding dissent by not partitioning the TAC. It can also concentrate effort over a short period of time within a confined geographic area. This facilitates monitoring and surveillance. It also raises the rebuttable presumption that vessels fishing in the area after the TAC is reached are engaged in IUU fishing. The disadvantages of this management strategy, also known as "the race to fish", are that it tends to lead to excess capacity, economic inefficiencies, misreporting, and high grading (discarding small or damaged fish); requires a robust monitoring and surveillance system to ensure reporting integrity; and as Balton notes, may force vessels to fish in hazardous weather conditions.[106]

Many RFMOs have in recent years endeavoured to develop criteria for allocating fishing opportunities to both established members and new members to achieve transparency and equity, to manage expectations, and to ensure the conservation of the stocks. The traditional criteria of "adjacency" and "historic performance" created the perverse situation in which new entrants would be induced to fish (without an agreed allocation) prior to joining the RFMO to establish a fishing record, or fishing States with long-established records of substantial historic catches would be perversely rewarded for having overfished in the past, often over and above the TAC in contravention of the RFMO measures.

While recognizing the need for balancing the rights and interests of long-standing member States with the rights and interests of new entrants, the task of allocating and redistributing fishing opportunities is particularly problematic when fisheries have been fully exploited or overexploited. Nonetheless, there is now a generally recognized need for identifying additional criteria that will take into consideration the factors provided in Article 11 of UNFSA. However, the adoption of "non-exhaustive, non-prioritised and non-weighted lists of allocation criteria"[107] by some RFMOs may not facilitate the decision-making process or ability of fishing States and companies to predict their catch allowances. There is a recognized urgent need for clear, predictable criteria for making allocation decisions. Such criteria may facilitate the settlement of allocation disputes.

8.5.5.1 NAFO

In 1997, NAFO established the Working Group on Allocation of Fishing Rights to Contracting Parties of NAFO and Chartering of Vessels between Contracting Parties to examine allocation practices. This group has not yet produced criteria for allocations to Contracting Parties, and the allocation and redistribution of quota remain a major source of contention among most members. In 1999, NAFO addressed the issue of fishing opportunities for new entrants by passing a resolution stipulating that stocks were fully allocated and that fishing opportunities would likely be limited to new fisheries and stocks that had not been allocated.

8.5.5.2 ICCAT

Having experienced a significant increase in membership from distant-water fishing nations[108] as well as interest in participating in the Atlantic tuna fishery from adjacent coastal States,[109] many of whom are new-entrant,[110] developing coastal States without a record of substantial past performance,[111] ICCAT reviewed its practice of allocating quotas based on historical catch for those species subject to ICCAT TAC controls. ICCAT introduced TACs and a sharing arrangement in 1998 for Western Atlantic bluefin tuna and in 1997 for South Atlantic swordfish. Following a period of considerable debate, confrontation, and unilateral quotas, in November 2001 ICCAT adopted Criteria for the Allocation of Fishing Possibilities, comprising 15 factors and categories. "The approach of gradual change to minimize economic disruption was recognized and the guiding concept [became]… 'fair and equitable, ensuring opportunities to all'. However, the Allocation Working Group could not agree on quantitative weightings or formulae linked to these criteria, and it was left to the ICCAT Special Panels to decide how to apply the criteria on a stock-by-stock basis as and when sharing agreements are (re) negotiated".[112] Cooperating Non-Contracting Parties to ICCAT are also entitled to allocations using historical catch as the main criterion. While this allocation-criteria scheme is a demonstrable effort to improve decision making, the ICCAT program is generally viewed as requiring further refinement given the ongoing dissent over the process, the use of the objection procedure, and the inability to reach agreement on quotas for some stocks.[113]

8.5.5.3 NEAFC

In 2003, NEAFC agreed that stocks under its jurisdiction were, and presumably still are, fully allocated and that fishing opportunities for new members would likely to be limited to new fisheries. New Contracting Parties to the NEAFC Convention will be entitled to participate on the same basis as existing Contracting Parties in future allocations of those stocks that are unregulated at the time of their application. Additionally, new Contracting Parties that were previously Cooperating Non-Contracting Parties may request part of any relevant cooperative quota. All allocations in NEAFC are considered on a case-by-case basis. However, the inability to agree on allocations for blue whiting led NEAFC to adopt a fast-track dispute resolution procedure in 2004.

8.5.6 Transparency

Transparency, openness, and participation, three principles embodied in Agenda 21 and the Rio Declaration, are the *sine qua non* of contemporary decision making and good governance. Therefore UNFSA and other treaties[114] include express provisions on transparency.[115] It has been argued that nontransparent decision making, policies, and management bodies, which have excluded nongovernmental organizations (NGOs) and interested citizens from participation, are partly to blame for current problems in fisheries.[116] As a result, "following UNCED, the requirement for transparency in the information and decision-making systems dramatically increased"[117] worldwide. Since 1995, all RFMOs have begun to introduce rules to increase transparency, especially with respect to participation of NGOs and intergovernmental organizations (IGOs). Further, post-UNFSA RFMO treaties include specific provisions on transparency. In the SEAFO Convention, for example, the Commission is required to provide for transparency in all activities of the organization as well as allow or invite observers.[118] The WCPFC Agreement contains two articles devoted to transparency and cooperation with other organizations to foster wider participation in the Commission's activities and decision-making process.[119]

Communicating knowledge and information transparently, effectively, and rapidly is central to EAF and modern governance. Taking advantage of the many benefits of information technology, all RFMOs have established publicly accessible Web sites as the main means for disseminating information on a worldwide basis. However, there is considerable variation in the amount of substantive content in these Web sites; most offer the complete minutes or proceedings of the latest meetings, subcommittee reports, documents, and other outputs. A few (BSDH, CCAMLR, CCSBT, IATTC, NAFO)[120] also allow access to scientific data and reports, either freely or through a restricted Web site with more stringent terms of access and use. This dependence on internet technology may, however, still disadvantage States and entities without the adequate electronic infrastructure. The amount of information available from RFMO Web sites implies that all users have access to the same technology. Otherwise, users must depend on more traditional, more time-consuming, and perhaps more expensive means of acquiring information.

All RFMOs have allowed their meetings to be attended by observers from IGOs, NGOs, and other RFMOs, although such participation is usually limited to giving opening statements, providing materials to participants, and observing the meeting. In a majority of cases, an invitation is required. In some instances, a member may veto participation of any observer. NGO observers are typically excluded from some events during RFMO meetings (such as "heads of delegation" sessions) at which critical decisions are often made.[121] In the case of the Central Bering Sea Donut Hole Agreement, observer rules do not address attendance by NGOs, and only observers from regional economic integration organizations (REIOs) and IGOs may attend meetings of the Annual Conference. Industry representatives and environmental organizations attend meetings of the BSDH as members of a party's delegation.

Meetings of most RFMOs are normally open to participation by Non-Contracting Parties that have vessels fishing in the convention area. Only BSDH, GFCM, and SEAFO do not expressly allow Non-Contracting Parties to attend their meetings, although these RFMOs may deal with such Non-Contracting Parties separately.

However, even with these encouraging developments, the idea of transparency should not be limited to a procedural transparency that allows for broader participation by diverse groups of stakeholders. There is also the matter of substantive transparency, where the focus is upon the clarity and accessibility of the underlying reasons and justifications for decisions. Increasing transparency in a substantive sense means ensuring that stakeholders are afforded access to not only the stated reasons for a decision, but also the information upon which the decision is made.

All RFMO treaties in the post-UNFSA period have adopted provisions to accommodate the procedural aspects of transparency, thereby facilitating the participation of the public, IGOs, NGOs, and other interested parties. The development and implementation of RFMO mechanisms for enhancing substantive transparency are still evolving. An example of a recent procedural change within NAFO that has improved the degree of transparency in decision making is the introduction of a Record of Decisions agreed by the NAFO Fisheries Commission. In 2004, the Executive Secretary of NAFO was asked to serve as the Rapporteur for the Annual General Meeting and to prepare and distribute a list of decisions including the TACs and quota table to all Contracting Parties prior to the close of the meeting. This enabled the Fisheries Commission to circulate and adopt an agreed Record of Decisions.

Transparent decision making is important in all cases, especially when there is a disparity in the levels of power, scientific expertise, or development among the various member States. Both procedural and substantive transparency are required. Otherwise, decision making may be cynically viewed as an empty and ceremonial practice that allows "rubber-stamp" approvals of technical and complex proposals and recommendations. Transparency of RFMO processes enables States to act on equal terms. At the same time, transparency allows non-State representatives of civil society and epistemic communities to contribute to the process and be informed. Sustainable fisheries management is no longer the sole concern of member States or fishing countries or fishing companies, but has become a societal concern worldwide and a fundamental objective of sustainable development. Sustainable fishing remains one of the most pressing environmental issues of our time. Accordingly, "the demand of society for more transparency is growing rapidly, and acceptable solutions must be found".[122] The inclusion of transparency in UNFSA formalizes this fundamental shift in governance and the fisheries management paradigm.

8.5.7 Compatibility

Chairman Nandan in his concluding address to the Conference in August 1995 stated that "One of the cornerstones of the Agreement is the provision to ensure compatibility of conservation and management measures throughout the range of stocks. In this sense, the scope of the Agreement is broad enough to embrace resources as a whole, while fully respecting the different jurisdictional responsibilities. No one should escape from the conservation and management principles of the Agreement".[123] The provisions on compatibility defined under Article 7 of UNFSA seek to reconcile the authority of coastal and flag States over straddling fish stocks and highly migratory fish stocks, and to remedy the inadequate guidance provided by LOSC and customary international law. The challenge, Professor McDorman points out, is deciphering what compatibility actually implies for management and the extent to which a coastal State may expect that measures it adopts in its EEZs are to be reflected in the measures adopted by an RFMO with jurisdiction over the adjacent high seas.[124] The question that prominently arises is, as between areas under national jurisdiction and an adjacent high seas management and conservation regime, which regime should prevail over or influence the other.

8.5.7.1 The 1982 LOSC

As Burke notes, "While the 1982 treaty addresses this [issue] in a general fashion, both for straddling and highly migratory stocks, it does not spell out how the balance between coastal and fishing states is to be worked out, other than by the direction to cooperate".[125] Under LOSC, coastal States have the exclusive authority to prescribe regulations for living marine resources occurring within areas of national jurisdiction; that is to say that coastal States exercise sovereign rights over fishing operations in their EEZ.[126] Outside areas of national jurisdiction the flag State exercises jurisdiction over vessels entitled to fly its flag. Fishing on the high seas is subject to the rights, duties, and interests of coastal States as provided in LOSC Article 116, and to agreement or arrangement. This division of regulatory jurisdiction gives rise to an issue of coordination between the two management regimes — the EEZ and the high seas. RFMO convention areas often cover vast expanses of ocean space that include areas both within and beyond national jurisdiction, but some argue that the regulatory jurisdiction extends only to the areas beyond national jurisdiction unless expressly agreed otherwise by international agreement or by a third-party tribunal. Balton, however, has commented that with respect to highly migratory fish stocks, both in law and practice, RFMO regulatory areas cover both the high seas and the EEZs. TAC, gear restrictions, closed areas, capacity limits, etc. are not restricted to international waters. The situation he explains is different for straddling stock regimes, and the distinction made under LOSC between Articles 63(2) and 64 is maintained in UNFSA Article 7.[127] Others like Molenaar assert that with respect to straddling fish stocks as per Article 63(2) of LOSC, it is a presumption that the regulatory area does not comprise national jurisdiction. He emphasizes that it is important to distinguish between the convention area and the regulatory area, especially in the cases of the NAFO Convention and the NEAFC Convention.[128] In the latter case, a coastal State can give NEAFC competence to deal with that coastal State's maritime zones. In CCAMLR, the situation is reversed, in that although the Convention applies throughout the convention area, coastal States like France and South Africa that have sub-Antarctic islands have the option to ensure that CCAMLR Conservation Measures are not applicable to the maritime zones around those sub-Antarctic islands.

The drafters of LOSC recognized the need for some means of ensuring coherence and consistency between the measures taken by the coastal State within the EEZ and those taken in the high seas area adjacent to the EEZ. This was necessary not only for effective management but also to ensure that the freedom of fishing on the high seas would not undermine the special interests, rights, and duties of coastal States within the EEZ.[129]

To address this, Articles 63(2) and 64 of LOSC lay the basis for two vital principles applicable, respectively, to straddling fish stocks and highly migratory fish stocks. The first, the "consistency" principle embodied in Article 63(2), provides that the coastal State and the State fishing for straddling fish stocks on the high seas "shall seek, either directly or through appropriate subregional or regional organizations, to agree upon the measures necessary for the conservation of *these stocks in the adjacent area*" (emphasis added). The second principle is that of "compatibility", contained in Article 64(1), which obliges States fishing for highly migratory species to "co-operate directly or through appropriate international organizations with a view to ensuring conservation and promoting the objective of optimum utilization of such species *throughout the region, both within and beyond the exclusive economic zone*" (emphasis added).[130] The application of these principles on the high seas must take into consideration Article 116 of LOSC, which qualifies the recognized right of States to engage in fishing in the high seas subject to the "rights and duties as well as interests of coastal States provided for, *inter alia*, in article 63, paragraph 2, and articles 64 to 67".[131]

Article 63(2) has been widely interpreted to mean that the prescriptive competence for these stocks while occurring within the EEZ remains with the coastal State, and that the objective of enacting this provision was to ensure that these straddling fish stocks would be conserved in the adjacent high seas area by the adoption of consistent measures that would not undermine the coastal State's management regime.[132] However, in the case of Article 64(1), the specific obligations to cooperate[133] and establish consistent or compatible measures *through* RFMOs also provide the basis for reasonably arguing that the RFMO could then exercise its own prescriptive competence over highly migratory stocks even though they occurred within an EEZ. Thus, under LOSC there is a subtle distinction,[134] in that while consistency demands that the high seas conservation and management regime must necessarily flow from and be consistent with the regime applicable in the adjacent EEZ, in the case of compatibility for highly migratory fish stocks, the conservation and management regime established through an RFMO flows over the high seas and into the EEZs of adjacent States.

These interpretations have been the subject of some debate since 1982.[135] Some commentators have argued that it is not clear that Articles 63(2) and 64(1) represent different principles, considering that both clauses are reproduced in Article 7 of UNFSA under the heading "compatibility" and that the main distinction between them is the use of different wording ("in the adjacent area" vs. "both within and beyond the exclusive economic zone"). Balton indicates that for more than a decade the authority for RFMOs charged with managing highly migratory fish stocks (i.e., CCSBT, IATTC, ICCAT, and IOTC) to prescribe fishery rules that apply in the EEZ as well as the high seas has become accepted practice. He further opines that while WCPFC is a relatively new organization, it will inevitably follow the same approach to ensure the sustainable management of the highly migratory fish stocks in question throughout their geographic range.[136]

8.5.7.2 UNFSA

Article 7[137] of UNFSA responds to the inadequate assignment of prescriptive jurisdiction for straddling fish stocks and highly migratory species in LOSC. It requires compatibility of measures established on the high seas by RFMOs with those adopted by the coastal State(s) in areas under national jurisdiction to allow for coherent and consistent management throughout the geographic range of the stocks. Cooperation is required for successful conservation and management of straddling fish stocks and highly migratory fish stocks, as by definition, no State has authority over the entire stock throughout its geographic range. UNFSA was intended to address the issue of coherent management through the adoption of the compatibility principle, "whereby measures to be adopted for straddling — *and* highly migratory — fish stocks in respect of both high seas areas and areas under national jurisdiction are to be *compatible*" (emphasis added).[138] The compatibility of conservation and management measures throughout the range of the stocks is a critical principle of UNFSA, which allows it to be broad enough to embrace the resources as a whole while respecting the different jurisdictional responsibilities allocated between States and the RFMOs.[139]

Burke considers the scheme to be "ingenious",[140] as it leaves largely intact the basic LOSC principles governing authority over these stocks within the EEZ and on the high seas. Article 7(1) reflects the language of Articles 63(2) and 64 of LOSC and restates the distinction between straddling fish stocks and highly migratory fish stocks. For straddling fish stocks, the coastal State continues to have competence to prescribe and enforce within its EEZ, while competence to prescribe for high seas fishing outside the EEZ is by international agreement. In the case of highly migratory fish stocks, coastal States continue to have the responsibility for enforcing and prescribing within their EEZs; however, a coastal State's jurisdiction to prescribe is subject to international agreement of the RFMO where the coastal State is a member of the RFMO.[141] Under UNFSA, both coastal States and high seas fishing States "have obligations that burden their authority to act".[142] In keeping with the principles of biological unity and the need for coherent management over the entire migratory range of the fish stocks, the international conservation and management standards under UNFSA Articles 5, 6, and 7, together with Annexes I and II, apply to fisheries *both* within national jurisdiction and beyond. In addition to the improved standards, *no State* (coastal or high seas fishing State) under Article 7(2)(f) shall take measures that will result in any harmful impact on the living marine resources as a whole. And no flag-State initiative or RFMO regulation can undermine the effectiveness of the conservation and management measures for the same stocks adopted and applied by the adjacent coastal State in accordance with Article 61 of LOSC. Thus, the responsibilities of coastal States under LOSC have been "further elaborated and reinforced"[143] by UNFSA, as have the obligations of flag States and RFMOs to ensure that stocks are managed responsibly and coherently throughout their range.

To ensure compatibility of management regimes inside EEZs and on the high seas, conservation and management measures are to be agreed upon within a reasonable period of time by high seas fishing States and coastal States, directly or through RFMOs.[144] UNFSA provides a number of factors that States must take into consideration in arriving at compatible measures, in addition to biological unity of the stocks. These factors include relationships among stock distribution, the geographic particularities of the region, and the extent to which stocks occur and are taken in areas under national jurisdiction (now frequently referred to as the concept of zonal attachment); and the respective dependence of the coastal and high seas fishing States on the particular fish stock.[145]

8.5.7.3 Pre-UNFSA treaties for the conservation and management of straddling fish stocks

Two pre-UNFSA RFMO treaties for straddling fish stocks — NAFO and NEAFC — include the consistency principle, to ensure that coastal-State rights, duties, and interests in managing the resource within the EEZ are not adversely impacted by measures taken on the high seas. For example, in the case of NAFO, in considering proposals for joint action by the States Parties:

> ...the Commission shall seek to ensure consistency between:
>
> a) any proposal that applies to a stock or group of stocks occurring both within the Regulatory Area and within an area under the fisheries jurisdiction of a coastal State, or any proposal that would have an effect through species interrelationships on a stock or group of stocks occurring in whole or in part within an area under the fisheries jurisdiction of a coastal State; and
>
> b) any measures or decisions taken by the coastal State for the management and conservation of that stock or group of stocks with respect to fishing activities conducted within the area under its fisheries jurisdiction.
>
> The appropriate coastal State and the Commission shall accordingly promote the coordination of such proposals, measures and decisions. Each coastal State shall keep the Commission informed of its measures and decisions for the purpose of this Article.[146]

Under the NAFO framework, consistency has been interpreted to mean that measures taken in the regulatory area are to be consistent with, and will not adversely impact, the EEZ regime of the coastal States concerned. The issue of compatibility of conservation measures was first discussed in NAFO in 1999 regarding the management of a straddling fish stock within the EEZ of Canada, where the high seas fishing nations "expressed concern over the fact that the stock had become subject to conflicting conservation and management measures...".[147] Canada responded that the measures adopted within its EEZ were consistent with its rights and obligations and it had not put the sustainability of the stock at risk.

The NEAFC Convention of 1980 similarly requires the Commission to ensure consistency as follows:

> 5.2. The Commission in the exercise of its functions under paragraph 1 shall seek to ensure consistency between: (a) any recommendation that applies to a stock or group of stocks occurring both within an area under the fisheries jurisdiction of a Contracting Party and beyond, or any recommendation that would have an effect through species inter-relationships on a stock or group of stocks occurring in whole or in part within an area under the fisheries jurisdiction of a Contracting Party, and (b) any measures and decisions taken by such Contracting Party for the management and conservation of that stock or group of stocks with respect to fisheries conducted within the area under its fisheries jurisdiction. The appropriate Contracting Party and the Commission shall accordingly promote the coordination of such recommendations, measures and decisions.[148]

NEAFC has adopted a two-tiered system to achieve consistency of measures between the EEZ regimes and that of NEAFC for most straddling fish stocks occurring in its convention area. Within the high seas enclave known as the Loop Hole, which is "officially" under the competence of NEAFC, cod and capelin are managed through the Joint Norwegian-Russian Fisheries Commission as well as by international agreements with Iceland and the European Union (EU). All other species are in principle under the competence of NEAFC. Under this system, the coastal States meet prior to the NEAFC Annual Meeting to establish a global TAC, consisting of a TAC for their EEZs and a TAC for the NEAFC Regulatory Area. The TAC for coastal States must then be allocated through a sharing arrangement. The TAC assigned to NEAFC for distribution among member States is decided at the Annual Meeting. In recent years States have not been able to reach agreement on compatible measures for blue whiting; nonetheless, for 2006 an allocation for blue whiting (but not for Atlanto-Scandian Spring Spawning herring) has been agreed to within NEAFC. A fast-track dispute resolution scheme, as called for in UNFSA, has been established by NEAFC to resolve these matters.

These articles and the practice of these two RFMOs imply that the EEZ regulatory regimes of the coastal State take precedence over measures for the high seas, and that the RFMOs' measures, actions, and activities must not undermine the conservation efforts of the former.

CCAMLR, however, adopted a different approach. The 1980 Statement of the Chairman of CCAMLR[149] indicates that measures adopted by consensus within CCAMLR "also apply to waters adjacent to the islands within the area to which [CCAMLR] applies over which the existence of State sovereignty is recognized by all Contracting Parties".[150] "This would not prevent that [coastal] State from promulgating national measures that were more strict than the Commission's measures or which dealt with other matters".[151] In practice, the concordance of CCAMLR measures primarily applies to Australia, France, South Africa, and the United Kingdom, who, for the most part, exercise jurisdiction on the basis of the Chairman's Statement.[152] While the United Kingdom does not invoke the Chairman's Statement with respect to the South Sandwich and South Georgia Islands, it has implemented conservation measures that are considered compatible. France and Australia also have islands within the Southern Ocean and exercise their coastal-State rights to prescribe who will fish within their respective EEZs. France, however, has set the TAC within its EEZ at levels significantly above those agreed within CCAMLR. Australia, for its part, has declared that if IUU fishing in the CCAMLR area continues, it will eliminate a legal fishery around the Herd and McDonald Islands within five years.

The Central Bering Sea Donut Hole Convention does not mention consistency or compatibility, nor does it mention the need to take into account measures established by coastal States in the adjacent EEZs. However, all decisions at the Annual Meeting are reached by consensus, and provisions for compatibility are mentioned in the appended Record of Discussion. Part B1(d) of the non-binding Record recognizes that the Bering Sea is a large marine ecosystem, and as pollock occur in the EEZs of Russia and the United States as well as on the high seas of the convention area, effective conservation "cannot be ensured unless both of such measures are based upon the best scientific information available and fully compatible with each other". The parties also noted that in adopting conservation measures for pollock within the convention area, all parties should take into account their compatibility with measures adopted within the coastal-State zones as well as their effect on ecologically related species throughout their range.

It is noteworthy that this approach and these formulations of the relationship between coastal and high seas fishing States for the management of straddling fish stocks within the Donut Hole were adopted in 1993 only a few weeks prior to the second negotiating session of UNFSA. The compatibility provisions were a major issue under negotiation at that time.

8.5.7.4 Post-UNFSA treaties for the conservation and management of straddling fish stocks and highly migratory fish stocks

Post-UNFSA RFMO treaties for straddling fish stocks seem to imply a fusion of the consistency and compatibility principles. For example, Article 19 of the constitutive agreement establishing SEAFO calls for the compatibility of measures in a manner consistent with Article 61 and 119 of LOSC[153] to ensure harmonization of measures in the convention area and in areas under national jurisdiction, and does not seem to distinguish between waters within and beyond national jurisdiction. The agreement establishing SIOFA, which is not yet adopted, expressly states that among the objectives of the parties is "to ensure that conservation and management measures for straddling fish stocks in waters under national jurisdiction and measures adopted under this Agreement are compatible".[154]

In the case of the tuna RFMOs, the issue of compatibility is generally interpreted to mean that RFMO conservation and management measures are to be applied within areas under national jurisdiction and on the high seas. For example, ICCAT's regulatory Recommendations (which are binding) are to be adopted and implemented by the Contracting Parties for fishing throughout the Atlantic Ocean, within the EEZs as well as on the high seas.[155] This is intended to provide for compatibility of measures, even though the ICCAT text does not expressly provide for it. The same is true for the adoption and implementation by members of IATTC's binding Resolutions. Article 5 of the IATTC - Antigua Convention provides that all IATTC management measures for tuna and tuna-like species apply equally inside and outside zones of national jurisdiction. Similarly, the CCSBT Agreement provides for the TAC and quota measures for southern bluefin tuna to apply throughout the regulatory area within the EEZs and high seas. IOTC has not formally addressed the issue of compatibility, and in practice no distinction is made between the high seas areas and those under national jurisdiction in the management of the Indian Ocean tuna stocks.

The provisions on compatibility[156] in the WCPFC Convention are the same as those of UNFSA, providing that "conservation and management measures established for the high seas and those adopted for areas under national jurisdiction shall be compatible…". WCPFC emphasizes the need for regional cooperation to ensure the adoption of compatible measures throughout the geographic range of the fish stocks. However, it remains to be determined how compatibility is interpreted and applied in the context of the South Pacific Island States.

It is the extent of coastal-State interests in high seas resources, on the one hand, and the assertion of high seas fishing interests in the exploitation and conservation of stocks under national jurisdiction, on the other, that have been a central source of friction in the development of many RFMO regimes. This issue continues to preoccupy fisheries management.

8.5.7.5 Interpreting compatibility

According to Burke, the decisive standard in determining compatibility of measures pertaining to fish stocks is whether the conservation and management measures of the coastal State within its EEZ are undermined by the measures applied to the high seas.[157] He argues that this is because the premise underlying the provision on compatibility is that "coastal States will continue to have exclusive authority to regulate fisheries within the EEZ, and that international agreement is necessary for regulations outside",[158] and that the compatibility principle expressed in UNFSA indicates that the regulations for the high seas should "be compatible with the substantive regulations adopted and applied by the coastal State for the inside portion of the straddling stock".[159] But then, he also agrees that there are other factors expressly mentioned to be taken into account[160] that might call for adjustments or accommodations in the degree of compatibility with coastal-State regulations.[161]

Orebech et al., on the other hand, argue that although UNFSA does not provide authority for an RFMO to encroach upon the jurisdiction of a coastal State, nothing prevents the geographical competency of an RFMO from being extended to include the EEZ. They perceive two approaches to achieving compatible measures. UNFSA Article 7(1) favours a "bottom-up approach", where the coastal State determines the conservation and management measures for the stocks within the EEZ and thereby essentially establishes the measures for the entire stock considering that comparable measures should be applied in the high seas. This interpretation is consistent with the differential treatment reflected in Articles 63 and 64 of LOSC. Article 7(2), however, favours a "top-down approach", where the members of the RFMO have a collective responsibility for determining the measures for the entire resource. They argue that these provisions allow each RFMO to choose either approach at its discretion.[162]

Oude Elferink further argues that there does not exist *ipso facto* precedence for either measures applicable to areas under national jurisdiction or measures applicable to high seas in determining compatible measures. He notes that the balancing of these measures always depends on the circumstances of the specific case, including the contents of the measures concerned. He emphasizes that the ultimate objective is in "according precedence to measures that ensure sustainable management and conservation over measures that do not".[163]

The compatibility provisions of UNFSA are still viewed as "exceedingly delicate",[164] manifesting "a difference in emphasis if not a spectrum of views".[165] Hayashi notes that these provisions were the result of difficult negotiations, prompted by coastal States who sought to obtain the international community's endorsement of their policy of applying the regimes they adopted for their own EEZs to foreign vessels operating in the adjacent high seas areas.[166] Most coastal States agreed that activities in the high seas should not have an adverse impact on the living resources within coastal-State jurisdiction, but distant-water fishing nations resisted what they saw as a renewed attempt to restrict the freedom of high seas fishing through creeping jurisdiction.[167]

Nonetheless, it is clear that the intention of the drafters of UNFSA was to ensure compatibility between the conservation and management provisions applied within areas of national jurisdiction and those applied on the high seas, in keeping with the need for integration, so that the stocks in question would be managed coherently and consistently throughout their geographic range. Applying the PA and EAF to the management of such stocks would otherwise not be possible. Perhaps, what needs to be emphasized is that ultimately, the drafters of UNFSA intended to set out detailed international minimum standards for the conservation and management of these two types of stocks.[168] Although it is true, particularly in the case of straddling fish stocks, that coastal States tend to be more stringent and protective of their EEZ regimes given that the high seas regime is still considered inadequately regulated and subject to little control, there are also many circumstances in which the coastal-State regime has been found wanting. As a result, the trend toward according prescriptive authority to RFMOs acting upon decisions typically reached through the consensus of States may be changing the way in which coastal States regulate their EEZs and flag States exercise control over their fishing vessels in high seas areas.

The brilliance of Article 7 (and the UNFSA regime) is that it downplays the vexing issue of dominance of prescriptive authority by either the coastal State[169] or the RFMO – flag State[170] that arises from LOSC. Instead it emphasizes that the overall objective of the regime is to be coherent in accordance with detailed minimum standards that are to apply both within and beyond EEZs. Accordingly, it is the responsibility of all States concerned with the particular fishery to cooperate in deciding on a consistent management strategy to restore and sustain the stocks throughout their range based on detailed standards.

Compatibility does not prevent the coastal State or the RFMO from promulgating stricter conservation measures within its area of competence. The question is how is compatibility to be determined so that the exclusive prerogatives of coastal States over their EEZs are not affected. It is arguable that one way to resolve this issue is to regard the minimum standards set out in UNFSA as placing an onus on proponents of the "less stringent regime" to prove that it is implementing these standards in good faith. Thus the burden of proof would rest with the regime or party contesting a regulation or a more stringent standard to demonstrate why noncompatible measures will not result in serious or irreversible harm to the fish stock in question or to the living marine resources as a whole. An RFMO with a less stringent regime than the adjacent coastal State would be required to demonstrate that its management strategy would not adversely impact the effectiveness or undermine the measures adopted by the coastal State for the same stocks in its EEZ. In the case where a coastal State is unable to meet the standards of the adjacent RFMO, particularly in the case of developing coastal States, it would be the obligation of the international community to assist such States to meet their responsibilities for the conservation and management of straddling fish stocks and highly migratory fish stocks to achieve coherent management of fish stocks, as well as to ensure that developing coastal States are able to realize the benefits afforded to them under LOSC and UNFSA. "The Agreement clearly recognizes the needs of developing States and provides for certain forms of cooperation and the provision of special assistance in order to enable those States to fulfil their obligations".[171] The ability of all States to take action at the national level is imperative to give the Agreement practical effect.

Article 7 provides a procedure in the event that States fail to agree on compatible measures. Paragraph (3) requires States to "make every effort" to agree on measures "within a reasonable period of time", and if no agreement is reached within a reasonable time frame, any of the States concerned pursuant to paragraph (4) can bring the matter to compulsory and binding dispute settlement. The Agreement further provides that pending an agreement on compatible conservation and management measures, the States concerned "in a spirit of understanding and cooperation shall make every effort to enter into provisional arrangements of a practical nature".

In proposing this formula for achieving compatibility between conservation and management measures within and outside the EEZ, Chairman Nandan's text stated that "until the dispute settlement process is ended, States shall continue to *observe the provisions herein, and in relevant minimum international standards and otherwise act in a manner consistent with the duties imposed on States under the Convention"*, and where the coastal State or States have adopted measures for the conservation and management of the stock(s), "States fishing on the high seas shall observe conservation and management measures equivalent in effect to the measures applying" in the EEZ. On the other hand, if measures have been agreed in respect of the high seas, in the absence of national measures for the EEZ, the coastal States "shall observe measures equivalent in effect to those agreed in respect of the same stock(s) in the high seas".[172] (emphasis added).

A coastal State party to UNFSA may launch a challenge against any RFMO member also party to UNFSA resulting in compulsory, binding procedures with respect to a measure that undermines measures it has established for the same stock in its EEZ. Pursuant to LOSC Article 297.3, however, a coastal State is not obliged to submit to such procedures in respect of the management of fisheries in its EEZ. This introduces an asymmetry in the application of the compatibility provisions of UNFSA, which results in the inability of members of an RFMO to impose compatibility of any RFMO measures on the adjacent coastal State.

No such barrier exists where a coastal State seeks compatibility of RFMO measures with its own, as it has sovereign rights to introduce measures within its EEZ. It can therefore be argued that Article 7 of UNFSA essentially obliges States fishing on the high seas to agree on measures that are compatible or more stringent than those taken by the coastal State for the same stock or be open to a legal challenge entailing compulsory, binding procedures.[173]

8.5.7.6 Inter-RFMO cooperation and compatibility

Cooperative arrangements and compatible measures are not only necessary between the EEZ regimes and those on the adjacent high seas but also among RFMOs to ensure coherence of measures throughout the geographic range of stocks. Many of the tuna RFMO constitutive agreements expressly provide for inter-RFMO cooperation. Southern bluefin tuna migrates through the convention areas of CCSBT, IATTC, and WCPFC. The IATTC and WCPFC convention areas overlap. There is mixing between the Western Atlantic bluefin stock and the Eastern Atlantic bluefin stock under the competence of ICCAT, and the Eastern Atlantic stock is also found within GFCM. GFCM, in practice, adopts the measures adopted within ICCAT. Straddling fish stocks also are now known to traverse large oceanic regions and convention areas. Oceanic redfish are found in both the NAFO and NEAFC regulatory areas. Horse mackerel, a straddling fish stock unregulated on the high seas of the Southern Pacific, has a migration pattern that extends from Chile and Peru to New Zealand. The initiative to establish a South Pacific RFMO will presumably embrace this distribution pattern. Such an RFMO would need to adopt measures that would be compatible with those of the regional coastal States. Conservation and management of the Patagonian toothfish fishery will require collaboration among CCAMLR, SEAFO, and SIOFA. Compatible measures will need to be established for high seas resources managed by SIOFA and those under the jurisdiction of the adjacent littoral States within the newly created FAO Article VI regional fishery body (SWIOFC). Thus there is a definitive requirement for inter-RFMO and inter-RFB cooperation to ensure the adoption of compatible measures throughout the geographic range of the fish stocks. In an effort to advance inter-RFMO cooperation and explore the possibility of adopting common practices, Japan will host the inaugural meeting of the five tuna RFMOs in Kobe, Japan, in January 2007.

8.6 Dispute settlement

UNFSA takes into account that no matter how elaborate decision-making mechanisms may be, there is always the possibility of dispute. Even consensus-based decision making may lead to a situation where consensus simply cannot be reached, and the participating States may either be left in a deadlock or encouraged to break the impasse by resorting to dispute settlement. For this reason, UNFSA promotes the peaceful settlement of disputes.[174] It reiterates that States have the obligation to prevent disputes through agreement on efficient, expeditious, or stronger decision-making procedures (e.g., decision rules)[175] and to settle by peaceful means any disputes that do arise.[176] These provisions enhance the role and effectiveness of RFMOs and recognize the logic that better decisions result in fewer disputes to be settled elsewhere.[177] A novel provision in UNFSA allows for the informal settlement of technical disputes, which are encouraged to be referred to ad hoc expert panels that shall try to resolve the dispute expeditiously without recourse to binding dispute settlement.[178]

The UNFSA dispute settlement provisions incorporate by reference the procedures for the settlement of disputes set out in Part XV of LOSC. One commentator asserts that the entry into force of UNFSA effectively allows the incorporation of Part XV of LOSC into subregional and regional fisheries management agreements, overriding prior dispute settlement clauses of the latter.[179] Professor McDorman counters that this view is inconsistent with the wording in LOSC and the Sourhern Bluefin Tuna Case.[180]

Obviously, RFMO members wishing to rely on the UNFSA dispute settlement procedures must be Parties to the Agreement, not merely members of the subregional or regional arrangement.[181] SEAFO and NEAFC address this problem with respect to their member States by providing that in case disputes are not referred for resolution or resolved within a reasonable time through the noncompulsory dispute settlement mechanisms provided for, they may be submitted for compulsory settlement entailing binding decisions in accordance with either Part XV of LOSC or Part VIII of UNFSA (where straddling fish stocks are concerned), and the relevant parts of these treaties shall be applied whether or not the parties to the dispute are also parties to the Convention or Agreement.[182] In resolving the dispute, the tribunal is authorized to apply the relevant provisions of LOSC, UNFSA, and compatible "generally accepted standards for the conservation and management of living marine resources and other rules of international law" to ensure the conservation of the fish stocks concerned.[183] For its part, WCPFC incorporates Part VIII of UNFSA by reference and allows for the application of its provisions *mutatis mutandis* to all the member States regardless of whether or not they are also Parties to UNFSA.[184]

UNFSA further recognizes that seeking agreement for decisions or resorting to dispute settlement over issues both take considerable time, during which the stability and effectiveness of the management regime may be placed in jeopardy. To preserve the status quo pending resolution of the dispute, UNFSA encourages States to establish provisional arrangements of a practical nature mutually agreed upon, or in case they are unable to do so, to submit the dispute to a judicial body for the purpose of obtaining provisional measures.[185] The International Tribunal for the Law of the Sea (ITLOS) has exercised this jurisdiction with respect to high seas fisheries management in the Southern Bluefin Tuna Case, a dispute that arose as a result of failure to reach agreement on catch quotas under the CCSBT. ITLOS issued provisional measures pending resolution of the dispute by the Annex VII Tribunal that was subsequently established.[186] The provisional measure issued by ITLOS set the quotas of the parties at their levels in the recent past, on the reason that this was the prudent and cautionary recourse to prevent further deterioration of the tuna resource.[187]

Although it involved the application of provisions of LOSC rather than those of UNFSA, the Southern Bluefin Tuna Case is an important example of resort to international judicial process. This is not without its problems, though. The LOSC Annex VII Tribunal for the Southern Bluefin Tuna Case subsequently ruled that by the terms of the dispute settlement provisions in the CCSBT, the parties were precluded from subjecting their dispute to LOSC Part XV, Section 2 procedures in accordance with LOSC Article 281(1)[188] and from accessing UNFSA dispute settlement procedures.[189] Some commentators have criticized the decision, fearing that the case will become a precedent for subsequent disputes arising under both RFMO agreements and UNFSA and LOSC. Commentators have acknowledged that the Annex VII Tribunal left the door open for a different outcome in subsequent cases, by hinting that the result may have been different had UNFSA already entered into force at the time of the Tribunal's decision.[190] Clearly, RFMO members who wish to maintain recourse to LOSC or UNFSA compulsory procedures entailing binding decisions must take care in drafting RFMO dispute settlement procedures not to fall within the scope of LOSC Article 281(1).

To date, only NEAFC has developed internal procedures for obtaining provisional measures. However, since NEAFC dispute settlement provisions incorporate Part XV of LOSC and Part VIII of UNFSA, provisional measures may also be prescribed pursuant to those provisions. LOSC Part XV provisional measures are also available in SEAFO and WCPFC.

The Central Bering Sea Donut Hole Convention has "default" provisions that serve as provisional measures. These can be triggered when consensus cannot be reached on the annual harvest level, in which case the fallback formulas annexed to the agreement determine the harvest levels.[191] The absence of express provisions for provisional measures does not prevent participants that are parties to either LOSC or UNFSA from resorting to the provisions related to provisional measures in those instruments. One disadvantage of a "default" formula such as this is that while it may serve to preserve the status quo in terms of catch allocation, it is not very responsive to the broader needs of conservation, particularly if stock scarcity characterizes the context of the inability to reach consensus.

With the adoption of amendments providing for dispute settlement procedures in the treaties of two RFMOs (IATTC, NEAFC),[192] only three RFMOs remain without express dispute settlement provisions (BSDH, ICCAT, NAFO). Eight RFMOs (CCAMLR, CCSBT, GFCM, IATTC, IOTC, NEAFC, SEAFO, WCPFC) contain internal dispute resolution provisions,[193] including procedures for the establishment of an arbitration panel supplemented by resort to LOSC or UNFSA procedures.[194] The WCPFC, as described in the section on decision making, additionally contains what may be considered a unique dispute prevention mechanism, through the appointment of conciliators who may reconcile differences between parties over a matter to be put to a vote, if it becomes apparent early on that the measure would be objected to.[195]

8.7 Conclusion

The past decade has seen significant changes in the institutional mechanisms for marine fisheries management. Preexisting RFMOs have sought to meet the expectations of the international community for more effective governance by increasingly adopting the provision of UNFSA and other international instruments, and new RFMOs have been established with improved or innovative decision-making and dispute resolution mechanisms. These changes have highlighted the role of science, precaution, and the ecosystem in the management of marine fisheries resources and are meant to lead to better and more efficient governance of the shared riches of the sea. There appears to be a certain willingness of States within RFMOs to improve the management of straddling fish stocks and highly migratory fish stocks by increasingly incorporating the "words" to modernize regional institutional frameworks and in establishing new agreements; what remains to be seen is whether these words are manifested in the actions of these member States not only in meeting rooms but also at sea and in port.

The current trend in RFMOs is to rely on consensus-based decision making even though many constitutive agreements prescribe voting procedures. Using consensus as a procedural device has its advantages in that it avoids marginalizing a minority viewpoint and can enhance compliance through collective agreement. However, the consensus approach has also been criticized for resulting in a protracted decision process, which often leads to paralysis to the detriment of making timely and effective management decisions. In the post-UNFSA era, RFMOs have endeavoured to address these shortcomings by adopting international minimum standards that serve as the lowest common denominator or benchmark for decisions. These standards and consensual applications are reinforced by preagreed decision rules, precautionary reference points, and criteria for setting TACs and determining quotas. Where consensus cannot be reached in a timely fashion, the process now in some RFMOs defaults to a voting procedure or to a fast-track dispute settlement process.

Objection procedures or opting-out provisions have also been known to frustrate the RFMO decision-making process and render decisions ineffective. To strengthen RFMO decision making while respecting the consensual nature of international relations and sovereignty of individual States, the use of the objection procedures is being qualified in a number of RFMOs. Conditions have been introduced limiting the circumstances under which an objection can be lodged: the objecting State is required to provide reasons or show cause and must demonstrate that post-opting-out behaviour is consistent with or will not undermine the effectiveness of the RFMO conservation measure.

The issue of compatibility is an important element in the cooperative management of straddling fish stocks and highly migratory fish stocks. It remains an often contentious aspect of decision making in some RFMOs that are open to further consideration about its application and interpretation in balancing the rights and interests of States over the range of the stocks. Other thorny and sometimes intractable issues include decisions pertaining to participation in international fisheries and the allocation of the allowed catch. At a time when many stocks have been depleted to historically low levels, there is continued pressure for access to the fisheries, redistribution of fishing opportunities, and increased participation by those with a "real interest". The establishment by all RFMOs of a fair and transparent process with clear criteria could improve the determination of allocations as well as the rules regarding new entrants. These issues require immediate attention and, to some extent, can be addressed by dispute settlement processes.

Dispute settlement is an essential component of good governance and achieving good order of the oceans. The inability to reconcile dissenting views and competing interests as well as to resolve seemingly intractable conflicts often undermines RFMO efforts to achieve sustainable international fisheries. UNFSA explicitly provides for the application of LOSC provisional measures and dispute settlement options to RFMO management disputes, including provisions for compulsory and binding dispute resolution where efforts to resolve the dispute have failed. UNFSA's emphasis on dispute settlement has directly encouraged RFMOs to develop dispute settlement mechanisms where none previously existed. Both UNFSA State Parties and flag States who are not party to the Agreement but are members of RFMOs have demonstrated active interest in the last few years in improving decision making, restricting use of opt-out procedures, and resorting to dispute resolution mechanisms to resolve disagreements. The expansion of dispute settlement modes available to RFMO members, whether they may be technical, procedural, or substantive in nature, is a welcome development and will undoubtedly contribute to the improved effectiveness of RFMO management. By submitting to a timely and fair dispute settlement process, States give effect to their dual duties to cooperate and to conserve and cannot use the inability to agree as a reason for postponing or failing to take conservation and management action.

Many challenges remain in the actual implementation of conservation and management measures worldwide. With civil society's increasing scrutiny and pressure on the fishing industry, States and RFMOs to be transparent and accountable for decisions and operations related to restoring and sustaining fish stocks, we will most likely see more evolutionary developments in the high seas management regime. It must be emphasized that while UNFSA strengthened RFMOs' role and management responsibilities, RFMOs' individual and collective strength and ability to make the best decisions possible and to establish effective management regimes are dependent on the commitment of their respective member States and other fishing interests to cooperate and fulfill their responsibilities both on the high seas and within areas under national jurisdiction.

For conservation and management decisions to be effective, there needs to be compliance with the overarching international legal regime as well as with regional agreements. Noncompliance whether by authorized vessels or by those of nonmembers exacerbates the problem and poses a systemic threat to achieving sustainable fisheries. The next chapter addresses monitoring, control, and surveillance (MCS), and enforcement.

Endnotes

[1] The precautionary approach to fisheries and the ecosystem approach to fisheries are more fully discussed in Chaps. 6 and 7, respectively. Transparency and compatibility are discussed further below in this chapter at 8.5.6 and 8.5.7.

[2] Decision making in RFMOs includes "formal" decision-making and objection procedures, as well as the secondary aspects with respect to directions, factors, considerations, and principles that are involved in the elaboration of specific decisions. See T. McDorman, "Implementing Existing Tools: Turning Words into Actions – Decision-Making Processes of Regional Fisheries Management Organizations (RFMOs)" (2005) 20 *International Journal of Marine and Coastal Law* 423–457 at 426.

[3] T. McDorman, at 440

[4] See S. Yee, "The Responsibility Of States Members Of An International Organization For Its Conduct As A Result Of Membership Or Their Normal Conduct Associated With Membership" in M. Ragazzi, ed., *International Responsibility Today: Essays in Memory of Oscar Schachter* (Leiden: Martinus Nijhoff Publishers, 2005) 435–454.

[5] A. Sydnes, "Regional Fishery Organizations: How and Why Organizational Diversity Matters" (2001) 32 *Ocean Development and International Law* 349–372 at 363.

[6] *Vienna Convention on the Law of Treaties*, done at Vienna, Austria, 23 May 1969, entered into force 27 January 1980, 1155 UNTS 331, Article 2. See also H. Schiffman, "Reservations in Marine Environmental Treaties: Practical Observations and Legal Limitations" (2005) 26 *Whittier Law Review* 1003–1024.

[7] Vienna Convention Art. 26; see also W. Burke, "Compatibility and Precaution in the 1995 Straddling Stock Agreement" in H.N. Scheiber, ed., *Law of the Sea: The Common Heritage and Emerging Challenges* (The Hague: Martinus Nijhoff Publishers, 2000) 105–126 at footnote 15 where Burke states, "Any time a fishing State accepts regulations promulgated pursuant to an international fisheries agreement, it accepts restrictions on its authority to choose, which is the essence of its sovereign rights".

[8] M.S. McDougal referring to misleading usage of doctrines of peaceful coexistence and sovereignty in "Perspectives for an International Law of Human Dignity" in M.S. McDougal et al., eds., *Studies in World Public Order* (New Haven, Conn.: Yale University Press, 1960) at 988. J.E. Nijman, *The Concept of International Legal Personality: An Inquiry Into The History And Theory Of International Law* (The Hague: T.M.C. Asser Press, 2004), writes about Myres McDougal and the theory of international law developed in the 1950s and 1960s at Yale Law School advocating a world public order of human dignity. As Nijman explains, McDougal "rejected the realist conception of law as 'a static body of rigid rules.' He submitted that law was 'not merely a 'limitation' upon power but… also a conscious, creative instrument in promoting both order and other values...' " (at p. 325). She continues that it was McDougal's view that "this planet needs a world public order that will structure and direct worldwide social human interaction — the 'world social process' — and that is actively involved in the pursuit and distribution of resources and values. The concept of 'process' here is used to indicate the continuous interaction among people involved in decision-making, i.e., in 'mak[ing] and enforce[ing] choices that affect the distribution of values among [the participants].' '[W]orld power processes' are those global social processes in which decision-making is controlled by power. *Power*: '…at the highest level of abstraction, …may be most usefully defined as *participation* in the making of decisions that are attended by sanctions.' The possibility of imposing sanctions indicates effective control" (at p. 326). And at p. 330 she continues to explain "In McDougal's view, the traditional doctrine of sovereignty tended to be used by proponents of (Soviet) totalitarianism and of the principle of non-intervention. Sovereignty was a basic but technical concept, which was mostly used by scholars who favour 'coexistence' and the preservation of the contemporary international situation. McDougal considered the doctrine 'peaceful coexistence' *misleading*, because it defined and applied fundamental concepts of traditional international law, *inter alia*, sovereignty, in such a way that they were made to serve 'the purposes not of peaceful cooperation, but rather of a temporizing tactic designed to obfuscate opponents while poising them for ultimate destruction.' Sovereignty was a key concept in traditional theory, and as such considered ill-suited to a theory on the contemporary global situation. A new meaning was required, that was not merely exclusive (like the traditional doctrine prescribed) but also inclusive: in McDougal's opinion, sovereignty also meant *responsibility*". Again quoting McDougal, Nijman continues " 'the concept of "sovereignty"… will be interpreted as embracing only that competence which remains to States after due account is taken of their obligations under international law,' McDougal pointed out, 'and not as a license for freedom from all restraint by shared authority.' " (at pp. 330–331). Nijman further writes that sovereignty in McDougal's view is interpreted as "a balance between the *exclusive* competence of a state over its territory, both formal and effective, and the *internal* competence of the international community, which limits the formal authority of the state in conformity with its obligations under international law" (at p. 331) (footnotes excluded).

[9] By requiring that all decisions pertaining to the RFMO management strategy be consistent with the minimum international standards elaborated under UNFSA (including the application of the precautionary approach, the requirement to set precautionary reference points and preagreed controls (Annex II, para. 4), the recognition of MSY as a limit and not a management target, the use of the best available science, and the consideration of ecosystem impacts and the biological unity of the stocks) member States cannot be permitted to dilute or undermine the overall management regime established by such norms by only agreeing to be bound to decisions reached by consensus or which do not meet minimum international conservation standards.

[10] W. Burke, "Compatibility and Precaution in the 1995 Straddling Stock Agreement" at 111

[11] W. Burke, "Compatibility and Precaution in the 1995 Straddling Stock Agreement" at 116.

[12] W. Burke, "Compatibility and Precaution in the 1995 Straddling Stock Agreement" at 106.

[13] M. Lodge, "Review of Factors of Unsustainability in Fisheries: Relationship to International Fisheries Instruments", Report and Documentation of the International Workshop on Factors of Unsustainability and Overexploitation in Fisheries, Bangkok, Thailand, 4–8 February, 2002, FAO Fisheries Report No. 672 (Rome: FAO, 2002) 91–109 at 105–106.

[14] See also T. McDorman, at 426. He adds that the decision-making process itself must also be one that attracts States to become members of the RFMO.

[15] Commentators have noted that there is a need to underscore the difference between "efficient" decision-making processes, where decisions are made efficiently, from "effective" decision making, where the "desired outcome", i.e., one that is consistent with UNFSA and contemporary international legal framework, is achieved. Ted McDorman, personal communication, January 2006, and Erik Molenaar, personal communication, January 2006.

[16] See T. McDorman, at 425–426.

[17] The Twenty-Sixth Session of COFI (2005) recognized the need for RFMO performance assessments but there was not consensus on conducting independent assessments (rather than by the RFMOs themselves), Report note 76 at para. 112. UNGA Resolution 59/25 paras. 70–71 in relation to paras. 66–71 ensures that performance of States, RFMOs, and arrangements is monitored and contains a commitment by UNGA to assess the adequacy of performance. UNGA Resolution 61/105 para. 73 urges States through their participation in RFMOs to undertake on an urgent basis performance reviews. The High Seas Task Force (HSTF) on IUU fishing directed its Secretariat in March 2005 "to conduct an assessment of the performance of high seas RFMOs against objective criteria (to be developed by the Secretariat) based on the standards established by relevant international agreements". High Seas Task Force, "First Meeting of the High Seas Task Force, Summary of Outcomes", Meeting of the High Seas Task Force, Paris, France, 9 March 2005, HSTF/10, at 4.. HSTF subsequently issued the report "Closing the Net: Stopping Illegal Fishing on the High Seas", Final Report of the Ministerially-Led Task Force on IUU Fishing on the High Seas, 2006, calling for several initiatives including the need for strengthening self-evaluation of RFMOs and encouraging a regular, independent review process of RFMO performance based on a comprehensive "model" of best practices, at 73, 98–99. See also S. Garcia and R.J. Grainger, "Gloom and Doom? The Future of Marine Capture Fisheries" (2005) 360 *Philosophical Transactions of the Royal Society Series B*, published online 29 January 2005, 21–46 at 33, where they state "RFMOs are being held to greater accountability by the international community. The process of sub-regionalization of the RFMOs, aimed at higher relevance and performance is not always supported by adequate resources and coordination".

[18] BSDH established pursuant to the *Convention on the Conservation and Management of Pollock Resources in the Central Bering Sea*, done at Washington, D.C., 16 June 1994, entered into force 8 December 1995, TIAS 11465, at Arts. III and IV; hereinafter referred to as the Central Bering Sea Donut Hole Convention.

[19] *Agreement for the Establishment of the Indian Ocean Tuna Commission*, done at Rome, Italy, 25 November 1993, entered into force 27 March 1996, 1927 UNTS 330; [1995] *OJ L* 236/25, Art. IX.2 (IOTC).

[20] *Convention for the Conservation of Southern Bluefin Tuna*, done at Canberra, Australia, 10 May 1993, entered into force 20 May 1994, 1994 ATS 16, Art 8.3 (CCSBT).

[21] See *Convention on Future Multilateral Cooperation in the Northwest Atlantic Fisheries*, as amended, Ottawa, Ont., 24 October 1978, entered into force 1 January 1979, 1135 UNTS 369, Art. III (NAFO).

[22] NAFO Art. XI.6.

[23] See IATTC online <http://www.iattc.org/HomeENG.htm> (accessed 26 August 2005) and <http://www.iccat.es/organization.htm> (accessed 26 August 2005); see NEAFC online <http://www.neafc.org/about/about_governance.htm> (accessed 26 August 2005); see CCAMLR online <http://www.ccamlr.org/pu/e/gen-intro.htm> (accessed 26 August 2005); see IOTC online <http://www.iotc.org/English/info/comstruct.php> (accessed 26 August 2005); see SEAFO online <http://www.seafo.org/> (accessed 26 August 2005); see WCPFC online <http://www.wcpfc.org/> (accessed 26 August 2005).

[24] *Agreement for the Establishment of a General Fisheries Commission for the Mediterranean*, as amended, done Rome, Italy, 24 September 1949, entered into force 20 February 1952, FAO online <http://www.fao.org/fi/body/rfb/GFCM/gfcm_basic.htm> (accessed 21 October 2005), Art. II.2 (GFCM); *International Convention for the Conservation of Atlantic Tunas*, as amended, done at Rio de Janeiro, Brazil, 4 May 1966, entered into force 21 March 1969, 673 UNTS 63; TIAS 6767; 6 ILM 293, Art. III.3 (ICCAT); NEAFC Art. 5.1; *Convention on the Conservation of Antarctic Marine Living Resources*, done at Canberra, Australia, 20 May 1980, entered into force 7 April 1982, ATS 1982 No. 9; 19 ILM 841, at Art. XII.2 (CCAMLR); Central Bering Sea Donut Hole Convention, Art. V.3; *Convention on the Conservation and Management of Fishery Resources in the South-East Atlantic Ocean*, done at Windhoek, Namibia, 20 April 2001, entered into force 30 April 2003, SEAFO online <http://www.seafo.org/> (accessed 21 October 2005), Art. 17 (SEAFO).

[25] *Convention for the Establishment of an Inter-American Tropical Tuna Commission*, as amended, done at Washington, D.C., 31 May 1949, entered into force 3 March 1950, TIAS 2044, (IATTC); SEAFO Art. 32.2.

[26] WCPFC Art. 20.2.

[27] NAFO Arts. V.2, XIV.2, and XXI.2.

[28] GFCM Arts. V.1 and XII; ICCAT Art. 2.1(b); NEAFC Arts. 5.1, 8.1 (a "qualified majority"), and 9; IOTC Art. IX.1.

[29] NAFO Art. XXI.2; NEAFC Art. 19.2; IOTC Art. XX.1.

[30] WCPFC Art. 20.

[31] IATTC Art. I.8; CCSBT Art. 7. The need for unanimity contributed to the stalemate in CCSBT on the allocation of southern bluefin tuna, which led to the ITLOS case between Australia and New Zealand and Japan.

[32] Central Bering Sea Donut Hole Convention, Arts. XII.5 and XVI.4.

[33] IATTC, under the Antigua Convention (*Convention for the Strengthening of the Inter-American Tropical Tuna Commission established by the 1949 Convention between the United States of America and the Republic of Costa Rica*, "Antigua Convention"), done at Washington, D.C., 14 November 2003, IATTC online <http://www.iattc.org/PDFFiles2/Antigua_Convention_Jun_2003.pdf> (accessed 21 October 2005)), also expressly uses consensus in Arts. IX and XX, and VII.1.l; CCAMLR Art. 12; Central Bering Sea Donut Hole Convention Art. V.2, SEAFO Art. 17.1; WCPFC Art. 10.4.

[34] Again IATTC, under the Antigua Convention Art. IX.3(a), expressly uses consensus; GFCM Art. IX; CCAMLR Arts. XII.1 and XIX.1; IOTC Art. XIII.2; WCPFC Art. 18.1.

[35] CCSBT Art. 9.2(d).

[36] Antigua Convention Art. I.5.

[37] WCPFC Art. 20.1.

[38] See discussion by M. Lodge, "Review of Factors of Unsustainability in Fisheries: Relationship to International Fisheries Instruments" at 105–106.

[39] M. Lodge, "Review of Factors of Unsustainability in Fisheries: Relationship to International Fisheries Instruments" at 105.

[40] See literature on consensus-based decision making that concludes consensus is more likely to be effective where the geopolitical dynamic of the organization is conducive to agreement, i.e., when the States are on good terms with each other and share a commonality of interests and perspectives on the issue, and when there is a high degree of trust and cooperation. Consensus-based decision making is therefore not only a procedural device, but an indicator and instrument of relations between and among States.

[41] T. McDorman, at 429.

[42] See UNFSA Arts. 3.3, 5(b), and 11(f), and Part VII .

[43] GFCM Art. IX; IOTC Art. XIII.2; WCPFC Art. 20.2.

[44] Antigua Convention Art. IX.

[45] See J. Beer-Gabel and V. Lestang, *Les commissions de pêche et leur droit* (Brussels: Université de Bruxelles, 2003).

[46] This device has been particularly vexing in NAFO and ICCAT.

[47] T. McDorman, at 429.

[48] T. McDorman, at 429.

[49] See BSDH, CCSBT, IATTC, WCPFC.

[50] CCSBT Art. 7.

[51] Central Bering Sea Donut Hole Convention Arts. V.2, VII, and VIII.

[52] David Balton, personal communication, May 2007.

[53] CCAMLR Art. IX(c) and (d); GFCM Art. V; ICCAT Art. VIII; IOTC Art. IX; NAFO Art. XII; NEAFC Art. 12; SEAFO Art. 23.

[54] Denzil Miller, personal communication, March 2005.

[55] GFCM, ICCAT, IOTC, NAFO, NEAFC.

[56] WCPFC Art. 20.4.

[57] WCPFC Art. 20.6.

[58] WCPFC Art. 20.9.

[59] SEAFO Art. 23.1.c to 23.1.g. See also T. McDorman, at 452.

[60] NEAFC, "Annex K — Amendment of the Convention on Dispute Settlement", Report of the 23rd Annual Meeting of the North-east Atlantic Fisheries Commission, 8–12 November 2004. NEAFC online <http://www.neafc.org/reports/annual-meeting/docs/full_reports/23neafc_annualmeeting_annexes_2004.pdf> (accessed 25 August 2005) (hereafter NEAFC 23 Annex K).

[61] This formulation was originally proposed by Canada and developed in the NAFO Dispute Settlement Process Working Group but consensus was not reached to adopt this procedure. Nadia Bouffard, personal communication, November 2005.

[62] These concepts of identifying conditions under which an objection can be made and reversing the onus to the opposing State to show cause could also be applied to qualify the use of objection procedures or opting-out clauses in other RFMOs.

[63] UNFSA Art. 10(a)

[64] UNFSA Art. 18.

[65] Rosemary Rayfuse, personal communication, September 2005, suggests that if it is accepted that there is an obligation for Parties to UNFSA to either comply with the measures adopted by RFMOs or to refrain from fishing, when parties object to a conservation and management measure that has been adopted by the RFMO and subsequently refuse to comply with the measure, then arguably the objecting State is required to refrain from engaging in the activity to which the adopted measure refers.

[66] For a general discussion of rule-based decision-making see F. Schauer, "Rule-Based Decision-Making in the Development of Legal Institutions in Political Environments", Lecture Series of the Office of the Prosecutor (The Hague: ICC-OTP, 2005). Available at <http://www.icc-cpi.int/library/organs/otp/Schauer_presentation.pdf> (accessed April 2008).

[67] UNFSA Art. 5(b), (c) and (g).

[68] S.M. Garcia, "Fishery Science and Decision Making: Dire Straights to Sustainability" (2005) 76(2) *Bulletin of Marine Science* 171–196 at 192.

[69] For a discussion on the role of science and decision making see S. Garcia, "Fishery Science and Decision-Making: Dire Straights to Sustainability".

[70] T. McDorman, at 435–436.

[71] S. Garcia, "Fishery Science and Decision Making: Dire Straights to Sustainability", at 174.

[72] S. Levin, "Science and Sustainability" (1993) 3 *Ecological Applications* 550–589.

[73] S. Garcia, "Fishery Science and Decision Making: Dire Straights to Sustainability" at 176. For further discussion on preagreed decision rules see section 8.5.4 and Chap. 5. Improving transparency in scientific and technical information, how analysis is undertaken, and how scientific recommendations are made can help managers clarify the issues, the margins of error, and the risk-averse analyses required.

[74] P. Harreomës, D. Gee, M. MacGarvin, A. Stirling, J. Keys, B. Wynne, and S. Guedes Vaz, eds., *Late Lessons from Early Warnings: The Precautionary Principle 1896–2000*, Environmental Issue Report No. 22 (Copenhagen: European Environmental Agency, 2001) at 169.

[75] P. Harremoës, D. Gee, M. MacGarvin, A. Stirling, J. Keys, B. Wynne, and S. Guedes Vaz, at 170.

[76] See GFCM online <http://www.fao.org/fi/body/rfb/GFCM/gfcm_home.htm> (accessed 26 August 2005); ICCAT online <http://www.iccat.es/> (accessed 26 August 2005); NAFO online <http://www.nafo.ca/About/FRAMES/AbFrMand.html> (accessed 26 August 2005); NEAFC online <http://www.neafc.org/> (accessed 26 August 2005); CCAMLR online <http://www.ccamlr.org/pu/e/gen-intro.htm> (accessed 26 August 2005); IOTC online <http://www.iotc.org/English/index.php> (accessed 26 August 2005).

[77] For futher discussion see Chap. 6, particularly at 6.4.1.1.

[78] IATTC under the Antigua Convention Art. IV; SEAFO Arts. 3(b) and 7; WCPFC, Art. 6. See generally Chap. 6 at 6.4.

[79] CCAMLR Art. II.3(a), (b) and (c); CCSBT Art. 9.2; IATTC (Antigua Convention) Art. VII.1(f); SEAFO Art. 3(c), 3(d), and 3(e); WCPFC Arts. 5(b), 5(d), 5(e), 5(f), 8.2(a), and 10.1(c).

[80] S. Nandan, "Statement of the Chairman Upon the Adoption of the Agreement on the Implementation of the Provisions of the United Nations Convention on the Law of the Sea of 10 December 1982 Relating to the Conservation and Management of Straddling Fish Stocks and Highly Migratory Fish Stocks" in J.-P. Levy and G. Schram, eds., *United Nations Conference on Straddling Fish Stocks and Highly Migratory Fish Stocks: Selected Documents* (The Hague, Boston, London: Martinus Nijhoff Publishers, 1995), 749–754.

[81] It should also be noted that there is also a model whereby the coastal State provides all scientific advice (e.g., the Russian Federation in the "Peanut Hole" Sea of Okhotsk under Art. 16 arrangement of UNFSA for enclaves surrounded by one EEZ).

[82] NEAFC Art. 14.

[83] WCPFC Art. 13.

[84] As Garcia notes, "Government scientists usually cannot express themselves freely. Part of the data they have access to are considered confidential by industry. From a professional ethics point of view, it is probably not considered appropriate to divulge the fact that a manager, faced with a series of options carefully crafted by his scientific analysts and advisers, has chosen an option which is politically reckless". S. Garcia, "Fishery Science and Decision-Making: Dire Straights to Sustainability" at 181 (footnote excluded).

[85] Lee Kimball, personal communications, May 2005. E. Meltzer, International Fisheries Governance: A New Paradigm, paper presented to ICES/SCOR Symposium, Ecosystem Effects of Fishing, March 1999, Montpelier, France, on file with author.

[86] See discussion in S. Garcia, "Fishery Science and Decision-Making: Dire Straights to Sustainability".

[87] P. Mace, "A New Role for MSY in Single-Species and Ecosystem Approaches to Fisheries Stock Assessment and Management" (2001) 2 *Fish and Fisheries* 2–32 at 27.

[88] FAO, "Fisheries Management 2. The Ecosystem Approach to Fisheries", FAO Technical Guideline for Responsible Fisheries No. 4 Suppl. 2 (Rome: FAO, 2003).

[89] Central Bering Sea Donut Hole Convention, at Annex 1, Part 1 (c) and (d).

[90] In the first set of criteria, krill is considered in a single-species context with a threshold biomass to maintain given fishing effort. The second set of criteria takes into account the role of krill in the food chain. The risk that biomass levels drop below predetermined thresholds is evaluated on a 20-year time scale. Depending on model inputs, the lower of the two levels is selected to be included in the calculation of krill yield and subsequent allocation of krill to the fishery. A. Constable, W. de la Mare, D. Agnew, I. Everson, and D. Miller, "Managing Fisheries to Conserve the Antarctic Living Marine Ecosystem: Practical Implementation of the Convention on the Conservation of Antarctic Living Marine Resources (CCAMLR)" (2000) 57 *ICES Journal of Marine Sciences* 778–791.

[91] F. Serchuk, D. Rivard, J. Casey, and R. Mayo, *Report of the Ad Hoc Working Group on the NAFO Scientific Council on the Precautionary Approach*, NAFO SCS Doc. 97/ 12 (Dartmouth, N.S.: NAFO, 1997).

[92] See C. Hedley, E. Molenaar, and A. Oude Elferink, *The Implications of the UN Fish Stocks Agreement (New York, 1995) for Regional Fisheries Organisations and International Fisheries Management* (Brussels: European Parliament, 2003) at 27–36.

[93] M.W. Lodge and S. Nandan, "Some Suggestions Towards Better Implementation of the United Nations Agreement on Straddling Fish Stocks and Highly Migratory Fish Stocks of 1995" (2005) 20 *International Journal of Marine and Coastal Law* 345–379 at 374.

[94] D. Butterworth and A. Penney, "Allocation in high seas fisheries: avoiding meltdown" in A.I.L. Payne, C.M. O'Brien, and S.I. Rogers, eds., *Management of Shared Fish Stocks* (Oxford: Blackwell Publishing, 2004) 165–189 at 166. This paper contains an excellent discussion on the need for RFMOs to address the dangers of fishing nations failing to reach a rapid resolution on setting TACs and allocations.

[95] CCSBT, ICCAT, NAFO, and more recently NEAFC have suffered from an inability to effectively and responsibly determine national allocations. The conflict over setting TACs and (or) allocations and the resulting nonsustainable fishing practices have characterized NAFO since 1986 with the entry of new members to the EU. These problems have led some commentators to suggest that NAFO has been ineffectual since that time. In the following 15 years, the EU consistently set unilateral quotas for its fleets at considerably higher levels than those prescribed by NAFO (12 times higher in 1989). This culminated in 1992 with the establishment of a moratorium on cod fishing in the Canadian EEZ and the 1992 Agreement between Canada and the EU. However, following EU dissatisfaction with its allocation of Greenland halibut in 1995, the EU once again unilaterally set a quota almost six times greater than that agreed by NAFO. In April 1995 Canada and the EU reached an agreement (following the controversial seizing of the Spanish vessel "The Estai") that was subsequently adopted by NAFO, which resolved the cooperative arrangement for sharing Greenland halibut and also contained improved measures for monitoring and enforcement. The issue of Greenland halibut is still contentious, and despite a recovery plan, the stock remains a source of conservation concern. The same is true for 10 other species, including cod, which have been under moratoria since 1992 and have not recovered to a point that would permit the fishery to reopen. For a discussion of the history and inadequacies with the decision-making and dispute resolution processes in NAFO see P.G.G. Davies, "The EC/Canadian Fisheries Dispute in the Northwest Atlantic" (1995) 44 *International and Comparative Law Quarterly* 927–939.

[96] It is also understood that to have sustainable fishing, decisions regarding TAC (and quota allocation) must be responsible and sustainable, and vessels must comply with the conservation measures if they are to be effective. In the end, it is the acceptance and compliance with conservation and management measures by vessels and the ability and willingness of States to regulate and control these vessels and their masters that will result in sustainable fishing.

[97] M.W. Lodge and S. Nandan, "Some Suggestions Towards Better Implementation of the United Nations Agreement on Straddling Fish Stocks and Highly Migratory Fish Stocks of 1995" at 374 (footnotes excluded).

[98] NAFO Art. XI.4; SEAFO Art. 20.1(b); WCPFC Arts. 5(b), 10.3(b), and 10.3(c).

[99] NAFO Art. XI.4; WCPFC Arts. 8.2(b) and 10.3(g).

[100] NAFO Art. XI.4; SEAFO Art. 20.1(e); WCPFC Art. 10.3(e).

[101] NAFO Art. XI.4; SEAFO Art. 20.1(d).

[102] SEAFO Art. 20.1(c).

[103] The WCPFC Convention contains an extensive list of other possible criteria or factors that may be taken into account in making decisions about allocation, such as special requirements of developing States and small-island developing States (Arts. 5.b, 10.3.d, and 10.3.j), the interests of artisanal and subsistence fishers (Art. 5.h), the record of compliance with conservation measures (Art. 10.3.f), special circumstances regarding the coastal State's EEZ (Art. 10.3.h), and geographic factors (Art. 10.3.i). IOTC Art. V.2 (b) and (d) and SEAFO Art. 20.1(d) provide for consideration of interests of developing coastal States.

[104] In 2001, ICCAT agreed on a TAC for southern albacore tuna but when agreement could not be reached on national quotas using the 2001 Allocation Criteria, ICCAT resorted to an Olympic fishery (see ICCAT Recommendation 01-02). Again in 2002, a TAC was established but no agreement could be reached on national quotas for southern albacore, so ICCAT members agreed to a binding Recommendation for a "sharing mechanism". This mechanism is a variant of an Olympic fishery whereby there is a competitive fishery (i.e., without allocation) after the TAC is set until 80% of the TAC has been caught. At that point the parties are notified to enter into multilateral discussions to decide on the allocation of the remaining 20% of the TAC.

[105] See C. Hedley, E. Molenaar, and A. Oude Elferink, at 30.

[106] David Balton, personal communication, May 2007.

[107] C. Hedley, E. Molenaar, and A. Oude Elferink, at 34.

[108] As noted earlier, in keeping with the requirement under UNFSA that all States fishing in the convention area have a duty to cooperate and conserve straddling fish stocks and highly migratory fish stocks so as not to undermine the management regime of the relevant RFMO, many RFMOs have encouraged the participation of nonmembers and fishing entities fishing in their area of competence. This, combined with the use of registers and positive and negative lists by many RFMOs, including ICCAT, has resulted in vessels flagged in States not authorized to fish in the convention area being classified as IUU.

[109] The Icelandic position on membership of ICCAT is of some note in the context of the question of potential new membership rights. Iceland had, in fact, indicated to ICCAT that it was considering membership, although the extent of tuna or other highly migratory stocks in the Icelandic EEZ was uncertain. However, Iceland eventually decided against membership on the grounds that its low track record would be used by ICCAT against it in setting quotas. R. Barston, "The Law of the Sea and Regional Fisheries Organisations" (1999) 14(3) *International Journal of Marine and Coastal Law* 333–352.

[110] See Art. 11 UNFSA.

[111] The need for introducing a TAC and sharing arrangement for South Atlantic swordfish stocks in 1996 precipitated the call to develop a new allocation scheme within ICCAT. Brazil called for new criteria that would be consistent with Art. 11 of UNFSA rather than using solely past performance as the basis for developing the sharing arrangement. The following year (1997) Brazil hosted an intersessional meeting of ICCAT Panel 4, which is responsible for swordfish management recommendations, at which time Brazil reiterated its position that UNFSA Art. 11 factors should also be considered. However, no agreement was reached and from 1998 to 2000 the allocation or sharing arrangement for South Atlantic swordfish was negotiated in camera by Heads of Delegation. This resulted in no allocations for some Southwest Atlantic coastal States who had small historic catches. Led by Brazil, these Sates considered that the status quo was unacceptable and successfully advocated a complete review of the ICCAT quota-allocation processes. This effort resulted in the 1998 establishment of the ICCAT Ad Hoc Working Group on Allocation Criteria mandated to develop recommendations on revised ICCAT allocation procedures. The Allocation Working Group met four times prior to the 2001 ICCAT Commission meeting. Brazil, Venezuela, and South Africa led the "Group of 16" primarily developing coastal States, who proposed that UNFSA Art. 11 be taken into account in redistribution and allocation. The EU and Japan, on the other hand, asserted the position that any new allocation process could not apply to stocks that had already been allocated (i.e., bluefin and swordfish) and that historic catches remain the primary criterion for new allocations.

[112] D. Butterworth and A. Penney, at 173, including quote from ICCAT Working Group on Allocation of Fishing Opportunities presented in Annex 8 of 2001 ICCAT Annual Report, which resulted in the 2001 ICCAT Criteria for Allocation of Fishing Opportunities.

[113] D. Butterworth and A. Penney, at 176, observe that ICCAT's adoption of a broad suite of nonquantified criteria is of little help.

[114] For a discussion on the implementation of transparency in RFMOs, UNFSA, CBD, and the Climate Change Convention see G. Wiser, "Transparency in 21st Century Fisheries Management: Options for Public Participation to Enhance Conservation and Management of International Fish Stocks" (2001) 4 *Journal of International Wildlife Law and Policy* 95–129.

[115] J. Swan, "Decision Making in Regional Fisheries Bodies and Arrangements: the Evolving Role of RFBs and International Agreement on Decision Making Processes", FAO Fisheries Circular No. 995 (Rome: FAO, 2004).

[116] G. Hewison, "The Precautionary Approach to Fisheries Management: An Environmental Perspective" (1996) 11 *International Journal of Marine and Coastal Law* 301–332 at 329.

[117] S. Garcia and D. Staples, "Sustainability Reference Systems and Indicators for Responsible Marine Capture Fisheries: A Review of Concepts and Elements for a Set of Guidelines" (2000) 51(5) *Marine and Freshwater Research* 385–426 at 418.

[118] SEAFO Art. 8.8, 8.9, and 8.10.

[119] WCPFC Arts. 21 and 22.

[120] See IATTC online; NAFO online; CCAMLR online; CCSBT online <http://www.ccsbt.org/> (accessed 26 August 2005); Alaskan Fishery Science Center online <http://www.afsc.noaa.gov/refm/cbs/> (accessed 26 August 2005).

[121] David Balton, personal communication, May 2007.

[122] S. Garcia, "Fishery Science and Decision Making: Dire Straights to Sustainability" at 181.

[123] S. Nandan, "Statement of the Chairman Upon the Adoption of the Agreement on the Implementation of the Provisions of the United Nations Convention on the Law of the Sea of 10 December 1982 Relating to the Conservation and Management of Straddling Fish Stocks and Highly Migratory Fish Stocks" at 751.

[124] T. McDorman, at 437–438

[125] W. Burke, "Compatibility and Precaution in the 1995 Straddling Stock Agreement" at 108.

[126] In some instances under the LOSC regime, the coastal State is obliged to cooperate with other States prior to taking action, but the final decision rests solely on the coastal State. See Arts. 62–67.

[127] Ambassador David Balton, personal communication, May 2007.

[128] Dr. Erik Jaap Molenaar, Senior Research Associate, Netherlands Institute of Law of the Sea, Utrecht, personal communication, December 2005.

[129] A. Tahindro, "Conservation and Management of Transboundary Fish Stocks: Comments in Light of the Adoption of the 1995 Agreement for the Conservation and Management of Straddling Fish Stocks and Highly Migratory Fish Stocks" (1997) 28 *Ocean Development and International Law* 1–58 at 15.

[130] LOSC Art. 64(1).

[131] LOSC Art. 116(b).

[132] Coastal States generally take the view that they have a special or superior interest in the stocks beyond their EEZ given the adverse impact of uncontrolled high seas fishing on domestic management efforts and coastal fishing communities. This view is protected and reinforced by the recognition of coastal States' interests in their EEZs and adjacent high seas as indicated in Arts. 63, 64, and 116 of LOSC. As noted in the commentary in M. Nordquist, N. Grandy, S. Nandan, and S. Rosenne, eds., *United Nations Convention on the Law of the Sea 1982: A Commentary, Vol II* (Leiden: Martinus Nijhoff Publishers, 1993) at 644–646, during the last months of negotiating the 1982 LOSC, coastal States proposed a number of provisions that would have accorded a certain precedence of measures adopted by coastal States to regulate straddling fish stocks in their EEZ over measures for the same stock adopted on the high seas by RFMOs. But since none of these were adopted, coastal States were left with what is in the Convention and were required to implement these measures in some reasonable manner.

[133] Art. 63(2) points to the obligation "to seek to agree", and Art. 118 calls on States to "cooperate with each other". States are enjoined to cooperate for the management of Art. 64 stocks.

[134] Professors Miles and Burke note that "[i]f Article 116 is to be effective, it may need to be interpreted to authorize the coastal State to secure its superior right by prescribing conservation measures with which high seas fishing states are obliged to comply. The rights of the coastal state, expressly made superior to the high seas fishing state, would otherwise be empty and the high seas state would have no meaningful obligation different from its obligation to any other state". E. Miles and W. Burke, "Pressures on the United Nations Convention on the Law of the Sea of 1982 Arising from New Fisheries Conflicts: The Problem of Straddling Stocks" (1989) 20 *Ocean Development and International Law* 343–362 at 352. High seas fishing nations have rejected the argument that coastal States were accorded special rights with respect to straddling fish stocks or highly migratory fish stocks.

[135] High seas nations generally perceive the coastal States' position as creeping jurisdiction and a further effort to enclose the seas and diminish flag-State freedom. In recent years, the position has also emerged that overfishing within EEZs will adversely impact the high seas fishery.

[136] Ambassador David Balton, personal communications, May 2007.

[137] For further analysis of Art. 7 of the UNFSA, see A. Oude Elferink, "The Determination of Compatible Conservation and Management Measures for Straddling Fish Stocks and Highly Migratory Fish Stocks" (2001) 5 *Max Planck Yearbook of United Nations Law* 551–607.

[138] R. Rayfuse, *Non-Flag State Enforcement in High Seas Fisheries* at 206.

[139] S. Nandan, "Statement of the Chairman Upon the Adoption of the Agreement on the Implementation of the Provisions of the United Nations Convention on the Law of the Sea of 10 December 1982 Relating to the Conservation and Management of Straddling Fish Stocks and Highly Migratory Fish Stocks" at 751. Also, as Chairman Nandan reported in 1995 to the American Society of International Law, the formulation of Art. 7 was one of the critical issues of the UNFSA negotiations, and the formula expressed the agreement of States that the approach to the two areas ought to be coherent and consistent, but within the rights and duties of States under the 1982 LOSC, so that the exclusive prerogatives of coastal States over their EEZ were not affected. S. Nandan, "Conservation and Management of Straddling Fish Stocks and Highly Migratory Fish Stocks under the Law of the Sea Convention" (1995) 89 *American Society of International Law Proceedings* 452–455 at 452.

[140] W. Burke, "Compatibility and Precaution in the 1995 Straddling Stock Agreement" at 111.

[141] Until 1990 the US interpretation of Art. 64 of LOSC was that coastal States did not enjoy exclusive management authority over tuna within their EEZs. For further discussion, see W. Burke, "Highly Migratory Species in the New Law of the Sea" (1984) 14 *Ocean Development and International Law* 273–314.

[142] W. Burke, "Compatibility and Precaution in the 1995 Straddling Stock Agreement" at 111.

[143] S. Nandan, "Statement of the Chairman Upon the Adoption of the Agreement on the Implementation of the Provisions of the United Nations Convention on the Law of the Sea of 10 December 1982 Relating to the Conservation and Management of Straddling Fish Stocks and Highly Migratory Fish Stocks" at 751.

[144] UNFSA Art. 7.3; UNFSA Art 7(2)(e).

[145] UNFSA Art. 7(2)(d) and (e)

[146] NAFO Art. XI.3.

[147] C. Hedley, E. Molenaar, and A. Oude Elferink, *The Implications of the UN Fish Stocks Agreement (New York, 1995) for Regional Fisheries Organizations and International Fisheries Management* at 18.

[148] NEAFC Art. 5.2.

[149] Statement by the Chairman of the Conference on the Conservation of Antarctic Marine Living Resources. This statement appears as an annex to CCAMLR. Full text available at CCAMLR online <http://www.ccamlr.org/pu/e/e_pubs/bd/pt1.pdf> (accessed 26 August 2005).

[150] See Statement by the Chairman of the Conference on the Conservation of Antarctic Marine Living Resources, para 5. In the absence of consensus, the coastal State can introduce any measure it deems appropriate.

[151] C. Hedley, E. Molenaar, and A. Oude Elferink, *The Implications of the UN Fish Stocks Agreement (New York, 1995) for Regional Fisheries Organizations and International Fisheries Management* at 18.

[152] For the last three years, there has been a debate within CCAMLR on whether certain countries (notably France) are rigorously implementing seabird migration measures within CCAMLR.

[153] SEAFO Art. 19.2.

[154] Draft SIOFA Art. 6.1(g) (copy on file with author).

[155] The position of Iceland with respect to ICCAT illustrates the need for clarity on the meaning of compatible measures elaborated in UNFSA. Iceland was not a member of ICCAT for many years; it wished to exert its sovereign rights over the part of the resource found within its EEZ to demonstrate a historical catch record prior to joining ICCAT, as it was concerned that it would otherwise be allocated an inappropriate quota to fish tuna within its EEZ.

[156] WCPFC Art. 8.1.

[157] W. Burke, "Compatibility and Precaution in the 1995 Straddling Stock Agreement" at 114.

[158] W. Burke, "Compatibility and Precaution in the 1995 Straddling Stock Agreement" at 114.

[159] W. Burke, "Compatibility and Precaution in the 1995 Straddling Stock Agreement" at 114.

[160] Burke identifies, for example, Art. 7(2)(d), which mentions the biological unity of the stocks; other biological characteristics; relationships among stock distribution, the geographic particularities of the region, and the extent to which stocks occur; Art. 7(2)(e), which highlights the "respective dependence" of the coastal and high seas fishing States on the stocks concerned; and Art. 7(2)(f), which calls for the measures to not result in harmful impacts on the living marine resources as a whole. W. Burke "Compatibility and Precaution in the 1995 Straddling Stock Agreement" at 114–115.

[161] W. Burke, "Compatibility and Precaution in the 1995 Straddling Stock Agreement" at 115.

[162] P. Orebech, K. Sigurjonsson, and T. McDorman, "The 1995 United Nations Straddling Fish Stocks and Highly Migratory Fish Stocks Agreement: Management, Enforcement and Dispute Settlement" (1998) 13 *International Journal of Marine and Coastal Law* 119–141.

[163] A. Oude Elferink, "The Impact of Article 7(2) of the Fish Stocks Agreement on the Formulation of Conservation & Management Measures for Straddling Fish Stocks and Highly Migratory Fish Stocks", FAO Legal Papers Online No. 4 (Rome: FAO, August 1999). Available at <http://www.fao.org/Legal/prs-ol/years/1999/list99.htm> (accessed 18 October 2005).

[164] Amos Donohue, personal communication, January 2005.

[165] Amos Donohue, personal communication, January 2005.

[166] M. Hayashi, "The Straddling and Highly Migratory Fish Stocks Agreement" in E. Hey, ed., *Developments in International Fisheries Law* (The Hague: Kluwer Law International, 1999) 55–83 at 62.

[167] M. Hayashi, "The Straddling and Highly Migratory Fish Stocks Agreement" at 63

[168] S. Nandan, "Statement of the Chairman Upon the Adoption of the Agreement on the Implementation of the Provisions of the United Nations Convention on the Law of the Sea of 10 December 1982 Relating to the Conservation and Management of Straddling Fish Stocks and Highly Migratory Fish Stocks" at 750.

[169] In the case of straddling fish stocks, coastal States and some jurists argue that coastal States under Art. 116 enjoy superior rights and duties and have a special interest in living resources beyond national jurisdiction. Coastal States generally take the view that they have a special or superior interest in the stocks beyond their EEZ given the adverse impact of uncontrolled high seas fishing on domestic management efforts and coastal fishing communities. This view is protected and reinforced by the recognition of coastal States' interests in their EEZs and adjacent high seas, as indicated in Arts. 63, 64, and 116 of LOSC.

[170] See discussion of the opposing views in M. Hayashi, "The Role of the United Nations in Managing the World's Fisheries" in G. Blake et al., eds., *The Peaceful Management of Transboundary Resources* (Leiden: Martinus Nijhoff Publishers, 1995) 373–393 at 388–389. High seas nations perceive the coastal States' position as creeping jurisdiction and a further effort to enclose the seas and diminish flag-State freedom. Some also argue that it would be unequal and unjust to give coastal States preferential or superior rights after acquiring through the LOSC EEZ regime a significant portion of the oceans resources. In recent years, the position has also emerged that overfishing within EEZs will adversely impact the high seas fishery.

[171] S. Nandan, "Statement of the Chairman Upon the Adoption of the Agreement on the Implementation of the Provisions of the United Nations Convention on the Law of the Sea of 10 December 1982 Relating to the Conservation and Management of Straddling Fish Stocks and Highly Migratory Fish Stocks" at 751. See also UNFSA Art. 25 (1)(b).

[172] UN 164/13/Rev.1: para 8. At the end of the second negotiating session, Chairman Nandan prepared a Revised Negotiating Text, which the Conference considered at the third session. UNFSA was adopted in August 1995. This text is found in J.-P. Levy and G. Schram, eds., *United Nations Conference on Straddling Fish Stocks and Highly Migratory Fish Stocks: Selected Documents* (The Hague, Boston, London: Martinus Nijhoff Publishers, 1995) paras. 8(b) and (c). at 441–442. While Nandan uses *minimum international standards* in this instance, this book uses the term "international minimum standards" in keeping with UNFSA Arts. 5(b) and 10(c).

[173] Amos Donohue, Senior Counselor International Law Section, Canadian Department of Justice, Ottawa, personal communication, December 2004.

[174] UNFSA Part VIII.

[175] UNFSA Art. 28.

[176] UNFSA Art. 27.

[177] A. Boyle, "Some Problems of Compulsory Jurisdiction and Settlement of Disputes Relating to Straddling Fish Stocks" (1999) 14 *International Journal of Marine and Coastal Law* 1–25 at 21.

[178] Art. 29; A. Boyle, at 21.

[179] Ted McDorman, personal communication, February 2008.

[180] See J. Peel, "A Paper Umbrella Which Dissolves in the Rain? The Future for Resolving Fisheries Disputes Under UNCLOS in the Aftermath of the Southern Bluefin Tuna Arbitration" (2002) 3 *Melbourne Journal of International Law* 53–78.

[181] See T. Stephens, "A Paper Umbrella Which Dissolves in the Rain? Implications of the Southern Bluefin Tuna Case for the Compulsory Resolution of Disputes Concerning the Marine Environment under the 1982 LOS Convention" (2001) 6(3–4) *Asia Pacific Journal of Environmental Law* 297–318; also see T. Stephens, "The Limits of International Adjudication in International Environmental Law: Another Perspective in the Southern Bluefin Tuna Case" (2004) 19(2) *International Journal of Marine and Coastal Law* 177–197; F. Orrego Vicuña, "From the 1893 Bering Sea Fur-Seals Case to the 1999 Southern Bluefin Tuna Cases: A Century of Efforts at Conservation of Living Resources of the High Seas", (2000) 10 *Yearbook of International Environmental Law 1999* 40–47.

[182] SEAFO Art. 24.4; NEAFC 23 Annex K, at 29 para. 5.

[183] SEAFO Art. 24.5; NEAFC 23 Annex K, at 29 para. 6.

[184] WCPFC Art. 31.

[185] UNFSA Arts. 7.5 and 31.

[186] See Southern Bluefin Tuna Cases (*New Zealand* v. *Japan*, *Australia* v. *Japan/Provisional Measures*) (1999) 38 *International Legal Materials* 1624.

[187] See G. Eiriksson, "Settlement of Disputes II: International Tribunal for the Law of the Sea" DOALOS/UNITAR Briefing on Developments in Ocean Affairs and the Law of the Sea 20 Years After the Conclusion of the United Nations Convention on the Law of the Sea, 25–26 September 2002 (New York: United Nations, 2002) at 6.

[188] Award, at 104. "The *Southern Bluefin Tuna* Award of 4 August 2000 marked the first instance of application of compulsory arbitration under Part XV, Section 2 of the 1982 UN Law of the Sea Convention and of the institution of an Arbitral Tribunal under Annex VII of the LOS Convention. The purpose of the Tribunal was to hear the merits of a dispute between Japan, on the one hand, and Australia and New Zealand, on the other hand, concerning fishing under the Convention on the Conservation of Southern Bluefin Tuna. The dispute had previously been subject to a provisional measures hearing before ITLOS. The Arbitral Tribunal ultimately declined jurisdiction, however, placing emphasis on the fundamental principle of consent", available at <http://72.14.205.104/search?q=cache:Z4aSO6LxoH4J:www.intfish.net/cases/fisheries/sbt2/index.htm+Southern+Bluefin+Tuna+AwaRD&hl=en&ct=clnk&cd=1&gl=ca> (accessed March 2008.) See also B. Kwiatkowska, "The *Australia and New Zealand v. Japan Southern Bluefin Tuna (Jurisdiction and Admissibility)* Award of the First LOSC Annex VII Arbitral Tribunal", OceanLaw On-Line Paper No. 17, March 2003 (revised), available at <http://www.oceanlaw.net/ops/17.htm> (accessed April 2006); D. Devine, "Compulsory Dispute Settlement in UNCLOS Undermined? Southern Bluefin Tuna Case: Australia and New Zealand v Japan" (2000) 25 *South African Yearbook of International Law* 97–112; J. Firestone and T. Polacheck, "The Effectiveness of the UN Convention on the Law of the Sea in Resolving International Fisheries Disputes: The Southern Bluefin Tuna Case" in H. Bryner, eds., *Science and Politics in the International Environment* (Oxford: Rowman & Littlefield Publishers, Inc., 2004).

[189] Award, at 16.

[190] See T. Stephens, "A Paper Umbrella Which Dissolves in the Rain? Implications of the Southern Bluefin Tuna Case for the Compulsory Resolution of Disputes Concerning the Marine Environment under the 1982 LOS Convention"; and J. Peel, "A Paper Umbrella Which Dissolves in the Rain? The Future for Resolving Fisheries Disputes under UNCLOS in the Aftermath of the Southern Bluefin Tuna Arbitration".

[191] Central Bering Sea Donut Hole Convention Arts. VII.2 and VIII.2.

[192] IATTC under the Antigua Convention Art. XXV; NEAFC 23 Annex K.

[193] IATTC under the Antigua Convention Art. XXV; GFCM Art. XVII; NEAFC 23 Annex K; CCAMLR Art. XXV; CCSBT Art. 16; IOTC Art. XXIII; SEAFO Art. 24; WCPFC Part IX.

[194] This is in the cases of CCAMLR, CCSBT, GFCM, and IOTC.

[195] WCPFC Art. 20.4 and 20.6.

Chapter 9

The MCS and Enforcement Regime

*By Evelyne Meltzer and Rosemary Rayfuse**

9.1 Introduction

The ability of a regional fisheries management organization (RFMO) to effectively regulate legal fishing activities and control illegal, unreported, and unregulated (IUU) fishing in its area of jurisdiction depends not only on its competence to adopt conservation and management measures, but also on its ability to achieve compliance with these measures[1]. Ensuring cooperation and compliance with regional management regimes is essential to manage for sustainability. However, effective regional governance requires that competing and often widely divergent interests be reconciled. This can give rise to a fundamental tension between the sovereignty of individual member States (and nonmember States) and the autonomous legal personality of the RFMO represented by the collective will of member States.

Regrettably, a major, and all too frequent, obstacle to the effectiveness of the efforts of RFMOs is noncompliance by member States, regional economic integration organizations (REIOs), and fishing entities with agreed conservation and management measures. The obligation on members or Contracting Parties of RFMOs to comply with conservation and management measures adopted by RFMOs may be modified through application of decision-making procedures under the constitutive treaty of the RFMO. Where RFMO measures are recommendatory only, no legal obligation to comply exists. Similarly, the existence of "opt-out" or objection procedures may affect the binding nature of a measure with respect to the objecting State and can undermine the management regime if conservation standards are compromised. A further impediment to the efficacy of RFMO measures is third-party noncompliance.

The United Nation Fish Stock Agreement (UNFSA)[2] seeks to address these impediments to effective RFMO regulation and sustainable development by elaborating on the duty to cooperate in the conservation and management of straddling fish stocks and highly migratory fish stocks embodied in Articles 116 to 119 of the United Nations Convention on the Law of the Sea (LOSC).[3] UNFSA reinforces the duty to cooperate by requiring that all States that are party to UNFSA, both members and nonmembers of RFMOs, comply with the measures adopted by RFMOs or refrain from fishing.[4] States that do not agree to apply the conservation and management measures adopted by RFMOs are still under the duty to cooperate, which includes the requirement that they must not authorize their vessels to fish for straddling fish stocks or highly migratory fish stocks that are subject to those measures.[5] States are obliged to take measures necessary to ensure their vessels comply with RFMO measures and do not engage in any activity that undermines the effectiveness of those measures.[6] These duties are also found in the preamble and provisions of the Agreement to Promote Compliance with International Conservation and Management Measures by Fishing Vessels on the High Seas (FAO Compliance Agreement).[7]

Accordingly, those States and entities that are party to either UNFSA or the FAO Compliance Agreement (or both) have undertaken to comply with RFMO management schemes or to refrain from fishing. While not all States have ratified or acceded to UNFSA, most of the major fishing States (as well as the European Union (EU)) are now party to UNFSA or the FAO Compliance Agreement or to both. Since the adoption of UNFSA, concerted efforts have been made to increase participation in RFMOs. All States and entities fishing in, or having a "real interest" in, a fishery over which an RFMO has competence have been strongly encouraged to join the relevant RFMO as members or cooperating nonmembers in accordance with UNFSA and the FAO Code of Conduct for Responsible Fisheries.

This chapter provides an overview of the comprehensive, practical, and innovative measures prescribed for under UNFSA to address monitoring, control, and surveillance

---

* This chapter was written with Dr. Rosemary Rayfuse, Professor, Faculty of Law, University of New South Wales, Sydney, Australia. Passages of this chapter are reproductions of sections from R. Rayfuse, *Non-Flag States Enforcement in High Seas Fisheries* (Leiden: Martinus Nijhoff Publishers, 2004) and R. Rayfuse, "To Our Children's Children's Children: From Promoting to Achieving Compliance in High Seas Fisheries" (2005) 20(3) *International Journal of Marine and Coastal Law* 509–532. These passages are reproduced with permission of Brill. The authors gratefully acknowledge the research assistance and contribution of Godwin Djokoto, Don Aldous, and Hugh Williamson in early drafts of this chapter.

(MCS) and enforcement. The chapter begins with a brief general discussion of MCS. Against this conceptual background, the chapter then identifies and analyses the core MCS measures required by UNFSA. This is followed by an examination of the practice of RFMOs in implementing these measures. A number of MCS and enforcement challenges pertaining to fisheries associated with straddling fish stocks and highly migratory fish stocks are discussed. These include developing effective frameworks for the discharge of flag-State responsibility; regulating the activities of nonparties to global and regional fisheries instruments; coordinating and establishing compatible MCS schemes for RFMOs, particularly those with overlapping or adjacent geographic areas of competence; developing an effective and transparent framework for the global exchange of MCS-related information on fishing vessels; and achieving widespread use of port-State and trade-related measures to enforce regional fisheries management measures.

9.2 International legal framework for MCS and enforcement

Under the 1982 LOSC, high seas fishing is open to all States, with some limited exceptions for certain species (such as anadromous stocks).[8] The LOSC imposes a general duty on States to cooperate in the conservation and management of the living resources of the high seas.[9] Flag States have exclusive jurisdiction over their vessels on the high seas, except where otherwise provided for by the LOSC or under customary international law.[10] While the LOSC envisages cooperative management of straddling fish stocks and highly migratory fish stocks by appropriate subregional or regional organizations, there is no clear enforcement power that permits nonflag States to deal directly with noncomplying vessels on the high seas.

Strengthening the LOSC framework for MCS and enforcement was recognized as a necessary component to improving the high seas fishing regime under UNFSA. To that end, it was necessary to elaborate the responsibilities of flag States vis-à-vis their vessels and to articulate clearly the content of the obligation to cooperate.

Flag-State responsibilities are now set out for their respective Parties in the FAO Compliance Agreement and UNFSA. Both instruments oblige flag States to ensure, through means such as vessel registers and fishing authorizations, that their vessels do not engage in activities that undermine the effectiveness of international conservation and management measures, including those adopted by RFMOs.[11] The FAO Compliance Agreement requires each Party to take enforcement measures against any of its vessels that act in contravention of the Compliance Agreement. This requires making contravention of the Compliance Agreement an offence under national law with provision for sanctions of sufficient gravity to effect compliance and deprive offenders of the benefits accruing from their illegal activities. For serious offences, sanctions must include refusal, suspension, or withdrawal of the vessel's authorization to fish on the high seas.[12]

UNFSA further elaborates flag-State responsibilities over vessels registered or licensed to fish on the high seas. Flag States are required, *inter alia*, to

> (1) establish and enforce licensing schemes and regulations for high seas fishing;
>
> (2) establish a national register of vessels authorized to fish on the high seas;
>
> (3) establish requirements and verification programs for marking vessels and gear, recording and reporting vessel position, and verifying catch of target and nontarget species;
>
> (4) adopt MSC measures, including national, regional, and global inspection and enforcement schemes; observer programs; and the implementation of vessel monitoring systems (VMS);
>
> (5) regulate high seas transshipment to ensure that the effectiveness of conservation and management measures is not undermined, and regulate fishing activities to ensure compliance with regional, subregional, and global measures.

In addition, where a subregional, regional, or global system of MCS exists, flag States must ensure that their measures are compatible with that system.[13]

UNFSA also seeks to ensure compliance not just with its own provisions but with measures adopted by RFMOs. Flag States are obliged to take enforcement action against their vessels for noncompliance with RFMO measures, irrespective of where violations occur, and must conduct immediate and full investigations of alleged violations and report on the progress and outcome of those investigations to the relevant RFMO and the State alleging the violation. Flag States must require their vessels to provide information on the vessel's fishing activities to the investigating authorities. Where sufficient evidence of a violation exists, flag States are obliged to refer the case to their authorities with a view to instituting proceedings without delay and to ensuring that further fishing activities are not carried out until any outstanding sanctions are complied with. Investigations and judicial proceedings are to be carried out expeditiously, and sanctions imposed are to be adequate in severity to be effective in securing compliance and discouraging violations. Sanctions shall also deprive offenders of the benefits accruing from their illegal activities. Flag States are obliged to ensure that masters and other officers of their fishing vessels risk refusal, withdrawal, or suspension of their licenses if they fail to comply with the flag State's obligations.[14]

UNFSA further provides that in fulfilling their obligation to cooperate through subregional or regional fisheries management organizations or arrangements, States shall "(h) establish appropriate cooperative mechanisms for effective monitoring, control, surveillance and enforcement".[15]

UNFSA then provides guidance about what these "appropriate cooperative mechanisms" must, at a minimum, include. Members of RFMOs are to exchange information on fishing activities of vessels of nonmembers and are to take measures consistent with international law and UNFSA to deter the activities of those vessels that undermine the effectiveness of RFMOs.[16] All States, both flag and nonflag, are obliged to cooperate either directly or through RFMOs to ensure compliance with and enforcement of RFMO measures. To this end, flag States may request the assistance of nonflag States, which are then obliged to endeavour to meet the reasonable requests of the flag States in investigating alleged violations of RFMO measures. States are also to assist each other in identifying vessels reported to have engaged in activities that undermine the effectiveness of RFMO measures and, to the extent permitted by their national laws, are to establish arrangements for making evidence relating to violations available to prosecuting authorities in other States.[17]

Recognizing the need for supplementing flag-State jurisdiction to ensure the long-term conservation and sustainable use of straddling fish stocks and highly migratory fish stocks, UNFSA also introduced a mechanism for dealing with situations in which the flag State is either temporarily or permanently unwilling or unable to act. UNFSA recognizes that on the high seas, members of an RFMO may take action in accordance with international law to deter vessels that have engaged in activities that undermine the effectiveness or otherwise violate the measures established by the RFMO until such time as the flag State takes appropriate action.[18] Precisely what deterrence action might be is not spelled out. However, UNFSA provides procedures for at-sea boarding and inspection by members of RFMOs of vessels belonging to any State Party to UNFSA, regardless of whether they are also members of the RFMO, to ensure compliance with that RFMO's measures. In addition, where there are clear grounds for believing a violation of the RFMO's measures has occurred, coastal-State members may board and inspect such vessels if they enter an area under national jurisdiction during the same fishing trip for violations committed in the regulatory area.[19] States are to establish, through RFMOs, boarding and inspection procedures that are consistent with those set out in UNFSA. Where they have not done so, the UNFSA procedures apply by default, although members of an RFMO can limit the application of these procedures among themselves where they have adopted an alternative mechanism that otherwise effectively discharges the obligation to ensure compliance.[20]

Under Articles 21 and 22 of UNFSA, a flag State has three working days to respond following an inspection of its vessel by another State in which a serious violation has been discovered and must either investigate and take enforcement action itself or authorize the inspecting State to do so. If the flag State fails to respond, the inspecting State may retain control of the vessel and take it to port pending further investigation and pending resumption, by the flag State, of control of the vessel.[21]

UNFSA also stipulates that port States have both a right and a duty to take measures on a nondiscriminatory basis, to promote the effectiveness of RFMOs when foreign vessels are voluntarily within their ports. Such measures include inspection of documents, fishing gear, and catches. Port States may also adopt regulations to prohibit landings and transshipments where it is established that the catch has been taken on the high seas in violation of RFMO measures.[22]

These treaties have been expanded on and supplemented by several international non legally binding instruments that further specify flag-State responsibilities regarding the conservation and management of the living resources of the high seas, notably the FAO Code of Conduct for Responsible Fisheries[23] and the FAO International Plan of Action to Prevent, Deter and Eliminate Illegal, Unreported and Unregulated Fishing (IPOA-IUU).[24] IUU fishing is broadly defined as fishing activities that are inconsistent with or in contravention of the management or conservation measures in force in a particular fishery or within the area of competence of an RFMO.[25] The IPOA-IUU calls on all States to take measures to ensure that their nationals do not engage in or support IUU fishing and sets out a range of "tools" to be adopted by States both individually and within RFMOs to achieve this. The IPOA-IUU also calls upon States to adopt national "plans of action" to combat IUU fishing. As of 30 March 2006 few States have done so.

9.3 What are MCS and enforcement mechanisms?

Although often used interchangeably, measures to promote compliance and enforcement are not coextensive; however, they are both necessary elements of a functioning and effective legal system. This section provides a brief overview of the definition of these and related terms.

9.3.1 MCS

In the context of international law, compliance may be defined as the conformance by a State with its international obligations.[26] Measures to promote compliance are therefore measures designed to encourage that conformance. These measures, as noted earlier, are generally referred to in the fisheries context as monitoring, control, and surveillance (MCS).

MCS serves two purposes: first, to obtain the basic data necessary for proper fisheries management, i.e., who is fishing where, how, when, for what, and in what quantity; and second, to deter infringements and to ensure that measures or regulations adopted are being complied with.[27] MCS measures are among the major operational tools of modern fisheries management and enforcement[28] and encompass a suite of programs including reporting obligations, observation, at-sea inspection, VMS, vessel registers, in-port inspections, and trade-related measures.

9.3.1.1 Monitoring

Monitoring relates to the first objective of MCS and is defined as "the continuous requirement for the measurement of fishing effort characteristics and resource yields".[29] It includes the collection, measurement, and analysis of catch, species composition, fishing effort, bycatch, discards, and area of operations. This function requires access to information supplied by fishers and the establishment of the statistical infrastructure to support fisheries management. Scientific, environmental, and economic data and information are gathered from sources such as vessel registries, licensing programs, observer reports, inspection reports, logbooks, fishery catch reports, air surveillance sorties, and satellites. Effective monitoring for conventional fisheries management requires computers, software programs, and the associated expertise. The collection, compilation, and analysis of accurate, complete, and timely information are essential for determining stock status and ecosystem implications. Reliable information is required to make science-based management decisions supporting sustainable and responsible fisheries management. The information collected on where and how the fishery is conducted is also used to plan efficient and effective control measures.

Typically a variety of tools are used in monitoring a fishery, and the appropriate mix of tools will vary depending on the characteristics and requirements of the fishery. RFMOs have developed their respective programs based on the particular dynamics, history, structure, and function of the organization, and the capacities of its members. Accordingly, RFMOs have not responded in a uniform fashion to the monitoring challenge.

9.3.1.2 Control

Control refers to the regulatory conditions under which the exploitation of the resources may be conducted[30] and includes the legislative and regulatory framework, procedures, and other arrangements established nationally, subregionally, or regionally. The objective of a control program is to deter noncompliance. Deterrence is given effect through legal power and agreements; patrol presence; threat of and imposition of appropriate sanctions and penalties; and the embarrassment or shame associated with being caught for illicit behavior. To be an effective deterrent, the risk of noncompliance, i.e., being caught and suffering a loss, needs to be greater than the benefit gained from disregarding the fishing rules.[31] The threat of effective judicial and nonjudicial sanctions for violations is an important element in modifying behavior and promoting compliance. The effectiveness of these control mechanisms is often dependent on the rigor and degree of compatibility of national legislation, awareness and capacity of national justice systems, and the participation of port States.

Control activities available to RFMOs in international fisheries law include observer programs, satellite-based VMS, air and sea surveillance, boarding and inspection schemes, in-port inspection, and trade information and catch documentation schemes.

9.3.1.3 Surveillance

Within the MCS context, surveillance is defined as "the degree and type of observations required to maintain compliance with the regulatory controls imposed in fishing activities".[32] This requires observing and overseeing fishing operations to ensure compliance and to detect and prevent infringement[33] through the use of patrol vessels, aircraft, satellites, or onboard observers mandated to carry out a compliance monitoring function. Given that surveillance programs are capital intensive, most coastal States will multitask and coordinate the deployment of vessels and aircraft with other services such as the national coast guard or military, both to maximize the use of existing platforms and to share the operating costs, especially the fuel expense.[34] RFMOs such as the Commission for the Conservation of Antarctic Marine Living Resources (CCAMLR), the Northwest Atlantic Fisheries Organization (NAFO), and the North East Atlantic Fisheries Commission (NEAFC) typically collaborate with member States and flag States to implement the surveillance program.

9.3.2 *Enforcement*

Enforcement consists of compelling restraint of behavior in a way that promotes conformity with prescribed rules and makes adherence to the rule more attractive than violation. Enforcement therefore aims both to terminate and to punish noncompliance as a means of deterrence and includes measures such as arrest, prosecution, and punishment.[35] After a violation has been detected or where there is reasonable cause to suspect noncompliance, there is typically an investigation and, where appropriate, detention, prosecution, or punishment for the violation. These elements form part of the enforcement function.[36] While effective enforcement itself serves as a deterrent, enforcement is sometimes distinguished from control in that it includes the policing of suspected offenders.

In the high seas fisheries context, enforcement jurisdiction resides primarily in the flag State. This is premised on the assumption that the flag State will effectively control its vessels in accordance with its international obligations. While there is a presumption that States will comply with their international legal obligations, in practice this is not always the case.[37] Regrettably, the history of continuing noncompliance on the high seas by "flags of noncompliance" and other flag States has resulted in initiatives such as the High Seas Task Force (HSTF) calling for practical measures to secure improved compliance and effective enforcement.

9.4 Implementation of MCS and enforcement measures in RFMOs

This section reviews the range of MCS and enforcement measures that have been adopted by RFMOs to assist them in ensuring that flag States effectively exercise their responsibilities to enforce compliance with RFMO measures. This review will adhere, as far as possible, to the framework of the MCS and enforcement provisions of UNFSA, looking first at flag-State responsibilities under Article 18, then at the requirements of international cooperation in MCS and enforcement in Articles 19–23.

9.4.1 *Fishing vessel registry*

UNFSA Article 18(3)(a)–(c) requires flag States to authorize and licence vessels to fish on the high seas, to maintain a national record of vessels so authorized, and to enable access to the information contained in that record by directly interested States. Though this provision (Article 18(3)(c)) is directed at flag States, since the adoption of UNFSA, most RFMOs have adopted measures requiring members to licence and authorize their vessels to fish on the high seas and have established regional registers of fishing vessels authorized to fish in their respective areas of competence.[38] These regional registers or records of vessels come in many different forms and have led to emergence of the so-called "negative" and "positive" lists. These lists reinforce and provide the basis upon which MCS measures can be imposed and enforcement action taken.

9.4.1.1 Negative lists

The first lists were negative lists, consisting of vessels that were believed, on the basis of information and intelligence supplied by various sources, to be engaged in IUU fishing in contravention of the rules prescribed by RFMOs. The concept was introduced in the International Commission for the Conservation of Atlantic Tunas (ICCAT) in 1999 and included a list of nonmember vessels that were observed fishing, transshipping, or offloading catch in the convention area. In the ensuing years, the Japanese government and industry have collaborated on this initiative through the Organization for the Promotion of Responsible Tuna Fisheries, which has as its objectives the elimination of IUU fishing, the reduction of excess fishing capacity, and the decrease of longline bycatch. CCAMLR, the General Fisheries Commission for the Mediterranean (GFCM), the Inter-American Tropical Tuna Commission (IATTC), ICCAT, the Indian Ocean Tuna Commission (IOTC), NAFO, and NEAFC have established IUU negative vessel lists. These negative lists of vessels are posted on some of the respective RFMO Web sites, and the information is typically shared with members as well as among RFMOs.

Although negative lists are effective in identifying unauthorized vessels and deterring noncompliance, they have some limitations. An IUU vessel, once identified, can simply change its name and register to avoid detection. This places a significant burden on flag States to determine the registration history of a vessel before granting their flag to it. For States willing to combat IUU fishing, this effort may be worthwhile. However, for those unwilling or, for whatever reason, unable to do so, the risk of registration of known IUU vessels remains.

In one notable incident, in 1999, the Panamanian flagged *Camouco*, which had previously been known by four other names, was arrested in French waters around the Crozet Islands in the Southern Ocean. The master was subsequently convicted and the owners fined, but not before Panama had made an application to the International Tribunal for the Law of the Sea (ITLOS) for the prompt release of the vessel.[39] Following proceedings in the court in Reunion, the vessel was renamed *Arvisa 1* and reflagged to Uruguay. In January 2002, the vessel was sighted, with no visible identification, well within the CCAMLR Convention Area by the Australian research vessel *Aurora Australis.* She claimed to be the *Kambott*, flying the Mauritian flag, but later proved to be the *Arvisa I*, flagged to Uruguay. In July 2003, the same vessel was apprehended by the French navy in Kerguelen waters but by this time she had temporarily reflagged to the Netherlands Antilles and was renamed the *Eternal*.[40]

This familiar high seas scenario with reflagging of vessels, also known as the free-rider problem associated with flags of noncompliance, has perennially undermined conservation strategies and regional MCS efforts. Numerous initiatives, including those of the Ministerially-led High Seas Task Force on IUU Fishing, have condemned the irresponsible use of "open registries" and have called for a "genuine link" to be established between the flag State and the registered fishing vessel. This genuine link recognizes that any right to fish on the high seas has concomitant duties on the part of the flag State to ensure their flag vessels and nationals conserve and cooperate in fishing responsibly[41]. The world community is not only endeavouring to hold flag States accountable for the actions of their flag vessels and nationals but is also pressuring flag States, port States, and RFMOs to impose adequate sanctions for noncompliance. The effectiveness of negative lists has increased in recent years, as a broad suite of sanctions have been imposed by RFMOs for IUU fishing, ports have denied access for IUU vessels or prohibited the landing of IUU fish, and certain flag States have begun to deregister IUU fishing vessels. For example in IATTC, vessels on the negative list are subject to a range of measures against them by Contracting Parties, including denial of landing, import, and transshipment; refusal of flag; and denial of chartering arrangements.[42]

9.4.1.2 Positive lists

While the purpose of a negative list is to identify vessels engaged in IUU fishing, a positive list constitutes the official register of fishing vessels authorized to fish in a regulatory area.[43] Vessel information is submitted by States and entered on the RFMO register or record.[44] The concept of registers for fisheries purposes was first introduced by the Pacific Islands Forum Fisheries Agency (FFA) in the 1980s and proved to be a simple and powerful tool for addressing illegal fishing. Under the minimum terms and conditions of FFA, a fishing vessel is required to provide full details and to be in "good standing".[45] Good standing status may be revoked if the vessel is found to have committed a serious offence. It then cannot be licensed to fish in the exclusive economic zone (EEZ) of any FFA member.[46]

IATTC was the first RFMO to apply the concept of the positive list to high seas fishing by way of binding Resolution in June 2000. In 2002, ICCAT also established a list of vessels authorized to fish ICCAT species[47] to complement and buttress its negative list.[48] CCAMLR, the Commission for the Conservation of Southern Bluefin Tuna (CCSBT), GFCM, IATTC, ICCAT, IOTC, NAFO, NEAFC, and the Western and Central Pacific Fisheries Commission (WCPFC) have all established vessel records or registries with "lists of vessels authorized to fish" in the area, or for the particular stock as in the case of CCSBT.

Seven RFMOs (CCAMLR, GFCM, IATTC, ICCAT, IOTC, NAFO, and NEAFC) have established both a record of known IUU vessels and a record or register of authorized fishing vessels. These negative and positive lists have considerably strengthened RFMOs' ability to regulate and control fishing in their area of competence and have facilitated flag-State control. As in the case of the negative list, most RFMOs have imposed sanctions for those fishing vessels not on the positive list that are presumed to have been fishing without authorization. Unlisted vessels in the IOTC area are prohibited from fishing for, retaining, or transshipping tuna or tuna-like species.

9.4.1.3 Challenges

The criteria for a vessel to be listed vary considerably from one RFMO to another, and there may be several lists within a single RFMO. In some cases the positive and negative list applies to vessels of all gear types (e.g., IATTC). The list may apply to a specific gear type for Non-Contracting Parties.[49] In other cases the vessel registry system applies only to authorized fishing vessels greater than 24 m.[50] This policy is based on the premise that only vessels of a certain size are able to fish on the high seas. Vessels under 24 m capable of fishing on the high seas are increasingly being designed and constructed, and as a result, several RFMOs now maintain a list of authorized vessels of all sizes (e.g., CCSBT amended its list in October 2004 to include smaller vessels).

Another weakness with most of the regional vessel registers, with the possible exception of CCAMLR, is that to a large extent RFMOs rely on the flag State to provide accurate information on the vessel and do not independently verify the information. IATTC requires that duly qualified and recognized surveyors provide information on purse-seine vessels included in its register. Moreover, the data and information compiled in the registers are not always widely accessible. For example, the CCAMLR IUU Vessel List has, until recently, been accessible only to members through a secure Web site because of confidentiality and other concerns. At the 2005 CCAMLR Annual Meeting, it was decided to make this information more widely available.[51] In addition, lists compiled within one regional framework are not necessarily available even to other management agencies.[52] The problem is further aggravated by differences and inconsistencies in data requirements that render comparisons between vessels and the establishment of linkages between movements of vessels from one register to another almost impossible. Another significant limitation of the effectiveness of vessel registration under the UNFSA provisions is that in many flag States, the links between the maritime administrations, which are responsible for vessel registration, and the fisheries administration, which issues the authorization to fish, are inadequate or absent.

A centralized global register of all high seas fishing vessels could be a powerful tool for law enforcement. This global information system could provide transparent and unbiased data regarding, *inter alia*, ownership and control of high seas fishing vessels.[53] To be effective, "such a register would need to have the capacity to issue unique, traceable identifiers to individual vessels and provide definitive, unique and verifiable information on each and every high seas fishing vessel, and the nature and extent of its fishing authorizations. In theory at least, the existence of such a record, when combined with enhanced monitoring, control and surveillance (including, for example, mandatory port State inspections combined with a requirement to provide VMS data or participate in centralized VMS), could operate as a powerful disincentive to the renaming and reflagging of IUU vessels because the true characteristics of the vessel and the commercial entity behind it would be made easier to trace".[54] Entry in good standing on the global register would become a mandatory requirement for any vessel fishing on the high seas. "The status of a vessel on the database could itself provide a strong basis for unilateral or regionally coordinated action (e.g., prohibition on import of product caught by a non-registered vessel, denial of port access). Ultimately, applying the precedent of Articles 21 and 22 of UNFSA, if there is a watertight system of global registration for fishing vessels, it could be argued that lack of registration creates a *prima facie* right for other States to take action against an unregistered vessel on the high seas".[55]

To address the issue of overcapacity, an alternative would be to require dual authorizations from both the flag State and the RFMO, the latter being based on catch capacity in the individual fisheries concerned. Only vessels with dual authorization would be permitted to fish and only in the fishery for which they were authorized. Again, the register could be used to establish a presumption that any vessel either not on the list or participating in an unauthorized fishery is engaged in IUU fishing. This presumption of IUU fishing gives rise to a right for other States to take action against such vessels.[56]

9.4.2 Data collection and reporting

Reporting and collecting data are needed for improved stock assessments, management plans, fisheries models, monitoring and control, and enforcement. Therefore, the importance of data collection, verification, and analysis within a structured database cannot be overstated. They are critical element of an effective MCS regime.

UNFSA Articles 10 and 14, together with Annex I, provide the basis for improved standards for reporting catch. FAO, with input from RFMOs, has developed protocols for data collection and standardized data formats. UNFSA Annex I requires that the information collected by RFMOs include catch of target and nontarget species, fishing effort data, and vessel particulars, including activity on the high seas and within EEZs.

The constitutive instruments of all the UNFSA-related RFMOs have a provision for, at minimum, the mandatory submission of fishery catch data on an annual basis. NAFO and NEAFC require monthly reporting by members, although some fisheries require vessels to report weekly.[57] In NAFO there is also a pilot project on observers, satellite tracing, and electronic reporting, which includes obligations for daily reporting.[58] In some of its fisheries, CCAMLR requires five-day or ten-day effort reporting. In this way, the Secretariat is able to ascertain when TACs have been met and when the fishery concerned should be closed. RFMOs also require reporting of fish bycatch, incidental catch of other species, fishing effort, transshipments, and infringements.

It is an essential requirement for the successful implementation of the management measures developed by RFMOs that all participating States, REIOs, or entities (i.e., including cooperating nonmembers) fulfill their obligation by providing all necessary data to the Commission in a timely manner. Inability or failure to fulfill commitments to provide data in time for analysis has impeded the stock assessment process. For example, in ICCAT the provision of "Task II" data[59] related to fish size is often not forthcoming from members, and the failure to provide "Task I" data related to catch in time for inclusion in stock assessment analysis has compromised the work of the Commission. To address this data collecting, reporting, and assessment challenge, ICCAT adopted a measure at its 2005 Annual Meeting that places the onus on members as well as Cooperating Non-Contracting Parties to demonstrate why reporting requirements were not met and provide plans for corrective action.

Since the coming into force of UNFSA, all RFMOs have introduced measures to strengthen data collection and reporting. Where gaps in the data supplied exist, RFMOs have begun to identify the reasons for gaps and take steps to complete the data sets through diplomatic measures or assistance programs.[60]

9.4.3 Trade information and market-related measures

RFMO data-reporting obligations only apply to members. For management purposes, where nonmembers engage in a fishery, data on their catches are also crucial to an accurate assessment of stock status and management. RFMOs have sought to address this by making provision of data a condition of obtaining the status of Cooperating Non-Contracting Party, which may, in turn, be a condition of access to quota. An alternative method of obtaining information on the catch of both members and nonmembers while simultaneously verifying the amount and nature of catch is through the use of market-based statistical document programs, as called for in UNFSA Article 18(3)(f). These programs assist not only in obtaining scientific data for management purposes but also in identifying potential IUU catch and helping to ensure it does not have a market.

ICCAT was the first RFMO to develop a mandatory statistical document to accompany fish shipments from exporter to market in a member nation.[61] This was done in response to a concern about the high level of bluefin tuna catch in the North Atlantic and was considered effective, since a high percentage of the global bluefin market is imported into a single ICCAT member, Japan. The Bluefin Statistical Document Program was adopted in 1994. This was followed by a similar program for bigeye tuna and swordfish in 2002.[62] In 2002, IOTC adopted the Bigeye Tuna Statistical Document Program modeled on the ICCAT programs.[63] Similarly, in 2003, CCSBT adopted the Southern Bluefin Tuna Statistical Document Program, which imposes a requirement for mandatory validation of each export by an official of the flag State. The program calls for submission of quarterly import information deficiencies to the CCSBT Secretariat.[64] IATTC adopted the Bigeye Statistical Documentation Program in June 2003.[65] CCAMLR has a catch documentation scheme (CDS) for Patagonian toothfish designed to track the landing and flow of fish.

These schemes generally require landings and imports of certain species into RFMO members, from any source, to be accompanied by documentation. Members (participating States, REIOs, entities) then submit the consolidated information, from which a record of the amount of catch taken by all States fishing for the species can be built up and an estimate of nonmember catch made. This estimate can then be figured into management decisions on a precautionary basis.

Documentation of landings and trade has also become a useful tool in separating fish and fish products taken by legal fishing from those taken by IUU fishing and in providing evidence of which vessels, and hence which States, are engaged in IUU activities. ICCAT has adopted the approach of having member States make diplomatic demarches to the nonmember States identified, and if no corrective action is taken, member States are required to refrain from importing specified fish or fish products from the identified State. In ICCAT, the first flag States found to be engaging in IUU activities were Belize, Cambodia, Honduras, and Saint Vincent and the Grenadines, who were subsequently notified in 1998 that a ban on bigeye imports from these States would be implemented in markets of ICCAT members.[66] This ban was later lifted for Saint Vincent and the Grenadines in 2001, and Belize and Honduras in 2002, because of significant progress made by these States and reported to ICCAT.[67] In 2005,

ICCAT maintained trade-restrictive measures on Bolivia (2002)[68] and Georgia (2003),[69] and Chinese Taipei was given one year to comply with ICCAT conservation measures before the trade-measure resolution (03-15) would be triggered.

In 2004, ICCAT introduced further measures to track imports of fish products into member States by requesting member States who import, or where landings take place, to collect information concerning landings and report annually to the Compliance Committee. There, analysis of import figures is conducted to verify origin of fish and compliance with bans in place. Also, in 2004, ICCAT prohibited landings of IUU fish and adopted the same approach of notification of parties and adoption of nondiscriminatory trade measures and recommendations against nonmembers as those that apply to member States.[70]

Provision for a similar approach has been made in CCSBT, which in 2000 adopted an action plan that included the possibility of imposing trade sanctions against nonmembers that continue to import IUU southern bluefin tuna.[71] Likewise, in 2003, IOTC adopted a voluntary measure requesting members that import tuna or in whose ports tuna are landed to collect as much information as possible, including area of catch, from vessels fishing in the convention area, and to forward the information to the IOTC on an annual basis. The IOTC Compliance Committee plans to maintain a list of both member- and nonmember-State vessels that fish in contravention of IOTC measures and apply a notification process of reminder of obligations, followed, if necessary, by imposition of nondiscriminatory trade measures.[72]

In IATTC all bigeye imported by Contracting Parties must be accompanied by the requisite statistical documents, which must be validated by the flag State. Documents are not required for purse seiners and bait boats delivering bigeye directly to canneries. In 2005, IATTC adopted a trade-measure Resolution requiring all parties and cooperating nonparties importing or landing bigeye to submit the data to IATTC. IATTC maintains a list of all States and fishing entities that have not complied with or have not cooperated with IATTC measures. The Commission can adopt trade-restrictive measures where other measures have proven to be ineffective in securing compliance. Given the innovative nature of such measures and concerns about their compatibility with international trade law, IATTC has called on its members to take trade action only when all other actions have failed to encourage compliance.[73]

CCAMLR's CDS is a slightly different approach. CCAMLR adopted the scheme in 2000 for *Dissostichus* (toothfish) species.[74] Generally, the CDS is a catch certification system that seeks to prevent the landing, transshipment, import, export, or reexport of toothfish at ports, or the transshipment to party vessels by both party and nonparty fishing vessels, unless the catch is accompanied by a valid *Dissostichus* catch document (DCD). Fish not accompanied by documentation cannot be landed, transshipped, imported, or exported. CCAMLR maintains a specific database pertaining to toothfish, which includes data on catches taken by vessels from nonmember States but landed in member States. The validity of the DCD is dependent upon several checks. These include the following: the vessel master is required to certify the origin of the catch, including whether or not it comes from within the CCAMLR Convention Area, and whether it was caught in a manner consistent with CCAMLR Conservation Measures; the port-State "landing and importation" authorities are required to certify whether the catch is consistent with details provided in the DCD; and if appropriate, the flag State may be required to verify the authenticity of the content of the DCD. In the event of reexports, where fish is exported from the country of landing, the subsequent importing State is to examine the relevant documentation. To supplement the DCD, CCAMLR has also examined the use of trade sanctions, and at its 2005 meeting, the Commission decided to send letters to a number of Non-Contracting Parties with a threat to impose such measures.[75]

Slightly different again is the approach adopted by NAFO in 2005, which requires that all fish products caught in the NAFO Regulatory Area (NRA) be labeled as such with species identification.[76] The usefulness of these programs in providing more detailed information on which more effective management measures can be based is well-known. However, these programs generate large amounts of documentation and are labour intensive. A possible solution to these problems is to adopt, as has been attempted by CCAMLR, an electronic version of the DCD.[77] For the season 2005–2006 CCAMLR operated a voluntary electronic Web-based CDS. CCAMLR's program is further reinforced by that of a key importer: the United States. The United States recently introduced a mandatory electronic CDS and requires since 2006 electronic DCDs as a condition of import.[78]

These statistical, trade, and catch documentation programs are, however, open to abuse through misreporting. The provision of false information may not be remedied by electronic reporting. The efficacy of such programs therefore depends on the willingness and ability of member States to fully implement them, to report the data received to the RFMO, and to monitor and, if necessary, verify through other means the accuracy of the information submitted. In this respect global harmonization and standardization of the reporting requirements under the various schemes have been suggested as a solution for improving both their implementation and their usefulness.[79]

The use of multilaterally agreed, nondiscriminatory trade measures to promote compliance with RFMO conservation and management measures and reinforce the statistical and catch documentation schemes is an innovative and important development in MCS and enforcement. Article 33(2) of UNFSA, although it does not directly mention trade measures, provides that "State Parties shall take measures consistent with this Agreement and international law to deter the activities of vessels flying the flag of non-parties which undermine the effective implementation of this Agreement". The IPOA to combat IUU fishing expands on UNFSA by making specific provision for the use of multilateral trade measures by RFMOs to enforce regional conservation and management measures. As noted above, several RFMOs (CCAMLR, CCSBT, IATTC (for bigeye), ICCAT, IOTC, NAFO, and NEAFC) have developed and implemented wide-ranging trade-related measures to address IUU fishing.

Measures taken by RFMOs include mandatory prohibition of imports of fish from nonmembers whose vessels undermine the effectiveness of agreed conservation measures (CCAMLR CDS); prohibition of imports from members who exceed their catch limits; prohibition of imports from nonmembers who fail to take corrective action; use of CDSs to determine the origin of fish and whether the fish are harvested in a manner consistent with the RFMOs' management measures; use of other statistical documentation programs that trace all fish and fish products entering international trade; provision of catch documents to vessels, including export (or reexport) validation by the exporting State; and marking or labeling of fish products to indicate species, production method (e.g., caught at sea, inland waters, or farmed), and catch or production area.

The efficacy of schemes for the imposition of trade-restrictive measures, however, needs to be further strengthened. The current processes designed to implement these measures are typically cumbersome and lengthy, and involve (1) identification; (2) a request to rectify; (3) further assessment at yet another meeting of the RFMO; and (4) a threat to impose such measures. Given the annual nature of most RFMO meetings, this process can take up to two or three years, during which time the IUU fishing continues. Then, when sanctions are imminent, the vessel can merely reflag to a nondesignated nonmember State. This approach, to date, appears to have encouraged the practice of flag and port hopping. Timely and extensive implementation of this approach would render these measures a more effective device for ensuring flag-State compliance.

9.4.4 Regulation of transshipment

UNFSA Article 18(3)(h) requires flag States to take measures to regulate "transshipment on the high seas to ensure that the effectiveness of conservation and management measures is not undermined". Regulation of transshipments provides a unique opportunity to confirm the catch of a fishing vessel either to verify the catch information supplied by the vessel or to collect catch data where no catch records exist. Regulation of transshipment also provides an opportunity to disrupt access to markets for IUU fish.

Several RFMOs have adopted regulations or bans on transshipment and landings. The NEAFC Scheme to Promote Compliance by Non-Contracting Party Vessels includes a ban on transshipment between vessels flying the flag of a Contracting Party or those flying the flag of a nonparty.[80] The IOTC Scheme to Promote Compliance by Non-Contracting Parties requires a mandatory prior authorization for all transshipments by members, whether between member vessels or between member and nonmember vessels. ICCAT also bans at-sea transshipments between ICCAT members and nonmembers. However, such measures were difficult to monitor or enforce without "on the water MCS". In 2005, in a landmark decision, ICCAT adopted a binding Recommendation to establish a regional observer program to be implemented not later than 2007. This program requires the presence of observers during at-sea transshipments involving large-scale tuna longline fishing vessels. GFCM,[81]which currently has no observer program, adopted in February 2006 the ICCAT transshipment Recommendation, calling for the establishment of a GFCM regional observer program. NAFO requires mandatory inspections on transshipments for all vessels having fished in the NRA. IATTC has adopted several legally binding Resolutions to control transshipments. The WCPFC Convention requires that transshipment at sea in the convention area beyond areas of national jurisdiction is to take place only in accordance with specified terms and conditions[82] and procedures to be established by the Commission.[83] The Commission has not yet established any such conditions. The WCPFC Convention also prohibits transshipment at sea by purse-seine vessels operating in the convention area.[84]

Existing RFMO transshipment practices are, however, limited in that they apply only to transshipments involving one or more member-State vessels and need to be reinforced to assist with the problem of transshipments between nonmember vessels. IUU fish can therefore still enter the market, either through multiple transshipments or by blending with legitimate catch.

Some have suggested that all RFMOs should move to ban, where practical, at-sea transshipments.[85] Requiring all fishing vessels to land their catch and be subject to in-port inspection would improve MCS efficacy. In-port inspections are essential for verifying the catch records supplied by the flag State to the RFMO and are particularly important in the tuna RFMOs where there is no at-sea inspection scheme. Alternatively, given the economic efficiencies involved in using transshipment vessels, RFMOs could establish a positive list for transshipment vessels. This development would assist in the regulation of transshipment outside the regulatory area as well.

9.4.5 Observer programs

UNFSA Article 18(3)(g)(ii) requires the implementation of national observer programs and subregional and regional observer programs in which the flag State is a participant, including requirements for such vessels to permit access by observers from other States to carry out the functions agreed under the programs. Debate persists about whether this requirement is fulfilled by the adoption of national programs that merely report to the relevant RFMO on a periodic basis or whether a regionally administered program is required. Many argue that a regionally administered observer program provides a better and more timely opportunity to gain specific knowledge from fishing operations in a standard format common to all members. Another debate persists about the role and function of observers, whether part of national, subregional, or regional observer programs. While the nature and scope of the programs vary considerably, most RFMOs responsible for straddling fish stocks with the exception of NEAFC have an observer program. RFMOs responsible for highly migratory fish stocks have limited observer programs.

9.4.5.1 Role of the observer

Traditionally there are two functions for an observer: scientific and compliance. A long-standing debate exists over whether the addition of the compliance function impedes the observer in his or her ability to fulfill the scientific function effectively. Specifically, observers have noted a lower degree of cooperation with scientific data collection on board the vessel when compliance issues are raised. Observers may be on board vessels for long periods of time, so a cooperative climate is essential for the retrieval and transmission of necessary information. Captains and crews can make it difficult for observers to perform their functions if the observers report infractions. Observers have been beaten, bribed, kept confined to their cabin, and otherwise mistreated and misled.[86] Such interference with the duties of the observers patently undermines their effectiveness.[87] On the other hand, where observers perform only scientific functions, an important monitoring opportunity is lost. The observer with both mandates may confront a challenge in fulfilling the compliance function without impacting negatively on his or her ability to collect the scientific data required.

Therefore, in some regional observer programs the observer undertakes solely a scientific function, while in others the observer program is also compliance oriented. The CCAMLR Scheme of International Scientific Observation is an example of a program for purely scientific observation, which is complemented by the CCAMLR System of Inspection for compliance-monitoring purposes. The NAFO Observer Program, on the other hand, is designed for the observer to provide both functions to ensure compliance with NAFO conservation and management measures. Observers under the IATTC-related Agreement on International Dolphin Conservation Program (AIDCP) fulfill both a scientific and a compliance-monitoring role in that once they verify that the dolphin mortality limit for a given vessel has been reached, all fishing on dolphin-associated tuna must cease. IATTC observers also document violations for both AIDCP and IATTC.

An observer is usually a civilian employee or contractor, who is generally embarked on the fishing vessel for the duration of the fishing voyage and observes or monitors the ongoing activity. Unlike an inspector, a fisheries observer usually has no official powers in regard to offences committed by the vessel, other than to make records and reports. In some jurisdictions, their testimony may not be admissible as evidence in legal proceedings related to a fisheries violation. It is important to distinguish an inspector from an observer, even one fulfilling an element of the compliance function.

9.4.5.2 Types of observer programs

RFMOs have adopted several approaches to the design and implementation of observer programs. Observer programs may be nationally operated with nationally trained observers operating either under national standards, as in ICCAT, IOTC, or NAFO, or under regionally agreed standards, as in CCSBT and the South East Atlantic Fisheries Organisation (SEAFO) (although this program is not yet in place). Provision may also be made for national observers to be placed on vessels of other flag States by mutual agreement, either reciprocally, as in SEAFO, or upon request, as in NAFO and the Central Bering Sea Donut Hole Convention (BSDH, under the Convention and not yet in practice because of moratoria since 1994). Observer programs may also be regionally administered by RFMOs. In this case, observers may be trained either nationally or internationally under an RFMO training program as in CCAMLR or WCPFC (though the latter program is not yet in place). The observers are then dispatched as RFMO observers and report directly to the flag State or the RFMO Secretariat or both. A minimum of 50% of observers under the AIDCP must be IATTC appointed, trained, and certified with the remainder coming from national observer programs. The use of non-national observers trained, certified, and deployed under the authority of the RFMO is designed to circumvent the potential for abuse and noncompliance present in nationally operated schemes.[88]

Data from national or regional observers may be submitted either to flag States alone or to both the flag State and the RFMO. In the CCSBT, data are maintained by members and not by the Secretariat. In NAFO, observer reports are made to the flag State and copied to the NAFO Secretariat. The IATTC and AIDCP require at-sea reporting by observers and weekly data submission to the IATTC Secretariat. Under the BSDH Convention, where no centralized secretariat exists, observers are mandated to report any observed violation to the flag State and the observer State. In CCAMLR, where observation is limited to the scientific function and is carried out by independent regional scientific observers, data are submitted directly to the CCAMLR database.

There is benefit to be derived from a centrally based regional observer program, since the issues of reporting standards and consistent training may be more easily addressed. Also a centrally based regional program is thought to be a more efficient method for delivering program objectives in terms of coverage levels and ensuring data quality. However, the logistics of running such an international observer program presents its own set of challenges. These include the high cost of foreign travel for observers at the beginning and the end of trips at sea; making last-minute complex travel arrangements; the provision of periodic training programs for recruits; and the potential negative impact of recruiting observers for the regional program on the national observer programs of RFMO members. This is recognized in the WCPFC Convention, which makes provision for an independent observer program that will be carried out in collaboration with other national and regional observer programs.[89]

9.4.5.3 Observer coverage levels

Setting and maintaining observer coverage level targets are considered by some to be essential in achieving representative data from the fishery and a meaningful level of compliance. These targets may vary, e.g., as provided for in SEAFO, according to fishing gear type, species targeted, or specific area fished. Some RFMOs require 100% observer coverage for particular fisheries.[90] Others require 100% coverage for all fisheries.[91] Lower levels of observer coverage are required in other RFMOs. Both CCSBT and IOTC call for 10% coverage under national programs. Coverage targets have not yet been set in ICCAT, SEAFO, and WCPFC. The MCS scheme in WCPFC is to be further developed by the Technical and Compliance Committee, which first met in December 2005. Although no targets have been set in ICCAT, the 2003 Ad Hoc Data Workshop concluded that comprehensive at-sea observer data were essential. As noted above, in 2005, ICCAT adopted a binding Recommendation for the establishment of a regional observer program by 2007, which will require the presence of observers on all large-scale tuna longline fishing vessels during at-sea transshipments. GFCM did not have an observer program in March 2006, but adopted a Recommendation similar to that of ICCAT for the purposes of monitoring at-sea transshipment in February 2006.

9.4.5.4 Impact of observer program

The impact of observer programs in ensuring compliance with management and conservation measures cannot be easily measured. In NAFO, where observers fulfill both the scientific and compliance function, there has been a significant decline in infringements since the introduction of the 100% observer coverage requirement in 1998.[92] Similarly, in AIDCP, which requires 100% coverage on large purse-seine vessels, with 70% of observers provided by IATTC and the remainder by Contracting Parties, the significant reduction of dolphin bycatch associated with yellowfin tuna is noteworthy (see Chapter 7 on ecosystem approach). While this increase in compliance with the conservation measures may be attributed to a combination of factors, the NAFO and AIDCP experiences suggest that observer programs, if well implemented, can provide the requisite data and serve as a deterrent to flag-State violations. Nonetheless, the use of observer programs as an MCS tool — while possibly less expensive than aerial surveillance and patrol vessels — remains controversial. In NAFO, for example, debate persists about the independence, or lack thereof, of observers, and their ability to communicate with inspectors and provide probative evidence in compliance-related proceedings.

Some RFMOs have considered introducing electronic logbooks to strengthen their MCS program, which could replace or supplement onboard observers. NAFO recently introduced a pilot project on observer, satellite tracking, and electronic reporting to test a near-real-time electronic reporting system by observers and vessel masters. Some contend the current technology is not as precise and will not have the same deterrent effect as onboard observers.[93]

9.4.6 Aerial surveillance

The traditional method deployed for surveillance and detection of illegal fishing activity is a system of routine air patrols coordinated with patrol vessels carrying enforcement officers. Aerial surveillance is usually provided by military aircraft on a multitasked mission or, in some instances, by civilian aircraft specifically chartered and equipped for maritime surveillance. Since these resources are expensive and not available to all national governments, RFMOs have endeavored to use aircraft from the armed forces and other government departments of member States for conducting surveillance in their convention areas. For example, the Canadian government assigns aircraft to the joint inspection and surveillance scheme to provide air surveillance in the NAFO zone, independently of Canadian national surveillance activities.[94] NEAFC has a virtually identical provision.[95] FFA has had coordinated aerial surveillance activities in the Western and Central Pacific for over 20 years and would be well positioned to continue this role within the WCPFC. In such a case, this would be the first regional aerial surveillance program directed by an RFMO. Although there is no air surveillance program within IATTC, there has been a sighting and reporting agreement for vessels operating in the convention area since 2004. Other RFMOs have no coordinated surveillance program, but may use member-State aircraft for the transportation of personnel to inspect vessels.

Aerial surveillance provides information to assist in monitoring the general level of fishing activity and assists in the identification and investigation of possible infringements. Air patrols serve as a major detection and investigation tool for unauthorized fishing and transshipment activity, as they enable the collection of photographic and observational evidence. Recent developments in technology allow photographic evidence to be procured at night as well.

As an MCS mechanism, aerial surveillance serves as a deterrent to illegal activity. Its usefulness is indicated by the success of the Parties to the North Pacific Anadromous Fish Commission (NPAFC) in working together to provide combined and coordinated military and coastguard surveillance and enforcement operations to deter and prevent illegal fishing operations in the North Pacific.[96] The enforcement practice in the NPAFC may serve as a model for other areas.

The advantages of air surveillance are that large areas can be covered and inspection platforms can be effectively and efficiently directed into position. For coastal States, surveillance aircraft can also be used during a hot pursuit to ensure continuous pursuit of a vessel fleeing the EEZ. RFMOs are also in a position to receive reports of aircraft surveillance missions, including reports of infringements, and to request information concerning the infringement from the flag State. Of particular note, aerial surveillance provides the investigative and evidentiary support necessary to assist flag States in prosecution and punishment of offenders.

Aerial surveillance does, however, have its limitations. Aircraft operations may be impeded by inclement weather or nonavailability of aircraft and air crews. They are constrained by distance and fuel needs and are unable to operate around the clock. Aerial surveillance is also extremely expensive — Canada commits in excess of CDN \$6 million annually to aerial surveillance in the NAFO area — making it difficult, if not impossible, for developing States to adopt aerial surveillance in areas such as the South Atlantic and Indian Ocean.

9.4.7 Satellite-based vessel monitoring systems

UNFSA Article 18(3)(g)(iii) requires flag States to develop and implement vessel monitoring systems (VMS), "including, as appropriate, satellite transmitter systems, in accordance with any national programmes and those which have been subregionally, regionally or globally agreed among the States concerned". A satellite-based VMS enables fisheries managers and flag States to track the movement of vessels and provides information for surveillance. VMS, as part of a global MCS strategy, is designed to reinforce effective fisheries management and vessel safety.[97]

VMS provides near-real-time position monitoring of all vessels fitted with transmitters or transceivers (often incorrectly referred to as transponders).[98] This information is usually sufficient to enable fisheries managers to target conventional air and surface surveillance assets more effectively. In addition, VMS information can be required by a CDS that tracks landings and transshipments to verify details of time and location of catch.[99] VMS can be highly effective at tracking vessels,[100] thereby increasing the chance of detecting of IUU fishing and making IUU fishing less attractive. Nonetheless, many notorious IUU fishing vessels may attempt to exploit loopholes in the regulations and not carry transreceivers. The application of VMS remains a key role in any MCS scheme.

VMS can be divided into three categories: nationally operated, dual, and fully centralized systems. In a nationally operated system, VMS information is relayed to a national fisheries monitoring centre (FMC), and the flag State is responsible for its analysis and use. However, in a nationally operated system, VMS data are not shared with the RFMO. In a dual system, VMS data are relayed to the national FMC, which then relays it to the RFMO secretariat after some delay. In a fully centralized system, the VMS is operated by the RFMO, which receives the data directly from the vessels in near-real time and is able to transmit information directly to inspection platforms. The flag State may also receive the data simultaneously.

Recent attempts at improving MCS have led a number of States[101] to adopt widespread and, in some cases, compulsory use of VMS for vessels fishing within their EEZ or flying their flag or both. RFMOs may either encourage or require member States to impose VMS requirements on their vessels fishing for some or all of the regulated species on the high seas. VMS initiatives are underway in a number of RFMOs — CCAMLR, IATTC, ICCAT, NAFO, NEAFC, and SEAFO (as of 1 April 2006) have adopted measures relating to VMS schemes. The WCPFC Convention provides that all vessels on the high seas will be equipped with VMS, and that VMS data will be sent directly to the Commission. WCPFC recently endorsed the establishment of a satellite VMS as an integral part of its MCS program. Whether this will be a fully centralized or dual system remains to be seen. The BSDH Convention also provides for VMS, though there is no fishing in the area because of the long-standing moratorium on pollock. A few of these RFMOs have had fully functional VMS for several years and have continually improved the respective systems and reporting functions. Despite occasional technical problems, satellite VMS is a powerful MCS tool, and there is enormous potential for expanded use of this technology for data gathering (e.g., monitoring vessel movements and collecting catch data) as well as electronic surveillance.

9.4.7.1 Various schemes

The actual operation or implementation of the schemes by RFMOs varies significantly. Some of these VMS schemes are nonlegally binding, while others call for VMS to be implemented on only a percentage of vessels fishing in the relevant area. Still others call for VMS in certain fisheries only. The frequency of submitting position reports from fishing vessels ranges from whenever available in ICCAT, to transmission every six hours in IOTC, to every four hours for finfish fisheries in CCAMLR, to every two hours in case of vessels fishing in the NEAFC and NAFO regulatory areas. Most of these schemes are nationally operated, although some are dual systems.

The IOTC in 2002 adopted in a nonbinding resolution a pilot program to implement nationally operated VMS schemes to be applied to 10% of vessels larger than 24 m.[102] The pilot program will be reviewed at the 2006 meeting. In 2000 and 2001, IATTC's Compliance Working Group discussed the possibility of introducing compulsory VMS, but concluded that it was not urgent, as there was 100% observer coverage on large purse-seine vessels. Nevertheless, in 2004, IATTC adopted a VMS scheme requiring, “where possible”, Contracting Parties to establish VMS by January 2005.[103] Progress reports on implementation were submitted to the Director of the Commission by 31 May 2005, and discussion is continuing on the implementation of the scheme and the uses to which VMS information might be put.

In 2004 ICCAT resolved that all vessels over 24 m in length that fish within its convention area must carry VMS by 1 November 2005. The ICCAT scheme is nationally operated with no requirement for members to submit VMS data to the Commission.[104] Similarly, all fishing vessels authorized to fish southern bluefin tuna under the competence of CCSBT must now carry VMS, and the data are reported to the flag State only. GFCM does not have a VMS scheme, but VMS is still under consideration and will be discussed at the 2007 meeting.

Since January 2001, NAFO has required all Contracting Parties vessels fishing in the NRA to be equipped with VMS. NAFO has established a comprehensive scheme of VMS messaging, which creates an electronic logbook transmission system from the vessel directly to national FMCs. These messages are then forwarded to Contracting Parties with an inspection presence in the NRA (Canada, Denmark on behalf of Greenland, and the EU).[105]

To curb tampering and to verify falsification of VMS data, the NAFO system requires a dual-reporting system, whereby vessels report to national authorities, who then convey the information to the Secretariat.

A dual scheme has also been in operation in CCAMLR, which has a mandatory requirement for VMS on all fisheries except krill fisheries.[106] VMS data are forwarded in near-real time to the CCAMLR Secretariat through the flag State or, if the flag State so desires, simultaneously to the flag State and the Secretariat. This is similar to the VMS scheme in operation in NEAFC, which requires all vessels fishing in the NEAFC Convention Area to be fitted with VMS.[107] The system is fully automated and reports, via the Secretariat, to members with an inspection presence in the area. The frequency with which position reports are made from fishing vessels has increased from six hours to two hours, and members with an inspection presence are supplied with real-time information about ongoing fishing activities 24 hours a day, 365 days a year.

As with observers, debate persists about whether a wholly nationally operated VMS scheme sufficiently meets the requirements of UNFSA or whether some rationalization of VMS technology and centralization of its administration within the RFMO are both preferable and obligatory. Harmonization of technology among members and centralization of reporting functions can reduce costs and improve efficiency in information exchange and data use. In terms of institutional efficiency and efficacy, some argue the preferred model for RFMOs is a centralized VMS with all members adopting compatible technology. Difficulties have arisen, however, because of the selection of differing technologies (i.e., there is the one-way system (Argos), the two-way system (INMARSAT), and two other bidirectional systems using the Orbcomm satellite or the Iridium satellite — these bidirectional systems now have global coverage and can be programmed from land after the fishing vessel is at sea). Once a VMS satellite technology is adopted, States are typically disinclined to replace their equipment given the economic outlay. Use of incompatible hardware, software, and other standards in national VMS programs compromises the possibility of effective use on a regional basis. A considerable improvement could be achieved if present technologies were able to "communicate" among each other, thereby allowing an exchange of information among the various member States who may have different VMS providers. Synergies could also be achieved if VMS strategies developed by individual States were compatible. One notable example of such an initiative is the North Atlantic Format program deployed in NEAFC and NAFO. Implementation of this standardized format among all RFMOs in the North Atlantic has greatly improved the efficacy of VMS in the region. Adoption of a unified format globally could result in important cost savings for fisheries authorities and owners of fishing vessels and improvements in the MCS capabilities of RFMOs and their members.

The efficacy of VMS as an MCS tool also depends on the frequency at which data are submitted or the number of times the onboard transceiver is "polled". Experience in NEAFC indicates that frequent reporting may be necessary to adequately monitor fishing activity. Debate persists, however, about the efficacy and necessity of frequent position reports, particularly where such reporting may inconvenience legitimate fishing operations where polling is not automatic.

VMS poses technical as well as institutional challenges. Technical problems can arise and require ongoing resolution on a daily basis so that the best use is made of this surveillance tool. More fundamentally, the efficacy of VMS technology may be undermined by natural causes or by human intervention, known as "spoofing". Some systems can simply be switched off. As well, the reliability of VMS transmission may be compromised where the system allows transmission of catch and effort data from the fishing vessel to the monitoring agency, but requires input of the information by the vessel operator.[108] As a result, vessel operators can manipulate VMS to transmit false positions while fishing out of season or in prohibited areas. This can be achieved by tampering with the onboard "black box", which transmits the signal; cloning the onboard communications terminal so that the surrogate gives out false information; and interfering with the outgoing signal from the blue box or the incoming positioning signal from the GPS.[109]

As a result of spoofing, technological improvements and the deployment of other means of verifying VMS and other data are still required. For example, the Orbcomm system is reportedly equipped with a number of operational parameters that make it difficult to spoof.[110] These parameters can be programmed into the Iridium system as well to circumvent spoofing. Additionally, both the Orbcomm and Iridium transceivers can be linked to sensors that can permit personnel onshore to monitor activities on the fishing vessel (e.g., a pressure sensor on the hydraulic system would indicate whether the vessel was pulling in or paying out nets; temperature sensors in the hold would indicate location of product; door sensors would indicate when hatch cover was removed or a hold opened).

The adoption of VMS by developing countries poses a key institutional challenge. To date, adoption of VMS technology, as well as participation in international gatherings and discussions on MCS and VMS, remains limited because of technical and financial constraints. Human capacity building is necessary to ensure that the required level of competence in designing, installing, and operating VMS is achieved. While much of the fisheries sector in the developing world is less technologically advanced, increasingly this is not the case — many fleet segments and vessels are modern and use highly sophisticated technology. Providing training and transferring technology to help developing States deploy VMS are important areas for development assistance.

While satellite-based VMS is an important tool, it can only monitor activities of vessels fitted with the requisite equipment. This equipment can be altered or moved to skew results. VMS does not provide evidence of the manner in which fishing is carried out or confirm the type of gear used or other effects of fishing activities on the wider environment. While considerable potential to expand the applications and uses of VMS technology exists, currently other MCS methodologies discussed in this chapter are still viewed as necessary to detect, deter, and prevent noncompliance with RFMOs' conservation and management regime.[111]

9.4.7.2 Vessel detection systems

Given the limitations of aerial surveillance and VMS, efforts are now being directed towards the development of satellite-based vessel detection systems (VDS).[112] VDS uses the remote sensing capabilities of satellites, whereas VMS uses the communication capability of satellites. There are currently two satellites available for this service and interest is growing. VDS uses technology called synthetic aperture radar (SAR), which is able to detect vessels at sea under all conditions, including darkness and cloud cover. The aim of the VDS is to determine the number of fishing vessels in the area in question and to provide their position. VDS signals can also be checked against VMS data to determine if IUU fishing is taking place.

VDS is not a replacement for VMS, it detects all vessels, whereas VMS detects only those vessels with functioning automatic location communicators. Nor are VDS and VMS absolute replacements for aerial surveillance but serve, rather, to direct air surveillance to areas and particular vessels in question. The application of these technologies is an important advancement in the identification and combating of IUU fishing.

A project was recently initiated to improve observation of fishing activities in the enormous area under the competence of CCAMLR. The use of SAR images will improve the detection of vessels and the capability of distinguishing them from icebergs.[113] Radar satellites have the possibility to store data on board or to send data via a relay satellite link; thus they can be valuable assets for monitoring compliance. Accordingly, there is considerable interest in investigating the capability to remotely map vessel positions in Antarctic and near-Antarctic waters.

Under a project by the Norwegian Defence Research Establishment, there will be sampling of some ENVISAT satellite images over Bouvet Island and southwards to distinguish between icebergs and vessels. The "Feasibility of ENVISAT Ship Detection for Fisheries Enforcement in the South Atlantic/Antarctica" is an ENVISAT-AO (Announcement of Opportunity) project. The European ENVISAT satellite is able to deliver information in a timely manner using Artemis or a solid-state recorder.

The project's goal is to demonstrate ENVISAT ASAR (Advanced Synthetic Aperture Radar) instrument's suitability for ship detection and the possibility for obtaining information quickly from the areas of interest, and to quantify the risk of false alarms caused by Antarctic icebergs. The project assesses the limitations and capability of ENVISAT, and it does not include operational use for real control or enforcement operations until thought to be successful. Databases on iceberg observations in Antarctic waters already exist, and icebergs can easily be detected at large incidence angles. A Canadian study analysing the difference between icebergs in co- and cross-polarized data was recently completed to improve vessel detection capability.[114]

9.4.8 At-sea boarding and inspection

UNFSA Article 18(3)(g)(i) calls for "the implementation of national inspection schemes and subregional and regional schemes for cooperation in enforcement pursuant to articles 21 and 22, including requirements for such vessels to permit access by duly authorized inspectors from other States". Within their EEZ, the sovereign right of coastal States over living marine resources provides the legal basis for inspection of their own and other flagged vessels. On the high seas, the right to inspect arises from other rules of international law, including bilateral arrangements or regional agreements, such as those establishing RFMOs, and the provisions of UNFSA.

Inspection means the examination of a vessel, gear, and catch for possible violations of the conservation and management measures adopted by an RFMO. The purpose of an at-sea inspection scheme is to establish a set of agreed upon conditions governing the powers and responsibilities of parties during an inspection at sea. The scheme is intended to provide a clear understanding of the respective roles of the vessel's crew and the boarding party during the inspection of the vessel's records, fishing equipment, and catch on board. The objective of the inspection is to verify that the vessel operators and owners are complying with international rules relating to data collection as well as with conservation and management measures. Inspection of a vessel at sea (or in port) also provides an opportunity to collect, first hand, cogent and accurate evidence, including interviews, documents, and photographs, which can be used in any future legal action that may arise as a result of the inspection. In this respect, the evidentiary value of at-sea inspections exceeds that of VMS and other MCS options.

An inspector is most often a fisheries or coast-guard officer of the State who is embarked on a patrol vessel or another official platform that carries out surveillance and enforcement duties.[115] National inspection schemes are intended to apply to all vessels and within all areas falling under the jurisdiction of a State. In their home State, inspectors generally possess powers of arrest, search, and seizure for offences occurring within areas under national jurisdiction and on board flag vessels, including on the high seas. However, geography and cost often render it difficult, if not impossible, for flag States to carry out inspections on the high seas. While Japan, for example, has, in the past, sent inspection vessels to monitor the activities of Japanese vessels fishing for tuna in the Mediterranean, such long-range deployment of inspection platforms by flag States is rare.[116]

9.4.8.1 Reciprocal boarding and inspection

Reciprocal boarding and inspection can be used to alleviate both practical and financial constraints and to assuage doubts about the efficacy of national inspections. Thus, Article 21(1) of UNFSA provides that State Parties to UNFSA who are members of or participants in an RFMO or arrangement are authorized to board and inspect fishing vessels flying the flag of any other State Party regardless of whether it is a member of or participant in the RFMO or arrangement in question. It has been argued that this, in effect, means that a State Party to UNFSA may become legally bound to apply conservation measures adopted under a regional agreement to which it is not a party,[117] and the provision has been labelled a "unique and far-reaching exception to the flag State principle"[118] and "the beginning of a new era in high seas fisheries enforcement".[119] However, nothing in this provision represents a departure from established principles of international law including the fundamental principles of *pacta tertiis*, according to which States are not bound to the provisions of a treaty unless they have in one way or another consented to them.[120] States Parties to UNFSA have accepted the obligation for vessels flying their flag to comply with RFMO measures or to refrain from fishing. Where such vessels fail to do either, UNFSA provides this mechanism to allow other UNFSA Parties to take action. As a matter of treaty law, there is nothing unique or far-reaching in this provision. It represents merely an agreement among State Parties to UNFSA on a *modus operandi* for nonflag-State boarding and inspection.

With respect to the application of UNFSA's boarding and inspection regime, Article 21(1) makes clear that the exercise of boarding and inspection rights between UNFSA Parties is limited to waters that fall within the geographical competence of RFMOs. In other words, no UNFSA boarding can be undertaken by one State within the EEZ of another State[121] or in high seas areas that are not governed by a relevant RFMO. In addition, action may be undertaken only by a member of the relevant RFMO and only for the purpose of ensuring compliance with conservation and management measures adopted by that RFMO. UNFSA Parties who are not members of the relevant RFMO may not invoke Article 21 and 22 against other UNFSA Parties who are. Additionally, only fishing vessels flying the flag of a State Party to UNFSA can be subjected to these enforcement measures, whether or not that State is also a member of the relevant RFMO. As a matter of treaty law, UNFSA can have no third-party application. This does not preclude, however, the development of rules of customary international law reflecting the rules in UNFSA that will be binding on all States, including nonparties to UNFSA.

9.4.8.2 Articles 21 and 22 procedures

The basic procedures for boarding and inspection under UNFSA are set out in Article 22, including the duties of the inspectors, the scope of the inspection, and the obligation of the flag State to ensure that masters of their vessels comply with the inspection requirements and, where they do not, to suspend the vessel's authorization to fish and order it to return to port immediately. Inspectors have the authority to inspect the vessel, its licence, gear, equipment, records, facilities, fish and fish products, and any relevant documents necessary to verify compliance with the relevant conservation and management measures. Inspectors are to avoid the use of force except to the degree necessary to ensure their safety and where they are obstructed in the execution of their duties. The degree of force used must not, however, exceed that reasonably required in the circumstances. Flag State Parties to UNFSA are to ensure that their vessels accept, facilitate the safe conduct of, and cooperate with the inspectors in the performance of their duties. If a master refuses to accept a boarding and inspection then, subject to safety considerations, the flag State must order the master to submit. In cases of noncompliance, the flag State is required to suspend the vessel's authorization to fish and order the vessel to return immediately to port.

To improve regional application of these measures, Article 21(2) goes on to require the establishment through RFMOs of nondiscriminatory procedures for boarding and inspection of vessels fishing for straddling fish stocks and highly migratory fish stocks. These procedures are to be consistent with the provisions of Articles 21 and 22 of UNFSA. Where such UNFSA-consistent RFMO procedures have not been established within two years of the adoption of UNFSA, in other words by August 1997, then Article 21(3) provides that the UNFSA procedures apply by default.

In addition, pursuant to Article 21(15), where an RFMO has established an "alternative mechanism which effectively discharges the obligation" under UNFSA to ensure compliance with that organization's conservation and management measures, RFMO members may agree to limit the application of Article 21(1) between themselves in respect of that particular high seas area. However, where the UNFSA default procedures apply pursuant to Article 21(3), it would appear that any RFMO member who has not become a party to any such agreement *inter se* may apply the UNFSA boarding and inspection procedures in the regulatory area to any vessel flying the flag of an UNFSA State Party.

9.4.8.3 At-sea boarding and inspection schemes

At-sea boarding and inspection schemes have been adopted by most of the RFMOs having competence for straddling fish stocks, namely BSDH Convention,[122] CCAMLR,[123]NAFO,[124] and NEAFC.[125]In the cases of BSDH and NAFO these schemes predate UNFSA. The post-UNFSA SEAFO Convention provides for the adoption of a scheme,[126] although as of May 2006 this had not occurred.

To date, most RFMOs responsible for highly migratory fish stocks have not adopted at-sea inspection schemes. ICCAT adopted a boarding and inspection scheme in 1975 but has never implemented it. The issue is being revisited by ICCAT and its Working Group on the Development of Integrated Monitoring Methods. IOTC has similarly not pursued the inclusion of at-sea boarding and inspection in its control and enforcement scheme. CCSBT, GFCM, nor IATTC have plans to introduce an at-sea boarding and inspection regime.

The notable exception is WCPFC. The WCPFC Convention makes specific provision in Article 26 for a boarding and inspection scheme and specifically provides that in the event that the Commission is unable to agree on boarding and inspection procedures or alternative measures within two years from the date of the entry into force of UNFSA, the UNFSA regime will apply. The WCPFC Convention entered into force on 19 June 2004. At the December 2005 Annual Meeting of the WCPFC it was agreed that Articles 21 and 22 of UNFSA would come into effect on 19 June 2006. WCPFC members agreed to form a working group to finalize modalities for implementation prior to the 2006 Annual Meeting and not to act on the provisions of UNFSA Articles 21 and 22 until that time. Accordingly, as yet, there is no evidence of the UNFSA provisions having been specifically invoked to justify nonflag at-sea boarding and inspection in any RFMO. It remains to be seen what happens in the WCPFC.

9.4.8.3.1 NAFO[127]

At-sea boarding and inspection has a long history in the NAFO Convention Area. Given that NAFO has one of the most developed and long standing boarding and inspection schemes of all the RFMOs, a more detailed description is provided herein. The NAFO Scheme, while well developed and the most detailed, is nonetheless not viewed by all States as being consistent with UNFSA.[128]

In 1970, NAFO's predecessor, ICNAF, adopted the Scheme of Joint International Enforcement of the Fishery Regulations in the Convention Area, which came into force in 1971. The ICNAF Scheme was similar to that adopted by NEAFC in 1967. By 1977 the extension of coastal-State jurisdiction was affecting the ICNAF Scheme, and discussions commenced on possible amendments to make it more compatible with the reduction in its area of application.[129] An extensive range of amendments were adopted by both ICNAF and NAFO, and in 1979 they came into force as the new NAFO Scheme of Joint International Enforcement.[130] However, this scheme met with limited success because of noncompliance with conservation measures and use of objection procedures. A series of difficult negotiations followed and resulted in the adoption in 1988 of a revised scheme. Although this revised scheme resulted in the EU sending an inspection vessel to the NRA for the first time, apparently this did little to assuage Canadian concerns about its integrity or effectiveness.[131]

In 1991, NAFO adopted the Guidelines for the Coordination and Optimization of Inspection and Control in the NRA.[132] These guidelines provided for an exchange of information among member States. Names and identification of inspectors and inspection vessels, together with their estimated times of arrival in and departure from the NRA, were to be notified to the Secretariat and other members with an inspection presence in the area. The Secretariat, in turn, was to provide the inspection vessel entering the area with a list of sightings and boardings including dates and position, which had been conducted in the previous 10-day period, together with any other relevant information. Inspection vessels were then to remain in contact on a daily basis, if possible, to continue to exchange information about sightings and boardings. Where possible, members were to exchange inspectors to develop a consistent approach to inspection and control in the NRA. Reports of inspection activity were to be prepared annually, outlining details of sightings, boardings, and apparent infringements.

In 1992 some modifications were adopted to provide for air surveillance and the introduction of a limited national observer pilot program.[133] In the Canada – European Community (EC) Agreement of December 1992, the parties undertook to work together to improve surveillance and control in NAFO. However, because of continuing disputes over the objectivity of inspection activities and accusations of ineffective flag-State follow-up and sanction for infringements, the scheme's efficacy remained limited. Amendments adopted in 1995 pursuant to the Canada–EC agreement[134] settling the *Estai* affair[135] were intended to rectify this. These related to increasing the effectiveness of at-sea inspections by allocation of inspection vessels and objective distribution of inspections among members; specific inspection procedures; advance notification of apparent infringements; reporting on suspected illegal practices; follow-up inspections and possible recall to port for enforcement action on infringements; and reporting and other requirements regarding disposition and follow-up of apparent infringements.[136] Consequential amendments have been made in subsequent years to provide further detail to the scheme and increase its effectiveness. The NAFO Scheme appears now to be the most detailed in existence.

Like its forbears, the NAFO Scheme of Joint International Inspection and Surveillance provides for reciprocal at-sea boarding and inspection between NAFO members.[137] It does not, however, provide for nonflag prosecution and sanction. This is still left to the flag State, although a number of provisions have been included to provide verifiable incentives for flag States to do so. Inspection under the scheme applies to all fishing vessels that are or have been engaged in fishing activities in the NRA. It also applies to all vessels equipped for fish processing that are or have been engaged in fish transferring operations in the NRA and to all transport vessels that are actually engaged in fish transferring operations. Inspection is carried out by designated inspectors operating from designated inspection platforms, and parties are at liberty to agree to placement of non-national inspectors on their inspection platforms. To ensure that the identity of inspectors and platforms is known, names of designated inspectors and vessels and call signs of assigned aircraft must be notified annually to the Secretariat. Identity documents, which must be carried and produced by all inspectors upon boarding a vessel, are then issued by the Secretariat. In addition, details relating to the commencement and termination of inspection activities are to be reported, and during the time they are on NAFO inspection duty, inspection platforms may only conduct NAFO-related activities. They may not conduct national inspections at the same time and may not enforce laws and regulations related to waters under the jurisdiction of their designating authority except in respect of their own flagged vessels.

The NAFO Scheme attempts to ensure prompt flag-State reaction and to avoid misunderstandings in the communication chain. Members must designate specific authorities to receive notices of apparent infringements and must even specify the means by which they will receive and respond to these communications. This information is to be provided to the Secretary, who acts as a clearing house in the case of requests for contact information by members. In addition, whenever a member party has more than 15 vessels engaged in fishing operations or the processing or transfer of fish from the NRA, it must ensure that it has an inspector or other designated authority present either in the NRA or in the territory of a coastal member to receive and respond without delay to notices of apparent infringements. This threshold number of vessels has been a matter of dispute in NAFO since its earliest days, with proposals ranging between 10 and 20.

Detailed requirements are set out relating to signals, inspection pennants, radio call signs, and radio communications. In addition, provisions are included relating to maintenance of safe boarding conditions, facilitation of boarding, and masters' duties once signal has been given that an inspection party intends to board. There is, however, no doctrine of absolute liability in the case of accident or injury during boarding operations. Presumably more general principles of liability for negligence do, however, apply to discourage the potential for injury or harm to inspectors. This is important given the requirement that no more than two inspectors, who must be unarmed, may board a vessel for the purposes of a nonflag inspection. Indeed, the use of arms in relation to inspections is strictly prohibited, although it is not clear whether this applies only to personal fire arms or also to force used against a vessel to stop it from fleeing.

Inspection time is not to exceed three hours or until the net is hauled and the net and catch have been examined, whichever is longer. In cases of apparent infringement this time limit does not apply. The usual provision with respect to conducting inspections in a manner that minimizes inconvenience and interference with catch and vessel is included, as are lengthy and specific provisions detailing the activities that inspectors may engage in when examining catch, nets, or other gear and any relevant documents deemed necessary. The objective is for the inspector to prepare an estimate of catch on board, which is then cross-checked with logbook entries, and any difference between the two is recorded in the inspection report.

In recognition of the effect of the objection procedure, inspectors may only inspect and enquire on the basis of measures to which the flag State of the vessel has not objected. The objection procedure therefore has the potential to undermine the efficacy of both the inspection scheme and conservation and management measures. Nevertheless, in the interest of ensuring that information is gained for scientific and management purposes even if quotas have been objected to, information on catch by species and area taken is to be summarized from the logbooks. In this way, information on total catches in the NRA can be obtained while maintaining the integrity of the objection procedure.

Inspectors may ask the master for any required assistance, and masters are obliged to facilitate their work. Resistance to an inspector or failure to comply with his or her directions is to be treated by the flag State of the inspected vessel as resistance to a national inspector. In addition, where boarding has not been enabled after proper signals have been given, the inspector is to report this as an apparent infringement as soon as possible to any flag State inspector in the vicinity or to the designated authority of the flag State. A detailed report of the incident, including information on the nature of the signal given, date, time, vessel position, sea conditions, wind, visibility, icing conditions, and so on, is also to be sent to the Secretariat.

Once the inspection is complete, a copy of the inspection report is to be given to the master. If no infringement is found, the original is transmitted to the designated authority of the flag State within 30 days of inspection. Copies of every report are also to be forwarded to the Secretariat. If an apparent infringement has been found, the inspector must note it in the report, sign the entry, and obtain the countersignature of the master. A notation of the details of the apparent infringement must also be entered in the vessel's logbook. Copies of all documentation and photos taken are to be given to the master and attached to the report sent to the flag State. Details of apparent infringements must be communicated to the designated flag-State authorities within one working day following the inspection where possible. In addition, a statement constituting advance notification of the apparent infringement must be sent to the flag State, with a copy to the Secretariat within 24 hours. Any notes or comments and observations must be accompanied by a statement citing relevant NAFO Measures, describing the practices observed, and substantiating their grounds for inspection. This must also be sent within 24 hours to the flag State of the inspected vessel and the Secretariat. The original inspection report identifying the apparent infringement or difference between recorded and estimated catches, together with all supporting documentation, must then be sent to the inspected vessel's flag-State authorities within 10 days after the inspection vessel returns to port.

Where an apparent infringement concerns fishing in a closed area or with prohibited gear or after quota attainment, inspectors must immediately attempt to communicate with an inspector of the flag State known to be in the vicinity or with the designated flag-State authority for the purposes of receiving further instructions. At the inspector's request, the master must cease violative fishing operations while the inspector completes the inspection, and if unable to communicate with the flag State within a reasonable time, the inspector must leave the vessel, which is then free to resume its activities. However, if contact is made and the flag State consents, the inspector may remain on board, and the master cannot resume fishing until the inspector is satisfied the infringement will not reoccur.

Flag States notified of apparent infringements that have been committed by their vessels are obliged to receive and consider the evidence provided by the inspecting State and to promptly conduct necessary investigations to obtain the evidence required. Whenever possible they are to board the vessel involved themselves. In particular, where differences between recorded and estimated catch are noted, the flag State shall, whenever possible, board the vessel and in any event cooperate with the inspectors to ensure proper preparation and presentation of evidence and shall investigate to determine appropriate follow-up. They must take immediate judicial or administrative action, as though the violations occurred within national waters, and they must cooperate with the inspecting State to ensure evidence of apparent infringements is prepared in a form that facilitates judicial action.

More definitive action is required in the case of certain apparent infringements. If an inspector cites a vessel for having committed "to a serious extent" an apparent infringement relating to misreporting of catches, mesh-size violations, hail-system violations, interference with VMS, preventing an inspector from carrying out her or his duties, or directed fishing for moratoria stocks or other stocks for which fishing is prohibited, then the inspector must immediately notify the Secretary, who must, in turn, notify other Contracting Parties with an inspection presence in the area. Notably, no definition of "serious extent" is given. In any event, the flag State must then ensure that the vessel is inspected within 72 hours — either by a flag-State inspector or by one authorized by the flag State. The original inspector may take all measures necessary to secure the evidence and may remain on the vessel until such time as he or she has been able to provide information to the authorized flag-State inspector.

Where justified, the flag State or its inspector may require the vessel to proceed to a designated port (chosen by the master) for inspection under flag-State authority and in the presence of a NAFO inspector from any other member that wishes to participate. In addition, subject to the consent of the flag State, if a vessel is obliged to proceed to port, then an inspector from another member may board the vessel and remain on board as it proceeds to port and be present during the inspection. Flag States that do not call a contravening vessel to port must provide due justification for failing to do so to the Secretary, who will make it available to any member on request.

The NAFO Scheme also contains provisions relating to "surveillance". When sighting information does not correspond to information provided by the hail or VMS systems, inspectors shall photograph and record the details of position, etc., verify them with the flag State, and within 72 hours, complete the surveillance report, which is to be forwarded to the flag State or its designated authority and to the Secretary. On receipt of a surveillance report, a flag State must take prompt action to receive and consider it and, wherever possible, board the vessel concerned and conduct any further investigations necessary to determine appropriate follow-up action. In this way, discrepancies in reported vessel locations can be identified, and the temptation to tamper with VMS or to otherwise provide false information is somewhat minimized.

The flag State of an inspected vessel is obliged to act on nonflag inspection reports as though they were their own. They are also obliged to give them appropriate evidentiary weight and to further cooperate to facilitate judicial or other proceedings arising from an inspection report. Nevertheless, ultimate discretion with respect to prosecution and sanction of apparent infringements rests with the flag State. The scheme does, however, provide a mechanism in the case of dispute between the inspecting and flag State Parties over issues such as failure or refusal to investigate or prosecute. When the inspecting and flag States disagree over the interpretation or application of the scheme, they shall first consult in an attempt to resolve the issue. If not successful, then the issue may be referred to a Standing Committee on International Control (STACTIC) special meeting, and a report of the STACTIC deliberations will be drawn up and given to the Fisheries Commission. There is no indication of what the Fisheries Commission is to do with the report or what other actions might be taken to resolve the issue, but clearly it is hoped that public scrutiny of this sort will suffice to bring the parties together to ensure no future dispute on the same issue. This mechanism has, however, never been used.

With respect to Non-Contracting Parties, the NAFO Non-Contracting Party Scheme[138] does not specifically provide for boarding and inspection of Non-Contracting Party vessels. Rather, the possibility is provided for by implication. Article 8 of the scheme provides that if a Non-Contracting Party vessel that has been sighted and reported as having engaged in fishing activities in the NRA consents to be boarded by NAFO inspectors, the finding of the inspectors shall be transmitted to the Secretariat. The Secretariat will, in turn, forward the information to all Contracting Parties within one day and to the flag State of the boarded vessel as soon as possible. The boarded vessel is to be provided with a copy of the inspector's findings.

With respect to stateless vessels, in 1999 NAFO adopted a measure stipulating that where there are reasonable grounds for suspecting that a vessel which has been sighted fishing in the NRA is stateless, a NAFO member may board and inspect the vessel. Where the evidence so warrants, the member can take such action as may be appropriate in accordance with international law. Members are to examine the appropriateness of their domestic measures relating to the exercise of jurisdiction over such vessels and provide annual reports to NAFO, including negative returns, on the actions they have taken to implement these requirements against Non-Contracting Party and stateless vessels.

9.4.8.3.2 CCAMLR[139]

The CCAMLR Convention specifically requires establishment of an at-sea observation and inspection system. The CCAMLR System of Observation and Inspection (OIS) was adopted by the Commission in 1988[140] and came into force for the 1989–1990 season.

The OIS is a nationally operated international system that provides for nonflag boarding and inspection. Inspectors are appointed by and report to their own governments, which, in turn, report to the Commission. All members of the Commission can designate inspectors,[141] who must be nationals of the designating State[142] and who must speak the language of the flag State of the vessels on which they carry out their activities.[143] The Commission maintains a register of designated inspectors[144] and of commercial and scientific vessels that will be operating in the convention area,[145] all of which are subject to the system.[146]

Inspection can be carried out from vessels of the designating States or from on board vessels subject to the inspection.[147] In the latter case, arrangements for scheduling inspections are a matter between the flag States and the designating State.[148] While carrying out their duties, inspectors are subject solely to the jurisdiction of the designating States.[149] Vessels carrying inspectors fly a special flag to indicate they are on board and carrying out their duties in accordance with this system.[150]

When requested to do so, all commercial and scientific vessels must promptly arrange for transfer of the inspector on board. When coming aboard, inspectors must show their approved identity documents. While on board, inspectors have the authority to inspect catch, nets, and other fishing gear, as well as harvesting and scientific activities, and must be given access to records and reports of catch and location data. They may take photographs, document violations, and tag gear that appears to have been used in contravention of conservation measures, and they must be given appropriate assistance and have access to communications equipment. They are, however, required to carry out their duties with a minimum of interference and inconvenience.[151]

Upon completion of their duties, inspectors provide the master of the vessel with a certificate of inspection and a notice of any violations observed. The master must acknowledge receipt of the notification and is entitled to comment on it in writing.[152] The flag State is obliged to take necessary steps to prosecute and impose sanctions for interference by its nationals with an inspector or for their failure to comply with reasonable requirements.[153]

Inspectors are required to compile detailed reports on their activities. The reports are provided to the designating State, which, in turn, makes the reports available to the Commission. The reports are also provided to the flag State, which is entitled to comment on them prior to their consideration by the Commission.[154] Prosecutions and imposition of sanctions for violation of conservation and other measures adopted by the Commission are the responsibility of the flag State, which is obliged to report the measures taken to the Commission.[155] However, no recourse exists against flag States for failure or refusal to prosecute.

Over the years, a number of improvements have been made to the OIS. To improve the Commission's ability to gauge the effectiveness of the OIS, since 1993 members have been asked to supply information about the actual number of inspectors deployed, the duration of their trips, and the areas covered, in addition to the numbers of designated inspectors and inspections actually conducted.[156] Compliance with this request has, however, not been overwhelming. Time frames have also been established for transmission of inspection reports to the designating member, the flag State, and the Commission, and for providing comments thereon.

A series of amendments made from 1994 to 1997 to Articles VII and VIII of the OIS provide an intersessional procedure for notifying members of infractions and actions taken.[157] Completed inspection forms are to be provided by the inspector to the designating member "not later than 15 days of his/her arrival in port".[158] "Not later than 15 days from its reception" the designating member must forward a copy of the inspection form to the Executive Secretary, who shall forward it to the flag State "not later than seven days from its receipt".[159] Fifteen days "after the transmission" of the inspection form to the flag States, the Executive Secretary shall submit that form to members "together with any comments or observations" received from the flag State.

Article IX, as amended in 1996, provides that supplementary reports or information prepared by inspectors or any report on refusal to stop or facilitate inspection or other interference with inspection, prepared in accordance with Article VII, shall likewise be provided by the designating members to the Executive Secretary and then to the flag States for comment and observation. Within 15 days of receipt it shall also be transmitted to members. Comments and observations received from the flag State are also to be transmitted to the members.[160]

Although not originally envisaged, amendments made in 1995 provide for the use of videotaped evidence. In addition, copies of any video footage as well as duplicate copies of any photographs taken are to be provided along with a copy of the inspection to both the designating member and the Executive Secretary, who shall then forward a copy to the flag State of the inspected vessel.

To overcome the problem of vessels seeking to evade inspection by claiming merely to be in transit through the convention area, in 1995 the OIS was further amended to confirm that the right of inspection extends to any fishing or fisheries research vessel registered in the territory of a member and present in the convention area.[161] Agreement on a definition of "fishing" was also achieved in 1995. The definition, which was further amended in 1996 and is now contained in Article X of the OIS, provides a list of indicators from which a rebuttable presumption can be established that fishing has occurred or is about to occur.[162] It is then up to the flag State, on the basis of the inspector's report or other evidence and the domestic law of the State, to determine whether a violation has occurred and, if so, what sanction will be imposed.[163]

Other amendments added to the system in 1997 reflect various requirements set out in UNFSA. Information about licences and permits is to be supplied to the Commission within seven days of the date of issue. The Commission must circulate the information to all parties within a further seven days. The OIS has also been further fine-tuned by amendments relating, *inter alia*, to time frames for reporting names of inspectors, lists of licenced vessels, and information on infringement follow-up, to the Commission.[164] New provisions added in 1997 also oblige flag States to prosecute and impose sanctions where evidence of violation exists. Sanctions imposed must be sufficiently severe as to ensure compliance with conservation measures and to deprive offenders of any economic benefit accruing from their illegal activities. Flag States are to report at least once a year to the Commission on prosecutions and sanctions and the reasons for any failure to prosecute or to secure conviction, and they must ensure that vessels do not resume fishing in the CCAMLR area until any sanctions imposed have been complied with.[165] However, beyond this public scrutiny, no provisions exist to specify what should happen should flag States fail to prosecute or sanction or stop vessels from fishing. Ultimate discretion still rests with the flag State.

9.4.9 Issues arising from UNFSA Articles 21–22

Three fundamental issues have been raised concerning the implementation of UNFSA Articles 21–22. The first issue is whether a pre-UNFSA RFMO that has not yet adopted a boarding and inspection scheme is now required to adopt the UNFSA scheme. The second issue concerns the extent to which pre- or post-UNFSA schemes must be consistent with UNFSA to satisfy Article 21(2), and whether UNFSA represents a compulsory minimum or maximum standard. Related to this is the third issue, namely what constitutes an "alternative mechanism" to "effectively discharge" the obligation to ensure compliance with the conservation and management measures established by the RFMO for the purposes of application of Article 21(15).

9.4.9.1 Obligation to adopt a boarding and inspection scheme

With respect to the first issue, as a matter of treaty law, UNFSA is binding only on its Parties. In an RFMO where all members are party to UNFSA and no boarding and inspection scheme currently exists, the UNFSA provisions must be applied as a minimum. In RFMOs where none of the States are party to UNFSA, adoption of the UNFSA regime would not be obligatory.

The more difficult scenario arises in RFMOs with "mixed" membership, that is, where some members are party to UNFSA and some are not. For State Parties to UNFSA, the regime has set out that there is a minimum they can accept to comply with their international obligations. Further, in any particular convention area, any UNFSA Party may, pursuant to Articles 21 and 22, board fishing vessels entitled to fly the flag of any other UNFSA Party. States that are not party to UNFSA are not so bound. This can result in paralysis within an RFMO and prevent the adoption of a boarding and inspection regime. This would appear to be the situation in the tuna RFMOs, where only WCPFC has adopted provisions for the implementation of a boarding and inspection scheme. In each of the other tuna organizations, objections to the adoption of such a scheme have related primarily to cost and efficacy in the context of their specific fisheries, rather than to absence of legal obligation. Arguably, the position of the tuna RFMOs is better interpreted as an attempted application of Article 21(15), leaving UNFSA Parties to apply Articles 21 and 22 among themselves.[166]

It is also arguable that the obligation to adopt a boarding and inspection scheme (unless an equally effective alternative is adopted) now rests on all members of all

RFMOs as part of the duty to cooperate. Numerous international instruments — both legally binding and non legally binding — call upon States to improve MCS measures and prevent, deter, and eliminate IUU fishing. Failure to adopt an equally effective scheme could therefore result in one of two situations. Between UNFSA Parties, be they members of the RFMO or not, the UNFSA provisions will apply. Between members of the RFMO who are UNFSA Parties and members who are not, it may be argued that these members may well be in a breach of their obligation to cooperate. The rectification of such a breach will lie in the adoption of a boarding and inspection scheme that is consistent with the UNFSA scheme.

9.4.9.2 The question of consistency

With respect to the second issue, the first question that arises is whether members of an RFMO who are also Parties to UNFSA can use the UNFSA provisions or are confined to using the existing RFMO regime. One view suggests that where there is an existing RFMO scheme, then it will apply vis-à-vis all members. Another view suggests that the UNFSA procedures are always available between UNFSA members. In effect, this latter view appears to establish a two-tiered boarding and inspection regime within an RFMO: members who are party to UNFSA being liable to boarding and inspection under the UNFSA provisions, and members who are not party to UNFSA being subject to boarding and inspection under the RFMO regime. This may be precisely what the negotiators of UNFSA intended. However, to date within RFMOs with boarding and inspection regimes, no UNFSA Party has yet relied on the provisions of UNFSA to board and inspect the vessels of another UNFSA Party — whether member or nonmember of the RFMO.[167] In other words, RFMO members who are party to UNFSA have heretofore not used the opportunity of enforcing the RFMO measures against nonmembers who are also UNFSA Parties.

Arguably, some UNFSA Parties may now be precluded by passage of time from relying on its provisions against other RFMO members that are Parties to UNFSA. The contrary argument is that in the absence of an express declaration by an RFMO that its boarding and inspection regime is UNFSA consistent, it is still open to UNFSA Parties to apply Article 21 and 22 procedures. The difficulty here is that RFMO members may or may not address the applicability of UNFSA for a range of reasons, and even where they do, given the opposing side of the argument, agreement is unlikely. Resolution of this debate is a viable issue for dispute settlement.

Additionally, this issue of compliance by UNFSA Parties with their UNFSA obligations is linked to the issue of consistency. What level of consistency as specified in Article 21(2) is required, how and by whom is this to be determined, and is some declaration of consistency necessary to oust the operation of UNFSA in favour of the RFMO scheme as between its parties? The issue of consistency has arisen within NAFO and in other RFMOs whose enforcement regime predates UNFSA. It has been the position of several States that a "specific statement" would have to be issued, based on an explicit decision by the RFMO to "set aside" the UNFSA provisions in favour of the RFMO procedure, and that in the absence of such a statement, the UNFSA regime would apply between RFMO members who are also party to UNFSA.

In most legal systems, *de minimus* claims will be rejected. In other words, where small variations of minimal importance and effect exist between RFMO schemes and the UNFSA regime, it is unlikely that any complaint regarding lack of consistency will be decisive. However, potentially significant differences exist between the RFMO schemes described above and the UNFSA scheme. For example, UNFSA and most RFMO schemes use a list of "serious violations", or in NAFO's case "serious infringements",[168] the detection of which triggers a more stringent level of response by both the inspecting and flag States. While there are some similar, and indeed identical, provisions in the lists of UNFSA and the RFMOs, some lists are sufficiently different to cause difficulties, especially in the situation where items on the UNFSA list are not included in the RFMO list. Table 9.1 indicates, by way of example, the level of difference among the UNFSA, NAFO, and NEAFC schemes.

As is evident from Table 9.1, both NAFO and NEAFC have a comparable list of "serious infringements". However, a close examination reveals that the NEAFC list duplicates most of the UNFSA list, while the NAFO list, though similar, has several additional specific offences and does not include the offence of "concealing or tampering with, or disposing of evidence". Nor is there an offence for "multiple violations which together constitute a serious disregard of conservation and management measures". It is open to debate whether the lack of consistent (i.e., updated or UNFSA compliant) definitions would allow an UNFSA Party to take action against the vessel of another UNFSA Party for a "serious violation" under the Fish Stock Agreement even if this was not a "serious infringement" under the respective regional arrangement.

A more obvious difference exists regarding an important development in the law found in UNFSA. Article 21(8)

Table 9.1. Comparison among Article 11 of UNFSA and provisions in the NAFO and NEAFC schemes (prepared by H. Williamson).

UNFSA Art. 11 — "serious violations"	NEAFC Art. 20 — "serious infringements"	NAFO Art. 3 —"serious infringements"
(a) fishing without a valid licence, authorization, or permit issued by the flag State in accordance with Art. 18, para. 3(a)	(a) fishing without a valid authorization issued by the flag Contracting Party	(f) fishing without a valid authorization issued by the flag Contracting Party (Art. 14)
(b) failing to maintain accurate records of catch and catch-related data, as required by the relevant RFMO or RFMA, or serious misreporting of catch, contrary to the catch reporting requirements	(d) serious misrecording of catches	(g) misrecording of catches (Art. 20)
(c) fishing in a closed area, fishing during a closed season, or fishing without or after attainment of a quota established by the relevant RFMO or RFMA	(b) fishing without, or after attainment of, a quota	(d) fishing in a closed area or with gear prohibited in a specific area (Art. 12)
(d) directed fishing for a stock that is subject to a moratorium or for which fishing is prohibited	(g) directed fishing for a stock that is subject to a moratorium or for which fishing is prohibited	(b) directed fishing for a stock that is subject to a moratorium or for which fishing is prohibited (Art. 9)
(e) using prohibited fishing gear	(c) use of prohibited fishing gear	(d) fishing in a closed area or with gear prohibited in a specific area (Art. 12) (e) mesh-size violations (Art. 10)
(f) falsifying or concealing the markings, identity, or registration of a fishing vessel	(h) falsifying or concealing the markings, identity, or registration of a fishing vessel	
(g) concealing, tampering with, or disposing of evidence relating to an investigation	(i) concealing, tampering with, or disposing of evidence relating to an investigation	
(h) multiple violations that together constitute a serious disregard of conservation and management measures	(j) multiple violations that together constitute a serious disregard of conservation and management measures	
(i) such other violations as may be specified in procedures established by the relevant RFMO or RFMA.	See below for additional violations specific to NEAFC	See below for additional violations specific to NAFO
	(f) preventing an inspector from carrying out his or her duties	(j) preventing inspectors or observers from carrying out their duties (Arts. 24 and 30)
	(e) repeated failure to comply with the provisions of Arts. 9 (VMS) 10 (communication of catches), and 12 (communication with Secretariat)	(h) interference with the satellite monitoring system (Art. 22)
	same as (e) above	(i) catch-communication violations (Art. 23)
		(a) fishing on an "others" quota without prior notification to the Executive Secretary, or more than seven working days after the notification by the Executive Secretary that fishing under an "others" quota for that stock or species was closed (Art. 3).
		(c) directed fishing for stocks or species after the date on which the Contracting Party for the inspected vessel has notified the Executive Secretary that vessels of that party will cease a directed fishery for those stocks or species (Art. 3)

allows an inspecting State to bring a vessel into port in the event a serious violation is found and the flag State has either failed to respond or failed to take action within three working days. The NAFO and NEAFC schemes[169] reserve this right to the flag State, making no provision for situations in which the flag State is unwilling or unable to comply. The CCAMLR scheme also makes no mention of the issue.

Debate persists about which State is entitled to make the determination that a flag State is not fulfilling its obligations as required under UNFSA, therefore opening the door to the inspecting State taking the specific enforcement actions allowed under UNFSA Article 22. One position is that the determination rests solely with the flag State and that no vessel may be brought to port in any circumstances without flag-State consent. The contrary position is that the inspecting State must be able to make that determination; otherwise the provisions of Article 22 would be meaningless and never actionable.

Reading down UNFSA to meet the standards of RFMO regimes simply nullifies UNFSA and may be a breach, by the relevant State, of its obligations thereunder, including the obligation of good faith set out in UNFSA Article 34. Accordingly, an alternative interpretation of UNFSA Article 21(1)–(3) might be that an UNFSA Party must elect whether to enforce an existing inconsistent scheme against Parties to UNFSA until the existing scheme is updated to be consistent with UNFSA Articles 21 and 22 or, based on its assessment of inconsistency and failure of RFMO members to comply with Articles 21(2) and (3), fulfill the requirements of Article 21(4) and apply UNFSA Articles 21 and 22 procedures against UNFSA Parties. The existing RFMO scheme will continue to apply against members that are not UNFSA Parties. Any RFMO member that disagrees with the assessment of the inspecting State that the existing scheme is inconsistent with UNFSA may, using the dispute settlement procedures of Part VIII of UNFSA, challenge the actions of the inspecting State where those actions exceed what is provided in the existing scheme. Such a challenge could also include the issue of whether it is for the inspecting State to make the determination that a flag State is not fulfilling its obligations as required under UNFSA.

The issue then arises as to the status of the "declaration" made by the EU upon ratification of UNFSA, which, in effect, does not accept a reading of Article 21(8) that would allow an inspecting State to bring a vessel inspected on the high seas to port without flag-State consent. While declarations are permitted, Article 42 makes it clear that reservations to UNFSA are not. The question is whether the declaration made by the EU is, in fact, a reservation that seeks to restrict and modify the operation of Article 21(8). If this declaration is, in fact, a reservation, the EU's ratification of UNFSA may be subject to the interpretation that allows, in certain circumstances, the inspecting State to unilaterally bring a vessel to port.[170] Resolution of this issue may require the use of international dispute settlement mechanisms.

9.4.9.3 Use of force[171]

Another major area of difference relates to the provisions on the use of force. UNFSA Article 22(1)(f) provides that inspecting States are to ensure that their inspectors avoid the use of force except when and to the degree necessary to ensure their safety and where they are obstructed in the execution of their duties. The degree of force used shall not exceed that reasonably required in the circumstances. Disagreement persists about whether this clause limits the use of force to personal self defence or applies in cases where vessels obstruct inspectors and attempt to flee. In these cases, can force be used to stop or disable the vessel so that the inspection can take place?

The prohibition of the threat or use of force embodied in Article 2(4) of the UN Charter is well recognized in international law.[172] However, even allowing for a broad interpretation of the "threat or use of force", which is generally accepted as including only "armed force",[173] it is arguable that using proportionate force to board and inspect vessels on the high seas for the purposes of monitoring and enforcing compliance with conservation and management measures established by RFMOs will not violate that prohibition.[174]

Article 2(4) only applies to threats or use of force "against the territorial integrity or political independence of any state, or in any other manner inconsistent with the Purposes of the United Nations". Within contemporary international law, boarding and inspection are not considered a forcible trespass on the territory of the flag State. Neither can they violate the flag State's political independence, as the purpose of boarding and inspection is to induce a vessel and its flag State to comply with their accepted international obligations. Likewise, boarding and inspection pursuant to the provisions of UNFSA pose no threat to international peace,[175] whereas insufficient curtailment of IUU fishing activities or escalatory reactions to boarding and inspection may.[176]

Even acknowledging that the term use of force is wider in meaning than "armed conflict",[177] the prevailing view is that it is only armed force that is prohibited.[178] International law does not prohibit what are effectively "police actions" — even where these actions may involve the use of military hardware such as warships — although it may prescribe certain limits on the manner in which these police actions may be conducted. This is acknowledged in LOSC itself, which provides for precisely the sort of police actions envisioned here to be taken against pirate vessels and those suspected of being engaged in the slave trade, unauthorized broadcasting vessels, or stateless vessels.[179] Neither does the use of military vessels alter the nature of the operation. Indeed, LOSC *requires* these police-type enforcement operations to be only carried out by authorized instruments of the State, which includes warships or military aircraft or other duly authorized ships or aircraft clearly marked and identifiable as being on government service.[180] LOSC cannot be taken to authorize the use of force in violation of the peremptory norm contained in the UN Charter prohibiting it. Accordingly, it must be concluded that boarding and inspection cannot be considered tantamount to the threat or use of force contrary to the purposes of the UN Charter.

UNFSA Article 21(18) merely codifies existing customary international law on the use of force at sea. Indeed, there are many examples of the use of force on the high seas to overcome obstruction consisting of refusal to submit to boarding and inspection. Regular police action is conducted on the high seas by, for example, the United States in respect of vessels suspected of illicit drug trafficking.[181] Likewise, members of the NPAFC have, on a number of occasions, boarded, inspected, and even arrested and prosecuted nonflag fishing vessels suspected of engaging in illegal driftnet fishing in the North Pacific Ocean.[182] In several cases warning shots and even disabling fire have been used.[183] This practice appears to confirm the customary position that international law does not prohibit resort to force to overcome obstruction. However, as established in the *I'm Alone*[184] and *Red Crusader*[185] cases and in LOSC,[186] international law does place certain limits on the degree of force that will be considered appropriate. If those limits are exceeded, the enforcing State will be liable to the flag State for any loss or damage caused.

Accordingly, an RFMO boarding regime that prohibits not only all use of force but also the right to bear personal side arms for self-defence appears inconsistent with UNFSA Article 22(1)(f) and the rest of the UNFSA regime that is designed to strengthen regional schemes for boarding and inspection. It would also be inconsistent with the decisions of ITLOS in the M/V *Saiga* (No. 2) Case (Saint Vincent and the Grenadines v. Guinea).

Monitoring cooperation, ensuring compliance, and controlling all vessels fishing within the convention areas of the respective RFMOs are imperative if RFMOs are to restore depleted fish stocks, avoid overfishing, and address ecosystem considerations. A contemporary inspection program designed in accordance with international standards as provided for in UNFSA is central to achieving these conservation objectives. Though there may be reasonable cause to suspect an infringement of the regional conservation regime and warrant boarding and inspection, some States oppose boarding and inspection fearing possible abuse of power and excessive force by inspectors from other flag States. Providing a well-trained, highly regarded team of inspectors who are to follow an agreed protocol, together with fair, timely legal recourse should excessive authority occur or be alleged to have occurred, may allay some of these concerns.

Compensation schemes could be developed within RFMOs against which monetary claims could be made should physical harm occur. The funds for these schemes could be obtained from fines or other sanctions imposed on vessels committing violations.[187]

9.4.9.4 Alternative mechanisms for the purposes of Article 21(15)

The third issue related to the notion of consistency is what constitutes an "alternative mechanism" to "effectively discharge" the obligation to ensure compliance with the conservation and management measures established by an RFMO for the purposes of application of Article 21(15).

In all of the examples discussed above, it is arguable that the RFMO schemes, while inconsistent, effectively discharge flag States' obligations to ensure compliance. The EU has sought to interpret the NAFO Joint Scheme of International Inspection and Surveillance as an alternative enforcement scheme within the meaning of UNFSA Article 21(15). Similarly, the argument has been made by members of tuna RFMOs that the various measures they have introduced, including trade-related measures, are effective alternative mechanisms and that there is no need to adopt boarding and inspection schemes.

No express guidance is given in UNFSA for determining what constitutes an effective discharge of the obligation to ensure compliance with conservation and management measures. Arguably, the question can be put to an objective test: whether the measures adopted have been effective in ensuring compliance by fishing vessels with applicable conservation and management measures. The continued prevalence of IUU fishing, conducted by both members and nonmembers of RFMOs (including Parties to UNFSA), would seem to suggest that the test has not been met, and therefore the claimed alternative mechanisms do not effectively discharge the obligation to ensure compliance. Moreover, pursuant to Article 21(15), where an effective alternative mechanism for ensuring compliance has been established, UNFSA Parties that are members of the RFMO may agree to apply the alternative mechanism between themselves and thereby derogate from Article 21(1). It follows that those UNFSA Parties that do not agree to do so could apply the boarding and inspection procedures in Articles 21 and 22 to other UNFSA Parties in the RFMO. These procedures would continue to apply, in any event, to vessels of other UNFSA Parties that are not members of the RFMO. Disputes between members of the RFMO that agree to derogate from Article 21 and 22 and those that do not could, presumably, be the subject of dispute settlement procedures in Part VIII of UNFSA.

9.4.9.5 Other issues of application

Other applications and practical issues arise from UNFSA Articles 21–22. For example, UNFSA provides that inspections are to be carried out by "duly authorized inspectors". Authorization is the responsibility of the State. However, some RFMOs with boarding and inspection schemes require "dual" authorization for inspectors to operate within their area of competence. CCAMLR, for example, requires inspectors to be designated by it when operating in the CCAMLR area. NEAFC and NAFO both require nationally authorized inspectors to be issued with RFMO documentation. In each of these RFMOs, members provide lists of nationally authorized inspectors, trainees, designated inspection vessels, and (where relevant) aircraft types to the Secretariat, which issues the appropriate credentials. In NAFO, these inspectors are then forbidden from enforcing their own national fisheries laws against other members' vessels, whereas in NEAFC there is no specific prohibition in the scheme against the enforcement of national laws against the vessels of other member States. The use of "dual-hatting" by Australian inspectors in the CCAMLR area has given rise, in the past, to some concern.

The issue of dual authorizations gives rise to the possibility of potentially absurd circumstances for RFMOs should an inspector operating under one RFMO mandate discover violations of measures adopted by another RFMO. Indeed, the issue arose in the CCAMLR Convention Area in 2005 in respect of fishing for southern bluefin tuna (which is under the competence of CCSBT). The same scenario may arise in NAFO where the NRA overlaps with the ICCAT Convention Area. Incidental catches of ICCAT-regulated tuna and swordfish species can occur in the NAFO regulated ground fishery. In this respect, simultaneous competence to act under the mandate of all RFMOs of the geographical area in which the inspection is being carried out, as long as communicated to the master upon boarding, as required by UNFSA Article 22(1)(a), would seem desirable. Nothing in UNFSA precludes this possibility. Article 21(4) merely requires that all States whose vessels fish on the high seas should be informed of the form of identification to be used by the authorized inspectors and that all vessels used for boarding and inspection be clearly marked and identifiable as being on government service.

Differences of opinion and practice also exist with respect to the gathering and use of evidence obtained during at-sea boarding and inspection. Many RFMOs restrict the sources of intelligence used for the purpose of directing inspection activity. More effective compliance could be gained if intelligence information was not only obtained from designated inspection platforms,[188] but also from a broad range of sources including trade, industry, VMS, airborne surveillance, and other shipping activities. This would appear to be consistent with the practice that has been adopted in the NPAFC. In addition, while evidence obtained from national inspections is admissible in national courts, evidence from inspections carried out by non-national inspectors operating under an RFMO scheme is either not admissible or is given little weight in proceedings in some flag States. This may undermine the effectiveness of any boarding and inspection regime.

Rationalization and better integration of RFMO schemes with national inspection agencies have been suggested as means of avoiding practical difficulties in the implementation of UNFSA and the enforcement of RFMO measures. The International MCS Network (a post-UNFSA initiative) currently provides an international forum for individuals involved in national and regional MCS programs to coordinate and improve the effectiveness of the surveillance service. A proposal to expand the network is a key recommendation of the HSTF.[189] In this respect, the International MCS Network could be tasked with the adoption of a set of minimum terms and conditions for MCS programs for RFMOs and assigned an audit function whereby it would be responsible for auditing all RFMO schemes to ensure consistency with the minimum terms and conditions.

9.4.10 Port-state control schemes

UNFSA Article 23 provides the legal basis for port States other than the flag State to conduct in-port inspections and exercise other measures of control. Article 23 stipulates that a port State has the right and the duty to take nondiscriminatory measures "to promote the effectiveness of subregional, regional and global conservation and management measures".[190] A port State may, among other things, "inspect documents, fishing gear and catch on board fishing vessels when such vessels are voluntarily within its ports or at its offshore terminals".[191] Port States may "prohibit landings and transshipment where it is established that the catch has been taken in a manner that undermines the effectiveness of subregional, regional or global conservation and management measures".[192]

Port States are in a unique position: while the vessel is in port, it comes under local jurisdiction, shifting the balance of control from the flag State to the port State. In most cases, the vessel must request permission to enter port and will be subject to inspection for customs, immigration, and other matters including fisheries. The powers of a port State may even be more important where a vessel wishes to transship fish, since the entire transaction may be monitored by inspectors. Port inspections during off-loading may be critical to the determination of the identity and quantity of catch. The data gathered during a port inspection can provide a good estimate of target species taken by the vessel and, combined with observer data from a vessel fishing under similar conditions, may provide estimates of the incidental catch of nontarget species. Port inspections are also far less expensive to conduct than inspections at sea. Mandatory inspections of all landings (off-loaded catch) would not only improve catch statistics but greatly improve compliance.

9.4.10.1 ICCAT[193]

RFMOs are taking steps to introduce requirements for their members to implement port-State control measures. In 1997, ICCAT adopted a revised port inspection scheme[194] that provides for inspection by Contracting Party authorities who will monitor compliance with ICCAT conservation and management measures at their own ports, without discrimination. Inspections are to include examination of fish, gear, fish samples, and relevant documents, including logbooks and cargo manifests. Where an apparent violation has been detected, the flag State is to notify ICCAT of actions taken to address it. ICCAT and its members have also recognized the need to control port access between Contracting and Non-Contracting Parties. In 1998 the Commission adopted a recommendation banning landings and in-port transshipments by Non-Contracting Party vessels in Contracting Party ports where port inspection reveals that fish on board were caught in the convention area or in violation of ICCAT conservation and management measures.[195]

9.4.10.2 IOTC[196]

In 2001, IOTC adopted a Non-Contracting Party scheme, which includes elements of port-State control.[197] Non-Contracting Party vessels observed to be fishing in contravention of IOTC conservation or management measures are to be reported immediately to the flag State. The vessel making the observation, in turn, is to notify the flag State of the observed vessel and the Secretariat, which will pass the information along to other Contracting Parties. Any vessel so sighted is presumed to be undermining IOTC conservation and management measures. When any such vessel voluntarily enters a port of a Contracting Party, it is to be inspected and be denied landing or transshipping privileges unless and until it is able to establish that any tuna or tuna-like species found on board were caught outside the IOTC area or in compliance with IOTC conservation and management measures. Information on all such inspections and any subsequent action is to be communicated to the relevant flag State and all members of the Commission via the Secretariat.

In 2002, IOTC also established a port inspection scheme.[198] The resolution provides that Contracting Parties may inspect fishing vessels voluntarily in their ports. They are to do so in a manner that ensures minimum interference and inconvenience and avoids degradation of fish quality. Contracting Parties are also to adopt regulations to prohibit landings and transshipments by Non-Contracting Party vessels where it has been established that the catch consists of species regulated by the IOTC and has been taken in a manner that undermines the effectiveness of IOTC conservation and management measures. Where evidence exists of a violation of such measures the port State is to draw this to the attention of both the flag State and the Commission, providing full documentation, including any record of inspection. The flag State is then to provide the Commission with details of any actions it has taken in respect of the matter. Although inspections should be carried out on a nondiscriminatory basis, in its first phase, inspection priority should be given to vessels of Non-Contracting Parties.

9.4.10.3 Other RFMOs

The FAO Model Scheme on Port State Measures by the FAO Committee on Fisheries (COFI) endorsed in March 2005 provides a useful guide for RFMOs, as it describes basic and minimum measures for subsequent action to be taken either through the adoption of regional Memoranda of Understanding, through RFMOs, or at the level of individual port States. The FAO Model Scheme includes information to be required by a port State prior to allowing access to a foreign fishing vessel, port inspection procedures of such a vessel, result indicators of port inspections, elements of training programs for port-State inspectors, and an outline of an information system on port-State inspections.[199]

In 2005, NEAFC agreed to establish a system to strengthen port-State control, and efforts are ongoing to establish a scheme similar to the FAO model. NEAFC's Permanent Committee on the Control and Enforcement (PECCOE) (established in 2000 to advise the Commission) held meetings in April 2006, and NEAFC finalized its port-State control scheme in 2006. With the establishment of the provisional A list and the permanent B list of IUU vessels, Contracting Parties are prohibited from authorizing landings in their ports of catch from listed vessels. NEAFC reports that the effectiveness of these port-State measures has greatly increased. CCAMLR Conservation Measure 147/XIX (1998) promotes cooperation between CCAMLR Parties to ensure port-State inspection and the prevention of toothfish landings by both CCAMLR and non-CCAMLR vessels. Since 2003, CCAMLR has implemented mandatory inspection of all toothfish vessels.[200] NAFO requires its members to perform port inspections on all Non-Contracting Party vessels that have been sighted engaging in fishing activities in the NRA. The port State shall not allow the vessel to land or tranship any fish until the inspection takes place. Landings and transshipments are prohibited in cases where the inspection reveals that the vessel has onboard stocks regulated by NAFO, unless the vessel demonstrates that the species were caught outside the NRA or in a manner consistent with NAFO conservation and management measures.[201] The SEAFO Convention also provides for in-port vessel inspection, imposing an obligation on all contracting port States, *inter alia*, to inspect documents, fishing gear, and catch on board fishing vessels when such vessels are voluntarily in their ports or offshore terminals.[202] Finally, the WCPFC Convention also establishes a port inspection scheme under which a port State may prohibit landings and transshipment if catch was taken in a manner that undermines conservation measures.[203]

9.4.10.4 Challenges to implementing port-State controls

There are several challenges to the adoption and effective implementation of port-State controls by RFMOs and their members. First, even assuming their uniform adoption and application by all members, an assumption that is not yet borne out in practice, these measures can only be required of RFMO members. Landings can therefore be directed through nonmember ports or what have come to be known as "ports of convenience". Second, disagreement persists about whether all vessels should be inspected or only those that have been cited for fishing in violation of RFMO measures. Additionally, in some RFMOs, port inspection of member's vessels is merely discretionary, whereas that of nonmember's vessels is compulsory. As Rayfuse notes, "Quite apart from the apparent discriminatory effect of this, the reality is that many port States lack the technical, administrative and financial resources necessary to fully implement such schemes. Moreover, such schemes are of limited use in dealing with fisheries processing and transshipping which is carried out at sea. They are also of no use when vessels can engage in port-hopping to off-load in non-participating port States".[204]

Further developments of port-State measures could be pursued with particular reference to the Model Scheme on Port State Measures to Combat IUU Fishing adopted by the FAO Technical Consultation to Review Port State Measures to Combat IUU Fishing in September 2004.[205] RFMOs could also take a leading role in promoting universal acceptance of the concept of the "responsible port State" proposed by the HSTF.[206] Some of the specific port-State measures the HSTF recommended RFMOs adopt include (1) demanding mandatory port-State control; (2) subjecting all fishing vessels, both national and foreign, to liability to in-port inspection; and (3) identifying ports of convenience and ensuring that these port States cooperate with the RFMO's port-State control regime.[207]

9.4.11 Enforcement measures

UNFSA Article 19 requires flag States to effectively discharge their flag-State responsibilities, including enforcing the measures adopted by RFMOs of which the flag State is a member and imposing effective sanctions that are adequate in severity to secure compliance, discourage violations, and "deprive offenders of the benefits accruing from their illegal activities". UNFSA Article 20 obliges States to cooperate directly or through RFMOs to ensure compliance with the measures adopted by RFMOs. To this end, flag States are entitled to seek and obtain assistance from other States in the conduct of investigations, which may be carried out either by the flag State directly or through the relevant RFMO. Information on the progress and outcome of investigations is to be provided to all interested or affected States. States are to assist each other in identifying vessels that have violated RFMO measures and are to make arrangements for the provision of information and evidence relating to these violations. Flag States must investigate the activities of their vessels where a coastal State in whose waters the vessels have been fishing requests this. UNFSA Parties who are RFMO members may also take action against vessels that have "engaged in activities which undermine the effectiveness of or otherwise violate" RFMO measures "until such time as appropriate action is taken by the flag State". What constitutes appropriate action by the flag State is the subject of considerable ongoing debate.

RFMOs have increasingly acted to articulate the obligations on flag States with respect to enforcement and to monitor flag States' compliance with their obligation to enforce RFMO measures. In some cases, RFMOs are taking steps to adopt measures to penalize flag States that do not comply with their obligation to enforce compliance by their vessels.

Measures adopted by RFMOs are increasingly UNFSA consistent in requiring members to adopt and implement legislation criminalizing IUU fishing or fishing in contravention of authorizations or licenses and to impose penalties that are adequate in severity to secure compliance, discourage violations, and deprive offenders of the benefits accruing from their illegal activities. However, in the absence of harmonized enforcement and penalty standards, reliance on national implementation means that a patchwork of differing and not necessarily compatible administrative and judicial measures and (or) penalties can cause confusion for both fishers and for national authorities trying to implement them. This can lead to disagreements within RFMOs about the relative efficacy of implementation and enforcement efforts.[208] For example, one of the more contentious issues in NAFO has been the implementation of effective penalties for noncompliance or offences committed by fishing vessels within the NRA.

Since prosecution and punishment are flag-State duties, effective enforcement becomes a matter of national implementation. In general, the ultimate penalty for noncompliance is a finding of guilt by a competent administrative tribunal or court, and the imposition of an appropriate penalty. This could be a fine, loss of quota or licence, confiscation of catch or vessel, or the imprisonment of the fishing crew. For this to occur, several key elements must be in place:

(1) The enforcing State must have the appropriate legal instruments, laws, and regulations in place to make the proscribed activity an offence and to provide a judicial body with the jurisdiction to try the offence. Rules of procedure must be in place to allow for the submission of evidence of the offence gathered by the appropriate investigating authority. This can be problematic if the investigating authorities and tribunals are of different nationalities and located in different States.

(2) The offence must be detected and evidence gathered in a manner acceptable to the tribunal that will hear the case. Different legal standards of proof, and the admissibility of certain types of evidence, make successful prosecutions of offences extremely difficult. In NAFO, for example, Canada will accept the testimony of fisheries observers as evidence in judicial proceedings. In the EU, however, some member States do not accept fisheries observers as competent witnesses.

(3) Integrity of the evidence must be maintained. For example, in NAFO, in the case of non-EU States, the vessel will usually be ordered into the nearest (Canadian) port where it will be boarded, and a thorough investigation will be conducted. In the case of the EU, the vessel is ordered back to its home port in Europe. According to NAFO reports, a number of violators were later cleared when no evidence was found on their return to Europe to corroborate a citation for an apparent infringement. This, of course, occurs after a voyage of many weeks and thousands of miles out of sight of investigating personnel. This problem is not unique to NAFO and has been cited as an issue in other RFMOs.

(4) Finally, an appropriate punishment should be imposed. A general rule is that for the control mechanisms to be effective, the economic consequences of noncompliance (i.e., the punishment) must be sufficiently stringent or costly to constitute an effective deterrent. This presupposes that the flag State follows up with the case.

One of the problems with IUU fishing is that vessel owners will often flag in a State that has no interest in or no effective means of prosecuting high seas fisheries offences or one that imposes penalties that are so low that they are viewed as a minor cost of doing business. Some States have adopted "administrative fines", "compounding", or the "ticketing of offences",[209] a procedure where offenders are permitted to admit their offence and pay a fine without going to court. These offenders are able to quickly, and sometimes discretely, pay their fines without incurring public censure. Another problem is that a vessel of a certain flag State may engage in illegal activities well beyond the reach of the authorities of its flag State. In such circumstances, the flag State may request the assistance of the States adjacent to the suspected illegal activity but is dependent on the will and capability of those States to take effective action.

It has been suggested that RFMOs should adopt a range of uniform standards for dealing with infractions, and a regionally consistent level of fines or penalties to be applied in cases of violation of RFMO measures, which members would then implement in their domestic legislation.[210] Such standards would remove the financial incentive for noncompliance, and the problem of chronic offenders seeking the cheapest RFMO member in which to flag and thereby pay the smallest fines. This would also serve to increase transparency and ensure the consistent application of fair, equitable, and nondiscriminatory penalties.

As a first step, the HSTF has suggested the establishment of a global database and clearing house for information on sanctions applied for IUU fishing, to be compiled from publicly available information. This database would serve to heighten awareness of the inadequacy of many penalties applied. It would also draw attention at the RFMO and national levels to the need for greater consistency in administrative and penal sanctions for IUU fishing.[211] This is another area in which interregional cooperation among RFMOs could be established. Agreements among RFMOs to enforce each other's fines or sanctions for vessels from both member Sates and nonmember States within their ports could help prevent the avoidance of sanctions by vessels merely leaving the jurisdiction.

With respect to monitoring flag-State compliance, RFMO members are increasingly being obliged to report on the legislation and regulations they have adopted to implement RFMO measures and on the steps they have taken to ensure and enforce compliance by their vessels with those measures, including reporting on reasons for noncompliance and on any sanctions imposed. Annual review of these reports is generally carried out by a "compliance" committee, which reports a summary of its findings and recommends measures for improving compliance to the plenary.[212] The efficacy of this review process rests, however, both on the submission and on the content of these reports. With respect to submission, although reports may be required, they may not be compelled. Currently, no sanction, other than public shaming, appears to exist for a failure to report. To address this, some RFMOs have been resorting to measures such as reliance on NGO reports and information received from other member States to assess the accuracy of reports and the relative merits of follow-up action taken by the State involved.

RFMOs have also adopted a range of measures requiring their members to take action against stateless, presumed stateless, and nonmember vessels that are believed to have been fishing in a manner that undermines the effectiveness of the RFMO measures. These include measures relating to the reporting of vessel sightings, reports of transshipment activities, reports of results from trade information and catch documentation schemes, and the obligation to inform nonmember flag States when their vessels have been sighted fishing in the RFMO area. In RFMOs with at-sea boarding and inspection regimes, members are even increasingly being encouraged to seek the consent of nonmembers to board and inspect vessels for the purposes of ascertaining compliance with RFMO measures. These enforcement measures adopt a reverse burden of proof, holding that mere presence in the RFMO areas is presumed to be IUU fishing unless and until the vessel or its flag State is able to prove otherwise. The intended effect of these measures is to stop members not only from engaging in IUU fishing themselves, but also from allowing their vessels and ports to transship or engage in any other relations with IUU vessels. By relegating IUU vessels to the status of pariahs, RFMOs are hoping to render these vessels' activities both unprofitable and unsustainable.

Vessels flying the flag of an UNFSA Party that are fishing in an RFMO area may be subject to boarding and inspection where warranted to ensure compliance with RFMO conservation and management measures.[213] In the event that there are clear grounds to believe that a violation has occurred, the inspecting State shall secure evidence and notify the flag State,[214] which may respond itself or authorize the inspecting State to act. Should the flag State fail to respond or to take action, the investigating State may take additional actions, including the gathering of additional evidence, ordering the vessel into port,[215] and arguably, prosecuting the offence itself.[216]

Several RFMOs have adopted specific procedures for dealing with nonmember vessels. The NEAFC Non-Contracting Party Scheme[217]calls for immediate notification of sightings to the Secretariat, which notifies other members if a nonmember vessel is sighted within the regulatory area. This is followed by a request for boarding at sea and encouragement "to take appropriate actions". Slightly more forceful is the requirement for any member to undertake an investigation if the vessel enters into any member-State port, as well as a prohibition on off-loading, transshipping, or any joint fishing operations. Further steps involve the Secretariat notifying the flag State, as well as other RFMOs, of the presumed IUU activity. NAFO has similar provisions.[218] Under both NAFO and NEAFC rules, any nonmember vessel spotted fishing in the regulatory area is "presumed to be undermining the effectiveness of the Conservation and Enforcement Measures".

The major problem, which has been recognized internationally,[219] is how to deal with vessels flagged in States that are neither party to UNFSA nor members of or participants in the relevant RFMO. While the issue of how to address IUU fishing by such vessels has been the subject of considerable effort by the UN, FAO, the Organization for Economic Cooperation and Development (OECD), HSTF, and environmental NGOs, to date there has been no satisfactory solution.

UNFSA Article 33 calls upon parties to take steps to deter the activities of the vessels of nonparties that undermine the effective implementation of the Agreement. Measures adopted by RFMOs, including procedures for reporting sightings, complaints to flag States, "naming and shaming", port denial, transshipment bans, and positive and negative lists, have, to date, been of limited success. One possibility, which has been used on several occasions, is boarding and inspection or the right of visit with respect to vessels having no valid flag or multiple registrations.[220] This is particularly useful in situations in which the flag State revokes registry when contacted by the inspecting State or in cases where a vessel has acquired multiple flags. In both cases the vessel becomes a "ship without nationality"[221] and may be boarded and inspected in accordance with Article 110 of LOSC.[222]

The question remains as to what steps RFMOs can take where reporting by member States identifies a failure to "effectively" implement and enforce. In other words, what measures can be taken against a member State that fails to comply with its obligations to control and penalize? There has been some criticism of the lack of enthusiasm of some RFMO members for prosecuting their own nationals based on what have been perceived to be lengthy delays in bringing matters to adjudication, and in the lack of transparency in both the process and the eventual resolution of the case.

Some members of ICCAT, CCAMLR, and other RFMOs have criticized others in RFMO meetings and in diplomatic demarches. The power of public shaming or ridicule is used in some RFMOs to attempt to influence recalcitrant members to mend their ways. A more determined approach is that adopted in ICCAT, where trade-restrictive measures that are usually applied against nonmembers have also been recommended against noncompliant members, although their effectiveness has been compromised by the use of the objection procedure.[223]

Pursuant to UNFSA, where a serious violation of RFMO measures has been detected and the flag State has failed to respond or to take action within three days, the inspecting State may take the vessel into port and must notify the flag State.[224] Even though the provision avoids the use of the expression "effectively", this is nonflag seizure and arrest.[225] It is not clear what further action the inspecting State may then take, although it would be unreasonable to assume that the "inspecting State ought simply to release the vessel in the face of inactivity on the part of the flag State".[226] In fact, it would appear that there is nothing in UNFSA that prohibits prosecution by the inspecting State in such situations.[227]

It has been suggested that RFMOs might take a more determined approach to ensuring members effectively prosecute and penalize IUU fishing through the adoption of measures specifying the consequences of noncompliance. For example, where a flag State fails to exercise control over its vessels, other members may be authorized to take enforcement action, including at-sea arrest, detention, and confiscation of vessel and (or) catch.[228] In situations of ongoing flag-State failure evidenced by a pattern of conduct where, over a period of time, the flag State has failed to adequately investigate, prosecute, or punish, RFMOs might consider applying disincentives that go beyond public shaming and restrictive trade measures. Remedies might lie in the reduction or denial of quota allocation, as adopted by ICCAT against Chinese Taipei (Taiwan) at its 2005 meeting, or in the temporary banning of vessels entitled to fly the flag of a member from RFMO waters.[229] UNFSA Parties may also consider a legal challenge pursuant to Article 30 of the Fish Stocks Agreement. These measures would require clear articulation of the criteria by which a failure will be judged (i.e., what constitutes an acceptable investigation), the time period over which failure is to be assessed, and any applicable exculpatory circumstances. Since such determinations are always likely to be contentious, an effective mechanism would also be required for settling disputes relating to the application of these RFMO measures.

9.5 Conclusion

The objective of MCS measures is to ascertain what and how much is being caught, where, how, and by whom, not only to provide the necessary data for stock assessment and broader ecosystem considerations but also to determine the extent to which RFMO conservation and management measures are being respected. Obviously, no one measure holds all the answers to dealing with the pernicious problems of noncompliance and IUU fishing. Instead, a comprehensive and integrated program providing real-time, accurate, meaningful, and verifiable information in a cost-effective and nondiscriminatory manner is required to achieve this goal.[230] To be useful, MCS must be accompanied by effective enforcement. This requires appropriate prosecution with punishment for a broad spectrum of violations to serve not only as a deterrent but also to restructure economic incentives so that problematic behavior is modified in favour of conservation.

It is clear that the best conservation standards agreed to by States within RFMOs will fail without compliance by those involved in fishing operations and responsible enforcement. A robust MCS and enforcement regime acknowledging the shared responsibility for enforcing measures to conserve and manage living resources harvested on the high seas is critical to achieving good order of the oceans for present and future generations. UNFSA introduced a range of MCS and enforcement measures at the national, subregional, and regional levels as well as further delineated the responsibilities of flag States and port States to strengthen the effectiveness of RFMOs in regulating international fisheries. In many instances RFMOs have adopted measures that are consistent with UNFSA. However, much more needs to be done to counter IUU fishing and to ensure compliance by all authorized participants both within and beyond areas under national jurisdiction. Compliance failures and unreliable data remain ongoing challenges for the international community, reinforcing the need for the wide application of the precautionary approach.

The coordinated, consistent, and rigorous implementation and further development of the MCS and enforcement regime provided for under UNFSA are essential to ensure the sustainability of straddling fish stocks and highly migratory fish stocks, nontarget species, marine ecosystems, and critical habitats. In this respect, widespread adoption of the recommendations in the Final Report of the HSTF[231] in the context of IUU fishing on the high seas can support the practical implementation of the MCS and enforcement provisions of UNFSA. Of particular relevance are recommendations to strengthen the International MCS Network; promote broader participation in UNFSA and the FAO Compliance Agreements; expand the use of certification and trade measures; and reinforce the role of port States, particularly the adoption and implementation of the FAO Model Port State Scheme by RFMOs.

Some of the UNFSA MCS and enforcement provisions pose difficulties for some States and have delayed ratification in certain cases. Nonetheless, all States are still obligated to cooperate with RFMOs. Adherence to the regulatory regime is essential both for authorized fishing interests and for those vessels that are not authorized to operate in fisheries within the convention areas of the respective RFMOs. Noncompliance undermines the efforts of RFMOs to restore depleted stocks, avoid overfishing, and control harm to the marine ecosystem from fishing activities. Given the precarious state of many stocks and international fisheries, noncompliance continually thwarts the quest for sustainability, increases the likelihood of serious or irreparable harm, and increases the specter of an ocean of species at risk. There is an urgent need to actively engage industry in the science and management of international fisheries to heighten global awareness of nonsustainable practices, improve decision making, and increase compliance. Compliance by all fishing interests with conservation measures decided in accordance with international minimum standards would result in enormous progress in achieving sustainable development of international fisheries.

Endnotes

[1] For a comprehensive analysis of this issue, see Marine Resources Assessment Group Ltd. (MRAG), "Review of Impacts of Illegal, Unreported and Unregulated Fishing on Developing Countries". Final Report prepared for UK Department of International Development (DFID) (London: MRAG, 2005); High Seas Task Force 2006, "Closing the Net: Stopping Illegal Fishing on the High Seas", Final Report of the Ministerially-led Task Force on IUU Fishing on the High Seas, Governments of Australia, Canada, Chile, Namibia, New Zealand, and the United Kingdom, WWF, IUCN, and the Earth Institute at Columbia University; OECD, *Fish Piracy: Combating Illegal, Unreported and Unregulated Fishing* (Paris: OECD, 2004); K. Bray, "A Global Review of Illegal, Unreported and Unregulated (IUU) Fishing", FAO – Government of Australia workshop on IUU fishing, Sydney 2000, Document AUS: IUU/2000/6. 2000; K. Rigg, R. Parmentier, D. Currie, and the Varda Group, "Halting IUU Fishing: Enforcing International Fisheries Agreements", AGR/FI/IUU/RD (2004)1 (Paris: OECD, 2004), available at <http://www.oecd.org/dataoecd/34/39/31426729.PDF> (accessed January 2005); D. Balton, "IUU Fishing and State Control Over Nationals", AGR/FI/IUU/RD (2004)2 (Paris: OECD, 2004), available at <https://www.oecd.org/dataoecd/58/28/31426742.PDF> (accessed February 2005); C. Joyner, "Compliance with and Enforcement of International Fisheries Law" in E. Hey, ed., *Developments in International Fisheries Law* (The Hague: Kluwer Law International, 1999) 327–355; D. Balton, "The Compliance Agreement" in E. Hey, ed., *Developments in International Fisheries Law* (The Hague: Kluwer Law International, 1999) 31–53; M. Gianni and W. Simpson, *The Changing Nature of High Seas Fishing: How Flags of Convenience Provide Cover for Illegal, Unreported and Unregulated Fishing* (Australian Department of Agriculture, Fisheries and Forestry, International Transport Workers' Federation, and WWF International, 2005); C.C. Schmidt, "Addressing Illegal, Unreported and Unregulated (IUU) Fishing", Prepared for the International Fisheries Compliance 2004 Conference, Brussels, Belgium, 29–30 September 2004, available at <http://www.oecd.org/dataoecd/0/21/34029751.pdf> (accessed March 2005).

[2] *Agreement for the Implementation of the Provisions of the United Nations Convention on the Law of the Sea of 10 December 1982 Relating to the Conservation and Management of Straddling Fish Stocks and Highly Migratory Fish Stocks*, done at New York, 4 December 1995, entered into force 11 December 2001, 2167 UNTS 88; (1995) 34 ILM 1542. Referred to as UNFSA or the Fish Stock Agreement.

[3] *United Nations Convention on the Law of the Sea*, done at Montego Bay, Jamaica, 10 December 1982, entered into force 16 November 1994, 33 ILM 1309. Referred to as LOSC.

[4] Although not expressed precisely in those terms, this is the combined effect of a number of UNFSA provisions, particularly UNFSA Art. 8. See D. Balton, "Strengthening the Law of the Sea: The New Agreement on Straddling Stocks and Highly Migratory Fish Stocks" (1996) 27 *Ocean Development and International Law* 125–151 at 138.

[5] UNFSA Art. 17(2).

[6] UNFSA Art. 18.

[7] *Agreement to Promote Compliance with International Conservation and Management Measures by Fishing Vessels on the High Seas*, done at Rome, Italy, 24 November 1993, entered into force April 2003, (1994) 33 ILM 968. Also known as the Flagging Agreement. Hereinafter referred to as the FAO Compliance Agreement. The FAO Compliance Agreement forms an integral part of the FAO *Code of Conduct for Responsible Fisheries*, Rome: FAO 1995. See Part III of this book for reproduced version of text of these Agreements. FAO Compliance Agreement Arts. III–VI.

[8] LOSC Arts. 87(1) and 86.

[9] LOSC Arts. 118–120.

[10] LOSC Art. 92(1).

[11] FAO Compliance Agreement Arts. III and IV, and UNFSA Art. 18.

[12] FAO Compliance Agreement Art. III(8).

[13] UNFSA Art. 18.

[14] UNFSA Art. 19.

[15] UNFSA Art. 10(h).

[16] UNFSA Art. 17(4).

[17] UNFSA Art. 20.

[18] UNFSA Art. 20(7).

[19] UNFSA Art. 21(14). This section extends the powers of a coastal State to taking action against a vessel that has committed a "serious violation" outside the coastal State's area of jurisdiction, provided that the vessel has come under the coastal State's legal jurisdiction, either voluntarily or as a result of another legal mechanism.

[20] UNFSA Arts. 21–22.

[21] UNFSA Art. 21(6)–(8).

[22] UNFSA Art. 23.

[23] Adopted by consensus at the 28th Session of the FAO Conference on 31 October 1995. The Code has several articles addressing MCS issues. See in particular, Art. 6 paras. 10–12, Art. 7.7 (Implementation), and Art. 8.2 (Flag State Duties).

[24] FAO International Plan of Action to Prevent, Deter, and Eliminate Illegal, Unreported and Unregulated Fishing adopted by consensus at the 25th Session of COFI on 2 March 2001 and endorsed by the 125th Session of the FAO Council on 23 June 2001. Available via FIGIS from <http://www.fao.org/DOCREP/003/y1224e00.HTM> (accessed March 2008).

[25] A formal definition of IUU fishing may be found in IPOA-IUU para. 3. IUU fishing can occur within EEZs as well as on the high seas. The focus here is on measures to address IUU fishing on the high seas because within EEZs coastal States have full authority and the necessary legal tools (if not the resources) to combat the problem.

[26] R. Rayfuse, "To Our Children's Children's Children: From Promoting to Achieving Compliance in High Seas Fisheries" 20(3) *International Journal of Marine and Coastal Law* (2005) 509–532. This article was written as a background paper for the Conference on the Governance of High Seas Fisheries and the UN Fish Stocks Agreement — Moving from Words to Action, St. John's, N.L., 1–5 May 2005.

[27] FAO Fisheries Department's Expert Consultation on Monitoring, Control and Surveillance, 1981. See also A.J. Armstrong, *Development of International/Regional Co-Operation in Fisheries Monitoring, Control and Surveillance* (London: Commonwealth Secretariat, 1991) at 1.

[28] P. Flewwelling, C. Cullinan, D. Balton, R.P. Sautter, and J.E. Reynolds, "Recent Trends in Monitoring, Control, and Surveillance Systems for Capture Fisheries", FAO Fisheries Technical Paper No. 415 (Rome: FAO, 2001) at 7.

[29] P. Flewwelling, C. Cullinan, D. Balton, R.P. Sautter, and J.E. Reynolds, FAO Fisheries Technical Paper No. 415 at 9.

[30] A.J. Armstrong, at 1.

[31] Contemporary compliance theory suggests that approximately 90% of a population will obey the law if they perceive it as fair and imposed by a legitimate authority, and 10% of the population will only obey the law if they have a real fear of sanction. However, if the 90% see the 10% getting away with it, they will cease obeying the law, especially if they see the violators making economic gains at their expense. This is why there is a need for effective and visible enforcement and meaningful sanctions. If the punishment is to be effective, it must be severe enough to deter the offending activity and not merely be written off by the offenders as a cost of doing business. See J.R. Nielsen and C. Mathiesen, "Important Factors Influencing Rule Compliance Lessons from Denmark" (2003) 27 *Marine Policy* 409–416. For a thorough analysis of this issue see Marine Resources Assessment Group Ltd. (MRAG), "Review of Impacts of Illegal, Unreported and Unregulated Fishing on Developing" at 48.

[32] A.J. Armstrong, at 7.

[33] A.J. Armstrong, at 7.

[34] In addition to the capital outlay required for surveillance, another reason for coordination is the continuous overlapping of areas of responsibility and jurisdiction in ocean space with other agencies, departments, or sectors. Environmental protection, customs, and maritime security authorities, to name a few, are also concerned with the location and activities of vessels within the areas of concern of national or regional resource management bodies. In fact, one of the critical problems facing all marine-related sectors is the need for harmonizing their requirements for reporting and other responses. For example, a fishing vessel seeking entry into a port may be required to send various reports to separate authorities concerning fishing and catch specifics, customs and cargo particulars, environmental protection status, ship safety, and, more recently, ship security where required under the International Ship and Port Facility Security Code (ISPS Code) adopted within the International Maritime Organization (IMO) or similar national legislation. There are certain advantages, however, to establishing cooperative and information sharing arrangements with other maritime sectors. Currently, the concerns of maritime security are paramount for many national governments. As a result, a fishing vessel seeking entry into port may be required to report all its previous at-sea ship-to-ship contacts, in compliance with the ISPS Code or similar regulations. This requirement would prove to be a useful source of information for tracking compliance with RFMO requirements, including the at-sea transshipment of fish. While the ISPS Code, as applied in most States, exempts fishing vessels, yachts, and similar small vessels, "mother ships" or other larger commercial vessels involved in servicing the high seas fishing fleet may well be subject to ISPS compliance with the Code. This would necessitate their keeping records of ports visited and ship-to-ship contacts while at sea (cf. ISPS Code Part B, para. 438).

[35] R. Rayfuse, "To Our Children's Children's Children: From Promoting to Achieving Compliance in High Seas Fisheries" at 510.

[36] R. Rayfuse, "To Our Children's Children's Children: From Promoting to Achieving Compliance in High Seas Fisheries" at 510.

[37] R. Rayfuse, "To Our Children's Children's Children: From Promoting to Achieving Compliance in High Seas Fisheries" at 510.

[38] See, e.g., IATTC Resolution C-03-02 on the Establishment of a List of Longline Fishing Vessels over 24 Meters (LSTLFVs) Authorized to Operate in the Eastern Pacific Ocean (2003); ICCAT Recommendation 02-22 Concerning the Establishment of an ICCAT Record of Vessels over 24 Meters Authorized to Operate in the Convention Area (2002); CCAMLR Conservation Measure (CM) 119/XVI, Requirement for Contracting Parties to Licence their Flag Vessels in the Convention Area (1998) as amended in CM 10-02 (2004); NEAFC Scheme of Control and Enforcement in respect of Fishing Vessels in Areas beyond the Limits of National Fisheries Jurisdiction in the Convention Area (1998); and IOTC Resolution 02/05 Concerning the Establishment of an IOTC Record of Vessels over 24 Meters Authorized to Operate in the IOTC Area (2002).

[39] ITLOS Case No. 5, Panama v. France, 7 February 2000.

[40] See Coalition of Legal Toothfish Operators (COLTO) Web site, <http://www.colto.org>, and ITLOS Judgment (Panama v. France). Another good illustration is that of the *Candor*. This vessel was previously named *Arosa Cuarto* (1989) and *Pescamex* III (1998) and owned by Compania Internacional de Pesca S.A. It was renamed *Cisne Azul* in January 1999, then renamed again to the *Viking* and flagged to the Seychelles. The vessel renamed to the *Inca* and was deregistered by Belize in April 2004. She is currently named *Condor*, and her flag State is Togo. See <http://www.colto.org>.

[41] LOSC Arts. 116–119.

[42] R. Rayfuse, "To Our Children's Children's Children: From Promoting to Achieving Compliance in High Seas Fisheries" at 516.

[43] "What is the Positive List", now available at <http://www.oprt.or.jp/eng/e_positivelist.htmll> (accessed March 2008) at 1.

[44] "What is the Positive List", now available at <http://www.oprt.or.jp/eng/e_positivelist.htmll> (accessed March 2008) at 1.

[45] For a discussion on the early implementation of registers and minimum terms and conditions for access, see M. Lodge, "Minimum Terms and Conditions of Access: Responsible Fisheries Management Measures in the South Pacific Region" (1992) 16 *Marine Policy* 277–305.

[46] Greenpeace, "Saving Our Seas from Piracy: A New Era in Fisheries Management", available at <http://weblog.greenpeace.org/pacific/background/new_era.html> (accessed September 2005). See E. Meltzer, "Global Overview of Straddling Fish Stocks and Highly Migratory Fish Stocks: Nonsustainable Nature of High Seas Fishing" (1994) 24 *Ocean development and International Law* 255–344 at 305–312 for a historical overview of the management of the tuna fisheries in the Western and South Pacific following the establishment of

the South Pacific Forum Fisheries Agency (FFA) in 1979, including the 1982 Nauru Agreement Concerning Cooperation in the Management of Fisheries of Common Interest that stipulated fishing licences would be issued only to those foreign vessels in good standing on the Regional Register of Foreign Fishing Vessels maintained by FFA. The 1990 review of the minimum terms and conditions of fisheries access led to a Second Implementing Arrangement. The 1990 Nauru Minute of Agreement was followed by the Niue Treaty on Cooperation in Fisheries Surveillance and Law Enforcement, which entered into force 20 May 1993.

[47] ICCAT Recommendation 02-22 Concerning the Establishment of an ICCAT Record of Vessels Over 24 Meters Authorized to Operate in the Convention Area.

[48] ICCAT Recommendation 02-03 to Establish a List of Vessels Presumed to have Carried out IUU Fishing Activities in the ICCAT Convention Area.

[49] ICCAT and IATTC have a positive longline list applicable to large longline vessels of Non-Contracting Parties.

[50] In 2002 IOTC established a positive list of vessels more than 24 m authorized to fish in the IOTC Regulatory Area.

[51] See CCAMLR Conservation Measures 10-06 (2005) and 10-07 (2005).

[52] O. Stokke and D. Vidas, "Regulating IUU fishing or combating IUU operations" in OECD, *Fish Piracy: Combating Illegal, Unreported and Unregulated Fishing* (Paris: OECD, 2004) at 19.

[53] See "Conference on the Governance of High Seas Fisheries and the UN Fish Agreement: Moving from Words to Action" (2005) 20 *International Journal of Marine and Coastal Law* 605–629 at 618.

[54] High Seas Task Force, "High Seas Governance", Meeting of the High Seas Task Force, Paris, France, 9 March 2005, HSTF/09. Available at <http://www.high-seas.org> (accessed August 2005).

[55] High Seas Task Force, "High Seas Governance", Meeting of the High Seas Task Force, Paris, France, 9 March 2005, HSTF/09. Available at <http://www.high-seas.org> (accessed August 2005).

[56] R. Rayfuse, "To Our Children's Children's Children: From Promoting to Achieving Compliance in High Seas Fisheries" at 528.

[57] See NEAFC Scheme Art. 10, and NAFO CEM Art. 23.

[58] See NAFO CEM Art. 52.

[59] ICCAT maintains several databases that are prioritized for reporting purposes. These include Task I: Best ICCAT estimates of catches; Task II a: Catch & Effort data; Task II b: Size composition of samples. ICCAT Field Manual at 4. Available at <http://www.iccat.es/pubs_FieldManual.htm> (accessed September 2005).

[60] For instance, see ICCAT, "Report of the 18th Regular Meeting of the Commission, Dublin, Ireland, 17–24 November 2003" in *Report for the Biennial Period, 2002–2003, Part II (2003)*. Available at <http://www.iccat.es/pubs_biennial.htm> (accessed March 2008).

[61] ICCAT Recommendation 1992-01.

[62] ICCAT Recommendations 2001-21 and 2001-22.

[63] IOTC Recommendation 01/06.

[64] CCSBT Southern Bluefin Tuna Statistical Document Program, October 2003, updated November 2006. Available at <http://www.ccsbt.org/docs/pdf/about_the_commission_/trade_information_schemes.pdf> (accessed March 2008)

[65] IATTC Resolution on Bigeye Tuna Statistical Document Program. C-03-01. Available at <http://www.iattc.org> (accessed August 2005).

[66] ICCAT Recommendation 1996-11.

[67] ICCAT Recommendation 2002-16.

[68] ICCAT Recommendation 2002-17.

[69] ICCAT Recommendation 2003-18.

[70] ICCAT Resolution 2003-15.

[71] CCSBT Action Plan, March 2000.

[72] IOTC Recommendation 03/05.

[73] IATTC Resolution C-05-04.

[74] CCAMLR Toothfish Catch Documentation Scheme, CCAMLR Conservation Measure 10-05 (2005), available at <http://www.ccamlr.org> (accessed September 2006). See D. Agnew, "The Illegal and Unregulated Fishery for Toothfish in the Southern Ocean and the CCAMLR Documentation Scheme" (1999) 24 *Marine Policy* 361–374.

[75] CCAMLR Conservation Measures 10-06 (2005) and 10-07 (2005) refer to the possibility of taking trade-related measures.

[76] NAFO, NAFO Conservation and Enforcement Measures. Articles relevant to 3M shrimp fisheries, NAFO FC Doc. 06/12, Serial 5206. FC Working Paper 06/24 Article 19 Product Labelling Requirements. Annex 6: Divisions 3LM Shrimp New Management Measures. No longer available online; hard copies on file with Evelyne Meltzer and at NAFO, Dartmouth, N.S.

[77] CCAMLR Resolution 21/XXIII.

[78] CCAMLR, *Report of the Twenty-Third Meeting of the Commission*, CCAMLR-XXIII, Hobart, Australia, 25 October – 5 November 2004, at para. 7.5.

[79] FAO, "Report of the Expert Consultation on Data Formats and Procedures for Monitoring Control and Surveillance, Bergen, Norway, 25–27 October 2004", FAO Fisheries Report No. 761 (Rome: FAO, 2005).

[80] See NEAFC Scheme Art. 6.

[81] GFCM at <http://www.fao.org/fi/body/rfb/GFCM/gfcm_home.htm> (accessed April 2006).

[82] WCPFC Convention Annex III.

[83] WCPFC Convention Art. 29(4).

[84] WCPFC Convention Art. 29(5).

[85] See Greenpeace, 10 April 2006, "The Binar 4 — A Case Study of Pirate Fishing Links between Europe and West Africa", available at <http://www.greenpeace.org/raw/content/international/press/reports/the-binar-4-a-case-study-of.pdf> (accessed May 2006).

[86] Anonymous personal communication, January 2005, March 2005, June 2005.
[87] R. Rayfuse, "To Our Children's Children's Children: From Promoting to Achieving Compliance in High Seas Fisheries" at 519.
[88] R. Rayfuse, "To Our Children's Children's Children: From Promoting to Achieving Compliance in High Seas Fisheries" at 518.
[89] WCPFC Convention Art. 28(4).
[90] CCAMLR requires observers on all fisheries other than krill, and the AIDCP and IATTC require observers on purse seiners over 363 tonnes.
[91] E.g., NAFO on all vessels fishing in the NAFO Regulatory Area, SEAFO in the SEAFO Convention Area, and under the BSDH Convention on all vessels fishing in the Donut Hole.
[92] Anonymous personal communication to Evelyne Meltzer, December 2005.
[93] R. Rayfuse, "To Our Children's Children's Children: From Promoting to Achieving Compliance in High Seas Fisheries" at 517.
[94] NEAFC CEM Art. 26.
[95] NEAFC Scheme Art. 14.
[96] For a detailed review of the NPAFC enforcement regime see R. Rayfuse, *Non-Flag State Enforcement in High Seas Fisheries* (Leiden: Martinus Nijhoff Publishers, 2004), Chap. 4.
[97] See Navigs S.A.R.L., "Fishing Vessel Monitoring Systems: Past, Present and Future", prepared for the High Seas Task Force, October 2005. Available at <http://www.high-seas.org/docs/hstf_vms_final1.pdf> (accessed February 2006)
[98] FAO, "Fishing Operations 1. Vessel Monitoring Systems", FAO Technical Guidelines for Responsible Fisheries No. 1, Suppl. 1 (Rome: FAO, 1998) at 8. Available at <http://www.fao.org/docrep/003/w9633e/w9633e00.HTM> (accessed February 2006).
[99] cf. CCAMLR CM 10-05 (2005) para. 15.
[100] cf. CCAMLR CM 10-05 (2005) para. 15.
[101] See E. Molenaar and M. Tsamenyi, "Satellite-Based Vessel Monitoring Systems: International Legal Aspects and Developments in State Practice" FAO Papers Online No. 7, April 2000. Available at <http://www.fao.org/legal/prs-ol/lpo7.pdf> (accessed August 2005).
[102] OTC Resolution 02/02, which was evaluated in 2005.
[103] IATTC Resolution C-04-06.
[104] ICCAT Recommendations 03-14 and 04-11.
[105] NAFO, "NAFO Conservation and Enforcement Measures. Chap. 7 — Electronic Reporting, Satellite Tracking and Observers", Art. 55–60. Available at <http://www.nafo.int/fisheries/frames/regulations.html> (accessed March 2008).
[106] CCAMLR Conservation Measure 10-04 (2005), para. 15.
[107] NEAFC Scheme of Control and Enforcement (1998) Arts. 8–12.
[108] FAO Technical Guidelines for Responsible Fisheries No. 1, Suppl. 1.
[109] For an exhaustive discussion of the phenomenon of tampering and interfering with VMS operations, see FAO Technical Guidelines for Responsible Fisheries No. 1 Suppl. 1 at 22–27.
[110] These parameters include codes in the output message to indicate that any of the following have occurred: loss of main power where the unit is disconnected from the vessel and placed on a raft or other floating object; reconnection of main power indicating a lapse; loss of connection to the GPS antenna indicating a broken or cut cable; loss of GPS signal indicating that the antenna has been covered preventing the vessel position to be determined. Greg Aitkins, CarteNav, Halifax, N.S., personal communication by e-mail to Evelyne Meltzer, January 2006.
[111] R. Rayfuse, "To Our Children's Children's Children: From Promoting to Achieving Compliance in High Seas Fisheries" at 519 and 526–528. See also Greenpeace, "Saving Our Seas from Piracy: A New Era in Fisheries Management", available at <http://weblog.greenpeace.org/pacific/background/new_era.html> (accessed May 2006); Navigs S.A.R.L., "Fishing Vessel Monitoring Systems: Past, Present and Future", prepared for the High Seas Task Force, October 2005. Available at <http://www.high-seas.org/docs/hstf_vms_final1.pdf> (accessed February 2006).
[112] FAO, "Making Monitoring Control and Surveillance of Fishing Vessels More Effective", COFI 25th Session , Rome, Italy, 24–28 February 2003, COFI/2003/4 at para. 13.
[113] The following description of the project was provided through a personal communication by e-mail to Evelyne Meltzer from Terje Løbach, April 2006.
[114] The ocean gives stronger reflection from vertical polarization than from horizontal polarization. Thus, it is better to use horizontal polarization for ship detection. That way, the reflections from the ocean are suppressed and the ships are clearer in the radar images. Based on ERS and RADARSAT experience, the HH-polarization in the outer Image Mode beams are preferred for ship detection. However, signatures from ships can be confused with those from icebergs. Thus, alternating polarization signatures have been obtained over Antarctic icebergs to see if the co-polarization/cross-polarization ratios deviate significantly from those of ships. Polarimetric signatures of ships have been studied by other projects in European waters. Terje Løbach, personal communication to Evelyne Meltzer, April 2006.
[115] This section deals only with at-sea inspections. See 9.4.10 below for a discussion on port-State inspections.
[116] R. Rayfuse, *Non-Flag State Enforcement in High Seas Fisheries*, at 196.
[117] B. Vukas and D. Vidas, "Flags of Convenience and High Seas Fishing: The Emergence of a Legal Framework" in O. Stokke, ed., *Governing High Seas Fisheries: The Interplay of Global and Regional Regimes* (Oxford: Oxford University Press, 2001) 53–90 at 76.
[118] M. Hayashi, "The 1995 UN Fish Stocks Agreement and the Law of the Sea" in D. Vidas and W. Østreng, eds., *Order for the Oceans* at *the Turn of the Century* (Dordrecht: Kluwer Law International, 1999) 37–53 at 43.
[119] "The implications of this regime are significant. Among the parties to the Convention, only the United States and possible the Russian Federation are likely to maintain fisheries enforcement vessels in or near the Doughnut Hole. Accordingly, only the coastal states will be in a position to board and inspect fishing vessels in the Doughnut Hole on a regular basis, If a vessel from another party is found to have committed a serious violation, the right of the coastal states to have their boarding ream remain on board the vessel should, in most cases, force an immediate end to the violation and, by virtue of the evidence collected by the boarding team, may lead to more vigorous

prosecution…these provisions became the model for Articles 21 and 22 or the Fish Stocks Agreement and are a part of what appears to represent the beginning of a new era in high seas fisheries enforcement". D. Balton, "The Bering Sea Donut Hole Convention: Regional Solution, Global Implications" in O. Stokke, ed., *Governing High Seas Fisheries: The Interplay of Global and Regional Regimes* (Oxford: Oxford University Press, 2001) 143–177 at 165.

[120] R. Rayfuse, "The United Nations Agreement on Straddling and Highly Migratory Fish Stocks as an Objective Regime: A Case of Wishful Thinking?" (1999) 20 *Australian Yearbook of International Law* 253–278.

[121] See, however, UNFSA Art. 21(14), which provides for the application of the UNFSA boarding and inspection regime in cases where there are clear grounds for believing that a fishing vessel flying the flag of another State Party has engaged in any activity contrary to RFMO conservation and management measures and the vessel has subsequently, during the same fishing trip, entered into an area under the national jurisdiction of the inspecting State Party.

[122] BSDH Convention Art. XI.

[123] CCAMLR, "Establishment of a System of Observation and Inspection, Article XXIV of the Convention" in *Report of the Seventh Meeting of the Commission*, Hobart, Australia, 24 October – 4 November 1988, paras. 124–125. The consolidated text of the scheme is available at <http://www.ccamlr.org> (accessed August 2005).

[124] NAFO, "NAFO Conservation and Enforcement Measures. Chap. 4 — Joint Inspection and Surveillance Scheme", Arts. 25–38. Available at <http://www.nafo.int/fisheries/frames/regulations.html> (accessed 27 May 2007).

[125] NEAFC, "Control and Enforcement Scheme, Report of the Seventeenth Annual Meeting of NEAFC" (1998) para. 49 and Annex E. The full text of the scheme is available at <http://www.neafc.org/> (accessed August 2005).

[126] SEAFO Convention Art. 16(3)(b).

[127] This section is reproduced from R. Rayfuse, *Non-Flag State Enforcement in High Seas Fisheries*, 234–243 (with copy-edits for house style and consistency).

[128] Nadia Bouffard, Canadian Department of Fisheries and Oceans, e-mail communication to Evelyne Meltzer, 28 January 2005. In the words of the Head of the EU delegation, John Spencer, at the 2004 NAFO Annual Meeting: "Canada has made clear that it views the NAFO Scheme as not being consistent with the Agreement [UNFSA] and that on such grounds, Canada reserves the right to apply the agreement vis-à-vis any other NAFO Party that has ratified it, even if it entails actions that do not respect NAFO provisions. In particular, this might imply the right of the inspection state under certain circumstances to bring a vessel to port, even in the absence of flag State consent. We have made it quite clear that the European Community does not agree with such an interpretation of the UN Fish Stocks Agreement, which amounts to negating the power of Regional Fisheries Organizations, such as NAFO to implement what their status already implies: a regional, multilateral approach to fishery regulation and control based on international cooperation. Such interpretations further result in nullifying NAFO rules and creating discriminatory treatment among the membership".

[129] ICNAF Annual Proceedings, Vol. 27, 1976–1977, at 57.

[130] ICNAF Annual Proceedings, Vol. 29, 1978–1979, at 24 and 31–33, and NAFO, *Annual Report 1979*, Vol. 1 (Dartmouth, N.S.: NAFO, 1979) at 60–61 and 70–71.

[131] For a description and critique of the amendments, see E. Wiseman and R. Steinbock, "Compliance" in M. Nordquist and J. Norton Moore, eds., *Current Fisheries Issues and the Food and Agriculture Organization of the United Nations* (Leiden: Martinus Nijhof Publishers, 2000) at 399–403

[132] Proposal 1/91 of the Fisheries Commission adopted on 12 April 1991 and approved in principle on 21 June 1991. Reproduced in NAFO Conservation and Enforcement Measures NAFO/FC Doc. 02/9 (2002) at 6.

[133] Interestingly, these included a certificate of harvest origin scheme similar to the catch documentation schemes now adopted in other RFMOs. Report of the STACTIC Working Group in Improvements to Inspection and Control in the NAFO Regulatory Area, July 1991, and Report of the Standing Committee on International Control at the 13th Annual Meeting (1991), Meeting Proceedings 1991, at 105 and 232, respectively.

[134] Canada – European Community Agreed Minute on the Conservation and Management of Fish Stocks, 34 ILM 1260 (1995).

[135] For a description of this matter see R. Rayfuse, *Non-Flag State Enforcement in High Seas Fisheries* at 230–232 and accompanying footnotes.

[136] Report of the Special Meeting of STACTIC (May 1995), Report of the Special Meeting of the Fisheries Commission and its Subsidiary Body (STACTIC) (May 1995), and Report of the 17th Annual Meeting of the Fisheries Commission and its Subsidiary Body (STACTIC) (1995), Meeting Proceedings 1995, at 47, 79, and 183, respectively.

[137] The consolidated version of the scheme is in NAFO, "NAFO Conservation and Enforcement Measures", NAFO FC Doc. 02/09 (2002), Serial No. 4624 at 31–77. No longer available online; was accessed September 2005.

[138] The text of the scheme is reproduced as Annex 4 to the Report of the Standing Committee on Fishing Activities of Non-Contracting Parties in the Regulatory Area (STACFAC), Report of the 19th Annual Meeting if the General Council, 1997 Meeting Proceedings, at 265–267.

[139] This section is reproduced from R. Rayfuse, *Non-Flag State Enforcement in High Seas Fisheries*, 272–276 (with copy-edits for house style and consistency).

[140] CCAMLR, *Report of the Seventh Meeting of the Commission*, Hobart, Australia, 24 October – 4 November 1988, paras. 124–125. The consolidated text of the scheme is available at <http://www.ccamlr.org> (accessed September 2005).

[141] CCAMLR OIS Art. I.

[142] CCAMLR OIS Art. I(c).

[143] CCAMLR OIS Art. I(d).

[144] CCAMLR OIS Art. II.

[145] CCAMLR OIS Art. IV.

[146] CCAMLR OIS Art. V.
[147] CCAMLR OIS Art. III.
[148] CCAMLR OIS Art. III(c).
[149] CCAMLR OIS Art. I(c).
[150] CCAMLR OIS Art. III(b).
[151] CCAMLR OIS Arts. V and VI.
[152] CCAMLR OIS Art. VIII.
[153] CCAMLR OIS Art. VII.
[154] CCAMLR OIS Art. IX.
[155] CCAMLR OIS Art. X. The OIS is intended to be an international system providing for the inspection of a fishing vessel operating under the flag of one member State by inspectors nominated by another member State. This does not, however, preclude national inspections.
[156] CCAMLR, *Report of the Twelfth Meeting of the Commission*, CCAMLR-XII, Hobart, Australia, 25 October – 5 November 1993, para. 6.5.
[157] CCAMLR, "Intersessional Procedure for Notifying CCAMLR Members of Infractions and Sanctions Imposed by Flag States" Appendix III in *Report of the Standing Committee on Observation and Inspection*, Hobart, Australia, 26 October – 4 November 1994; and CCAMLR, *Report of the Thirteenth Meeting of the Commission*, CCAMLR-XIII, Hobart, Australia, 26 October – 4 November 1994, para. 5.36.
[158] CCAMLR OIS Art. VIII(d) as amended, CCAMLR, *Report of the Sixteenth Meeting of the Commission*, CCAMLR-XVI, Hobart, Australia, 27 October – 7 November 1997, para. 8.14.
[159] CCAMLR OIS Art. VIII(e) as amended, CCAMLR, *Report of the Sixteenth Meeting of the Commission*, CCAMLR-XVI, Hobart, Australia, 27 October – 7 November 1997, para. 8.14.
[160] CCAMLR, *Report of the Standing Committee on Observation and Inspection*, Hobart, Australia, 21 October – 1 November 1996, paras. 1.69–1.72; and CCAMLR, *Report of the Fifteenth Meeting of the Commission*, CCAMLR XV, Hobart, Australia, 21 October – 1 November 1996, para. 7.24.
[161] CCAMLR OIS Art. III as amended now reads: "In order to verify compliance with Conservation Measures adopted under the Convention, inspectors designated by members shall be entitled to board a fishing or fisheries research vessel in the area to which the Convention applies to determine whether the vessel is, or has been, engaged in scientific research, or harvesting of marine living resources". CCAMLR, *Report of the Fourteenth Meeting of the Commission*, CCAMLR XIV, Hobart, Australia, 24 October – 4 November 1995, para. 7.24. Available at <http://www.ccamlr.org/pu/E/e_pubs/cr/95/all.pdf> (accessed March 2005)
[162] CCAMLR, *Report of the Fourteenth Meeting of the Commission*, CCAMLR XIV, Hobart, Australia, 24 October – 4 November 1995, para. 7.26; and CCAMLR, *Report of the Fifteenth Meeting of the Commission*, CCAMLR XV, Hobart, Australia, 21 October – 1 November 1996, para. 7.24.
[163] CCAMLR, *Report of the Standing Committee on Observation and Inspection*, Hobart, Australia, 24 October – 4 November 1995, para. 2.18.
[164] CCAMLR, *Report of the Eighteenth Meeting of the Commission*, CCAMLR XVIII, Hobart, Australia, 25 October – 5 November 1999, paras. 8.15–8.30.
[165] CCAMLR OIS Arts. XII, XII, and XIV. CCAMLR, *Report of the Sixteenth Meeting of the Commission*, CCAMLR XVI, Hobart, Australia, 27 October – 7 November 1997, para. 8.14.
[166] R. Rayfuse, *Non-Flag State Enforcement in High Seas Fisheries* at 343
[167] In 2004 amendments to Canada's Coastal Fisheries Protection Regulations were approved, enabling Canada to apply UNFSA enforcement procedures to all UNFSA Parties with fleets in the NRA. See <http://www.dfo-mpo.gc.ca/overfishing-surpeche/issue_e.htm> (accessed September 2005).
[168] NAFO Scheme Art. 3.32.
[169] NAFO Scheme Art. 32.4–32.6, and NEAFC Scheme Art. 20.4–20.6.
[170] The consequences of invalid reservations to treaties are still a matter of debate in international law. cf, the decisions of Lauterpacht J. in the *Certain Norwegian Loans* case, [1957] ICJ Rep. 9 and the *Interhandel* case, [1959] ICJ Rep. 6, and the work of the International Law Commission (ILC) on the issue of reservations to treaties. The latest report of the ILC is found in Chap. X of ILC, *Report on the Work of its Fifty-Seventh Session (2 May – 3 June and 11 July – 5 August 2005)*, ICL Rep. (A/60/10). Available at <http://untreaty.un.org/ilc/reports/2005/2005report.htm> (accessed September 2005).
[171] For a more thorough analysis of this issue see R. Rayfuse, "Countermeasures and High Seas Fisheries Enforcement" (2004) 51(1) *Netherlands International Law Review* 41–76, from which this summary is drawn.
[172] UN Charter Art. 2(4).
[173] Simma notes that the concept of "use of force" is wider than armed attack but that it still requires the use of "armed" force. B. Simma, ed., *The Charter of the United Nations: A Commentary*, 2nd ed., Vol. 1 (Oxford: Oxford University Press, 2002) at 117 and 796.
[174] See R. Rayfuse, "Countermeasures and High Seas Fisheries Enforcement" at 73–75.
[175] M. Lodge and S. Nandan, "Some Suggestions Towards Better Implementation of the United Nations Agreement on Straddling Fish Stocks and Highly Migratory Fish Stocks of 1995" (2005) 20 *International Journal of Marine and Coastal Law* 345–380 at 351, and at 363 note, "One of the most contentious achievements of UNFSA was to provide for the possibility of boarding and inspection of fishing vessels on the high seas by states other than the flag state. It is the first global instrument to establish such a procedure and in the process broke with several hundred years of international law. Importantly, however, the power to board and inspect is strictly circumscribed by

safeguards designed to preserve at all times the right of the flag state to assert its primary jurisdiction, and indeed to encourage it to do so, and to prevent abuse as well as risk to human life and safety".

[176] At the FAO COFI Ministerial Meeting held in Rome, Italy, March 2005, the actions of those on board IUU vessels where likened to actions of pirates and criminals.

[177] B. Simma, at 791.

[178] B. Simma, at 117.

[179] LOSC Arts. 101–110.

[180] LOSC Arts. 107, 110(4), 110(5), and 111(5).

[181] See the "Discussions of Panel IV: Caribbean Drug Challenges" in M. Nordquist and J. Norton Moore, eds., *Oceans Policy: New Institutions, Challenges and Opportunities* (Leiden: Martinus Nijhoff Publishers, 1999) at 173–236.

[182] For a comprehensive analysis of the state practice referred to here, see R. Rayfuse, *Non-Flag State Enforcement in High Seas Fisheries*.

[183] In one case, the Russian Border Guard ended a five-day hot pursuit of a Chinese vessel found illegally driftnet fishing within its EEZ by firing disabling shots that killed two crew members and wounded three others. See US Coast Guard, "1998 Report to the Committee on Enforcement of the North Pacific Anadromous Fish Commission", NPAFC Doc. 354.

[184] UN Rep. Vol III (1935) 1609.

[185] The Red Crusader case (1962) 35 ILR 485. For a discussion of the effect of these cases, see D.P. O'Connell, *The International Law of the Sea*, Vol. II (Oxford: Clarendon Press, 1984) at 1072–1074.

[186] LOSC Art. 106 (however this refers specifically to piracy).

[187] R. Rayfuse, "To Our Children's Children's Children: From Promoting to Achieving Compliance in High Seas Fisheries" at 528.

[188] R. Rayfuse, "To Our Children's Children's Children: From Promoting to Achieving Compliance in High Seas Fisheries" at 528.

[189] High Seas Task Force, "Better High Seas Monitoring, Control and Surveillance — An Improved Network", Meeting of the High Seas Task Force, Paris, France, 9 March 2005, HSTF/04. Available at <http://www.high-seas.org> (accessed August 2005).

[190] UNFSA Art. 23(1).

[191] UNFSA Art. 23(2).

[192] UNFSA Art. 23.3.

[193] This section is reproduced from R. Rayfuse, *Non-Flag State Enforcement in High Seas Fisheries* at 177–178 (with copy-edits for house style and consistency).

[194] ICCAT Recommendation 97-10 for a Revised ICCAT Port Inspection Scheme.

[195] IOTC Recommendation 98-11 Concerning the Ban on Landings and Transhipments of Vessels from Non-Contracting Parties Identified as Having Committed a Serious Infringement.

[196] This section is reproduced from R. Rayfuse, *Non-Flag State Enforcement in High Seas Fisheries* at 190–191 (with copy-edits for house style and consistency).

[197] IOTC Resolution 01/03 Establishing a Scheme to Promote Compliance by Non-Contracting Party Vessels with Resolutions Established by IOTC.

[198] IOTC Resolution 02/01 Relating to the Establishment of an IOTC Programme of Inspection in Port.

[199] T. Løbach, "Measures to be Adopted by the Port State in Combating IUU Fishing", Document AUS:IUU/2000/15 (Rome: FAO, 2000). Available at <http://www.fao.org> and <www.intfish.net/docs/2004/fao_portstate.pdf> (accessed August 2005).

[200] CCAMLR Conservation Measure 10-03 (2005).

[201] NAFO CEM Chap. V, Arts. 39–40.

[202] SEAFO Convention Art. 15(2).

[203] WCPFC Convention Arts. 27 and 29.

[204] R. Rayfuse, "To Our Children's Children's Children: From Promoting to Achieving Compliance in High Seas Fisheries" at 521.

[205] FAO, "Report of the Technical Consultation to Review Port State Measures to Combat Illegal, Unreported and Unregulated Fishing. Rome, 31 August – 2 September 2004", FAO Fisheries Report No. 759 (Rome: FAO, 2004). See also R. Rayfuse, "To Our Children's Children's Children: From Promoting to Achieving Compliance in High Seas Fisheries" at 527.

[206] High Seas Task Force, "Promoting Responsible Port States", Meeting of the High Seas Task Force, Paris, France, 9 March 2005, HSTF/06. Available at <http://www.high-seas.org> (accessed August 2005).

[207] High Seas Task Force, "Promoting Responsible Port States" at 16.

[208] R. Rayfuse, "To Our Children's Children's Children: From Promoting to Achieving Compliance in High Seas Fisheries" at 523.

[209] P. Cacaud, M. Kuruc, and M. Spreij, "Administrative Sanctions in Fisheries Law", FAO Legislative Study No. 82 (Rome: FAO, 2003).

[210] P. Cacaud, M. Kuruc, and M. Spreij, FAO Legislative Study No. 82.

[211] High Seas Task Force, "High Seas Governance", Meeting of the High Seas Task Force, Paris, France, 9 March 2005, HSTF/09. Available at <http://www.high-seas.org> (accessed September 2005).

[212] See, e.g., the terms of reference for and reports of the CCAMLR Standing Committee on Implementation and Compliance (SCIC), the NAFO Standing Committee on International Control (STACTIC), the NEAFC Permanent Committee on Control and Enforcement (PECCOE), the NPAFC Committee on Enforcement (ENFO), the IOTC Technical and Compliance Committee, the ICCAT Compliance Committee, the WCPFC Technical and Compliance Committee, and the IATTC Permanent Working Group on Compliance.

[213] UNFSA Art. 21(1).

[214] UNFSA Art. 21(5).

[215] UNFSA Art. 21(8).

[216] R. Rayfuse, "To Our Children's Children's Children: From Promoting to Achieving Compliance in High Seas Fisheries" at 516; R. Rayfuse *Non-Flag State Enforcement in High Seas Fisheries* at 331; and F. Orrego Vicuña, *The Changing International Law of High Seas Fisheries* (Cambridge: Cambridge University Press, 1999) at 253.

[217] The text of the Scheme is available at <http://www.neafc.org> (accessed September 2005).

[218] NAFO Scheme Arts. 40–44.

[219] See S. Nandan, "Current Fisheries Governance" prepared for *Conference on the Governance of High Seas Fisheries and the United Nations Fish Agreement — Moving from Words to Action*, St. John's, N.L., 1–5 May 2005. Available at <http://www.dfo-mpo.gc.ca/fgc-cgp/documents/presentations/nandan_e.pdf> (accessed August 2005).

[220] R. Rayfuse, *Non-Flag State Enforcement in High Seas Fisheries*, refers to a number of such instances, cf. 126–136.

[221] LOSC Art. 92(2).

[222] The extent to which such vessels are unprotected and the actions that may be taken against them following a right of visit remain unclear in international law at the present. The IPOA-IUU adopted a circular approach. Para. 20 specifies that "States should take measures consistent with international law in relation to vessels without nationality on the high seas involved in IUU fishing". For a discussion of the significant overuse of terms such as "consistent with international law" in the IPOA-IUU, see W. Edeson "The International Plan of Action on Illegal, Unreported and Unregulated Fishing: The Legal Context of a Non-Legally Binding Instrument" (2001) 16 *International Journal of Marine and Coastal Law* 603–623 at 614.

[223] R. Rayfuse, *Non-Flag State Enforcement in High Seas Fisheries* at 162.

[224] UNFSA Art. 21(8).

[225] R. Rayfuse, *Non-Flag State Enforcement in High Seas Fisheries* at 331.

[226] F. Orrego Vicuña, at 254.

[227] R. Rayfuse, *Non-Flag State Enforcement in High Seas Fisheries* at 331.

[228] R. Rayfuse, "Countermeasures and High Seas Fisheries Enforcement" at 41

[229] R. Rayfuse, "To Our Children's Children's Children: From Promoting to Achieving Compliance in High Seas Fisheries" at 529.

[230] R. Rayfuse, "To Our Children's Children's Children: From Promoting to Achieving Compliance in High Seas Fisheries" at 528.

[231] High Seas Task Force 2006, "Closing the Net: Stopping Illegal Fishing on the High Seas", Final Report of the Ministerially-led Task Force on IUU Fishing on the High Seas, Governments of Australia, Canada, Chile, Namibia, New Zealand, and the United Kingdom, WWF, IUCN, and the Earth Institute at Columbia University.

Afterword*

The 2006 United Nations Fish Stocks Agreement Review Conference and beyond

United Nations Fish Stocks Agreement Review Conference (22–26 May 2006)

The Review Conference, held in New York from 22 to 26 May 2006, was the first opportunity for State Parties to the United Nations Fish Stocks Agreement (UNFSA) and non-State Parties to formally review the implementation and application of the Agreement since it had entered into force in 2001.

The Review Conference, as called for by Article 36 of UNFSA, had the overall objective of determining whether the Agreement had been effective in securing the conservation and management of straddling fish stocks and highly migratory fish stocks. To this end, the Conference reviewed and assessed the implementation and application of the provisions of the Agreement, based on the four pillars on which the Agreement is built (i.e., conservation and management of stocks; mechanisms for international cooperation; monitoring, control, and surveillance; compliance and enforcement; developing States and nonparties) and proposed means of strengthening the substance and methods of implementation of those provisions to better address any continuing problems in the conservation and management of those stocks.

The final report[1] of the Review Conference constitutes a political document, adopted by consensus of the participants, including State and non-State Parties, which contains comprehensive measures to strengthen the implementation of the Agreement and includes, *inter alia*, commitments to urgently strengthen the mandates of regional fisheries management organizations (RFMOs) to implement modern approaches to fisheries (including the introduction of performance reviews); integrate ecosystem considerations in fisheries management; reduce the world's fishing capacity to levels commensurate with the sustainability of fish stocks; develop a legally binding instrument on minimum standards for port-State measures and a comprehensive global register of fishing vessels; expand assistance to developing countries; and continue a dialogue to address ongoing concerns raised by non-State parties.

Consistent with the provisions of the Agreement, an important outcome of the Conference in terms of governance of the high seas was the recognition of RFMOs as primary mechanisms for international cooperation in conserving and managing fish stocks. The Conference also proposed that new RFMOS should be established where needed and that interim measures be agreed upon until permanent organizations are in place.

In sum, the Review Conference has not only renewed interests in the Agreement—14 States have joined since May 2006[2] (see Table 10.1). Its outcomes have also provided an action plan of sorts to lead the way to improving international high seas governance. The developments and efforts outlined below provide but a broad-brush sketch of the changes in train in the international fisheries world. In this respect, the Review Conference has proven to be a catalyst for change in high seas fisheries governance.

New international developments and efforts since the Review Conference

Conservation and, Management of stocks

Since the May 2006 Review Conference, many States have been actively involved in the process of establishing new RFMOs. For example, in July 2006, the Agreement for the establishment of the South Indian Ocean Fisheries Arrangement (SIOFA) was signed by the Comoros, France, Kenya, Mozambique, New Zealand, the Seychelles, and the European Community, and incorporates many of the provisions found in UNFSA.

Since 2006, Australia, Chile, and New Zealand have been leading a process to develop a new RFMO to address

* Original version by Terje Løbach, Norway in 2007, with updates by DFO - Nathalie Lavoie of Canada in 2008.

gaps in the international conservation and management of non-highly migratory fisheries and protection of biodiversity in high seas areas of the South Pacific Ocean. The gap extends from the most eastern part of the South Indian Ocean through the Pacific towards the exclusive economic zones of South America. The Sixth International Consultation on the Establishment of the South Pacific RFMO (SPRFMO) is planned from 6 to 10 October 2008, in Canberra, Australia.[3]

Mechanisms for international cooperation and nonmembers

At the Review Conference, RFMOs were also encouraged to strengthen their mandates and to enhance interorganization cooperation with the objective of implementing modern approaches to fisheries management, including addressing participatory rights, establishing mechanisms to promote the participation of nonmembers, ensuring that State behaviour following opt-out of RFMO measures is constrained, establishing clear processes for dispute resolution, and improving transparency.

In August 2007, Chatham House published the final report of the independent high-level panel mandated to develop a model RFMO for improved governance, entitled Recommended Best Practices for Regional Fisheries Management Organizations. This initiative, which was based on a recommendation from the Ministerially-led High Seas Task Force on Illegal, Unreported and Unregulated fishing on the high seas, included a comprehensive assessment of best practices of RFMOs worldwide and the provisions of international instruments. The report has been tabled at several international meetings as a means of highlighting its role as a practical tool to which RFMOs can turn and refer to in their efforts towards modernization.

Some concrete coordination initiatives have also take place among RFMOs. Indeed, in January 2007, Japan hosted a joint meeting of the five RFMOs that manage high seas tuna stocks, in Kobe. Participants developed a course of action and agreed on follow-up actions to be taken by the five tuna RFMOs and their members. It was the first time that all the tuna RFMOs joined forces to find collaborative ways to rebuild tuna stocks and prevent further depletion.

In June 2007, at an extraordinary meeting of the North East Atlantic Fisheries Commission (NEAFC), the Commission considered the Panel report of the first Performance Review and agreed on some actions based on its recommendations. NEAFC has also amended its Convention to strengthen its mandate to implement modern approaches to fisheries management and to introduce a fast-track dispute settlement mechanism.

Similarly, at its 2007 Annual Meeting, the Contracting Parties to the Northwest Atlantic Fisheries Organization (NAFO) agreed on outstanding NAFO reform issues and adopted the new NAFO Convention. The NAFO reform process was first launched in 2005, following the St. John's Conference on High Seas Overfishing in May 2005, with a view to ensuring that NAFO's management mandate is consistent with modern standards of international fisheries management, including those espoused by UNFSA.

Integrate ecosystem considerations in fisheries management

Concrete actions were also taken by some RFMOs (NAFO, NEAFC, and the South East Atlantic Fisheries Organisation (SEAFO)) to implement ecosystem considerations in the management of fisheries, ranging from a ban on shark finning, the protection of sea turtles, the reduction of mortality of seabirds, to the prohibition of ghost fishing.

In that context, it is important to underline the adoption of the United Nations General Assembly Resolution 61/105 on the protection of vulnerable marine ecosystems (VMEs) from the effects of fishing. The Resolution, which calls on flag States and RFMOs as well as RFMOs in development, to take action by 31 December 2008, constitutes an important regime shift in the management of high seas fisheries. In accordance with the Resolution, fishing activities must be assessed for significant adverse impacts to VMEs, which could either lead to fishing not being authorized to proceed, or to imposing appropriate mitigation measures as a condition to the fishing activity, where necessary. The United Nations will review progress of implementation of the resolution in the fall of 2009.

RFMOs have responded to international concerns about threats to sensitive marine ecosystems and have taken steps to implement the commitments found in Resolution 61/105. For instance, the developing SPRFMO has paved the way to a set of interim measures to protect VMEs; these measures are being further detailed for practical implementation by participating States.

In May 2008, NAFO's Contracting Parties successfully adopted new measures to map current fishing areas outside of Canada's 200-mile limit, and to provide for assessment of the impact of current fishing practices on VMEs. Precautionary protocols will also be put in place to protect areas that have not been recently fished. This builds on existing VME measures adopted by NAFO, including the closure of four seamounts to commercial fishing in 2006 and the creation of a coral protection zone in 2007.

In November 2007, the NEAFC Commission agreed to maintain the closure of five sensitive marine areas, until the end of 2008, and to extend the boundaries of several closed areas on the Hatton and Rockall Banks to provide better protection for cold-water corals.

Monitoring, control, and surveillance and compliance and enforcement

The Review Conference also emphasized that effective compliance with and enforcement of agreed conservation and management measures, supported by effective monitoring, control, and surveillance, are critical to achieving the long-term conservation and sustainable use of straddling fish stocks and highly migratory fish stocks. The Review Conference drew attention to a number of actions to be taken, many of which are found in the Food and Agriculture Organization's (FAO) International Plan of Action on Illegal, Unreported, and Unregulated Fishing (IPOA-IUU).

The conference participants further agreed to develop appropriate processes to assess the performance of flag States, to develop regional guidelines for fisheries sanctions, and to consider the use of multilaterally agreed trade measures. Finally the Conference participants agreed to cooperate with FAO to develop a comprehensive global register of fishing vessels and to ensure that all vessels fishing on the high seas carry vessel monitoring systems. On the issue of flag-State performance, in preparation for the work called for by FAO-COFI in 2007, and to develop a better global understanding of obligations with respect to flag-State responsibilities, Canada (with support from the European Commission and Iceland) hosted an expert workshop in March 2008 to develop criteria for assessing flag-State performance and identifying possible actions in the event of noncompliance. At this workshop, international experts focussed their discussions on criteria to assess performance of flag-State responsibilities and possible actions against vessels in the event of noncompliance. In addition, considerations for helping developing countries to meet their flag-State commitments were also discussed. Work is underway to develop a guidance document for future FAO work in this area.

The Review Conference also called for a legally binding instrument on minimum standards for port-State measures, which would build on the 2005 FAO Model Scheme on Port State Measures to Combat IUU Fishing and the IPOA-IUU. The United Nations General Assembly repeated this call in both the 2006 and 2007 Sustainable Fisheries Resolutions.

While some RFMOs have already introduced certain port-State duties, a comprehensive scheme with a wider application is needed, since not all port States are members of an RFMO; and there are regions where RFMOs are unlikely to be established, where the current RFMOs deal only with specific species, and where port control might involve more than one RFMO.

Following the FAO expert consultation that took place in September 2007, the FAO has also convened a Technical Consultation for 23-27 June 2008, to finalize the draft instrument. FAO-COFI is working towards a presentation of the final version of the text at its next session in 2009.

As well, RFMOs such as NAFO, NEAFC, the Commission for the Conservation of Antarctic Marine Living Resources (CCAMLR), the Inter-American Tropical Tuna Commission (IATTC), the Indian Ocean Tuna Commission (IOTC), SEAFO, the Commission for the Conservation of Southern Bluefin Tuna (CCSBT), and the Western and Central Pacific Tuna Commission (WCPFC) have introduced or are about to introduce mandatory vessel monitoring systems for vessel operating within their areas of competence. RFMOs have also recognized that enhanced international cooperation is needed regarding transshipments at sea, where monitoring is extremely difficult. Regulating transshipments has become an important tool in the fight against IUU fishing in areas under the jurisdiction of RFMOs, and for collecting and verifying data. Many IUU operators tranship at sea as a way of avoiding detection. By transferring catches onto reefers, IUU fishing vessels avoid entering ports for landing their fish. On reefers, IUU catches are often laundered by mixing them with legally caught fish. In November 2006, NEAFC adopted a new Scheme on Control and Enforcement, based partly on the 2005 FAO Model Scheme on Port State Measures to Combat IUU Fishing. Some RFMOs (the International Commission for the Conservation of Tunas (ICCAT), NAFO, NEAFC, SEAFO) have also put in place specific programs concerning at-sea transshipping operations.

Furthermore, several RFMOs (CCAMLR, IATTC, ICCAT, IOTC, NAFO, NEAFC, SEAFO) have established specific schemes in an attempt to combat IUU fishing by nonparties, which include lists of vessels involved in such activities in areas under their competence, the so-called "negative" or "black lists".

It is important to note the commonality of actions and concerted efforts by RFMOs in respect of vessels on these lists. For instance, RFMO parties must take all necessary measures against listed vessels to (1) prohibit the issuance of a licence to fish in waters under their jurisdiction, (2) refuse to grant their flag, (3) deny port access, (4) prohibit chartering, and (5) ensure that fishing, support, refuel, and cargo vessels and mother ships flying their flag do not participate in any transshipment or joint fishing operations, support, or resupply. In addition there are specific measures concerning trade, as parties to RFMOs must also encourage importers, transporters, and other sectors concerned to refrain from transaction and transshipment of any species caught by black-listed vessels. All RFMOs have established a policy of transmitting their lists to other RFMOs, and these lists are posted on their respective Web sites. NAFO and NEAFC have gone even further in their cooperation efforts, as vessel blacklisted by NAFO will also formally be listed by NEAFC, and vice versa. CCSBT has agreed to specific actions against their own members who are involved in IUU fishing.

Developing States and nonparties

Assistance to developing States for the realization of their rights and fulfilment of their obligations under UNFSA is crucial for the implementation of the Agreement as a whole. UNFSA acknowledges that lack of or limited capacity in many developing States is a serious impediment to its implementation. UNFSA emphasizes the need for building capacity and providing technical assistance to developing countries, including financial assistance, assistance relating to human resource development, technical assistance, transfer of technology, and advisory and consultative services.

At the Review Conference, some States reiterated their concerns over specific provisions, namely those related to the implementation of Article 7 (compatibility of conservation and management measures) and particularly to Article 21 (subregional and regional cooperation in enforcement), as impediments that have prevented them from becoming Parties to UNFSA. UNFSA Parties, for their part, have reiterated that the agreement represents a carefully achieved balance of interests with appropriate safeguards

There have been new developments, since the Review Conference, in respect of concerns raised by nonparties. The Informal Consultations of State Parties (ICSP) meeting provides the only forum for State Parties, non-State Parties, and observers to discuss successes and collective challenges with regard to the ratification and implementation of UNFSA.

While ICSP-6 held in the Spring of 2007 did not allow for in-depth exchanges on Review Conference commitments, calling into question for some the need for annual meetings, ICSP-7 meeting proved to be constructive and rejuvenated the Parties' interest in having annual meetings. It renewed a substantive exchange of ideas on UNFSA implementation and ways to promote further ratification and accession to the Agreement. Clarification was sought by several State Parties on the types of impediments non-States Parties face in ratifying the Agreement, as possible solutions will vary depending upon the cause. Three informal categories of impediments were identified: (1) lack of capacity; (2) policy issues, including technical and legal differences and interpretations; and (3) issues of a "political" nature. There was an expressed willingness to discuss these issues in more detail at the next States Parties meeting in 2009 and planning for the meeting is underway

There was also an acknowledgement by developed countries that when it comes to helping developing countries build their capacity to ratify and implement UNFSA, there is a need for more concrete action. Multiple sources of funding and capacity-building assistance are available (beyond the UNFSA Trust Fund) but these are not well known.

The success of the Seventh Informal Meeting of States Parties to UNFSA reaffirmed UNFSA's continuing relevance and its undeniable role as a vehicle of modern high seas governance.

All in all, this renewed spirit of cooperation among participants augurs well for the resumption of the Review Conference, now planned for 2010.

Endnotes

[1] A/CONF.210/2006/15

[2] Slovenia, Estonia, Japan, Trinidad and Tobago, Niue, Bulgaria, Latvia, Lithuania, Czech Republic, Romania, Republic of Korea, Palau, Oman, Hungary

[3] The SPRFMO is the first of many RFMOs to adopt measures to implement the provisions of the United Nations Resolution 61/105 dealing with the protection of Vulnerable Marine Ecosystems from the effects of fishing.

Table 10.1. Parties to UNFSA and FAO Compliance Agreement as of July 1, 2008

State or Entity	UNFSA	Compliance Agreement
Albania		■
Angola		■
Argentina		■
Australia	■	■
Austria	■	*
Bahamas	■	
Barbados	■	■
Belgium	■	*
Belize	■	■
Benin		■
Brazil	■	
Bulgaria	■	*
Canada	■	■
Cape Verde		■
Chile		■
Cook Islands	■	■
Costa Rica	■	
Cyprus	■	■
Czech Republic	■	*
Denmark	■	*
Egypt		■
Estonia	■	*
EC[3]	■	■
Fiji	■	
Finland	■	*
France	■	*
Georgia		■
Germany	■	*
Ghana		■
Greece	■	*
Guinea	■	
Hungary	■	*
Iceland	■	
India	■	
Iran	■	
Ireland	■	*
Italy	■	*
Japan	■	■
Kiribati	■	
Korea	■	■
Kenya	■	
Latvia	■	*
Liberia	■	
Lithuania	■	*
Luxembourg	■	*
Madagascar		■
Maldives	■	
Malta	■	*
Marshall Islands	■	
Mauritius	■	■
Mexico		■
Micronesia (FSM)	■	
Monaco	■	
Morocco		■
Myanmar		■
Namibia	■	■
Nauru	■	
Netherlands	■	*
New Zealand	■	■
Niue	■	
Norway	■	■
Oman	■	■
Palau	■	
Papua New Guinea	■	
Peru		■
Poland	■	*
Portugal	■	*
Romania	■	*
Russian Federation	■	
St. Kitts & Nevis		■
St. Lucia	■	■
Samoa	■	
Senegal	■	
Seychelles	■	■
Slovakia		*
Slovenia	■	*
Solomon Islands	■	
South Africa	■	
Spain	■	*
Sri Lanka	■	
Sweden	■	■
Syria		■
Tanzania		■
Tonga	■	
Trinidad and Tobago	■	
Ukraine	■	
United Kingdom	■	*
USA	■	■
Uruguay	■	■
Total	**71**	**36 (61)***

[1] UNFSA: Agreement for the Implementation of the Provisions of the Convention Relating to the Conservation and Management of Straddling Fish Stocks and Highly Migratory Fish Stocks. Source: United Nations Oceans and Law of the Sea, Division of Ocean Affairs and Law of the Sea. http://www.un.org/Depts/los/reference_files/status2008.pdf, 1 July 2008.

[2] FAO Compliance Agreement to Promote Compliance with International Conservation and Management Measures by Fishing Vessels on the High Seas. Source: FAO http://www.fao.org/Legal/treaties/012s-e.htm, 1 July 2008.

[3] EC accepted FAO Compliance Agreement in 1996 on behalf of all EU member States, which now include Austria, Belgium, Bulgaria, Cyprus, Czech Republic, Denmark, Estonia, Finland, France, Germany, Greece, Hungary, Ireland, Italy, Latvia, Lithuania, Luxembourg, Malta, Netherlands, Poland, Portugal, Romania, Slovakia, Slovenia, Spain, Sweden, and United Kingdom. Cyprus and Sweden accepted the FAO Compliance Agreement prior to joining the EC and are hence listed separately.

■ Indicates that the Parties have acceded or ratified these legal agreements.

* Indicates Parties to FAO Compliance Agreement under EC.

Evelyne Meltzer: 07/2008

Part III:
Matrices and Maps

Matrices and Maps of the RFMOs

The material in Part III consists of a set of matrices for RFMOs with competence for straddling fish stocks (Northwest Atlantic Fisheries Organization (NAFO), North East Atlantic Fisheries Commission (NEAFC), Southeast Atlantic Fisheries Organization (SEAFO), Commission for the Conservation of Antarctic Marine Living Resources (CCAMLR), and the Central Bering Sea Donut Hole Arrangement (BSDH)*, and for highly migratory fish stocks (International Commission for the Conservation of Atlantic Tunas (ICCAT), General Fisheries Commission for the Mediterranean (GFCM), Inter-American Tropical Tuna Commission (IATTC), Commission for the Conservation of Southern Bluefin Tuna (CCSBT), Indian Ocean Tuna Commission (IOTC) and Western and Central Pacific Fisheries Commission (WCPFC).

The matrices provide an overview of each RFMO and include information on:

- RFMO structure which covers RFMO establishment (entry into force), area of competence, key objectives, organizational structure and the scientific regime;
- Fisheries for each RFMO, including key species and gear, stock status, bycatch and the state of implementation of the precautionary approach and ecosystem approach to fisheries;
- Participation in RFMOs with sections on membership, membership categories and related participation criteria as well as transparency;
- Monitoring, control, and surveillance (MCS) mechanisms including data collection and reporting requirements, observer programs, satellite-based vessel monitoring systems (VMS), enforcement measures, and catch documentation schemes and trade measures;
- Governance efforts including decision making, dispute settlement mechanisms, inter-RFMO cooperation and finally, key challenges facing the respective RFMOs.

The matrices appear as follows : NAFO, NEAFC, SEAFO, CCAMLR, BSDH, ICCAT, GFCM, IATTC, WCPFC, CCSBT AND IOTC.

Maps

Part III includes a set of maps illustrating the geographical area of application covered by the respective RFMO constitutive agreements to manage a given fish stock or stocks. Additionally there are four maps illustrating some other areas where arrangements for straddling fish stocks exist or are proposed (Sea of Okhotsk "Peanut Hole", Barents Sea "Loop Hole", South Tasman Rise Arrangement, SIOFA as proposed). The overview map identifying existing or proposed RFMOs with competence for straddling fish stocks is found in the Summary (Figure S.1), and the overview map depicting the six RFMOs with competence for highly migratory fish stocks (tuna and tuna-like) is also found in the Summary (Figure S.2). The maps in Part III appear as follows: NAFO, NEAFC, SEAFO, CCAMLR, BSDH, Sea of Okhotsk "Peanut Hole", Barents Sea "Loop Hole", South Tasman Rise Arrangement, SIOFA, ICCAT, GFCM, IATTC, WCPFC, CCSBT and IOTC.

A list of acronyms is included in the front of this book to facilitate the use of the matrices, maps and figures.

* BSDH is an RFMA. For the purposes of this work, the term RFMO generally includes RFMA.

Northwest Atlantic Fisheries Organization (NAFO)

Evelyne Meltzer and Susanna D. Fuller: 05/2006

	NAFO	Area of competence	Objectives	Organizational structure	Scientific regime
RFMO structure	**Northwest Atlantic Fisheries Organization** **Headquarters:** Dartmouth, N.S., Canada. **Convention:** Convention on Future of Multilateral Cooperation in the Northwest Atlantic Fisheries (NAFO Convention). **Signed:** 24 October 1978, Ottawa, Canada. **In force:** 1 January 1979. **Web site:** www.nafo.int	The Northwest Atlantic Ocean. North of 35ºN and west of a line extending north from 35ºN and 42ºW to 59ºN, then west to 44ºW, and then north to the coast of Greenland, and the waters of the Gulf of St. Lawrence, Davis Strait and Baffin Bay south of 78º10'N.	To contribute through consultation and cooperation to the optimum utilization, rational management, and conservation of the fishery resources of the NAFO Convention Area (NCA). The NAFO Regulatory Area (NRA) is the high seas component of NCA.	General Council (GC) has 2 standing committees: Finance and Administration (STACFAD) and Non-Contracting Party Fishing Activities (STACFAC). Scientific Council (SC) has 4 standing committees: Fisheries Science (STACFIS), Publications (STACPUB), Research Coordination (STACREC), and, Fisheries Environment (STACFEN). Fisheries Commission (FC) has 1 standing committee: International Control (STACTIC). Secretariat (Executive Secretary).	SC is composed of scientists of CPs and has 4 standing committees: STACFIS, STACPUB, STACREC, and STACFEN. SC acts upon annual requests from FC and coastal States for advice on stock assessments. Work of SC forms foundation upon which FC determines management measures for NRA. FC also reviews information and scientific advice from the SC, establishes TACs for different species and national quota allocations, and establishes conservation measures.
	Key species and gear	**Stock status**	**Bycatch**	**Ecosystem approach (EA)**	**Precautionary approach (PA)**
The fishery, EA, and PA	**Straddling stocks:** cod, Greenland halibut, redfish, skates, American plaice, yellowtail flounder, white hake, witch flounder, capelin, squid, shrimp. **Discrete stocks:** 3M cod, 3M American plaice, 3M redfish, and 3M shrimp. **Gear:** bottom trawling (including otter and shrimp trawls), midwater trawls, longlines, and gill nets.	**Collapsed:** 9 stocks under moratoria. Since 1993: 3NO capelin, 3L cod. Since 1995: 3NO cod, 3M cod, 3LNO American plaice, 3M American plaice, 3NO witch flounder, 3L witch flounder. Since 1998: 3LN redfish. **Status uncertain:** 3LNO thorny skates, 30 redfish (no TAC), 3NO white hake (current TAC unsustainable), 3LMNO Greenland halibut (15-year rebuilding plan, 2004 catches exceeded TAC by 27%). **Recovered:** 3LNO yellowtail (moratorium lifted in 1998, in 2005 B_{lim} set at 30% B_{MSY}). **Other:** 3M shrimp (under effort regulation, 2005 biomass > B_{lim}), 3M redfish, (spawning biomass increasing as of 2005).	All nontarget species. Bycatch mitigation measures include gear mesh size, minimum fish size, time–area closures, use of sorting grates in shrimp fisheries, and percentage bycatch limits for contents of vessel holds and on a per-set (haul) basis (lower bycatch limits for moratoria species). In 2005, adopted measures banning shark finning as well as transshipment and landing of shark fins in NCA.	NAFO included EA in 2005 workplan. Following 2005 meeting, Contracting Parties (CPs) requested to collect data on seamounts in NCA. Northwest Atlantic Ecosystem Workshop planned for 2006. NAFO fisheries generally managed on an annual stock-by-stock basis. As of 2005, newly regulated species have multiyear (3 year) TACs. Since 2003, Greenland halibut under (15 year) rebuilding plan — poor progress, as 2004 TAC exceeded.	NAFO established PA Working Group in 1997. In 2004, FC adopted PA framework *in toto* and adopted separate proposal to apply PA framework initially to 3LNO yellowtail flounder and 3M shrimp. FC requested that additional stocks be considered under PA framework in the future.
	Members – Contracting Parties*	**Cooperating nonmembers**	**Membership**	**Participation criteria**	**Transparency**
Participation	Bulgaria, ***Canada***, Cuba, **Denmark** (Faroe Islands and Greenland), ***EU***, **France** (St. Pierre-et-Miquelon), **Iceland**, *Japan*, *Republic of Korea*, ***Norway***, **Russian Federation**, **Ukraine**, and ***USA*** (13 members).	There is no cooperating status for Non-Contracting Parties (NCPs).	Membership in NAFO GC is open to each CP. Membership in FC is only open to CPs who participate in NRA fisheries or who provide satisfactory evidence that they expect to participate in such fisheries. Membership is reviewed annually by GC. 60% of annual individual CP contribution to budget based on nominal catches in NCA.	1999 Resolution clarifies stocks are fully allocated and fishing opportunities for new members are likely to be limited to new fisheries, shared quotas, and stocks currently not allocated. NAFO annually sets aside an allocation of regulated species called "other" in part for States not included in list of CP allocations or for exceptional arrangements. Art. XIX provides for CPs to draw the attention of (i.e., contact) NCPs regarding fishing activities of nationals that may adversely effect NAFO conservation and management objectives. NCP vessels engaged in fishing in the NRA are presumed to be IUU and undermining effectiveness of NAFO measures.	Publicly accessible Web site. NAFO proceedings, scientific documents, and reports are available online. Meetings are open to all IGOs upon notification. NGOs and industry have participated as members of CP delegations. SC may invite NGOs and others. To date, one NGO has participated in SC meeting. Procedures were agreed in 2002 facilitating NGO participation as observers at FC plenary meetings with 100 days prior notice. NGOs may make oral statements upon invitation of Chair and distribute materials through Secretariat. Media policy prepared in 2004 to allow coverage of opening session.

*Bold type indicates that member has ratified UNFSA; italic type indicates that member has ratified the FAO Compliance Agreement.

	Data collecting and reporting	Observer program	Vessel monitoring system (VMS)	Enforcement	Catch documentation – trade schemes documentation
Monitoring, control, and surveillance	Data collection by NAFO via CPs includes catch and effort data; reports from port inspection, at-sea inspection and observer program; VMS messages; and aerial surveillance data. Data from CPs scientific surveys is used by members of SC in developing stock assessments and other scientific advice.	Since 1998, 100% observer coverage on all CP vessels fishing in NRA is required (for monitoring and compliance purposes). Reports made to flag State and copied to NAFO Secretariat. NAFO has introduced Pilot Project on Observer, Satellite Tracking and Electronic Reporting to test real-time electronic reporting system by observers and vessel masters.	Since 2001, CPs fishing in NRA must be equipped with VMS. Vessels transmit every 2 hours automatic positional and other reports to national Fisheries Monitoring Centres (FMC), which, in turn, forward to NAFO Secretariat. NAFO and NEAFC are developing North Atlantic Format (NAF) for harmonized, comprehensive electronic messaging transmitted through VMS. As of 2005, NAF Web site available.	NAFO Scheme of Joint International Inspection and Surveillance applies to all vessels fishing in NRA. NAFO requires CPs to perform in-port inspections on vessels having fished in NRA. CPs shall deny port access and transshipment for noncompliance pending vessel inspection. Vessel must demonstrate that species were caught outside NRA or consistent with NAFO rules. In 1997, adopted Scheme to Promote Compliance by NCPs; presumption that NCP vessels fishing in NRA are undermining conservation regime and are IUU. In 2005, NAFO created IUU list for NCP vessels. All NCP vessels in NRA can be inspected with prior consent of vessel – flag State ("courtesy boarding"). Information on NCP vessels is immediately communicated to NAFO Secretariat, all CPs and to flag State. Secretariat sends information on IUU activity to other RFMOs. There is a Canadian air surveillance program dedicated to NAFO surveillance.	As of 2005, all processed fish products from NRA must be labelled as caught in NRA with species and product category identified. Other than labelling, there is no formal catch or trade documentation scheme.
	Decision making	**Dispute settlement mechanism**	**Cooperative efforts**	**Efforts to Strengthen RFMO**	**Challenges**
Other governance efforts and challenges	Decisions are reached by consensus whenever possible; otherwise by majority vote (last vote 2002). Each CP has one vote. CPs in arrears for a period of more than 2 years cannot vote. Objection procedure enables CPs to formally object up to 60 days following decision.	NAFO does not have a dispute settlement mechanism.	NEAFC, NAFO, CCAMLR, and other RFBs cooperating to expand NAF (common data communications standard and other fisheries data — electronic logbooks). Has Memorandum of Understanding with ICES for joint stock assessments and other scientific matters. Cooperates with NEAFC on oceanic redfish management. Secretariat attends RFB Network since inception in 1999 and meets regularly with Executive Secretaries of North Atlantic RFMOs. NAFO is a partner in FIRMS to report and share information on status and trends of fishery resources (posted on FIGIS). NAFO is long-standing member of CWP. Cooperating with other RFMOs and FAO on IUU NCP vessel listing.	Overhaul of conservation and enforcement measures. Improved Web site for public access to information. Increased institutional transparency with 2002 provisions for NGO participation. Development of EA included in 2005 work plan. Established IUU NCP vessel list in 2005. Beginning in 2006, voluntary submission of data collected from seamounts. In 2004, adopted PA framework with initial application for 3LNO yellowtail flounder and 3M shrimp; also adopted multiyear TACs for newly regulated species; first RFMO to introduce management measures for elasmobranchs (skates regulated in 2005). Since 2003, Greenland halibut rebuilding plan — TAC exceeded in 2004; rebuilding prospects poor. In 2004, prepared first official compliance report. Implemented VMS and 100% observer coverage for NRA. At 2005 Annual Meeting adopted EU–Canada proposal for NAFO reform and established Working Group on Reform (first meeting in April 2006).	Avoiding overfishing. Respecting preagreed TACs. Rebuilding and maintaining stocks, particularly given poor recovery of 9 moratoria stocks and continued overfishing of Greenland halibut. Reducing and mitigating bycatch (urgent concern for incidental catch of moratoria species). Developing EA. Broadening scope to include marine biodiversity and habitat conservation concerns. Amending objection provisions to ensure conservation regime respected. Improving CP compliance. Improving timely and effective follow-up by flag States to violations of NAFO measures. Introducing effective deterrents for noncompliance. Funding the increasing management, research, and enforcement costs. CPs in arrears. Establishing global standards for observers and inspectors. Increasing collaboration with other RFBs, IGOs, and NGOs to strengthen marine conservation and integrated management in Atlantic Ocean. Modernizing Convention to be consistent with UNFSA. Reducing and managing fishing capacity.

North East Atlantic Fisheries Commission (NEAFC)

Evelyne Meltzer and Susanna D. Fuller: 05/2006

	NEAFC	Area of competence	Objectives	Organizational structure	Scientific regime
RFMO Structure	**North East Atlantic Fisheries Commission** **Headquarters:** London, UK. **Convention:** The Convention on Future Multilateral Cooperation in North-East Atlantic Fisheries (NEAFC Convention). **Signed:** 18 November 1980, London, UK. **In force:** 17 March 1982. **Web site:** www.neafc.org	The Northeast (NE) Atlantic and Arctic Oceans. East of a line south of Cape Farewell — the southern tip of Greenland — at 42ºW, north of a line to the west of Cape Hatteras — the southern tip of Spain — at 36ºN, and west of a line touching the western tip of Novya Semlya at 51ºE. Three high seas areas constitute NEAFC Regulatory Area: The Irminger Sea – Reykjanes – Azores Area; the Norwegian Sea "Banana Hole"; and the Barents Sea "Loophole". NEAFC does not regulate demersal fisheries in the Loophole.	To promote conservation and optimum utilisation of fishery resources of convention area within a framework appropriate to regime of extended coastal-State jurisdiction over fisheries, and to encourage international cooperation and consultation with respect to these resources. *In 2005, Contracting Parties (CPs) agreed in principle to update Convention. Proposed objectives are to promote the long-term conservation and optimum utilisation of fishery resources of NE Atlantic area, safeguard marine ecosystems in which resources occur, and encourage international cooperation and consultation with respect to these resources.*	Commission; 3 permanent committees: Permanent Committee on Control and Enforcement (PECCOE); Finance and Administration Committee; and new 2005 Permanent Committee on Management and Science; 4 working groups: Working Group on the Future of NEAFC, Working Group on Deep-Sea Species, Working Group on Blue Whiting (not met since 2000), Advisory Group on Data Communications; Secretariat (Executive Secretary).	NEAFC receives scientific advice from ICES (Art. 14 of Convention). In 1999, NEAFC formalized cooperative arrangement under Memorandum of Understanding (MOU) for all stocks under NEAFC. New MOU (2004–2006) is in place. ICES is compensated for these services.
	Key species and gear	**Stock status**	**Bycatch**	**Ecosystem approach (EA)**	**Precautionary approach (PA)**
The fishery, EA, and PA	**Straddling stocks:** redfish (pelagic), blue whiting, mackerel, Atlanto-Scandian (Norwegian spring-spawning) herring, Rockall haddock. As of November 2004, several deep-sea species are also regulated by NEAFC, including blue ling, black scabbardfish, and orange roughy. **Gear:** purse seine, pelagic trawl, demersal trawl, bottom longline, and bottom gill nets.	**Status uncertain:** redfish, Rockall haddock. **Full reproductive capacity, may be harvested unsustainably:** blue whiting. **At risk of reduced reproductive capacity:** mackerel. **Full reproductive capacity:** Atlanto-Scandian (Norwegian spring-spawning) herring. **Generally unknown:** deep-sea species, but according to ICES, many are beyond safe biological limits.	In 2005, adopted Recommendation, effective February 2006, to temporarily prohibit use of gill nets, entangling nets, and trammel nets in depths greater than 200 m in regulatory area to reduce discards and ghost fishing.	Working Group on the Future of NEAFC is examining how to strengthen NEAFC's role in addressing overall ocean management. EA now permanent agenda item at annual meetings. Closed area adjacent to Rockall Bank to trawl fishing in 2001; implemented temporary freeze in deep-sea fisheries in regulatory area in 2003; closed 5 vulnerable habitats to demersal fishing gear for 2005–2007; agreed to 30% reduction in deep-sea fisheries effort for 2005 onwards following ICES advice. In 2005 CPs agreed in principle to update NEAFC Convention to be consistent with UNFSA for EA.	In 1996, requested that ICES include PA in advice provided to Commission. Annual ICES advice includes management recommendations on PA reference points. Long-term PA management plans exist for mackerel and Atlanto-Scandian herring and blue whiting. In 2004, decided to close 5 high seas areas for 2005–2007 period. In 2005, established ban on targeting basking shark and ban on gill nets below 200 m.
	Members – Contracting Parties*	**Cooperating nonmembers***	**Membership**	**Participation criteria**	**Transparency**
Participation	**Denmark** (Faroe Islands and Greenland), Estonia, ***EC***, **Iceland**, ***Norway***, Poland, and **Russian Federation** (7 members). (Latvia and Lithuania are part of the EC as of 1 May 2004. As of May 2006, Estonia and Poland will be included in the EC.)	***Canada***, *Japan*, and **New Zealand**. ***Belize*** applied in 2005 for Non-Contracting Party (NCP) status. Application under consideration.	NEAFC is not considered an open organization (new entrant applications have been denied). States may accede to the Convention (except an EC member State) if application is approved by 3/4 of CPs. In May 2003, Working Group on the Future of NEAFC agreed to develop guidelines for new members and allocations. The guidelines were adopted in 2003.	In November 2003, NEAFC agreed that stocks regulated by NEAFC are fully allocated and fishing opportunities for new members are likely to be limited to new fisheries. New CPs will participate on same basis as existing CPs in future allocation of stocks unregulated at time of application. In addition, new CPs that were previously Cooperating NCPs may request part of relevant cooperative quota. Allocations will be considered on case-by-case basis.	Publicly accessible Web site. In 2001, NEAFC adopted rules to allow NGOs to participate in meetings. Secretary reviews NGO applications and notifies CPs. If one or more CPs object in writing within 30 days, participation by NGO will be put to vote. NGOs who attend Commission meetings may make oral statement upon invitation of Chairman, distribute documents, and engage in other activities approved by Chairman. IGOs and NCPs are also invited to Annual Meeting.

*Bold type indicates that member has ratified UNFSA; italic type indicates that member has ratified the FAO Compliance Agreement.

Monitoring, control, and surveillance	Data collecting and reporting	Observer program	Vessel monitoring system (VMS)	Enforcement	Catch documentation – trade schemes documentation
	1999 Scheme of Control and Enforcement requires each CP to report to Secretary monthly catches of regulated species landed or transshipped. Data collection via CPs includes infringement data, port inspection, at-sea inspection and boarding, and VMS data. Scientific and survey data are collected by ICES. 2005 Recommendation requires CPs to provide information on management measures for deep-sea species in convention area, develop sampling programs for deep-sea fisheries, and submit logbook data via NEAFC to ICES.	NEAFC does not have an observer program.	Since January 2000, all vessels fishing outside EEZs require VMS. Secretariat supplies CPs with an inspection presence, with information about ongoing fishing activities (24 hours a day, 365 days a year). Frequency of position reports from fishing vessels has increased from 6 hours to 2 hours. VMS operating successfully. NEAFC and NAFO are leading development of North Atlantic Format (NAF) — a comprehensive table of message types that can be transmitted through VMS.	In 1999 implemented Scheme of Control and Enforcement for CPs, as well as Scheme to Promote NCP Compliance. In case of infringement, CP flag State is notified and vessel to be examined by its inspectors within 72 hours. Flag State where warranted can require vessel to proceed to port for further inspection. CPs are to ensure legal action taken and penalties imposed to remove economic benefit derived from infringements. Information on NCP vessels in regulatory area circulated by Secretariat to CPs and RFMOs. In 2000, established committee (PECCOE) to advise Commission. Adopted 2003 Resolution on actions against IUU NCP vessels, including vessel lists. In 2005 agreed to consider strengthening system of port-State control. Naming and shaming of IUU NCP vessels on provisional A list and permanent B list is deterring IUU landings in CP ports. Key flag of convenience has deflagged and delisted vessels.	CPs are to refuse landing of catches deemed to be taken in contravention of NEAFC management recommendations.

Other governance efforts and challenges	Decision making	Dispute settlement mechanism	Cooperative efforts	Efforts to strengthen RFMO	Challenges
	Decisions are made by simple majority or, where Convention requires qualified majority, by 2/3 majority of votes of all CPs present and casting affirmative or negative votes. Each CP has 1 vote. A quorum of 2/3 of CPs is required. In even split of votes on any matter subject to simple majority, the proposal is rejected. Recommendations become binding on date determined by Commission. In emergency, votes may be taken by post or other means of communication. Any CP may object to a recommendation (for management measures only) within 50 days of date of notification. Since 2004 (amended Convention) parties are required to provide written statement identifying reason for objection, their intentions, and alternative conservation and management measures. Management of all stocks discussed in plenary meeting. In 2005, because of lack of consensus at meeting on oceanic redfish, management measures agreed in subsequent postal vote. In recent years, no agreement on allocations for Atlanto-Scandian herring and blue whiting. Blue whiting quotas agreed for 2006. Herring allocations still unresolved.	At 2004 Annual Meeting, Commission unanimously adopted amendments to Convention establishing fast-track dispute settlement mechanism. Ratification is expected in due course; however, parties have agreed to immediately make use of this fast-track mechanism on a voluntary basis.	NEAFC sets TAC for oceanic redfish for both NAFO and NEAFC. NEAFC Secretariat actively participates in RFB Secretariats Network biannual meetings. NEAFC Secretariat initiated North Atlantic Regional Fisheries Management Organization (NARFMO) and has organized annual meetings since 2001; in an effort to develop NAF, invited NAFO and CCAMLR to meet to discuss common data communication and standards for VMS and electronic logbooks (with the objective to engage other RFMOs to use this format as a common standard globally); established Advisory Group for Data Communication in April 2005 open to all RFBs. Group met again in October 2005. NEAFC performance review process involves members appointed by FAO, UNDOALOS, and an independent scientific institution not involved in NEAFC area.	Reactivated NEAFC following UNFSA in 1995. Tasked working groups with modernizing NEAFC. Incorporated PA references. Set TAC and allocations for oceanic redfish and Atlanto-Scandian herring in 1997 and mackerel in 2000. Implemented Scheme of Control and Enforcement and NCP Scheme in 1999. Expanded Scheme in 2003 and introduced NCP IUU lists. Identification of IUU NCP vessels has resulted in decline of IUU landings in CP ports ,as CPs prohibit landings from IUU fishing. Amended objection procedure in 2004. Established dispute settlement mechanism. Adopted sophisticated VMS system. Discussing new port-State scheme, based on FAO model. Agreed to close 5 areas with vulnerable habitats to demersal fishing gear. Set up arrangement whereby NEAFC sets TAC and allocation for oceanic redfish for NAFO. Created a publicly accessible Web site. Working to accommodate new entrants. In 2005, CPs agreed to text modernizing Convention to include PA, EA, and biodiversity protection, and to apply amendments on voluntary basis, pending ratification. In 2005, agreed to conduct independent performance review to ensure NEAFC alignment with UNFSA and related instruments. Most CPs are State Parties to UNFSA.	Reaching agreements on TACs and management measures. Sustainable development of deep-sea stocks. Clarifying ICES advice on stock status for mackerel and redfish. Considering new entrants. Ongoing modernization of RFMO. Implementation of Convention amendments updating EA, PA, biodiversity provisions, and dispute settlement. Need for improving decision-making process to ensure coherent, compatible, and effective management throughout convention area (e.g., blue whiting allocations). Reviewing and possibly revising Convention to incorporate broader ocean management perspectives, including EA as agreed in principle by CPs at 2005 Annual Meeting.

South East Atlantic Fisheries Organisation (SEAFO)

Evelyne Meltzer and Susanna D. Fuller: 05/2006

RFMO Structure	**SEAFO**	**Area of competence**	**Objectives**	**Organizational structure**	**Scientific regime**
	South East Atlantic Fishery Organisation **Headquarters:** Walvis Bay, Namibia. **Convention:** Convention on the Conservation and Management of Fishery Resources in the South-East Atlantic Ocean (SEAFO Convention). **Signed:** 20 April 2001, Windhoek, Namibia. **In force:** 13 April 2003. (First meeting of Commission held March 2004 in Swakopmund, Namibia. Second meeting held October 2005 in Windhoek, Namibia.) **Web site:** www.seafo.org	The Southeast Atlantic Ocean. Beyond national jurisdiction in areas bounded by a line beginning at the outer limit of waters under national jurisdiction at 6ºS, then west along 6ºS to 10ºW, then north along 10ºW to the equator, then west to 20ºW, then south along 20ºW to 50ºS, then east along 50ºS to 30ºE, then north along 30ºE to the coast of the African continent. Northern limit of area under review to reflect inclusion of Cabinda, Angola.	To ensure the long-term conservation and sustainable use of fishery resources on the high seas, other than highly migratory stocks, found in the Southeast Atlantic Ocean beyond the limits of national jurisdiction taking into account other living marine resources and the protection of the marine environment. Includes discrete stocks.	Commission, Scientific Committee (SC), Compliance Committee (establishment to be addressed at 2006 meeting), Secretariat (Executive Secretary).	SC provides advice to Commission and is composed of one representative appointed by each Contracting Party (CP). SC may establish any subsidiary body and submit to Commission for approval. Provisional Working Group (PWG) established in 2005 to collect and analyse catch and environmental data.
The fishery, EA, and PA	**Key species and gear**	**Stock Status**	**Bycatch**	**Ecosystem approach (EA)**	**Precautionary approach (PA)**
	Straddling stocks: Alphonsino, armourhead, deep-sea red crab, orange roughy, Patagonian toothfish, sharks (blue and short finned mako), swordfish. **Discrete stocks:** deep-sea red crab **Gear:** Bottom trawl, purse-seine, and traps used in crab fisheries.	Stock status is generally unknown. Uncertainties exist on reported catches and stock assessments have not been undertaken for high seas fisheries. SC in process of gathering existing data, in cooperation with flag States and related organizations.	CPs are to take into account the impact of fishing operations on ecologically related species, including seabirds, cetaceans, seals ,and sea turtles (Art. 3). Deep-sea shrimp are caught as bycatch in trawl fishery.	Convention provides for adoption of measures to conserve nontarget species belonging to the same ecosystem as associated or dependent species; minimize harmful impacts on living marine resources; and protect marine biodiversity. First meeting of SC in 2005 addressed ecosystem considerations. SC to make recommendations to Commission on wider ecosystem impacts of fishing, including overexploitation, bycatch, dumping, gear effects, habitat destruction, and impacts on adjacent EEZ.	No formal adoption of PA to date. PA is included in Commission mandate. In implementing PA, the Commission must consider international best practices regarding its application, including Annex II of UNFSA and the FAO Code of Conduct. In 2005, SC recommended no increase beyond current levels of fishing for 2006 for all species.
Participation	**Members – Contracting Parties***	**Cooperating nonmembers**	**Membership**	**Participation criteria**	**Transparency**
	Parties: Angola, ***EC***, ***Nambia***, and ***Norway***. Nonmember signatories: **South Africa**, United Kingdom (on behalf of St. Helena and its dependencies of Tristan da Cuhna and Ascension Island), **Iceland**, *Republic of Korea*, and **USA**.	None at this time.	Membership is open to all States having an interest in the convention area as well as to States and REIOs who participated in the SEAFO Conference, or whose vessels fish for stocks covered by the Convention or had done so in the 4 years prior to adoption of the Convention. CPs are to request Non-Contracting Parties (NCPs) whose vessels fish in the area to cooperate fully by becoming Parties to the Convention.	Each CP contributes to the budget. CPs must take measures to ensure that its nationals and industries fishing in the convention area comply with Convention provisions. NCP vessels are to cooperate fully by agreeing to apply conservation and management measures. NCPs will enjoy benefits from participation in the fishery commensurate with their commitment to compliance.	Publicly accessible Web site. Meeting minutes are available. NGOs and IGOs may be invited to meetings of the Commission and other Committees as observers. The financial activities of the Organization are subject to an annual audit by independent auditors. Information can also be obtained directly from the Secretariat.

*Bold type indicates that member has ratified UNFSA; italic type indicates that member has ratified the FAO Compliance Agreement.

	Data collecting and reporting	Observer program	Vessel monitoring system (VMS)	Enforcement	Catch documentation – trade schemes documentation
Monitoring, control, and surveillance	CPs must collect and exchange scientific, technical, and statistical data and forward to SEAFO. Each party must also provide information concerning fishing activities, including fishing areas and fishing vessels to facilitate compilation of reliable catch and effort statistics. SC requested to develop log sheets and data forms for each fishery. In January 2006, sampling forms developed by SC made available to CPs.	Observer program is based on common standards, including placing with prior consent observers on a reciprocal basis on vessels flying the flag of another CP; an appropriate level of coverage for different types and sizes of fishing vessels and fishery research vessels; and ensuring the safety of vessels and observers. In the event that a port State suspects a violation by a CP, the port State will notify the flag State and the Commission. In 2005, adopted measure requiring all vessels operating in area and targeting species not under jurisdiction of other RFMOs to carry scientific observers as of 1 January 2006. SC is developing format for observer data; date to be transmitted to SEAFO Secretariat.	Convention requires near-to-real time reporting of vessel movements including by satellite surveillance. In 2005, adopted minimum standards for establishment of VMS. As of 1 April 2006 all vessels authorized to fish in area shall implement VMS, with data transmitted to flag State Fisheries Monitoring Centre (FMC). FMC required to submit data to SEAFO Secretariat within 24 hours of receipt.	Convention provides broad guidelines on enforcement, specifically for at-sea and in-port inspection, including boarding and inspection of vessels on a reciprocal basis. Commission is in process of establishing specific measures. In 2005, adopted measure on interim port-State control, with each CP to maintain effective system. Measure provides guidelines for port-State inspections. Information to be shared with other CPs and transmitted to Secretariat.	None at this time.
	Decision making	**Dispute settlement mechanism**	**Cooperative efforts**	**Efforts to strengthen RFMO**	**Challenges**
Other governance efforts and challenges	Decisions on all matters of substance are by consensus. Whether a matter is one of substance is treated as a matter of substance. Other decisions may be determined by a simple majority. Conservation and management measures adopted by the Commission become binding on all CPs 60 days after notice is given by the Secretariat. Objection procedure exists: if CP notifies Commission that it is unable to accept a measure, the measure is then not binding on that CP but remains binding on all other CPs.	Art. 24 of the Convention urges CPs to resolve disputes by negotiation, inquiry, mediation, conciliation, arbitration, judicial settlement, or other peaceful means. Technical disputes may be referred to an ad hoc expert panel established by the Commission who will resolve the dispute expeditiously without resorting to binding procedures. Where a dispute is not referred for settlement within a reasonable time, or where a dispute is not resolved, the dispute, at the request of one party to the dispute, may be submitted for a binding decision.	Convention provides that Commission will cooperate with FAO, IUCN, and other organizations on issues of mutual interest. Letters of introduction sent to NAFO, ICCAT, CCAMLR, NEAFC, CITES, and FAO. SEAFO Executive Secretary participated in 2005 biennial RFB meeting. In 2005, SC recommended to Commission that a formal relationship with the Benguela Current Large Marine Ecosystem (BCLME) Project be established. SC cooperating with numerous organizations to gather necessary data for SEAFO area, related ecosystem, and relevant fisheries. Executive Secretary visited NEAFC in 2005 to gather information on data handling and VMS.	First post-UNFSA RFMO for straddling stocks. SEAFO Convention in force in 2003; has competence over discrete stocks. Recognizes special requirements for developing States in the region. CPs to ensure effective control over nationals and industries fishing in SEAFO area. Created Web site. Established permanent Secretariat with Executive Secretary and staff. Opened permanent office in March 2005. Adopted Headquarters Agreement with Namibia. Cooperating with relevant organizations to collect required information on ecosystem and fisheries. In 2005, agreed to establish scientific observer scheme, satellite-based VMS, and interim port-State control measures. In 2006, Angola became a CP. Several signatories completing legislative process to ratify Convention. At 2005 Annual Meeting, informally discussed extent to which CPs are meeting their responsibilities within SEAFO.	Need for ensuring full participation (ratification, accession, etc.) by all States and entities having "real interest" in fishing in SEAFO area. Many coastal States in SEAFO area are developing nations that require assistance to fully participate in RFMO. No Cooperating NCP status. Other challenges include addressing IUU; obtaining reliable data for stock assessment and fisheries management; determining northern limit of convention area; increasing fisheries monitoring and data collection; and establishing and adopting complete MCS system.

Commission for the Conservation of Antarctic Marine Living Resources (CCAMLR)

Evelyne Meltzer and Susanna D. Fuller: 05/2006

RFMO structure	CCAMLR	Area of competence	Objectives	Organizational structure	Scientific regime
	Commission for the Conservation of Antarctic Marine Living Resources **Headquarters:** Hobart, Tasmania, Australia. **Convention:** Convention on the Conservation of Antarctic Marine Living Resources (CCAMLR Convention). **Signed:** 20 May 1980, Canberra, Australia. **In force:** 7 April 1982. **Web site:** www.ccamlr.org	The Southern Ocean surrounding Antarctica. which constitutes the area south of 60ºS and between that latitude and the Antarctic Convergence. The Antarctic Convergence is a line joining the following points along latitude and longitude lines: 50ºS, 0º; 50ºS, 30ºE; 45ºS, 30ºE; 45ºS, 80ºE; 55ºS, 80ºE; 55ºS, 150ºE; 60ºS, 150ºE; 60ºS, 50ºW; 50ºS, 50º W; 50ºS, 0º. CCAMLR area incorporates FAO Statistical Areas 48, 58, and 88.	To ensure the conservation, including rational use, of Antarctic living marine resources.	Commission, Standing Committee on Implementation and Compliance (SCIC), Standing Committee on Administration and Finance (SCAF), Scientific Committee (SC) and its working groups (WGs), and Secretariat (Executive Secretary).	SC is composed of scientific representatives and advisors from member nations. SC provides the Commission with a summary of discussion including rationale for findings and recommendations. SC establishes permanent WGs and recommends research programs, conservation, and other measures to Commission. WGs include Fish Stock Assessment (WG-FSA), Ecosystem Modeling and Management (WG-EMM), Incidental Mortality Associated with Fishing (WG-IMAF), and Subgroup on Acoustic Survey and Analysis Methods (SG-ASAM), and Ad Hoc Joint Assessment Group (JAG). The CCAMLR Ecosystem Monitoring Program (CEMP) monitors key life-history parameters of selected dependent species.

The fishery, EA, and PA	Key species and gear	Stock status	Bycatch	Ecosystem approach (EA)	Precautionary approach (PA)
	Species: Antarctic krill, Patagonian toothfish, Antarctic toothfish, sub-Antarctic lantern fish, mackerel icefish, sevenstar flying squid, Antarctic rock cod, and crabs. Currently, no active crab or squid fishery. **Gear:** pots (crab), bottom trawls, bottom longlines, squid jigs.	**Stock status assessed in some areas:** Antarctic Krill — all fisheries subject to precautionary TACs. **Depleted:** Patagonian Toothfish may be significantly depleted in Subareas 58.6 and 58.7 — many stocks regulated by precautionary TACs; mackerel icefish (Area 48) — highly variable recruitment, restricted fishery, TAC imposed around South Georgia (Subarea 48.3); Antarctic rock cod — depleted in Area 48 and 58, no directed fishery. **Stock status unknown:** Antarctic toothfish — full status unknown, regulated by precautionary TACs; sub-Antarctic lantern fish — no fishery since 1991–1992; sevenstar flying squid — precautionary TACs imposed.	Seabirds are significant bycatch issue. Seabird bycatch mitigation measures in place since 1992. Other bycatch species include Antarctic rock cods, ice fishes, skates, and rays. Bycatch considered by WG-FSA. Instituted bycatch Conservation Measures, including mesh-size regulation, bottom trawl prohibition around South Georgia, and bycatch of grenadiers, skates, and rays (counted collectively) limited to 5% of toothfish catch; also established WG-IMAF in 1994.	Krill, as prey species, is currently managed under EA. Art. II of Convention outlines EA principle setting benchmark for other fisheries organizations. Began CEMP in 1984. Management approach regulates human activities to avoid deleterious changes in Antarctic ecosystem, including bycatch reduction of seabirds and other nontarget species. 2005 SC MPA Workshop held in USA. At 2005 Annual Meeting, CCAMLR endorsed SC recommendation to establish harmonized regime across the Antarctic Treaty System (ATS). SC and relevant agencies to develop MPA network strategy. Next Workshop to be held 2007. 2005 Valdivia, Chile, Symposium called for CCAMLR policy on destructive fishing practices, including impacts of bottom trawling.	As of 2005, the only RFMO to have fully incorporated PA into stock assessment and decision making. All regulated fisheries in areas under CCAMLR jurisdiction are subject to precautionary catch limits. In addition, both krill and Patagonian toothfish fisheries are subject to (predetermined) decision rules. Though PA is not specified, Convention is first international fisheries instrument to outline the PA conservation principle. CCAMLR is pioneering efforts to manage marine ecosystem according to PA, and new fisheries shall not develop faster than Commission is able to evaluate their potential consequences.

Participation	Members*	Cooperating Parties*	Membership	Participation criteria	Transparency
	Members of the Commission: *Argentina*, ***Australia***, **Belgium**, **Brazil**, *Chile*, ***EC***, **France**, **Germany**, **India**, **Italy**, *Japan* , *Republic of Korea*, ***Namibia***, **New Zealand**, ***Norway***, Poland, **Russian Federation**, **South Africa**, **Spain**, ***Sweden***, **Ukraine**, **United Kingdom**, ***USA***, and ***Uruguay*** (24 members).	Acceding States Party to CCAMLR, but not members of the Commission: Bulgaria, ***Canada***, **Cook Islands**, **Finland**, **Greece**, ***Mauritius***, **Netherlands**, *Peru*, and Vanuatu (9). Members and Cooperating Parties are Contracting Parties (CPs). **Catch documentation scheme (CDS) Parties: Canada**, China, ***Mauritius***, *Peru*, ***Seychelles***, Singapore. CPs and CDS Parties require CDS documents prior to importing toothfish.	Any State or REIO can become a CCAMLR Party subject to criteria listed in Arts. XXIX and VI of CCAMLR Convention (including review by Commission and submission of written statement outlining reasons for application).	There are two categories of membership: (1) members of Commission who pay dues are involved in scientific research and (or) fishing subject to Conservation Measures, and have voting rights; and (2) acceding States or REIOs that are interested in research or harvesting activities, pay no dues, and agree to be bound by Convention. In addition, States who are nonparties to CCAMLR, including port States, can participate in the implementation of CDS (open and transparent process). In 2005, Commission adopted Resolution 24/XXIV on NCP Enhancement Program consistent with UNFSA.	Publicly accessible Web site that includes scientific data, Commission and SC proceedings, and member activity reports. Availability of unpublished documents subject to Rules for Access to Data. Participation of observers from IGOs and NGOs is subject to Rules of Procedure of the Commission and SC (NGOs may attend as observers of Commission meetings and the main meeting of SC with unanimous approval by members). Participating NGOs include ASOC and COLTO. Some meetings may be closed to observers, if requested by member. Participating IGOs and RFMOs include the CEP, FAO, CITES, IWC, IOC, IUCN, and CCSBT.

*Bold type indicates that member has ratified UNFSA; italic type indicates that member has ratified the FAO Compliance Agreement.

	Data collecting and reporting	Observer program	Vessel monitoring system (VMS)	Enforcement	Catch documentation – trade schemes documentation
Monitoring, control, and surveillance	Data collection by CCAMLR includes fishery catch and effort statistics; data collected by scientific observers on fish bycatch, incidental mortality of seabirds, and marine mammals; biological information and biomass estimates obtained during fishery-independent scientific surveys; biological information on dependent species collected as part of CEMP; CDS landing and trade statistics. All vessels participating in exploratory fisheries must complete research and tagging requirements.	CCAMLR Scheme of International Scientific Observation has 100% observer coverage on all finfish fisheries by independent regional scientific observers. In 2005, observers on 8 krill vessels. At 2005 Annual Meeting, UK proposal for mandatory 100% observer coverage on krill vessels (1-year trial period) not adopted; however, voluntary 100% observer coverage encouraged by most members. Data collected using standard format, submitted directly to CCAMLR, and used by the SC and its WGs.	VMS not required for krill fishery. In 2004, adopted centralized VMS for finfish fisheries with data transmitted every 4 hours to flag-State Fisheries Monitoring Centre (FMC) then communicated to Secretariat, or data can be transmitted on a voluntary basis directly from vessel to Secretariat. As of 2005, each vessel licensed by CCAMLR members for finfish fisheries is required to have satellite-linked VMS with continuous reporting. VMS position data are used by Secretariat to corroborate toothfish landings in CDS. Exploratory fisheries are subject to exception, whereby VMS reports are transmitted by flag State after vessel has left the area. VMS reporting standards follow North Atlantic Format.	Enforcement of CCAMLR measures is undertaken through system of observation and inspection adopted in 1998. This is a nationally operated scheme with CCAMLR designated inspectors. Where members designate inspectors, results of inspection are reported to CCAMLR. Vessels licensed to fish are included on list of regulated vessels (rebuttable preumption that those not on list are unregulated or IUU). CCAMLR annually estimates level of IUU fishing and has taken comprehensive measures to address IUU fishing, including VMS, observers, and CDS (binding on all members in 2000). As of 2003, there is also mandatory port inspection of all toothfish vessels. CPs following in-port inspection notify Secretariat of all vessels carrying toothfish denied port access, landing, or transshipment. Secretariat informs CPs and CDS Parties.	CCAMLR adopted CDS for Patagonian toothfish in 2000. Scheme is designed to determine whether toothfish are caught in compliance with Conservation Measures by tracking landings and trade flows (corroborated by mandatory VMS). Regulated high seas fishing in conformance with CDS protocol includes all necessary documentation pertaining to landings, imports, and transshipment of toothfish. Presumption exists that if documentation is not completed, fishing is IUU. To expedite and improve CDS, CPs and CDS Parties urged to use electronic CDS (E-CDS).
	Decision making	**Dispute settlement mechanism**	**Cooperative efforts**	**Efforts to strengthen RFMO**	**Challenges**
Other governance efforts and challenges	Decision making is consensus based; only members of Commission may participate in decisions. Though voting procedure exists (each member has 1 vote), there has not been vote in 23 years on any matter of substance (whether an issue is matter of substance is regarded as matter of substance). Members failing to pay contributions for 2 consecutive years lose right to participate in decision making during period of default. Resolutions are nonbinding. Decisions on Conservation Measures are binding 180 days after Commission notification. While objection procedure exists under Convention, full procedure has never been activated, and only 2 technical objections to a measure have occurred.	Under Art. XXV and in the event of a dispute, CPs must consult among themselves with view to resolution by negotiation, inquiry, mediation, conciliation, arbitration, judicial settlement, or other peaceful means. If dispute is not resolved, it must, with parties' consent, be referred to International Court of Justice or to arbitration. If dispute goes to arbitration, the tribunal must be constituted as provided in Annex to CCAMLR Convention.	Various levels of cooperation exist with Committee on Environmental Protection (CEP) of the Antarctic Treaty, FAO, IUCN, IWC, SCAR, IOC, the Pacific Community (formerly the South Pacific Commission), the FFA, CCSBT, IOTC, ACAP, and ICES. CDS collaboration with ICCAT, and IATTC. Seeks cooperation and info exchange with other organizations on IUU fishing. Actively involved with NEAFC and NAFO in the development of the NAF for common VMS reporting standards. Secretariat chairs and participates in RFB meeting held biennially on margin of COFI. Joined FIRMS. Joint CCAMLR–IWC workshop planned for 2006.	Continues to lead in operationalizing PA and EA. Continues to develop PA reference points as basis for decision rules. EA management of krill fishery. Held 2005 MPA workshop. Yearly symposium starting 2005. Introduced comprehensive reporting system through CDS and mandatory VMS for management of toothfish fishery. Established registry of vessels permitted to fish. Implemented seabird bycatch limits; if limit exceeded, vessel must cease to fish and leave area. Established independent regional scientific observer program mandatory for all finfish fisheries; regional inspection program; and in-port inspection program for toothfish. Has reduced IUU fishing through enforcement, CDS, and diplomatic efforts. Encourages use of E-CDS. Has increased CDS Parties. Adopted 2005 NCP Cooperation Enhancement Program to further reduce IUU fishing.	Avoiding, reducing, and eliminating overfishing and rebuilding depleted stocks. IUU fishing (particularly for toothfish) by CPs and NCPs. IUU fishing in adjacent Southern Indian Ocean area (proposed SIOFA may address this). Controlling toothfish transshipment. Assessing and enhancing compliance throughout CCAMLR area including within EEZs of coastal States. Extending observer program and VMS to krill fishery. Monitoring impacts of new krill harvesting method (pumping) and potential vessel and fleet-capacity increases. Improving compliance with seabird bycatch mitigation measures, including within CCSBT Area. Balancing data access with confidentiality and security concerns. Increasing numbers of CDS Parties including port States. Late payment of dues by members. Geopolitical and jurisdictional sensitivities.

Central Bering Sea Donut Hole

Evelyne Meltzer and Susanna D. Fuller: 05/2006

	BSDH	Area of competence	Objectives	Organizational structure	Scientific regime
RFMO structure	**Central Bering Sea Donut Hole** **Convention:** Convention on the Conservation and Management of Pollock Resources in the Central Bering Sea (Donut Hole Convention). **Signed:** 16 June 1994, in Washington, D.C., USA, by China, Republic of Korea, Russian Federation, and USA; Japan, 4 August 1994; and Poland, 25 August 1994. **In force:** 8 December 1995 for Russia Federation, Poland, China, and USA; 21 December 1995 for Japan; and 4 January 1996 for Republic of Korea. **Web site:** www.afsc.noaa.gov/refm/cbs/default.htm	The high seas area of the Bering Sea. Covers beyond 200 nautical miles from baselines from which breadth of the territorial sea of coastal States (USA and Russia) of Bering Sea is measured, except as otherwise provided in Convention. Activities related to scientific data collection may extend beyond convention area within Bering Sea.	To establish international regime for conservation, management, and optimum utilization of pollock resources in convention area; to restore and maintain pollock resources in Bering Sea at levels that will permit MSY; to cooperate in gathering and examining factual information concerning pollock and other living marine resources in Bering Sea; and to provide, if parties agree, a forum in which to consider establishment of necessary conservation and management measures for living marine resources other than pollock in convention area.	Convention does not provide for a Commission. Parties convene Annual Conference. The Convention established a Scientific and Technical Committee (STC) that meets prior to Annual Conference. 10th Meeting of Parties held September 2005 in Busan, Korea.	STC was established to compile, exchange, and analyze information on fisheries harvests, pollock, and other living marine resources covered by Convention. Parties cooperate in scientific research and exchange data on pollock resources, including trial fishing efforts.
	Key species and gear	**Stock status**	**Bycatch**	**Ecosystem approach (EA)**	**Precautionary approach (PA)**
The fishery, EA, and PA	**Straddling stock:** *Theragra chalcogramma* (common name is Walleye or Alaska pollock). **Gear:** pelagic–midwater trawls.	Moratorium on pollock fishery since 1993 in Donut Hole. USA subsequently ceased directed fishery within its EEZ in Aleutian Basin and Bogoslof area. At 2005 Annual Conference, STC confirmed continued low abundance of pollock stocks in Central Bering Sea. Estimated pollock biomass level in convention area in 2005 was 422 000 tonnes (t), substantially below levels that would trigger lifting moratorium, as described in Part I of Convention Annex (biomass less than or equal to 1.67×10^6 t, harvest level is 0). Occasional trial fisheries in past. No trial fisheries in 2005. Parties authorized Korea to conduct trial fishery in July and August 2006 to locate pollock concentrations.	There is concern over bycatch of Pacific salmon and Pacific halibut. If fishery reopens, STC will address issues pertaining to bycatch mitigation measures.	Convention pertains only to conservation and management of pollock. Other species would be considered if parties unanimously agree.	Though not specifically identified as "precautionary", the threshold formula in the Convention Annex is considered to represent precautionary reference points. The fishery is not to reopen unless threshold biomass of Aleutian Basin pollock is reached.
	Members – Contracting Parties*	**Cooperating nomembers**	**Membership**	**Participation criteria**	**Transparency**
Participation	*Japan*, China, *Republic of Korea*, Poland, **Russian Federation,** and ***USA*** are signatories to Convention (6 members).	None at this time.	Parties may, by unanimous agreement, invite other States to become Parties to the Convention.	Although a moratorium on commercial pollock has been in place since 1993, parties may conduct trial fishing using no more than two vessels in the convention area at any time. There are numerous provisions for such fishing to occur (e.g., information on trial fishing operations must be provided to other parties 2 weeks in advance of operations; trial fishing vessels must have a flag-State scientific observer on board and offer other parties an opportunity to place their own observers on board; and trial fishing vessels must follow measures adopted by parties regarding vessel monitoring systems, entry and transshipment notifications, and shipboard logs and records among others). Annual harvest levels and individual national quotas are established during Annual Conference.	USA developed Web site containing reports on Convention, information from annual conferences, workshop records, documents and data, and information on key contact persons from each party. Observers from REIOs and IGOs attend meetings. Current observer rules do not address attendance by NGOs. NGOs have generally not asked to attend annual conferences. Industry representatives and representatives of environmental organizations can attend these meetings as part of a party delegation. In addition, parties may, by unanimous agreement, invite representatives of any nonparties to participate as observers at annual conferences.

*Bold type indicates that member has ratified UNFSA; italic type indicates that member has ratified the FAO Compliance Agreement.

Monitoring, control, and surveillance	Data collecting and reporting	Observer program	Vessel monitoring system (VMS)	Enforcement	Catch documentation – trade schemes documentation
	Parties annually submit to STC: catch and effort statistics; information regarding time and area of fishing operations; and incidental taking of anadromous species or other marine living resources. Vessels must notify all parties of intention to enter Donut Hole to fish 48 hours in advance, provide location of any transshipments of fish 24 hours in advance, and report catch data regularly.	Under Convention, observer coverage is 100%. Each vessel fishing in Donut Hole must, upon request, accept an observer from another party. If observer is not available, flag State must place one of own observers aboard. Observers report findings to flag State and observer State.	VMS is required on all vessels. Location data are to be shared with all parties on real-time basis.	Parties are responsible for enforcing provisions of Convention. Each Party agrees in advance to boarding and inspection of vessels flying its flag by authorized officials from any other party. Flag-State party is notified promptly of alleged violations and shall take appropriate measures in accordance with national laws and regulations, including prompt investigation. Flag-State party must order fishing vessel to cease violations and leave convention area immediately. Enforcement regime allows for significant level of coverage by observers from nonflag States.	None at this time.

Other governance efforts and challenges	Decision making	Dispute settlement mechanism	Cooperative efforts	Efforts to strengthen RFMO	Challenges
	Decisions on matters of substance are made by consensus. A matter is deemed to be of substance if any party considers it to be of substance. Other decisions are made by a simple majority of votes of all parties casting affirmative or negative votes. Each party has 1 vote. Parties establish annual allowable harvest level for pollock in convention area; determine individual national pollock quota for each party; adopt appropriate conservation and management measures; establish plan of work for STC; discuss cooperative enforcement measures; and receive enforcement reports from each party. There is no opting-out procedure. Coastal States have agreed to moratorium on Aleutian Basin pollock fishing in EEZs as long as there is no fishing of stock in Donut Hole. This was agreed upon in non legally binding Record of Discussions adopted in conjunction with Convention. Annual discussions on fishing within EEZs are part of Record of Discussion.	When parties fail to reach consensus on annual harvest level, it is determined by fall-back formula in Part 1 of Annex to Convention, prohibiting directed fishing if biomass is equal to or less than 1.67×10^6 t.	USA State Department has invited North Pacific and Bering Sea Fisheries Advisory Body to advise the USA delegation.	Created publicly accessible Web site. Donut Hole Convention established moratorium on commercial pollock fishing in the Central Bering Sea. Moratorium maintained and respected since 1993. Convention established precautionary threshold (decision rule) for reopening pollock fishery. Convention provides for 100% observer coverage and nonflag-State boarding and inspection procedures to be agreed in advance. Enforcement in convention area facilitated by cooperation between USA Coast Guard and Russian Federal Security Service. Convention established cost-effective institutional arrangement. In June 2005, USA hosted Central Bering Sea Pollock Workshop on Allowable Harvest Level and Stock Identification as part of agreed work plan. Korea authorized to conduct trial fishery in July–August 2006 using 4 vessels, as part of efforts to increase data collection in absence of comprehensive survey of area.	Rebuilding pollock stocks. Maintaining moratorium until stock reaches agreed threshold. Expanding Convention objectives to implement broader EA. Ensuring enforcement and fisheries management regimes are effective in event that fishery reopens. Enhancing transparency by allowing NGO participation. Resolving USA – Russian Federation maritime boundary dispute.

International Commission for the Conservation of Atlantic Tunas (ICCAT)

Evelyne Meltzer and Susanna D. Fuller: 05/2006

	ICCAT	Area of competence	Objectives	Organizational structure	Scientific regime
RFMO structure	**International Commission for the Conservation of Atlantic Tunas (ICCAT)** **Headquarters:** Madrid, Spain. **Convention:** International Convention for the Conservation of Atlantic Tunas (ICCAT Convention). Amended by the 1984 Paris Protocol (in force 1997) and the 1992 Madrid Protocol (in force 2005). **Signed:** 14 May 1966, Rio de Janeiro, Brazil. **In Force:** 21 March 1969. **Web site:** www.iccat.int	The Atlantic Ocean, including adjacent seas. There is no precise definition for area. The longitude of 20°E is used for scientific and compliance purposes as the border with the Indian Ocean.	To maintain populations of tuna and tuna-like fishes found in the Atlantic Ocean at levels that permit the maximum sustainable catch for food and other purposes.	Commission; Council (now inactive); species panels (4); Standing Committee on Research and Statistics (SCRS); Standing Committee on Finance and Administration (STACFAD); Conservation and Management Measures Compliance Committee; Permanent Working Group for the Improvement of ICCAT Statistics and Conservation Measures (PWG); and various ad hoc working groups (WGs) (Integrated Monitoring and Control Measures, to Develop Integrated and Coordinated Atlantic Bluefin Tuna Management Strategies. Secretariat (Executive Secretary).	Small permanent scientific staff. SCRS consists of scientists from Contracting Parties and Cooperating Non-Contracting Parties, Entity, and Fishing Entity (CPCs). SCRS advises on conservation and management measures; addresses specific ICCAT requests; meets annually; produces annual report on stock status and research topics serving as scientific basis for ICCAT decisions; coordinates national research activities; and develops plans for cooperative research programs. SCRS has Sub-Committee on Statistics and Sub-Committee on Ecosystems (in 2006 Environment and Bycatch rolled into Ecosystems); species groups (sharks); and WGs (e.g., Stock Assessment Methods, Joint GFCM–ICCAT on Large Pelagic Fishes in the Mediterranean, on SCRS Organization) and ad hoc WG on SCRS Organization. Relies on CPCs for fishery-dependent data.

	Key species and gear	Stock status	Bycatch	Ecosystem approach (EA)	Precautionary approach (PA)
The fishery, EA, and PA	About 30 key highly migratory fish stocks including: yellowfin tuna, bigeye tuna, albacore tuna (North and South Atlantic, Mediterranean stocks), skipjack tuna, bonito, Atlantic bluefin tuna (East and West Atlantic stocks), swordfish (North and South Atlantic, Mediterranean), blue marlin and white marlin, shortfin mako, blue shark, and porbeagle. **Gear:** purse seine (36%), longline (30%), and bait boats major gears. The use of airplanes or helicopters supporting bluefin fishing operations in Mediterranean is prohibited. Since 1999, seasonal moratoria on FADs in Gulf of Guinea. Use of driftnets for large pelagic fisheries in Mediterranean prohibited.	**Above F_{MSY}** (overfishing): North Atlantic (NA) albacore, West Atlantic (WA) bluefin, blue marlin, white marlin. **Near F_{MSY}:** bigeye, NA swordfish, yellowfin. **Below F_{MSY}:** South Atlantic (SA) albacore. **Above B_{MSY}:** SA albacore, NA and SA blue shark. **Near B_{MSY}:** bigeye, NA swordfish, yellowfin. **Below B_{MSY}** (overfished): NA albacore, WA bluefin, blue marlin, white marlin. **Under recovery plan:** WA bluefin tuna (by 2018), NA swordfish (by 2009), blue marlin and white marlin (by 2006). **Uncertain:** skipjack, East Atlantic bluefin, sailfish, SA swordfish, Mediterranean swordfish, Mediterranean albacore (never assessed), NA shortfin mako, small tunas (never assessed).	Seabirds, turtles, sharks, and juveniles of target species. Bycatch Sub-Committee met yearly and in 2006 rolled into Ecosystems Sub-Committee. 2005 Tropical Species Workshop evaluated alternative to minimum size to reduce juvenile tuna bycatch. Mimimum size and time–area closure Recommendations for several species (yellowfin, bluefin, and swordfish). ICCAT encourages submission of bycatch and interaction statistics as well as development of NPOAs for sharks and seabirds. In 2005 mandatory reporting requirement for shark catch was adopted. 2005 resolution encourages CPCs to conduct research on circle hook use for sea turtle mortality reduction.	2005 Recommendation on bluefin tuna farming in cooperation with GFCM requires tracking farmed tuna, identification and record of tuna farming operations, and submission of data to Commission. 2005 Resolution on sargassum, requires data collection on sargassum given importance to juveniles and feeding grounds for highly migratory species. In 2006 SCRS will establish Ecosystem Sub-Committee, which includes EAF in terms of reference.	In 1997, SCRS created Ad Hoc Working Group on PA, which last met in conjunction with 2001 SCRS Plenary. In 2004, draft Recommendation on PA was introduced; not adopted. The ICCAT Convention implies that MSY is a target, not a limit.

	Members*	Cooperating Nonmembers*	Membership	Participation criteria	Transparency
Participation	Algeria, Angola, ***Barbados***, ***Belize***, **Brazil**, ***Canada***, Cape Verde, China, Côte d'Ivoire, Croatia, Equatorial Guinea, ***EC***, **France** (St. Pierre and Miquelon), Gabon, *Ghana*, Guatemala, Guinea Conakry, Honduras, **Iceland**, *Japan*, *Republic of Korea*, Libya, *Mexico*, *Morocco*, ***Namibia***, Nicaragua, ***Norway***, Panama, Philippines, **Russian Federation,** Sao Tomé and Principe, **Senegal**, **South Africa**, *Syria*, Trinidad and Tobago, Tunisia, Turkey, **United Kingdom** (Overseas Territories), ***USA***, **Uruguay**, Vanuatu, and Venezuela (42 members from 5 continents).	Created special status known as Cooperating Non-Contracting Party, Entity or Fishing Entity, i.e., Non-Contracting Cooperating Party (NCC): Guyana, **Netherlands Antilles**, and Chinese Taipei.	Membership in ICCAT is open to any State that is a member of the UN, specialized UN agency and through 1984 Paris Protocol (in force 1997) includes REIOs (i.e., EC). In 1994 established Cooperating Non-Contracting Party Status: revised in 1997 to be NCC (to include Chinese Taipei). Annually, ICCAT contacts all NCPs fishing species under ICCAT competence in area to urge them to become a new member or NCC. Since 1995, 23 new members have joined. Contracting Parties, Fishing Entities, and Cooperating Non-Contracting Parties are collectively referred to as CPCs.	CPs vote and contribute to budget. The revised contribution formula in force as of March 2005 reflects economic development status of CPs and will apply to 2006 budget. Flag States to maintain vessel records, control those authorized to fish in area, and ensure compliance with ICCAT measures. CPCs are entitled to allocations under 2001 ICCAT Allocation of Fishing Possibilities using historical catch as main criterion. Some members are also seeking access to tuna stocks within their EEZs based on adjacency. ICCAT controls access and fishing capacity by TACs, quotas, and vessel effort limitations.	Publicly accessible Web site. Meeting minutes and background papers are available on line. IGOs and NGOs can attend meetings (with 50 days prior notice) unless 1/3 of members object in writing 30 days prior to meeting. NGOs may make oral statements by invitation and distribute documents through Secretariat. Atlantic coastal State NCPs in area may attend meetings as observers. There is a participation fee for all observers.

*Bold type indicates that member has ratified UNFSA; italic type indicates that member has ratified the FAO Compliance Agreement.

	Data collecting and reporting	Observer program	Vessel monitoring system (VMS)	Enforcement	Catch documentation – trade schemes documentation
Monitoring, control, and, surveillance	Scientific data (primarily fishery-dependent) are collected by CPCs. SCRS relies on CPCs to submit current and complete data by 31 July prior to Annual Meeting. Data submitted to the SCRS must be separated into 2 groups: Task I (annual catch by species, gear, region, and flag); and Task II (monthly catch and fishing effort statistics for each species by small area, and size data). SCRS advice is based on these and tagging data. Tagging data are used for growth rate, movement, and abundance estimates. 1999 Resolution calls for provision of recreational fishery data. Tuna farming data required as of 2006. Standardized National Annual Reports, and therefore ICCAT Reporting Tables (used by the Compliance Committee) are to be submitted 1 month prior to Annual Meeting. Data in ICCAT Reporting Table (for monitoring) may differ from Task I data (scientific). Compliance Committee decisions are based on ICCAT Reporting Tables. Partial, late (or no) data are often submitted (penalty may be imposed), compromising stock assessments and scientific advice. 2005 Recommendations requires CPCs to provide explanations for reporting deficiencies and plans for corrective actions; PWG to identify data gaps affecting stock assessments.	Recommendation 04-01 introduced limited regional observer program requiring that there be observers on board at least 5% of longline vessels over 24 m fishing for bigeye. 2003 Ad Hoc Data Workshop concluded comprehensive at-sea observer data are essential. Observer program (and associated data) coverage since 1996 has ranged from nil to 100% depending on CPC. In 2005, regional observer program established for large-scale tuna longline fishing vessels (LSTLVs) for transshipment. All CPC vessels authorized to receive transshipments must have ICCAT observer on board during all transshipments in convention area. Observers appointed and deployed on carrier vessels by Secretariat. Standards and training have been developed. Observer program paid by CPCs for LSTLVs engaged in transshipment. No observer assigned to vessels where fees are unpaid, potentially compromising compliance scheme.	2003 Recommendation calling for CPCs to implement VMS no later than 1 November 2005 for commercial vessels >24 m. VMS data, when available, to be transmitted to flag State and then to ICCAT.	Convention obliges CPCs to establish high seas international enforcement system; currently no at-sea boarding or inspection. PWG monitors activities of NCPs and NCCs. Compliance Committee considers noncompliance by CPs. In cases of quota overruns, subject to Commission approved "payback" plans, CPs may transfer some quota to other CPs. In 2005, adopted Recommendation for monitoring at-sea and in-port transshipment by LSTLVs, including regional observer program. 1997 Port Inspection Scheme provides for inspections of flag and non-flag State vessels during off-loading and in-port transshipment. Not all CPCs implement Scheme. In 1999, established "negative" list of vessels believed to be IUU. Established in 2002 "positive" list of authorized fishing vessels building on 1998 and 2000 Recommendations. Vessels not recorded are deemed not authorized to fish, transship, or land tuna or tuna-like species. CPCs are to enforce compliance through domestic measures prohibiting landing or transshipment from unauthorized vessels. In cases of violation, report sent to flag State for follow-up action. Flag States failing to comply with ICCAT measures may be subject to trade restrictions and (or) having vessel removed from the positive list. At 2005 meeting, Chinese Taipei subject to significant quota cuts for bigeye and required to reduce capacity to 15 vessels (>24 m) to control bigeye fishery in ICCAT area.	ICCAT was first RFMO to introduce statistical documentation program (SDP) in 1994 for Atlantic bluefin, then extended to swordfish and bigeye. ICCAT now implements multilateral, transparent trade measures against parties who undermine effectiveness of conservation measures. 2003 resolution concerning trade measures replaced 1994 Bluefin Tuna Action Plan, 1995 Swordfish Action Plan, and 1998 IUU catches resolution to include CPCs and NCPs. In 2005, trade-restrictive measures maintained on Bolivia and Georgia, and Chinese Taipei given one year to comply before triggering trade-measure resolution (03-15). Several States have been identified under 2003 trade-measure resolution.

	Decision making	Dispute settlement mechanism	Cooperative Efforts	Efforts to strengthen RFMO	Challenges
Other governance efforts and challenges	Many decisions are made by consensus. Resolutions are nonbinding. Binding Recommendations are adopted by a simple majority vote with a quorum of 2/3 of CPs, and enter into force 6 months later. Each CP has 1 vote. Voting rights may be suspended if CP is in arrears for 2 years or more (voting rights have never been suspended, though many are in arrears). In 2004, requested CPs in arrears to present payment plan before 2005 Annual Meeting. There is a detailed objection procedure. CPs are not bound by Recommendations to which they have registered an objection. Resolution to amend objection procedure proposed (not adopted) requiring CPs to provide reasons for objection, and alternative conservation and management measures. Recommendations adopted by ICCAT apply throughout area (implemented domestically by CPCs for fishing inside and outside their EEZs).	No mechanism exists at this time.	Consistent with Art. XI of Convention, ICCAT cooperates with FAO (e.g., partner in FIRMS, endorsed IPOAs and Compliance Agreement, and is a member of CWP), and coordinates and exchanges data with other organizations (e.g., Joint GFCM–ICCAT Ad Hoc Working Group on Large Pelagic Fishes in the Mediterranean, SBT management with CCSBT, seabird bycatch with CCAMLR, shark data collection and assessment with ICES). Secretariat works closely and meets annually with other tuna RFMOs; participates in RFB biennial meetings; is part of efforts to establish global vessel register for tuna RFMOs; hosts workshops (e.g., 2004 Bigeye Tuna Review, and in 2005 Reduced Mortality of Juvenile Tuna); and assists CPCs to meet data collection, quality assurance, and reporting obligations.	Amended Convention to include REIOs. Created CPC Status to include Chinese Taipei and other NCPs. Increased ICCAT membership of many former NCPs fishing in area. Adopted in 2001 Criteria for Allocation of Fishing Possibilities (first RFMO) to address new entrants and limit capacity. Pioneered use of SDP and continual refinement of trade measures to address noncompliance of vessels undermining conservation regime. Established list of vessels permitted to fish and list of IUU fishing vessels. Implemented time–area closures throughout area. Established rebuilding plans for key species (e.g., NA swordfish, WA bluefin, and marlin). Strengthened CPC data reporting requirements, accountability, and corrective action. Increased transparency at Commission meetings. Reformed budget contribution formula with Madrid Protocol. Established special fund to provide capacity building as needed. In 2006, will establish regional observer program for transshipment and develop workplan to strengthen and modernize ICCAT in line with international legal instruments.	Avoiding overfishing and preventing further decline of target species. Rebuilding overfished and depleted stocks. Reducing bycatch, addressing gear issues, and dealing with overcapacity. Setting PA limits, given data inadequacies, IUU fishing, and status of many key stocks. Strengthening data quality and collection. Managing access, allocation, and redistribution expectations of large membership. Amending objection procedure. Controlling IUU fishing and transshipments. Implementing regional observer program for transshipments. Lacking at-sea boarding or inspection scheme. Strengthening port-State inspection scheme. Increasing MCS by Atlantic coastal States within EEZs. Increasing NCP coastal-State participation in ICCAT. Increasing and standardizing observer coverage throughout area. Standardizing and integrating catch and trade measures with other RMFOs, port States, and CPCs. Ensuring payment of budget contributions. Modernizing Convention and RFMO practice.

General Fisheries Commission for the Mediterranean (GFCM)

Evelyne Meltzer and Susanna D. Fuller: 05/2006

RFMO structure	GFCM	Area of competence	Objectives	Organizational structure	Scientific regime
	General Fisheries Commission for the Mediterranean **Headquarters:** Rome, Italy. **Convention:** Agreement for the Establishment of a General Fisheries Council for the Mediterranean. (FAO Article XIV Fisheries Body). **Signed:** 24 September 1949, Rome, Italy. **In force:** 20 February 1952. 1997 Agreement changed name to "Commission" and adopted amendment establishing autonomous budget, which came into force 29 April 2004. **Web site:** www.fao.org/fi/body/rfb/GFCM/gfcm_home.htm	The Mediterranean, the Black Sea, and connecting waters. This area coincides with FAO Statistical Area 37. Overlapping mandate with ICCAT for highly migratory fish stocks.	To promote development, conservation, and management of living marine resources; to formulate and recommend conservation measures; and to encourage training cooperative projects (includes aquaculture).	Commission, Scientific Advisory Committee (SAC) with 4 subcommittees, Committee on Aquaculture (CAQ) (4 networks). In 2006, agreed to Terms of Reference for Compliance Committee. Ad-hoc technical panels and working groups (WG). Secretariat (Secretary) (FAO provided Secretariat until 2005).	SAC was established in 1997, meets annually, and provides advice to Commission. Advice on aquaculture is provided by CAQ. SAC subcommittees (SC) include SC Stock Assessment (SCSA). Stock assessments are done for subareas. Assessment of large pelagic species is done in cooperation with ICCAT. SC Statistics and Information (SCSI), SC Economics and Social Sciences (SCESS), SC Marine Environment and Ecosystem (SCMEE). Several WGs, including WG Operational Units; GFCM–ICCAT WG Sustainable Tuna Farming; and GFCM–ICCAT WG Large Pelagics. Permanent Working Group on Stock Assessment Methodology (PWSAM) — established in 2005, first meeting in March 2006. Frequently organizes scientific workshops, e.g., Measurement of Fishing Effort (2005) and Gear Selectivity (2005). CAQ has 4 networks: Social, Economic and Legal Aspects of Aquaculture (SELAM); Technology and Aquaculture (TECAM); Environment and Aquaculture (EAM); and Information Systems for the Promotion of Aquaculture (SIPAM).
The fishery, EA, and PA	**Key species and gear**	**Stock status**	**Bycatch**	**Ecosystem approach (EA)**	**Precautionary approach (PA)**
	Transboundary stocks: hake, red mullet, striped mullet, blue and red shrimp, Norway lobster, anchovy, sardine, dolphinfish. **Highly migratory fish stocks:** Eastern Atlantic bluefin tuna, albacore, bonito, swordfish. **Gear:** bottom trawl, dredge, purse seine, surface longline, gill net, artisanal gear.	**Highly migratory fish stocks:** Above B_{MSY}: Eastern Atlantic bluefin. Declining: swordfish. **Demersal:** many stocks fully exploited.	Sharks and juveniles of target species. Mitigation measures include gear restrictions for certain vessels, bycatch limits, minimum size for landed fish, and measures to limit fishing mortality. 2004 ICCAT Recommendation prohibits catch of Atlantic bluefin <10 kg in Mediterranean. GFCM adheres to ICCAT Recommendation on shark bycatch including 2005 amendment requiring Contracting Parties (CPs) to report on implementation of mitigation measures. Protocol for mitigating shark bycatch included in Mediterranean Large Elasmobranchs Monitoring (MEDLEM) program.	SCMEE mandate includes EA and reported on ecosystem effects of fishing in 2004. Created ad-hoc WG on EA in 2005. 2005 Workshop on Ecosystem Approach to Fisheries included essential and sensitive habitats, ecosystem indicators, gear selectivity, bycatch of vulnerable and protected species. In 2005, Commission prohibited towed dredges and trawl fisheries below 1000 m protecting deepwater habitat. In 2006, applied these gear prohibitions to specific areas to protect cold seeps, corals, and seamounts; and established closed season (1 Jan. – 14 Aug.) for dolphinfish fisheries on FADs (SAC to assess effect of measure in 2010). In 2006, GFCM endorsed 2005 technical guidelines from GFCM–ICCAT WG Sustainable Tuna Farming for capture, transport, transfer, farming, harvesting, and marketing of bluefin; and invited CAQ to expand guidelines as required through EAM.	PA implementation is addressed in Art. III.2 of the 1997 amended Agreement. SAC is in the process of defining reference points for specific stocks.

	Members – Contracting Parties*	Cooperating Nonmembers	Membership	Participation criteria	Transparency
Participation	Albania, Algeria, Bulgaria, Croatia, ***Cyprus***, *Egypt*, ***EC***, **France**, **Greece**, Israel, **Italy**, *Japan*, Lebanon, Libya, **Malta**, **Monaco**, *Morocco*, Romania, Serbia and Montenegro, Slovenia, **Spain**, *Syria*, Tunisia, and Turkey (24 members).	None at this time. 2006 Recommendation establishes Cooperating Non-Contracting Party (Cooperating NCP) status and encourages States fishing in area to become CPs or Cooperating NCPs. Status Cooperating NCPs reviewed annually.	Membership is open to member nations and associate members of the FAO. Other States that are members of the UN or its specialized agencies may be admitted as members by a 2/3 majority. 1997 GFCM Agreement provides for membership of REIOs and States or associate members whose vessels engage in fishing in region for stocks covered by Agreement.	Members must contribute to budget and have voting privileges. Each member is represented on the Commission by one delegate and advisors. Members are obligated to report and submit required data.	Publicly accessible Web site. Plenary meetings of Commission are held in public unless otherwise decided by the Commission. Russian Federation has permanent observer status. GFCM grants observer status for these plenary and committee (SAC and CAQ) meetings to IGOs and NGOs based on the GFCM Rules of Procedure (now being amended). Medisamak, the Mediterranean fishing association (an industry organization), and as of 2006 CPIS (Confederation internationale de la pêche sportive) are official observers. All other Commission meetings are open to scientific experts, IGOs, and NGOs.
	Data collecting and reporting	**Observer program**	**Vessel monitoring system (VMS)**	**Enforcement**	**Catch documentation – trade schemes documentation**
Monitoring, control, and surveillance	Data collected through GFCM–FAO regional projects and by CPs. Data include catch, effort, scientific trawl and acoustic survey data. Collaboration with Mediterranean International Trawl Survey Program (MEDITS) organized by EC. The Mediterranean Fishery Statistics Information System (MEDFISIS) serves Commission's statistical arm, reformatting and aggregating submitted data. Data on aquaculture activities submitted through the Information System for the Promotion of Aquaculture (SIPAM). In 2006, agreed to data confidentiality policy and procedures regarding electronic data transmission.	No regional observer program. In 2005, adopted general guidelines for the establishment of a control scheme that envisages an observer program and VMS, to be further examined and discussed at next GFCM session scheduled in January 2006.	No VMS at this time; however, VMS is currently under consideration.	No at sea boarding or inspection. In 2005, adopted binding Recommendation to establish register for vessels >15 m authorized to fish in area. In 2006, agreed to establish IUU vessel list. CPs and Cooperating NCPs to annually transmit to Secretariat list of NCP vessels presumed to be carrying out IUU fishing in area. Secretariat to compile and publish IUU list. As of 2006, GFCM cooperating with ICCAT on transshipment measures for large-scale longline fishing vessels. Enforcement essentially rests with CPs.	None in place at this time. As of 2006, Japan will prohibit imports of farmed bluefin tuna not in compliance with requirement under GFCM– ICCAT Recommendations.
	Decision making	**Dispute settlement mechanism**	**Cooperative efforts**	**Efforts to strengthen RFMO**	**Challenges**
Other governance efforts and challenges	Management decisions are typically taken by consensus. Other decisions, when taken by vote, require a majority. A majority of the membership constitutes quorum. Each member has 1 vote. Whenever vote occurs, there is a declaration of competency made by the EC and its member States. Commission may adopt and amend its own rules of procedure, by a 2/3 majority, provided consistent with General Rules of FAO. Only Recommendations adopted pursuant to Art. V are binding. Resoultions are nonbinding. An objection procedure exists whereby members submit objection within 120 days of notification of a Recommendation, after which period decisions become binding.	Art. XVII provides a mechanism for the establishment of a committee to which disputes may be referred. The committee is composed of 1 member of each of the parties to the dispute and an independent chairman chosen by members of the committee. If the dispute is not settled at the committee level, it shall be referred to the International Court of Justice. To date, Art. XVII establishing a committee has never been invoked.	Collaborates with ICCAT, e.g., GFCM–ICCAT WG on Large Pelagics has met annually since 1990; GFCM–ICCAT WG on Sustainable Tuna Farming (to define guidelines for bluefin tuna farming). Supported by subregional projects, e.g., Responsible Fisheries in the Adriatic Sea (ADRIAMED); Advice, Technical Support and Establishment of Cooperation Networks to Facilitate Coordination to Support Fisheries Management in the Western and Central Mediterranean (COPEMED); and Assessment and Monitoring of the Fishery Resources and the Ecosystems in the Straits of Sicily (MedSudMed). Incorporates fisheries data into FIRMS as FAO partner. Cooperates with Medisamak, exchanging information and attending meetings. Partnerships with IGOs, e.g., UNEP, IUCN, CIHEAM, ACCOBAMS, and tuna RFMOs.	Amended Agreement in 1997 to establish autonomous budget (in force 2004). In 2005, agreed to establish independent Commission headquarters in Rome, Italy. In 2005, established ad hoc WG on EA. In 2005, closed area below 1000 m to protect deepwater fish (habitat and spawning grounds). In 2006, prohibited towed dredges and bottom trawls from sensitive areas and established closed season for fishing on FADs. Strengthened participation in SCSA and WGs in 2004. Increased participation of less-developed countries in projects. Assessed PA, EA, and IUU for Commission work plan. Established vessel register. In 2006, created IUU vessel list. In 2006, reestablished Environment and Aquaculture (EAM) under CAQ. In 2006, adopted recommendation establishing Cooperating NCP status. Cooperating with ICCAT to control transshipment. Conducted independent evaluation in 2003 of SAC and CAQ.	Avoiding overfishing. Achieving sustainable fisheries. Strengthening capacity of Secretariat to fulfill RFMO functions and responsibilities. Controlling IUU fishing. Ensuring timely and reliable data on catch and effort are submitted to Secretariat. Ensuring continued active participation in all SCs and WGs. Effective implementation and enforcement of decisions and measures by member States. Increasing membership of States with "real interest" (including all Black Sea States). Addressing management of small-scale fisheries. Regulating tuna farming.

*Bold type indicates that member has ratified UNFSA; italic type indicates that member has ratified the FAO Compliance Agreement.

Inter-American Tropical Tuna Commission (IATTC)

Evelyne Meltzer and Susanna D. Fuller: 05/2006

RFMO structure

IATTC	Area of competence	Objectives	Organizational structure	Scientific regime
Inter-American Tropical Tuna Commission (IATTC) **Headquarters:** La Jolla, California, USA. **Convention:** Convention for the Establishment of an Inter-American Tropical Tuna Commission (IATTC Convention). 1999 Protocol to Permit REIO Membership (not in force). **Signed:** 31 May 1949, Washington, D.C., USA. **In force:** 3 March 1950. Associated with 1997 Agreement on the International Dolphin Conservation Program (AIDCP) in force in 1999. *Amended by 2003 Convention for the Strengthening of the Inter-American Tropical Tuna Convention (Antigua Convention) (closed for signature 31 December 2004 not yet in force).* **Web site:** www.iattc.org	Generally considered to be the Eastern Pacific Ocean (EPO). *Precisely defined under Antigua Convention to be: along the 50ºN parallel from the coast of North America to the intersection with 150ºW, and from that line to the intersection with 50ºS, and from that line to its intersection with the coast of South America (extends the notional IATTC boundaries by 10º both N and S).*	To maintain populations of yellowfin and skipjack tuna as well as other species taken by tuna vessels in EPO and to cooperate in gathering and interpreting data to facilitate management of stocks at levels permitting MSY year after year. *Antigua Convention: to ensure long-term conservation and sustainable use of tunas and tuna-like species, and other species taken by tuna-fishing vessels in the EPO, in accordance with relevant rules of international law.* AIDCP key objectives are to progressively reduce and eliminate incidental dolphin mortalities in tuna purse-seine fishery in area; seek ecologically sound means of capturing large yellowfin tunas not in association with dolphins; ensure long-term sustainability in area; avoid, reduce, and minimize bycatch and discards of juvenile tunas and nontarget species.	IATTC: Commission (Work Plan includes Tuna–Billfish Program and Tuna–Dolphin Program); National Sections; Permanent Working Group on Fleet Capacity; Permanent Working Group on Compliance; Joint Working Group on Fishing by Non-Parties; Working Group on Bycatch (WG-Bycatch); Working Group on Stock Assessments (WG-SA); WG on Limit Reference Points (WG-LRP); permanent independent Scientific Staff; and Secretariat (Director). AIDCP: International Review Panel (IRP); Scientific Advisory Board; Tuna Tracking WG; WG to Promote and Publicize the AIDCP Dolphin-Safe Certification Program; WG on Financing and Vessel Assessments; Joint WG on Fishing by Non-Parties. IATTC provides Secretariat for AIDCP. Working Group on Limit Reference Points (WG-LRP).	Large permanent independent scientific staff with offices in major fishing ports. WG-Bycatch, WG-LRP, and WG-SA. Director and scientific staff provide advice to Commission after review by WG-SA. IATTC collaborates on stock assessment with Secretariat of the Pacific Community (SPC) (for bigeye and billfish) and Interim Scientific Committee for Tuna and Tuna-Like Species in the North Pacific (ISC); with EC and Chile for swordfish pursuant to Southeast Pacific Swordfish Arrangement (SPSA). All members have equal access to scientific information. *Antigua Convention includes a Scientific Advisory Committee to review research programs, stock assessments, research, and recommendations of scientific staff (Annex 4).*

The fishery, EA, and PA

Key species and gear	Stock status	Bycatch	Ecosystem approach (EA)	Precautionary approach (PA)
Highly migratory fish stocks: yellowfin tuna, bigeye tuna, albacore tuna, skipjack tuna, bonito tuna, Pacific bluefin tuna, sailfish, and billfishes including marlin and swordfish. **Gear:** purse seine, longline, pole-and-line, trolling, and others.	**Above B_{MSY}:** skipjack ($F < F_{MSY}$, stock abundant), southern albacore (catches < MSY), striped marlin, swordfish. **At AMSY:** yellowfin. **Below B_{MSY}:** bigeye (2003 biomass at lowest levels since 1975). **Fully exploited:** northern albacore ($F > F_{MSY}$). **Status uncertain:** blue marlin (appears to be fully exploited), black marlin or sailfish (no recent assessment). (yellowfin and bigeye purse-seine fisheries are managed by closed seasons; bigeye longline fishery managed by quota limits).	Dolphins, turtles, seabirds, sharks, other nontarget species, and juvenile target species. 2004 Resolution on bycatch requires CPCs to reduce incidental mortality of juvenile tuna, release unharmed nontarget species, and reduce turtle bycatch. Under 2004 Resolution, adopted 3-year program to mitigate turtle bycatch. In 1999, AIDCP replaced voluntary La Jolla Agreement to provide measures to mitigate the effect of purse-seining on dolphin stocks. 2005 Resolution on seabird bycatch requires CPCs to report bycatch data and inform Commission on status of NPOAs. 2005 Resolution on sharks requires CPCs to collect data for stock assessments, reduce shark finning, and limit fins to >5% of total landed catch.	EA is envisaged in Art. II. Advice to Commission now includes information on ecosystem effects of fishing. For yellowfin and bigeye, Commission adopted Resolutions to reduce bycatch of target and nontarget species in purse-seine and longline fisheries, improve longline technology, develop ecosystem models for tropical EPO, and investigate how habitat affects juvenile tuna. *EA is envisaged in Art. VII of Antigua Convention.*	Since 1980s, has included precaution in absence of information and adaptive management approach when assessing impacts of expanded fisheries on stocks. Interprets MSY as limit reference point. If catches for target species reach MSY limit, management measures are imposed. WG-LRP was established to suggest PA limits and targets. Fleet capacity was limited to precautionary level by Resolution (2002). *PA is included in Art. IV of Antigua Convention consistent with UNFSA.*

Participation

Contracting Parties*	Cooperating nonmembers*	Membership	Participation criteria	Transparency
IATTC: **Costa Rica**, Ecuador, El Salvador, **France**, Guatemala, *Japan*, *Korea*, *Mexico*, Nicaragua, Panama, *Peru*, **Spain**, ***USA***, Vanuatu, Venezuela. AIDCP: **Costa Rica**, Ecuador, El Salvador, ***EC***, Guatemala, Honduras, *Mexico*, Nicaragua, Panama, *Peru*, ***USA***, Vanuatu, and Venezuela. Bolivia and Columbia are applying the AIDCP provisionally.	IATTC Cooperating Non-Contracting Parties (Cooperating NCPs) or cooperating fishing entities: ***Canada***, China, Chinese Taipei, the ***EC***, and Honduras. Contracting Parties, Cooperating Non-Contracting Parties, fishing entities, and REIOs are collectively referred to as CPCs.	Membership is open to States (not REIOs) whose nationals participate in fisheries in area upon CPs' unanimous consent. Members pay dues, participate in data collection, and research. States may become Cooperating NCPs to Convention as well as to AIDCP. *Antigua Convention is open to all coastal States and fishing nations, including Chinese Taipei and REIOs, and does not require unanimous approval for new membership.*	Each CP may establish Advisory Committee (AC) for its National Section; AC may attend non-Executive sessions and address sessions at Chair's discretion. Each CP is member of all WGs. Cooperating NCPs fishing entities participate in meetings as observers, must report fisheries statistics and research programs in area, comply with all conservation measures and resolutions, and inform IATTC of compliance measures (VMS, observer programs, inspections). Status is reviewed annually. Purse-seine capacity Resolution prohibits new vessels without corresponding reduction of existing fleet, and new entrants to the purse-seine fishery must make arrangement with existing participants to enter.	Publicly accessible Web site. Meeting minutes, reports, and scientific information available online. NCPs, IGOs, NGOs, and owners of tuna vessels fishing in EPO under jurisdiction of any CP can participate as observers with 120 days prior notice and upon unanimous decision of parties. Observers may present orally at Commission meetings subject to Chair's discretion, provided no member objects. Chair must give prior approval for circulation of documents by observers. 2005 Resolution on Financing invites NGOs to make contribution to budget. *Under Antigua Convention, NGOs must give 50 days notice to participate in meetings and may attend provided less than 1/3 of members object.*

*Bold type indicates that member has ratified UNFSA; italic type indicates that member has ratified the FAO Compliance Agreement.

	Data collecting and reporting	Observer program	Vessel monitoring system (VMS)	Enforcement	Catch documentation – trade schemes documentation
Monitoring, control, and surveillance	Fishery-dependent data (from vessels, observers, managers, and processing facilities) are provided to IATTC scientific staff. 2003 Resolution on data provisions outlines specific data requirements to be submitted to the Commission. 2004 Resolution on catch reporting requires Director to report annual catches of Convention species by flag and gear type to parties by 30 June of following year. Purse-seine vessels required to report weekly to Commission. Tagging program in place to collect data on tuna populations.	Since 1994, 100% observer coverage on purse-seine vessels (> 363 tonnes) with at-sea reporting by observers and weekly data submission to Secretariat. No observers on longline and small purse-seine vessels. 70% of observers are employed by IATTC, the remainder by national observer programs. Observer effectiveness monitored through comparison of national and IATTC programs to identify possible discrepancies. AIDCP Parties may maintain own national observer programs (e.g., Colombia, Ecuador, EU, Mexico, and Venezuela), provided they collect and report information in same standard as IATTC observers. IATTC runs observer program on cost-recovery basis for AIDCP.	As of 1 January 2005 where possible, CPCs must establish VMS. Each CPC provided progress reports to Director by 31 May 2005. Based on CPC reports, at June 2005 Meeting, Commission discussed, but did not agree on, measures for enhanced use of VMS for compliance and data collection, e.g., a more comprehensive and timely data reporting system could be established using VMS (potential pilot program under consideration with Vanuatu).	IATTC established register of tuna fishing vessels active in area (1999); register for large purse-seine vessels authorized to fish in area, including capacity limits and vessel replacement rules (2002); "positive" list of longline vessels >24 m authorized to fish in area (2003); and list of IUU vessels. In 2005, adopted measures prohibiting interaction with IUU listed vessels. There is also a sighting and reporting system for vessels operating in area (2004). Director notifies flag States of noncompliant vessels. Flag States then order vessel to withdraw from area. There is no at-sea inspection scheme for nonflag States. On-board observers report possible infractions, which are investigated by the flag States and reviewed by Compliance Committee. (IRP reviews AIDCP infractions.)	In 2003, introduced Bigeye Tuna Statistical Documentation Program (SDP). All bigeye imported into CPs must have statistical document, which must be validated by flag State. Documents are not required for purse seiners and baitboats delivering bigeye directly to canneries. Trade Measure Resolution adopted in 2005 requires CPCs importing– receiving landings to submit data to IATTC. CPCs and nonparties identified annually for noncompliance or failure to cooperate with IATTC measures. Commission can adopt trade-restrictive measures where other efforts ineffective in obtaining compliance. IATTC maintains list of those subject to trade-restrictive measures.

	Decision making	Dispute settlement mechanism	Cooperative efforts	Efforts to strengthen RFMO	Challenges
Other governance efforts and challenges	Convention provides for unanimous voting procedure. In practice, decisions, resolutions, recommendations, and publications are approved by consensus. There is no objection or opting-out procedure. All Resolutions are binding (recommendations are nonbinding). All management measures apply inside EEZ and on high seas, i.e., throughout the range of the Agreement area. AIDCP: All decisions are made by consensus. *Antigua Convention adopts consensus-based decision making, contains no opting-out clause, and includes provisions to contact and obtain consent from CPs not in attendance at annual meetings.*	There is no formal dispute settlement mechanism. *Art. XXV of Antigua Convention outlines dispute settlement: member may consult with another member to resolve a dispute related to the interpretation or application of the provisions of the Convention; disputes unable to be resolved by members may be referred by mutual consent to a nonbinding ad hoc expert panel.*	Since 2000 meets yearly with tuna RFMOs to discuss common concerns. Cooperates with other tuna RFMOs on vessel register, and big eye tuna review (2004); and ad hoc consultations on fleet capacity and effort reduction (N. albacore) with WCPFC. Consultation with CCAMLR on seabird bycatch (2005). Secretariat chaired RFB meeting for 2 years and is actively involved in FIRMS to provide comprehensive global reporting system. Collaborates on stock assessments and scientific research with other organizations and countries (e.g., SPC, ISC, SPSA, PICES, and CPPS), as well as oceanic information and training (CPPS). Secretariat monitors international initiatives and attends meetings of other RFMOs as appropriate. *Art. XXIV of Antigua Convention obligates Commission to cooperate where necessary with other fishery organizations–arrangements to achieve Convention objectives. Such arrangements are being discussed with WCPFC.*	Modernized 1949 Convention adopting Antigua Convention in 2003 (if ratified, would address most UNFSA provisions). Ratified and successfully implemented AIDCP. Improved transparency. Included ecosystem effects of fishery in science mandate and developed biological reference points. Considered MSY a catch limit, rather than target. Established time–area closures and 2005 catch limits for bigeye. Implemented capacity reduction Resolution for purse-seine fleet (capacity frozen at 2002 levels). In 2004, established 3-year program to reduce sea turtle bycatch and mitigate impact of tuna fishing. In 2005 established measures to mitigate seabird bycatch and reduce shark finning. In 2005, introduced trade measures and further improved compliance and enforcement framework including vessel sighting and reporting and prohibiting all interactions with IUU listed vessels, e.g., IPOA-IUU. Maintains vessel register for large purse-seine and large longline vessels and IUU vessel list. Improved NCP flag-State enforcement of unauthorized fishing in area. Established 2003 bigeye SDP. Expanded data collection–analysis through observer program, FIRMS, and tagging programs. Implemented at-sea observer reporting on large purse-seine vessels. Improved data provision by CPs. Inter-RMFO cooperation.	Obtaining necessary ratifications to bring Antigua Convention into force. IUU fishing for bigeye and other species. Avoiding overfishing and rebuilding bigeye stocks. Implementing PA (e.g., scientific staff recommended stricter bigeye fishery controls than adopted by Commission) and operationalizing EA. Mitigating seabird bycatch in longline fishery (particularly with extended Southern boundary under Antigua Convention). Implementing observer program on longline and small purse-seine vessels. Standardizing national observer programs. Reducing capacity for purse-seine and longline fleets commensurate with resource. Ensuring compliance with vessel replacement rules under regional vessel register. Coordinating IUU lists, statistical documents, and VMS data with other RFMOs. Monitoring flag-State action on violations. Budgetary constraints and payment difficulties for some States. Implementing 2005 Resolution on financing and funding formula. Several NCPs fishing in EPO derive benefit, but do not contribute to budget, and are therefore requested to make voluntary contributions. Achieving NCP participation. Geopolitical sensitivities.

Western and Central Pacific Fisheries Commission (WCPFC)

Evelyne Meltzer and Susanna D. Fuller: 05/2006

RFMO structure

WCPFC	Area of competence	Objectives	Organizational structure	Scientific regime
Western and Central Pacific Fisheries Commission **Headquarters:** Kolonia, Pohnpei State, Federated States of Micronesia. **Convention:** Convention on the Conservation and Management of Highly Migratory Fish Stocks in the Western and Central Pacific Ocean. **Signed:** 4 September 2000, Honolulu, Hawaii, USA. **In force:** 19 June 2004. (First Regular Session held December 2004. Second Regular Session held December 2005.) **Web site:** www.wcpfc.org	The Western and Central Pacific Ocean (WCPO). From the south coast of Australia south along 141°E to its intersection with 55°S, then east along 55°S to its intersection with 150°E, then south along 150°E to its intersection with 60°S, then east along 60°S to its intersection with 130°W, then north along 130°W to its intersection with 4°S, then west along 4°S to its intersection with 150°W, then north along 150°W.	To ensure, through effective management, the long-term conservation and sustainable use of highly migratory fish stocks (Annex 1 of LOSC and other species as the Commission may determine) in the WCPO in accordance with the LOSC and UNFSA.	Commission; Scientific Committee (SC). SC has six specialist working groups (SWG) — since 2004, established Ecosystem and Bycatch (EB-SWG), Methods (ME-SWG), Fishing Technology (FT-SWG), Statistics (ST-SWG), Biology (BI-SWG), and Stock Assessment (SA-SWG). Technical and Compliance Committee (TCC), Northern Committee (NC). Secretariat (Executive Director).	WCPFC will employ scientific and technical staff. The Secretariat of the Pacific Community (SPC) Oceanic Fisheries Program will be interim provider of science to SC, with provision to engage scientific experts as required. SC (first meeting in August 2005) composed of Contracting Party (CP) representation or others with demonstrated interest in WCPO tuna science. SC has established six SWGs, recommends research plans, and reviews stock assessments (which may be produced by independent experts) to Commission. SC can make recommendations as directed by Commission or on own initiative. SC will work with TCC on observer program standards and procedures.

The fishery, EA, and PA

Key species and gear	Stock status	Bycatch	Ecosystem approach (EA)	Precautionary approach (PA)
Highly migratory fish stocks: skipjack tuna, yellowfin tuna, albacore tuna, bigeye tuna. **Gear:** purse seine, longline, handline, pole and line, troll, artisanal gear.	**Above B_{MSY}:** skipjack, South Pacific albacore. **Nearly fully exploited:** bigeye (although $F_{CURRENT} > F_{MSY}$), yellowfin, and North Pacific albacore.	Seabirds, sharks, sea turtles, billfish, and juvenile target species (especially surrounding FADs). Adopted nonbinding resolutions for bycatch mitigation (for sea turtles, seabirds, and nontarget fish species) at 2005 meeting. CCMs are to collect and submit sea turtle catch data to Commission, reduce catch, and implement bycatch mitigation measures (for purse-seine and longline gear). Commission, in consultation with TCC, will consider seabird mitigation measures including those applied by other RFMOs. CCMs to avoid catching and, where possible, release unharmed nontarget fish species that would be discarded.	EA initially considered during the WCPFC Prep. Con. Art. V(d) outlines the responsibility of Commission in assessing fishing impacts, other human activities, and environmental factors on target stocks, nontarget species, and species belonging to the same ecosystem. EB-SWG to address EAF and take into account national and regional initiatives in the WCPO.	PA is adopted in Art. 5(c). Art. 6 outlines PA with reference to Annex II of UNFSA. Commission will examine interpretation and implementation of reference points.

Participation

Members*	Cooperating nonmembers	Membership	Participation criteria	Transparency
Australia, ***Canada***, China (not Hong Kong), **Cook Islands**, **Federated States of Micronesia**, **Fiji Islands**, **France** (with respect to French Polynesia, New Caledonia, and Wallis and Futuna), *Japan*, **Kiribati**, *Republic of Korea*, **Marshall Islands**, **Nauru**, ***New Zealand*** (including ***Tokelau***), Niue, **Palau**, **Papua New Guinea**, Philippines, **Samoa**, **Solomon Islands**, **Tonga**, Tuvalu, and Vanuatu. Convention allows for participation of fishing entities and REIOs. Fishing Entity of Taiwan (Chinese Taipei) agreed to be bound by Convention as of Nov. 2004. EC joined following invitation of Dec. 2004 (24 members).	Cooperating nonmembers (CNMs) are Indonesia and USA (as of December 2005)	Membership is open to participants in the negotiating process of the Convention. States or REIOs who wish to fish for highly migratory fish stocks in convention area may be invited, by consensus, to accede to the Convention. This approach restricts access, emphasizing initiative to accede lies with existing members, not with new applicants. Chinese Taipei is a member. Members and CNMs are collectively referred to as CCMs.	All members must contribute to the budget; the amount is to be based on an equal basic fee, the wealth of the country, and a variable fee depending on catch taken within EEZ and on high seas. CNMs must apply Convention principles and can make voluntary financial contributions.	Publicly accessible Web site. meeting minutes and background papers are available online. Subject to Commission rules and procedures, representatives from NCPs, IGOs, and NGOs may participate in Commission meetings and its subsidiary bodies as observers or otherwise as appropriate; have access to pertinent information subject to Commission rules and procedures; and are permitted to give oral presentations and distribute papers through the Secretariat.

*Bold type indicates that member has ratified UNFSA; italic type indicates that member has ratified the FAO Compliance Agreement.

	Data collecting and reporting	Observer program	Vessel monitoring system (VMS)	Enforcement	Catch documentation – trade schemes documentation
Monitoring, control, and surveillance	CCMs must provide data annually and as otherwise requested to Commission. Catch, effort, biological and other data, and information must be collected in accordance with Annex I of UNFSA.	Region-wide, independent observer program established under Art. 28(1) with guidelines specified in Annex III of the Convention. Commission will determine final details, including percent coverage. The program will be coordinated to the extent possible with other national or regional programs. Decided at first meeting in 2005 that Commission with TCC will develop regional observer program tailored to WCPO in 2006. Adopted at December 2004 meeting procedures to record fishing vessels and authorizations to fish, specifications for vessel marking, and procedures for CNMs.	Art. 24(8) requires all vessels fishing in the convention area to use VMS. CNMs must cooperate to ensure compatibility between national and high seas VMS. Options for VMS implementation and harmonization with coastal and island States considered by TCC (December 2005); to be further clarified in 2006.	Art. 26 establishes a boarding and inspection (B&I) procedure. If no procedure is in place within 2 years of entry into force, Arts. 21 and 22 of UNFSA will be applied. First meeting of TCC held in Dec. 2005 prior to 2nd Annual Meeting. As of 19 June 2006, UNFSA Arts. 21 and 22 will apply in the WCPFC Regulatory Area. However, B&I procedures will not be used until agreed at 3rd Annual Meeting. Art. 27 established port-State inspection procedure, which allows the port State to prohibit landings and transshipments if catch was taken through noncompliance. Art. 29 outlines procedures for both in-port and at-sea transshipment. Port-State schemes (based on 2005 FAO Model Scheme on Port State Measures to Combat IUU) to be developed in 2006. Transshipment procedures to be further developed by TCC in 2006 as a matter of priority.	There is no catch or trade documentation scheme at this time.
	Decision making	**Dispute settlement mechanism**	**Cooperative efforts**	**Efforts to strengthen RFMO**	**Challenges**
Other governance efforts and challenges	Decisions will be made by consensus wherever possible. Consensus means absence of any formal objection. In cases where consensus is not reached, decisions on questions of substance will be taken by a 3/4 majority overall and in each chamber (i.e., 3/4 majority of Pacific Islands Forum Fisheries Agency (FFA) members and 3/4 majority of nonmembers of the FFA). Such decisions will be binding. There is no objection procedure. A member who has voted against a decision or was absent when decision was taken may seek a review by a panel constituted as outlined in Annex II of the Convention. The results of the review panel determine whether the decision is revisited at a subsequent Commission meeting.	Art. 31 addresses the peaceful settlement of disputes and refers to Part VIII of the Convention. Provisions set out in Part VIII of Convention apply to any dispute between members of Commission, whether or not they are Parties to the Agreement.	Art. 22 provides for cooperation with FAO and other UN specialized agencies. Commission will act in cooperation with CCAMLR, CCSBT, IOTC, and IATTC with regard to related and similar objectives. WCPFC cooperates with regional programs, including FFA and SPC. Recognizes need for comprehensive global system to regulate at-sea and in-port transshipment, across all RFMOs.	Negotiated Convention and established WCPFC as RFMO largely consistent with UNFSA, completing RFMO coverage for highly migratory fish stocks. Convention includes PA and EA and provisions for regional, independent observer program. As of January 2006, all States, REIOs, and entities with major fishing interest participate in WCPFC as CCMs. Coastal and island States developing national tuna management plans, improving conservation of tuna stocks throughout geographic range. FFA considering EAF guidelines with possible pilot program. Options for conservation and management measures (catch–effort limits, time–area closures, fishing impacts, compatibility, and bycatch mitigation) outlined by TCC in December 2005.	Addressing conservation concerns for bigeye and yellowtail. Adopting and implementing conservation and management measures, including appropriate allocations and capacity reduction, throughout convention area. Reducing bycatch and implementing mitigation measures for seabirds, sharks, and turtles. Interpreting and implementing PA and **EAF**. Ensuring compatible management throughout range of stocks and determining respective roles of WCPFC and coastal and island States and the design of an effective implementation of MCS scheme (including boarding and inspection, port-State measures, and transshipment). Determining western boundary of convention area. Implementing vessel registry. IUU fishing. Developing mechanism for internal assessment and audit of RFMO.

Commission for Conservation of Southern Bluefin Tuna (CCSBT)

Evelyne Meltzer and Susanna D. Fuller: 05/2006

RFMO Structure	**CCSBT**	**Area of competence**	**Objectives**	**Organizational structure**	**Scientific regime**
	Commission for Conservation of Southern Bluefin Tuna **Headquarters:** Canberra, Australia. **Convention:** Convention for the Conservation of the Southern Bluefin Tuna (CCSBT Convention). **Signed:** 10 May 1993, Canberra, Australia. **In force:** 20 May 1994 **Web site:** www.ccsbt.org	The entire range of southern bluefin tuna (SBT), covering approximately 30ºS–50ºS in all oceans and the spawning ground south of Java, Indonesia. Where SBT geographic range overlaps with other RFMOs, CCSBT has primary competence.	To ensure, through appropriate management, conservation and optimum **use** of SBT. The Commission provides internationally recognized forum for other countries–entities to actively participate in SBT issues.	Commission, Scientific Committee (SC), Stock Assessment Group (SAG), Independent Advisory Panel. In 2005, agreed to activate Compliance Committee. Extended Commission and Extended Scientific Committee were created in 2001 (enabling Fishing Entity of Taiwan to participate). Secretariat (Executive Secretary).	SC advises Commission. SC has independent Chair and independent Advisory Panel. Stock assessment is conducted by SAG established to separate technical evaluation and advisory roles of SC. SAG has independent external Chair. Independent Advisory Panel was created to assist national scientists with stock assessment, provide support to process, and comment on members' papers. If members cannot agree on science, Advisory Panel will prepare independent advice. SC considers advice of SAG and advises Commission. No permanent scientific staff exist. Established extended SC to allow Fishing Entity of Taiwan to participate in scientific deliberations.
The fishery, EA, and PA	**Key species and gear**	**Stock status**	**Bycatch**	**Ecosystem approach (EA)**	**Precautionary approach (PA)**
	Highly migratory stocks: southern bluefin tuna. **Gear:** purse seine and longline.	SBT may have declined by up to 90% since 1960. In 2005, if no further reductions and continued failure to implement management procedure (CMP-2, preagreed formula adopted to set TAC), there is a possibility that spawning stock will decline to 0 tonnes by 2030. 2004 assessment concluded that catch rates at that time were more likely to result in further decline than stock increase. (Spawning grounds continue to be targeted within Indonesian EEZ.) Initial rebuilding plan to reach 1980 levels by 2020 was unachievable. 2005 adopted management procedure and quota reductions to ensure a 90% probability that spawning stock biomass in 2014 will not fall below 2004 level. (In 2005, market reviews indicate 10 000 tonnes SBT illegally caught).	Seabirds, sharks, rays, turtles, and juvenile tuna. Measures adopted in 1995 mandate use of streamer lines and bait dying to mitigate seabird bycatch. Identification guides for sharks and seabirds have been published by CCSBT. Ecologically Related Species Working Group (ERSWG) meets biennially. Members and cooperating nonmembers (CNMs) are required to report on implementation of required bycatch mitigation measures.	While EA is not expressly referred to in Convention, Commission has established ERSWG charged with reducing bycatch and evaluating effects on associated species. ERSWG implemented data collection by fishing vessels on bycatch species. Focus remains on managing target stock.	No formal adoption of PA to date. Stock assessments take into account unreported catch and juvenile mortality.
Participation	**Members – Contracting Parties***	**Cooperating nonmembers***	**Membership**	**Participation criteria**	**Transparency**
	Commission members: ***Australia***, **New Zealand**, *Republic of Korea*, *Japan*. Extended Commission: Fishing Entity of Taiwan and Commission members. Extended Commission created in 2002 (5 members of Extended Commission).	In 2003, Extended Commission decided to include status of CNM. Philippines is CNM. Negotiations to join CCSBT are ongoing with **South Africa** and Indonesia, as their nationals and vessels fish SBT. South Africa was offered a catch quota of 45 tonnes (approximately three times their current catch levels).	Membership is open to any State whose vessels engage in fishing for SBT or to any coastal State through whose EEZ or fishing zone SBT migrates. REIOs are not permitted to join. CNM status is reviewed annually based on adherence to conservation management activities of Commission. In 2004, Extended Commission considered policy options for admission of new entrants, including setting of catch limits. No decisions were taken. Engagement of new members or CNMs is encouraged to ensure that conservation measures are not undermined.	CNMs cannot vote, do not pay contributions but participate fully in discussions, although they may be excluded from discussion on some agenda items at discretion of Chair (no exclusions have occurred to date). CNMs are required to observe conservation and management measures of CCSBT including any catch limit agreed with the CCSBT. Vessels of CNMs are added to CCSBT list of vessels authorized to fish for SBT.	Publicly accessible Web site. Meeting minutes and catch data are available online. NGOs may attend meetings as observers and submit documents with 100 days notice and unanimous consent of parties. NGOs may also attend as members of party delegations. Reports and Commission documents are placed in public domain unless party asks for document to be kept confidential. IGOs and other organizations are invited. Opening stages of Commission meeting are open to public, but subsequent discussions of substance are limited to members, CNMs, and observers.

*Bold type indicates that member has ratified UNFSA; italic type indicates that member has ratified the FAO Compliance Agreement.

	Data collecting and reporting	Observer program	Vessel monitoring system (VMS)	Enforcement	Catch documentation – trade schemes documentation
Monitoring, control, and surveillance	Parties to Convention provide Commission with scientific information, catch and effort statistics, and other data relevant to conservation of SBT and ecologically related species. There is agreed data set required to be collected by members and CNMs, and maintained by CCSBT Secretariat. CCSBT Secretariat is implementing 5-year tagging program and maintains database on tag releases and recoveries. Monthly catch reporting agreed upon in 2005.	CCSBT has agreed observer program standards that members must implement for fleets catching SBT. Standards have target observer coverage of 10% and data set to be collected. Currently, observer data are maintained by members and not in data set managed by CCSBT Secretariat. International observer program (IOP) discussed in 2005. Extended Commission to prepare discussion document examining IOP rationale, other RFMO practice, and implementation logistics.	No centralized VMS scheme. All authorized fishing vessels are operating with VMS but report back to national administrations and not to CCSBT.	CCSBT established list of authorized fishing vessels, which, as of 1 July 2005, includes all vessels regardless of length. Obligatory for members to implement CCSBT conservation and management measures within EEZs and fleets fishing for SBT; no boarding and inspection scheme; compliance is sought through trade measures. In 2005, agreement to activate Compliance Committee for which Terms of Reference (TOR) were originally prepared in 1997, to monitor compliance of members, CNMs as well as Non-Contracting Parties (NCPs).	Trade Information Scheme (TIS) implemented in 2000. Members and CNMs require CCSBT statistical documents for all imports. Members and CNMs will not validate trade documents for, or accept imports from, large-scale vessels not on authorized list. States not in cooperation with CCSBT conservation and management measures can face restrictions, as outlined in 2000 Action Plan. Since Japan is member and primary market, trade control measures can be effective; however, no trade documentation required if fish are caught and consumed by same country. 2005 discussion on implementation of CDS instead of or in addition to existing TIS. In 2005, USA recognized trade and catch schemes of all RFMOs and will reject all product without valid trade documentation. In 2005, technical working group established to consider the implications of a catch documentation scheme (CDS) for CCSBT.
	Decision making	**Dispute settlement mechanism**	**Cooperative Efforts**	**Efforts to strengthen RFMO**	**Challenges**
Other governance efforts and challenges	Decisions are made by consensus of members present at Commission meeting. In case of vote, each member has one vote. Two-thirds of members constitutes a quorum. Commission's work is principally done within Extended Commission. Decision making by consensus resulting in deadlock and failure to agree on catch quota in 1998–2002. No agreement reached in 2005 for TAC reductions for 2006–2007. Failure to reach consensus has lead to adoption of model-based management procedure, with predetermined decision rules and data inputs, precluding opting out.	Art. XIV provides dispute resolution directions. Annex I includes instructions for creating Arbitral Tribunal if other dispute resolution mechanisms fail. Parties have submitted a dispute to ITLOS (and not resolved through CCSBT dispute settlement procedures).	Obligated under Art. 12 of Convention to cooperate with other RFMOs and IGOs. Cooperates with other tuna RFMOs regarding SBT. Tuna bodies met annually and participate in RFB biennial meetings. Effectiveness of RFMO improved through regular contact and shared data, trade statistics, and share their list of IUU vessels with IOTC and WCPFC. Has cooperation agreements with ICCAT, IOTC, and WCPFC and participates in meetings of other RFMOs when the agenda includes relevant issues. Cooperates with CCAMLR on seabird bycatch measures.	Created Extended Commission and Extended SC to allow participation by Fishing Entity of Taiwan. Created status of CNM. Developed authorized vessel list (registry). Since 2001, enhanced Secretariat's role to strengthen intersessional management including maintaining database, managing TIS, and negotiating with potential CNMs. Can control nationals under flag State (Art. 15.4). CCSBT is considering implementing catch monitoring scheme. Adopted preagreed management formula outlined in CMP-2. 2005 Activation of Compliance Committee.	SBT seriously overfished and at historically low levels. Current levels of catch nonsustainable. Urgent need for achieving stock rebuilding target, implementing management procedure (CMP-2), and reducing TAC for 2006–2007. IUU fishing. Increase in number of nonparties fishing SBT. Impact of harvesting of SBT on spawning grounds within Indonesian EEZ. Meeting targets for observer coverage of 10%. Data uncertainty, potentially corrupting important inputs on which SAG assesses and Commission sets quotas. Implementation of bycatch mitigation measures.

Indian Ocean Tuna Commission (IOTC) Evelyne Meltzer and Susanna D. Fuller: 05/2006

	IOTC	Area of competence	Objectives	Organizational structure	Scientific regime
RFMO structure	**Indian Ocean Tuna Commission** **Headquarters:** Victoria, Seychelles. **Agreement:** Agreement for the Establishment of the Indian Ocean Tuna Commission (IOTC Agreement) (FAO Art. XIV body). **Signed:** 25 November 25, Rome, Italy. **In force:** 27 March 1996. **Web site:** www.iotc.org	The Indian Ocean. Defined for the purpose of the Agreement as FAO Areas 51 and 57, and adjacent seas, north of the Antarctic Convergence. There is an overlap on the western boundary with WCPFC. In 1999, it was agreed that area of competence be extended to 20°E.	To promote cooperation among members with a view to ensuring, through appropriate management, the conservation and optimum utilisation of stocks covered by the Agreement and encouraging sustainable development of fisheries based on such stocks. IOTC has competency for SBT but, by agreement, CCSBT manages SBT in IOTC Convention Area.	Commission,;Scientific Committee (SC); Compliance Committee; Standing Committee on Administration and Finance; sucommissions (stock-specific, to be convened only if necessary); Working Party on Tropical Tunas (WPTT); Working Party on Tagging (WPT); Working Party on Billfish (WPB); Working Party on Temperate Tunas (WPTMT); Working Party on Neritic Tunas; Working Party on Bycatch (WPB) — met in 2005; Working Party on Methods (WPM); and in 2005 agreed to establish Working Party on Management Options. Secretariat (Executive Secretary).	SC is composed of representatives of Contracting Parties (CPs) and reviews work of WPs and advises Commission (and subcommissions, if established) on research, data collection, stock status, and management issues. SC meetings have been held conjointly with Commission meetings. Subcommissions will examine management options and recommend to Commuission appropriate management measures for particular stocks, WPs are meetings of scientists (in their individual capacity) who conduct stock assessment and proposes management recommendations.
	Key species and gear	**Stock status**	**Bycatch**	**Ecosystem approach (EA)**	**Precautionary approach (PA)**
The fishery, EA, and PA	**Highly migratory species:** yellowfin tuna, skipjack tuna, bigeye tuna, albacore tuna, longtail tuna, kawakawa, frigate tuna, bullet tuna, narrow-barred Spanish mackerel, Indo-Pacific king mackerel, Indo-Pacific blue marlin, black marlin, striped marlin, Indo-Pacific sailfish, swordfish (also southern bluefin tuna (SBT) but CCSBT has primary responsibility). **Gear:** purse seine, bait boat, gill net, troll line, handline, and trawl.	**Above MSY:** yellowfin (further increase in effort and catch above 2000 levels should be avoided, uncertainties in stock status), bigeye (possible overfishing occurring). **Stable:** skipjack, swordfish (current catch level unsustainable) possible overfishing in SW Indian Ocean). **Uncertain:** albacore (no reliable assessment, catch increases to consider PA if adopted). Assessments not available for other target species.	Sea turtles, sharks, seabirds, and juveniles of target species, particularly swordfish. Established WPB in 2002, first in-person meeting of WPB in July 2005. 2005 recommendation for sea turtle data collection and bycatch mitigation guidelines for purse-seine and longline vessels. In 2002, IOTC agreed to implement IPOA-Sharks. 2005 Resolution requires CPCs to collect catch data, fully use shark catch; ensure fins be <5% of total shark weight onboard; release live, to the extent possible, sharks caught in nontarget fisheries; increase gear selectivity; and identify shark nursery areas. 2005 seabird recommendation calls for CPCs to submit bycatch data, and report on NPOA-Seabirds implementation.	EA is pursued through requesting data on nontarget, associated, and dependent species (data are often not available). While IOTC is endeavouring to understand impacts on fisheries, organizational and research capacity of some CPs has been further compromised.	In 1999, SC invoked PA in its recommendation to reduce bigeye catches. IOTC has adopted PA recognizing the need for incorporating uncertainty in stock assessments.
	Members – Contracting Parties*	**Cooperating nonmembers***	**Membership**	**Participation criteria**	**Transparency**
Participation	***Australia***, China, Comoros, *Eritrea*, ***EC***, **France**, **Guinea**, **India**, **Iran**, **Japan**, **Kenya**, *Republic of Korea*, *Madagascar*, Malaysia, ***Mauritius***, Oman, Pakistan, Philippines, ***Seychelles***, **Sri Lanka**, Sudan, Thailand, **United Kingdom**, and Vanuatu. France, and United Kingdom are CPs on behalf of their Indian Ocean territories (24 members).	Cooperating Non-Contracting Parties (NCPs): Indonesia and **South Africa**, status renewed for both in 2005. 2005 application by **Belize** rejected. One Senegalese vessel on approved vessel registry; IOTC awaiting Cooperating NCP application.	Membership is open to members of FAO that are Indian Ocean coastal States or REIOs, or whose vessels fish in the IOTC Agreement Area. The Commission may (by a 2/3 vote) admit States that are members of the UN or one of its specialized agencies or the IAEA. IOTC encourages Cooperating NCPs (Resolution 03/02). Contracting Parties and Cooperating Non-Contracting Parties collectively referred to as CPCs.	CPs can vote and must contribute to budget. Subcommissions will be open to those CP coastal States lying on migratory path of stocks concerned in subcommission or are States whose vessels participate in fisheries of these stocks. To date, no subcommissions have been constituted, but possible management actions have been discussed in the context of the full Commission.	Publicly accessible Web site. Meeting minutes are published online. NCPs may attend meetings. IGOs and NGOs, upon request, may be invited to participate as observers. A list of observers must be submitted 30 days prior to meeting. The Fishing Entity of Taiwan attends as "invited experts".

*Bold type indicates that member has ratified UNFSA; italic type indicates that member has ratified the FAO Compliance Agreement.

Monitoring, control, and surveillance	Data collecting and reporting	Observer program	Vessel monitoring system (VMS)	Enforcement	Catch documentation – trade schemes documentation
	CPs are subject to mandatory statistical reporting and confidentiality procedures. Stock assessment is peer reviewed through species working parties. The Secretariat maintains a capability in stock assessment to ensure that parties without scientific capabilities have access to relevant information. Ongoing Indian Ocean tuna tagging project (IOTTP). As the Fishing Entity of Taiwan is not a FAO member, information from fishing vessels from Taiwan is supplied through the Japanese Organization for Promotion of Responsible Fisheries to China and then to the Commission.	IOTC has encouraged members to implement national observer programs, covering at least 10% vessels.	In 2002, passed binding Resolution to establish pilot program to implement VMS on 10% CP and NCP vessels >24 m. Pilot project to run 2-year schemes. Project to be reviewed in 2006. Information is to be sent to land-based Fisheries Monitoring Centre every 6 hours.	In 2002, established IOTC vessel record ("positive" list of authorized vessels >24 m). Since 2005, all vessels (including <24 m fishing outside EEZs) fishing in regulatory area must be on IOTC record. Unlisted vessels cannot fish for, retain, or transship tuna or tuna like-species. In 2002, established IUU vessel list. Since 1998 vessel registration and information exchange required for NCP vessels fishing in regulatory area (since 2005, includes vessels <24 m fishing outside EEZs). CPCs require prior authorization for at-sea or in-port transshipment, consistent with reported catch validating statistical documents. Measures addressing transshipment challenges and noncompliance (e.g., banning transshipment, observers on carrier vessels) to be further discussed at 2006 IOTC Meeting. There are currently no provisions for at-sea boarding and inspection. 2001 Resolution on Scheme of Control and Inspection for NCPs — phased implementation in place. 2002 Port Inspection Scheme binding on CPCs. Resolution in 2005 on in-port inspection requires landings data to be submitted annually to Commission by 1 July.	In 2001, implemented statistical documentation scheme (SDS) for bigeye (applies only to frozen bigeye). Statistical documents and prior authorization are required for at-sea or in-port transshipment by CPCs and NCPs. 2003 Resolution Concerning Trade Measures requires monitoring of noncompliance, identifying CPCs and NCPs in violation of IOTC measures, and applying trade-restrictive measures.
Other governance efforts and challenges	**Decision making**	**Dispute settlement mechanism**	**Cooperative efforts**	**Efforts to strengthen RFMO**	**Challenges**
	Decisions are generally reached through consensus. Each CP has one vote. Commission can adopt binding measures by 2/3 majority. CPs may object, in which case the decision is not binding on them. Resolutions are binding, and nonbinding recommendations can be adopted by majority.	Disputes, if not settled by Commission, will be referred for settlement to a conciliation procedure. If the dispute is not settled, it may be referred to the ICJ unless parties to the dispute agree to another method of settlement.	Cooperates with CCSBT in SBT management and of Indonesian catches. In 2001, established 5-year project with Overseas Fisheries Cooperation Foundation of Japan to improve data collection in regulatory area. Cooperates with various countries in the region in small-scale tuna tagging program funded by EC, Japan, and China. Has Memorandum of Understanding with Indian Ocean Commission (IOC) to conduct large-scale tropical tuna tagging program with EC funding. In 2004, joined FIRMS to report and share information on status and trends of fishery resources. IOTC to attend 2007 meeting of tuna RFBs.	Established IOTC as an Art. XIV FAO body. Improved enforcement through establishment of Compliance Committee to review annual report of compliance by CPCs. Established IOTC record for positive list of vessels and IUU list. Expanded MCS to vessels <24 m fishing on high seas. Established SDS. Ongoing efforts to control transshipment. Efforts to address bycatch, especially nontarget species through IPOAs and mitigation measures. Ongoing discussions to enable participation of Taiwan as fishing entity. Established WP for Management Options. Meeting to improve effectiveness and efficiency of IOTC to be held May 2006.	Avoiding and reducing overfishing (especially bigeye). Reducing and eliminating IUU fishing. Addressing transshipment. Reducing excess fishing capacity. Implementing bycatch mitigation measures. Developing and implementing EA and PA. Improving and expanding the SDS. Implementing at-sea and in-port NCP inspection schemes. Establishing effective mechanism for participation of Fishing Entity of Taiwan. Ensuring timely and accurate collection of scientific data from CPCs. Strengthening RFMO effectiveness. Addressing organizational capacity concerns of CPCs, given tsunami devastation. Administrative efficiency. Members in arrears.

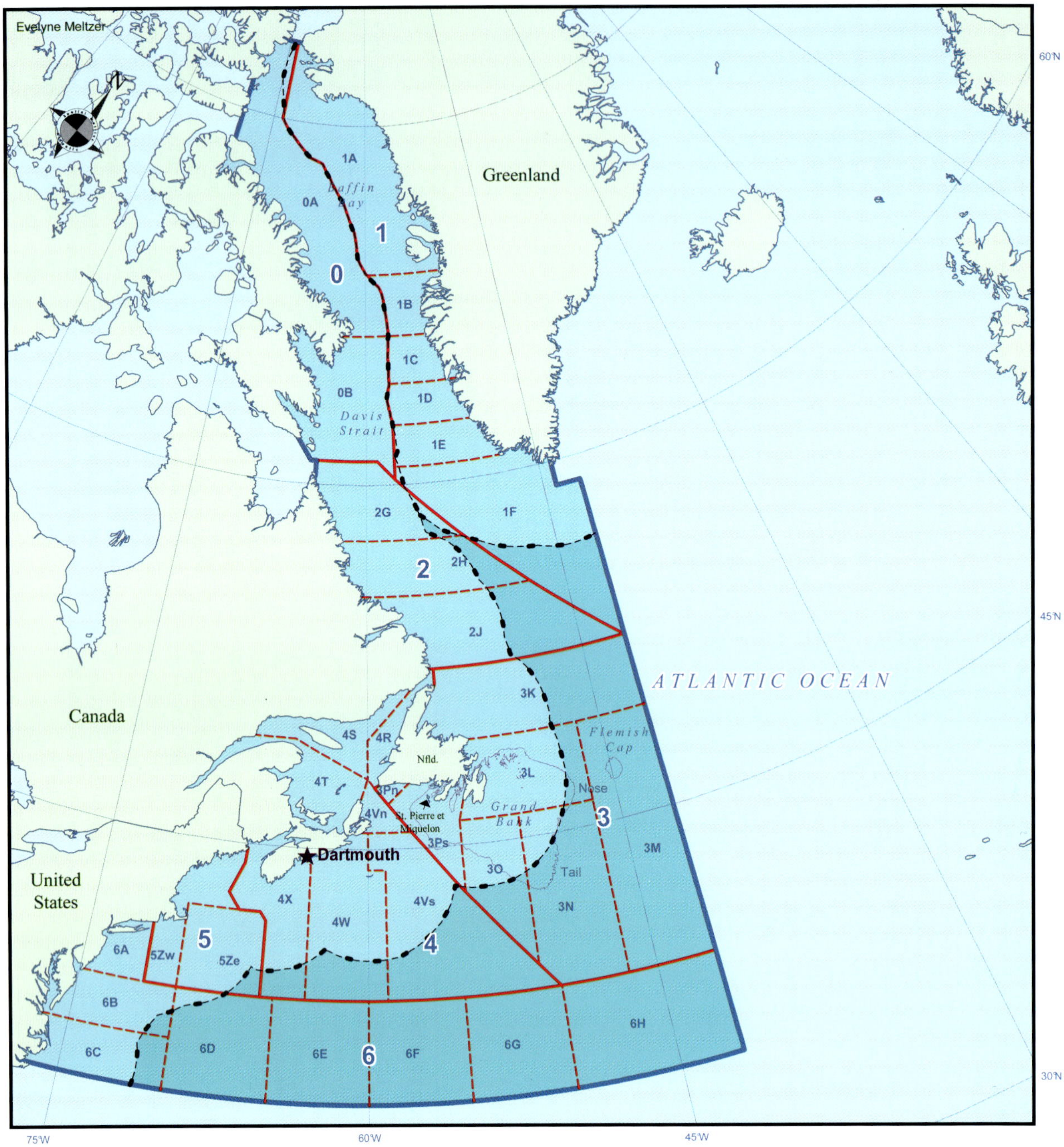

For illustration purposes only.
Map Projection : Lambert Conf. Conic
Standard Parallels : 49°N, 77°N
Central Meridian : 60°W

Evelyne Meltzer. 05/2006

Northwest Atlantic Fisheries Organization (NAFO)

★ Headquarters: Dartmouth, Nova Scotia, Canada

- RFMO Boundary NAFO Convention Area
- NAFO Regulatory Area
- 200 mile limit
- NAFO Scientific and Statistical Subareas
- NAFO Scientific and Statistical Divisions

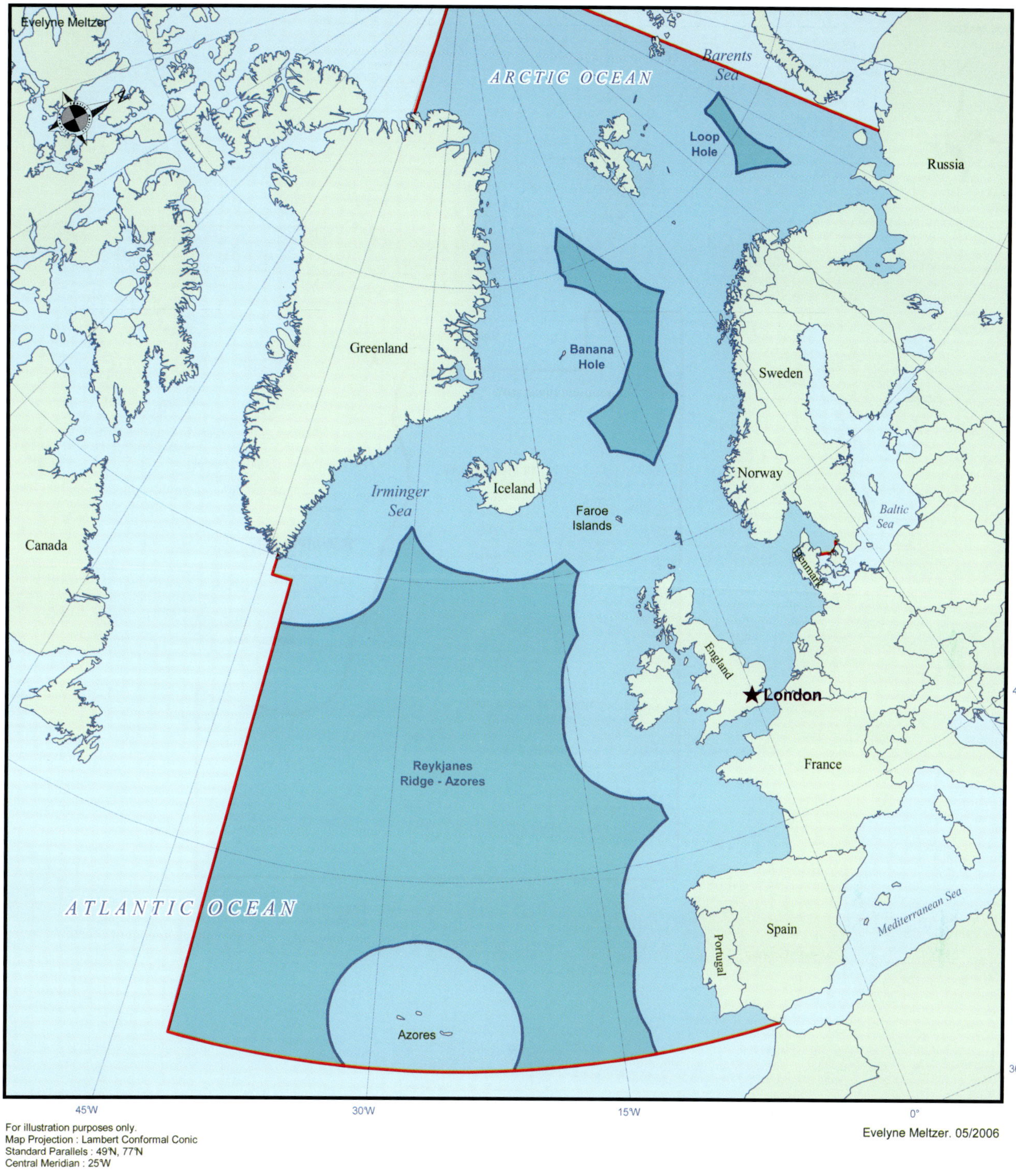

North East Atlantic Fisheries Commission (NEAFC)

RFMO Boundary

NEAFC Convention Area

NEAFC Regulatory Area

★ Headquarters: London, England

RFMO Boundary

South East Atlantic Fisheries Organisation (SEAFO)

★ Headquarters: Walvis Bay, Namibia

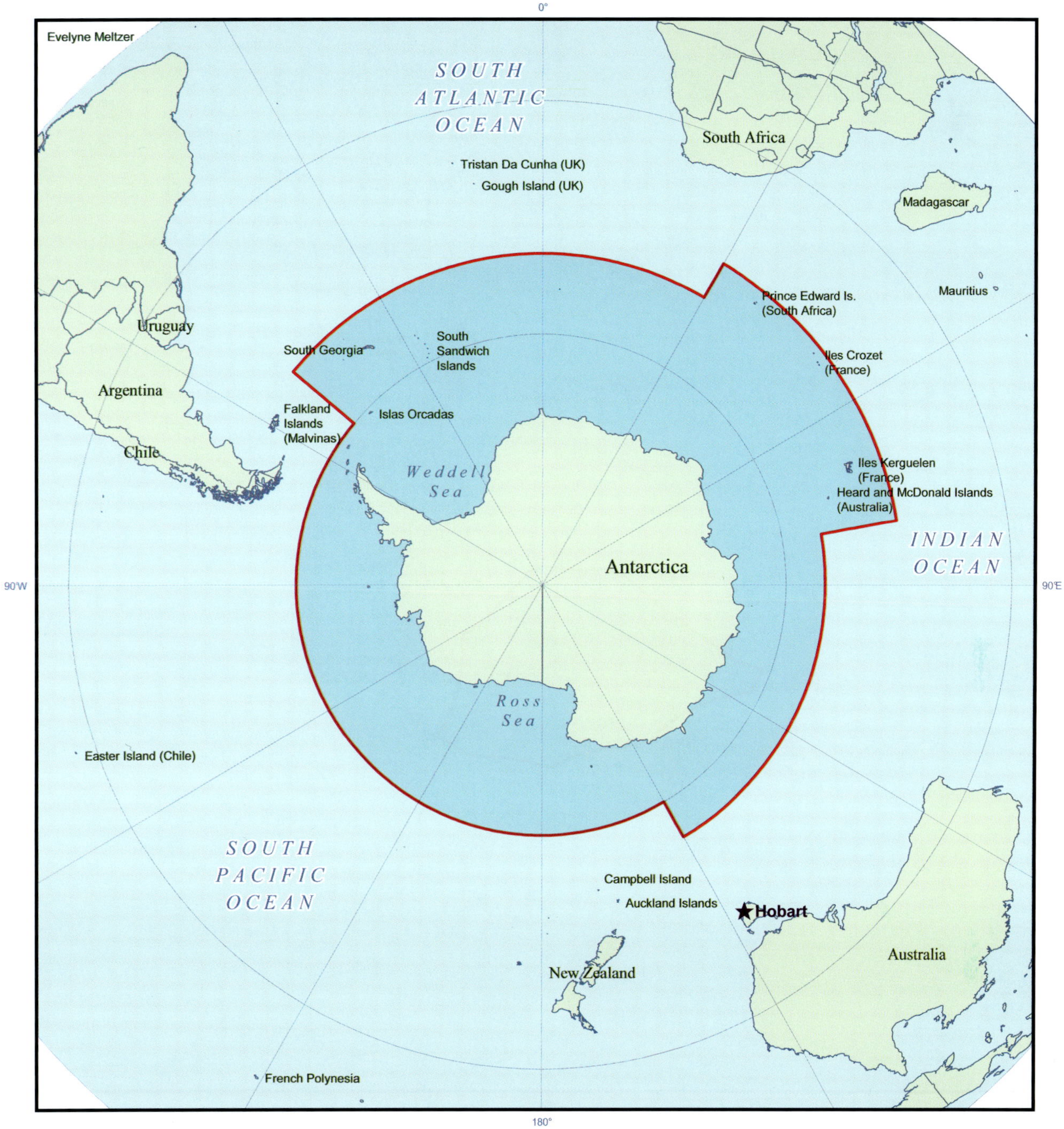

For illustration purposes only.
Map Projection : Lambert Azimuthal

Evelyne Meltzer. 05/2006

Commission for the Conservation of Antarctic Marine Living Resources (CCAMLR)

— RFMO Boundary

★ Headquarters: Hobart, Tasmania, Australia

The "Donut Hole" in the Central Bering Sea

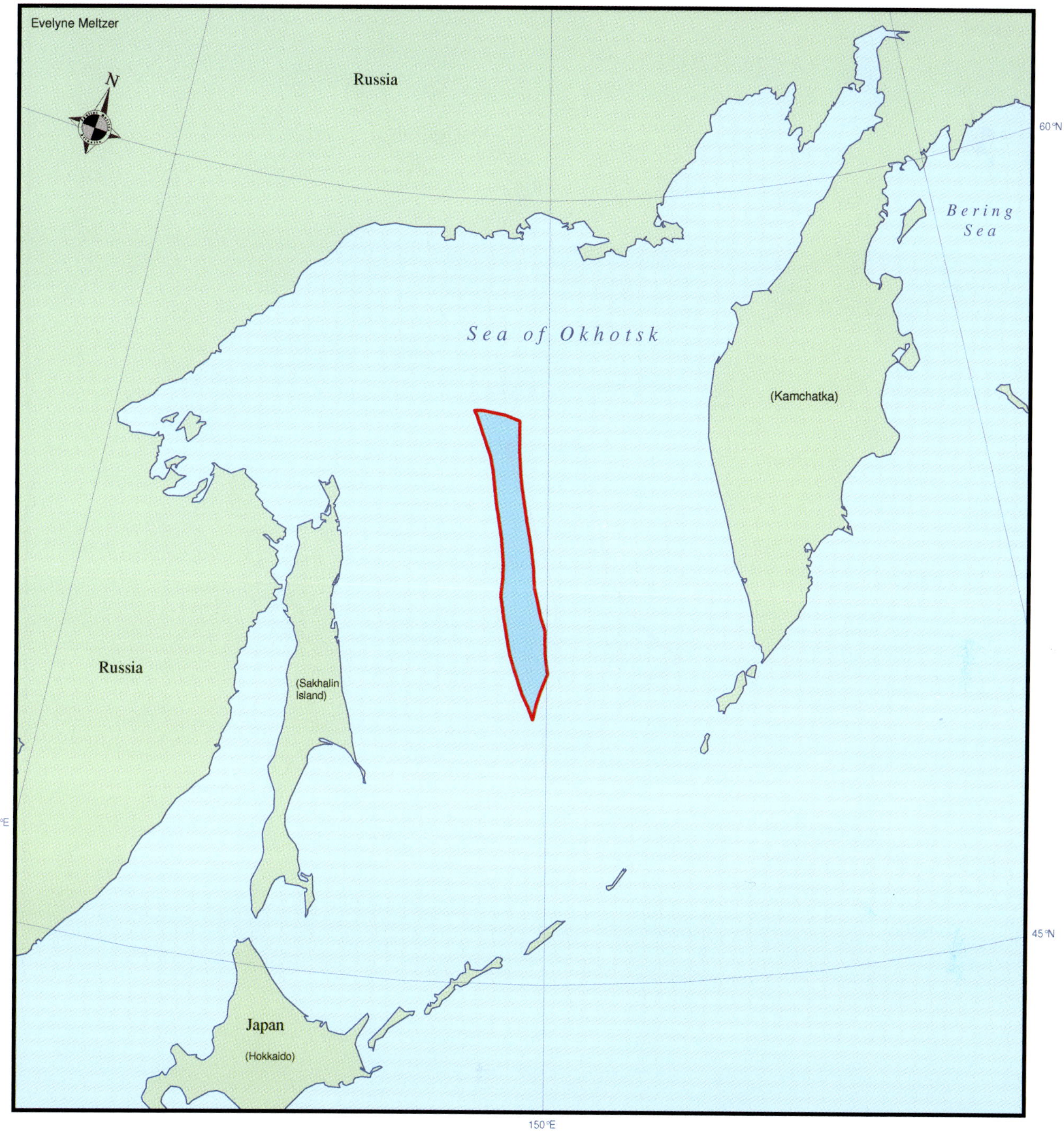

For illustration purposes only.
Map Proj. : Lambert Conformal Conic
Standard Parallels : 49°N, 77°N
Central Meridian : 150°E

Global Overview of Straddling and Highly Migratory Fish Stocks. Evelyne Meltzer: working copy 04/2005

The "Peanut Hole" in the Sea of Okhotsk

— Boundary

Area managed by Russia as UNFSA Article 16 Enclave:
currently a moratorium on fishing in Area

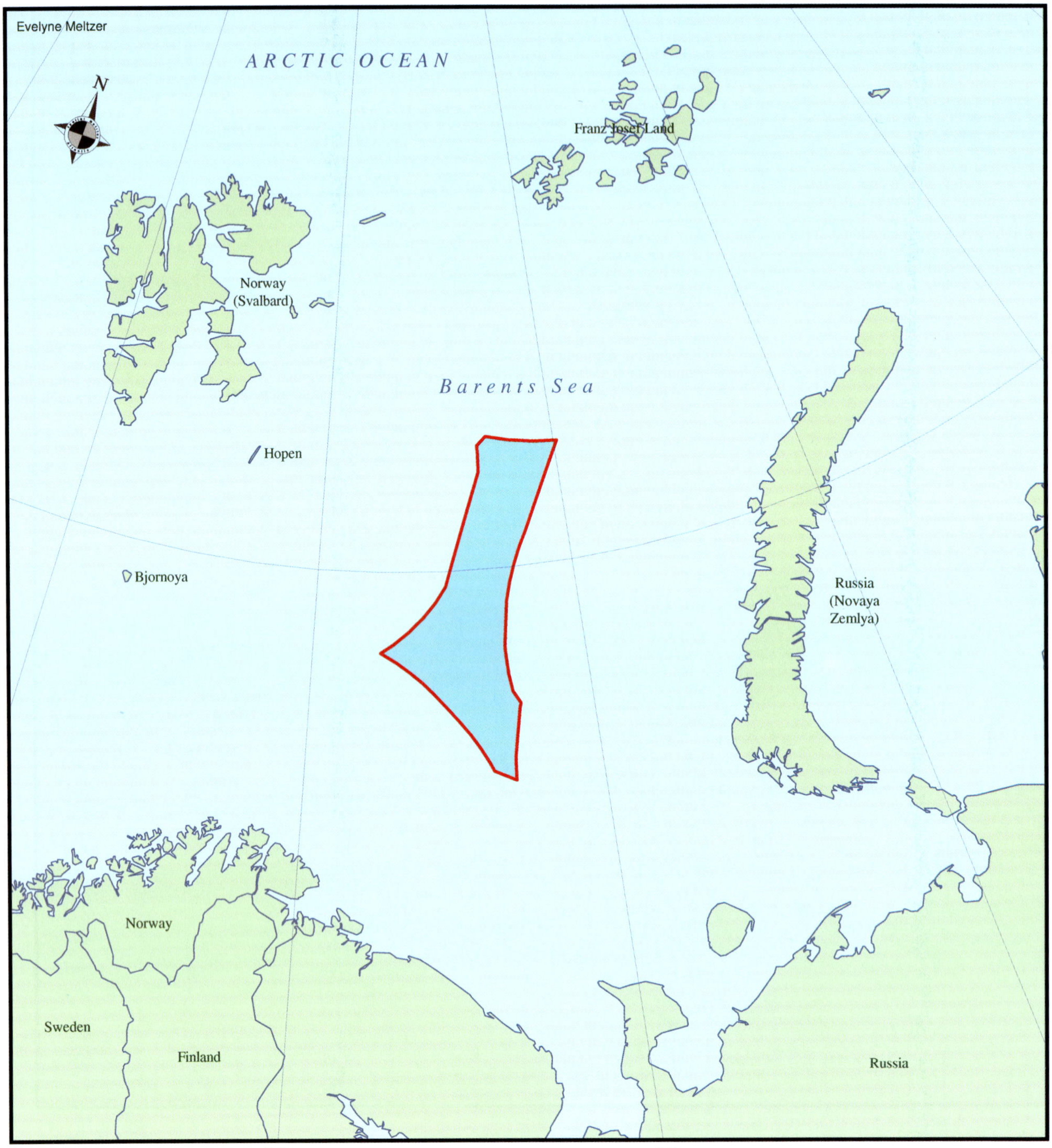

For illustration purposes only.
Map Proj. : Lambert Conformal Conic
Standard Parallels : 49°N, 77°N
Central Meridian : 35°W

Global Overview of Straddling and Highly Migratory Fish Stocks. Evelyne Meltzer: working copy 04/2005

The "Loop Hole"

Boundary

South Tasman Rise Orange Roughy Arrangement

— RFMA boundary

Southern Indian Ocean Fisheries Arrangement (proposed) (SIOFA)

— RFMA boundary

For illustration purposes only.
Map Projection : Mercator

Global Overview of Straddling and Highly Migratory Fish Stocks. Evelyne Meltzer: working copy 04/2005

— RFMO boundary

International Commission for the Conservation of Atlantic Tunas (ICCAT)

★ Headquarters: Madrid, Spain

General Fisheries Commission for the Mediterranean (GFCM)

(FAO Article XIV)

★ Headquarters: Rome, Italy

Inter-American Tropical Tuna Commission (IATTC)

★ Headquarters: La Jolla, California, US

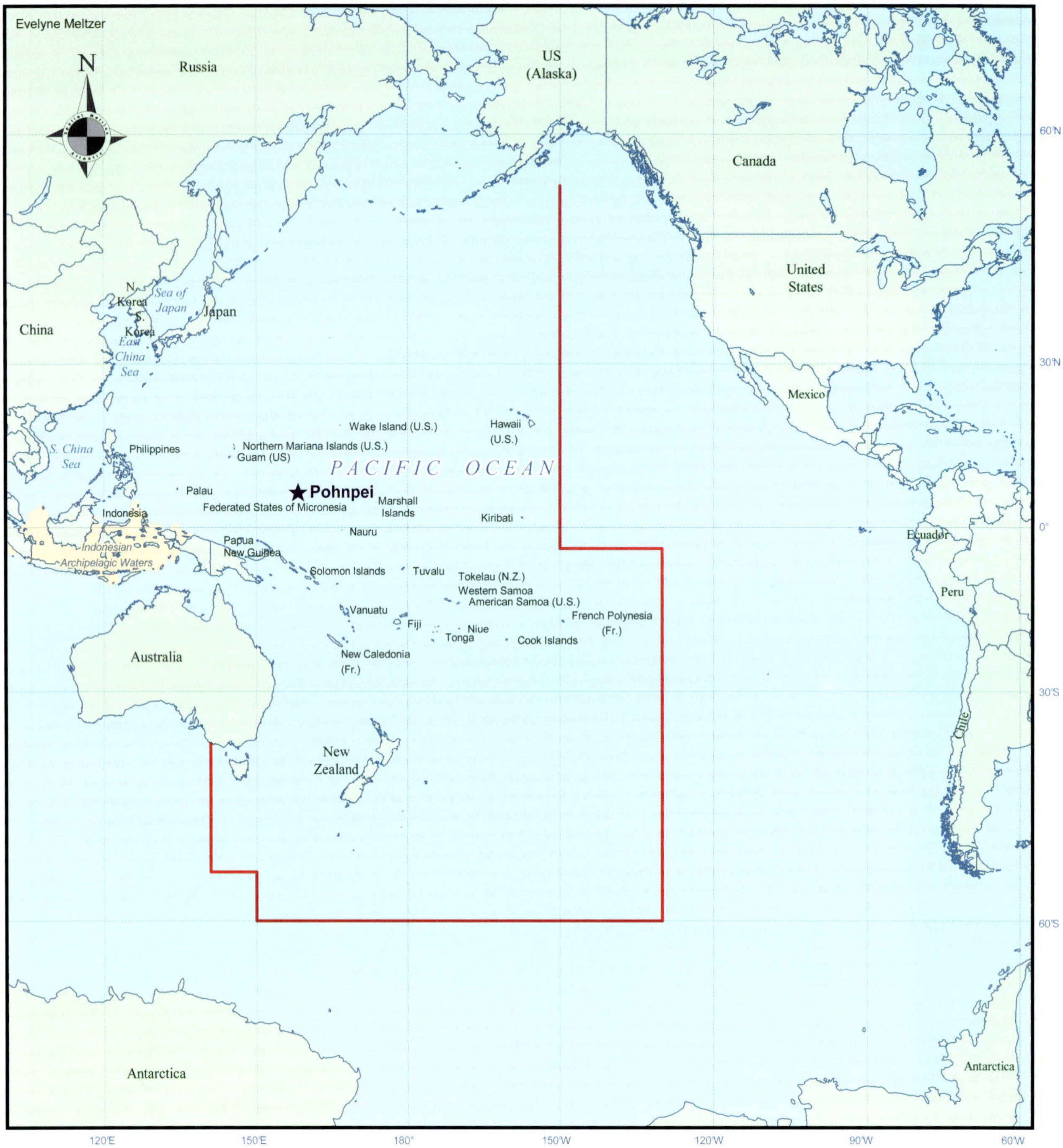

For illustration purposes only.
Map Projection : Mercator

Evelyne Meltzer. 05/2006

RFMO Boundary

Western and Central Pacific Fisheries Commission (WCPFC)

★ Headquarters: Pohnpei, Federated States of Micronesia

Note : Northern boundary and most of Western boundary for RFMO are not defined. WCPFC area includes all waters of the Pacific Ocean bounded by the WCPFC limits indicated in the South-west, South and East, and is not intended to include waters in South-East Asia which are not part of the Pacific Ocean; nor is it intended to include waters of the South China Sea.

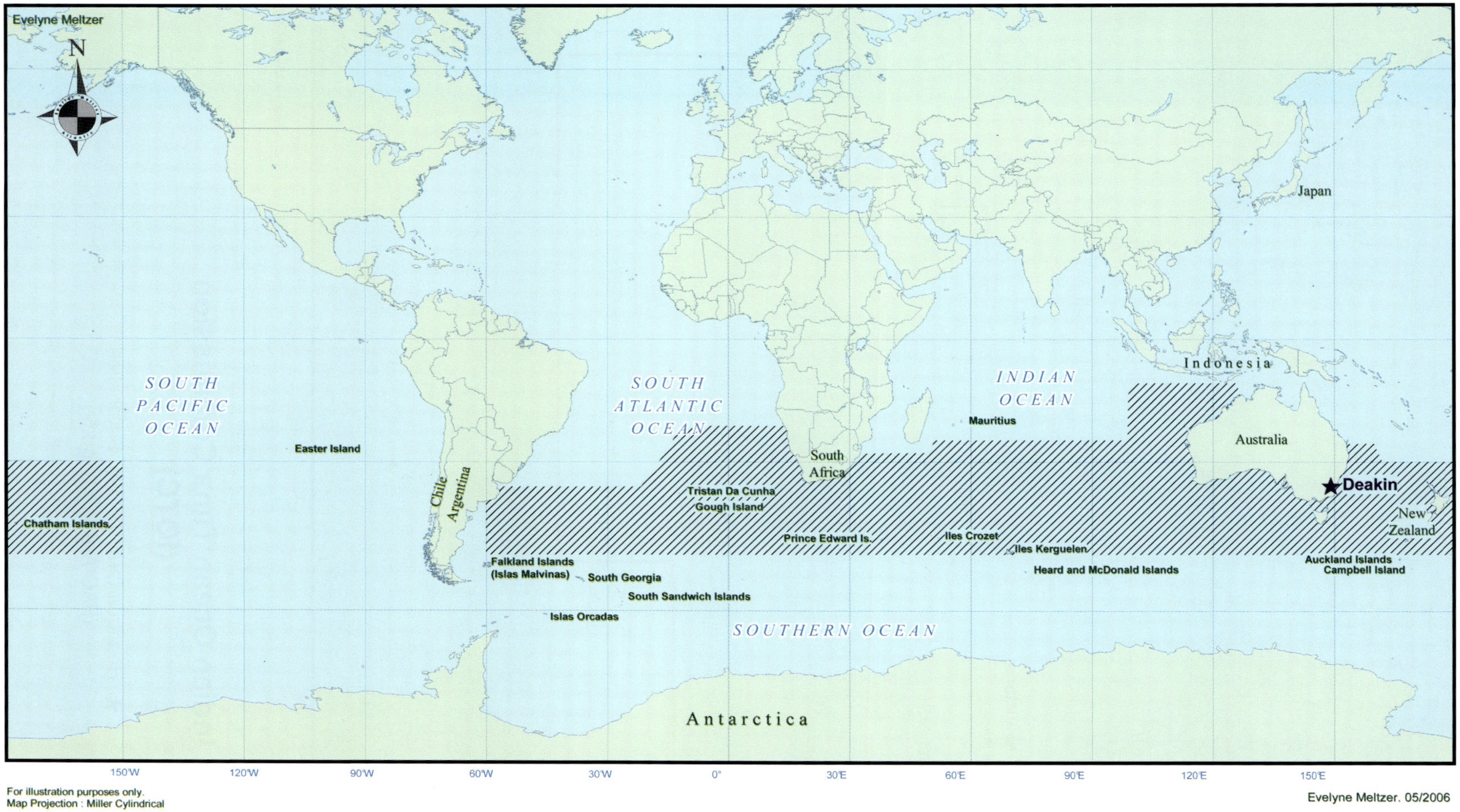

RFMO Area of Competence is throughout SBT geographical range

Approximate distribution of SBT

Commission for Conservation of Southern Bluefin Tuna (CCSBT)

★ Headquarters: Deakin (Canberra), Australia

Indian Ocean Tuna Commission (IOTC)

★ Headquarters: Victoria, Seychelles
(FAO Article XIV body)

Appendix I – UNFSA

UNITED NATIONS

A

General Assembly

Distr.
GENERAL

A/CONF.164/37
8 September 1995

ORIGINAL: ENGLISH

UNITED NATIONS CONFERENCE ON
STRADDLING FISH STOCKS AND
HIGHLY MIGRATORY FISH STOCKS
Sixth session
New York, 24 July-4 August 1995

AGREEMENT FOR THE IMPLEMENTATION OF THE PROVISIONS OF THE UNITED NATIONS CONVENTION ON THE LAW OF THE SEA OF 10 DECEMBER 1982 RELATING TO THE CONSERVATION AND MANAGEMENT OF STRADDLING FISH STOCKS AND HIGHLY MIGRATORY FISH STOCKS

AGREEMENT FOR THE IMPLEMENTATION OF THE PROVISIONS OF THE UNITED NATIONS CONVENTION ON THE LAW OF THE SEA OF 10 DECEMBER 1982 RELATING TO THE CONSERVATION AND MANAGEMENT OF STRADDLING FISH STOCKS AND HIGHLY MIGRATORY FISH STOCKS

The States Parties to this Agreement,

Recalling the relevant provisions of the United Nations Convention on the Law of the Sea of 10 December 1982,

Determined to ensure the long-term conservation and sustainable use of straddling fish stocks and highly migratory fish stocks,

Resolved to improve cooperation between States to that end,

Calling for more effective enforcement by flag States, port States and coastal States of the conservation and management measures adopted for such stocks,

Seeking to address in particular the problems identified in chapter 17, programme area C, of Agenda 21 adopted by the United Nations Conference on Environment and Development, namely, that the management of high seas fisheries is inadequate in many areas and that some resources are overutilized; noting that there are problems of unregulated fishing, over-capitalization, excessive fleet size, vessel reflagging to escape controls, insufficiently selective gear, unreliable databases and lack of sufficient cooperation between States,

Committing themselves to responsible fisheries,

Conscious of the need to avoid adverse impacts on the marine environment, preserve biodiversity, maintain the integrity of marine ecosystems and minimize the risk of long-term or irreversible effects of fishing operations,

Recognizing the need for specific assistance, including financial, scientific and technological assistance, in order that developing States can participate effectively in the conservation, management and sustainable use of straddling fish stocks and highly migratory fish stocks,

Convinced that an agreement for the implementation of the relevant provisions of the Convention would best serve these purposes and contribute to the maintenance of international peace and security,

Affirming that matters not regulated by the Convention or by this Agreement continue to be governed by the rules and principles of general international law,

Have agreed as follows:

PART I

GENERAL PROVISIONS

Article 1

Use of terms and scope

1. For the purposes of this Agreement:

(a) "Convention" means the United Nations Convention on the Law of the Sea of 10 December 1982;

(b) "conservation and management measures" means measures to conserve and manage one or more species of living marine resources that are adopted and applied consistent with the relevant rules of international law as reflected in the Convention and this Agreement;

(c) "fish" includes molluscs and crustaceans except those belonging to sedentary species as defined in article 77 of the Convention; and

(d) "arrangement" means a cooperative mechanism established in accordance with the Convention and this Agreement by two or more States for the purpose, *inter alia*, of establishing conservation and management measures in a subregion or region for one or more straddling fish stocks or highly migratory fish stocks.

2. (a) "States Parties" means States which have consented to be bound by this Agreement and for which the Agreement is in force.

(b) This Agreement applies *mutatis mutandis*:

(i) to any entity referred to in article 305, paragraph 1 (c), (d) and (e), of the Convention and

(ii) subject to article 47, to any entity referred to as an "international organization" in Annex IX, article 1, of the Convention

which becomes a Party to this Agreement, and to that extent "States Parties" refers to those entities.

3. This Agreement applies *mutatis mutandis* to other fishing entities whose vessels fish on the high seas.

Article 2

Objective

The objective of this Agreement is to ensure the long-term conservation and sustainable use of straddling fish stocks and highly migratory fish stocks through effective implementation of the relevant provisions of the Convention.

Article 3

Application

1. Unless otherwise provided, this Agreement applies to the conservation and management of straddling fish stocks and highly migratory fish stocks beyond areas under national jurisdiction, except that articles 6 and 7 apply also to the conservation and management of such stocks within areas under national jurisdiction, subject to the different legal regimes that apply within areas under national jurisdiction and in areas beyond national jurisdiction as provided for in the Convention.

2. In the exercise of its sovereign rights for the purpose of exploring and exploiting, conserving and managing straddling fish stocks and highly migratory fish stocks within areas under national jurisdiction, the coastal State shall apply *mutatis mutandis* the general principles enumerated in article 5.

3. States shall give due consideration to the respective capacities of developing States to apply articles 5, 6 and 7 within areas under national jurisdiction and their need for assistance as provided for in this Agreement. To this end, Part VII applies *mutatis mutandis* in respect of areas under national jurisdiction.

Article 4

Relationship between this Agreement and the Convention

Nothing in this Agreement shall prejudice the rights, jurisdiction and duties of States under the Convention. This Agreement shall be interpreted and applied in the context of and in a manner consistent with the Convention.

PART II

CONSERVATION AND MANAGEMENT OF STRADDLING FISH STOCKS AND HIGHLY MIGRATORY FISH STOCKS

Article 5

General principles

In order to conserve and manage straddling fish stocks and highly migratory fish stocks, coastal States and States fishing on the high seas shall, in giving effect to their duty to cooperate in accordance with the Convention:

(a) adopt measures to ensure long-term sustainability of straddling fish stocks and highly migratory fish stocks and promote the objective of their optimum utilization;

(b) ensure that such measures are based on the best scientific evidence available and are designed to maintain or restore stocks at levels capable of producing maximum sustainable yield, as qualified by relevant environmental and economic factors, including the special requirements of developing States, and taking into account fishing patterns, the interdependence of stocks and any generally recommended international minimum standards, whether subregional, regional or global;

(c) apply the precautionary approach in accordance with article 6;

(d) assess the impacts of fishing, other human activities and environmental factors on target stocks and species belonging to the same ecosystem or associated with or dependent upon the target stocks;

(e) adopt, where necessary, conservation and management measures for species belonging to the same ecosystem or associated with or dependent upon the target stocks, with a view to maintaining or restoring populations of such species above levels at which their reproduction may become seriously threatened;

(f) minimize pollution, waste, discards, catch by lost or abandoned gear, catch of non-target species, both fish and non-fish species, (hereinafter referred to as non-target species) and impacts on associated or dependent species, in particular endangered species, through measures including, to the extent practicable, the development and use of selective, environmentally safe and cost-effective fishing gear and techniques;

(g) protect biodiversity in the marine environment;

(h) take measures to prevent or eliminate overfishing and excess fishing capacity and to ensure that levels of fishing effort do not exceed those commensurate with the sustainable use of fishery resources;

(i) take into account the interests of artisanal and subsistence fishers;

(j) collect and share, in a timely manner, complete and accurate data concerning fishing activities on, inter alia, vessel position, catch of target and non-target species and fishing effort, as set out in Annex I, as well as information from national and international research programmes;

(k) promote and conduct scientific research and develop appropriate technologies in support of fishery conservation and management; and

(l) implement and enforce conservation and management measures through effective monitoring, control and surveillance.

Article 6

Application of the precautionary approach

1. States shall apply the precautionary approach widely to conservation, management and exploitation of straddling fish stocks and highly migratory fish stocks in order to protect the living marine resources and preserve the marine environment.

2. States shall be more cautious when information is uncertain, unreliable or inadequate. The absence of adequate scientific information shall not be used as a reason for postponing or failing to take conservation and management measures.

3. In implementing the precautionary approach, States shall:

(a) improve decision-making for fishery resource conservation and management by obtaining and sharing the best scientific information available and implementing improved techniques for dealing with risk and uncertainty;

(b) apply the guidelines set out in Annex II and determine, on the basis of the best scientific information available, stock-specific reference points and the action to be taken if they are exceeded;

(c) take into account, inter alia, uncertainties relating to the size and productivity of the stocks, reference points, stock condition in relation to such reference points, levels and distribution of fishing mortality and the impact of fishing activities on non-target and associated or dependent species, as well as existing and predicted oceanic, environmental and socio-economic conditions; and

(d) develop data collection and research programmes to assess the impact of fishing on non-target and associated or dependent species and their environment, and adopt plans which are necessary to ensure the conservation of such species and to protect habitats of special concern.

4. States shall take measures to ensure that, when reference points are approached, they will not be exceeded. In the event that they are exceeded, States shall, without delay, take the action determined under paragraph 3 (b) to restore the stocks.

5. Where the status of target stocks or non-target or associated or dependent species is of concern, States shall subject such stocks and species to enhanced monitoring in order to review their status and the efficacy of conservation and management measures. They shall revise those measures regularly in the light of new information.

6. For new or exploratory fisheries, States shall adopt as soon as possible cautious conservation and management measures, including, inter alia, catch limits and effort limits. Such measures shall remain in force until there are sufficient data to allow assessment of the impact of the fisheries on the long-term sustainability of the stocks, whereupon conservation and management measures based on that assessment shall be implemented. The latter measures shall, if appropriate, allow for the gradual development of the fisheries.

7. If a natural phenomenon has a significant adverse impact on the status of straddling fish stocks or highly migratory fish stocks, States shall adopt conservation and management measures on an emergency basis to ensure that fishing activity does not exacerbate such adverse impact. States shall also adopt such measures on an emergency basis where fishing activity presents a serious threat to the sustainability of such stocks. Measures taken on an emergency basis shall be temporary and shall be based on the best scientific evidence available.

Article 7

Compatibility of conservation and management measures

1. Without prejudice to the sovereign rights of coastal States for the purpose of exploring and exploiting, conserving and managing the living marine resources within areas under national jurisdiction as provided for in the Convention, and the right of all States for their nationals to engage in fishing on the high seas in accordance with the Convention:

(a) with respect to straddling fish stocks, the relevant coastal States and the States whose nationals fish for such stocks in the adjacent high seas area shall seek, either directly or through the appropriate mechanisms for cooperation provided for in Part III, to agree upon the measures necessary for the conservation of these stocks in the adjacent high seas area;

(b) with respect to highly migratory fish stocks, the relevant coastal States and other States whose nationals fish for such stocks in the region shall cooperate, either directly or through the appropriate mechanisms for cooperation provided for in Part III, with a view to ensuring conservation and promoting the objective of optimum utilization of such stocks throughout the region, both within and beyond the areas under national jurisdiction.

2. Conservation and management measures established for the high seas and those adopted for areas under national jurisdiction shall be compatible in order to ensure conservation and management of the straddling fish stocks and highly migratory fish stocks in their entirety. To this end, coastal States and States fishing on the high seas have a duty to cooperate for the purpose of achieving

compatible measures in respect of such stocks. In determining compatible conservation and management measures, States shall:

(a) take into account the conservation and management measures adopted and applied in accordance with article 61 of the Convention in respect of the same stocks by coastal States within areas under national jurisdiction and ensure that measures established in respect of such stocks for the high seas do not undermine the effectiveness of such measures;

(b) take into account previously agreed measures established and applied for the high seas in accordance with the Convention in respect of the same stocks by relevant coastal States and States fishing on the high seas;

(c) take into account previously agreed measures established and applied in accordance with the Convention in respect of the same stocks by a subregional or regional fisheries management organization or arrangement;

(d) take into account the biological unity and other biological characteristics of the stocks and the relationships between the distribution of the stocks, the fisheries and the geographical particularities of the region concerned, including the extent to which the stocks occur and are fished in areas under national jurisdiction;

(e) take into account the respective dependence of the coastal States and the States fishing on the high seas on the stocks concerned; and

(f) ensure that such measures do not result in harmful impact on the living marine resources as a whole.

3. In giving effect to their duty to cooperate, States shall make every effort to agree on compatible conservation and management measures within a reasonable period of time.

4. If no agreement can be reached within a reasonable period of time, any of the States concerned may invoke the procedures for the settlement of disputes provided for in Part VIII.

5. Pending agreement on compatible conservation and management measures, the States concerned, in a spirit of understanding and cooperation, shall make every effort to enter into provisional arrangements of a practical nature. In the event that they are unable to agree on such arrangements, any of the States concerned may, for the purpose of obtaining provisional measures, submit the dispute to a court or tribunal in accordance with the procedures for the settlement of disputes provided for in Part VIII.

6. Provisional arrangements or measures entered into or prescribed pursuant to paragraph 5 shall take into account the provisions of this Part, shall have due regard to the rights and obligations of all States concerned, shall not jeopardize or hamper the reaching of final agreement on compatible conservation and management measures and shall be without prejudice to the final outcome of any dispute settlement procedure.

7. Coastal States shall regularly inform States fishing on the high seas in the subregion or region, either directly or through appropriate subregional or regional fisheries management organizations or arrangements, or through other appropriate means, of the measures they have adopted for straddling fish stocks and highly migratory fish stocks within areas under their national jurisdiction.

8. States fishing on the high seas shall regularly inform other interested States, either directly or through appropriate subregional or regional fisheries management organizations or arrangements, or through other appropriate means, of the measures they have adopted for regulating the activities of vessels flying their flag which fish for such stocks on the high seas.

PART III

MECHANISMS FOR INTERNATIONAL COOPERATION CONCERNING STRADDLING FISH STOCKS AND HIGHLY MIGRATORY FISH STOCKS

Article 8

Cooperation for conservation and management

1. Coastal States and States fishing on the high seas shall, in accordance with the Convention, pursue cooperation in relation to straddling fish stocks and highly migratory fish stocks either directly or through appropriate subregional or regional fisheries management organizations or arrangements, taking into account the specific characteristics of the subregion or region, to ensure effective conservation and management of such stocks.

2. States shall enter into consultations in good faith and without delay, particularly where there is evidence that the straddling fish stocks and highly migratory fish stocks concerned may be under threat of over-exploitation or where a new fishery is being developed for such stocks. To this end, consultations may be initiated at the request of any interested State with a view to establishing appropriate arrangements to ensure conservation and management of the stocks. Pending agreement on such arrangements, States shall observe the provisions of this Agreement and shall act in good faith and with due regard to the rights, interests and duties of other States.

3. Where a subregional or regional fisheries management organization or arrangement has the competence to establish conservation and management measures for particular straddling fish stocks or highly migratory fish stocks, States fishing for the stocks on the high seas and relevant coastal States shall give effect to their duty to cooperate by becoming members of such organization or participants in such arrangement, or by agreeing to apply the conservation and management measures established by such organization or arrangement. States having a real interest in the fisheries concerned may become members of such organization or participants in such arrangement. The terms of participation in such organization or arrangement shall not preclude such States from membership or participation; nor shall they be applied in a manner which discriminates against any State or group of States having a real interest in the fisheries concerned.

4. Only those States which are members of such an organization or participants in such an arrangement, or which agree to apply the conservation and management measures established by such organization or arrangement, shall have access to the fishery resources to which those measures apply.

5. Where there is no subregional or regional fisheries management organization or arrangement to establish conservation and management measures for a particular straddling fish stock or highly migratory fish stock, relevant coastal States and States fishing on the high seas for such stock in the subregion or region shall cooperate to establish such an organization or enter into other appropriate arrangements to ensure conservation and management of such stock and shall participate in the work of the organization or arrangement.

6. Any State intending to propose that action be taken by an intergovernmental organization having competence with respect to living resources should, where such action would have a significant effect on conservation and management measures already established by a competent subregional or regional fisheries management organization or arrangement, consult through that organization or arrangement with its members or participants. To the extent practicable, such consultation should take place prior to the submission of the proposal to the intergovernmental organization.

Article 9

Subregional and regional fisheries management organizations and arrangements

1. In establishing subregional or regional fisheries management organizations or in entering into subregional or regional fisheries management arrangements for straddling fish stocks and highly migratory fish stocks, States shall agree, *inter alia*, on:

(a) the stocks to which conservation and management measures apply, taking into account the biological characteristics of the stocks concerned and the nature of the fisheries involved;

(b) the area of application, taking into account article 7, paragraph 1, and the characteristics of the subregion or region, including socio-economic, geographical and environmental factors;

(c) the relationship between the work of the new organization or arrangement and the role, objectives and operations of any relevant existing fisheries management organizations or arrangements; and

(d) the mechanisms by which the organization or arrangement will obtain scientific advice and review the status of the stocks, including, where appropriate, the establishment of a scientific advisory body.

2. States cooperating in the formation of a subregional or regional fisheries management organization or arrangement shall inform other States which they are

aware have a real interest in the work of the proposed organization or arrangement of such cooperation.

Article 10

Functions of subregional and regional fisheries management organizations and arrangements

In fulfilling their obligation to cooperate through subregional or regional fisheries management organizations or arrangements, States shall:

(a) agree on and comply with conservation and management measures to ensure the long-term sustainability of straddling fish stocks and highly migratory fish stocks;

(b) agree, as appropriate, on participatory rights such as allocations of allowable catch or levels of fishing effort;

(c) adopt and apply any generally recommended international minimum standards for the responsible conduct of fishing operations;

(d) obtain and evaluate scientific advice, review the status of the stocks and assess the impact of fishing on non-target and associated or dependent species;

(e) agree on standards for collection, reporting, verification and exchange of data on fisheries for the stocks;

(f) compile and disseminate accurate and complete statistical data, as described in Annex I, to ensure that the best scientific evidence is available, while maintaining confidentiality where appropriate;

(g) promote and conduct scientific assessments of the stocks and relevant research and disseminate the results thereof;

(h) establish appropriate cooperative mechanisms for effective monitoring, control, surveillance and enforcement;

(i) agree on means by which the fishing interests of new members of the organization or new participants in the arrangement will be accommodated;

(j) agree on decision-making procedures which facilitate the adoption of conservation and management measures in a timely and effective manner;

(k) promote the peaceful settlement of disputes in accordance with Part VIII;

(l) ensure the full cooperation of their relevant national agencies and industries in implementing the recommendations and decisions of the organization or arrangement; and

(m) give due publicity to the conservation and management measures established by the organization or arrangement.

Article 11

New members or participants

In determining the nature and extent of participatory rights for new members of a subregional or regional fisheries management organization, or for new participants in a subregional or regional fisheries management arrangement, States shall take into account, *inter alia*:

(a) the status of the straddling fish stocks and highly migratory fish stocks and the existing level of fishing effort in the fishery;

(b) the respective interests, fishing patterns and fishing practices of new and existing members or participants;

(c) the respective contributions of new and existing members or participants to conservation and management of the stocks, to the collection and provision of accurate data and to the conduct of scientific research on the stocks;

(d) the needs of coastal fishing communities which are dependent mainly on fishing for the stocks;

(e) the needs of coastal States whose economies are overwhelmingly dependent on the exploitation of living marine resources; and

(f) the interests of developing States from the subregion or region in whose areas of national jurisdiction the stocks also occur.

Article 12

Transparency in activities of subregional and regional fisheries management organizations and arrangements

1. States shall provide for transparency in the decision-making process and other activities of subregional and regional fisheries management organizations and arrangements.

2. Representatives from other intergovernmental organizations and representatives from non-governmental organizations concerned with straddling fish stocks and highly migratory fish stocks shall be afforded the opportunity to take part in meetings of subregional and regional fisheries management organizations and arrangements as observers or otherwise, as appropriate, in accordance with the procedures of the organization or arrangement concerned. Such procedures shall not be unduly restrictive in this respect. Such intergovernmental organizations and non-governmental organizations shall have

timely access to the records and reports of such organizations and arrangements, subject to the procedural rules on access to them.

Article 13

Strengthening of existing organizations and arrangements

States shall cooperate to strengthen existing subregional and regional fisheries management organizations and arrangements in order to improve their effectiveness in establishing and implementing conservation and management measures for straddling fish stocks and highly migratory fish stocks.

Article 14

Collection and provision of information and cooperation in scientific research

1. States shall ensure that fishing vessels flying their flag provide such information as may be necessary in order to fulfil their obligations under this Agreement. To this end, States shall in accordance with Annex I:

(a) collect and exchange scientific, technical and statistical data with respect to fisheries for straddling fish stocks and highly migratory fish stocks;

(b) ensure that data are collected in sufficient detail to facilitate effective stock assessment and are provided in a timely manner to fulfil the requirements of subregional or regional fisheries management organizations or arrangements; and

(c) take appropriate measures to verify the accuracy of such data.

2. States shall cooperate, either directly or through subregional or regional fisheries management organizations or arrangements:

(a) to agree on the specification of data and the format in which they are to be provided to such organizations or arrangements, taking into account the nature of the stocks and the fisheries for those stocks; and

(b) to develop and share analytical techniques and stock assessment methodologies to improve measures for the conservation and management of straddling fish stocks and highly migratory fish stocks.

3. Consistent with Part XIII of the Convention, States shall cooperate, either directly or through competent international organizations, to strengthen scientific research capacity in the field of fisheries and promote scientific research related to the conservation and management of straddling fish stocks and highly migratory fish stocks for the benefit of all. To this end, a State or the competent international organization conducting such research beyond areas under national jurisdiction shall actively promote the publication and

dissemination to any interested States of the results of that research and information relating to its objectives and methods and, to the extent practicable, shall facilitate the participation of scientists from those States in such research.

Article 15

Enclosed and semi-enclosed seas

In implementing this Agreement in an enclosed or semi-enclosed sea, States shall take into account the natural characteristics of that sea and shall also act in a manner consistent with Part IX of the Convention and other relevant provisions thereof.

Article 16

Areas of high seas surrounded entirely by an area under the national jurisdiction of a single State

1. States fishing for straddling fish stocks and highly migratory fish stocks in an area of the high seas surrounded entirely by an area under the national jurisdiction of a single State and the latter State shall cooperate to establish conservation and management measures in respect of those stocks in the high seas area. Having regard to the natural characteristics of the area, States shall pay special attention to the establishment of compatible conservation and management measures for such stocks pursuant to article 7. Measures taken in respect of the high seas shall take into account the rights, duties and interests of the coastal State under the Convention, shall be based on the best scientific evidence available and shall also take into account any conservation and management measures adopted and applied in respect of the same stocks in accordance with article 61 of the Convention by the coastal State in the area under national jurisdiction. States shall also agree on measures for monitoring, control, surveillance and enforcement to ensure compliance with the conservation and management measures in respect of the high seas.

2. Pursuant to article 8, States shall act in good faith and make every effort to agree without delay on conservation and management measures to be applied in the carrying out of fishing operations in the area referred to in paragraph 1. If, within a reasonable period of time, the fishing States concerned and the coastal State are unable to agree on such measures, they shall, having regard to paragraph 1, apply article 7, paragraphs 4, 5 and 6, relating to provisional arrangements or measures. Pending the establishment of such provisional arrangements or measures, the States concerned shall take measures in respect of vessels flying their flag in order that they not engage in fisheries which could undermine the stocks concerned.

PART IV

NON-MEMBERS AND NON-PARTICIPANTS

Article 17

Non-members of organizations and non-participants in arrangements

1. A State which is not a member of a subregional or regional fisheries management organization or is not a participant in a subregional or regional fisheries management arrangement, and which does not otherwise agree to apply the conservation and management measures established by such organization or arrangement, is not discharged from the obligation to cooperate, in accordance with the Convention and this Agreement, in the conservation and management of the relevant straddling fish stocks and highly migratory fish stocks.

2. Such State shall not authorize vessels flying its flag to engage in fishing operations for the straddling fish stocks or highly migratory fish stocks which are subject to the conservation and management measures established by such organization or arrangement.

3. States which are members of a subregional or regional fisheries management organization or participants in a subregional or regional fisheries management arrangement shall, individually or jointly, request the fishing entities referred to in article 1, paragraph 3, which have fishing vessels in the relevant area to cooperate fully with such organization or arrangement in implementing the conservation and management measures it has established, with a view to having such measures applied de facto as extensively as possible to fishing activities in the relevant area. Such fishing entities shall enjoy benefits from participation in the fishery commensurate with their commitment to comply with conservation and management measures in respect of the stocks.

4. States which are members of such organization or participants in such arrangement shall exchange information with respect to the activities of fishing vessels flying the flags of States which are neither members of the organization nor participants in the arrangement and which are engaged in fishing operations for the relevant stocks. They shall take measures consistent with this Agreement and international law to deter activities of such vessels which undermine the effectiveness of subregional or regional conservation and management measures.

PART V

DUTIES OF THE FLAG STATE

Article 18

Duties of the flag State

1. A State whose vessels fish on the high seas shall take such measures as may be necessary to ensure that vessels flying its flag comply with subregional and regional conservation and management measures and that such vessels do not engage in any activity which undermines the effectiveness of such measures.

2. A State shall authorize the use of vessels flying its flag for fishing on the high seas only where it is able to exercise effectively its responsibilities in respect of such vessels under the Convention and this Agreement.

3. Measures to be taken by a State in respect of vessels flying its flag shall include:

(a) control of such vessels on the high seas by means of fishing licences, authorizations or permits, in accordance with any applicable procedures agreed at the subregional, regional or global level;

(b) establishment of regulations:

(i) to apply terms and conditions to the licence, authorization or permit sufficient to fulfil any subregional, regional or global obligations of the flag State;

(ii) to prohibit fishing on the high seas by vessels which are not duly licensed or authorized to fish, or fishing on the high seas by vessels otherwise than in accordance with the terms and conditions of a licence, authorization or permit;

(iii) to require vessels fishing on the high seas to carry the licence, authorization or permit on board at all times and to produce it on demand for inspection by a duly authorized person; and

(iv) to ensure that vessels flying its flag do not conduct unauthorized fishing within areas under the national jurisdiction of other States;

(c) establishment of a national record of fishing vessels authorized to fish on the high seas and provision of access to the information contained in that record on request by directly interested States, taking into account any national laws of the flag State regarding the release of such information;

(d) requirements for marking of fishing vessels and fishing gear for identification in accordance with uniform and internationally recognizable vessel and gear marking systems, such as the Food and Agriculture Organization of the United Nations Standard Specifications for the Marking and Identification of Fishing Vessels;

(e) requirements for recording and timely reporting of vessel position, catch of target and non-target species, fishing effort and other relevant fisheries data in accordance with subregional, regional and global standards for collection of such data;

(f) requirements for verifying the catch of target and non-target species through such means as observer programmes, inspection schemes, unloading reports, supervision of transshipment and monitoring of landed catches and market statistics;

(g) monitoring, control and surveillance of such vessels, their fishing operations and related activities by, inter alia:

(i) the implementation of national inspection schemes and subregional and regional schemes for cooperation in enforcement pursuant to articles 21 and 22, including requirements for such vessels to permit access by duly authorized inspectors from other States;

(ii) the implementation of national observer programmes and subregional and regional observer programmes in which the flag State is a participant, including requirements for such vessels to permit access by observers from other States to carry out the functions agreed under the programmes; and

(iii) the development and implementation of vessel monitoring systems, including, as appropriate, satellite transmitter systems, in accordance with any national programmes and those which have been subregionally, regionally or globally agreed among the States concerned;

(h) regulation of transshipment on the high seas to ensure that the effectiveness of conservation and management measures is not undermined; and

(i) regulation of fishing activities to ensure compliance with subregional, regional or global measures, including those aimed at minimizing catches of non-target species.

4. Where there is a subregionally, regionally or globally agreed system of monitoring, control and surveillance in effect, States shall ensure that the measures they impose on vessels flying their flag are compatible with that system.

PART VI

COMPLIANCE AND ENFORCEMENT

Article 19

Compliance and enforcement by the flag State

1. A State shall ensure compliance by vessels flying its flag with subregional and regional conservation and management measures for straddling fish stocks and highly migratory fish stocks. To this end, that State shall:

(a) enforce such measures irrespective of where violations occur;

(b) investigate immediately and fully any alleged violation of subregional or regional conservation and management measures, which may include the physical inspection of the vessels concerned, and report promptly to the State alleging the violation and the relevant subregional or regional organization or arrangement on the progress and outcome of the investigation;

(c) require any vessel flying its flag to give information to the investigating authority regarding vessel position, catches, fishing gear, fishing operations and related activities in the area of an alleged violation;

(d) if satisfied that sufficient evidence is available in respect of an alleged violation, refer the case to its authorities with a view to instituting proceedings without delay in accordance with its laws and, where appropriate, detain the vessel concerned; and

(e) ensure that, where it has been established, in accordance with its laws, a vessel has been involved in the commission of a serious violation of such measures, the vessel does not engage in fishing operations on the high seas until such time as all outstanding sanctions imposed by the flag State in respect of the violation have been complied with.

2. All investigations and judicial proceedings shall be carried out expeditiously. Sanctions applicable in respect of violations shall be adequate in severity to be effective in securing compliance and to discourage violations wherever they occur and shall deprive offenders of the benefits accruing from their illegal activities. Measures applicable in respect of masters and other officers of fishing vessels shall include provisions which may permit, *inter alia*, refusal, withdrawal or suspension of authorizations to serve as masters or officers on such vessels.

Article 20

International cooperation in enforcement

1. States shall cooperate, either directly or through subregional or regional fisheries management organizations or arrangements, to ensure compliance with and enforcement of subregional and regional conservation and management measures for straddling fish stocks and highly migratory fish stocks.

2. A flag State conducting an investigation of an alleged violation of conservation and management measures for straddling fish stocks or highly migratory fish stocks may request the assistance of any other State whose cooperation may be useful in the conduct of that investigation. All States shall endeavour to meet reasonable requests made by a flag State in connection with such investigations.

3. A flag State may undertake such investigations directly, in cooperation with other interested States or through the relevant subregional or regional fisheries management organization or arrangement. Information on the progress and outcome of the investigations shall be provided to all States having an interest in, or affected by, the alleged violation.

4. States shall assist each other in identifying vessels reported to have engaged in activities undermining the effectiveness of subregional, regional or global conservation and management measures.

5. States shall, to the extent permitted by national laws and regulations, establish arrangements for making available to prosecuting authorities in other States evidence relating to alleged violations of such measures.

6. Where there are reasonable grounds for believing that a vessel on the high seas has been engaged in unauthorized fishing within an area under the jurisdiction of a coastal State, the flag State of that vessel, at the request of the coastal State concerned, shall immediately and fully investigate the matter. The flag State shall cooperate with the coastal State in taking appropriate enforcement action in such cases and may authorize the relevant authorities of the coastal State to board and inspect the vessel on the high seas. This paragraph is without prejudice to article 111 of the Convention.

7. States Parties which are members of a subregional or regional fisheries management organization or participants in a subregional or regional fisheries management arrangement may take action in accordance with international law, including through recourse to subregional or regional procedures established for this purpose, to deter vessels which have engaged in activities which undermine the effectiveness of or otherwise violate the conservation and management measures established by that organization or arrangement from fishing on the high seas in the subregion or region until such time as appropriate action is taken by the flag State.

Article 21

Subregional and regional cooperation in enforcement

1. In any high seas area covered by a subregional or regional fisheries management organization or arrangement, a State Party which is a member of such organization or a participant in such arrangement may, through its duly authorized inspectors, board and inspect, in accordance with paragraph 2, fishing vessels flying the flag of another State Party to this Agreement, whether or not such State Party is also a member of the organization or a participant in the arrangement, for the purpose of ensuring compliance with conservation and management measures for straddling fish stocks and highly migratory fish stocks established by that organization or arrangement.

2. States shall establish, through subregional or regional fisheries management organizations or arrangements, procedures for boarding and inspection pursuant to paragraph 1, as well as procedures to implement other provisions of this article. Such procedures shall be consistent with this article and the basic procedures set out in article 22 and shall not discriminate against non-members of the organization or non-participants in the arrangement. Boarding and inspection as well as any subsequent enforcement action shall be conducted in accordance with such procedures. States shall give due publicity to procedures established pursuant to this paragraph.

3. If, within two years of the adoption of this Agreement, any organization or arrangement has not established such procedures, boarding and inspection pursuant to paragraph 1, as well as any subsequent enforcement action, shall, pending the establishment of such procedures, be conducted in accordance with this article and the basic procedures set out in article 22.

4. Prior to taking action under this article, inspecting States shall, either directly or through the relevant subregional or regional fisheries management organization or arrangement, inform all States whose vessels fish on the high seas in the subregion or region of the form of identification issued to their duly authorized inspectors. The vessels used for boarding and inspection shall be clearly marked and identifiable as being on government service. At the time of becoming a Party to this Agreement, a State shall designate an appropriate authority to receive notifications pursuant to this article and shall give due publicity of such designation through the relevant subregional or regional fisheries management organization or arrangement.

5. Where, following a boarding and inspection, there are clear grounds for believing that a vessel has engaged in any activity contrary to the conservation and management measures referred to in paragraph 1, the inspecting State shall, where appropriate, secure evidence and shall promptly notify the flag State of the alleged violation.

6. The flag State shall respond to the notification referred to in paragraph 5 within three working days of its receipt, or such other period as may be prescribed in procedures established in accordance with paragraph 2, and shall either:

(a) fulfil, without delay, its obligations under article 19 to investigate and, if evidence so warrants, take enforcement action with respect to the vessel, in which case it shall promptly inform the inspecting State of the results of the investigation and of any enforcement action taken; or

(b) authorize the inspecting State to investigate.

7. Where the flag State authorizes the inspecting State to investigate an alleged violation, the inspecting State shall, without delay, communicate the results of that investigation to the flag State. The flag State shall, if evidence so warrants, fulfil its obligations to take enforcement action with respect to the vessel. Alternatively, the flag State may authorize the inspecting State to take such enforcement action as the flag State may specify with respect to the vessel, consistent with the rights and obligations of the flag State under this Agreement.

8. Where, following boarding and inspection, there are clear grounds for believing that a vessel has committed a serious violation, and the flag State has either failed to respond or failed to take action as required under paragraphs 6 or 7, the inspectors may remain on board and secure evidence and may require the master to assist in further investigation including, where appropriate, by bringing the vessel without delay to the nearest appropriate port, or to such other port as may be specified in procedures established in accordance with paragraph 2. The inspecting State shall immediately inform the flag State of the name of the port to which the vessel is to proceed. The inspecting State and the flag State and, as appropriate, the port State shall take all necessary steps to ensure the well-being of the crew regardless of their nationality.

9. The inspecting State shall inform the flag State and the relevant organization or the participants in the relevant arrangement of the results of any further investigation.

10. The inspecting State shall require its inspectors to observe generally accepted international regulations, procedures and practices relating to the safety of the vessel and the crew, minimize interference with fishing operations and, to the extent practicable, avoid action which would adversely affect the quality of the catch on board. The inspecting State shall ensure that boarding and inspection is not conducted in a manner that would constitute harassment of any fishing vessel.

11. For the purposes of this article, a serious violation means:

(a) fishing without a valid licence, authorization or permit issued by the flag State in accordance with article 18, paragraph 3 (a);

(b) failing to maintain accurate records of catch and catch-related data, as required by the relevant subregional or regional fisheries management organization or arrangement, or serious misreporting of catch, contrary to the catch reporting requirements of such organization or arrangement;

(c) fishing in a closed area, fishing during a closed season or fishing without, or after attainment of, a quota established by the relevant subregional or regional fisheries management organization or arrangement;

(d) directed fishing for a stock which is subject to a moratorium or for which fishing is prohibited;

(e) using prohibited fishing gear;

(f) falsifying or concealing the markings, identity or registration of a fishing vessel;

(g) concealing, tampering with or disposing of evidence relating to an investigation;

(h) multiple violations which together constitute a serious disregard of conservation and management measures; or

(i) such other violations as may be specified in procedures established by the relevant subregional or regional fisheries management organization or arrangement.

12. Notwithstanding the other provisions of this article, the flag State may, at any time, take action to fulfil its obligations under article 19 with respect to an alleged violation. Where the vessel is under the direction of the inspecting State, the inspecting State shall, at the request of the flag State, release the vessel to the flag State along with full information on the progress and outcome of its investigation.

13. This article is without prejudice to the right of the flag State to take any measures, including proceedings to impose penalties, according to its laws.

14. This article applies mutatis mutandis to boarding and inspection by a State Party which is a member of a subregional or regional fisheries management organization or a participant in a subregional or regional fisheries management arrangement and which has clear grounds for believing that a fishing vessel flying the flag of another State Party has engaged in any activity contrary to relevant conservation and management measures referred to in paragraph 1 in the high seas area covered by such organization or arrangement, and such vessel has subsequently, during the same fishing trip, entered into an area under the national jurisdiction of the inspecting State.

15. Where a subregional or regional fisheries management organization or arrangement has established an alternative mechanism which effectively discharges the obligation under this Agreement of its members or participants to ensure compliance with the conservation and management measures established by the organization or arrangement, members of such organization or participants in such arrangement may agree to limit the application of paragraph 1 as between themselves in respect of the conservation and management measures which have been established in the relevant high seas area.

16. Action taken by States other than the flag State in respect of vessels having engaged in activities contrary to subregional or regional conservation and management measures shall be proportionate to the seriousness of the violation.

17. Where there are reasonable grounds for suspecting that a fishing vessel on the high seas is without nationality, a State may board and inspect the vessel. Where evidence so warrants, the State may take such action as may be appropriate in accordance with international law.

18. States shall be liable for damage or loss attributable to them arising from action taken pursuant to this article when such action is unlawful or exceeds that reasonably required in the light of available information to implement the provisions of this article.

Article 22

Basic procedures for boarding and inspection pursuant to article 21

1. The inspecting State shall ensure that its duly authorized inspectors:

(a) present credentials to the master of the vessel and produce a copy of the text of the relevant conservation and management measures or rules and regulations in force in the high seas area in question pursuant to those measures;

(b) initiate notice to the flag State at the time of the boarding and inspection;

(c) do not interfere with the master's ability to communicate with the authorities of the flag State during the boarding and inspection;

(d) provide a copy of a report on the boarding and inspection to the master and to the authorities of the flag State, noting therein any objection or statement which the master wishes to have included in the report;

(e) promptly leave the vessel following completion of the inspection if they find no evidence of a serious violation; and

(f) avoid the use of force except when and to the degree necessary to ensure the safety of the inspectors and where the inspectors are obstructed in the execution of their duties. The degree of force used shall not exceed that reasonably required in the circumstances.

2. The duly authorized inspectors of an inspecting State shall have the authority to inspect the vessel, its licence, gear, equipment, records, facilities, fish and fish products and any relevant documents necessary to verify compliance with the relevant conservation and management measures.

3. The flag State shall ensure that vessel masters:

(a) accept and facilitate prompt and safe boarding by the inspectors;

(b) cooperate with and assist in the inspection of the vessel conducted pursuant to these procedures;

(c) do not obstruct, intimidate or interfere with the inspectors in the performance of their duties;

(d) allow the inspectors to communicate with the authorities of the flag State and the inspecting State during the boarding and inspection;

(e) provide reasonable facilities, including, where appropriate, food and accommodation, to the inspectors; and

(f) facilitate safe disembarkation by the inspectors.

4. In the event that the master of a vessel refuses to accept boarding and inspection in accordance with this article and article 21, the flag State shall, except in circumstances where, in accordance with generally accepted international regulations, procedures and practices relating to safety at sea, it is necessary to delay the boarding and inspection, direct the master of the vessel to submit immediately to boarding and inspection and, if the master does not comply with such direction, shall suspend the vessel's authorization to fish and order the vessel to return immediately to port. The flag State shall advise the inspecting State of the action it has taken when the circumstances referred to in this paragraph arise.

Article 23

Measures taken by a port State

1. A port State has the right and the duty to take measures, in accordance with international law, to promote the effectiveness of subregional, regional and global conservation and management measures. When taking such measures a port State shall not discriminate in form or in fact against the vessels of any State.

2. A port State may, inter alia, inspect documents, fishing gear and catch on board fishing vessels, when such vessels are voluntarily in its ports or at its offshore terminals.

3. States may adopt regulations empowering the relevant national authorities to prohibit landings and transshipments where it has been established that the catch has been taken in a manner which undermines the effectiveness of subregional, regional or global conservation and management measures on the high seas.

4. Nothing in this article affects the exercise by States of their sovereignty over ports in their territory in accordance with international law.

PART VII

REQUIREMENTS OF DEVELOPING STATES

Article 24

Recognition of the special requirements of developing States

1. States shall give full recognition to the special requirements of developing States in relation to conservation and management of straddling fish stocks and highly migratory fish stocks and development of fisheries for such stocks. To this end, States shall, either directly or through the United Nations Development Programme, the Food and Agriculture Organization of the United Nations and other specialized agencies, the Global Environment Facility, the Commission on Sustainable Development and other appropriate international and regional organizations and bodies, provide assistance to developing States.

2. In giving effect to the duty to cooperate in the establishment of conservation and management measures for straddling fish stocks and highly migratory fish stocks, States shall take into account the special requirements of developing States, in particular:

(a) the vulnerability of developing States which are dependent on the exploitation of living marine resources, including for meeting the nutritional requirements of their populations or parts thereof;

(b) the need to avoid adverse impacts on, and ensure access to fisheries by, subsistence, small-scale and artisanal fishers and women fishworkers, as well as indigenous people in developing States, particularly small island developing States; and

(c) the need to ensure that such measures do not result in transferring, directly or indirectly, a disproportionate burden of conservation action onto developing States.

Article 25

Forms of cooperation with developing States

1. States shall cooperate, either directly or through subregional, regional or global organizations:

(a) to enhance the ability of developing States, in particular the least-developed among them and small island developing States, to conserve and manage straddling fish stocks and highly migratory fish stocks and to develop their own fisheries for such stocks;

(b) to assist developing States, in particular the least-developed among them and small island developing States, to enable them to participate in high seas fisheries for such stocks, including facilitating access to such fisheries subject to articles 5 and 11; and

(c) to facilitate the participation of developing States in subregional and regional fisheries management organizations and arrangements.

2. Cooperation with developing States for the purposes set out in this article shall include the provision of financial assistance, assistance relating to human resources development, technical assistance, transfer of technology, including through joint venture arrangements, and advisory and consultative services.

3. Such assistance shall, inter alia, be directed specifically towards:

(a) improved conservation and management of straddling fish stocks and highly migratory fish stocks through collection, reporting, verification, exchange and analysis of fisheries data and related information;

(b) stock assessment and scientific research; and

(c) monitoring, control, surveillance, compliance and enforcement, including training and capacity-building at the local level, development and funding of national and regional observer programmes and access to technology and equipment.

Article 26

Special assistance in the implementation of this Agreement

1. States shall cooperate to establish special funds to assist developing States in the implementation of this Agreement, including assisting developing States to meet the costs involved in any proceedings for the settlement of disputes to which they may be parties.

2. States and international organizations should assist developing States in establishing new subregional or regional fisheries management organizations or arrangements, or in strengthening existing organizations or arrangements, for the conservation and management of straddling fish stocks and highly migratory fish stocks.

PART VIII

PEACEFUL SETTLEMENT OF DISPUTES

Article 27

Obligation to settle disputes by peaceful means

States have the obligation to settle their disputes by negotiation, inquiry, mediation, conciliation, arbitration, judicial settlement, resort to regional agencies or arrangements, or other peaceful means of their own choice.

Article 28

Prevention of disputes

States shall cooperate in order to prevent disputes. To this end, States shall agree on efficient and expeditious decision-making procedures within subregional and regional fisheries management organizations and arrangements and shall strengthen existing decision-making procedures as necessary.

Article 29

Disputes of a technical nature

Where a dispute concerns a matter of a technical nature, the States concerned may refer the dispute to an ad hoc expert panel established by them. The panel shall confer with the States concerned and shall endeavour to resolve the dispute expeditiously without recourse to binding procedures for the settlement of disputes.

Article 30

Procedures for the settlement of disputes

1. The provisions relating to the settlement of disputes set out in Part XV of the Convention apply *mutatis mutandis* to any dispute between States Parties to this Agreement concerning the interpretation or application of this Agreement, whether or not they are also Parties to the Convention.

2. The provisions relating to the settlement of disputes set out in Part XV of the Convention apply *mutatis mutandis* to any dispute between States Parties to this Agreement concerning the interpretation or application of a subregional, regional or global fisheries agreement relating to straddling fish stocks or highly migratory fish stocks to which they are parties, including any dispute concerning the conservation and management of such stocks, whether or not they are also Parties to the Convention.

3. Any procedure accepted by a State Party to this Agreement and the Convention pursuant to article 287 of the Convention shall apply to the settlement of disputes under this Part, unless that State Party, when signing, ratifying or acceding to this Agreement, or at any time thereafter, has accepted another procedure pursuant to article 287 for the settlement of disputes under this Part.

4. A State Party to this Agreement which is not a Party to the Convention, when signing, ratifying or acceding to this Agreement, or at any time thereafter, shall be free to choose, by means of a written declaration, one or more of the means set out in article 287, paragraph 1, of the Convention for the settlement of disputes under this Part. Article 287 shall apply to such a declaration, as well as to any dispute to which such State is a party which is not covered by a declaration in force. For the purposes of conciliation and

arbitration in accordance with Annexes V, VII and VIII to the Convention, such State shall be entitled to nominate conciliators, arbitrators and experts to be included in the lists referred to in Annex V, article 2, Annex VII, article 2, and Annex VIII, article 2, for the settlement of disputes under this Part.

5. Any court or tribunal to which a dispute has been submitted under this Part shall apply the relevant provisions of the Convention, of this Agreement and of any relevant subregional, regional or global fisheries agreement, as well as generally accepted standards for the conservation and management of living marine resources and other rules of international law not incompatible with the Convention, with a view to ensuring the conservation of the straddling fish stocks and highly migratory fish stocks concerned.

Article 31

Provisional measures

1. Pending the settlement of a dispute in accordance with this Part, the parties to the dispute shall make every effort to enter into provisional arrangements of a practical nature.

2. Without prejudice to article 290 of the Convention, the court or tribunal to which the dispute has been submitted under this Part may prescribe any provisional measures which it considers appropriate under the circumstances to preserve the respective rights of the parties to the dispute or to prevent damage to the stocks in question, as well as in the circumstances referred to in article 7, paragraph 5, and article 16, paragraph 2.

3. A State Party to this Agreement which is not a Party to the Convention may declare that, notwithstanding article 290, paragraph 5, of the Convention, the International Tribunal for the Law of the Sea shall not be entitled to prescribe, modify or revoke provisional measures without the agreement of such State.

Article 32

Limitations on applicability of procedures for the settlement of disputes

Article 297, paragraph 3, of the Convention applies also to this Agreement.

PART IX

NON-PARTIES TO THIS AGREEMENT

Article 33

Non-parties to this Agreement

1. States Parties shall encourage non-parties to this Agreement to become parties thereto and to adopt laws and regulations consistent with its provisions.

2. States Parties shall take measures consistent with this Agreement and international law to deter the activities of vessels flying the flag of non-parties which undermine the effective implementation of this Agreement.

PART X

GOOD FAITH AND ABUSE OF RIGHTS

Article 34

Good faith and abuse of rights

States Parties shall fulfil in good faith the obligations assumed under this Agreement and shall exercise the rights recognized in this Agreement in a manner which would not constitute an abuse of right.

Part XI

RESPONSIBILITY AND LIABILITY

Article 35

Responsibility and liability

States Parties are liable in accordance with international law for damage or loss attributable to them in regard to this Agreement.

PART XII

REVIEW CONFERENCE

Article 36

Review conference

1. Four years after the date of entry into force of this Agreement, the Secretary-General of the United Nations shall convene a conference with a view

to assessing the effectiveness of this Agreement in securing the conservation and management of straddling fish stocks and highly migratory fish stocks. The Secretary-General shall invite to the conference all States Parties and those States and entities which are entitled to become parties to this Agreement as well as those intergovernmental and non-governmental organizations entitled to participate as observers.

2. The conference shall review and assess the adequacy of the provisions of this Agreement and, if necessary, propose means of strengthening the substance and methods of implementation of those provisions in order better to address any continuing problems in the conservation and management of straddling fish stocks and highly migratory fish stocks.

PART XIII

FINAL PROVISIONS

Article 37

Signature

This Agreement shall be open for signature by all States and the other entities referred to in article 1, paragraph 2(b), and shall remain open for signature at United Nations Headquarters for twelve months from the fourth of December 1995.

Article 38

Ratification

This Agreement is subject to ratification by States and the other entities referred to in article 1, paragraph 2(b). The instruments of ratification shall be deposited with the Secretary-General of the United Nations.

Article 39

Accession

This Agreement shall remain open for accession by States and the other entities referred to in article 1, paragraph 2(b). The instruments of accession shall be deposited with the Secretary-General of the United Nations.

Article 40

Entry into force

1. This Agreement shall enter into force 30 days after the date of deposit of the thirtieth instrument of ratification or accession.

2. For each State or entity which ratifies the Agreement or accedes thereto after the deposit of the thirtieth instrument of ratification or accession, this Agreement shall enter into force on the thirtieth day following the deposit of its instrument of ratification or accession.

Article 41

Provisional application

1. This Agreement shall be applied provisionally by a State or entity which consents to its provisional application by so notifying the depositary in writing. Such provisional application shall become effective from the date of receipt of the notification.

2. Provisional application by a State or entity shall terminate upon the entry into force of this Agreement for that State or entity or upon notification by that State or entity to the depositary in writing of its intention to terminate provisional application.

Article 42

Reservations and exceptions

No reservations or exceptions may be made to this Agreement.

Article 43

Declarations and statements

Article 42 does not preclude a State or entity, when signing, ratifying or acceding to this Agreement, from making declarations or statements, however phrased or named, with a view, *inter alia*, to the harmonization of its laws and regulations with the provisions of this Agreement, provided that such declarations or statements do not purport to exclude or to modify the legal effect of the provisions of this Agreement in their application to that State or entity.

Article 44

Relation to other agreements

1. This Agreement shall not alter the rights and obligations of States Parties which arise from other agreements compatible with this Agreement and which do not affect the enjoyment by other States Parties of their rights or the performance of their obligations under this Agreement.

2. Two or more States Parties may conclude agreements modifying or suspending the operation of provisions of this Agreement, applicable solely to the

relations between them, provided that such agreements do not relate to a provision derogation from which is incompatible with the effective execution of the object and purpose of this Agreement, and provided further that such agreements shall not affect the application of the basic principles embodied herein, and that the provisions of such agreements do not affect the enjoyment by other States Parties of their rights or the performance of their obligations under this Agreement.

3. States Parties intending to conclude an agreement referred to in paragraph 2 shall notify the other States Parties through the depositary of this Agreement of their intention to conclude the agreement and of the modification or suspension for which it provides.

Article 45

Amendment

1. A State Party may, by written communication addressed to the Secretary-General of the United Nations, propose amendments to this Agreement and request the convening of a conference to consider such proposed amendments. The Secretary-General shall circulate such communication to all States Parties. If, within six months from the date of the circulation of the communication, not less than one half of the States Parties reply favourably to the request, the Secretary-General shall convene the conference.

2. The decision-making procedure applicable at the amendment conference convened pursuant to paragraph 1 shall be the same as that applicable at the United Nations Conference on Straddling Fish Stocks and Highly Migratory Fish Stocks, unless otherwise decided by the conference. The conference should make every effort to reach agreement on any amendments by way of consensus and there should be no voting on them until all efforts at consensus have been exhausted.

3. Once adopted, amendments to this Agreement shall be open for signature at United Nations Headquarters by States Parties for twelve months from the date of adoption, unless otherwise provided in the amendment itself.

4. Articles 38, 39, 47 and 50 apply to all amendments to this Agreement.

5. Amendments to this Agreement shall enter into force for the States Parties ratifying or acceding to them on the thirtieth day following the deposit of instruments of ratification or accession by two thirds of the States Parties. Thereafter, for each State Party ratifying or acceding to an amendment after the deposit of the required number of such instruments, the amendment shall enter into force on the thirtieth day following the deposit of its instrument of ratification or accession.

6. An amendment may provide that a smaller or a larger number of ratifications or accessions shall be required for its entry into force than are required by this article.

7. A State which becomes a Party to this Agreement after the entry into force of amendments in accordance with paragraph 5 shall, failing an expression of a different intention by that State:

(a) be considered as a Party to this Agreement as so amended; and

(b) be considered as a Party to the unamended Agreement in relation to any State Party not bound by the amendment.

Article 46

Denunciation

1. A State Party may, by written notification addressed to the Secretary-General of the United Nations, denounce this Agreement and may indicate its reasons. Failure to indicate reasons shall not affect the validity of the denunciation. The denunciation shall take effect one year after the date of receipt of the notification, unless the notification specifies a later date.

2. The denunciation shall not in any way affect the duty of any State Party to fulfil any obligation embodied in this Agreement to which it would be subject under international law independently of this Agreement.

Article 47

Participation by international organizations

1. In cases where an international organization referred to in Annex IX, article 1, of the Convention does not have competence over all the matters governed by this Agreement, Annex IX to the Convention shall apply *mutatis mutandis* to participation by such international organization in this Agreement, except that the following provisions of that Annex shall not apply:

(a) article 2, first sentence; and

(b) article 3, paragraph 1.

2. In cases where an international organization referred to in Annex IX, article 1, of the Convention has competence over all the matters governed by this Agreement, the following provisions shall apply to participation by such international organization in this Agreement:

(a) at the time of signature or accession, such international organization shall make a declaration stating:

(i) that it has competence over all the matters governed by this Agreement;

(ii) that, for this reason, its member States shall not become States Parties, except in respect of their territories for which the international organization has no responsibility; and

(iii) that it accepts the rights and obligations of States under this Agreement;

(b) participation of such an international organization shall in no case confer any rights under this Agreement on member States of the international organization;

(c) in the event of a conflict between the obligations of an international organization under this Agreement and its obligations under the agreement establishing the international organization or any acts relating to it, the obligations under this Agreement shall prevail.

Article 48

Annexes

1. The Annexes form an integral part of this Agreement and, unless expressly provided otherwise, a reference to this Agreement or to one of its Parts includes a reference to the Annexes relating thereto.

2. The Annexes may be revised from time to time by States Parties. Such revisions shall be based on scientific and technical considerations. Notwithstanding the provisions of article 45, if a revision to an Annex is adopted by consensus at a meeting of States Parties, it shall be incorporated in this Agreement and shall take effect from the date of its adoption or from such other date as may be specified in the revision. If a revision to an Annex is not adopted by consensus at such a meeting, the amendment procedures set out in article 45 shall apply.

Article 49

Depositary

The Secretary-General of the United Nations shall be the depositary of this Agreement and any amendments or revisions thereto.

Article 50

Authentic texts

The Arabic, Chinese, English, French, Russian and Spanish texts of this Agreement are equally authentic.

IN WITNESS WHEREOF, the undersigned Plenipotentiaries, being duly authorized thereto, have signed this Agreement.

OPENED FOR SIGNATURE at New York, this fourth day of December, one thousand nine hundred and ninety-five, in a single original, in the Arabic, Chinese, English, French, Russian and Spanish languages.

ANNEX I

STANDARD REQUIREMENTS FOR THE COLLECTION AND SHARING OF DATA

Article 1

General principles

1. The timely collection, compilation and analysis of data are fundamental to the effective conservation and management of straddling fish stocks and highly migratory fish stocks. To this end, data from fisheries for these stocks on the high seas and those in areas under national jurisdiction are required and should be collected and compiled in such a way as to enable statistically meaningful analysis for the purposes of fishery resource conservation and management. These data include catch and fishing effort statistics and other fishery-related information, such as vessel-related and other data for standardizing fishing effort. Data collected should also include information on non-target and associated or dependent species. All data should be verified to ensure accuracy. Confidentiality of non-aggregated data shall be maintained. The dissemination of such data shall be subject to the terms on which they have been provided.

2. Assistance, including training as well as financial and technical assistance, shall be provided to developing States in order to build capacity in the field of conservation and management of living marine resources. Assistance should focus on enhancing capacity to implement data collection and verification, observer programmes, data analysis and research projects supporting stock assessments. The fullest possible involvement of developing State scientists and managers in conservation and management of straddling fish stocks and highly migratory fish stocks should be promoted.

Article 2

Principles of data collection, compilation and exchange

The following general principles should be considered in defining the parameters for collection, compilation and exchange of data from fishing operations for straddling fish stocks and highly migratory fish stocks:

(a) States should ensure that data are collected from vessels flying their flag on fishing activities according to the operational characteristics of each fishing method (e.g., each individual tow for trawl, each set for long-line and purse-seine, each school fished for pole-and-line and each day fished for troll) and in sufficient detail to facilitate effective stock assessment;

(b) States should ensure that fishery data are verified through an appropriate system;

(c) States should compile fishery-related and other supporting scientific data and provide them in an agreed format and in a timely manner to the relevant subregional or regional fisheries management organization or arrangement where one exists. Otherwise, States should cooperate to exchange data either directly or through such other cooperative mechanisms as may be agreed among them;

(d) States should agree, within the framework of subregional or regional fisheries management organizations or arrangements, or otherwise, on the specification of data and the format in which they are to be provided, in accordance with this Annex and taking into account the nature of the stocks and the fisheries for those stocks in the region. Such organizations or arrangements should request non-members or non-participants to provide data concerning relevant fishing activities by vessels flying their flag;

(e) such organizations or arrangements shall compile data and make them available in a timely manner and in an agreed format to all interested States under the terms and conditions established by the organization or arrangement; and

(f) scientists of the flag State and from the relevant subregional or regional fisheries management organization or arrangement should analyse the data separately or jointly, as appropriate.

Article 3

Basic fishery data

1. States shall collect and make available to the relevant subregional or regional fisheries management organization or arrangement the following types of data in sufficient detail to facilitate effective stock assessment in accordance with agreed procedures:

(a) time series of catch and effort statistics by fishery and fleet;

(b) total catch in number, nominal weight, or both, by species (both target and non-target) as is appropriate to each fishery. [Nominal weight is defined by the Food and Agriculture Organization of the United Nations as the live-weight equivalent of the landings];

(c) discard statistics, including estimates where necessary, reported as number or nominal weight by species, as is appropriate to each fishery;

(d) effort statistics appropriate to each fishing method; and

(e) fishing location, date and time fished and other statistics on fishing operations as appropriate.

2. States shall also collect where appropriate and provide to the relevant subregional or regional fisheries management organization or arrangement information to support stock assessment, including:

(a) composition of the catch according to length, weight and sex;

(b) other biological information supporting stock assessments, such as information on age, growth, recruitment, distribution and stock identity; and

(c) other relevant research, including surveys of abundance, biomass surveys, hydro-acoustic surveys, research on environmental factors affecting stock abundance, and oceanographic and ecological studies.

Article 4

Vessel data and information

1. States should collect the following types of vessel-related data for standardizing fleet composition and vessel fishing power and for converting between different measures of effort in the analysis of catch and effort data:

(a) vessel identification, flag and port of registry;

(b) vessel type;

(c) vessel specifications (e.g., material of construction, date built, registered length, gross registered tonnage, power of main engines, hold capacity and catch storage methods); and

(d) fishing gear description (e.g., types, gear specifications and quantity).

2. The flag State will collect the following information:

(a) navigation and position fixing aids;

(b) communication equipment and international radio call sign; and

(c) crew size.

Article 5

Reporting

A State shall ensure that vessels flying its flag send to its national fisheries administration and, where agreed, to the relevant subregional or regional fisheries management organization or arrangement, logbook data on catch and effort, including data on fishing operations on the high seas, at sufficiently frequent intervals to meet national requirements and regional and international obligations. Such data shall be transmitted, where necessary, by radio, telex, facsimile or satellite transmission or by other means.

Article 6

Data verification

States or, as appropriate, subregional or regional fisheries management organizations or arrangements should establish mechanisms for verifying fishery data, such as:

(a) position verification through vessel monitoring systems;

(b) scientific observer programmes to monitor catch, effort, catch composition (target and non-target) and other details of fishing operations;

(c) vessel trip, landing and transshipment reports; and

(d) port sampling.

Article 7

Data exchange

1. Data collected by flag States must be shared with other flag States and relevant coastal States through appropriate subregional or regional fisheries management organizations or arrangements. Such organizations or arrangements shall compile data and make them available in a timely manner and in an agreed format to all interested States under the terms and conditions established by the organization or arrangement, while maintaining confidentiality of non-aggregated data, and should, to the extent feasible, develop database systems which provide efficient access to data.

2. At the global level, collection and dissemination of data should be effected through the Food and Agriculture Organization of the United Nations. Where a subregional or regional fisheries management organization or arrangement does not exist, that organization may also do the same at the subregional or regional level by arrangement with the States concerned.

ANNEX II

GUIDELINES FOR THE APPLICATION OF PRECAUTIONARY REFERENCE POINTS IN CONSERVATION AND MANAGEMENT OF STRADDLING FISH STOCKS AND HIGHLY MIGRATORY FISH STOCKS

1. A precautionary reference point is an estimated value derived through an agreed scientific procedure, which corresponds to the state of the resource and of the fishery, and which can be used as a guide for fisheries management.

2. Two types of precautionary reference points should be used: conservation, or limit, reference points and management, or target, reference points. Limit reference points set boundaries which are intended to constrain harvesting within safe biological limits within which the stocks can produce maximum sustainable yield. Target reference points are intended to meet management objectives.

3. Precautionary reference points should be stock-specific to account, <u>inter alia</u>, for the reproductive capacity, the resilience of each stock and the characteristics of fisheries exploiting the stock, as well as other sources of mortality and major sources of uncertainty.

4. Management strategies shall seek to maintain or restore populations of harvested stocks, and where necessary associated or dependent species, at levels consistent with previously agreed precautionary reference points. Such reference points shall be used to trigger pre-agreed conservation and management action. Management strategies shall include measures which can be implemented when precautionary reference points are approached.

5. Fishery management strategies shall ensure that the risk of exceeding limit reference points is very low. If a stock falls below a limit reference point or is at risk of falling below such a reference point, conservation and management action should be initiated to facilitate stock recovery. Fishery management strategies shall ensure that target reference points are not exceeded on average.

6. When information for determining reference points for a fishery is poor or absent, provisional reference points shall be set. Provisional reference points may be established by analogy to similar and better-known stocks. In such situations, the fishery shall be subject to enhanced monitoring so as to enable revision of provisional reference points as improved information becomes available.

7. The fishing mortality rate which generates maximum sustainable yield should be regarded as a minimum standard for limit reference points. For stocks which are not overfished, fishery management strategies shall ensure that fishing mortality does not exceed that which corresponds to maximum sustainable yield, and that the biomass does not fall below a predefined threshold. For overfished stocks, the biomass which would produce maximum sustainable yield can serve as a rebuilding target.

Appendix 2 – FAO Compliance Agreement

AGREEMENT TO PROMOTE COMPLIANCE WITH INTERNATIONAL CONSERVATION AND MANAGEMENT MEASURES BY FISHING VESSELS ON THE HIGH SEAS

PREAMBLE

The Parties to this Agreement,

Recognizing that all States have the right for their nationals to engage in fishing on the high seas, subject to the relevant rules of international law, as reflected in the United Nations Convention on the Law of the Sea,

Further recognizing that, under international law as reflected in the United Nations Convention on the Law of the Sea, all States have the duty to take, or to cooperate with other States in taking, such measures for their respective nationals as may be necessary for the conservation of the living resources of the high seas,

Acknowledging the right and interest of all States to develop their fishing sectors in accordance with their national policies, and the need to promote cooperation with developing countries to enhance their capabilities to fulfil their obligations under this Agreement,

Recalling that Agenda 21, adopted by the United Nations Conference on Environment and Development, calls upon States to take effective action, consistent with international law, to deter reflagging of vessels by their nationals as a means of avoiding compliance with applicable conservation and management rules for fishing activities on the high seas,

Further recalling that the Declaration of Cancun, adopted by the International Conference on Responsible Fishing, also calls on States to take action in this respect,

Bearing in mind that under Agenda 21, States commit themselves to the conservation and sustainable use of marine living resources on the high seas,

Calling upon States which do not participate in global, regional or subregional fisheries organizations or arrangements to join or, as appropriate, to enter into understandings with such organizations or with parties to such organizations or arrangements with a view to achieving compliance with international conservation and management measures,

Conscious of the duties of every State to exercise effectively its jurisdiction and control over vessels flying its flag, including fishing vessels and vessels engaged in the transhipment of fish,

Mindful that the practice of flagging or reflagging fishing vessels as a means of avoiding compliance with international conservation and management measures for living marine resources, and the failure of flag States to fulfil their responsibilities with respect to fishing vessels entitled to fly their flag, are among the factors that seriously undermine the effectiveness of such measures,

Realizing that the objective of this Agreement can be achieved through specifying flag States' responsibility in respect of fishing vessels entitled to fly their flags and operating on the high seas, including the authorization by the flag State of such operations, as well as through strengthened international cooperation and increased transparency through the exchange of information on high seas fishing,

Noting that this Agreement will form an integral part of the International Code of Conduct for Responsible Fishing called for in the Declaration of Cancun,

Desiring to conclude an international agreement within the framework of the Food and Agriculture Organization of the United Nations, hereinafter referred to as FAO, under Article XIV of the FAO Constitution,

Have agreed as follows:

Article I

DEFINITIONS

For the purposes of this Agreement:

(a) "fishing vessel" means any vessel used or intended for use for the purposes of the commercial exploitation of living marine resources, including mother ships and any other vessels directly engaged in such fishing operations;

(b) "international conservation and management measures" means measures to conserve or manage one or more species of living marine resources that are adopted and applied in accordance with the relevant rules of international law as reflected in the 1982 United Nations Convention on the Law of the Sea. Such measures may be adopted either by global, regional or subregional fisheries organizations, subject to the rights and obligations of their members, or by treaties or other international agreements;

(c) "length" means

(i) for any fishing vessel built after 18 July 1982, 96 percent of the total length on a waterline at 85 percent of the least moulded depth measured from the top of the keel, or the length from the foreside of the stem to the axis of the rudder stock on that waterline, if that be greater. In ships designed with a rake of keel the waterline on which this length is measured shall be parallel to the designed waterline;

(ii) for any fishing vessel built before 18 July 1982, registered length as entered on the national register or other record of vessels;

3

(d) "record of fishing vessels" means a record of fishing vessels in which are recorded pertinent details of the fishing vessel. It may constitute a separate record for fishing vessels or form part of a general record of vessels;

(e) "regional economic integration organization" means a regional economic integration organization to which its Member States have transferred competence over matters covered by this Agreement, including the authority to make decisions binding on its Member States in respect of those matters;

(f) "vessels entitled to fly its flag" and "vessels entitled to fly the flag of a State", includes vessels entitled to fly the flag of a Member State of a regional economic integration organization.

Article II

APPLICATION

1. Subject to the following paragraphs of this Article, this Agreement shall apply to all fishing vessels that are used or intended for fishing on the high seas.

2. A Party may exempt fishing vessels of less than 24 metres in length entitled to fly its flag from the application of this Agreement unless the Party determines that such an exemption would undermine the object and purpose of this Agreement, provided that such exemptions:

(a) shall not be granted in respect of fishing vessels operating in fishing regions referred to in paragraph 3 below, other than fishing vessels that are entitled to fly the flag of a coastal State of that fishing region; and

(b) shall not apply to the obligations undertaken by a Party under paragraph 1 of Article III, or paragraph 7 of Article VI of this Agreement.

3. Without prejudice to the provisions of paragraph 2 above, in any fishing region where bordering coastal States have not yet declared exclusive economic zones, or equivalent zones of national jurisdiction over fisheries, such coastal States as are Parties to this Agreement may agree, either directly or through appropriate regional fisheries organizations, to establish a minimum length of fishing vessels below which this Agreement shall not apply in respect of fishing vessels flying the flag of any such coastal State and operating exclusively in such fishing region.

Article III

FLAG STATE RESPONSIBILITY

1. (a) Each Party shall take such measures as may be necessary to ensure that fishing vessels entitled to fly its flag do not engage in any activity that

undermines the effectiveness of international conservation and management measures.

(b) In the event that a Party has, pursuant to paragraph 2 of Article II, granted an exemption for fishing vessels of less than 24 metres in length entitled to fly its flag from the application of other provisions of this Agreement, such Party shall nevertheless take effective measures in respect of any such fishing vessel that undermines the effectiveness of international conservation and management measures. These measures shall be such as to ensure that the fishing vessel ceases to engage in activities that undermine the effectiveness of the international conservation and management measures.

2. In particular, no Party shall allow any fishing vessel entitled to fly its flag to be used for fishing on the high seas unless it has been authorized to be so used by the appropriate authority or authorities of that Party. A fishing vessel so authorized shall fish in accordance with the conditions of the authorization.

3. No Party shall authorize any fishing vessel entitled to fly its flag to be used for fishing on the high seas unless the Party is satisfied that it is able, taking into account the links that exist between it and the fishing vessel concerned, to exercise effectively its responsibilities under this Agreement in respect of that fishing vessel.

4. Where a fishing vessel that has been authorized to be used for fishing on the high seas by a Party ceases to be entitled to fly the flag of that Party, the authorization to fish on the high seas shall be deemed to have been cancelled.

5. (a) No Party shall authorize any fishing vessel previously registered in the territory of another Party that has undermined the effectiveness of international conservation and management measures to be used for fishing on the high seas, unless it is satisfied that

(i) any period of suspension by another Party of an authorization for such fishing vessel to be used for fishing on the high seas has expired; and

(ii) no authorization for such fishing vessel to be used for fishing on the high seas has been withdrawn by another Party within the last three years.

(b) The provisions of subparagraph (a) above shall also apply in respect of fishing vessels previously registered in the territory of a State which is not a Party to this Agreement, provided that sufficient information is available to the Party concerned on the circumstances in which the authorization to fish was suspended or withdrawn.

(c) The provisions of subparagraphs (a) and (b) shall not apply where the ownership of the fishing vessel has subsequently changed, and the new owner has provided sufficient evidence demonstrating that the previous owner or operator has no further legal, beneficial or financial interest in, or control of, the fishing vessel.

(d) Notwithstanding the provisions of subparagraphs (a) and (b) above, a Party may authorize a fishing vessel, to which those subparagraphs would otherwise apply, to be used for fishing on the high seas, where the Party concerned, after having taken into account all relevant facts, including the circumstances in which the fishing authorization has been withdrawn by the other Party or State, has determined that to grant an authorization to use the vessel for fishing on the high seas would not undermine the object and purpose of this Agreement.

6. Each Party shall ensure that all fishing vessels entitled to fly its flag that it has entered in the record maintained under Article IV are marked in such a way that they can be readily identified in accordance with generally accepted standards, such as the FAO Standard Specifications for the Marking and Identification of Fishing Vessels.

7. Each Party shall ensure that each fishing vessel entitled to fly its flag shall provide it with such information on its operations as may be necessary to enable the Party to fulfil its obligations under this Agreement, including in particular information pertaining to the area of its fishing operations and to its catches and landings.

8. Each Party shall take enforcement measures in respect of fishing vessels entitled to fly its flag which act in contravention of the provisions of this Agreement, including, where appropriate, making the contravention of such provisions an offence under national legislation. Sanctions applicable in respect of such contraventions shall be of sufficient gravity as to be effective in securing compliance with the requirements of this Agreement and to deprive offenders of the benefits accruing from their illegal activities. Such sanctions shall, for serious offences, include refusal, suspension or withdrawal of the authorization to fish on the high seas.

Article IV

RECORDS OF FISHING VESSELS

Each Party shall, for the purposes of this Agreement, maintain a record of fishing vessels entitled to fly its flag and authorized to be used for fishing on the high seas, and shall take such measures as may be necessary to ensure that all such fishing vessels are entered in that record.

Article V

INTERNATIONAL COOPERATION

1. The Parties shall cooperate as appropriate in the implementation of this Agreement, and shall, in particular, exchange information, including evidentiary material, relating to activities of fishing vessels in order to assist the flag State in identifying those fishing vessels flying its flag reported to have engaged in activities undermining international conservation and management measures, so as to fulfil its obligations under Article III.

2. When a fishing vessel is voluntarily in the port of a Party other than its flag State, that Party, where it has reasonable grounds for believing that the fishing vessel has been used for an activity that undermines the effectiveness of international conservation and management measures, shall promptly notify the flag State accordingly. Parties may make arrangements regarding the undertaking by port States of such investigatory measures as may be considered necessary to establish whether the fishing vessel has indeed been used contrary to the provisions of this Agreement.

3. The Parties shall, when and as appropriate, enter into cooperative agreements or arrangements of mutual assistance on a global, regional, subregional or bilateral basis so as to promote the achievement of the objectives of this Agreement.

Article VI

EXCHANGE OF INFORMATION

1. Each Party shall make readily available to FAO the following information with respect to each fishing vessel entered in the record required to be maintained under Article IV:

(a) name of fishing vessel, registration number, previous names (if known), and port of registry;

(b) previous flag (if any);

(c) International Radio Call Sign (if any);

(d) name and address of owner or owners;

(e) where and when built;

(f) type of vessel;

(g) length.

2. Each Party shall, to the extent practicable, make available to FAO the following additional information with respect to each fishing vessel entered in the record required to be maintained under Article IV:

(a) name and address of operator (manager) or operators (managers) (if any);

(b) type of fishing method or methods;

(c) moulded depth;

(d) beam;

(e) gross register tonnage;

(f) power of main engine or engines.

3. Each Party shall promptly notify to FAO any modifications to the information listed in paragraphs 1 and 2 of this Article.

4. FAO shall circulate periodically the information provided under paragraphs 1, 2, and 3 of this Article to all Parties, and, on request, individually to any Party. FAO shall also, subject to any restrictions imposed by the Party concerned regarding the distribution of information, provide such information on request individually to any global, regional or subregional fisheries organization.

5. Each Party shall also promptly inform FAO of -

(a) any additions to the record;

(b) any deletions from the record by reason of -

(i) the voluntary relinquishment or non-renewal of the fishing authorization by the fishing vessel owner or operator;

(ii) the withdrawal of the fishing authorization issued in respect of the fishing vessel under paragraph 8 of Article III;

(iii) the fact that the fishing vessel concerned is no longer entitled to fly its flag;

(iv) the scrapping, decommissioning or loss of the fishing vessel concerned; or

(v) any other reason.

6. Where information is given to FAO under paragraph 5(b) above, the Party concerned shall specify which of the reasons listed in that paragraph is applicable.

7. Each Party shall inform FAO of

(a) any exemption it has granted under paragraph 2 of Article II, the number and type of fishing vessel involved and the geographical areas in which such fishing vessels operate; and

(b) any agreement reached under paragraph 3 of Article II.

8. (a) Each Party shall report promptly to FAO all relevant information regarding any activities of fishing vessels flying its flag that undermine the effectiveness of international conservation and management measures, including the identity of the fishing vessel or vessels involved and measures imposed by the Party in respect of such activities. Reports on measures imposed by a Party may be subject to such limitations as may be required by national legislation with respect to confidentiality, including, in particular, confidentiality regarding measures that are not yet final.

(b) Each Party, where it has reasonable grounds to believe that a fishing vessel not entitled to fly its flag has engaged in any activity that undermines the effectiveness of international conservation and management measures, shall draw this to the attention of the flag State concerned and may, as appropriate, draw it to the attention of FAO. It shall provide the flag State with full supporting evidence and may provide FAO with a summary of such evidence. FAO shall not circulate such information until such time as the flag State has had an opportunity to comment on the allegation and evidence submitted, or to object as the case may be.

9. Each Party shall inform FAO of any cases where the Party, pursuant to paragraph 5(d) of Article III, has granted an authorization notwithstanding the provisions of paragraph 5(a) or 5(b) of Article III. The information shall include pertinent data permitting the identification of the fishing vessel and the owner or operator and, as appropriate, any other information relevant to the Party's decision.

10. FAO shall circulate promptly the information provided under paragraphs 5, 6, 7, 8 and 9 of this Article to all Parties, and, on request, individually to any Party. FAO shall also, subject to any restrictions imposed by the Party concerned regarding the distribution of information, provide such information promptly on request individually to any global, regional or subregional fisheries organization.

11. The Parties shall exchange information relating to the implementation of this Agreement, including through FAO and other appropriate global, regional and subregional fisheries organizations.

Article VII

COOPERATION WITH DEVELOPING COUNTRIES

The Parties shall cooperate, at a global, regional, subregional or bilateral level, and, as appropriate, with the support of FAO and other international or regional organizations, to provide assistance, including technical assistance, to Parties that are developing countries in order to assist them in fulfilling their obligations under this Agreement.

Article VIII

NON-PARTIES

1. The Parties shall encourage any State not party to this Agreement to accept this Agreement and shall encourage any non-Party to adopt laws and regulations consistent with the provisions of this Agreement.

2. The Parties shall cooperate in a manner consistent with this Agreement and with international law to the end that fishing vessels entitled to fly the flags of non-Parties do not engage in activities that undermine the effectiveness of international conservation and management measures.

3. The Parties shall exchange information amongst themselves, either directly or through FAO, with respect to activities of fishing vessels flying the flags of non-Parties that undermine the effectiveness of international conservation and management measures.

Article IX

SETTLEMENT OF DISPUTES

1. Any Party may seek consultations with any other Party or Parties on any dispute with regard to the interpretation or application of the provisions of this Agreement with a view to reaching a mutually satisfactory solution as soon as possible.

2. In the event that the dispute is not resolved through these consultations within a reasonable period of time, the Parties in question shall consult among themselves as soon as possible with a view to having the dispute settled by negotiation, inquiry, mediation, conciliation, arbitration, judicial settlement or other peaceful means of their own choice.

3. Any dispute of this character not so resolved shall, with the consent of all Parties to the dispute, be referred for settlement to the International Court of Justice, to the International Tribunal for the Law of the Sea upon entry into force of the 1982 United Nations Convention on the Law of the Sea or to arbitration. In the case of failure to reach agreement on referral to the International Court of Justice, to the International Tribunal for the Law of the Sea or to arbitration, the Parties shall continue to consult and cooperate with a view to reaching

settlement of the dispute in accordance with the rules of international law relating to the conservation of living marine resources.

Article X

ACCEPTANCE

1. This Agreement shall be open to acceptance by any Member or Associate Member of FAO, and to any non-member State that is a Member of the United Nations, or of any of the specialized agencies of the United Nations or of the International Atomic Energy Agency.

2. Acceptance of this Agreement shall be effected by the deposit of an instrument of acceptance with the Director-General of FAO, hereinafter referred to as the Director-General.

3. The Director-General shall inform all Parties, all Members and Associate Members of FAO and the Secretary-General of the United Nations of all instruments of acceptance received.

4. When a regional economic integration organization becomes a Party to this Agreement, such regional economic integration organization shall, in accordance with the provisions of Article II.7 of the FAO Constitution, as appropriate, notify such modifications or clarifications to its declaration of competence submitted under Article II.5 of the FAO Constitution as may be necessary in light of its acceptance of this Agreement. Any Party to this Agreement may, at any time, request a regional economic integration organization that is a Party to this Agreement to provide information as to which, as between the regional economic integration organization and its Member States, is responsible for the implementation of any particular matter covered by this Agreement. The regional economic integration organization shall provide this information within a reasonable time.

Article XI

ENTRY INTO FORCE

1. This Agreement shall enter into force as from the date of receipt by the Director-General of the twenty-fifth instrument of acceptance.

2. For the purpose of this Article, an instrument deposited by a regional economic integration organization shall not be counted as additional to those deposited by Member States of such an organization.

Article XII

RESERVATIONS

Acceptance of this Agreement may be made subject to reservations which shall become effective only upon unanimous acceptance by all Parties to this Agreement. The Director-General shall notify forthwith all Parties of any reservation. Parties not having replied within three months from the date of the notification shall be deemed to have accepted the reservation. Failing such acceptance, the State or regional economic integration organization making the reservation shall not become a Party to this Agreement.

Article XIII

AMENDMENTS

1. Any proposal by a Party for the amendment of this Agreement shall be communicated to the Director-General.

2. Any proposed amendment of this Agreement received by the Director-General from a Party shall be presented to a regular or special session of the Conference for approval and, if the amendment involves important technical changes or imposes additional obligations on the Parties, it shall be considered by an advisory committee of specialists convened by FAO prior to the Conference.

3. Notice of any proposed amendment of this Agreement shall be transmitted to the Parties by the Director-General not later than the time when the agenda of the session of the Conference at which the matter is to be considered is dispatched.

4. Any such proposed amendment of this Agreement shall require the approval of the Conference and shall come into force as from the thirtieth day after acceptance by two-thirds of the Parties. Amendments involving new obligations for Parties, however, shall come into force in respect of each Party only on acceptance by it and as from the thirtieth day after such acceptance. Any amendment shall be deemed to involve new obligations for Parties unless the Conference. in approving the amendment, decides otherwise by consensus.

5. The instruments of acceptance of amendments involving new obligations shall be deposited with the Director-General, who shall inform all Parties of the receipt of acceptance and the entry into force of amendments.

6. For the purpose of this Article, an instrument deposited by a regional economic integration organization shall not be counted as additional to those deposited by Member States of such an organization.

Article XIV

WITHDRAWAL

Any Party may withdraw from this Agreement at any time after the expiry of two years from the date upon which the Agreement entered into force with respect to that Party, by giving written notice of such withdrawal to the Director-General who shall immediately inform all the Parties and the Members and Associate Members of FAO of such withdrawal. Withdrawal shall become effective at the end of the calendar year following that in which the notice of withdrawal has been received by the Director-General.

Article XV

DUTIES OF THE DEPOSITARY

The Director-General shall be the Depositary of this Agreement. The Depositary shall:

(a) send certified copies of this Agreement to each Member and Associate Member of FAO and to such non-member States as may become Party to this Agreement;

(b) arrange for the registration of this Agreement, upon its entry into force, with the Secretariat of the United Nations in accordance with Article 102 of the Charter of the United Nations;

(c) inform each Member and Associate Member of FAO and any non-member States as may become Party to this Agreement of:

  (i) instruments of acceptance deposited in accordance with Article X;

  (ii) the date of entry into force of this Agreement in accordance with Article XI;

  (iii) proposals for and the entry into force of amendments to this Agreement in accordance with Article XIII;

  (iv) withdrawals from this Agreement pursuant to Article XIV.

Article XVI

AUTHENTIC TEXTS

The Arabic, Chinese, English, French, and Spanish texts of this Agreement are equally authentic.

Appendix 3 –
FAO Code Of Conduct For Responsible Fisheries

CODE OF CONDUCT

FOR

RESPONSIBLE FISHERIES

FOOD AND AGRICULTURE ORGANIZATION OF THE UNITED NATIONS
ROME, 1995

Reprinted 1996, 2000

Bibliographic reference:
FAO.
Code of Conduct for Responsible Fisheries
Rome, FAO. 1995. 41 p.

ISBN 92-5-103834-1

CONTENTS

		Page
PREFACE		v
INTRODUCTION		1
Article 1 -	Nature and Scope of the Code	1
Article 2 -	Objectives of the Code	2
Article 3 -	Relationship with other International Instruments	3
Article 4 -	Implementation, Monitoring and Updating	3
Article 5 -	Special Requirements of Developing Countries	4
Article 6 -	General Principles	4
Article 7 -	Fisheries Management	8
Article 8 -	Fishing Operations	16
Article 9 -	Aquaculture Development	23
Article 10 -	Integration of Fisheries into Coastal Area Management	26
Article 11 -	Post-harvest Practices and Trade	28
Article 12 -	Fisheries Research	32
Annex 1	BACKGROUND TO THE ORIGIN AND ELABORATION OF THE CODE	35
Annex 2	RESOLUTION	40

PREFACE

From ancient times, fishing has been a major source of food for humanity and a provider of employment and economic benefits to those engaged in this activity. The wealth of aquatic resources was assumed to be an unlimited gift of nature. However, with increased knowledge and the dynamic development of fisheries after the second world war, this myth has faded in face of the realization that aquatic resources, although renewable, are not infinite and need to be properly managed, if their contribution to the nutritional, economic and social well-being of the growing world's population is to be sustained.

The widespread introduction in the mid-seventies of exclusive economic zones (EEZs) and the adoption in 1982, after long deliberations, of the United Nations Convention on the Law of the Sea provided a new framework for the better management of marine resources. The new legal regime of the ocean gave coastal States rights and responsibilities for the management and use of fishery resources within their EEZs which embrace some 90 percent of the world's marine fisheries. Such extended national jurisdiction was a necessary but insufficient step toward the efficient management and sustainable development of fisheries. Many coastal States continued to face serious challenges as, lacking, experience and financial and physical resources, they sought to extract greater benefits from the fisheries within their EEZs.

In recent years, world fisheries have become a market-driven, dynamically developing sector of the food industry and coastal States have striven to take advantage of their new opportunities by investing in modern fishing fleets and processing factories in response to growing international demand for fish and fishery products. By the late 1980s it became clear, however, that fisheries resources could no longer sustain such rapid and often uncontrolled exploitation and development, and that new approaches to fisheries management embracing conservation and environmental considerations were urgently needed. The situation was aggravated by the realization that unregulated fisheries on the high seas, in some cases involving straddling and highly migratory fish species, which occur within and outside EEZs, were becoming a matter of increasing concern.

The Committee on Fisheries (COFI) at its Nineteenth Session in March 1991 called for the development of new concepts which would lead to responsible, sustained fisheries. Subsequently, the International Conference on Responsible Fishing, held in 1992 in Cancún (Mexico) further requested. FAO to prepare an international Code of Conduct to address these concerns. The outcome of this Conference, particularly the Declaration of Cancún, was an important contribution to the 1992 United Nations Conference on Environment and Development (UNCED), in particular its Agenda 21. Subsequently, the United

Nations Conference on Straddling Fish Stocks and Highly Migratory Fish Stocks was convened, to which FAO provided important technical back-up. In November 1993, the Agreement to Promote Compliance with International Conservation and Management Measures by Fishing Vessels on the High Seas was adopted at the Twenty-seventh Session of the FAO Conference.

Noting these and other important developments in world fisheries, the FAO Governing Bodies recommended the formulation of a global Code of Conduct for Responsible Fisheries which would be consistent with these instruments and, in a non-mandatory manner, establish principles and standards applicable to tile conservation, management and development of all fisheries. The Code, which was unanimously adopted on 31 October 1995 by the FAO Conference, provides a necessary framework for national and international efforts to ensure sustainable exploitation of aquatic living resources in harmony with the environment.

FAO, in accordance with its mandate, is fully committed to assisting Member States, particularly developing countries, in the efficient implementation of the Code of Conduct for Responsible Fisheries and will report to the United Nations community on the progress achieved and further action required.

INTRODUCTION

Fisheries, including aquaculture, provide a vital source of food, employment, recreation, trade and economic well-being for people throughout the world, both for present and future generations and should therefore be conducted in a responsible manner. This Code sets out principles and international standards of behaviour for responsible practices with a view to ensuring the effective conservation, management and development of living aquatic resources, with due respect for the ecosystem and biodiversity. The Code recognizes the nutritional, economic, social, environmental and cultural importance of fisheries and the interests of all those concerned with the fishery sector. The Code takes into account the biological characteristics of the resources and their environment and the interests of consumers and other users. States and all those involved in fisheries are encouraged to apply the Code and give effect to it.

ARTICLE 1 - NATURE AND SCOPE OF THE CODE

1.1 This Code is voluntary. However, certain parts of it are based on relevant rules of international law, including those reflected in the United Nations Convention on the Law of the Sea of 10 December 1982[1]. The Code also contains provisions that may be or have already been given binding effect by means of other obligatory legal instruments amongst the Parties, such as the Agreement to Promote Compliance with International Conservation and Management Measures by Fishing Vessels on the High Seas, 1993, which, according to FAO Conference resolution 15/93, paragraph 3, forms an integral part of the Code.

1.2 The Code is global in scope, and is directed toward members and non-members of FAO, fishing entities, sub regional, regional and global organizations, whether governmental or non-governmental, and all persons concerned with the conservation of fishery resources and management and development of fisheries, such as fishers, those engaged in processing and marketing of fish and fishery products and other users of the aquatic environment in relation to fisheries.

1.3 The Code provides principles and standards applicable to the conservation, management and development of all fisheries. It also covers the capture, processing and trade of fish and fishery products, fishing operations, aquaculture, fisheries research and the integration of fisheries into coastal area management.

---

[1] References in this Code; to the United Nations Convention on the Law of the Sea, 1982, or to other international agreements do not prejudice the position of any State with respect to signature, ratification or accession to the Convention or with respect to such other agreements.

1.4 In this Code, the reference to States includes the European Community in matters within its competence, and the term fisheries applies equally to capture fisheries and aquaculture.

ARTICLE 2 - OBJECTIVES OF THE CODE

The objectives of the Code are to:

a) establish principles, in accordance with the relevant rules of international law, for responsible fishing and fisheries activities, taking into account all their relevant biological, technological, economic, social, environmental and commercial aspects;

b) establish principles and criteria for the elaboration and implementation of national policies for responsible conservation of fisheries resources and fisheries management and development;

c) serve as an instrument of reference to help States to establish or to improve the legal and institutional framework required for the exercise of responsible fisheries and in the formulation and implementation of appropriate measures;

d) provide guidance which may be used where appropriate in the formulation and implementation of international agreements and other legal instruments, both binding and voluntary;

e) facilitate and promote technical, financial and other cooperation in conservation of fisheries resources and fisheries management and development;

f) promote the contribution of fisheries to food security and food quality, giving priority to the nutritional needs of local communities;

g) promote protection of living aquatic resources and their environments and coastal areas;

h) promote the trade of fish and fishery products in conformity with relevant international rules and avoid the use of measures that constitute hidden barriers to such trade;

i) promote research on fisheries as well as on associated ecosystems and relevant environmental factors; and

j) provide standards of conduct for all persons involved in the fisheries sector.

ARTICLE 3 - RELATIONSHIP WITH OTHER INTERNATIONAL INSTRUMENTS

3.1 The Code is to be interpreted and applied in conformity with the relevant rules of international law, as reflected in the United Nations Convention on the Law of the Sea, 1982. Nothing in this Code prejudices the rights, jurisdiction and duties of States under international law as reflected in the Convention.

3.2 The Code is also to be interpreted and applied:

a) in a manner consistent with the relevant provisions of the Agreement for the Implementation of the Provisions of the United Nations Convention on the Law of the Sea of 10 December 1982 relating to the Conservation and Management of Straddling Fish Stocks and Highly Migratory Fish Stocks;

b) in accordance with other applicable rules of international law, including the respective obligations of States pursuant to international agreements to which they are party; and

c) in the light of the 1992 Declaration of Cancún, the 1992 Rio Declaration on Environment and Development, and Agenda 21 adopted by the United Nations Conference on Environment and Development (UNCED), in particular Chapter 17 of Agenda 21, and other relevant declarations and international instruments.

ARTICLE 4 - IMPLEMENTATION, MONITORING AND UPDATING

4.1 All members and non-members of FAO, fishing entities and relevant sub-regional, regional and global organizations, whether governmental or non-governmental, and all persons concerned with the conservation, management and utilization of fisheries resources and trade in fish and fishery products should collaborate in the fulfillment and implementation of the objectives and principles contained in this Code.

4.2 FAO, in accordance with its role within the United Nations system, will monitor the application and implementation of the Code and its effects on fisheries and the Secretariat will report accordingly to the Committee on Fisheries (COFI).

All States, whether members or non-members of FAO, as well as relevant international organizations, whether governmental or non-governmental should actively cooperate with FAO in this work.

4.3 FAO, through its competent bodies, may revise the Code, taking into account developments in fisheries as well as reports to COFI on the implementation of the Code.

4.4 States and international organizations, whether governmental or non-governmental, should promote the understanding of the Code among those involved in fisheries, including, where practicable, by the introduction of schemes which would promote voluntary acceptance of the Code and its effective application.

ARTICLE 5 - SPECIAL REQUIREMENTS OF DEVELOPING COUNTRIES

5.1 The capacity of developing countries to implement the recommendations of this Code should be duly taken into account.

5.2 In order to achieve the objectives of this Code and to support its effective implementation, countries, relevant international' organizations, whether governmental or non-governmental, and financial institutions should give full recognition to the special circumstances and requirements of developing countries, including in particular the least-developed among them, and small island developing countries. States, relevant intergovernmental and non-governmental organizations and financial institutions should work for the adoption of measures to address the needs of developing countries, especially in the areas of financial and technical assistance, technology transfer, training and scientific cooperation and in enhancing their ability to develop their own fisheries as well as to participate in high seas fisheries, including access to such fisheries.

ARTICLE 6 - GENERAL PRINCIPLES

6.1 States and users of living aquatic resources should conserve aquatic ecosystems. The right to fish carries with it the obligation to do so in a responsible manner so as to ensure effective conservation and management of the living aquatic resources.

6.2 Fisheries management should promote the maintenance of the quality, diversity and availability of fishery resources in sufficient quantities for present and future generations in the context of food security, poverty alleviation and sustainable development. Management measures should not only ensure the conservation of target species but also of species belonging to the same ecosystem or associated with or dependent upon the target species.

6.3 States should prevent over fishing and excess fishing capacity and should implement management measures to ensure that fishing effort is commensurate with the productive capacity of the fishery resources and their sustainable utilization. States should take measures to rehabilitate populations as far as possible and when appropriate.

6.4 Conservation and management decisions for fisheries should be based on the best scientific evidence available, also taking into account traditional knowledge of the resources and their habitat, as well as relevant environmental, economic and social factors. States should assign priority to undertake research and data collection in order to improve scientific and technical knowledge of fisheries including their interaction with the ecosystem. In recognizing the transboundary nature of many aquatic ecosystems, States should encourage bilateral and multilateral cooperation in research, as appropriate.

6.5 States and subregional and regional fisheries management organizations should apply a precautionary approach widely to conservation, management and exploitation of living aquatic resources in order to protect them and preserve the aquatic environment, taking account of the best scientific evidence available. The absence of adequate scientific information should not be used as a reason for postponing or failing to take measures to conserve target species, associated or dependent species and non-target species and their environment.

6.6 Selective and environmentally safe fishing gear and practices should be further developed and applied, to the extent practicable, in order to maintain biodiversity and to conserve the population structure and aquatic ecosystems and protect fish quality. Where proper selective and environmentally safe fishing gear and practices exist, they should be recognized and accorded a priority in establishing conservation arid management measures for fisheries. States and users of aquatic ecosystems should minimize waste, catch of non-target species, both fish and non-fish species, and impacts on associated or dependent species.

6.7 The harvesting, handling, processing and distribution of fish and fishery products should be carried out in a manner which will maintain the nutritional value, quality and safety of the products, reduce waste and minimize negative impacts on the environment.

6

6.8 All critical fisheries habitats in marine and fresh water ecosystems, such as wetlands, mangroves, reefs, lagoons, nursery and spawning areas, should be protected and rehabilitated as far as possible and where necessary. Particular effort should be made to protect such habitats from destruction, degradation, pollution and other significant impacts resulting from human activities that threaten the health and viability of the fishery resources.

6.9 States should ensure that their fisheries interests, including the need for conservation of the resources, are taken into account in the multiple uses of the coastal zone and are integrated into coastal area management, planning and development.

6.10 Within their respective competences and in accordance with international law, including within the framework of subregional or regional fisheries conservation and management organizations or arrangements, States should ensure compliance with and enforcement of conservation and management measures and establish effective mechanisms, as appropriate, to monitor and control the activities of fishing vessels and fishing support vessels.

6.11 States authorizing fishing and fishing support vessels to fly their flags should exercise effective control over those vessels so as to ensure the proper application of this Code. They should ensure that the activities of such vessels do not undermine the effectiveness of conservation and management measures taken in accordance with international law and adopted at the national, subregional, regional or global levels. States should also ensure that vessels flying their flags fulfill their obligations concerning the collection and provision of data relating to their fishing activities.

6.12 States should, within their respective competences and in accordance with international law, cooperate at subregional, regional and global levels through fisheries management organizations, other international agreements or other arrangements to promote conservation and management, ensure responsible fishing and ensure effective conservation and protection of living aquatic resources throughout their range of distribution, taking into account the need for compatible measures in areas within and beyond national jurisdiction.

6.13 States should, to the extent permitted by national laws and regulations, ensure that decision making processes are transparent and achieve timely solutions to urgent matters. States, in accordance with appropriate procedures, should facilitate consultation and the effective participation of industry, fishworkers, environmental and other interested organizations in decision–making with respect to the development of laws and policies related to fisheries management, development, international lending and aid.

6.14 International trade in fish and fishery products should be conducted in accordance with the principles, rights and obligations established in the World Trade Organization (WTO) Agreement and other relevant international agreements. States should ensure that their policies, programmes and practices related to trade in fish and fishery products do not result in obstacles to this trade, environmental degradation or negative social, including nutritional, impacts.

6.15 States should cooperate in order to prevent disputes. All disputes relating to fishing activities and practices should be resolved in a timely, peaceful and cooperative manner, in accordance with applicable international agreements or as may otherwise be agreed between the parties. Pending settlement of a dispute, the States concerned should make every effort to enter into provisional arrangements of a practical nature which should be without prejudice to the final outcome of any dispute settlement procedure.

6.16 States, recognising the paramount importance to fishers and fishfarmers of understanding the conservation and management of the fishery resources on which they depend, should promote awareness of responsible fisheries through education and training. They should ensure that fishers and fishfarmers are involved in the policy formulation and implementation process, also with a view to facilitating the implementation of the Code.

6.17 States should ensure that fishing facilities and equipment as well as all fisheries activities allow for safe, healthy and fair working and living conditions and meet internationally agreed standards adopted by relevant international organizations.

6.18 Recognizing the important contributions of artisanal and small-scale fisheries to employment, income and food security, States should appropriately protect the rights of fishers and fishworkers, particularly those engaged in subsistence, small-scale and artisanal fisheries, to a secure and just livelihood, as well as preferential access, where appropriate, to traditional fishing grounds and resources in the waters under their national jurisdiction.

6.19 States should consider aquaculture, including culture-based fisheries, as a means to promote diversification of income and diet. In so doing, States should ensure that resources are used responsibly and adverse impacts on the environment and on local communities are minimized.

8

ARTICLE 7 - FISHERIES MANAGEMENT

7.1 General

7.1.1 States and all those engaged in fisheries management should, through an appropriate policy, legal and institutional framework, adopt measures for the longterm conservation and sustainable use of fisheries resources. Conservation and management measures, whether at local, national, subregional or regional levels, should be based on the best scientific evidence available and be designed to ensure the long-term sustainability of fishery resources at levels which promote the objective of their optimum utilization and maintain their availability for present and future generations; short-term considerations should not compromise these objectives.

7.1.2 Within areas under national jurisdiction, States should seek to identify relevant domestic parties having a legitimate interest in the use and management of fisheries resources and establish arrangements for consulting them to gain their collaboration in achieving responsible fisheries.

7.1.3 For transboundary fish stocks, straddling fish stocks, highly migratory fish stocks and high seas fish stocks, where these are exploited by two or more States, the States concerned, including the relevant coastal States in the case of straddling and highly migratory stocks, should cooperate to ensure effective conservation and management of the resources. This should be achieved, where appropriate, through the establishment of a bilateral, subregional or regional fisheries organization or arrangement.

7.1.4 A subregional or regional fisheries management organization or arrangement should include representatives of States in whose jurisdictions the resources occur, as well as representatives from States which have a real interest in the fisheries or the resources outside national jurisdictions. Where a subregional or regional fisheries management organization or arrangement exists and has the competence to establish conservation and management measures, those States should cooperate by becoming a member of such organization or a participant in such arrangement, and actively participate in its work.

7.1.5 A State which is not a member of a subregional or regional fisheries management organization or is not a participant in a subregional or regional fisheries management arrangement should nevertheless cooperate, in accordance with relevant international agreements and international law, in the conservation and management of the relevant fisheries resources by giving effect to any conservation and management measures adopted by such organization or arrangement.

7.1.6 Representatives from relevant organizations, both governmental and non-governmental, concerned with fisheries should be afforded the opportunity to take part in meetings of subregional and regional fisheries management organizations and arrangements as observers or otherwise, as appropriate, in accordance with the procedures of the organization or arrangement concerned. Such representative should be given timely access to the records and reports of such meetings, subject to the procedural rules on access to them.

7.1.7 States should establish, within their respective competences and capacities. effective mechanisms for fisheries monitoring, surveillance, control and enforcement to ensure compliance with their conservation and management measures, as well as those adopted by subregional or regional organizations or arrangements.

7.1.8 States should take measures to prevent or eliminate excess fishing capacity and should ensure that levels of fishing effort are commensurate with the sustainable use of fishery resources as a means of ensuring the effectiveness of conservation and management measures.

7.1.9 States and subregional or regional fisheries management organizations and arrangements should ensure transparency in the mechanisms for fisheries management and in the related decision-making process.

7.1.10 States and subregional or regional fisheries management organizations and arrangements should give due publicity to conservation and management measures and ensure that laws, regulations and other legal rules governing their implementation are effectively disseminated. The bases and purposes of such measures should be explained to users of the resource in order to facilitate their application and thus gain increased support in the implementation of such measures.

7.2 Management objectives

7.2.1 Recognizing that long-term sustainable use of fisheries resources is the overriding objective of conservation and management, States and subregional or regional fisheries management organizations and arrangements should, *inter alia,* adopt appropriate measures, based on the best scientific evidence available, which are designed to maintain or restore stocks at levels capable of producing maximum sustainable yield, as qualified by relevant environmental and economic factors, including the special requirements of developing countries.

7.2.2 Such measures should provide *inter alia* that:

a) excess fishing capacity is avoided and exploitation of the stocks remains economically viable;

b) the economic conditions under which fishing industries operate promote responsible fisheries;

c) the interests of fishers, including those engaged in subsistence, small-scale and artisanal fisheries, are taken into account;

d) biodiversity of aquatic habitats and ecosystems is conserved and endangered species are protected;

e) depleted stocks are allowed to recover or, where appropriate, are actively restored;

f) adverse environmental impacts on the resources from human activities are assessed and, where appropriate, corrected; and

g) pollution, waste, discards, catch by lost or abandoned gear, catch of non-target species, both fish and non-fish species, and impacts on associated or dependent species are minimized, through measures including, to the extent practicable, the development and use of selective, environmentally safe and cost-effective fishing gear and techniques.

7.2.3 States should assess the impacts of environmental factors on target stocks and species belonging to the same ecosystem or associated with or dependent upon the target stocks, and assess the relationship among the populations in the ecosystem.

7.3 Management framework and procedures

7.3.1 To be effective, fisheries management should be concerned with the whole stock unit over its entire area of distribution and take into account previously agreed management measures established and applied in the same region, all removals and the biological unity and other biological characteristics of the stock. The best scientific evidence available should be used to determine, *inter alia*, the area of distribution of the resource and the area through which it migrates during its life cycle.

7.3.2 In order to conserve and manage transboundary fish stocks, straddling fish stocks, highly migratory fish stocks and high seas fish stocks throughout their range, conservation and management measures established for such stocks in accordance with the respective competences of relevant States or, where appropriate, through subregional and regional fisheries management organizations and arrangements, should be compatible. Compatibility should be achieved in a manner consistent with the rights, competences and interests of the States concerned.

7.3.3 Long-term management objectives should be translated into management actions, formulated as a fishery management plan or other management framework.

7.3.4 States and, where appropriate, subregional or regional fisheries management organizations and arrangements should foster and promote international cooperation and coordination in all matters related to fisheries, including information gathering and exchange, fisheries research, management and development.

7.3.5 States seeking to take any action through a non-fishery organization which may affect the conservation and management measures taken by a competent subregional or regional fisheries management organization or arrangement should consult with the latter, in advance to the extent practicable, and take its views into account.

7.4 Data gathering and management advice

7.4.1 When considering the adoption of conservation and management measures, the best scientific evidence available should be taken into account in order to evaluate the current state of the fishery resources and the possible impact of the proposed measures on the resources.

7.4.2 Research in support of fishery conservation and management should be promoted, including research on the resources and on the effects of climatic, environmental and socio-economic factors. The results of such research should be disseminated to interested parties.

7.4.3 Studies should be promoted which provide an understanding of the costs, benefits and effects of alternative management options designed to rationalize fishing, in particular, options relating to excess fishing capacity and excessive levels of fishing effort.

7.4.4 States should ensure that timely, complete and reliable statistics on catch and fishing effort are collected and maintained in accordance with applicable international standards and practices and in sufficient detail to allow sound statistical analysis. Such data should be updated regularly and verified through an appropriate system. States should compile and disseminate such data in a manner consistent with any applicable confidentiality requirements.

7.4.5 In order to ensure sustainable management of fisheries and to enable social and economic objectives to be achieved, sufficient knowledge of social, economic and institutional factors should be developed through data gathering, analysis and research.

7.4.6 States should compile fishery-related and other supporting scientific data relating to fish stocks covered by subregional or regional fisheries management organizations or arrangements in an internationally agreed format and provide them in a timely manner to the organization or arrangement. In cases of stocks which occur in the jurisdiction of more than one State and for which there is no such organization or arrangement, the States concerned should agree on a mechanism for cooperation to compile and exchange such data.

7.4.7 Subregional or regional fisheries management organizations or arrangements should compile data and make them available, in a manner consistent with any applicable confidentiality requirements, in a timely manner and in an agreed format to all members of these organizations and other interested parties in accordance with agreed procedures.

7.5 Precautionary approach

7.5.1 States should apply the precautionary approach widely to conservation, management and exploitation of living aquatic resources in order to protect them and preserve the aquatic environment. The absence of adequate scientific information should not be used as a reason for postponing or failing to take conservation and management measures.

7.5.2 In implementing the precautionary approach, States should take into account, *inter* alia, uncertainties relating to the size and productivity of the stocks, reference points, stock condition in relation to such reference points, levels and distribution of fishing mortality and the impact of fishing activities, including discards, on non-target and associated or dependent species as well as environmental and socio-economic conditions.

7.5.3 States and subregional or regional fisheries management organizations and arrangements should, on the basis of the best scientific evidence available, *inter alia*, determine:

a) stock specific target reference points, and, at the same time, the action to be taken if they are exceeded; and

b) stock specific limit reference points and, at the same time, the action to be taken if they are exceeded; when a limit reference point is approached, measures should be taken to ensure that it will not be exceeded.

7.5.4 In the case of new or exploratory fisheries, States should adopt as soon as possible cautious conservation and management measures, including, *inter alia*, catch limits and effort limits. Such measures should remain in force until there are sufficient data to allow assessment of the impact of the fisheries on the long-term sustainability of the stocks, whereupon conservation and management measures based on that assessment should be implemented. The latter measures should, if appropriate, allow for the gradual development of the fisheries.

7.5.5 If a natural phenomenon has a significant adverse impact on the status of living aquatic resources, States should adopt conservation and management measures on an emergency basis to ensure that fishing activity does not exacerbate such adverse impact. States should also adopt such measures on an emergency basis where fishing activity presents a serious threat to the sustainability of such resources. Measures taken on an emergency basis should be temporary and should be based on the best scientific evidence available.

7.6 Management measures

7.6.1 States should ensure that the level of fishing permitted is commensurate with the state of fisheries resources.

7.6.2 States should adopt measures to ensure that no vessel be allowed to fish unless so authorized, in a manner consistent with international law for the high seas or in conformity with national legislation within areas of national jurisdiction.

7.6.3 Where excess fishing capacity exists, mechanisms should be established to reduce capacity to levels commensurate with the sustainable use of fisheries resources so as to ensure that fishers operate under economic conditions that promote responsible fisheries. Such mechanisms should include monitoring the capacity of fishing fleets.

7.6.4 The performance of all existing fishing gear, methods and practices should be examined and measures taken to ensure that fishing gear, methods and practices which are not consistent with responsible fishing are phased out and replaced with more acceptable alternatives. In this process, particular attention should be given to the impact of such measures on fishing communities, including their ability to exploit the resource.

7.6.5 States and fisheries management organizations and arrangements should regulate fishing in such a way as to avoid the risk of conflict among fishers using different vessels, gear and fishing methods.

7.6.6 When deciding on the use, conservation and management of fisheries resources, due recognition should be given, as appropriate, in accordance with national laws and regulations, to the traditional practices, needs and interests of indigenous people and local fishing communities which are highly dependent on fishery resources for their livelihood.

7.6.7 In the evaluation of alternative conservation and management measures, their cost-effectiveness and social impact should be considered.

7.6.8 The efficacy of conservation and management measures and their possible interactions should be kept under continuous review. Such measures should, as appropriate, be revised or abolished in the light of new information.

7.6.9 States should take appropriate measures to minimize waste, discards, catch by lost or abandoned gear, catch of non-target species, both fish and non-fish species, and negative impacts on associated or dependent species, in particular endangered species. Where appropriate, such measures may include technical measures related to fish size, mesh size or gear, discards, closed seasons and areas and zones reserved for selected fisheries, particularly artisanal fisheries. Such measures should be applied, where appropriate, to protect juveniles and spawners. States and sub-regional or regional fisheries management organizations and arrangements should promote, to the extent practicable, the development and use of selective, environ-mentally safe and cost effective gear and techniques.

7.6.10 States and subregional and regional fisheries management organizations and arrangements, in the framework of their respective competences, should introduce measures for depleted resources and those resources threatened with depletion that facilitate the sustained recovery of such stocks. They should make every effort to ensure that resources and habitats critical to the well-being of such resources which have been adversely affected by fishing or other human activities are restored.

7.7 Implementation

7.7.1 States should ensure that an effective legal and administrative framework at the local and national level, as appropriate, is established for fisheries resource conservation and fisheries management.

7.7.2 States should ensure that laws and regulations provide for sanctions applicable in respect of violations which are adequate in severity to be effective, including sanctions which allow for the refusal, withdrawal or suspension of authorizations to fish in the event of non-compliance with conservation and management measures in force.

7.7.3 States, in conformity with their national laws, should implement effective fisheries monitoring, control, surveillance and law enforcement measures including, where appropriate, observer programmes, inspection schemes and vessel monitoring systems. Such measures should be promoted and, where appropriate, implemented by subregional or regional fisheries management organizations and arrangements in accordance with procedures agreed by such organizations or arrangements.

7.7.4 States and subregional or regional fisheries management organizations and arrangements, as appropriate, should agree on the means by which the activities of such organizations and arrangements will be financed, bearing in mind, *inter alia*, the relative benefits derived from the fishery and the differing capacities of countries to provide financial and other contributions. Where appropriate, and when possible, such organizations and arrangements should aim to recover the costs of fisheries conservation, management and research.

7.7.5 States which are members of or participants in subregional or regional fisheries management organizations or arrangements should implement internationally agreed measures adopted in the framework of such organizations or arrangements and consistent with international law to deter the activities of vessels flying the flag of non-members or non-participants which engage in activities which undermine the effectiveness of conservation and management measures established by such organizations or arrangements.

7.8 Financial institutions

7.8.1 Without prejudice to relevant international agreements, States should encourage banks and financial institutions not to require, as a condition of a loan or mortgage, fishing vessels or fishing support vessels to be flagged in a jurisdiction other than that of the State of beneficial ownership where such a requirement would have the effect of increasing the likelihood of non-compliance with international conservation and management measures.

ARTICLE 8 - FISHING OPERATIONS

8.1 Duties of all States

8.1.1 States should ensure that only fishing operations allowed by them are conducted within waters under their jurisdiction and that these operations are carried out in a responsible manner.

8.1.2 States should maintain a record, updated at regular intervals, on all authorizations to fish issued by them.

8.1.3 States should maintain, in accordance with recognized international standards and practices, statistical data, updated at regular intervals, on all fishing operations allowed by tem.

8.1.4 States should, in accordance with international law, within the framework of subregional or regional fisheries management organizations or arrangements, cooperate to establish systems for monitoring, control, surveillance and enforcement of applicable measures with respect to fishing operations and related activities in waters outside their national jurisdiction.

8.1.5 States should ensure that health and safety standards are adopted for everyone employed in fishing operations. Such standards should be not less than the minimum requirements of relevant international agreements on conditions of work and service.

8.1.6 States should make arrangements individually, together with other States or with the appropriate international organization to integrate fishing operations into maritime search and rescue systems.

8.1.7 States should enhance through education and training programmes the education and skills of fishers and, where appropriate, their professional qualifications. Such programmes should take into account agreed international standards and guidelines.

8.1.8 States should, as appropriate, maintain records of fishers which should, whenever possible, contain information on their service and qualifications, including certificates of competency, in accordance with their national laws.

8.1.9 States should ensure that measures applicable in respect of masters and other officers charged with an offence relating to the operation of fishing vessels should include provisions which may permit, *inter alia,* refusal, withdrawal or suspension of authorizations to serve as masters or officers of a fishing vessel.

8.1.10 States, with the assistance of relevant international organizations, should endeavour to ensure through education and training that all those engaged in fishing operations be given information on the most important provisions of this Code, as well as provisions of relevant international conventions and applicable environmental and other standards that are essential to ensure responsible fishing operations.

8.2 Flag State duties

8.2.1 Flag States should maintain records of fishing vessels entitled to fly their flag and authorized to be used for fishing and should indicate in such records details of the vessels, their ownership and authorization to fish.

8.2.2 Flag States should ensure that no fishing vessels entitled to fly their flag fish on the high seas or in waters under the jurisdiction of other States unless such vessels have been issued with a Certificate of Registry and have been authorized to fish by the competent authorities. Such vessels should carry on board the Certificate of Registry and their authorization to fish.

8.2.3 Fishing vessels authorized to fish on the high seas or in waters under the jurisdiction of a State other than the flag State, should be marked in accordance with uniform and internationally recognizable vessel marking systems such as the FAO Standard Specifications and Guidelines for Marking and Identification of Fishing Vessels.

8.2.4 Fishing gear should be marked in accordance with national legislation in order, that the owner of the gear can be identified. Gear marking requirements should take into account uniform and internationally recognizable gear marking systems.

8.2.5 Flag States should ensure compliance with appropriate safety requirements for fishing vessels and fishers in accordance with international conventions, internationally agreed codes of practice and voluntary guidelines. States should adopt appropriate safety requirements for all small vessels not covered by such international conventions, codes of practice or voluntary guidelines.

8.2.6 States not party to the Agreement to Promote Compliance with International Conservation and Management Measures by Vessels Fishing in the High Seas should be encouraged to accept the Agreement and to adopt laws and regulations consistent with the provisions of the Agreement.

8.2.7 Flag States should take enforcement measures in respect of fishing vessels entitled to fly their flag which have been found by them to have contravened applicable conservation and management measures, including, where appropriate, making the contravention of such measures an offence under national legislation. Sanctions applicable in respect of violations should be adequate in severity to be effective in securing compliance and to discourage violations wherever they occur and should deprive offenders of the benefits accruing from their illegal activities. Such sanctions may, for serious violations, include provisions for the refusal, withdrawal or suspension of the authorization to fish.

8.2.8 Flag States should promote access to insurance coverage by owners and charterers of fishing vessels. Owners or charterers of fishing vessels should carry sufficient insurance cover to protect the crew of such vessels and their interests, to indemnify third parties against loss or damage and to protect their own interests.

8.2.9 Flag States should ensure that crew members are entitled to repatriation, taking account of the principles laid down in the "Repatriation of Seafarers Convention (Revised), 1987, (No.166)".

8.2.10 In the event of an accident to a fishing vessel or persons on board a fishing vessel, the flag State of the fishing vessel concerned should provide details of the accident to the State of any foreign national on board the vessel involved in the accident. Such information should also, where practicable, be communicated to the International Maritime Organization.

8.3 Port State duties

8.3.1 Port States should take, through procedures established in their national legislation, in accordance with international law, including applicable international agreements or arrangements, such measures as are necessary to achieve and to assist other States in achieving the objectives of this Code, and should make known to other States details of regulations and measures they have established for this purpose. When taking such measures a port State should not discriminate in form or in fact against the vessels of any other State.

8.3.2 Port States should provide such assistance to flag States as is appropriate, in accordance with the national laws of the port State and international law, when a fishing vessel is voluntarily in a port or at an offshore terminal of the port State and the flag State of the vessel requests the port State for assistance in respect of non-compliance with subregional, regional or global conservation and management measures or with internationally agreed minimum standards for the prevention, of pollution and for safety, health and conditions of work on board fishing vessels.

8.4 Fishing operations

8.4.1 States should ensure that fishing is conducted with due regard to the safety of human life and the International Maritime Organization International Regulations for Preventing Collisions at Sea, as well as International Maritime Organization requirements relating to the organization of marine traffic, protection of the marine environment and the prevention of damage to or loss of fishing gear.

8.4.2 States should prohibit dynamiting, poisoning and other comparable destructive fishing practices.

8.4.3 States should make every effort to ensure that documentation with regard to fishing operations, retained catch of fish and non-fish species and, as regards discards, the information required for stock assessment as decided by relevant management bodies, is collected and forwarded systematically to those bodies. States should, as far as possible, establish programmes, such as observer and inspection schemes, in order to promote compliance with applicable measures.

8.4.4 States should promote the adoption of appropriate technology, taking into account economic conditions, for the best use and care of the retained catch.

8.4.5 States, with relevant groups from industry, should encourage the development and implementation of technologies and operational methods that reduce discards. The use of fishing gear and practices that lead to the discarding of catch should be discouraged and the use of fishing gear and practices that increase survival rates of escaping fish should be promoted.

8.4.6 States should cooperate to develop and apply technologies, materials and operational methods that minimize the loss of fishing gear and the ghost fishing effects of lost or abandoned fishing gear.

8.4.7 States should ensure that assessments of the implications of habitat disturbance are carried out prior to the introduction on a commercial scale of new fishing gear, methods and operations to an area.

8.4.8 Research on the environmental and social impacts of fishing gear and, in particular, on the impact of such gear on biodiversity and coastal fishing communities should be promoted.

8.5 Fishing gear selectivity

8.5.1 States should require that fishing gear, methods and practices, to the extent practicable, are sufficiently selective so as to minimize waste, discards, catch of non-target species, both fish and non-fish species, and impacts on associated or dependent species and that the intent of related regulations is not circumvented by technical devices. In this regard, fishers should cooperate in the development of .selective fishing gear and methods. States should ensure that information on new developments and requirements is made available to all fishers.

8.5.2 In order to improve selectivity, States should, when drawing up their laws and regulations, take into account the range of selective fishing gear, methods and strategies available to the industry.

8.5.3 States and relevant institutions should collaborate in developing standard methodologies for research into fishing gear selectivity, fishing methods and strategies.

8.5.4 International cooperation should be encouraged with respect to research programmes for fishing gear selectivity, and fishing methods and strategies, dissemination of the results of such research programmes and the transfer of technology.

8.6 Energy optimization

8.6.1 States should promote the development of appropriate standards and guidelines which would lead to the more efficient use of energy in harvesting and postharvest activities within the fisheries sector.

8.6.2 States should promote the development and transfer of technology in relation to energy optimization within the fisheries sector and, in particular, encourage owners, charterers and managers of fishing vessels to fit energy optimization devices to their vessels.

8.7 Protection of the aquatic environment

8.7.1 States should introduce and enforce laws and regulations based on the International Convention for the Prevention of Pollution from Ships, 1973, as modified by the Protocol of 1978 relating there to (MARPOL 73/78).

8.7.2 Owners, charterers and managers of fishing vessels should ensure that their vessels are fitted with appropriate equipment as required by MARPOL 73/78 and should consider fitting a shipboard compactor or incinerator to relevant classes of vessels in order to treat garbage and other shipboard wastes generated during the vessel's normal service.

8.7.3 Owners, charterers and managers of fishing vessels should minimize the taking aboard of potential garbage through proper provisioning practices.

8.7.4 The crew of fishing vessels should be conversant with proper shipboard procedures in order to ensure discharges do not exceed the levels set by MARPOL. 73/78. Such procedures should, as a minimum, include the disposal of oily waste and the handling and storage of shipboard garbage.

8.8 Protection of the atmosphere

8.8.1 States should adopt relevant standards and guidelines which would include provisions for the reduction of dangerous substances in exhaust gas emissions.

8.8.2 Owners, charterers and managers of fishing vessels should ensure that their vessels are fitted with equipment to reduce emissions of ozone depleting substances. The responsible crew members of fishing vessels should be conversant with the proper running and maintenance of machinery on board.

8.8.3 Competent authorities should make provision for the phasing out of the use of chlorofluorocarbons (CFCs) and transitional substances such as hydrochlorofluorocarbons (HCFCs) in the refrigeration systems of fishing vessels and should ensure that the shipbuilding industry and those engaged in the fishing industry are informed of and comply with such provisions.

8.8.4 Owners or managers of fishing vessels should take appropriate action to refit existing vessels with alternative refrigerants to CFCs and HCFCs and alternatives to Halons in fire fighting installations. Such alternatives should be used in specifications for all new fishing vessels.

8.8.5 States and owners, charterers and managers of fishing vessels as well as fishers should follow international guidelines for the disposal of CFCs, HCFCs and Halons.

8.9 Harbours and landing places for fishing vessels

8.9.1 States should take into account, inter alia, the following in the design and construction of harbours and landing places:

a) safe havens for fishing vessels and adequate servicing facilities for vessels, vendors and buyers are provided;

b) adequate freshwater supplies and sanitation arrangements should be provided;

c) waste disposal systems should be introduced, including for the disposal of oil, oily water and fishing gear;

d) pollution from fisheries activities and external sources should be minimized; and

e) arrangements should be made to combat the effects of erosion and siltation.

8.9.2 States should establish an institutional framework for the selection or improvement of sites for harbours for fishing vessels which allows for consultation among the authorities responsible for coastal area management.

8.10 Abandonment of structures and other materials

8.10.1 States should ensure that the standards and guidelines for the removal of redundant offshore structures issued by the International Maritime Organization are followed. States should also ensure that the competent fisheries authorities are consulted prior to decisions being made on the abandonment of structures and other materials by the relevant authorities.

8.11 Artificial reef's and fish aggregation devices

8.11.1 States, where appropriate, should develop policies for increasing stock populations and enhancing fishing opportunities through the use of artificial structures, placed with due regard to the safety of navigation, on or above the seabed or at the surface. Research into the use of such structures, including the impacts on living marine resources and the environment, should be promoted.

8.11.2 States should ensure that, when selecting the materials to be used in the creation of artificial reefs as well as when selecting the geographical location of such artificial reefs, the provisions of relevant international conventions concerning the environment and safety of navigation are observed.

8.11.3 States should, within the framework of coastal area management plan, establish management systems for artificial reefs and fish aggregation devices. Such management systems should require approval for the construction and deployment of such reefs and devices and should take into account the interests of fishers, including artisanal and subsistence fishers.

8.11.4 States should ensure that the authorities responsible for maintaining cartographic records and charts for the purpose of navigation, as well as relevant environmental authorities, are informed prior to the placement or removal of artificial reefs or fish aggregation devices.

ARTICLE 9 - AQUACULTURE DEVELOPMENT

9.1 Responsible development of aquaculture, including culture-based fisheries, in areas under national jurisdiction

9.1.1 States should establish, maintain and develop an appropriate legal and admi-nistrative framework which facilitates the development of responsible aquaculture.

9.1.2 States should promote responsible development and management of aquaculture, including an advance evaluation of the effects of aquaculture development on genetic diversity and ecosystem integrity, based on the best available scientific information.

9.1.3 States should produce and regularly update aquaculture development strategies and plans, as required, to ensure that aquaculture development is ecologically sustainable and to allow the rational use of resources shared by aquaculture and other activities.

9.1.4 States should ensure that the livelihoods of local communities, and their access to fishing grounds, are not negatively affected by aquaculture developments.

9.1.5 States should establish effective procedures specific to aquaculture to undertake appropriate environmental assessment and monitoring with the aim of minimizing adverse ecological changes and related economic and social consequences resulting from water extraction, land use, discharge of effluents, use of drugs and chemicals, and other aquaculture activities.

9.2 Responsible development of aquaculture including culture-based fisheries within transboundary aquatic ecosystems

9.2.1 States should protect transboundary aquatic ecosystems by supporting responsible aquaculture practices within their national jurisdiction and by cooperation in the promotion of sustainable aquaculture practices.

9.2.2 States should, with due respect to their neighbouring States and in accordance with. international law, ensure responsible choice of species, siting and management of aquaculture activities which could affect transboundary aquatic ecosystems.

9.2.3 States should consult with their neighbouring States, as appropriate, before .introducing non-indigenous species into transboundary aquatic ecosystems.

9.2.4 States should establish appropriate mechanisms, such as databases and information networks to collect, share and disseminate data related to their aquaculture activities to facilitate cooperation on planning for aquaculture development at the national, subregional, regional and global level.

9.2.5 States should cooperate in the development of appropriate mechanisms, when required, to monitor the impacts of inputs used in aquaculture.

9.3 Use of aquatic genetic resources for the purposes of aquaculture including culture-based fisheries

9.3.1 States should conserve genetic diversity and maintain integrity of aquatic communities and ecosystems by appropriate management. In particular, efforts should be undertaken to minimize the harmful effects of introducing non-native species or genetically altered stocks used for aquaculture including culture-based fisheries into waters, especially where there is a significant potential for the spread of such non-native species or genetically altered stocks into waters under the jurisdiction of other States as well as waters under the jurisdiction of the State of origin. States should, whenever possible, promote steps to minimize adverse genetic, Disease and other effects of escaped farmed fish on wild stocks.

9.3.2 States should cooperate in the elaboration, adoption and implementation of international codes of practice and procedures for introductions and transfers of aquatic organisms.

9.3.3 States should, in order to minimize risks of disease transfer and other adverse effects on wild and cultured stocks, encourage adoption of appropriate practices in the genetic improvement of broodstocks, the introduction of non-native species, and in the production, sale and transport of eggs, larvae or fry, broodstock or other live materials. States should facilitate the preparation and implementation of appropriate national codes of practice and procedures to this effect.

9.3.4 States should promote the use of appropriate procedures for the selection of broodstock and the production of eggs, larvae and fry.

9.3.5 States should, where appropriate, promote research and, when feasible, the development of culture techniques for endangered species to protect, rehabilitate and enhance their stocks, taking into account the critical need to conserve genetic diversity of endangered species.

9.4 Responsible aquaculture at the production level

9.4.1 States should promote responsible aquaculture practices in support of rural communities, producer organizations and fish farmers.

9.4.2 States should promote active participation of fishfarmers and their communities in the development of responsible aquaculture management practices.

9.4.3 States should promote efforts which improve selection and use of appropriate feeds, feed additives and fertilizers, including manures.

9.4.4 States should promote effective farm and fish health management practices favouring hygienic measures and vaccines. Safe, effective and minimal use of therapeutants, hormones and drugs, antibiotics and other disease control chemicals should be ensured.

9.4.5 States should regulate the use of chemical inputs in aquaculture which are hazardous to human health and the environment.

9.4.6 States should require that the disposal of wastes such as offal, sludge, dead or diseased fish, excess veterinary drugs and other hazardous chemical inputs does not constitute a hazard to human health and the environment.

9.4.7 States should ensure the food safety of aquaculture products and promote efforts which maintain product quality and improve their value through particular care before and during harvesting and on-site processing and in storage and transport of the products.

ARTICLE 10 - INTEGRATION OF FISHERIES INTO COASTAL AREA MANAGEMENT

10.1 Institutional framework

10.1.1 States should ensure that an appropriate policy, legal and institutional framework is adopted to achieve the sustainable and integrated use of the resources, taking into account the fragility of coastal ecosystems and the finite nature of their natural resources and the needs of coastal communities.

10.1.2 In view of the multiple uses of the coastal area, States should ensure that representatives of the fisheries sector and fishing communities are consulted in the decision-making processes and involved in other activities related to coastal area management planning and development.

10.1.3 States should develop, as appropriate, institutional and legal frameworks in order to determine the possible uses of coastal resources and to govern access to them taking into account the rights of coastal fishing communities and their customary practices to the extent compatible with sustainable development.

10.1.4 States should facilitate the adoption of fisheries practices that avoid conflict among fisheries resources users and between them and other users of the coastal area.

10.1.5 States should promote the establishment of procedures and mechanisms at the appropriate administrative level to settle conflicts which arise within the fisheries sector and between fisheries resource users and other users of the coastal area.

10.2 Policy measures

10.2.1 States should promote the creation of public awareness of the need for the protection and management of coastal resources and the participation in the management process by those affected.

10.2.2 In order to assist decision-making on the allocation and use of coastal resources, States should promote the assessment of their respective value taking into account economic, social and cultural factors.

10.2.3 In setting policies for the management of coastal areas, States should take due account of the risks and uncertainties involved.

10.2.4 States, in accordance with their capacities, should establish or promote the establishment of systems to monitor the coastal environment as part of the coastal management process using physical, chemical, biological, economic and .social parameters. .

10.2.5 States should promote multidisciplinary research in support of coastal area management, in particular on its environmental, biological, economic, social, legal and institutional aspects.

10.3 Regional cooperation

10.3.1 States with neighbouring coastal areas should cooperate with one another to facilitate the sustainable use of coastal resources and the conservation of the environment.

10.3.2 In the case of activities that may have an adverse transboundary environmental effect on coastal areas, States should:

a) provide timely information and, if possible, prior notification to potentially affected States;

b) consult with those States as early as possible.

10.3.3 States should cooperate at the subregional and regional level in order to improve coastal area management.

10.4 Implementation

10.4.1 States should establish mechanisms for cooperation and coordination among national authorities involved in planning, development, conservation and management of coastal areas.

10.4.2 States should ensure that the authority or authorities representing the fisheries sector in the coastal management process have the appropriate technical capacities and financial resources.

ARTICLE 11 - POST-HARVEST PRACTICES AND TRADE

11.1 Responsible fish utilization

11.1.1 States should adopt appropriate measures to ensure the right of consumers to safe, wholesome and unadulterated fish and fishery products.

11.1.2 States should establish and maintain effective national safety and quality assurance systems to protect consumer health and prevent commercial fraud.

11.1.3 States should set minimum standards for safety and quality assurance and make sure that these standards are effectively applied throughout the industry. They should promote the implementation of quality standards agreed within the context of the FAO/WHO Codex Alimentarius Commission and other relevant organizations or arrangements.

11.1.4 States should cooperate to achieve harmonization, or mutual recognition, or both, of national sanitary measures and certification programmes as appropriate and explore possibilities for the establishment of mutually recognized control and certification agencies.

11.1.5 States should give due consideration to the economic and social role of the post-harvest fisheries sector when formulating national policies for the sustainable development and utilization of fishery resources.

11.1.6 States and relevant organizations should sponsor research in fish technology and quality assurance and support projects to improve post-harvest handling of fish, taking into account the economic, social, environmental and nutritional impact of such projects.

11.1.7 States, noting the existence of different production methods, should through cooperation and by facilitating the development and transfer of appropriate technologies, ensure that processing, transporting and storage methods are environmentally sound.

11.1.8 States should encourage those involved in fish processing, distribution and marketing to:

a) reduce post-harvest losses and waste;

b) improve the use of by-catch to the extent that this is consistent with responsible fisheries management practices; and

c) use the resources, especially water and energy, in particular wood, in an environmentally sound manner.

11.1.9 States should encourage the use of fish for human, consumption and promote consumption of fish whenever appropriate.

11.1.10 States should cooperate in order to facilitate the production of value-added products by developing countries.

11.1.11 States should ensure that international and domestic trade in fish and fishery products accords with sound conservation and management practices through improving the identification of the origin of fish and fishery products traded.

11.1.12 States should ensure that environmental effects of post-harvest activities are considered in the development of related laws, regulations and policies without creating any market distortions.

11.2 Responsible international trade

11.2.1 The provisions of this Code should be interpreted and applied in accordance with the principles, rights and obligations established in the World Trade Organization (WTO) Agreement.

11.2.2 International trade in fish and fishery products should not compromise the sustainable development of fisheries and responsible utilization of living aquatic resources.

11.2.3 States should ensure that measures affecting international trade in fish and fishery products are transparent, based, when applicable, on scientific evidence, and are in accordance with internationally agreed rules.

11.2.4 Fish trade measures adopted by States to protect human or animal life or health, the interests of consumers or the environment, should not be discriminatory and should be in accordance with internationally agreed trade rules, in particular the principles, rights and obligations established in the Agreement on the Application of Sanitary and Phytosanitary Measures and the Agreement on Technical Barriers to Trade of the WTO.

11.2.5 States should further liberalize trade in fish and fishery products and eliminate barriers and distortions to trade such as duties, quotas and non-tariff barriers in accordance with the principles, rights and obligations of the WTO Agreement.

11.2.6 States should not directly or indirectly create unnecessary or hidden barriers to trade which limit the consumer's freedom of choice of supplier or that restrict market access.

11.2.7 States should not condition access to markets to access to resources. This principle does not preclude the possibility of fishing agreements between States which include provisions referring to access to resources, trade and access to markets, transfer of technology, scientific research, training and other relevant elements.

11.2.8 States should not link access to markets to the purchase of specific technology or sale of other products.

11.2.9 States should cooperate in complying with relevant international agreements regulating trade in endangered species.

11.2.10 States should develop international agreements for trade in live speci-mens where there is a risk of environmental damage in importing or exporting States.

11.2.11 States should cooperate to promote adherence to, and effective implementation of relevant international standards for trade in fish and fishery products and living aquatic resource conservation.

11.2.12 States should not undermine conservation measures for living aquatic resources in order to gain trade or investment benefits.

11.2.13 States should cooperate. to develop internationally acceptable rules or standards for trade in fish and fishery products in accordance with the principles, rights, and obligations established in the WTO Agreement.

11.2.14 States should cooperate with each other and actively participate in relevant regional and multilateral fora, such as the WTO in order to ensure equitable, nondiscriminatory trade in fish and fishery products as well as wide adherence to multilaterally agreed fishery conservation measures.

11.2.15 States, aid agencies, multilateral development banks and other relevant international organizations should ensure that their policies and practices related to the promotion of international fish trade and export production do not result in environmental degradation or adversely impact the nutritional rights and needs of people for whom fish is critical to their health and well-being and for whom other comparable sources of food are not readily available or affordable.

11.3 Laws and regulations relating to fish trade

11.3.1 Laws, regulations and administrative procedures applicable to international trade in fish and fishery products should be transparent, as simple as possible, comprehensible and, when appropriate, based on scientific evidence.

11.3.2 States, in accordance with their national laws, should facilitate appropriate consultation with and participation of industry as well as environmental and consumer groups in the development and implementation of laws and regulations related to trade in fish and fishery products.

11.3.3 States should simplify their laws, regulations and administrative procedures applicable to trade in fish and fishery products without jeopardizing their effectiveness.

11.3.4 When a State introduces changes to its legal requirernents affecting trade in fish and fishery products with other States, sufficient information and time should be given to allow the States and producers affected to introduce, as appropriate, the changes needed in their processes and procedures. In this connection, consultation with affected States on the time frame for implementation of the changes would be desirable. Due consideration should be given to requests from developing countries for temporary derogations from obligations.

11.3.5 States should periodically review laws and regulations applicable to international trade in fish and fishery products in order to determine whether the conditions which gave rise to their introduction continue to exist.

11.3.6 States should harmonize as far as possible the standards applicable to international trade in fish and fishery products in accordance with relevant internationally recognized provisions.

11.3.7 States should collect, disseminate and exchange timely, accurate and pertinent statistical information on international trade in fish and fishery products through relevant national institutions and international organizations.

11.3.8 States should promptly notify interested States, WTO and other appropriate international organizations on the development of and changes to laws, regulations and administrative procedures applicable to international trade in fish and fishery products.

ARTICLE 12 - FISHERIES RESEARCH

12.1 States should recognize that responsible fisheries requires the availability of a sound .scientific basis to assist fisheries managers and other interested parties in making decisions. Therefore, States should ensure that appropriate research is conducted into} all aspects of fisheries including biology, ecology, technology, environmental science, economics, social science, aquaculture and nutritional science. States should -ensure the availability of research facilities and provide appropriate training, staffing and institution building to conduct the research, taking into account the special needs of developing countries.

12.2 States should establish an appropriate institutional framework to determine the applied research which is required and its proper use.

12.3 States should ensure that data generated by research are analyzed, that the results of such analyses are published, respecting confidentiality where appropriate, and distributed in a timely and readily understood fashion, in order that the best scientific evidence is made available as a contribution to fisheries conservation, management and development. In the absence of adequate .scientific information, appropriate research should be initiated as soon as possible.

12.4 States should collect reliable and accurate data which are required to assess the status of fisheries and ecosystems, including data on bycatch, discards and waste. Where appropriate, this data .should be provided, at an appropriate time and level of aggregation, to relevant States and subregional, regional and global fisheries organizations.

12.5 States should he able to monitor and assess the state of the stocks under their jurisdiction, including the impacts of ecosystem changes resulting from fishing pressure, pollution or habitat alteration. They should also establish the research capacity necessary to assess the effects of' climate or environment change on fish stocks and aquatic ecosystems.

12.6 States should support and strengthen national research capabilities to meet acknowledged scientific standards.

12.7 States, as appropriate in cooperation with relevant international organisations, should encourage research to ensure optimum utilization of fishery resources and stimulate the research required to support national policies related to fish as food

12.8 States should conduct research into, and monitor, human food supplies from aquatic sources and the environment from which they are taken and ensure that .here is no adverse health impact on consumers. The results of such research should be made publicly available.

12.9 States should ensure that the economic, social, marketing and institutional aspects of fisheries are adequately researched and that comparable data are generated for ongoing monitoring, analysis and policy formulation.

12.10 States should carry out studies on the selectivity of fishing gear, the environmental impact of fishing gear on target species and on the behaviour of target and non-target species in relation to such fishing gear as an aid for management decisions and with a view to minimizing non-utilized catches as well as safeguarding the biodiversity of ecosystems and the aquatic habitat.

12.11 States should ensure that before the commercial introduction of new types of gear, a scientific evaluation of their impact on the fisheries and ecosystems where they will be used should be undertaken. The effects of such gear introductions should be monitored.

12.12 States should investigate and document traditional fisheries knowledge and technologies, in particular those applied to small-scale fisheries, in order to assess their application to sustainable fisheries conservation, management and development.

12.13 States should promote the use of research results as a basis for the setting of management objectives, reference points and performance criteria, as well as for ensuring adequate linkage, between applied research and fisheries management,

12.14 States conducting scientific research activities in waters under the jurisdiction of another State should ensure that their vessels comply with the laws and regulations of that State and international law.

12.15 States should promote the adoption of uniform guidelines governing fisheries research conducted on the high seas.

12.16 States should, where appropriate, support the establishment of mechanisms, including, *inter alia*, the adoption of uniform guidelines, to facilitate research at the subregional or regional level and should encourage the sharing of the results of such research with other regions.

12.17 States, either directly or with the support of relevant international organizations, should develop collaborative technical and research programmes to improve understanding of the biology, environment and status of transboundary aquatic stocks.

12.18 States and relevant international organizations should promote and enhance the research capacities of developing countries, inter alia, in the areas of data collection and analysis, information, science and technology, human resource development anti provision of research facilities, in order for them to participate effectively in the conservation, management and sustainable use of living aquatic resources.

12.19 Competent rotor national organizations should, where appropriate, render technical and financial support to States upon request and when engaged in research investigations aimed at evaluating stocks which have been previously unfished or very lightly fished.

12.20 Relevant technical and financial international organizations should, upon request, support States in their research efforts, devoting special attention to developing countries, in particular the least developed among them and small island developing countries.

Annex 1

BACKGROUND TO THE ORIGIN AND ELABORATION OF THE CODE

1. This annex describes the process of elaboration and negotiation of the Code, which led to its submission for adoption to the Twenty-eighth Session of the FAO Conference. It has peen felt useful to annex this section as a reference to the origin and the development of the Code and thus reflect the interest generated and the spirit of compromise of all the parties involved in its elaboration. It is hoped that this will contribute to the promotion of the commitment necessary for its implementation

2. At various international fora, concern had 'long been expressed regarding the clear signs of over-exploitation of important fish stocks, damage to ecosystems, economic losses, and issues affecting fish trade - all of which threatened the long-term sustainability of fisheries and, in turn, harmed the contribution of fisheries to food supply. In discussing the current state arid prospects of world fisheries, the Nineteenth Session of the FAO Committee on Fisheries (COFI), held in March 19111, recommended that FAO should develop the concept of responsible fisheries and elaborate a Code of Conduct to this end.

3. Subsequently, the Government of Mexico, in collaboration with FAO, organized an International Conference oil Responsible Fishing in Cancún, in May 1992. The Declaration of Cancún endorsed at that Conference further developed the concept of responsible fisheries, stating that "this concept encompasses the sustainable utilization of fisheries resources to harmony with the environment; the use of capture and aquaculture practices which are not harmful to ecosystems, resources or their quality; the incorporation of added value to such products through transformation processes meeting the required sanitary standards; the conduct of commercial practices so as to provide consumers access to good quality products".

4. The Cancún Declaration was brought to the attention of the UNCED Rio Summit in June 1992, which supported the preparation of a Code of Conduct for Responsible Fisheries, The FAO Technical Consultation on High Seas Fishing, held in September 1992, further recommended the elaboration of a Code to address the issues regarding high seas fisheries.

5. The one hundred and Second Session of the FAO Council, held in November 1992, discussed the elaboration of the Code, recommending that priority be given to high seas issues and requested that proposals for the Code be presented to the 1993 session of the Committee on Fisheries.

6. The Twentieth Session of COFI, held in March 1993, examined general principles for such a Code, including the elaboration of guidelines and endorsed a timeframe for the further elaboration of the Code. It also requested FAO to prepare, on a "fast track" basis, as part of the Code, proposals to prevent reflagging of fishing vessels which affect conservation and management measures on the high seas.

7. The further development of the Code of Conduct for Responsible Fisheries was accordingly carried out in consultation and collaboration with relevant United Nations Agencies and other international organizations including non-governmental organizations.

8. In pursuance of the instructions of the FAO Governing Bodies, the draft Code was formulated in such a way as to be consistent with the 1982 United Nations Convention on the Law of the Sea, taking into account the 1992 Declaration of Cancun, the 1992 Rio Declaration and the provisions of Agenda 21 of UNCED, the conclusions and recommendations of the 1992 FAO Technical Consultation on High Seas Fishing, the Strategy endorsed by the 1984 FAO World Conference on Fisheries Management and Development, and other relevant instruments including the outcome of the then ongoing United Nations Conference on Straddling Fish Stocks and Highly Migratory Fish Stocks which, in August 1995, adopted an Agreement for the Implementation of the Provisions of the United Nations Convention on the Law of the Sea of 10 December 1982 Concerning Straddling Fish Stocks and Highly Migratory Fish Stocks.

9. The FAO Conference, at its Twenty-seventh Session in November 1993, adopted the Agreement to Promote Compliance with International Conservation and Management Measures by Fishing Vessels on the High Seas and recommended that the General Principles of the Code of Conduct for Responsible Fisheries be prepared on a "fast track" in order to orientate formulation of thematic articles. Accordingly, a draft text of the General Principles was reviewed by an informal Working Group of Government nominated experts, which met in Rome in February 1994. A revised draft was widely circulated to all FAO Members and Associate Members as well as intergovernmental and non-governmental organizations. Comments received on the second version of the General Principles were incorporated in the draft Code together with proposals for an alternative text. This document was also the subject of' informal consultation with non-governmental organizations on the occasion of the Fourth Session of the United Nations Conference on Straddling Fish Stocks and Highly Migratory Fish Stocks, held in August 1994 in New York.

10. In order to facilitate consideration of the full text of the draft Code, the Director-General proposed to the Council at its Hundred and Sixth Session in

June 1994, that a Technical Consultation on the Code of Conduct for Responsible Fishing be organized, open to all FAO Members, interested non-members, intergovernmental and non-governmental organizations, in order to provide an opportunity for the widest involvement of all concerned parties at an early stage of its elaboration.

11. This Technical Consultation took place in Rome from 26 September to 5 October 1994 and a draft for the entire Code and a first draft of technical guidelines to support most of the Thematic Articles of the Code were presented. Following a thorough review of all the Articles of the complete draft Code of Conduct, an Alternative Secretariat Draft was then prepared on the basis of comments made during the discussions in plenary and specific drafting changes submitted in writing during the Consultation.

12. The Consultation was able to review also in detail an alternative draft for three of the six Thematic Articles of the Code, i.e., Article 9 "Integration of Fisheries into Coastal Area Management", Article 6 "Fisheries Management", Article 7 "Fishing Operations", except for those principles which were likely to be affected by the outcome of the ongoing UN Conference on Straddling Fish Stocks and Highly Migratory Fish Stocks. A short Administrative Report was prepared and presented to the FAO Council and to COFI.

13. The Technical Consultation proposed to the Council at its Hundred and Seventh Session, 15-24 November 1994, that the final wording of those principles dealing mainly with high seas issues be left in abeyance pending the outcome of the UN Conference. The Council generally endorsed the proposed procedure, noting that following discussions at the next session of COFI, a final draft of the Code would be submitted to the FAO Council in June 1995 which would then decide upon the necessity for a Technical Committee to meet in parallel to that Session of the Council in order to elaborate further the detailed provisions of the Code if required.

14. Based upon the substantial comments and detailed suggestions received at the Technical Consultation, the Secretariat elaborated a revised draft of the Code of Conduct for Responsible Fisheries, which was submitted to the Twenty-first Session of the Committee on Fisheries, held from 10 to 15 March 1995.

15. The Committee on Fisheries was also informed that the UN Conference was expected to conclude its work in August 1995. It was proposed that principles left in abeyance in the draft text of the Code could then be reconciled with the language agreed upon at the UN Conference in accordance with a mechanism to be decided upon by the Committee and the Council, before

submission of the complete Code for its adoption at the Twenty-eighth Session of the FAO Conference in October 1995.

16. The Committee was informed of the various steps the Secretariat had undertaken in preparing the draft Code of Conduct. The Committee established an open-ended Working Group in order to review the draft text of the Code. The Working Group, which met from 10 to 14 March 1995, undertook a detailed revision of the draft Code in continuation of the work carried out by the Technical Consultation. It completed and approved the text of Articles 8 to 11. In view of the time constraints, the Working Group provided directives to the Secretariat to redraft Articles 1 to 5. It was also recommended that the elements of research and cooperation as well as aquaculture be included in Article 5, General Principles, to reflect issues developed in the Thematic Articles of the Code.

17. The Committee supported the proposal endorsed by the Hundred and Seventh Session of the Council oil mechanism; to finalize the Code. The final wording of those principles dealing mainly With issues concerning straddling fish stocks and highly migratory fish stocks, ,which formed only a small part of the Code, should be re-examined in the light of the outcome of the UN Conference. The Group also recommended that once agreement was reached on the substance, it would be necessary to harmonize legal, technical and idiomatic aspects of the Code, in order to facilitate its final approval.

18. The Report of tile open-ended Working Group was presented to a Ministerial Meeting on Fisheries, held on 14 and 15 March 1995, in conjunction with the COFI Session. The Rome Consensus oil World Fisheries emanating from this meeting urged that "Governments and international organizations take prompt action to complete the International Code of Conduct for Responsible Fisheries with a view to submitting the final text to the FAO Conference in October 1995"

19. The Hundred and Eighth Session of the Council was presented with a revised version of the Code of Conduct. The Council established an open-ended Technical Committee, which held its First Session from 5 to 9 June 1995, with a broad regional representation of members and observers. A number of intergovernmental and non-governmental organizations also participated.

20. The Council was informed by the Technical Committee that it had undertaken a thorough review of Articles 1 to 5 including the Introduction. It had also examined, amended and approved Articles 8 to 11. The Council was also informed that the Committee had started the revision of Article 6.

21. The Council approved the work carried out by the Technical Committee and endorsed its recommendation for a Second Session to he held from 25 to 29 September 1995 to complete the revision of the Code once the Secretariat had harmonized the text linguistically and juridically, taking into account the outcome of the UN Conference on Straddling Fish Stocks and Highly Migratory Fish Stocks.

22. A revised version of the Code as approved by the Open-ended Technical Committee at its First Session (5-9 June 1995) and endorsed by the One Hundred and Eighth Session of the Council was issued, both as a Conference document (C 95/20) and as a working paper for the Second Session of the Technical Committee. Elements pending agreement were clearly identified.

23. In order to facilitate the finalization of the entire Code, the Secretariat prepared the document "Secretariat Proposals for Article 6, Fisheries Management, and Article 7, Fishing Operations, of the Code of Conduct for Responsible Fisheries", taking into account the Agreement relating to the Conservation and Management of Straddling Fish Stocks and Highly Migratory Fish Stocks, adopted by the UN Conference in August 1995. The Secretariat also completed proposals for the harmonization of the text on legal and linguistic aspects and made this available to the Committee in three languages for the session (English, French and Spanish).

24. A Second Session of the Open-ended Technical Committee of the Council met from 25 to 29 September 1995, with a wide representation of regions and interested organizations. The Committee, working in a full spirit of collaboration, successfully concluded its mandate, finalizing and endorsing all Articles and the Code as a whole. The Technical Committee agreed that the negotiations of the text of the Code were finalized. An Open-ended Informal Group on Language Harmonization held an additional session and, together with the Secretariat, completed the harmonization on the basis of the text as adopter: at the closing session. The Technical Committee instructed tile Secretariat to already submit the finalized version as a revised Conference document to the Hundred and Ninth Session of the Council and to the Twenty-eighth Session of the Conference for its adoption. The Council endorsed the Code of Conduct as finalized by the Technical Committee. The Secretariat was requested to prepare the required draft resolution for the Conference, including also a call on countries to ratify, as a matter of urgency, the Compliance Agreement adopted at the last session of the Conference. The Twenty-eighth Session of the Conference adopted on 31 October 1995, by consensus, the Code of Conduct for Responsible Fisheries and the respective Resolution shown in Annex 2.

40

Annex 2

RESOLUTION

THE CONFERENCE

Recognizing the vital role of fisheries in world food security, and economic and social development, as well as the need to ensure the sustainability of the living aquatic resources and their environment for present and future generations,

Recalling that the Committee on Fisheries on 19 March 1991 recommended the development of the concept of responsible fishing and the possible formulation of an instrument on the matter.

Considering that the Declaration of Cancún, which emanated from the Inter-national Conference on Responsible Fisheries of May 1992, organized by the Government of Mexico in collaboration with FAO, had called for the preparation of a Code of Conduct on Responsible Fisheries.

Bearing in mind that with the entry into force of the United Nations Convention on the Law of the Sea, 1982, and the adoption of the Agreement for the implementation of the provisions of the United Nations Convention on the Law of the Sea of 10 December 1982 Relating to the Conservation and Management of Straddling Fish Stocks and Highly Migratory Fish Stocks, as was anticipated in the 1992 Rio Declaration and the provisions of Agenda 21 of UNCED, there is an increased need for subregional and regional cooperation, and that significant responsibilities are placed upon FAO in accordance with its mandate,

Recalling further that the Conference in 1993 adopted the FAO Agreement to Promote Compliance with International Conservation and Management Measures by Fishing Vessels on the high Seas, and that this Agreement would constitute an integral part of the Code of Conduct,

Noting with satisfaction that FAO, in accordance with the decisions of its Governing Bodies, had organized a series of technical meetings to formulate the Code of Conduct and that these meetings have resulted in agreement being reached on the text of the Code of Conduct for Responsible Fisheries,

Acknowledging that the Rome Consensus on World Fisheries, which emanated from the Ministerial Meeting on Fisheries of 14-15 March 1995, urged governments and international organizations to respond effectively to the current fisheries situation, *inter alia*, by completing the Code of Conduct for Responsible

Fisheries and to consider adopting the Agreement to Promote Compliance with International Conservation and Management Measures by Fishing Vessels on the High Seas:

1. Decides to adopt the Code of Conduct for Responsible Fisheries;

2. Calls on States, International Organizations, whether Governmental or Non-Governmental, and all those involved in fisheries to collaborate in the fulfillment and implementation of the objectives and principles contained in this Code;

3 Urges that special requirements of developing countries be taken into account in implementing the provisions of this Code;

4. Requests FAO to make provision in the Programme of Work and Budget for providing advice to developing countries in implementing this Code and for the elaboration of art Interregional Assistance Programme for external assistance aimed at supporting implementation of the Code;

5. Further requests FAO, in collaboration with members and interested relevant organizations, to elaborate, as appropriate, technical guidelines in support of the implementation of the Code;

6. Calls upon FAO to monitor and report on the implementation of the Code and its effects on fisheries, including action taken under other instruments and resolutions by UN organizations and, in particular, the resolutions adopted by the General Assembly to give effect to the Conference on Straddling Fish Stocks and Highly Migratory Fish Stocks leading to the Agreement for the implementation of the provisions of the United Nations Convention on the Law of the Sea of 10 December 1982 Relating to the Conservation and Management of Straddling Fish Stocks and Highly Migratory Fish Stocks;

7. Urges FAO to strengthen Regional Fisheries Bodies in order to deal more effectively with fisheries conservation and management issues in support of subregional, regional and global cooperation and coordination in fisheries.

Selected Bibliography

Abdul, H.S., "Development and Management of Malaysian Marine Fisheries: Technical Conservation Measures" (1995) 19 *Marine Policy* 115.

Agnew, D.J., "The CCAMLR Ecosystem Monitoring Programme" (1997) 9(3) *Antarctic Science* 235.

Agnew, D.J., "The Illegal and Unregulated Fishery for Toothfish in the Southern Ocean, and the CCAMLR Catch Documentation Scheme" (2000) 24 *Marine Policy* 361.

Agnew, D.J., and C.T. Barnes, "Economic Aspects and Drivers of IUU Fishing: Building a Framework" in K. Gray et al., eds., *Fish Piracy: Combatting Illegal, Unreported and Unregulated Fishing* (Paris: OECD, 2004) 169.

Agüero, M., "Implementation of the Provisions of the UN Fish Stocks Agreement: Conditions for Success – The Case of Latin America (The Galapagos Agreement)" in IDDRA Ltd., ed., *Policy Research: Options for Strengthening National, Sub-regional and Regional Institutions and Policies to Better Address Developing Countries' Needs* (Portsmouth: IDDRA Ltd., 2003). Available at <http://www.onefish.org/cds_upload/1051272053537_UNFSAAppendix6Aguero.doc> (accessed 4 March 2008).

Akaha, T., "Bilateral Fisheries Relations in the Seas of Japan and Okhotsk: A Catalyst for Cooperation or Seed of Conflict?" in E. Mann Borgese, N. Ginsburg, and J.R. Morgan, eds., *Ocean Yearbook 11* (Chicago: University of Chicago Press, 1994) 384.

Akiba, O., "International Law of the Sea: The Legality of Canadian Seizure of the Spanish Trawler (Estai)" (1997) 37 *Natural Resources Journal* 809.

Akiwumi, P., and T. Melvasalo, "UNEP's Regional Seas Programme: Approach, Experience and Future Plans" (1998) 22 *Marine Policy* 229.

Alcock, F., "Bargaining, Uncertainty, and Property Rights in Fisheries" (2002) 54 *World Politics* 437.

Alder, J., and G. Lugten, "Frozen Fish Block: How Committed Are North Atlantic States to Accountability, Conservation and Management of Fisheries?" (2002) 26 *Marine Policy* 345.

Alexander, L., "New Trends in Regionalism" in E. Mann Borgese, N. Ginsburg, and J. Morgan, eds., *Ocean Yearbook 11* (Chicago: University of Chicago Press, 1994) 1.

Alvarez, J.E., "Why Nations Behave" (1998) 19 *Michigan Journal of International Law* 303.

Anand, R.P., "The Politics of a New Legal Order for Fisheries" (1982) 11 *Ocean Development and International Law* 265.

Anderson, D., "The Straddling Stocks Agreement of 1995 – An Initial Assessment" (1996) 45 *International and Comparative Law Quarterly* 463.

Anderson, D., "Port States and Environmental Protection" in A. Boyle and D. Freestone, eds., *International Law and Sustainable Development: Past Achievements and Future Challenges* (Oxford: Oxford University Press, 1999) 325.

Ando, N., E. McWhinney, and R. Wolfrum, eds., *Liber Amicorum Judge Shigeru Oda* Vol. II (The Hague: Kluwer Law International, 2002).

Andrade, A.S., B. Ortiz-von Halle, and N. Naranjo Cuvi, eds., *La Pesca y El Comercio De Bacalao De Profundidad, Dissostichus Eleginoides: En America Del Sur, Una Perspectiva Regional* (Quito: TRAFFIC America del Sur, 2002).

Andresen, S., "The Convention for the Conservation of Antarctic Marine Living Resources (CCAMLR): Improved Procedures but Lacking Results" in E.L. Miles et al., eds., *Environmental Regime Effectiveness: Confronting Theory with Evidence* (Cambridge: The MIT Press, 2001) 405.

Andresen, S., and T. Skodvin, "Non-State Influence in the International Whaling Commission 1970–1990" (2003) 3 *Global Environmental Politics* 61.

Andrews-Chouicha, E., and K. Gray (prepared and edited), *Why Fish Piracy Persists: The Economics of Illegal, Unreported and Unregulated Fishing* (Paris: OECD, 2005).

Anferova, E., M. Vetema, and R. Hannesson, "Fish Quota Auctions in the Russian Far East: A Failed Experiment" (2005) 29 *Marine Policy* 47.

Applebaum, B., "The Straddling Stocks Problem: The Northwest Atlantic Situation, International Law and Options for Coastal State Action" in A.H.A. Soons, ed., *The Implementation of the Law of the Sea Convention Through International Institutions* (Honolulu, Hawaii: Law of the Sea Institute, 1990) 282.

Applebaum, B., and A. Donohue, "The Role of Regional Fisheries Management Organisations" in E. Hey, ed., *Developments in International Fisheries Law* (The Hague: Kluwer Law International, 1999) 217.

Aps, R., "Management of Shared Baltic Fishery Resources" in A.I.L. Payne, C.M. O'Brien, and S.I. Rogers, eds., *Management of Shared Fish Stocks* (Oxford: Blackwell Publishing Ltd., 2004) 190.

Aqorau, T., "Tuna Fisheries Management in the Western and Central Pacific Ocean: A Critical Analysis of the Convention for the Conservation and Management of Highly Migratory Fish Stocks in the Western Central Pacific Ocean and Its Implications for the Pacific Island States" (2001) 16 *International Journal of Marine and Coastal Law* 379.

Aqorau, T., and A. Bergin, "Ocean Governance in the Western Pacific Purse Seine Fishery — The Palau Arrangement" (1997) 21 *Marine Policy* 173.

Aqorau, T., and A. Bergin, "The UN Fish Stocks Agreement: A New Era for International Cooperation to Conserve Tuna in the Central Western Pacific" (1998) 29 *Ocean Development and International Law* 21.

Arbo, P., and B. Hersoug, "The Globalization of the Fishing Industry and the Case of Finnmark" (1997) 21 *Marine Policy* 121.

Ardia, D.S., "Does the Emperor Have No Clothes? Enforcement of International Laws Protecting the Environment" (1998) 19 *Michigan Journal of International Law* 497.

Arland, K., and T. Bjorndal, "Fisheries Management in Norway — an Overview" (2002) 26 *Marine Policy* 307.

Armas-Pfirter, M.F., "Straddling Stocks and Highly Migratory Stocks in Latin American Practice and Legislation: New Perspectives in Light of Current International Negotiations" (1995) 26 *Ocean Development and International Law* 93.

Armstrong, A.J., *Development of International/Regional Co-Operation in Fisheries Monitoring, Control and Surveillance* (London: Commonwealth Secretariat, 1991).

Armstrong, C.W., "Sharing a Fish Resource: Bargaining Theoretical Analysis of an Applied Allocation Rule" (1998) 22 *Marine Policy* 119.

Arnason, R., "Property Rights Quality and Economic Efficiency of Fisheries Management Regimes: Some Basic Results" in T. Bjorndal, D. Gordon, R. Arnason, and U.R. Sumaila, eds., *Advances in Fisheries Economics: Festschrift in Honour of Professor Gordon R. Munro* (Oxford: Blackwell Publishing Ltd., 2007).

Asia–Pacific Economic Cooperation (APEC), *Study into the Nature and Extent of Subsidies in the Fisheries Sector of APEC Members Economies* (Singapore: APEC Secretariat, 2000).

Baird, R., "Fishing in the Southern Ocean: The Development of Fisheries and the Role of CCAMLR in Their Management" (1997) 16 *University of Tasmania Law Review* 160.

Balton, D.A., "Strengthening the Law of the Sea: The New Agreement on Straddling Stocks and Highly Migratory Fish Stocks" (1996) 27 *Ocean Development and International Law* 125.

Balton, D.A., "The Compliance Agreement" in E. Hey, ed., *Developments in International Fisheries Law* (The Hague: Kluwer Law International, 1999) 31.

Balton, D.A., "Making New Rules Work: Implementation of the Global Fisheries Instruments" in M.H. Nordquist and J.N. Moore, eds., *Current Fisheries Issues and the Food and Agriculture Organization of the United Nations* (The Hague: Kluwer Law International, 2000) 107.

Balton, D.A., "The Bering Sea Doughnut Hole Convention: Regional Solution, Global Implications" in O.S. Stokke, ed., *Governing High Seas Fisheries: The Interplay of Global and Regional Regimes* (Oxford: Oxford University Press, 2001) 143.

Balton, D.A., "Dealing with the 'Bad Actors' of Ocean Fisheries". Paper presented to the OECD Workshop on Illegal, Unreported and Unregulated (IUU) Fishing Activities, Paris, France, 19–20 April 2004, AGR/FI/IUU/RD(2004)3 (Paris: OECD, 2004).

Balton, D.A., "Global Review of Illegal Unreported and Unregulated Fishing Issues: What's the Problem?". Paper presented to the OECD Workshop on Illegal, Unreported and Unregulated (IUU) Fishing Activities, Paris, France, 19–20 April 2004, AGR/FI/IUU(2004)16 (Paris: OECD, 2004).

Balton, D.A., "IUU Fishing and State Control Over Nationals". Paper presented to the OECD Workshop on Illegal, Unreported and Unregulated (IUU) Fishing Activities, Paris, France, 19–20 April 2004, AGR/FI/IUU/RD(2004)2 (Paris: OECD, 2004). Available at <http://www.oecd.org/dataoecd/58/28/31426742.PDF> (accessed February 2005).

Barria, P.M., M.P. Donoso, and J.R. Azocar, *Investigación Situación Pesquería Recursos Altamente Migratorios 2002: Informe De Diffusion, Programa De Sequimiento Del Estado De Situacion De Las Prinicpales Pesquerias Nacionales* (Valparaiso: Instituto de Fomento Pesquero, 2003).

Barston, R., "United Nations Conference on Straddling and Highly Migratory Fish Stocks" (1995) 19 *Marine Policy* 159.

Barston, R., "The Law of the Sea and Regional Fisheries Organisations" (1999) 14 *International Journal of Marine and Coastal Law* 333.

Barton, J.A., D.J. Agnew, and L.W. Purchase, "The Southwest Atlantic: Achievements of Bilateral Management and the Case for a Multilateral Arrangement" in A.I.L. Payne, C.M. O'Brien, and S.I. Rogers, eds., *Management of Shared Fish Stocks* (Oxford: Blackwell Publishing Ltd., 2004).

Bascompte, J., C.J. Melián, and E. Sala, "Interaction Strength Combinations and the Overfishing of a Marine Food Web" (2005) 102 *Proceedings of the National Academy of Sciences* 5443.

Baxter, R.R., "International Law in 'Her Infinite Variety'" (1980) 29 *International and Comparative Law Quarterly* 549.

Beer-Gabel, J., and V. Lestang, *Les Commissions de pêche et leur droit : La conservation et la gestion des ressources marines vivantes* (Brussels: Université de Bruxelles, 2003).

Beesley, J.A., and M. Rowe, '*Canada and Spain: A Conservation Dispute' LOS Leider* (Honolulu, Hawaii: Law of the Sea Institute, 1995).

Benitah, M., "Ongoing WTO Negotiations on Fisheries Subsidies" (June 2004) *ASIL Insights*. Available at <http://www.asil.org/insights/insigh136.htm> (accessed 28 February 2005).

Bergin, A., "Albatross and Longlining: Managing Seabird Bycatch" (1997) 21 *Marine Policy* 63.

Berkes, F., J. Golding, and C. Folke, eds., *Navigating Social and Ecological Systems: Building Resilience for Complexity and Change* (Cambridge: Cambridge University Press, 2003).

Beslier, S., "Enforcement and Surveillance: What Are Our Technical Capacities and How Much Are We Willing to Pay?", AGR/FI/IUU(2004)19 (Paris: OECD, 2004).

Beverton, R.J.H., and S.J. Holt, *On the Dynamics of Exploited Fish Populations* (Caldwell, N.J.: Blackburn Press, 2004).

Bialek, D., "Sink or Swim: Measures Under International Law for the Conservation of the Patagonian Toothfish in the Southern Ocean" (2003) 34 *Ocean Development and International Law* 105.

Bilder, R., "Beyond Compliance: Helping Nations Cooperate" in D. Shelton, ed., *Commitment and Compliance: The Role of Non-Binding Norms in the International Legal System* (New York: Oxford University Press, 2000) 65.

Birnie, P., "Are Twentieth-Century Marine Conservations Adaptable to Twenty-First Century Goals and Principles?" (1997) 12 *International Journal of Marine and Coastal Law* Part I, 307; Part II, 488.

Birnie, P., "New Approaches to Ensuring Compliance at Sea: The FAO Agreement to Promote Compliance with International Conservation and Management Vessels by Fishing Vessels on the High Seas" (1999) 8 *Review of European Community and International Environmental Law* 48.

Birnie, P., and A. Boyle, *International Law and Environment*, 2nd. ed. (Oxford: Oxford University Press, 2002).

Bisiaux, A., P. Nuengsigkapian, and C. Salpin, "A Brief History of Oceans and the Law of the Sea and Open-Ended Informal Consultative Process" (2003) 25 *Earth Negotiations Bulletin* 1.

Bjørndal, T., et al., "The Management of High Seas Fisheries" (2000) 94 *Annals of Operations Research* 183.

Blanco-Bazán, A., "The IMO Guidelines on Particular Sensitive Sea Areas (PSSAs): Their Possible Application to the Protection of Underwater Cultural Heritage" (1996) 20 *Marine Policy* 343.

Bodansky, D., "Scientific Uncertainty and the Precautionary Principle" (1991) 33 *Environment* 1.

Bodansky, D., "Customary (and Not So Customary) International Environmental Law" (1995) 3 *Indiana Journal of Global Legal Studies* 105.

Bohnsack, J.A., "Incorporating No-Take Marine Reserves in Precautionary Management and Stock Assessment" in V.R. Restrepo, ed., *Proceedings of the 5th National NMFS Stock Assessment Workshop: Providing Scientific Advice to Implement the Precautionary Approach Under the Magnuson–Stevens Fishery Conservation and Management Act*, U.S. Dep. Commer. NOAA Technical Memorandum NMFS-F/SPO-40 (Silver Springs, Md.: U.S. Department of Commerce, NOAA, 1999) 7.

Borgese, E.M., *Ocean Governance and the United Nations* (Halifax, N.S.: Centre for Foreign Policy Studies, Dalhousie University, 1995).

Bothe, M., "Legal and Non-Legal Norms—A Meaningful Distinction in International Relations?" (1980) XI *Netherlands Yearbook of International Law* 65.

Boude, J.-P., J. Boncoeur, and D. Bailly, "Regulating the Access to Fisheries: Learning From European Experiences" (2001) 25 *Marine Policy* 313.

Bours, H., M. Gianni, and D. Mather, *Pirate Fishing Plundering the Oceans* (Amsterdam: Greenpeace, 2001).

Boustany, A.M., et al., "Satellite Tagging: Expanded Niche for White Sharks" (2002) 415 *Nature* 35.

Boyle, A.E., "International Law and the Protection of the Global Atmosphere: Concepts, Categories and Principles" in R.R. Churchill and D. Freestone, eds., *International Law and Global Climate Change* (The Hague: Kluwer Law International, 1991) 7.

Boyle, A.E., "Some Reflections on the Relationship of Treaties and Soft Law" (1999) 48 *International and Comparative Law Quarterly* 901.

Boyle, A.E., "Problems of Compulsory Jurisdiction and Settlement of Disputes Relating to Straddling Fish Stocks" in O.S. Stokke, ed., *Governing High Seas Fisheries: The Interplay of Global and Regional Regimes* (Oxford: Oxford University Press, 2001) 91.

Boyle, A.E., and D. Freestone, ed., *International Law and Sustainable Development: Past Achievements and Future Challenges* (Oxford, New York: Oxford University Press, 1999).

Bradshaw, M., and O. Tully, "Use of Private Access Rights in Fisheries: Effective Management Through Public Transferability" (2004) 28 *Marine Policy* 411.

Bray, K., "A Global Review of Illegal, Unreported and Unregulated (IUU) Fishing", FAO – Government of Australia Workshop on IUU Fishing, Sydney 2000, Document AUS: IUU/2000/6 (Rome: FAO, 2000).

Bray, K., "Illegal, Unreported, and Unregulated (IUU) Fishing" in M.H. Nordquist and J.N. Moore, eds., *Current Fisheries Issues and the Food and Agriculture Organization of the United Nations* (The Hague: Kluwer Law International, 2000) 115.

Bręthes, J.-C., "The Canadian Atlantic Groundfish Experience and the Constraints to the Conservation of Fisheries Resources: A Perspective" (1998) 39 *Ocean & Coastal Management* 135.

Brierley, A., and K. Reid, "Kingdom of Krill" (1999) 2182 *New Scientist* 37.

Brock, J., "Poaching Vessels: CCAMLR Clamps Down", (17 November 2004) South Atlantic Remote Territories Media Association (SARTMA). Available at <http://www.sartma.com/cat_3_1.html> (accessed 2 March 2005).

Browman, H.I., et al., "Perspectives on Ecosystem-Based Approaches to the Management of Marine Resources" (2004) 274 *Marine Ecology Progress Series* 269.

Brown, D.N., and R.S. Pomeroy, "Co-Management of Caribbean Community (CARICOM) Fisheries" (1999) 23 *Marine Policy* 549.

Brown, E.D., "The Law of the Sea at the Millennium" in H.D. Smith, ed., *The Oceans: Key Issues in Marine Affairs* (Dordrecht: Kluwer Academic Publishers, 2004) 25.

Brown, J., "An Account of the Dolphin-Safe Tuna Issue in the UK" (2005) 29 *Marine Policy* 39.

Brown, K., "Truce in the Salmon War: Alternatives for the Pacific Salmon Treaty" (1999) 74 *Washington Law Review* 605.

Brubaker, R.D., *The Russian Arctic Strait* (Leiden: Martinus Nijhoff Publishers, 2004).

Burke, W.T., "Recent Decisions on International Fisheries Governance" (1998) 6 *Willamette Journal of International Law and Dispute Resolution* 15.

Burke, W.T., "Highly Migratory Species in the New Law of the Sea" (1984) 14 *Ocean Development and International Law* 273.

Burke, W.T., "Highly Migratory Species in the New Law of the Sea" (1989) 16 *Ocean Development and International Law* 285.

Burke, W.T., "Regulation of Driftnet Fishing on the High Seas and the New International Law of the Sea" (1990) 3 *Georgetown International Environmental Law Review* 265.

Burke, W.T., "UNCED and the Oceans" (1993) 17 *Marine Policy* 519.

Burke, W.T., *The New International Law of Fisheries: UNCLOS and Beyond* (Oxford: Clarendon Press, 1994).

Burke, W.T., "Some Comments on High Seas Fishing and International Law" in R. St. J. Macdonald, ed., *Essays in Honour of Wang Tieya* (Dordrecht: Martinus Nijhoff Publishers, 1994) 103.

Burke, W.T., "Importance of the 1982 UN Convention on the Law of the Sea and Its Future Development" (1996) 27 *Ocean Development and International Law* 1.

Burke, W.T., "Compatibility and Precaution in the 1995 Straddling Stock Agreement" in H.N. Scheiber, ed., *Law of the Sea: The Common Heritage and Emerging Challenges* (The Hague: Martinus Nijhoff Publishers, 2000) 105.

Burke, W.T., "Evolution in the Fisheries Provisions of UNCLOS" in N. Ando et al., eds., *Liber Amicorum Judge Shigeru Oda*, Vol. 2 (The Hague: Kluwer Law International, 2002) 1355.

Burke, W.T., M. Freeberg, and E.L. Miles, "United Nations Resolutions on Driftnet Fishing: An Unsustainable Precedent for High Seas and Coastal Fisheries Management" (1994) 25 *Ocean Development and International Law* 127.

Busch, J.M., *La Pesca En Mar Presencial* [The Fishing in the Actual Sea] [in Spanish]. Seminar, Universidad Gabriela Mistral, 2000 [unpublished].

Busch, W.-D.N., B.L. Brown, and G.F. Mayer, eds., *Strategic Guidance for Implementing an Ecosystem-Based Approach to Fisheries Management* (Silver Spring, Md.: United States Department of Commerce, National Oceanic and Atmospheric Administration, NMFS, 2003).

Butterworth, D., and A.J. Penney, "Allocation in High Seas Fisheries: Avoiding Meltdown" (2004) *Management of Shared Fish Stocks* 165.

Caddy, J.F., "Fisheries Management in the Twenty-First Century: Will New Paradigms Apply?" (1999) 9 *Fish Biology and Fisheries* 1.

Caddy, J.F., "Deciding on Precautionary Management Measures of a Stock Based on a Suite of Limit Reference Points (LRPs) as a Basis for a Multi-LRP Harvest Law" (1998) 32 *NAFO Scientific Council Studies* 55.

Caddy, J.F., and D.J. Agnew, "An Overview of Recent Global Experience with Recovery Plans for Depleted Marine Resources and Suggested Guidelines for Recovery Planning" (2004) 14 *Reviews in Fish Biology and Fisheries* 43.

Caddy, J.F., and K.L. Cochrane, "A Review of Fisheries Management Past and Present and Some Future Perspectives for the Third Millennium" (2001) 44 *Ocean & Coastal Management* 653.

Caddy, J.F., and R. Mahon, "Reference Points for Fisheries Management", FAO Fisheries Technical Paper No. 347 (Rome: FAO, 1995).

Caddy, J.F., and R. McGarvery, "Targets or Limits for the Management of Fisheries?" (1996) 16 *North American Journal of Fisheries Management* 479.

Caddy, J., and J. Seijo, "This Is More Difficult Than We Thought! The Responsibility of Scientists, Managers and Stakeholders to Mitigate the Unsustainability of Marine Fisheries" (2005) 360 *Philosophical Transactions of the Royal Society: Biological Sciences. Series **B*** 59.

Calderón, M., "The Theory of the Actual Sea, New Contribution of Chile to International Law of the Sea", Comment on *International Law of the Sea* (1998) Institute of Studies of the Pacific.

Cameron, J., and J. Abouchar, "The Precautionary Principle: A Fundamental Principle of Law and Policy for the Protection of the Global Environment" (1991) 14 *Boston College International Law Review* 1.

Campbell, D., D. Brown, and T. Battaglene, "Individual Transferable Catch Quotas: Australian Experience in the Southern Bluefin Tuna Fishery" (2000) 24 *Marine Policy* 109.

Campbell, H.F., "Managing Tuna Fisheries: A New Strategy for the Western and Central Pacific Ocean" (2000) 24 *Marine Policy* 159.

Campbell, H., S.F. Herrick Jr., and D. Squires, "The Role of Research in Fisheries Management: The Conservation of Dolphins in the Eastern Tropical Pacific and the Exploitation of Southern Bluefin Tuna in the Southern Ocean" (2000) 31 *Ocean Development and International Law* 347.

Canfield, J., "Recent Developments in Bering Sea Fisheries Conservation and Management" (1993) 24 *Ocean Development and International Law* 257.

Caron, D.D., and H.N. Scheiber, eds., *Bringing New Law to Ocean Waters* (Leiden: Martinus Nijhoff Publishers, 2004).

Carr, C.J., "Recent Developments in Compliance and Enforcement for International Fisheries" (1997) 24 *Ecology Law Quarterly* 847.

Carr, J., and M. Gianni, "High Seas Fisheries, Large-Scale Drift Nets, and the Law of the Sea" in in J. M. Van Dyke, D. Zaelke and G. Hewison, eds., *Freedom for the Seas in the 21st Century—Ocean Governance and Environmental Harmony* (Washington, D.C.: Island Press, 1993).

Catchpole, J.L., L.J. Frid, and T.S. Gray, "Discards in North Sea Fisheries: Causes, Consequences and Solutions" (2005) 29 *Marine Policy* 421.

Centre for Environment, Fisheries and Aquaculture Science (CEFAS), "Overfishing and the ICES Precautionary Approach to Management" (Lowestoft: CEFAS, n.d.). Available at <http://www.cefas.co.uk/data/fisheries-information/management-and-advice/overfishing-and-the-ices-precautionary-approach-to-management.aspx> (accessed 21 October 2005).

Chakalall, B., R. Mahon, and P. McConney, "Current Issues in Fisheries Governance in the Caribbean Community (CARICOM)" (1998) 22 *Marine Policy* 29.

Chee, C.I., "Consistency and Compatibility in Conservation and Management on Straddling Fish Stocks and Highly Migratory Fish Stocks Between the EEZ and the Adjacent High Seas" (1995) 40 *The Korean Journal of International Law* 171.

Chinkin, C.M., "The Challenge of Soft Law: Development and Change in International Law" (1989) 38 *International and Comparative Law Quarterly* 850.

Chircop, A., "Participation in Marine Regionalism: An Appraisal in a Mediterranean Context" in E.M. Borgese, N. Ginsburg, and J.R. Morgan, eds., *Ocean Yearbook 7* (Chicago: University of Chicago Press, 1989) 402.

Christensen, V., and R. Mahon, "Researchable Issues Relevant for Management of Large Marine Ecosystems, with Special Reference to the Caribbean". Available at <http://www.ecopath.org/modules/Publications/ViewPublications/publicat/lme_carib.html> (accessed 10 January 2006).

Christie, D.R., "The Conservation and Management of Stocks Located Solely within the Exclusive Economic Zone" in E. Hey, ed., *Developments in International Fisheries Law* (The Hague: Kluwer Law International, 1999) 395.

Christopher, M., "Toward a Rational Harvest: The United Nations Agreement on Straddling Fish Stocks and Highly Migratory Species" (1996) 5 *Minnesota Journal of Global Trade* 357.

Churchill, R.R., "Falkland Islands: Maritime Jurisdiction and Co-operative Arrangements with Argentina" (1997) 46 *International and Comparative Law Quarterly* 463.

Churchill, R., "Legal Uncertainties in International High Seas Fisheries Management" (1998) 37 *Fisheries Research* 225.

Churchill, R.R., "The Barents Sea Loophole Agreement: A 'Coastal State' Solution to a Straddling Stock Problem" (1999) 14 *International Journal of Marine and Coastal Law* 467.

Churchill, R., "Managing Straddling Fish Stocks in the North-East Atlantic: A Multiplicity of Instruments and Regime Linkages — But How Effective a Management?" in O.S. Stokke, ed., *Governing High Seas Fisheries: The Interplay of Global and Regional Regimes* (Oxford: Oxford University Press, 2001) 235.

Churchill, R.R., and C. Hedley, *The Meaning of the "Genuine Link" Requirement in Relation to the Nationality of Ships*. A study prepared for the International Transport Workers' Federation, 2000.

Churchill, R.R., and A.V. Lowe, *The Law of the Sea*, 3rd ed. (Manchester: Manchester University Press, 1999).

Cicin-Sain, B., and R.W. Knecht, eds., *Integrated Coastal and Ocean Management: Concepts and Practices* (Washington, D.C.: Island Press, 1998).

Clark, B.C., and A.D. Hemmings, "Problems and Prospects for the Convention on the Conservation of Antarctic Marine Living Resources Twenty Years On" (2001) 4 *Journal of International Wildlife Law & Policy* 47.

Clark, C.W., "Fisheries Bioeconomics: Why Is it so Wildly Misunderstood?" (2006) 48 *Population Ecology* 95.

Clark, C.W., "Worldwide Crisis in Fisheries: The Use and Misuse of Models" (Cambridge: Cambridge University Press, 2006).

Clark, C.W., G. Munro, and U.R. Sumaila, "Subsidies, Buybacks, and Sustainable Fisheries" (2005) 50 *Journal of Environmental Economics and Management* 47–58.

Clerveaux, W., "Towards a Better Understanding of Conflict Management in Tropical Fisheries: Evidence From Ghana, Bangladesh and the Caribbean" (2001) 25 *Marine Policy* 365.

Clingan, T.A., Jr., "Mar Presencial (The Presential Sea): Deja Vu All Over Again? — A Response to Francisco Orrego Vicuña" (1993) 24 *Ocean Development and International Law* 93.

Cochrane, K.L., and D.J. Doulman, "The Rising Tide of Fisheries Instruments and the Struggle to Keep Afloat" (2005) 360(1453) *Philosophical Transactions of the Royal Society: Biological Sciences. Series **B*** 77.

Cockcroft, A.C., and A.I.L. Payne, "A Cautious Fisheries Management Policy in South Africa: The Fisheries for Rock Lobster" (1999) 23 *Marine Policy* 587.

Cole, H., "Contemporary Challenges: Globalisation, Global Interconnectedness and That 'There Are *Not* Plenty More Fish in the Sea'. Fisheries Governance and Globalisation: Is There a Relationship?" (2003) 46 *Ocean & Coastal Management* 77.

Collie, J.S., and H. Gislason, "Biological Reference Points for Fish Stocks in a Multispecies Context" (1999) 58 *Canadian Journal of Fisheries and Aquatic Sciences* 2167.

Colmenares, N.A., and J.J. Escobar, "Ocean and Coastal Issues and Policy Responses in the Caribbean" (2002) 45 *Ocean & Coastal Management* 905.

Colson, D.A., and P. Hoyle, "Satisfying the Procedural Prerequisites to the Compulsory Dispute Settlement Mechanisms of the 1982 Law of the Sea Convention: Did the *Southern Bluefin Tuna* Tribunal Get It Right?" (2003) 34 *Ocean Development and International Law* 59.

Commission on Global Governance, *Our Global Neighbourhood: The Report of the Commission on Global Governance* (Oxford: Oxford University Press, 1995).

Commission for the Conservation of Antarctic Marine Living Resources (CCAMLR), "Intersessional Procedure for Notifying CCAMLR Members of Infractions and Sanctions Imposed by Flag States" Appendix III in *Report of the Standing Committee on Observation and Inspection*, Hobart, Australia, 26 October – 4 November 1994.

Commission for the Conservation of Antarctic Marine Living Resources (CCAMLR), "Report of the Working Group on Fish Stock Assessment" Annex 5 in *the Report of the Seventeenth Meeting of the Scientific Committee*, Hobart, Australia, 26–30 October, 1998.

Commission for the Conservation of Antarctic Marine Living Resources (CCAMLR), "Report of the Working Group on Fish Stock Assessment" Annex 5 in *the Report of the Eighteenth Meeting of the Scientific Committee*, Hobart, Australia, 25–29 October, 1999.

Commission for the Conservation of Antarctic Marine Living Resources (CCAMLR), "Report of the Working Group on Fish Stock Assessment" Annex 5 in *the Report of the Nineteenth Meeting of the Scientific Committee*, Hobart, Australia, 23–27 October, 2000.

Commission for the Conservation of Antarctic Marine Living Resources (CCAMLR), 2000. *Statistical Bulletin* Volume 12. http://www.ccamlr.org/English/e_pubs/e_stat_bull/e_stat_bull_toc.htm. April 2000.

Commission for the Conservation of Antarctic Marine Living Resources (CCAMLR), *CCAMLR XXIII — Meeting Report — Antarctic Southern Ocean Coalition*, 25 October – 5 November 2004, Hobart, Australia.

Commission for the Conservation of Southern Bluefin Tuna (CCSBT), *Report of the Resumed Third Annual Meeting*, Canberra, Australia, 18–22 February 1997.

Commission for the Conservation of Southern Bluefin Tuna (CCSBT), *Report of the 1997 Scientific Committee Meeting*, Canberra, Australia, 28 July – 8 August 1997.

Commission for the Conservation of Southern Bluefin Tuna (CCSBT), *Report of the Fifth Meeting of the Ecologically Related Species Working Group*, Wellington, New Zealand, 2–5 February 2004.

Constable, A.J., "CCAMLR Ecosystem Monitoring and Management: Future Work" (2002) 9 *CCAMLR Science* 233.

Constable, A.J.,"The Status of Antarctic Fisheries Research" in J. Jabour-Green and M. Haward, eds., *The Antarctic: Past, Present and Future*, Antarctic CRC Research Report No 28 (Hobart: University of Tasmania, 2002) 71.

Constable, A.J., and W.K. de la Mare, "Bias in the Estimation of Krill Yield From Using a Deterministic Formulation of the Median Unexploited Spawning Biomass" (1998) 5 *CCAMLR Science* 1.

Constable, A.J., and W.K. de la Mare, "A Generalized Yield Model for Evaluating Yield and the Long Term Status of Fish Stocks Under Conditions of Uncertainty" (1996) 3 *CCAMLR Science* 31.

Constable, A.J., et al., "Managing Fisheries to Conserve the Antarctic Marine Ecosystem: Practical Implementation of the Convention on the Conservation of Antarctic Marine Living

Resources (CCAMLR)" (2000) 57 *ICES Journal of Marine Science* 778.

Consultative Group on Flag State Implementation, *Oceans and Law of the Sea: Report of the Secretary General*, Fifty-ninth Session, A/59/63 (New York: United Nations General Assembly, 1995).

Copeland, B.R., and M.S. Taylor, "Trade, Tragedy, and the Commons", NBER Working Paper No. 10836 (Cambridge, Mass.: National Bureau of Economic Research, 2004).

Cordonnery, L., "A Note on the 2000 Convention for the Conservation and Management of Tuna in the Western and Central Pacific Ocean" (2002) 33 *Ocean Development and International Law* 1.

Co-Sponsors Australia, Chile and New Zealand, "Proposed Elements of a Legal Framework". Paper presented to the First International Meeting on the Establishment of the South Pacific Regional Fisheries Management Organization, Wellington, New Zealand, 14–17 February 2006. Meeting Doc. SP/01/Inf3. Available at <http//:www.southpacificrfmo.org/Event.first-meeting/?operation=documents> (accessed April 2006).

Costanza, R. et. al., "Principles for Sustainable Governance of the Oceans" (1998) 10 *Science* 198.

Couper, A.D., and H.D. Smith, "The Development of Fishermen-Based Policies" (1997) 21 *Marine Policy* 111.

Cox, A., and C.-C. Schmidt, "Subsidies in the OECD Fisheries Sector: A Review of Recent Analysis and Future Directions". Paper presented to the FAO Expert Consultation on Identifying, Assessing and Reporting on Subsidies in the Fishing Industry, Rome, Italy, 3–6 December 2002 (Paris: OECD, 2002).

Cripps, S., "What Data Do NGOs Have?". Paper presented to the OECD Workshop on Illegal, Unreported and Unregulated (IUU) Fishing Activities, Paris, France, 19–20 April 2004. Available at <http://www.oecd.org/dataoecd/1/24/31593530.ppt#348,1,OECD Workshop on IUU activities Paris, 19-20 April 2004> (accessed March 2006).

Croxall, J.P., and P.N. Tranthan, "The Southern Ocean: A Model System for Conserving Resources" in L.K. Glover and S.A. Earle, eds., *Defying Ocean's End* (Washington, D.C.: Island Press, 2004) 71.

Cury, P.M., and V. Christensen, "Quantitative Ecosystem Indicators for Fisheries Management" (2005) 62(3) *ICES Journal of Marine Science* 307.

Dalton, J.G., "The Chilean Mar Presencial: A Harmless Concept or a Dangerous Precedent?" (1993) 8(3) *International Journal of Marine and Coastal Law* 397.

Davidse, W.P., L.V. McEwan, and N. Vestergaard, "Property Rights in Fishing: From State Property Towards Private Property? A Case Study of Three EU Countries" (1999) 23 *Marine Policy* 537.

Davies, P.G.G.,"The EC/Canadian Fisheries Dispute in the Northwest Atlantic" (1995) 44 *International and Comparative Law Quarterly* 927.

Dean, J., "Trade Policy and the Environment: Developing Country Concerns". Manuscript prepared for OECD Workshop on Trade, Environment and Development (Paris: OECD, 1991).

Deep Sea Conservation Coalition, "A Net with Holes: The Regional Fisheries Management System" Deep Sea Conservation Coalition Policy Paper (Deep Sea Conservation Coalition, 2004).

de Fontaubert, A.C., "The Politics of Negotiation at the United Nations Conference on Straddling Fish Stocks and Highly Migratory Fish Stocks" (1995) 29 *Ocean and Coastal Management* 79.

de Fontaubert, A.C., D.R. Downes, and T.S. Agardy, *Biodiversity in the Seas: Implementing the Convention on Biological Diversity in Marine and Coastal Habitats* (Gland: IUCN, 1996).

de Fontaubert, C., "The Status of Natural Resources on the High-Seas: Legal and Political Considerations" in WWF–IUCN, eds., *The Status of Natural Resources on the High-Seas* (Gland: WWF–IUCN, 2001).

de Fontaubert, C., et al., *Achieving Sustainable Fisheries, Implementing the New International Legal Regime* (Gland: IUCN, 2003).

De Klemm, C., "Fisheries Conservation and Management and the Conservation of Marine Biological Diversity" in E. Hey, ed., *Developments in International Fisheries Law* (The Hague: Kluwer Law International, 1999) 423.

de la Mare, W.K., "Some Recent Developments in the Management of Marine Living Resources" in R. Floyd, A. Sheppard, and P. Barro, eds., *Frontiers of Population Ecology* (Melbourne: CSIRO Publishing, 1996) 599.

de Leiva Moreno, J.I., and J. Majkowski, "Status of the Tuna Stocks of the World" in W.H. Bayliff, J.I. de Leiva Moreno, and J. Majkowski, eds., *Second Meeting of the Technical Advisory Committee of the FAO Project: Management of Tuna Fishing Capacity: Conservation and Socio-Economics*, FAO Fisheries Proceedings No. 2 (Rome: FAO, 2005).

DeLone, E., "Improving the Management of Atlantic Tuna: The Duty to Strengthen ICCAT in Light of the 1995 Straddling Stocks Agreement" (1998) 6 *New York University Environmental Law Journal* 656.

Department of State, *Foreign Relations of the United States, (1955–1957)* Vol. X (Washington, D.C.: Department of State, 1958).

de Reynier, Y.L., "Evolving Principles of International Fisheries Law and the North Pacific Anadromous Fish Commission" (1998) 29 *Ocean Development and International Law* 147.

DeSombre, E.R., *Flagging Standards: Globalization and Environmental, Safety, and Labor Regulations at Sea* (Cambridge, Mass.: MIT Press, 2006).

Dietz, T., E. Ostrom, and P. Stern, "The Struggle to Govern the Commons" (2003) 302 *Science* 1907.

Doulman, D.J., "Structure and Process of the 1993–1995 United Nation Conference on Straddling Fish Stocks and Highly Migratory Fish Stocks" FAO Fisheries Circular No. 898 FID/C898 (Rome: FAO, 1995).

Doulman, D.J., *The Code of Conduct for Responsible Fisheries: The Requirement for Structural Change and Adjustment in the Fisheries Sector* (Rome: FAO, 1998).

Doulman, D.J., "Events Leading to the Elaboration of an International Plan of Action to Combat Illegal, Unreported and Unregulated Fishing". Paper presented to the International Conference on Fisheries Monitoring, Control and Surveillance, Brussels, Belgium, 24–27 October 2000.

Doulman, D.J., "Commentary. Some Observations on Developments in World Fisheries" (2003) 8 *Economic Perspectives*.

Downes, D., and B. Van Dyke, *Fisheries Conservation and Trade Rules: Ensuring That Trade Law Promotes Sustainable Fisheries* (Washington D.C.: Center for International Environmental Law and Greenpeace, 1998).

Dulvy, N.K., et al., "Comparison of Threat and Exploitation Status in Northeast Atlantic Marine Populations" (2005) 42 *Journal of Applied Ecology* 883.

Dulvy, N.K., Y. Sadovy, and J.D. Reynolds, "Extinction Vulnerability in Marine Populations" (2003) 4 *Fish and Fisheries* 25.

Dunlap, W.V., "Straddling Stocks in the Barents Sea Loophole" (1996–1997) 4 *Boundary and Security Bulletin* 79.

Dzidzornu, D.M., "Coastal State Obligations and Powers Respecting EEZ Environmental Protection Under Part XII of the UNCLOS: A Descriptive Analysis" (1997) 8 *Colorado Journal of International Environmental Law and Policy* 286.

Easty, D.C., *Greening the GATT: Trade Environment and the Future* (Washington, D.C.: Institute of International Economics, 1994).

Ebbin, S.A., "Swimming Upstream: Institutional Dimensions of Asymmetrical Problems in Two Salmon Management Regimes" (2003) 27 *Marine Policy* 441.

Ebbin, S.A., A.H. Hoel, and A.K. Syndes, eds., *A Sea Change: The Exclusive Economic Zone and Governance Institutions for Living Marine Resources* (Dordrecht: Springer, 2005)

Edeson, W.R., "The Code of Conduct for Responsible Fisheries: An Introduction" (1996) 11 *International Journal of Marine and Coastal Law* 233.

Edeson, W.R., "Towards Long-Term Sustainable Use: Some Recent Developments in the Legal Regime of Fisheries" in A. Boyle and D. Freestone, eds., *International Law and Sustainable Development: Past Achievements and Future Challenges* (Oxford: Oxford University Press, 1999) 165.

Edeson, W.R., "Closing the Gap: The Role of 'Soft' International Instruments to Control Fishing" (1999) 20 *Australian Year Book of International Law* 83.

Edeson, W.R. "Tools to Address IUU Fishing: The Current Legal Situation". Background papers for the Expert Consultation on Illegal, Unreported and Unregulated Fishing Organized by the Government of Australia in Co-operation with FAO, Sydney, Australia, 15–19 May 2000, Document AUS:IUU/2000/8 (Rome: FAO, 2000). Available at <http://www.fao.org/documents/show_cdr.asp?url_file=/DOCREP/005/Y3274E/y3274e0a.htm> (accessed 24 October 2005).

Edeson, W.R., "Confusion Over the Use of UNCLOS, and References to Other Recent Agreements" (2000) 15 *International Journal of Marine and Coastal Law* 415.

Edeson, W.R., "The International Plan of Action on Illegal, Unreported, and Unregulated Fishing: The Legal Context of a Non-Legally Binding Instrument" (2001) 16 *International Journal of Marine and Coastal Law* 603.

Edeson, W.R., "Soft and Hard Law Aspects of Fisheries Issues: Some Recent Global and Regional Approaches" in M.H. Nordquist, J.N. Moore, and S. Mahmoudi, eds., *The Stockholm Declaration and the Law of the Marine Environment* (The Hague, New York: Martinus Nijoff Publishers, 2003) 165.

Edeson, W.R., D. Freestone, and E. Gudmundsdottir, *Legislating for Sustainable Fisheries: A Guide to Implementing the 1993 FAO Compliance Agreement and 1995 UN Fish Stocks Agreement*, The World Bank Law, Justice and Development Series (Washington, D.C.: The World Bank, 2001).

Eggert, H., "Towards an Integrated Sustainable Management of Fisheries" in M.K. Tolba, ed., *Our Fragile World: Challenges and Opportunities for Sustainable Development* (Oxford: EOLSS, 2001).

Eggert, H., and A. Elegård, "Fishery Control and Regulation Compliance: A Case for Co-Management in Swedish Commercial Fisheries" (2003) 27 *Marine Policy* 525.

Elias, O., "The Nature of the Subjective Element in Customary International Law" (1995) 44 *International and Comparative Law Quarterly* 501.

Ellis, J., "The Straddling Stocks Agreement and the Precautionary Principle as Interpretive Device and Rule of Law" (2001) 32 *Ocean Development and International Law* 289.

Erceg, D., "Deterring IUU Fishing Through State Control Over Nationals" (2006) 30 *Marine Policy* 173.

Escobar, I., "The Development and Enforcement of National Plans of Action to Prevent, Deter and Eliminate Illegal, Unreported and Unregulated Fishing: The Spanish Case". Paper presented to the OECD Workshop on Illegal, Unreported and Unregulated (IUU) Fishing Activities, Paris, France, 19–20 April 2004, AGR/FI/IUU(2004)20 (Paris, OECD, 2004).

Escobar, I., et al., "Key Observations and Findings by the Workshop Chairs". Paper presented to the OECD Workshop on Illegal, Unreported and Unregulated (IUU) Fishing Activities, Paris, France, 19–20 April 2004 (Paris: OECD, 2004).

Esmark, M., and N. Jensen, *The Barents Sea Cod: Last of the Large Cod Stock*, 1st ed. (Oslo: WWF Norway, 2004).

European Commission Directorate General Fisheries, *Some Principles and Some Data: European Distant Water Fishing Fleets* (2001). Available at <http://ec.europa.eu/fisheries/publications/facts/peche_en.pdf>.

European Union Delegation, "Presentation on Fishing Efforts and Compliance 2003–2004", 26th Annual Meeting Northwest Atlantic Fisheries Organization, Dartmouth, N.S., 13–17 September 2004.

Everson, I., "Consideration of Major Issues in Ecosystem Monitoring and Management" (2002) 9 *CCAMLR Science* 213.

Exel, M., "Working Together—What Industry Can Do to Help". Paper presented to the OECD Workshop on Illegal, Unreported and Unregulated (IUU) Fishing Activities, Paris, France, 19–20 April 2004, AGR/FI/IUU(2004)24/REV1 (Paris: OECD, 2004) in Gray, K. et al. (Ed.) (2004). *Fish piracy: combating illegal, unreported and unregulated fishing*. pp. 317-318

Fanta, E., *CCAMLR Report 2004: CCAMLR XXIII: Commission for the Conservation of Antarctic Marine Living Resources*, Hobart, Australia, 25 October – 5 November 2004 (Curitiba, Brazil: Universidade Federal do Paraná, 2004). Scientific Committee on Antarctic Research (SCAR). Available at <http://www.scar.org/researchgroups/lifescience/ccamlr/hobart04/fantaCCAMLR2004.pdf> (accessed 21 October 2005).

Fauteux, P., "The Canadian Legal Initiative on High Seas Fishing" (1993) 4 *Yearbook of International Environmental Law* 51.

Feeny, D., S. Hanna, and A.F. McEvoy, "Questioning the Assumptions of the 'Tragedy of the Commons' Model of Fisheries" (1996) 72 *Land Economics* 187.

Fitzpatrick, J., *Illegal, Unreported and Unregulated Fishing: Measures to Enhance the Capability of a Flag State to Exercise Effective Control Over a Fishing Vessel*, Document AUS:IUU/2000/11 (Rome: FAO, 2000).

Fletcher, W.J., "Frameworks for Managing Fisheries Resources in Australia Using Ecosystem Approaches: Do They Fit Together and Can They Be Useful?" (2006) 78 *Bulletin of Marine Science* 691.

Fletcher, W.J., *A Guide to Implementing Ecosystem Approach to Fisheries Management (EAFM) within the Western and Central Pacific Region* (Honiara: Forum Fisheries Agency, 2007).

Fogg, G.E., *A History of Antarctic Science* (Cambridge: Cambridge University Press, 1992).

Food and Agriculture Organization (FAO), *Marine Fisheries and the Law of the Sea: A Decade of Change* (Rome: FAO, 1992).

Food and Agriculture Organization (FAO), "Reference Points for Fisheries Management: Their Potential Application to Straddling Stocks and Highly Migratory Resources", FAO Fisheries Circular No. 864 (Rome: FAO, 1993).

Food and Agriculture Organization (FAO), *World Review of Highly Migratory Species and Straddling Stocks* (Rome: FAO, 1995).

Food and Agriculture Organization (FAO), "The Precautionary Approach to Fisheries with Reference to Straddling Fish Stocks and Highly Migratory Fish Stocks", UN Doc. A/CONF. 164/INF/8 (New York: UN, 1995).

Food and Agriculture Organization (FAO), "Precautionary Approach to Fisheries. Part 2: Scientific Papers", FAO Fisheries Technical Paper No. 350 Part 2 (Rome: FAO, 1996).

Food and Agriculture Organization (FAO), *Major Issues in World Fisheries*, Doc. COFI/97/2 (Rome: FAO COFI, 1997).

Food and Agriculture Organization (FAO), "An Appraisal of the Suitability of the CITES Criteria for Listing Commercially Exploited Aquatic Species", FAO Fisheries Circular No. 954 (Rome: FAO, 2000).

Food and Agriculture Organization (FAO), "International Plan of Action to Prevent, Deter and Eliminate Illegal, Unreported and Unregulated Fishing Approved by FAO Committee on Fisheries, 2 March 2001" (2001) 16 *International Journal of Marine and Coastal Law* 660.

Food and Agriculture Organization (FAO), "Fishing Vessels Operating Under Open Registers and the Exercise of Flag State Responsibilities: Information and Options", FAO Fisheries Circular No. 980 (Rome: FAO, 2002).

Food and Agriculture Organization (FAO), "Papers presented at the Norway–FAO Expert Consultation on the Management of Shared Fish Stocks, Bergen, Norway, 7–10 October 2002", FAO Fisheries Report No. 695, Suppl. (Rome: FAO, 2002).

Food and Agriculture Organization (FAO), "Report of the International Workshop on Factors Contributing to Unsustainability and Overexploitation in Fisheries, Bangkok, Thailand", FAO Fisheries Report No. 672 (Rome: FAO, 2002).

Food and Agriculture Organization (FAO), "Fisheries Management 2: The Ecosystem Approach to Fisheries", FAO Technical Guidelines for Responsible Fisheries No. 4. Suppl. 2 (Rome: FAO, 2003).

Food and Agriculture Organization (FAO), *Fisheries Subsidies: A Short Term and Long Term Work Programme for FAO* (Rome: FAO, 2004).

Food and Agriculture Organization (FAO), *Follow-Up to the Recommendations of the Twenty-Fifth Session of the Committee on Fisheries Rome, Italy, 24–28 February 2003* (Rome: FAO, 2004).

Food and Agriculture Organization (FAO), *A Global Project for the Study of Impacts of Fisheries Subsidies* (Rome: FAO, 2004).

Food and Agriculture Organization (FAO), *A Global Technical Initiative on Fisheries Subsidies* (Rome: FAO, 2004).

Food and Agriculture Organization (FAO), "Report of the Expert Consultation on Legal Issues Related to CITES and Commercially-Exploited Aquatic Species", FAO Fisheries Report. No. 746 (Rome: FAO, 2004).

Food and Agriculture Organization (FAO), "Report of the Technical Consultation on the Use of Subsidies in the Fisheries Sector", FAO Fisheries Report No. 752 (Rome: FAO, 2004).

Food and Agriculture Organization (FAO), *A Summary of Recent Works on Subsidies in the Fishing Sector* (Rome: FAO, 2004).

Food and Agriculture Organization (FAO), *Technical Consultation on the Use of Subsidies in the Fisheries Sector* (Rome: FAO, 2004).

Food and Agriculture Organization (FAO), *The State of World Fisheries and Aquaculture 2004* (Rome: FAO, 2005).

Food and Agriculture Organization (FAO), "Review of the State of World Marine Fishery Resources", FAO Fisheries Technical Paper No. 457 (Rome: FAO, 2005).

Food and Agriculture Organization (FAO), "FAO Fact Sheet for WTO Ministerial Conference in Cancun, Mexico". Available at <http://www.globefish.org/index.php?id=2112> (accessed September 2005).

Food and Agriculture Organization (FAO), *Guidelines for the Eco-labelling of Fish and Fishery Products from Marine Capture Fisheries* (Rome: FAO, 2005). Available at <ftp://ftp.fao.org/docrep/fao/008/a0116t/a0116t00.pdf> (accessed March 2006).

Food and Agriculture Organization (FAO), "Reports of the IOC/FAO/IOTC Symposium and Workshop to Strengthen Port State Measures in the Indian Ocean. Port Louis, Mauritius, 18–22 June 2007", FAO Fisheries Report. No. 844 (Rome: FAO, 2007).

Food and Agriculture Organization (FAO) Subregional Office for Southern and East Africa, "Report of the Third Intergovernmental Consultation on the Establishment of a Southwest Indian Ocean Fisheries Commission", FAO Fisheries Report No. 742 (Rome: FAO, 2004).

Food and Agriculture Organization (FAO) Subregional Office for Southern and East Africa, "Report of the Fourth Intergovernmental Consultation on the Establishment of a Southwest Indian Ocean Fisheries Commission, Mah, Seychelles, 13–26 July 2004", FAO Fisheries Report No. 776 (Rome: FAO, 2005).

Fortier, L.Y., "From Confrontation to Cooperation on the High Seas: Recent Developments in International Law Concerning the Conservation of Marine Resources" in N. Ando et al., eds., *Liber Amicorum Judge Shigeru Oda* Vol. 2 (The Hague: Kluwer Law International, 2002) 1377.

Franckx, E., "Pacta Tertiis and the Agreement for the Implementation of the Provisions of the United Nations Convention on the Law of the Sea of 10 December 1982 Relating to the Conservation and Management of Straddling Fish Stocks and Highly Migratory Fish Stocks" (2000) 8 *Tulane Journal of International Comparative Law* 49.

Frank, K.T., et al., "Trophic Cascades in a Formerly Cod Dominated Ecosystem" (2005) 308 *Science* 1621. Frank, K.T., B. Petrie, and N.L. Schackell, "The Ups and Downs of Trophic Control in Continental Shelf Ecosystems" (2007) 22 *Trends in Ecology and Evolution 236.*

Freestone, D., "The Precautionary Principle" in R. Churchill and D. Freestone, eds., *International Law and Global Climate Change* (London: Kluwer Law International, 1991) 21.

Freestone, D., *Requirements of Proof for Conservation in High Seas Fisheries.* Paper for FAO Legal Office (Rome: FAO, 1992).

Freestone, D., "The Effective Conservation and Management of High Seas Living Resources: Towards a New Regime?" (1994) 5 *Canterbury Law Review* 341.

Freestone, D., "Canada and the EU Reach Agreement to Settle the *Estai* Dispute" (1995) 10 *International Journal of Marine and Coastal Law* 397.

Freestone, D., "The Conservation of Marine Ecosystems Under International Law" in D. Bowman and C. Redgwell, eds., *International Law and the Conservation of Biological Diversity* (London: Kluwer Law International, 1996) 91.

Freestone, D., "Implementing Precaution Cautiously: The Precautionary Approach in the Straddling and Highly Migratory Fish Stocks Agreement" in E. Hey, ed., *Developments in International Fisheries Law* (The Hague: Kluwer Law International, 1999) 287.

Freestone, D., "International Fisheries Law Since Rio: The Continuing Rise of the Precautionary Principle" in A. Boyle and D. Freestone, eds., *International Law and Sustainable Development: Past Achievements and Future Challenges* (Oxford: Oxford University Press, 1999).

Freestone, D.R. Barnes, and D.M. Ong, eds., *The Law of the Sea: Progress and Prospects* (Oxford: Oxford University Press, 2006).

Freestone, D., and E. Hey, eds., *The Precautionary Principle and International Law: The Challenge of Implementation* (The Hague: Kluwer Law International, 1996).

Freestone, D., and Z. Makuch, "The New International Environmental Law of Fisheries: The 1995 Straddling Stocks Agreement" (1996) 7 *Yearbook of International Environmental Law* 3.

Frid, C., O. Paramor, and C. Scott, "Ecosystem-Based Fisheries Management: Progress in the NE Atlantic" (2005) 29 *Marine Policy* 461.

Gabriel, W.L., and P.M. Mace, "A Review of Biological Reference Points in the Context of the Precautionary Approach" in V.R. Restrepo, ed., *Proceedings of the 5th National NMFS Stock Assessment Workshop: Providing Scientific Advice to Implement the Precautionary Approach Under the Magnuson–Stevens Fishery Conservation and Management Act*, U.S. Dep. Commer. NOAA Technical Memorandum NMFS-F/SPO-40 (Silver Springs, Md.: U.S. Department of Commerce, NOAA, 1999) 34.

Gallagher, R., *Fishing Vessel Monitoring Systems: Past, Present and Future* (Paris: High Seas Task Force, 2005).

Garcia, S.M., "The Precautionary Approach to Fisheries with Reference to Straddling Fish Stocks and Highly Migratory Fish Stocks", FAO Fisheries Circular No. 871 (Rome: FAO, 1994).

Garcia, S.M., "The Precautionary Principle: Its Implications in Capture Fisheries Management" (1994) 22 *Ocean & Coastal Management* 99.

Garcia, S.M., "The Precautionary Approach to Fisheries and Its Implications for Fishery Research, Technology and Management: An Updated Review" in FAO, *Precautionary Approach to Fisheries. Part 2: Scientific Papers*, Fisheries Technical Paper No. 350 Part 2 (Rome: FAO, 1996) 1.

Garcia, S.M., *Report of the Meeting of the ICCAT Ad Hoc Working Group on the Precautionary Approach*, Miami, Fla., 13–14 May 1998 (Madrid: ICCAT, 1998).

Garcia, S.M., "The Precautionary Approach to Fisheries: Progress Review and Main Issues (1995–2000)" in M. Nordquist and J.N. Moore, eds., *Current Fisheries Issues and the Food and Agriculture Organization of the United Nations* (The Hague: Kluwer Law International, 2000) 470.

Garcia, S.M., "Fishery Science and Decision-Making: Dire Straights to Sustainability" (2005) 76(2) *Bulletin of Marine Science* 171.

Garcia, S.M., and K.L. Cochrane, "Ecosystem Approach to Fisheries: A Review of Implementation Guidelines" (2005) 62 *ICES Journal of Marine Science* 311.

Garcia, S.M., and I. de Leiva Moreno, "Global Overview of Marine Fisheries" in M. Sinclair and G. Valdimarsson, eds., *Responsible Fisheries in the Marine Ecosystem* (Rome: FAO, 2003) 40.

Garcia S.M., and D. Doulman, "FAO's Fisheries Programme and the Plan of Implementation of the World Summit on Sustainable Development" in S.A. Ebbin, A.H. Hoel, and A.K. Sydnes, eds., *A Sea Change: The Exclusive Economic Zone and Governance Institutions for Living Marine Resources* (Dordrecht: Springer, 2005) 169.

Garcia, S.M., and J.R. Grainger, "Gloom and Doom? The Future of Marine Capture Fisheries" (2005) 360 *Philosophical Transactions of the Royal Society: Biological Sciences. Series **B*** 21.

Garcia, S.M., and J. Majowski, "State of High Seas Resources" in T. Kuribayashi and E.L. Miles, eds., *The Law of the Sea in the 1990s: A Framework for Further International Cooperation* (Honolulu: Law of the Sea Institute, 1992).

Garcia, S.M., and C. Newton, "Current Situation and Prospects in World Capture Fisheries" in E. Pikitch, D. Huppert, and M. Sissenwine, eds., *Global Trends: Fisheries Management* (AFS Symposium 20) (Bethesda, Md.: American Fisheries Society Publication, 1997) 3.

Garcia, S.M., and D.J. Staples, "Sustainability Reference Systems and Indicators for Responsible Marine Capture Fisheries: A Review of Concepts and Elements for a Set of Guidelines" (2000) 51(5) *Marine and Freshwater Research* 385.

Garcia, S.M., et al. "The Ecosystem Approach to Fisheries. Issues, Terminology, Principles, Institutional Foundations, Implementation and Outlook", FAO Fisheries Technical Paper No. 443 (Rome: FAO, 2003).

Garcia Amador, F.V., *The Exploitation and Conservation of the Resources of the Sea : A Study of Contemporary International Law* (Leiden: Sythoff, 1959).

Garza-Gil, M.D., and M.M. Varela-Lafuente, "The Profitability of the Spanish Swordfish Fleet in the North Atlantic" (2005) 29 *Marine Policy* 533.

Garza-Gil, M.D., M.M. Varela-Lafuente, and C. Iglesias-Malvido, "Spain's North Atlantic Swordfish Fishery" (2003) 27 *Marine Policy* 31.

Gavaris, S., et al., *Review of Management Plan Conservation Strategies for Canadian Fisheries on Georges Bank: A Test of a Practical Ecosystem-Based Framework*, ICES CM 2005/BB:05 (Copenhagen: ICES, 2005).

General Fisheries Commission for the Mediterranean (GFCM), *Selected Global Issues in Fisheries of Relevance to GFCM*, GFCM 25th Session, Sliema, Malta, 12–15 September 2000 (Rome: GFCM, 2000). Available at <http://www.fao.org/DOCREP/MEETING/X7663E.HTM> (accessed 20 October 2005).

General Fisheries Commission for the Mediterranean (GFCM), *Report of the Workshop on Ecosystem Approach to Fisheries*, Salammbô, Tunisa, 7–9 September 2005, SAC8/2005/Dma.6 (Rome: GFCM, 2005).

Gezelius, S.S., "Limits to Externalisation: The EU NAFO Policy 1979–1999" (1999) 23 *Marine Policy* 147.

Gianni, M., "The Entry into Force of the 1995 UN Fish Stocks Agreement: An NGO Perspective", OceanLaw On-Line Paper No. 8 (January 2002). Available at <http://www.intfish.net/igifl/archive/ops/olp/papers/8.htm> (accessed August 2008).

Gianni, M., *High Seas Bottom Trawl Fisheries and Their Impacts on the Biodiversity of Vulnerable Deep-Sea Ecosystems: Options for International Action* (Gland: IUCN, 2004).

Gianni, M., "IUU Fishing and the Cost to Flag of Convenience Countries". Paper presented to the OECD Workshop on Illegal, Unreported and Unregulated (IUU) Fishing Activities, Paris, France, 19–20 April 2004, AGR/FI/IUU/RD(2004)25 (Paris: OECD, 2004).

Gianni, M., and W. Simpson, "Flags of Convenience, Transshipment, Re-Supply and at-Sea Infrastructure in Relation to IUU Fishing (Management Implications and Recommendations for International Action Arising From a Case Study Prepared for WWF)", AGR/FI/IUU(2004)22 (Paris: OECD, 2004).

Gianni, M., and W. Simpson, *The Changing Nature of High Seas Fishing: How Flags of Convenience Provide Cover for Illegal, Unreported and Unregulated Fishing* (Australian Department of Agriculture, Fisheries and Forestry, International Transport Workers' Federation, and WWF International, 2005). Available at <http://assets.panda.org/downloads/iiumr.pdf> (accessed March 2008).

Gjerde, K.M., ed., "Special Double Issue on High Seas Fisheries Governance: Moving from Words to Action" (2005) 20 *International Journal of Marine and Coastal Law*.

Gjerde, K.M., "Ecosystems and Biodiversity in Deep Waters and High Seas", UNEP Regional Seas Reports and Studies No. 178 (Gland: UNEP–IUCN, 2006).

Gjerde, K., and C. Breide, *Towards a Strategy for High Seas Marine Protected Areas: Proceedings of the IUCN, WCPA and WWF Experts Workshop on High Seas Marine Protected Areas*, Malaga, Spain, 15–17 January 2003 (Gland: IUCN, 2003).

Glavin, T., "Wanted: New Management: Many of Canada's Fisheries Are in Crisis, but New Forms of Management Suggest a More Sustainable Future May Be Possible" (2004) 30 *Alternatives Journal* 20.

Glover, L.K., and S.A. Earle, *Defying Ocean's End: An Agenda for Action* (Washington, D.C.: Island Press, 2004).

Goltz, J.K., "The Sea of Okhotsk Peanut Hole: How the United Nations Draft Agreement on Straddling Stocks Might Preserve the Pollock Fishery" (1995) 4 *Pacific Rim Law & Policy Journal* 443.

Gonçalves, M.E., "Concepts of Marine Region and the New Law of the Sea" (1979) 3 *Marine Policy* 255.

Goñi, R., "Ecosystem Effects of Marine Fisheries: An Overview" (1998) 40 *Ocean & Coastal Management* 37.

González-Laxe, F., "The Precautionary Principle in Fisheries Management" (2005) 29 *Marine Policy* 495.

Gordon, K., "Promoting Corporate Responsibility: The OECD Guidelines for Multinational Enterprises". Paper presented to the OECD Workshop on Illegal, Unreported and Unregulated (IUU) Fishing Activities, Paris, France, 19–20 April 2004, AGR/FI/IUU(2004)12 (Paris: OECD, 2004).

Gorez, B., "Experiences with Subsidies and Fisheries Management: The Case of EU–ACP Fisheries Access Agreements". Paper presented at the UNEP Workshop on Fisheries Subsidies and Sustainable Fisheries Management, Geneva, Switzerland, 26–27 April 2004.

Gorina-Ysern, M., "World Ocean Public Trust: High Seas Fisheries after Grotius — Towards a New Ocean Ethos?" (2004) 34 *Golden Gate University Law Review* 645.

Grafton, R.Q., D. Squires, and J.E. Kirkley, "Private Property Rights and Crises in World Fisheries: Turning the Tide?" (1996) 14 *Contemporary Economic Policy* 90.

Grainger, R.J.R., and S.M. Garcia, *Chronicles of Marine Fishery Landings (1950–1994): Trend Analysis and Fisheries Potential* (Rome: FAO, 1996).

Grant, A., ed., *Special Issue: Sustainability Indicators in Marine Capture Fisheries* (2002) 51 *Marine and Freshwater Research*.

Gray, K., F. Legg, and E. Andrews-Chouicha, eds., *Fish Piracy: Combating Illegal, Unreported and Unregulated Fishing* (Paris: OECD, 2004).

Gray, T., and J. Hatchard, "The 2002 Reform of the Common Fisheries Policy's System of Governance – Rhetoric or Reality" (2003) 27 *Marine Policy* 545.

Greenpeace, *Chang Xing: Case Studies in IUU Vessels No. 2* (Amsterdam: Greenpeace, 2005).

Greenpeace, "CCAMLR Governments Are Failing the Toothfish and Albatross". Available at <http://archive.greenpeace.org/oceans/piratefishing/ccamlr.html> (accessed March 2005).

Greenpeace, *The Secret Shame of the Anuva: Case Studies on IUU Vessels No. 1* (Amsterdam: Greenpeace, 2005).

Greenpeace, "Patagonian Toothfish (Antarctic sea bass)". Available at <http://archive.greenpeace.org/oceans/piratefishing/toothfish.html> (accessed March 2005).

Greenpeace, "The Binar 4 — A Case Study of Pirate Fishing Links between Europe and West Africa" (10 April 2006). Available at <http://www.greenpeace.org/raw/content/international/press/reports/the-binar-4-a-case-study-of.pdf> (accessed May 2006).

Greenpeace, "Saving Our Seas From Piracy: A New Era in Fisheries Management". Available at <http://weblog.greenpeace.org/pacific/background/new_era.html> (accessed April 2006).

Greenpeace, "Too Many Ships". Available at <http://www.greenpeace.org/australia/issues/overfishing/overview/causes/corporate> (accessed 8 March 2005).

Greenpeace Australia Pacific, "Legally binding UN fisheries agreement could be used to protect world's oceans", Media Release, 11 December 2001, available at <http://www.greenpeace.org/australia/news-and-events/media/releases/overfishing/legally-binding-un-fisheries-a> (accessed 11 December 2001).

Grotius, H., *The Freedom of the Seas or the Right Which Belongs to the Dutch to Take Part in the East Indian Trade (1604)*, Edited by Scott, J.B., trans. by R. van Deman Magoffinn, translation and revision of the text of 1633 (New York: Carnegie Endowment for International Peace, Oxford University Press, 1916).

Grynberg, R., "WTO Fisheries Subsidies Negotiations: Implications for Fisheries Access Arrangements and Sustainable Management" (2003) 27 *Marine Policy* 499.

Grzybowski, D.M., et al., "A Historical Perspective Leading Up to and Including the United Nations Conference on Straddling Fish Stocks and Highly Migratory Fish Stocks" (1995) 13 *Pace Environmental Law Review* 49.

Gudmundur, E. "Settlement of Disputes II: International Tribunal for the Law of the Sea. DOALOS/UNITAR Briefing on Developments in Ocean Affairs and the Law of the Sea". Paper presented to the 20 Years After the Conclusion of the United Nations Convention on the Law of the Sea, New York, United Nations (New York: UN, 2002).

Gulland, J.A., "The Management Regime for Living Resources" in C.C. Joyner and S.K. Chopra, eds., *The Antarctic Legal Regime* (Dordrecht: Martinus Nijhoff Publishers, 1989) 219.

Guyader, O., and O. Thébaud, "Distributional Issues in the Operation of Rights-Based Fisheries Management Systems" (2001) 25 *Marine Policy* 103.

Haas, P.M., "Introduction: Epistemic Communities and International Policy Coordination" (1992) 46 *International Organization* 1.

Halfar, J., and R.M. Fujita, "Precautionary Management of Deep-Sea Mining" (2002) 26 *Marine Policy* 103.

Hall-Spencer, J., V. Allain, and J.H. Fosså, "Trawling Damage to Northeast Atlantic Ancient Coral Reefs" (2002) 269 (1490) *Proceedings of the Royal Society: Biological Sciences. Series **B*** 507.

Hampton, J., et al., "Fisheries: Decline of Pacific Tuna Populations Exaggerated?" 434 (7037) (2005) *Nature* 1.

Hanafusa, K., and N. Yagi, "Effort of Elimination of IUU Large-Scale Tuna Longline Vessels". Paper presented to the OECD Workshop on Illegal, Unreported and Unregulated (IUU) Fishing Activities, Paris, France, 19–20 April 2004, AGR/FI/IUU(2004)17 (Paris: OECD, 2004).

Hanna, S.S., "User Participation and Fishery Management Performance within the Pacific Fishery Management Council" (1995) 28 *Ocean & Coastal Management* 23.

Hannesson, R., "Effects of Liberalizing Trade in Fish, Fishing Services and Investment in Fishing Vessels", OECD Papers Offprint No. 8 (Paris: OECD, 2001).

Hannesson, R., "The Development of Economic Institutions in World Fisheries". Background Paper World Development Report, Dynamic Development in a Sustainable World (Washington, D.C.: World Bank, 2003).

Hanson, A.J., "Measuring Progress Towards Sustainable Development" (2003) 46 *Ocean & Coastal Management* 381.

Harbitz, A.M.A., and M. Aschan, "A Two-Dimensional Geostatistic Method to Simulate the Precision of Abundance Estimates" (2003) 60 *Canadian Journal of Fisheries and Aquatic Sciences* 1539.

Hardin, G., "The Tragedy of the Commons" (1968) 162 *Science* 1243.

Hare, J., "Port State Control: Strong Medicine to Cure a Sick Industry" (1997) 26 *Georgia Journal of International and Comparative Law* 71.

Harremoës, P., et al., eds., "Late Lessons From Early Warnings: The Precautionary Principle 1896–2000", Environmental Issue Report No. 22 (Copenhagen: European Environmental Agency, 2001).

Harwood, J., and K. Stokes, "Coping with Uncertainty in Ecological Advice: Lessons From Fisheries" (2003) 18 *Trends in Ecology and Evolution* 617.

Hatcher, A., "Subsidies for European Fishing Fleets: The European Community's Structural Policy for Fisheries 1971–1999" (2000) 24 *Marine Policy* 129.

Hatcher, A., "Incentives for Investment in IUU Fishing Capacity". Paper presented to the OECD Workshop on Illegal, Unreported and Unregulated (IUU) Fishing Activities, Paris, France, 19–20 April 2004, AGR/FI/IUU(2004)4O (Paris: OECD, 2004).

Hatcher, A., et al., "'Quota-Hopping' and the Foreign Ownership of UK Fishing Vessels" (2002) 26 *Marine Policy* 1.

Haughton, M.O., "Fisheries Subsidy and the Role of Regional Fisheries Management Organisations: The Caribbean Experience". Paper presented at the UNEP Workshop on Fisheries Subsidies and Sustainable Fisheries Management, Geneva, Switzerland, 26–27 April 2004.

Haughton, M.O., et al., "Establishment of the Caribbean Regional Fisheries Mechanism" (2004) 28 *Marine Policy* 351.

Haward, M., and A. Bergin, "Taiwan's Distant Water Tuna Fisheries" (2000) 24 *Marine Policy* 33.

Hayashi, M., "United Nations Conference on Straddling Fish Stocks and Highly Migratory Fish Stocks: An Analysis of the 1993 Sessions" in E. Mann Borgese, N. Ginsburg, and J.R. Morgan, eds., *Ocean Yearbook 11* (Chicago: University of Chicago Press, 1994) 20.

Hayashi, M., "The 1995 Agreement on the Conservation and Management of Straddling and Highly Migratory Fish Stocks: Significance for the Law of the Sea Convention" (1995) 29 *Ocean & Coastal Management* 51.

Hayashi, M., "The Role of the United Nations in Managing the World's Fisheries" in G. Blake et al., eds., *The Peaceful Management of Transboundary Resources* (London: Graham & Trotman Ltd., 1995) 373.

Hayashi, M., "Enforcement by Non-Flag States on the High Seas Under the 1995 Agreement on Straddling and Highly Migratory Stocks" (1996) 9 *Georgetown International Environmental Law Review* 1.

Hayashi, M., "The 1995 UN Fish Stocks Agreement and the Law of the Sea" in D. Vidas and W. Østreng, eds., *Order for the Oceans at the Turn of the Century* (The Hague: Kluwer Law International, 1999) 37.

Hayashi, M., "The Straddling and Highly Migratory Fish Stocks Agreement" in E. Hey, ed., *Developments in International Fisheries Law* (The Hague: Kluwer Law International, 1999) 55.

Hayes, P., and L. Zarsky, "Environmental Issues and Regimes in Northeast Asia" (1994) 6 *International Environmental Affairs* 283.

Healey, M.C., and T. Hennessey, "The Paradox of Fairness: The Impact of Escalating Complexity on Fishery Management" (1998) 22 *Marine Policy* 109.

Heap, J.A., "Has CCAMLR Worked? Management Policies and Ecological Needs" in A. Jørgensen-Dahl and W. Østreng, ed., *The Antarctic Treaty System in World Politics* (London: St. Martin's Press, 1991) 43.

Hedly, C., ed., "Special Report: FAO International Plan of Action on IUU Fishing", OceanLaw On-line Paper, No. 1 (March 2001). Available at <http://www.intfish.net/ops/papers/1.htm>.

Hedley, C., "NEAFC: First Meeting of the Permanent Committee on Control and Enforcement (PECCOE)" (2001) *International Fisheries Bulletin*, Focus No. 8.

Hedley, C., "The South-East Fisheries Organization (SEAFO) Convention: An Initial Review" (2001) *International Fisheries Bulletin*, Focus No. 12.

Hedley, C., "The 1998 Agreement on the International Dolphin Conservation Program: Recent Developments in the Tuna–Dolphin Controversy in the Eastern Pacific Ocean" (2001) 32 *Ocean Development and International Law* 71.

Hedley, C., "Entry into Force of the United Nations Fish Stocks Agreement: An Initial Assessment" (2001) *International Fisheries Bulletin*, Focus No. 24.

Hedley C., "The European Community and the United Nations Fish Stock Agreement" (2002) *International Fisheries Bulletin*, Focus No. 1.

Hedley, C., "The Precautionary Principle and International Fisheries Law" (2004) 1 *International Fisheries Law and Policy Review* 1.

Hedley, C., E.J. Molenaar, and A.G. Oude Elferink, *The Implications of the UN Fish Stocks Agreement (New York, 1995) for Regional Fisheries Organisations and International Fisheries Management* (Brussels: European Parliament, 2003).

Henriksen, T., G. Hønneland, and A. Sydnes, *Law and Politics in Ocean Governance: the UN Fish Stocks Agreement and Regional Fisheries Management Regimes* (Leiden, Boston: Martinus Nijhoff Publishers, 2006).

Herr, R.A., "The International Regulation of Patagonian Toothfish: CCAMLR and High Seas Fisheries Management" in O.S. Stokke, ed., *Governing High Seas Fisheries: The Interplay of Global and Regional Regimes* (Oxford: Oxford University Press, 2001) 303.

Herrington, W.C., "The Convention on Fisheries and Conservation of Living Resources: Accomplishments of the 1958 Geneva Conference" in L.M. Alexander, ed., *The Law of the Sea: Offshore Boundaries and Zones* (Columbus, Ohio: Ohio State University Press, 1967) 26.

Hersoug, B., and P. Holm, "Change without Redistribution: An Institutional Perspective on South Africa's New Fisheries Policy" (2000) 24 *Marine Policy* 221.

Hewison, G.J., "The Legally Binding Nature of the Moratorium on Large-Scale High Seas Driftnet Fishing" (1994) 25 *Journal of Maritime Law and Commerce* 557.

Hewison, G. J., "The Precautionary Approach to Fisheries Management: An Environmental Perspective" (1996) 11 *International Journal of Marine and Coastal Law* 301.

Hey, E., *The Regime for the Exploitation of Transboundary Marine Fisheries Resources: The United Nations Law of the Sea Convention Cooperation between States* (Dordrecht: Martinus Nijhoff, 1989).

Hey, E., "The Precautionary Concept in Environmental Policy and Law: Institutionalizing Caution" (1992) 4 *Georgetown International Environmental Law Review* 303.

Hey, E., ed., *Developments in International Fisheries Law* (The Hague: Kluwer Law International, 1999).

High Seas Task Force, "High Seas Governance", Meeting of the High Seas Task Force, Paris, France, 9 March 2005, HSTF/09. Available at <http://www.high-seas.org> (accessed August 2005).

High Seas Task Force, "Promoting Responsible Port States", Meeting of the High Seas Task Force, Paris, France, 9 March 2005, HSTF/06. Available at <http://www.high-seas.org> (accessed August 2005).

High Seas Task Force, "Closing the Net: Stopping Illegal Fishing on the High Seas", Final Report of the Ministerially-led Task Force on IUU Fishing on the High Seas, 2006, Governments of Australia, Canada, Chile, Namibia, New Zealand, and the United Kingdom, WWF, IUCN, and the Earth Institute at Columbia University. Available at <http://www.high-seas.org/> (accessed March 2008).

Hilborn, R., J.M. Orensanz, and A. Parma, "Institutions, Incentives and the Future of Fisheries" (2005) 360 *Philosophical Transactions of the Royal Society: Biological Sciences. Series **B*** 47.

Hilborn, R., et al., "When Can Marine Reserves Improve Fisheries Management?" (2004) 47 *Ocean & Coastal Management* 197.

Hillgenberg, H., "A Fresh Look at Soft Law" (1999) 10 *European Journal of International Law* 499.

Hinds, L., "Crisis in Canada's Atlantic Sea Fisheries" (1995) 19 *Marine Policy* 271.

Hinds, L., "Policy Implications and Reconstruction of Marine Institutional Architectures: The Regional Indigenous Organization (RIO) Option" (2001) 25 *Marine Policy* 415.

Hinds, L., "Oceans Governance and the Implementation Gap" (2003) 27 *Marine Policy* 349.

Hinkel, R., *The Effectiveness of Regional Fisheries Management Regimes: An Application of International Relations Theory to Marine Policy* (Masters Thesis, University of Delaware, 2000) [unpublished].

Hoel, A.H., "Political Uncertainty in International Fisheries Management" (1998) 37 *Fisheries Research* 239.

Holland, G., "The Arctic Ocean – The Management of Change in the Northern Seas" (2002) 45 *Ocean & Coastal Management* 841.

Holt, T., et al., *Certification Report for South Georgia Patagonian Toothfish Longline Fishery* (Derby: Moody Marine Ltd., 2004).

Hønneland, G.B., *Coercive and Discursive Compliance Mechanisms in the Management of Natural Resources. A Case Study From the Barents Sea Fisheries* (Dordrecht: Kluwer Academic Publishers, 2000).

Hønneland, G.B., "Recent Global Agreements on High Seas Fisheries: Potential Effects on Fisherman Compliance" in O.S. Stokke, ed., *Governing High Seas Fisheries: The Interplay of Global and Regional Regimes* (Oxford: Oxford University Press, 2001) 121.

Hønneland, G.B., "Fisheries in the Svalbard Zone: Legality, Legitimacy and Compliance" in A.G. Oude Elferink and D.R. Rothwell, eds., *The Law of the Sea and Polar Marine Delimitation and Jurisdiction* (The Hague: Kluwer Law International, 2001) 317.

Hønneland, G.B., "Enforcement Co-Operation Between Norway and Russia in the Barents Sea Fisheries" (2000) 31 *Ocean Development and International Law* 249.

Hønneland, G.B., *Russian Fisheries Management: The Precautionary Approach in Theory and Practice* (Leiden: Martinus Nijhoff Publishers, 2004).

Hønneland, G.B., and A.-K. Jorgensen, "Implementing International Fisheries Agreements in Russia — Lessons From the Northern Basin" (2002) 26 *Marine Policy* 359.

Hoydal, K., "Report on a Conference on Fisheries Economics, 6–7 June 2002, Sphienberg Manor, Denmark". Paper presented to the Conference on Fisheries Economics, Sphienberg Manor, Denmark, 6–7 June 2002.

Hoydal, K., "IUU Fishing in the NEAFC Area: How Big Is the Problem and What Have We Done?" in K. Gray et al., eds., *Fish Piracy: Combating Illegal, Unreported and Unregulated Fishing* (Paris: OECD, 2004) 159.

Huber, M.E., et al., "Priority Problems Facing the Global Marine and Coastal Environment and Recommended Approaches to Their Solution" (2003) 46 *Ocean & Coastal Management* 479.

Hughes, T.P., D.R. Bellwood, C. Folke, R.S. Steneck, and J. Wilson, "New Paradigms for Supporting the Resilience of Marine Ecosystems" (2005) 20 *Trends in Ecology & Evolution* 380.

Hughey, K.F.D., R. Cullen, and G.N. Kerr, "Stakeholder Groups in Fisheries Management" (2000) 24 *Marine Policy* 119.

Hunt, C., "Management of the South Pacific Tuna Fishery" (1997) 21 *Marine Policy* 155.

Hunt, C., "Fiji's Fisheries: Their Contribution to Development and Their Future" (1999) 23 *Marine Policy* 571.

Huppert, D.D., "Auctions of IFQs as a Means to Share the 'Rent'" in T. Bjørndal, D.V. Gordon, R. Arnason, and U.R. Sumaila, eds., *Advances in Fisheries Economics: A Festschrift in Honour of Professor Gordon R. Munro* (London: Blackwell Publishing Ltd., 2007).

Hutchings, J.A., "Collapse and Recovery of Marine Fishes" (2000) 406 *Nature* 882.

Hutchings, J.A., "Conservation Biology of Marine Fishes: Perceptions and Caveats Regarding Assignment of Extinction Risk" (2001) 58 *Canadian Journal of Fisheries and Aquatic Sciences* 108.

Hutchings, J.A., and J.D. Reynolds, "Marine Fish Population Collapses: Consequences for Recovery and Extinction Risk" (2004) 54 *BioScience* 297.

Hyvarinen, J., E. Wall, and I. Lutchman, *NGO Oceans Network Campaign Handbook: The UN Fish Stock Agreement in 1998* (Boston, Mass.: NGO Oceans Network, nd).

Hyvarinen, J., E. Wall, and I. Lutchman, "The United Nations and Fisheries in 1998" (1998) 29 *Ocean Development and International Law* 323.

Hyvättinen, H., and M. Hildén, "Environmental Policies and Marine Engines – Effects on the Development and Adoption of Innovations" (2004) 28 *Marine Policy* 491.

Indian Ocean Tuna Commission (IOTC), *Report of the Fourth Session of the IOTC*, Kyoto, Japan, 13–16 December 1999, IOTC/S/04/99/R[E] (Victoria, Seychelles: IOTC, 2000).

Indian Ocean Tuna Commission (IOTC), Resolution 05/01 On Conservation and Management Measures for Bigeye Tuna. Available at <http://www.iotc.org/English/resolutions/reso_detail.php?reso=35> (accessed February 2006).

INFOFISH, "FISH INFOnetwork Market Report on Tuna" (March–April 2005) *Eurofish Magazine*.

Inter-American Tropical Tuna Commission (IATTC), "Tuna and Billfishes in the Eastern Pacific Ocean", IATTC Fishery Status Report No. 2 (La Jolla, Calif.: IATTC, 2004).

International Commission for the Conservation of Atlantic Tunas (ICCAT), "Bluefin Year Program" in *Report for Biennial Period 2004–2005 Part I (2004) — Vol. 2*. Available at <http://www.iccat.int/Documents/BienRep/REP_EN_04-05_I_2.pdf> (accessed March 2006).

International Commission for the Conservation of Atlantic Tunas (ICCAT), "North Atlantic swordfish stock assessment" in *Report for Biennial Period 2002–2003 Part 1 (2002) — Vol. 2* at 115-124. Available at <http://www.iccat.es/Documents/BienRep/REP_EN_02-03_I_2.pdf> (accessed 15 October 2005).

International Commission for the Conservation of Atlantic Tunas (ICCAT), "Report of the Sub-Committee on Environment" in *Report for Biennial Period 2004–2005 Part II (2005) — Vol. 2*. Available at <http://www.iccat.int/Documents/BienRep/REP_EN_04-05_II_2.pdf> (accessed March 2006).

International Commission for the Conservation of Atlantic Tunas (ICCAT), "Swordfish stock assessment" in *Report for Biennial Period 2004–2005 Part II (2005) — Vol. 2*. Available at <http://www.iccat.int/Documents/BienRep/REP_EN_04-05_II_2.pdf> (accessed March 2006).

International Council for the Exploration of the Sea (ICES), *Report of the Precautionary Approach to Fisheries Management*, Copenhagen, Denmark, 5–11 February 1997, ICES CM 1997/Assess: 7

International Council for the Exploration of the Sea (ICES), *Report of the Precautionary Approach to Fisheries Management*, Copenhagen, Denmark, 3–6 February 1998, ICES CM 1998/ACFM: 10.

International Council for the Exploration of the Sea (ICES), 2004, Advice to NEAFC on deep sea fisheries. Available at <http://www.neafc.org/fisheries/docs/acfm_reports-2004/deep-water-general-2004.pdf> (accessed 21 October 2005).

International Institute for Sustainable Development (IISD), "A Summary of the Proceedings of the United Nations Conference on Environment and Development 3–14 June 1992", Earth Negotiations Bulletin No. EMB:02:13 (Winnipeg, Man: IISD, 1992).

International Labour Organisation, "ILO Work Underway to Prepare a Comprehensive Standard on Work in the Fishing Sector". Paper presented to the OECD Workshop on Illegal, Unreported and Unregulated (IUU) Fishing Activities, Paris, France, 19–20 April 2004, AGR/FI/IUU/RD(2004)4 (Paris: OECD, 2004).

International Union for the Conservation of Nature (IUCN), "The Marine Protected Areas of Antarctica", IUCN News Release. Available at <http://www.iucn.org/en/news/archive/2001_2005/press/ccamlr.pdf> (accessed March 2006).

International Union for the Conservation of Nature (IUCN), *The IUCN Red List of Threatened Species*. Available at <http://www.iucnredlist.org/info/programme> (accessed March 2006).

International Union for the Conservation of Nature (IUCN), *Ten-Year High Seas Marine Protected Area Strategy: A Ten-Year Strategy to Promote the Development of a Global Representative System of High Seas Marine Protected Area Networks,* as agreed by Marine Theme Participants at the Vth IUCN World Parks Congress, Durban, South Africa, 8–17 September 2003.

International Whaling Commission (IWC),"Chairman's Report of the Thirty-Fourth Annual Meeting," (1983) 33 *Reports of the International Whaling Commission* 20.

Jackson, A., "Developments in the South East Atlantic, 1997–1999" in M.H. Nordquist and J.N. Moore, eds., *Current Fisheries and the Food Agriculture Organization of the United Nations* (The Hague: Kluwer Law International, 2000).

Jackson, A., "The Convention on the Conservation and Management of Fishery Resources in the South East Atlantic Ocean, 2001: An Introduction" (2002) 17 *International Journal of Marine and Coastal Law* 33.

Jackson, J.B.C., et al., "Historical Overfishing and the Recent Collapse of Coastal Ecosystems" (2001) 293 *Science* 629.

Jensen, C.L., "Reduction of the Fishing Capacity in 'Common Pool' Fisheries" (2002) 26 *Marine Policy* 155.

Jentoft, S., "The Community: A Missing Link of Fisheries Management" (2000) 24 *Marine Policy* 53.

Jentoft, S., "Legitimacy and Disappointment in Fisheries Management" (2000) 24 *Marine Policy* 141.

Jentoft, S., "Fisheries Co-Management As Empowerment" (2005) 29 *Marine Policy* 1.

Jentoft, S., and B. McCay, "User Participation in Fisheries Management: Lessons Drawn From International Experiences" (1995) 19 *Marine Policy* 227.

Johannesburg Summit 2002, "Treaty to Halt Over-fishing of Oceans Takes Effect" (11 December 2001). Available at <http://www.un.org/jsummit/html/whats_new/otherstories_fisheries_11dec.html> (accessed 26 January 2005).

Johnson, G., and S. Magick, "Fishing for Answers: Tuna Commission Establishment Not A Minute Too Soon" (1 November 2004) *Pacific Magazine*. Available at <http://www.pacificmagazine.net/issue/2004/11/01/fishing-for-answers> (8 March 2005).

Johnston, D.M., *The International Law of Fisheries: A Framework for Policy-Oriented Inquiries* (New Haven, Conn.: New Haven Press, 1987).

Johnston, D.M., "The Driftnetting Problem in the Pacific Ocean: Legal Considerations and Diplomatic Options" (1990) 21 *Ocean Development and International Law* 5.

Joseph, J., "The Tuna–Dolphin Controversy in the Eastern Pacific Ocean: Biological, Economic and Political Impacts" (1994) 25 *Ocean Development and International Law* 1.

Joyner, C.C. and A.A. von Gustedt, "The Turbot War of 1995: Lessons for the Law of the Sea" (1996) 11 *International Journal of Marine and Coastal Law* 425.

Joyner, C.C., *Governing the Frozen Commons: the Antarctic Regime and Environmental Protection* (Columbia, S.C.: University of South Carolina Press, 1998).

Joyner, C.C., "Compliance with and Enforcement of International Fisheries Law" in E. Hey, ed., *Developments in International Fisheries Law* (The Hague: Kluwer Law International, 1999) 327.

Joyner, C.C., "On the Borderline? Canadian Activism in the Grand Banks" in O.S. Stokke, ed., *Governing High Seas Fisheries: The Interplay of Global and Regional Regimes* (Oxford: Oxford University Press, 2001) 207.

Joyner, C.C., and Z. Tyler, "Marine Conservation versus International Free Trade: Reconciling Dolphins with Tuna and Sea Turtles with Shrimp" (2000) 31 *Ocean Development and International Law* 127.

Juda, L., *International Law and Ocean Use Management: The Evolution of Ocean Governance* (London: Routledge, 1996).

Juda, L., "The United Nations Agreement on Straddling and Highly Migratory Fish Stocks: Policy Problems in Implementation" in M. Nordquist, ed., *Implementing the 1982 Law of the Sea Convention* (Charlottesville, Va.: University of Virginia, Center for Oceans Law and Policy, 1996).

Juda, L., "The 1995 United Nations Agreement on Straddling Fish Stocks and Highly Migratory Fish Stocks: A Critique" (1997) 28 *Ocean Development and International Law* 147.

Juda, L., "The United Nations Fish Stocks Agreement" in O.S. Stokke and Ř.B. Thommessen, eds., *Yearbook of International Co-operation on Environment and Development 2001/2002* Vol. 2 (London: Earthscan Publications, 2001) 53.

Juda, L., "Rio Plus Ten: The Evolution of International Marine Fisheries Governance" (2002) Vol. 2 33 *Ocean Development and International Law* 109.

Juda, L., "Changing National Approaches to Ocean Governance: The United States, Canada, and Australia" (2003) 34 *Ocean Development and International Law* 161.

Kang, J.-S., "The United Nations Convention on the Law of the Sea and Fishery Relations Between Korea, Japan and China" (2003) 27 *Marine Policy* 111.

Karagiannakos, A., "Total Allowable Catch (TAC) and Quota Management System in the European Union" (1996) 20 *Marine Policy* 235.

Kawaley, I., "Implications of Exclusive Economic Zone Management and Regional Cooperation Between South Pacific Small Midocean Island Commonwealth Territories" (1999) 30 *Ocean Development and International Law* 333.

Kaye, S.M., "Implementing High Seas Biodiversity Conservation: Global Geopolitical Considerations" (2004) 28 *Marine Policy* 221.

Kaye, S.M., *International Fisheries Management* (The Hague: Kluwer Law International, 2001).

Kedziora, D.M., "Gunboat Diplomacy in the Northwest Atlantic: The 1995 Canada–EU Fishing Dispute and the United Nations Agreement on Straddling and High Migratory Fish Stocks" (1996/1997) 17 *Northwestern Journal of International Law and Business* 1132.

Kesteven, G.L., "MSY Revisited: A Realistic Approach to Fisheries Management and Administration" (1997) 21 *Marine Policy* 73.

Keyuan, Z., "Sino–Japanese Joint Fishery Management in the East China Sea" (2003) 27 *Marine Policy* 125.

Kim, S.P., "The UN Convention on the Law of the Sea and New Fisheries Agreements in North East Asia" (2003) 27 *Marine Policy* 97.

Kimball, L.A., "Whither International Institutional Arrangements to Support Ocean Law?" (1997) 36 *Columbian Journal of Transnational Law* 307.

Kimball, L.A., *International Ocean Governance: Using International Law and Organizations to Manage Marine Resources Sustainably* (Gland: IUCN–NOAA, 2001).

Kimball, L.A., "Deep-Sea Fisheries of the High Seas: The Management Impasse" (2004) 19 *International Journal of Marine and Coastal Law* 259.

Kingsbury, B., "The Concept of Compliance as a Function of Competing Conceptions of International Law" (1998) 19 *Michigan Journal of International Law* 345.

Kirkwood, G.P., and D.J. Agnew, "Deterring IUU Fishing" in A.I.L. Payne, C.M. O'Brien, and S.I. Rogers, eds., *Management of Shared Fish Stocks* (Oxford: Blackwell Publishing Ltd., 2004) 1.

Kirkwood, G.P., and A.J. Constable, "Integration of CPUE Data into Assessments Using the Generalized Yield Model" (2001) 8 *CCAMLR Science* 65.

Klabbers, J., "The Redundancy of Soft Law" (1996) 65 *Nordic Journal of International Law* 167.

Klabbers, J., "The Undesirability of Soft Law" (1998) 67 *Nordic Journal of International Law* 381.

Kuemlangen, B., "Legal Implications and Options for Regional Cooperative Arrangements for the Management of Coastal and Oceanic Large Pelagic Resources for CARICOM States" in R. Mahon and P.A. McConney, eds., *Management of Large Pelagic Fisheries in CARICOM Countries*, FAO Fisheries Technical Paper No. 464 (Rome: FAO, 2004).

Kwiatkowska, B., "High Seas Fisheries Regime: At a Point of No Return?" (1993) 8 *International Journal of Marine and Coastal Law* 327.

Kwiatkowska, B., "The *Southern Bluefin Tuna* Arbitral Tribunal Did Get It Right: A Commentary and Reply to the Article by David A. Colson and Dr. Peggy Hoyle" (2003) 34 *Ocean Development and International Law* 369.

Kwiatkowska, B., "Straddling and Migratory Fish Stocks in the New Law of the Sea: Reconciling Rights, Freedoms and Responsibilities" in R. St.J. Macdonald, ed., *Essays in Honour of Wang Tieya* (Dordrecht: Martinus Nijhoff Publishers, 1994).

Lack M. and G. Sant, "Patagonian Toothfish: Are Conservation and Management Measures Working?" 2001 *Traffic Bulletin* Vol. 19 No. 1.

Lack, M., K. Short, and A. Willock, *Managing Risk and Uncertainty in Deep-Sea Fisheries: Lessons from Orange Roughy* (NSW, Australia: TRAFFIC Oceania and WWF Australia, 2003).

Lack, M., *Antarctic Toothfish: An Analysis of Management, Catch and Trade* (Sydney: TRAFFIC Oceania, 2001).

Langley, A., J. Hampton, and P. Williams, "The Western and Central Pacific Tuna Fishery: 2002: Overview and Status of Stocks", Oceanic Fisheries Programme Tuna Fisheries Assessment Report No. 5 (Noumea: Secretariat of the Pacific Community, 2002).

Lapeyre, E.G., "The Maritime Front of the Río De La Plata as an Instrument for Binational Fisheries Management" (1999) 42 *Ocean & Coastal Management* 155.

Lapko, V.V., and V.I. Radchenko, "Sea of Okhotsk" in C.R.C. Sheppard, ed., *Seas at the Millennium: An Environmental Evaluation* (New York: Elsevier Publishing Company, 2000) Vol. II.

Large, P.A., et al., "Deep-Water Fisheries of the Northeast Atlantic: II Assessment and Management Approaches" (2003) 31 *Journal of Northwest Atlantic Fishery Science* 151.

Larkin, P., "An Epitaph for the Concept of MSY" (1977) 106 *Transactions of the American Fisheries Society* 1.

Larocque, E., *The Convention on the Conservation and Management of Highly Migratory Fish Stocks in the Western and Central Pacific Ocean: Can Tuna Promote Development of Pacific Island Nations* (Mānoa: University of Hawaii, 2003).

Laves, R.C., *The Falklands/Malvinas Case: Breaking the Deadlock in the Anglo-Argentine Sovereignty Dispute* (The Hague: Martinus Nijhoff Publishers, 2001).

Leadbitter, D., "Seafood Trade and Market Access: Threats and Opportunities" in A.G. Brown, ed., *Fish, Aquaculture and Food Security: Sustaining Fish as a Food Supply* (Canberra: Gippsland Aquaculture Industry Network, 2004).

Lem, A., "The WTO Doha Round and Fisheries: What's at Stake". FAO Fact Sheet for WTO Ministerial Conference in Cancun, Mexico (Rome: FAO, 2003).

Lem, A., "China, the WTO and World Fish Trade", Globefish Industry Issues (Rome: FAO, 2004). Available at at <http://www.globefish.org/?id=2123>.

Lévy, J.-P., and G.G. Schram, *United Nations Conference on Straddling Fish Stocks and Highly Migratory Fish Stocks — Selected Documents* (The Hague: Martinus Nijhoff Publishers, 1996).

Levin, S.A., "Forum: Science and Sustainability" (1993) 3(4) *Ecological Applications* 545.

Lewison, R.L., et al., "Understanding Impacts of Fisheries Bycatch on Marine Megafauna" (2004) 19(11) *Trends in Ecology and Evolution* 598.

Lindroos, M., "Restricted Coalitions in the Management of Regional Fisheries Organizations" (2004) 17 *Natural Resource Modeling* 1.

Link., J.S., "Ecological Considerations in Fisheries Management: When Does It Matter?" (2002) 27 *Fisheries* 10.

Løbach, T., "Measures to be Adopted by the Port State in Combating IUU Fishing", Document AUS:IUU/2000/15 (Rome: FAO, 2000). Available at <http://www.fao.org/docrep/005/Y3274E/y3274e0h.htm> (accessed August 2007).

Løbach, T., "Port State Measures". Paper presented to the OECD Workshop on Illegal, Unreported and Unregulated (IUU) Fishing Activities, Paris, France, 19–20 April 2004, AGR/FI/IUU(2004) 9 (Paris: OECD, 2004).

Løbach, T., "Advances in Port State Control Measures" in K. Gray et al., eds., *Fish Piracy: Combating Illegal, Unreported and Unregulated Fishing* (Paris: OECD, 2004) 291.

Lodge, M.W., "Minimum Terms and Conditions of Access: Responsible Fisheries Management Measures in the South Pacific Region" (1992) 16 *Marine Policy* 277.

Lodge, M.W., "The South Pacific Forum Fisheries Agency and Legal Aspects of Fisheries Monitoring, Control and Surveillance". Paper presented to the Report of a Regional Workshop on Fisheries Monitoring Control and Surveillance, Albion, Mauritius, 16–20 December 1996.

Lodge, M.W., "The Development of the Palau Arrangement for the Management of the Western Pacific Purse Seine Fishery" (1998) 22 *Marine Policy* 1.

Lodge, M.W., "The Fisheries Regimes of Enclosed and Semi-Enclosed Seas and High Seas Enclaves" in Hey, E. ed., *Developments in International Fisheries Law* (The Hague: Kluwer Law International, 1999) 193.

Lodge, M.W., and S.N. Nandan, "Some Suggestions Towards Better Implementation of the United Nations Agreement on Straddling Fish Stocks and Highly Migratory Fish Stocks of 1995" (2005) 20 *International Journal of Marine and Coastal Law* 345.

Lodge, M.W., et al., *Recommended Best Practices for Regional Fisheries Management Organizations: Report of an independent panel to develop a model for improved governance by Regional Fisheries Management Organizations* (London: Chatham House, 2007).

Lugten, G.L., *A Review of Measures Taken by Regional Marine Fishery Bodies to Address Contemporary Fishery Issues*, FAO Circular No. 940 (Rome: FAO, 1999).

Lutchman, I., "Have Subsidies Played a Role in IUU Fishing for Patagonian Toothfish in the Southern Oceans?" Paper presented at the UNEP Workshop on Fisheries Subsidies and Sustainable Fisheries Management, Geneva, Switzerland, 26–27 April 2004.

Lutchman, I., and D. Hoggarth, "Net Losses: Untying the Gordian Knot of Fishing Overcapacity", IUCN Marine Conservation and Development Report (Gland: IUCN - The World Conservation Union, 1999).

Macdonald, J.M., "Appreciating the Precautionary Principle as an Ethical Evolution in Ocean Management" (1995) 26 *Ocean Development and International Law* 255.

Macdonald, R. St. J., "Solidarity in the Practice and Discourse of Public International Law" (1996) VIII(II) *Pace International Law Review* 259.

Macdonald, R. St. J., "The Charter of the United Nations in Constitutional Perspective" (1999) 20 *Australian Year Book of International Law* 205.

Macdonald, R. St. J., and D.M. Johnston, eds., *Towards World Constitutionalism: Issues in the Legal Ordering of the World Community* (Leiden: Martinus Nijhoff Publishers, 2005).

Mace, P.M., "A New Role for MSY in Single-Species and Ecosystem Approaches to Fisheries Stock Assessment and Management" (2001) 2 *Fish and Fisheries* 2.

MacLennan, D.N., "Technology in the Capture Fisheries". Paper presented at the Kyoto Conference – Outcomes and Papers Presented. Available at <http://www.fao.org/docrep/006/AC442e/AC442e6.htm#> (accessed August 2005).

Madaria, E., and A.H. Vallega, *El Sector Pesquero Argentino: Informe General (Preliminar II)* (Buenos Aires: Universidad Catolica Argentina, 1999).

Mahon, R., and H.A. Oxenford, "Precautionary Assessment and Management of Dolphinfish in the Caribbean" (1999) 63(3-4) *Scientia Marina* 429.

Majkowski, J., "Global Fishery Resources of Tuna and Tuna-Like Species", FAO Fisheries Technical Paper. No. 483 (Rome: FAO, 2007).

Mangel, M., and P.S. Levin, "Regime, Phase and Paradigm Shifts: Making Community Ecology the Basic Science for Fisheries" (2005) 360 *Philosophical Transactions of the Royal Society: Biological Sciences. Series **B*** 95.

Marine Resources Assessment Group Ltd. (MRAG), "Review of Ecosystem-Bycatch Issues for the Western Central Pacific Region: Interim Advice to the Preparatory Conference for the Western and Central Pacific Fisheries Commission (WCPFC)". Prepared for the Preparatory Conference for the Western and Central Pacific Fisheries Commission (WCPFC), 12–18 November 2002, Manila, Philippines, WCPFC Prep Con/WP.9.

Marine Resources Assessment Group Ltd. (MRAG), "Review of Impacts of Illegal, Unreported and Unregulated Fishing on Developing Countries". Final Report prepared for UK Department of International Development (DFID) (London: MRAG, 2005).

Marr, S., "The Southern Bluefin Tuna Cases: The Precautionary Approach and Conservation and Management of Fish Resources" (2000) 11(4) *European Journal of International Law* 815.

Martin-Castex, B., "Potential Link Between IUU Fishing and the Status of Safety-Related International Instruments Applicable to Fishing Vessels and Fishers". Paper presented to the OECD Workshop on Illegal, Unreported and Unregulated (IUU) Fishing Activities, Paris, France, 19–20 April 2004, AGR/FI/IUU(2004)14 (Paris: OECD, 2004).

Matthew, S., "Whose Seas? Whose Freedom?" (1995) 12 *Samudra — Triannual Report of ICSF* 46.

Maunder, M., "Paradigm Shifts in Fisheries Stock Assessment: From Integrated Analysis to Bayesian Analysis and Back Again" (2003) 16(4) *Natural Resource Modeling* 465.

McCay, B.J., "Social and Ecological Implications of ITQs: An Overview" (1995) 28 *Ocean & Coastal Management* 3.

McConnell, M., "Darkening Confusion Upon Darkening Confusion: The Search for the Elusive Genuine Link" (1985) 16 *Journal of Maritime Law and Commerce* 365.

McDiarmid, B., M. Gotje, and K. Sack, *NAFO Case Study* (Amsterdam: Greenpeace, 2005). Available at <http://www.greenpeace.org/international/press/reports/NAFO-Case-Study> (accessed March 2006).

McDorman, T.L., "Stateless Fishing Vessels, International Law and the UN High Seas Fisheries Conference" (1994) 25 *Journal of Maritime Law and Commerce* 531.

McDorman, T.L., "Port State Enforcement: A Comment on Article 218 of the 1982 Law of the Sea Convention" (1997) 28(2) *Journal of Maritime Law and Commerce* 305.

McDorman, T.L., "Implementing Existing Tools: Turning Words into Actions – Decision-Making Processes of Regional Fisheries

Management Organizations (RFMOs)" (2005) 20 *International Journal of Marine and Coastal Law* 423.

McDougal, M., and W.T. Burke, *The Public Order of the Oceans: A Contemporary International Law of the Sea* (New Haven, Conn.: Yale University Press, 1962).

McGarvey, R., "Demand-Side Fishery Management: Integrating Two Forms of Input Control" (2003) 27 *Marine Policy* 207.

McKay, B., K. Mulvaney, and B. Thorne-Miller, "Danger At Sea: Our Changing Ocean", SeaWeb Reports. Available at <http://www.seaweb.org/resources/reports.php#danger> (accessed 26 January 2005).

McWhinnie, S., "The Tragedy of the Commons in International Fisheries: An Empirical Examination" (2004) [unpublished].

Meltzer, E., "Global Overview of Straddling and Highly Migratory Fish Stocks: The Nonsustainable Nature of High Seas Fisheries" (1994) 25 *Ocean Development and International Law* 255.

Meltzer, E., "Global Overview of Straddling and Highly Migratory Fish Stocks". Presented to the Conference on the Governance of High Seas Fisheries and the United Nations Fish Agreement, St. John's, N.L., 1–5 May 2005. Available at <http://www.dfo-mpo.gc.ca/fgc-cgp/documents/meltzer_e.htm> (accessed March 2006).

Meltzer, E., "Global Overview of Straddling and Highly Migratory Fish Stocks: Maps and Charts Detailing RFMO Coverage and Implementation" (2005) 20 *International Journal of Marine and Coastal Law* (Special Issue) 571.

Michaud, P., "Regional Fisheries Organisations: Interaction and Participations" (1996) 30 *Law of the Sea Institute Proceedings* 143.

Mikalsen, K.H., and S. Jentoft, "From User-Groups to Stakeholders? The Public Interest in Fisheries Management" (2001) 25 *Marine Policy* 281.

Milazzo, M., "Subsidies in World Fisheries: A Re-Examination", World Bank Technical Paper No. 406 (Washington, D.C.: World Bank, 1998).

Miles, E.L., "The Management of Marine Regions: The North Pacific" (1979) 6 *Ocean Development and International Law* vii.

Miles, E.L., "Concepts, Approaches, and Applications in Sea Use Planning and Management" (1989) 20 *Ocean Development and International Law* 213.

Miles, E.L., "Towards More Effective Management of High Seas Fisheries" (1993) III(1) *Asian Yearbook of International Law*.

Miles, E.L., "The Concept of Ocean Governance: Evolution Toward the 21st Century and the Principle of Sustainable Ocean Use" (1999) 27 *Coastal Management* 1.

Miles, E.L., and W.T. Burke, "Pressures on the United Nations Convention on the Law of the Sea of 1982 Arising from New Fisheries Conflicts: The Problem of Straddling Stocks" (1989) 20 *Ocean Development and International Law* 343.

Miles, E.L., et al., *Environmental Regime Effectiveness: Confronting Theory with Evidence* (Cambridge: The MIT Press, 2002).

Miller, D., "Management and Governance Conventions and Protocols — SEAFC, WCPFC and SADC" in R. Shotton, ed., *Deep Sea 2003: Conference on the Governance and Management of Deep-sea Fisheries. Part 1: Conference Reports*. Queenstown, New Zealand, 1–5 December 2003. FAO Fisheries Proceedings No. 3/1 (Rome: FAO, 2005) 596.

Miller, D.G.M., "Patagonian Toothfish: The Storm Gathers". Paper presented to the OECD Workshop on Illegal, Unreported and Unregulated (IUU) Fishing Activities, Paris, France, 19–20 April 2004, AGR/FI/IUU(2004)7 (Paris: OECD, 2004).

Miller, D.G.M., "What Role for RFMOs?". Paper presented to the OECD Workshop on Illegal, Unreported and Unregulated (IUU) Fishing Activities, Paris, France, 19–20 April 2004, AGR/FI/IUU(2004)18 (Paris: OECD, 2004).

Miller, D.G.M., and D. Agnew, "Krill Fisheries in the Southern Ocean" in I. Everson, ed., *Krill: Biology, Ecology and Fisheries* Vol. 6, Fish and Aquatic Resources Series 300 (London: Blackwell Science, 2000).

Miller, D.G.M., and E.J. Molenaar, "The SEAFC Convention: A Comparative Analysis in a Developing Coastal State Perspective" in A. Chircop, S. Coffen-Smout, and M. McConnell, eds., *Ocean Yearbook 20* (Ardsley, N.Y.: Transnational Press, 2006).

Miller, D.G.M., E.N. Sabourenkov, and D.C. Ramm, "Managing Antarctic Marine Living Resources: The CCAMLR Approach" (2004) 19 *International Journal of Marine and Coastal Law* 317.

Mirovitskaya, N.S., and J.C. Haney, "Fisheries Exploitation as a Threat to Environmental Security: The North Pacific Ocean" (1992) 16 *Marine Policy* 243.

Mitchell, C.L., "Fisheries Management in the Grand Banks, 1980–1992 and the Straddling Stock Issue" (1997) 21 *Marine Policy* 97.

Miyake, M.P., "A Review of the Fishing Capacity of the Longline Fleets of the World" in W.H. Bayliff, J.I. de Leiva Moreno, and J. Majkowski, eds., *Second Meeting of the Technical Advisory Committee of the FAO Project: Management of Tuna Fishing Capacity: Conservation and Socio-Economics*, FAO Fisheries Proceedings No. 2 (Rome: FAO, 2005).

Miyake, M.P., N. Miyabe, and H. Nakano, "Historical Trends of Tuna Catches in the World", FAO Fisheries Technical Paper No. 467 (Rome: FAO, 2004).

Molenaar, E.J., "The Concept of 'Real Interest' and Other Aspects of Co-Operation Through Regional Fisheries Management Mechanisms" (2000) 15 *International Journal of Marine and Coastal Law* 475.

Molenaar, E.J., "The South Tasman Rise Arrangement of 2000 and Other Initiatives on Management and Conservation of Orange Roughy" (2001) 16 *International Journal of Marine and Coastal Law* 77.

Molenaar, E.J., "CCAMLR and Southern Ocean Fisheries" (2001) 16 *International Journal of Marine and Coastal Law* 465.

Molenaar, E.J., "Ecosystem Based Fisheries Management, Commercial Fisheries, Marine Mammals and the 2001 Reykjavik Declaration in the Context of International Law" (2002) 17 *International Journal of Marine and Coastal Law* 561.

Molenaar, E.J., "Marine Fisheries in the Netherlands Antilles and Aruba in the Context of International Law" (2003) 18 *International Journal of Marine and Coastal Law* 127.

Molenaar, E.J., "Participation, Allocation and Unregulated Fishing: The Practice of Regional Fisheries Management Organisations" (2003) 18 *International Journal of Marine and Coastal Law* 457.

Molenaar, E.J., "Addressing Regulatory Gaps in High Seas Fisheries" (2005) 20 *International Journal of Marine and Coastal Law* 533.

Molenaar, E.J., "Global, Regional and Unilateral Approaches to Unregulated Deep-Sea Fisheries" in R. Shotton, ed., *Deep Sea 2003: Conference on the Governance and Management of Deep-sea Fisheries. Part 1: Conference Reports*. Queenstown, New Zealand, 1–5 December 2003, FAO Fisheries Proceedings No. 3/1 (Rome: FAO, 2005).

Molenaar, E.J. "Managing Biodiversity in Areas Beyond National Jurisdiction" in M.H. Nordquist, R. Long, T.H. Heidar, and J.N. Moore, eds., *Law, Science & Ocean Management* (London: Martinus Nijhoff Publishers, 2007) 625.

Molenaar, E.J., and M. Tsamenyi, "Satellite Based Vessel Monitoring Systems (VMS) for Fisheries Management — International Legal Aspects and Developments in State Practice", FAO Legal Paper On-Line No. 7 (Rome: FAO, 2000). Available at <http://www.fao.org/Legal/prs-ol/lpo7.pdf>.

Moore, G., "Enforcement without Force: New Concepts in Compliance Control for Foreign Fishing Operations Based on Regional Cooperation" (1993) 24 *Ocean Development and International Law* 197.

Moore, G., "FAO – The Food and Agriculture Organisation of the United Nations Compliance Agreement" (1995) 10 *International Journal of Marine and Coastal Law* 412.

Moore, G., "Chapter 5: The Code of Conduct for Responsible Fisheries" in E. Hey, ed., *Developments in International Fisheries Law* (The Hague: Kluwer Law International, 1999) 85.

Moore, J.A., "Deep-sea Finfish Fisheries: Lessons from History" (1999) 24 *Fisheries* 16.

Moore, J.N., "The Rule of Law in the Oceans" in M.H. Nordquist and J.N. Moore, eds., *Security Flashpoints: Oil, Islands, Sea Access and Military Confrontation* (The Hague: Martinus Nijhoff Publishers, 1998) 467.

Morato, T., and D. Pauly, eds., "Seamounts: Biodiversity and Fisheries 2004", Fisheries Centre Research Reports No. 12(5) (Vancouver, B.C.: The University of British Columbia, 2004).

Morling, P., "The Economics of Marine Protected Areas in the High Seas" (2004) 21 *The George Wright Forum* 49.

Moron, J., "Report on Management Measures for the European Tuna Purse Seine Fleet". Presented to the Fourteenth Meeting of the Standing Committee on Tuna and Billfish, Noumea, New Caledonia, 9–16 August 2001. Available at <http://www.spc.int/oceanfish/Html?SCTB/SCTB14/FT-9%20Moron-Management.pdf> (accessed August 2007).

Mossop, J., "High Seas Marine Protected Areas: Unfulfilled Promise or Impossible Premise?" in A. Costi and Y.-L. Sage, eds., *Droit de l'environnement dans le Pacifique: problématiques et perspectives croisées/Environmental Law in the Pacific: International and Comparative Perspectives* (Wellington: New Zealand Association of Comparative Law and Association de Legislation Comparée des Pays du Pacifique, 2005) 147.

Mossop, J., "Book Review: Dispute Settlement in the UN Convention on Law of the Sea" (2005) 36(4) *Victoria University of Wellington Law Review* 867.

Mullon, C., P. Fréon, and P. Cury, "The Dynamics of Collapse in Worldwide Fisheries" (2005) 6(2) *Fish and Fisheries* 111.

Munro, G.R., and U.R. Sumaila, "Subsidies and Their Potential Impact on the Management of the Ecosystems of the North Atlantic" in T. Pitcher, U.R. Sumaila, and D. Pauly, eds., *Fisheries Impacts on North Atlantic Ecosystems: Evaluations and Policy Explorations*, Fisheries Centre Research Reports No. 9(5) (Vancouver, B.C.: The University of British Columbia, 2004) 10.

Munro, G.R., A. Van Houtte, and R. Willmann, "The Conservation and Management of Shared Fish Stocks: Legal and Economic Aspects", FAO Fisheries Technical Paper No. 465 (Rome: FAO, 2004).

Musick, J.A., "Criteria Define Extinction Risk in Marine Fishes" (1999) 24 *American Fisheries Society Initiative* 6.

Myers, R.A., and B. Worm, "Rapid Worldwide Depletion of Predatory Fish Communities" (2003) 423 *Nature* 280.

Myers, R.A., and B. Worm, "Extinction, Survival or Recovery of Large Predatory Fishes" (2005) 360 *Philosophical Transactions of the Royal Society* 13.

Nakken, O., P. Sandberg, and S.I. Steinshamn, "Reference Points for Optimal Fish Stock Management: A Lesson to be Learned from the Northeast Arctic Cod Stock" (1996) 20 *Marine Policy* 447.

Nandan, S., "Statement of the Chairman, Ambassador Satya N. Nandan, on 4 August 1995, Upon the Adoption of the Agreement on the Implementation of the Provisions of the United Nations Convention on the Law of the Sea of 10 December 1982 Relating to the Conservation and Management of Straddling Fish Stocks and Highly Migratory Fish Stocks" Document A/CONF:164/35 in J.-P. Lévy and G.G. Schram, eds., *United Nations Conference on Straddling Fish Stocks and Highly Migratory Fish Stocks: Selected Documents* (The Hague: Martinus Nijhoff Publishers, 1996) 749.

Nandan, S.N., "Conservation and Management of Straddling Fish Stocks and Highly Migratory Fish Stocks Under the Convention of the Law of the Sea" (1995) 89 *American Society of International Law Proceedings* 452.

Nandan, S., "Statement of Satya N. Nandan, Secretary General of the International Seabed Authority", Comment on *Agenda Item 49: Oceans and the Law of the Sea, 59th Session of the General Assembly of the United Nations* (New York: United Nations, 2004).

Nandan, S., "Current Fisheries Governance". Prepared for *Conference on the Governance of High Seas Fisheries and the United Nations Fish Agreement — Moving from Words to Action*, St. John's, N.L., 1–5 May 2005. Available at <http://www.dfo-mpo.gc.ca/fgc-cgp/documents/presentations/nandan_e.pdf> (accessed September 2005).

National Environmental Trust, *Black Market for White Gold: The Illegal Trade in Chilean Sea Bass* (Washington, D.C.: National Environmental Trust, 2004).

National Fisheries Conservation Center, "Fishing Capacity". Available at <http://nfcc-fisheries.org/ir_io_fc.html> (accessed 28 February 2005).

National Marine Fisheries Service, *Ecosystem-Based Fishery Management. A Report to Congress by the Ecosystems Principles Advisory Panel* (Silver Spring, Md: NMFS - U.S. Department of Commerce, 1999).

National Oceanic and Atmospheric Administration (NOAA), "10th Annual Conference of the Parties to the Convention on the Conservation and Management of Pollock Resources in the Central Bering Sea, Busan, Korea, 6–8 September 2005". Available at <http://www.afsc.noaa.gov/refm/cbs/Docs/10th%20Annual%20Conference/10thS&T_FINAL.pdf> (accessed February 2006).

National Research Council, *Sustaining Marine Fisheries* (Washington, D.C.: National Academy Press, 1999).

Navigs S.A.R.L., "Fishing Vessel Monitoring Systems: Past, Present and Future", prepared for the High Seas Task Force, October 2005. Available at <http://www.high-seas.org/docs/hstf_vms_final1.pdf> (accessed October 2005).

Ndiaye, T.M., and R. Wolfrum, eds., *Law of the Sea, Environmental Law and Settlement of Disputes: Liber Amicorum Judge Thomas A. Mensah* (Leiden: Martinus Nijhoff Publishers, 2007).

Nicol, S., "CCAMLR's Approaches to Ecosystem Based Management". Paper presented to the Workshop on Ecosystem Based Management, Cairns, Australia, 16–20 June 2003.

Nielsen, J.R., and C. Mathiesen, "Important Factors Influencing Rule Compliance in Fisheries – Lessons From Denmark" (2003) 27 *Marine Policy* 409.

Nielsen, J.R., "An Analytical Framework for Studying: Compliance and Legitimacy in Fisheries Management" (2003) 27 *Marine Policy* 425.

Nijman, J.E., *The Concept of International Legal Personality: An Inquiry into the History and Theory of International Law* (The Hague: TMC Asser Press, 2004).

Nordquist, M.H., and J.N. Moore, eds., *Current Fisheries and the Food and Agriculture Organization of the United Nations* (The Hague: Kluwer Law International, 2000).

Nordquist, M.H., with S.N. Nandan and S. Rosenne, eds., *United Nations Convention on the Law of the Sea 1982: A Commentary* Vol. II (Dordrecht: Martinus Nijhoff Publishers, 1993).

Norse, E., ed., *Global Marine Biological Diversity: A Strategy for Building Conservation into Decision Making* (Washington, D.C.: Island Press, 1993).

North-East Atlantic Fisheries Commission (NEAFC), *Report on the 24th Annual Meeting of the North-East Atlantic Fisheries Commission, 14–18 November 2005* (London: NEAFC, 2005). Available at <http://www.neafc.org/reports/annual-meeting/docs/24neafc_annual_vol-1-main-report_2005.pdf> (accessed February 2006).

North-East Atlantic Fisheries Commission (NEAFC) Secretariat, "IUU Fishing in NEAFC: How Big Is the Problem and What Have We Done". Paper presented to the OECD Workshop on Illegal, Unreported and Unregulated (IUU) Fishing Activities, Paris, France, 19–20 April 2004, AGR/FI/IUU/RD(2004)5 (Paris: OECD, 2004).

North-East Atlantic Fisheries Commission (NEAFC), *Report of the 26th Annual Meeting of the North-East Atlantic Fisheries Commission, 12–16 November 2007* (London: NEAFC, 2007).

Northwest Atlantic Fisheries Organization (NAFO), *Report of the Joint Scientific Council and Fisheries Commission Working Group on Precautionary Approach*, NAFO/FC Doc 99/2 Serial No. N4051 (Dartmouth, N.S.: NAFO, 1999).

Northwest Atlantic Fisheries Organization (NAFO), *Report of the Joint Scientific Council and Fisheries Commission Working Group on Precautionary Approach*, NAFO/ FC Doc 00/2 Serial No. N4207 (Dartmouth, N.S.: NAFO, 2000).

Northwest Atlantic Fisheries Organization (NAFO), *Report of the General Council*, GC Doc. 04/5 (Dartmouth, N.S.: NAFO, 2004).

Northwest Atlantic Fisheries Organization (NAFO), "Fisheries Commission Request for Scientific Advice on Management in 2006 of Certain Stocks in Subareas 2, 3 and 4", NAFO 26th Annual Meeting, Dartmouth, N.S., 13–17 September 2004, Serial No. N5031, NAFO/FC Doc. 04/7. Available at <http://www.nafo.int/science/advice/2006/FCrequest.html> (accessed 21 October 2005).

Northwest Atlantic Fisheries Organization (NAFO), "NAFO 2005 shrimp stock assessment". Available at <http://www.nafo.ca/fisheries/fishery/species/shrimp.html> (accessed 3 October 2005).

Northwest Atlantic Fisheries Organization (NAFO), *Report of the Intersessional Meeting of the General Council*, Montréal, Quebec, 19–20 April 2007.

Northwest Atlantic Fisheries Organization (NAFO), *Annual Report 2007* (Dartmouth, N.S.: NAFO, 2007).

Northwest Atlantic Fisheries Organization (NAFO) Scientific Council, *Proposed NAFO Precautionary Approach Framework*, SCS Doc. 03/23 (Dartmouth, N.S.: NAFO, 2003).

Onestini, M., "Resource Impacts of Fisheries Subsidies: Is Effective Management Likely in the Presence of Subsidies?". Paper presented at the UNEP Workshop on Fisheries Subsidies and Sustainable Fisheries Management, Geneva, Switzerland, 26–27 April 2004.

Orbach, M.K., "Beyond the Freedom of the Seas: Ocean Policy for the Third Millennium". Paper presented to the Fourth Annual Roger Revelle Commemorative Lecture, Washington, D.C., 13 November 2002. Available at <http://www.tos.org.oceanography/issue_archive/issue_pdfs/16_1/16.1_orbach.pdf> (accessed June 2008).

Ørebech, P., and R.D. Brubaker, *Interim Report – Implications of GATS/E.U. Law and Law of the Sea for the Russian Northern Sea Route in the Barents Sea* (Workshop 2.2. Project funded by the European Community under the 'Competitive and Sustainable Growth' Programme, 1998–2002, 2004) Deliverable D2.1.2. Legal Status of the NSR, Intermediate Report for the Workshop 2.2.

Ørebech, P., K. Sigurjonsson, and T. McDorman, "The 1995 United Nations Straddling and Highly Migratory Fish Stocks Agreement: Management, Enforcement and Dispute Settlement" (1998) 13 *International Journal of Marine and Coastal Law* 119.

Orellana, M.A., "The Swordfish Dispute Between the EU and Chile at the ITLOS and the WTO" (2002) 71 *Nordic Journal of International Law* 55.

Organisation for Economic Co-operation and Development (OECD), *Synthesis Report for the Study on Economic Aspects of Management of Marine Living Resources by Ad Hoc Expert Group on Fisheries*, AGR/FI/EG (96)1 (Paris: OECD Fisheries Committee, 1996).

Organisation for Economic Co-operation and Development (OECD), *Transition to Responsible Fisheries, Economic and Policy Implications: Part 3, Government Financial Transfers and Resource Sustainability* (Paris: OECD, 2000).

Orrego Vicuña, F., "The Effectiveness of the Decision-Making Machinery of CCAMLR: An Assessment" in A. Jørgenson-Dahl and W. Østreng, eds., *The Antarctic Treaty System in World Politics* (Oslo: Macmillam – Fridtjof Nansen Institute, 1991).

Orrego Vicuña, F., "Towards an Effective Management of High Seas Fisheries and the Settlement of the Pending Issues of the Law of the Sea: The View of Developing Countries Ten Years After the

Signature of the Law of the Sea Convention" (1993) 26 *Law of the Sea Institute Proceedings* 415.

Orrego Vicuña, F., "Toward an Effective Management of High Seas Fisheries and the Settlement of the Pending Issues of the Law of the Sea" (1993) 24 *Ocean Development and International Law* 81.

Orrego Vicuña, F., *The Changing International Law of High Seas Fisheries* (Cambridge: Cambridge University Press, 1999).

Orrego Vicuña, F., "The International Law of High Seas Fisheries: From Freedom of Fishing to Sustainable Use" in O.S. Stokke, ed., *Governing High Seas Fisheries: The Interplay of Global and Regional Regimes* (Oxford: Oxford University Press, 2001) 23.

Orrego Vicuña, F., "The Law Governing High Seas Fisheries: In Search of New Principles" in A. Chircop and M.L. McConnell, eds., *Ocean Yearbook 18* (Chicago: University of Chicago Press, 2004) 383.

Ostrom, E., "Coping with Tragedies of the Commons" (1999) 2 *Annual Review of Political Science* 493.

Ostrom, E., *Governing the Commons. The Evolution of Institutions for Collective Action* (Cambridge: Cambridge University Press, 1990).

Oude Elferink, A.G., "Fisheries in the Sea of Okhotsk High Seas Enclave — The Russian Federation's Attempts at Coastal State Control" (1995) 10 *International Journal of Marine and Coastal Law* 1.

Oude Elferink, A.G., "The Impact of Article 7(2) of the Fish Stocks Agreement on the Formulation of Conservation and Management Measures for Straddling and Highly Migratory Fish Stocks", FAO Legal Paper Online No. 4 (Rome: FAO, 1999).

Oude Elferink, A.G., "The Determination of Compatible Conservation and Management Measures for Straddling Fish Stocks and Highly Migratory Fish Stocks" (2001) 5 *Max Planck Yearbook of United Nations Law* 551.

Oude Elferink, A.G., "The Sea of Okhotsk Peanut Hole: *De Facto* Extension of Coastal State Control" in O.S. Stokke, ed., *Governing High Seas Fisheries: The Interplay of Global and Regional Regimes* (Oxford: Oxford University Press, 2001) 179.

Oude Elferink, A.G., ed., *Stability and Change in the Law of the Sea: The Role of the LOS Convention* (Leiden: Martinus Nijhoff Publishers, 2005).

Oude Elferink, A.G., and D.R. Rothwell, eds., *Oceans Management in the 21st Century: Institutional Frameworks and Responses* (Leiden: Martinus Nijhoff Publishers, 2004).

Oxman, B., "The Duty to Respect Generally Accepted International Standards" (1991) 24 *Journal of International Law and Policy* 109.

Pacific Environment, *Sustaining the Bering: An International Conference for Collaboration (Conference Summary and Proceedings)*, Girdwood, Alaska, 1–4 April 2003 (San Francisco, Calif.: Pacific Environment, 2003).

Paolillo, F.H., "The Exclusive Economic Zone in Latin American Practice and Legislation" (1995) 26 *Ocean Development and International Law* 105.

Parkes, G., "Precautionary Fisheries Management: The CCAMLR Approach" (2000) 24 *Marine Policy* 83.

Parsons, S., "Ecosystem Considerations in Fisheries Management: Theory and Practice" (2005) 20 *International Journal of Marine and Coastal Law* 381.

Pauly, D., "Empty Nets: Fisheries Are in Decline Around the World As Fishing Technology Outstrips Ecological Capacity" (2004) 30 *Alternatives Journal* 8.

Pauly, D., and V. Christensen, "Primary Production Required to Sustain Global Fisheries" (1995) 374 *Nature* 255.

Pauly, D., and J. Maclean, *In a Perfect Ocean* (Washington, D.C.: Island Press, 2003).

Pauly, D., and M.-L. Palomares, "Fishing Down Marine Food Web: It is Far More Pervasive Than We Thought" (2005) 76 *Bulletin of Marine Science* 197.

Pauly, D., et al., "Fishing Down Marine Food Webs" (1998) 279 *Science* 860.

Pauly, D., et al., "Towards Sustainability in World Fisheries" (2002) 418 *Nature* 689.

Pautzke, C.G., *Russian Far East Fisheries Management*. North Pacific Fishery Management Council Report to Congress (Anchorage, Alaska: North Pacific Fishery Management Council, 1997).

Payne, A.I.L., C.M. O'Brien, and S.I. Rogers, eds., *Management of Shared Fish Stocks* (London: Blackwell Publishing Ltd., 2004).

Peel, J., "A Paper Umbrella Which Dissolves in the Rain? The Future for Resolving Fisheries Disputes Under UNCLOS in the Aftermath of the Southern Bluefin Tuna Arbitration" (2002) 3 *Melbourne Journal of International Law* 53.

Peña-Torres, J., R. Serra, and M. Basch, "Straddling Stocks in the High Seas: The U.N. 1995 Agreement and Its Implications for Chilean Fisheries" in T. Bjoerndal, G.R. Munro and R. Arnason, eds., *Proceedings from the Conference on the Management of Straddling Fish Stocks and Highly Migratory Fish Stocks, and the UN Agreement*. Fiskerioekonomiske sma-skrifter/Papers on Fisheries Economics No. 38 (Bergen: Norwegian School of Economics and Business Administration, Centre for Fisheries Economics, 1999).

Perez, J.G., *Why Bottom Trawling Has to Be Permitted? (Deep Sea Trawling and the Protection of Deep Sea Biodiversity and Ecosystems)*. Speech, IUCN World Conservation Congress, Bangkok, Thailand, 17–25 November 2004 [unpublished].

Petersen, E.H., "Economic Policy, Institutions and Fisheries Development in the Pacific" (2002) 26 *Marine Policy* 315.

Petersen, E.H., "The Catch in Trading Fishing Access for Foreign Aid" (2003) 27 *Marine Policy* 219.

Pfirter, F.M.A., "Straddling Stocks and Highly Migratory Stocks. Latin America Perspective and Legislation: New Perspectives in Light of Current International Negotiations" (1995) 26 *Ocean Development and International Law* 127.

Pikitch, E.K., et al., "Ecosystem-Based Fishery Management" (2004) 305 *Science* 346.

Plé, J.-P., "Responding to Non-Member Fishing in the Atlantic: The ICCAT and NAFO Experiences" in H.N. Scheiber, ed., *Law of the Sea: The Common Heritage and Emerging Challenges* (The Hague: Martinus Nijhoff Publishers, 2000) 197.

Polacheck, T., "Experimental Catches and the Precautionary Approach: The Southern Bluefin Tuna Dispute" (2002) 26 *Marine Policy* 283.

Pontecorvo, G., "Supply Side Uncertainty and the Management of Commercial Fisheries: Peruvian Anchovetta, an Illustration" (2001) 25 *Marine Policy* 169.

Pontecorvo, G., "Insularity of Scientific Disciplines and Uncertainty About Supply: The Two Keys to The Failure of Fisheries Management" (2003) 27 *Marine Policy* 69.

Porter, G., "Analyzing the Resource Impacts of Fisheries Subsidies: A Matrix Approach". Paper presented at the UNEP Workshop on Fisheries Subsidies and Sustainable Fisheries Management, Geneva, Switzerland, 26–27 April 2004.

Porter, G., *Fisheries Subsidies and Overfishing: Towards a Structured Discussion* (Geneva: UNEP, 1998).

Potter, B., "Improving Regimes to Manage Natural Resources: Lessons From an Example of United States-Republic of Korea Pollack Negotiations" in A. Richards, ed., *The Political Economy of International Environmental Cooperation* (San Diego, Calif.: Institute on Global Conflict and Cooperation, 1997) 27.

Pretes, M., and E. Petersen, "Rethinking Fisheries Policy in the Pacific" (2004) 28 *Marine Policy* 297.

Punt, A., and A. Smith, "The Gospel of Maximum Sustainable Yield in Fisheries Management: Birth, Crucifixion and Reincarnation" Chap. 3 in J. Reynolds, et al., eds., *Conservation of Exploited Species* (Cambridge: Cambridge University Press, 2001) 41.

Ram-Bidesi, V., and M. Tsamenyi, "Implications of the Tuna Management Regime for Domestic Industry Development in the Pacific Island States" (2004) 28 *Marine Policy* 383.

Rätz, H.-J., and C. Stransky, "Distribution Patterns and Selected Stock Parameters of Pelagic Redfish 'Oceanic *Sebastes mentella*' in the NAFO and NEAFC Convention Areas". Paper presented to the Northwest Atlantic Fisheries Organization Scientific Council Meeting (2001) Serial No. N4353 NAFO SCR Doc. 01/1.

Rayfuse, R.G., "Enforcement of High Seas Fisheries Agreements: Observation and Inspection under the Convention on the Conservation of Antarctic Marine Living Resources" (1998) 13(4) *International Journal of Marine and Coastal Law* 579.

Rayfuse, R.G., "To Our Children's Children's Children: From Promoting to Achieving Compliance in High Seas Fisheries" (2005) 20 *International Journal of Marine and Coastal Law* 509.

Rayfuse, R.G., "Canada and Regional Fisheries Organizations: Implementing the United Nations Fish Stocks Agreement" (2003) 34 *Ocean Development and International Law* 209.

Rayfuse, R.G., "Regulation and Enforcement in the Law of the Sea: Emerging Assertions of a Right to Non-Flag State Enforcement in the High Seas Fisheries and Disarmament Contexts" (2005) 24 *Australian Year Book of International Law* 181.

Rayfuse, R.G., "Countermeasures and High Seas Fisheries Enforcement" (2004) 51(1) LI *Netherlands International Law Review* 41.

Rayfuse, R.G., *Non-Flag State Enforcement in High Seas Fisheries* (Leiden: Martinus Nijhoff Publishers, 2004).

Reid, C., et al., "An Analysis of Fishing Capacity in the Western and Central Pacific Ocean Tuna Fishery and Management Implications" (2003) 27 *Marine Policy* 449.

Rengifo, A., "Protection of Marine Biodiversity: A New Generation of Fisheries Agreements" (1997) 6(3) *Review of European Community and International Environmental Law* 313.

Restrepo, V., ed., *Proceedings of the 5th National NMFS Stock Assessment Workshop: Providing Scientific Advice to Implement the Precautionary Approach Under the Magnuson–Stevens Fishery Conservation and Management Act*, U.S. Dep. Commer. NOAA Technical Memorandum NMFS-F/SPO-40 (Silver Springs, Md.: U.S. Department of Commerce, NOAA, 1999).

Restrepo, V., "Estimation of Unreported Catches by ICCAT". Paper presented to the OECD Workshop on Illegal, Unreported and Unregulated (IUU) Fishing Activities, Paris, France, 19–20 April 2004, AGR/FI/IUU(2004)21 (Paris: OECD, 2004).

Reynolds, J., N. Dulvy, N. Goodwin, and J. Hutchings, "Biology of Extinction Risk in Marine Fishes" (2005) 272 *Philosophical Transactions of the Royal Society: Biological Sciences. Series **B*** 2337.

Rice, J., "Is the Failure of Conventional Fisheries Management Making the Conservation Approach More Appealing, Offering a Way Out of Making Tough Decisions?" in J. Swan and D. Gréboval, eds., *Overcoming Factors of Unsustainability and Overexploitation in Fisheries: Selected Papers on Issues and Approaches*, FAO Fisheries Report No. 782 (Rome: FAO, 2005) 49.

Rice, J., "Will an Ecosystem Approach Mitigate the Factors of Unsustainability?" in J. Swan and D. Gréboval, eds., *Overcoming Factors of Unsustainability and Overexploitation in Fisheries: Selected Papers on Issues and Approaches*, FAO Fisheries Report No. 782 (Rome: FAO, 2005) 191.

Rice, J., and M.-J. Rochet, "A Framework for Selecting a Suite of Indicators for Ecosystem Fisheries" (2005) 62 *ICES Journal of Marine Science* 516.

Rice, J., S. Trujillo, S. Jennings, K. Hvlland, O. Hagstrom, A. Astudillo, and J. Jenson, "Guidance on the Application of the Ecosystem Approach to Management of Human Activities in the European Marine Environment", ICES Cooperative Research Report No. 273 (Copenhagen: ICES, 2005).

Richards, L.J., and J.-J. Maguire, "Recent International Agreements and the Precautionary Approach: New Directions for Fisheries Management Science" (1998) 55 *Canadian Journal of Fisheries and Aquatic Sciences* 1545.

Riddle, K.W., "Illegal, Unreported, and Unregulated Fishing: Is International Cooperation Contagious?" (2006) 37 *Ocean Development and International Law* 265.

Rieser, A., "International Fisheries Law, Overfishing and Marine Biodiversity" (1997) 9 *Georgetown International Environmental Law Review* 251.

Rigg, K., R. Parmentier, D. Currie, and the Varda Group, "Halting IUU Fishing: Enforcing International Fisheries Agreements", AGR/FI/IUU/RD (2004)1 (Paris: OECD, 2004). Available at <http://www.oecd.org/dataoecd/34/39/31426729.PDF> (accessed January 2005).

Roberts, C., J. Hawkins, and F. Gell, "The Role of Marine Reserves in Achieving Sustainable Fisheries" (2005) 360 *Philosophical Transactions of the Royal Society: Biological Sciences. Series **B*** 123.

Roberts, C.M., L. Mason, and P. Hawkins, *Roadmap to Recovery: A Global Network of Marine Reserves* (Amsterdam: Greenpeace International, 2006). Available at <http://oceans.greenpeace.org/en/documents-reports, roadmap-to-recovery>.

Rochet, M.-J., and J. Rice, "Do Explicit Criteria Help in Selecting Indicators for Ecosystem-Based Fisheries Management?" (2005) 62(3) *ICES Journal of Marine Science* 528.

Roger, A., "The Biology of *Lophelia pertusa* (Linneaus, 1758) and Deep-Water Reef-Forming Corals and Impacts From Human Activities" (1999) 84 *International Review of Hydrobiology* 315.

Roheim, C., and J. Sutinen, "Trade and Marketplace Measures to Promote Sustainable Fishing Practices", ICTSD Natural Resources, International Trade and Sustainable Development Series Issue Paper No. 3 (Geneva: International Centre for Trade and Sustainable Development, 2006).

Romano, C., "The Southern Bluefin Tuna Dispute: Hints of a World to Come…Like It or Not" (2001) 32 *Ocean Development and International Law* 313.

Rosenberg, A., "How Can Managers Use Precautionary Management Advice?"in V.R. Restrepo, ed., *Proceedings of the 5th National NMFS Stock Assessment Workshop: Providing Scientific Advice to Implement the Precautionary Approach Under the Magnuson–Stevens Fishery Conservation and Management Act*, U.S. Dep. Commer. NOAA Technical Memorandum NMFS-F/SPO-40 (Silver Springs, Md.: U.S. Department of Commerce, NOAA, 1999) 102.

Rosenberg, A.A., and C.B. Mogensen, *A Template for the Development of Plans to Recover Overfished Stocks* (Brussels: WWF European Policy Office, 1999).

Rothwell, D.R., "Building on the Strengths and Addressing the Challenges: The Role of Law of the Sea Institutions" (2004) 35 *Ocean Development and International Law* 131.

Rothwell, D.R., and D.L. VanderZwaag, eds., *Towards Principled Oceans Governance: Australian and Canadian Approaches and Challenges* (London: Routledge Press, 2006).

Russ, G.R., and D.C. Zeller, "From Mare Liberum to Mare Reservarum" (2003) 27 *Marine Policy* 75.

Sabourenkov, E.N., and D.G.M. Miller, "The Management of Transboundary Stocks of Toothfish, *Dissostichus* spp., Under the Convention on the Conservation of Antarctic Marine Living Resources" in A.I.L. Payne, C.M. O'Brien, and S.I. Rogers, eds., *Management of Shared Fish Stocks* (London: Blackwell Publishers, 2004) 68.

Sainsbury, K., "The Ecosystem Approach to Fisheries". Keynote presentation to the ICES 2005 Annual Science Conference, Aberdeen, Scotland, 20–24 September 2005. Available at <http://www.ices.dk/iceswork/asc/2005/KeithSainsbury_talk.pdf> (accessed 20 October 2005).

Sakagawa, G.T., "Are US Regulations on Tuna–Dolphin Fishing Driving US Seiners to Foreign Flag Registry?" (1991) 11 *North American Journal of Fisheries Management* 241.

Salmona, P., "The Regional Seas in the 21st Century: The Need for Data" (2002) 45 *Ocean & Coastal Management* 935.

Sano, H., "Are Private Initiatives a Possible Way Forward? Actions Taken by Private Stake Holders to Eliminate IUU Tuna Fishing Activities". Paper presented to the OECD Workshop on Illegal, Unreported and Unregulated (IUU) Fishing Activities, Paris, France, 19–20 April 2004, AGR/FI/IUU(2004)13 (Paris: OECD, 2004).

Sans, P., "Compliance with International Environmental Obligations: Existing International Legal Arrangements" in J. Cameron, J. Werksman, and P. Roderick, eds., *Improving Compliance with International Environmental Law* (London: Earthscan, 1996) 48.

Satia, B., "The World Summit on Sustainable Development and Fisheries" (2003) *International Fisheries Law and Policy Review* (Special Issue) 11.

Scandol, J.P., et al., "Ecosystem-Based Fisheries Management: An Australian Perspective" (2005) 18 *Aquatic Living Resources* 261.

Schaefer, M.B., "Some Recent Developments Concerning Fishing and Conservation of the Living Resources of the High Seas" (1970) 7 *San Diego Law Review* 371.

Scheiber, H.N., "Origins of the Abstention Doctrine in Ocean Law: Japanese–US Relations and Pacific Fisheries 1937–1958" (1989) 16 *Ecology Law Quarterly* 23.

Scheiber, H.N., ed., *Law of the Sea: The Common Heritage and Emerging Challenges*, Publications on Ocean Development Vol. 34 (The Hague: Martinus Nijhoff Publishers, 2000).

Schiffman, H., "The Precautionary Approach at the International Tribunal for the Law of the Sea: The Southern Bluefin Tuna Cases" (2005) 5 *International Journal of Global Environmental Issues* 78.

Schiffman, H.S., "Reservations in Marine Environmental Treaties: Practical Observations and Legal Limitations" (2005) 26 *Whittier Law Review* 1003.

Schmidt, C.-C., "Chapter 3: Measures in Place Against IUU Fishing Activities" in E. Andrews-Chouicha and K. Gray, eds., *Why Fish Piracy Persists: The Economics of IUU Fishing* (Paris: OECD, 2005) 107.

Schonberger, S.N., and J.J. Agar, *Argentina: Towards Rights Based Fisheries Management*, World Bank Departmental Working Paper (Washington, D.C.: World Bank, 2001).

Schrank, W.E., "Extended Fisheries Jurisdiction: Origins of the Current Crisis in Atlantic Canada's Fisheries" (1995) 19 *Marine Policy* 285.

Schrank, W.E., "The Newfoundland Fishery: Ten Years After the Moratorium" (2005) 29 *Marine Policy* 407.

Schrank, W.E., and W.R. Keithly, Jr., "Thalassorama: The Concept of Subsidies" (1999) 14 *Marine Resource Economics* 151.

Scott, S.V., "Australia's First Tuna Negotiations with Japan" (2000) 24 *Marine Policy* 309.

Seafood Industry Council, "Battling for New Zealand's Seafood Trade at the WTO" (September 2003) New Zealand Seafood Industry Council Ltd. [News Centre]. Available at <http://www.seafood.co.nz/040903>.

Serchuk, F.M., D. Rivard, J. Casey, and R. Mayo, *Report of the Ad Hoc Working Group of the NAFO Scientific Council on the Precautionary Approach*, NAFO SCS Doc. 97/12 (Dartmouth, N.S.: NAFO, 1997).

Serdy, A., "The Submerged Entry into Force of the UN Fish Stocks Agreement" (2001) *International Fisheries Bulletin* No. 22.

Serdy, A., "How Long Has the United Nations Fish Stocks Agreement Been in Force?" (2003) 34 *Ocean Development and International Law* 29.

Serdy, A., "One Fin, Two Fins, Red Fins, Bluefins: Some Problems of Nomenclature and Taxonomy Affecting Legal Instruments Governing Tuna and Other Highly Migratory Species" (2004) 28 *Marine Policy* 235.

Serdy, A., "Bringing Taiwan into the International Fisheries Fold: The Legal Personality of a Fishing Entity" (2005) LXXV *British Year Book of International Law* 183.

Seung, W.C., "GATTing a Green Trade Barrier: Eco-Labelling and the WTO Agreement on Technical Barriers to Trade" (1997) 31 *Journal of World Trade* 137.

Shabtai, R., *Practice and Methods of International Law* (London: Ocean Publications, 1984).

Shaeffer, M.B., "Some Aspects of the Dynamics of Populations Important to the Management of the Commercial Marine Fisheries" (1954) 1 *Inter-American Tropical Tuna Commission Bulletin* 25.

Shaw, S., "Models for Inter-Institutional Cooperation". Paper presented at the UNEP Workshop on Fisheries Subsidies and Sustainable Fisheries Management, Geneva, Switzerland, 26–27 April 2004.

Shelton, D., "Law, Non-Law and the Problem of 'Soft Law'" in S. Dinah, ed., *Commitment and Compliance: The Role of Non-Binding Norms in the International Legal System* (New York: Oxford University Press, 2000) 1.

Shelton, P., et al., *A Proposal for a More Flexible Framework for Implementing the Precautionary Approach on NAFO Stocks*, NAFO SCR Doc. 03/58 [revised] Serial No. N4876 (Dartmouth, N.S.: NAFO, 2003).

Sheppard, C., *Seas at the Millennium: An Environmental Evaluation* Vol. II (New York: Pergamon Publishing Company, 2000).

Shrank, W., "Introducing Fisheries Subsidies", FAO Fisheries Technical Paper No. 437 (Rome: FAO, 2003).

Sibert, J., and J. Hampton, "Mobility of Tropical Tunas and the Implications for Fisheries Management" (2003) 27 *Marine Policy* 87.

Simma, B., *The Charter of the United Nations: A Commentary* Vol. 1, 2nd ed. (Oxford: Oxford University Press, 2002).

Sinclair, M., and R. O'Boyle, "The Management of Human Impacts on the Benthic Communities of the Scotian Shelf: A Case Study on Ecosystem-Based Management", ICES CM 2005/BB:18 (Copenhagen: ICES, 2005).

Sinclair, M., and G. Valdimarsson, eds., *Responsible Fisheries in the Marine Ecosystem* (Wallingford: CABI International Publishing and FAO, 2003).

Sinclair, M., et al., "Responsible Fisheries in the Marine Ecosystem" (2002) 58 *Fisheries Research* 255.

Singh-Renton, S., R. Mahon, and P. McConney, "Small Caribbean (CARICOM) States Get Involved in Management of Shared Large Pelagic Species" (2003) 27 *Marine Policy* 39.

Skjærseth, J.B., "Protecting the North-East Atlantic: Enhancing Synergies by Institutional Interplay" (2006) 30 *Marine Policy* 157.

Small, C.J., *Regional Fisheries Management Organisations: Their Duties and Performance in Reducing Bycatch of Albatrosses and Other Species* (Cambridge: Birdlife International, 2005).

Smith, A.D.M., K.J. Sainsbury, and R.A. Stevens, "Implementing Effective Fisheries-Management Systems – Management Strategy Evaluation and the Australian Partnership Approach" (1999) 56 *ICES Journal of Marine Science* 967.

Smith, H.D., ed., *The Oceans: Key Issues in Marine Affairs* (Dordrecht: Kluwer Academic Publishers, 2004).

Smith, T.D., "United States Practice and the Bering Sea: Is It Consistent with a Norm of Ecosystem Management?" (1995) 1 *Ocean and Coastal Law Journal* 141.

Soares, M., *The Ocean Our Future: The Report of the Independent World Commission on the Oceans* (Cambridge: Cambridge University Press, 1998).

Song, Y.-H., "The Canada–European Union Turbot Dispute in the Northwest Atlantic: An Application of the Incident Approach" (1997) 28 *Ocean Development and International Law* 269.

South East Atlantic Fisheries Organisation (SEAFO), *Report of Meeting, First Session, 9–13 March 2004, Swakopmund, Namibia*. Available at <http://www.seafo.org/reports/Meeting_Report_final_1.pdf> (accessed 20 October 2005).

South East Atlantic Fisheries Organisation (SEAFO) Scientific Committee, *Report of the SEAFO Scientific Committee* (2005). Available at <http://www.seafo.org/Scientific%20Committee/reports/SC%20Report%202005%20Eng_final.pdf> (accessed February 2006).

Speer, L., and S. Chasis, "The Agreement on the Conservation and Management of Straddling and Highly Migratory Fish Stocks: An NGO Perspective" (1995) 29 *Ocean and Coastal Management* 71.

Steenblik, R.P., and G.R. Munro, *International Work on Fishing Subsidies — An Update.* Available at <http://www.oecd.org/dataoecd/3/16/1918004.pdf> (accessed March 2006).

Stenlund, P., "Lessons in Regional Cooperation From the Arctic" (2002) 45 *Ocean & Coastal Management* 835.

Stephens, F., *Seafood Solutions: A Chef's Guide to Ecologically Responsible Fish Procurement* (Boston, Mass.: Chefs Collaborative, 2000).

Stephens, T., "A Paper Umbrella Which Dissolves in the Rain? Implications of the *Southern Bluefin Tuna Case* for the Compulsory Resolution of Disputes Concerning the Marine Environment Under the 1982 LOS Convention" (2001) 6 *Asia Pacific Journal of Environmental Law* 297.

Stokke, O.S., "Managing Straddling Stocks: The Interplay of Global and Regional Regimes" (2000) 43 *Ocean & Coastal Management* 205.

Stokke, O.S., "Sub-Regional Cooperation and Protection of the Arctic Marine Environment: The Barents Sea" in D. Vidas, ed., *Protecting the Polar Marine Environment: Law and Policy for Pollution Prevention* (Cambridge: Cambridge University Press, 2000) 124.

Stokke, O.S., "The Loophole of the Barents Sea Fisheries Regime" in O.S. Stokke, ed., *Governing High Seas Fisheries: The Interplay of Global and Regional Regimes* (Oxford: Oxford University Press, 2001) 273.

Stokke, O.S., ed., *Governing High Seas Fisheries: The Interplay of Global and Regional Regimes* (Oxford: Oxford University Press, 2001).

Stokke, O.S., "Boolean Analysis, Mechanisms, and the Study of Regime Effectiveness" in A. Underdahl and O.R. Young, eds., *Regime Consequences: Methodological Challenges and Research Strategies* (Dordrecht: Kluwer Academic Publishers, 2004 89).

Stokke, O.S., and C. Coffey, "Institutional Interaction in High Seas Fisheries Management: An Inventory". Inventory paper (Lysaker and London: Fridtjof Nansen Institute and Institute for European Environmental Policy, 2001).

Stokke, O.S., and C. Coffey, "Precaution, ICES and the Common Fisheries Policy: A Study of Regime Interplay" (2004) 28 *Marine Policy* 117.

Stokke, O.S., and D. Vidas, "Regulating IUU Fishing or Combating IUU Operations?". Paper presented to the OECD Workshop on Illegal, Unreported and Unregulated (IUU) Fishing Activities, Paris, France, 19–20 April 2004, AGR/FI/IUU(2004)8 (Paris: OECD, 2004).

Stokke, O.S., L.G. Anderson, and N. Mirovitskaya, "The Barents Sea Fisheries" in O.R. Young, ed., *The Effectiveness of International Environmental Regimes: Causal Connections and Behavioral Mechanisms* (Cambridge, Mass.: MIT Press, 1999).

Stone, C.D., "Too Many Fishing Boats, Too Few Fish: Can Trade Laws Trim Subsidies and Restore the Balance in Global Fisheries?" (1997) 24 *Ecology Law Quarterly* 505.

Stone, C.D., D. Downes, and A.C. de Fontaubert, *Biodiversity, Trade, and the Fishing Sector: Case Study: West Africa.* (Gland, Switzerland: IUCN - The World Conservation Union, 1999).

Suárez de Vivero, J.L., and J.C. Rodríguez Mateos, "New Factors in Ocean Governance. From Economic to Security-Based Boundaries" (2004) 28 *Marine Policy* 185.

Sullivan, K.M., "Conflict in the Management of a Northwest Atlantic Transboundary Cod Stock" (1989) 13 *Marine Policy* 118.

Sullivan, M.S., "The Case in International Law for Canada's Extension of Fisheries Jurisdiction Beyond 200 Miles" (1997) 28 *Ocean Development and International Law* 203.

Sumaila, U.R., and H. Keith, "Regulating Fisheries Subsidies — A Role for RFMOs?" (2006) 10 *Bridges Monthly Review* 21 (Geneva: International Centre for Trade and Sustainable Development, 2006).

Sumaila, U.R., J. Alder, and H. Keith, "The Cost of Being Apprehended for Fishing Illegally: Empirical Evidence and Policy Implications" in K. Gray et al., eds., *Fish Piracy: Combating Illegal, Unreported and Unregulated Fishing* (Paris: OECD, 2004) 201.

Swan, J., "Regional Fishery Bodies and Governance: Issues, Actions and Future Directions", FAO Fisheries Circular No. 959 (Rome: FAO, 2000).

Swan, J., "International Action and Responses by Regional Fishery Bodies or Arrangements to Prevent, Deter and Eliminate Illegal, Unreported and Unregulated Fishing", FAO Fisheries Circular No. 996 (Rome: FAO, 2004).

Swan, J., "Decision Making in Regional Fishery Bodies or Arrangements: The Evolving Role of RFBs and International Agreement on Decision Making Processes", FAO Fisheries Circular No. 995 (Rome: FAO & SwanSea Oceans Environment Inc., 2004).

Sydnes, A.K., "Establishing a Regional Fisheries Management Organization for the Western and Central Pacific Tuna Fisheries" (2001) 44 *Ocean & Coastal Management* 787.

Sydnes, A.K., "New Regional Fisheries Management Regimes: Establishing the South East Atlantic Fisheries Organisation" (2001) 25 *Marine Policy* 353.

Sydnes, A.K., "Regional Fishery Organizations: How and Why Organizational Diversity Matters" (2001) 32 *Ocean Development and International Law* 349.

Sydnes, A.K., "Regional Fishery Organisations in Developing Regions: Adapting to Changes in International Fisheries Law" (2002) 26 *Marine Policy* 373.

Symes, D., "The European Pond: Who Actually Manages the Fisheries?" (1995) 27 *Ocean & Coastal Management* 29.

Tahindro, A., "Conservation and Management of Transboundary Stocks: Comments in Light of the Adoption of the 1995 Agreement for the Conservation and Management of Straddling Fish Stocks and Highly Migratory Fish Stocks" (1997) 28 *Ocean Development and International Law* 1.

Tarasofsky, R.G., *Regional Fisheries Organizations and the World Trade Organization: Compatibility or Conflict?* (Cambridge: TRAFFIC International, 2003).

Teece, D.R., "Global Overfishing and the Spanish–Canadian Turbot War: Can International Law Protect the High-Seas Environment?" (1997) 8 *Colorado Journal of International Environmental Law and Policy* 89.

Thébaud, O., "Transboundary Marine Fisheries Management. Recent Developments and Elements of Analysis" (1997) 21 *Marine Policy* 237.

The World Bank, *Governance and Development* (Washington, D.C.: The International Bank for Reconstruction and Development/The World Bank, 1992).

Thompson, A., "Canadian Foreign Policy and Straddling Stocks: Sustainability in an Interdependent World" (2000) 28 *Policy Studies Journal* 219.

Townsend, R.E., "Fisheries Self-Governance: Corporate or Cooperative Structures?" (1995) 19 *Marine Policy* 39.

Townsend, R.E., "Transferable Dynamic Stock Rights" (1995) 19 *Marine Policy* 153.

Tsamenyi, M., L. Manarangi-Trott, and S. Rajkumar, "The International Legal Regime for Fisheries Management". Paper presented to the UNEP Workshop on Fisheries Subsidies and Sustainable Fisheries Management, Geneva, Switzerland, 26–27 April 2004.

Tsuru, Y., "Rethinking the Principle of Abstention: The North Pacific and Beyond" (2004) 28 *Marine Policy* 541.

Tutangata, T., and M. Power, "The Regional Scale of Ocean Governance Regional Cooperation in the Pacific Islands" (2002) 45 *Ocean & Coastal Management* 873.

Tyagi, Y.K., "Towards UNCLOS IV?" (1993) 27 *Journal of World Trade* 143.

United Nations, "Agreement on the High Seas Fishing: An Update. Special Session of the General Assembly to Review and Appraise the Implementation of Agenda 21". Comment on *Agreement on the High Seas Fishing* (1997) DPI/1871/SD-February 1997, United Nations Department of Public Information.

United Nations Environment Program (UNEP), "Ecosystem-Based Management of Fisheries: Opportunities and Challenges for Coordination Between Marine Regional Fishery Bodies and Regional Seas Conventions", Regional Seas Reports and Studies No. 175 (Nairobi: UNEP, 2001).

United Nations Environment Program (UNEP), "Chairman's Summary". Paper presented to the Workshop on the Impacts of

Trade-Related Policies on Fisheries and Measures Required for the Sustainable Management, Geneva, 15 March 2002.

United Nations Environment Program (UNEP), *Integrated Assessment of Trade Liberalization and Trade-Related Policies: A Country Study on the Argentina Fisheries Sector* (New York: UNEP, 2002).

United Nations Environment Program (UNEP), *Management of Risks to the Biodiversity of Seamounts and Cold-Water Coral Communities Beyond National Jurisdiction. Seventh Meeting. Backgrounder for the Conference of the Parties to the Convention on Biological Diversity*, Kuala Lumpur, Malaysia, UNEP/CBD/COP/7/INF/25 (np: UNEP/CBD/COP, 2003).

United Nations Environment Program (UNEP), *Analyzing the Resource Impact of Fisheries Subsidies: A Matrix Approach* (Geneva: UNEP, 2004).

United Nations Environment Program (UNEP) Division of Technology - Industry and Economics, *Fisheries Subsidies and Sustainable Fisheries Management*, Workshop, Geneva, Switzerland, 26–27 April 2004 (Switzerland: UNEP, 2004).

U.S. Coast Guard, "1998 Report to the Committee on Enforcement of the North Pacific Anadromous Fish Commission", NPAFC Doc. 354.

Vaisman, A., *Trawling in the Mist: Industrial Fisheries in the Russian Part of the Bering Sea*, A TRAFFIC Network Report (Cambridge: TRAFFIC International, 2001).

Vallega, A., "Ocean Governance in Post-Modern Society — A Geographical Perspective" (2001) 25 *Marine Policy* 399.

Vallega, A., "The Regional Approach to the Ocean, the Ocean Regions, and Ocean Regionalisation: A Post-Modern Dilemma" (2002) 45 *Ocean & Coastal Management* 721.

Vallega, A., "The Regional Seas in the 21st Century: An Overview" (2002) 45 *Ocean & Coastal Management* 925.

VanderZwaag, D., "The Precautionary Principle and Marine Environmental Protection: Slippery Shores, Rough Seas, and Rising Normative Tides" (2002) 33 *Ocean Development and International Law* 165.

VanderZwaag, D., S. Fuller, and R. Myers, "Canada and the Precautionary Principle/Approach in Ocean and Coastal Management: Wading and Wandering in Tricky Currents" (2003) 34 *Ottawa Law Review* 117.

Van Dyke, J.M., "The Evolution and International Acceptance of the Precautionary Principle" in D.D. Caron and H.N. Scheiber, eds., *Bringing New Law to Ocean Waters* (Leiden: Brill, 2004) 357.

Van Dyke, J.M., D. Zaelke, and G. Hewison, eds., *Freedom for the Seas in the 21st Century: Ocean Governance and Environmental Harmony* (Washington, D.C.: Island Press, 1993).

Van Santen, G., and P. Muller Working Apart or Together: *The Case for a Common Approach to Management of the Tuna Resources in Exclusive Economic Zones of the Pacific Island Countries.* (Pacific Islands Discussion Paper Series, Number 10, The World Bank, Washington DC, 2000).

van Sittert, L., "'Those Who Cannot Remember the Past Are Condemned to Repeat It': Comparing Fisheries Reforms in South Africa" (2002) 26 *Marine Policy* 295.

Verhaag, M.A., "It Is Not Too Late: The Need for a Comprehensive International Treaty to Protect the Arctic Environment" (2003) 15 *Georgetown International Environmental Law Review* 555.

Vidas, D., and W. Østreng, *Order for the Oceans at the Turn of the Century* (The Hague, Boston: Kluwer Law International, 1999).

Vitalis, V., "Round Table on Sustainable Development: Preparing for the World Summit: Some Information About Sustainable Development", OECD Headquarters, Paris, France, 11 July 2002.

Vukas, B., and D. Vidas, "Flags of Convenience and High Seas Fishing: The Emergence of a Legal Framework" in O.S. Stokke, ed., *Governing High Seas Fisheries: The Interplay of Global and Regional Regimes* (Oxford: Oxford University Press, 2001) 53.

Walters, C., *Adaptive Management of Renewable Resources* (New York: Macmillan, 1986).

Walters, C., "Folly and Fantasy in the Analysis of Spatial Catch Rate Data" (2003) 60 *Canadian Journal of Fisheries and Aquatic Sciences* 1433.

Waluda, C.M., et al., "Throwing Light on Straddling Stocks of *Illex Argentinus*: Assessing Fishing Intensity with Satellite Imagery" (2002) 59 *Canadian Journal of Fisheries and Aquatic Sciences* 592.

Wang, H., "Ecosystem Management and Its Application to Large Marine Systems: Science, Law and Politics" (2004) *Ocean Development and International Law* 41.

Ward, P.K., B. Kearney, and N. Tsirbas, "Science Arrangements for the Regional Management of Tuna Fisheries" (2000) 24 *Marine Policy* 93.

Ward, T., D. Tarte, E. Hegerl, and K. Short, *Ecosystem-Based Management of Marine Capture Fisheries* (Sydney: WWF Australia, 2002).

Weber, P. "Net Loss: Fish, Jobs, and the Marine Environment", Worldwatch Institute Paper No. 120 (Washington, D.C.: Worldwatch Institute, 1994).

Westlund, L., "Guide for Identifying, Assessing and Reporting on Subsidies in the Fisheries Sector", FAO Fisheries Technical Paper No. 438 (Rome: FAO, 2004).

Westlund, L., "Apparent Historical Consumption and Future Demand for Fish and Fishery Products: Exploratory Calculation". Paper presented to the Kyoto Conference Outcomes and Papers Presented. Available at <http://www.fao.org/DOCREP/006/AC442e/AC442e13.htm>.

Whitlow, J., "The Social Dimension of IUU Fishing" in K. Gray, et al., eds., *Fish Piracy: Combating Illegal, Unreported and Unregulated* (Paris: OECD, 2004) 231.

Willock, A., *Uncharted Waters: Implementation Issues and Potential Benefits of Listing Toothfish in Appendix II of CITES* (Cambridge: TRAFFIC International, 2002).

Willock, A., "Using Trade and Market Information to Assess IUU Fishing Activities". Paper presented to the OECD Workshop on Illegal, Unreported and Unregulated (IUU) Fishing Activities, Paris, France, 19–20 April 2004, AGR/FI/IUU(2004)10 (Paris: OECD, 2004).

Wiseman, E., and R. Steinbock, "Compliance" in M.H. Nordquist and J.N. Moore, eds., *Current Fisheries Issues and the Food and Agriculture Organization of the United Nations* (The Hague: Martinus Nijhoff Publishers, 2000) 387.

Wiser, G.A., "Transparency in 21st Century Fisheries Management: Options for Public Participation to Enhance Conservation and Management of International Fish Stocks" (2001) 4 *Journal of International Wildlife Law and Policy* 95.

World Trade Organization (WTO), *Benefits of Eliminating Trade Distorting and Environmentally Damaging Subsidies in the Fisheries Sector: Submission by New Zealand* (Geneva: WTO, 1999).

World Trade Organization (WTO), "Ministerial Declaration". Paper presented to the WTO Ministerial Conference Fourth Session, Doha, Qatar, 9–14 November 2001, WT/MIN (01)/DEC/1.

World Trade Organization (WTO), "WT/CTE/W/236". Paper presented to the UNEP Workshop on Fisheries Subsidies and Sustainable Fisheries Management, Geneva, 26–27 April 2004.

World Trade Organization (WTO) Committee on Trade and Environment, *Subsidies in the Fisheries Sector: Update on Recent Work Conducted by New Zealand* (Geneva: WTO, 2000).

World Trade Organization (WTO) Committee on Trade and Environment, *Environmentally-Harmful and Trade-Distorting Subsidies in Fisheries: Communication From the United States* (Geneva: WTO, 2000).

World Trade Organization (WTO) Committee on Trade and Environment, *Fisheries Subsidies: Submission From New Zealand* (Geneva: WTO, 2002).

World Trade Organization (WTO) Committee on Trade and Environment, *MEA Information Session on Technical Assistance, Capacity Building and Enhancing Information Exchange* (Geneva: WTO, 2002).

World Trade Organization (WTO) Committee on Trade and Environment, *Update of FAO Activities Related to Fisheries* (Geneva: WTO, 2002).

World Trade Organization (WTO) Committee on Trade and Environment, *Analysis on the Relationship Between Fisheries Subsidies and Over-Exploitation of Fisheries Resources: Submission From Japan* (Geneva: WTO, 2003).

World Trade Organization (WTO) Committee on Trade and Environment, *UNEP Workshop on Fisheries Subsidies and Sustainable Fisheries Management, Geneva, 26–27 April 2004, Summary of the Chairs: Contribution by UNEP* (Geneva: WTO, 2004).

World Trade Organization (WTO) Negotiating Group on Rules, *The Doha Mandate to Address Fisheries Subsidies: Issues*, WTO TN/RL/W/3. Available at <http://docsonline.wto.org/DDFDocuments/t/tn/rl/W3.doc> (accessed 15 March 2006).

World Wildlife Fund (WWF), *The Footprint of Distant Water Fleets on World Fisheries* (Godalming, England: WWF, 1998).

World Wildlife Fund (WWF), "Massive Government Subsidies Bloat Fishing Industry and Violate World Trade Rules; WWF Calls for New International Fisheries Rules and Mechanisms" (9 September 1998) Common Dreams News Wire. Available at <http://www.commondreams.org/pressreleases/Sept98/090998h.htm>.

World Wildlife Fund (WWF), *Fishing Madness: 101 Reasons Why the CFP Needs Radical Reform* (Brussels: WWF, 2002).

World Wildlife Fund (WWF), *Turning the Tide on Fishing Subsidies: Can the World Trade Organization Play a Positive Role?* (Washington, D.C.: WWF, 2002).

World Wildlife Fund (WWF), *Conserving Marine Turtles on a Global Scale* 2nd ed. (Gland, Switzerland: WWF, 2004).

World Wildlife Fund (WWF), TRAFFIC, and IUCN, *Regulating International Trade in Commercial Marine Fisheries Products* (np: WWF, TRAFFIC, and IUCN, 2002).

Worm, B., et al., "Impacts of Biodiversity Loss on Ocean Ecosystem Services" (2006) 314 *Science* 787.

Wynhoven, U., "OECD Instruments and IUU Fishing". Paper presented to the OECD Workshop on Illegal, Unreported and Unregulated (IUU) Fishing Activities, Paris, France, 19–20 April 2004, AGR/FI/IUU(2004)1 (Paris: OECD, 2004).

Yee, S., "*The Responsibility of States Members of an International Organization for its Conduct as a Result of Membership or Their Conduct Associated with Membership*, in M. Ragassi (ed.), International Responsibiity Today: Essays in Memory of Oscar Schachter (Martinus Nijhoff Publishers) (2005) 435.

Young, O., and G. Oshrenko, eds., *Polar Politics: Creating International Environmental Regimes* (Ithaca, N.Y.: Cornell University Press, 1993).

Yturriaga, J.A. de, *The International Regime of Fisheries: From UNCLOS 1982 to the Presential Sea* (The Hague: Martinus Nijoff Publishers, 1997).

Zackrison, J.L., and J.E. Meason, "Chile, *Mar Presencial*, and the Law of the Sea" (Summer 1997) *Naval War College Review* 65.

Index

abundance, 8, 44, 77–78, 80, 82–87, 90–91, 99, 115, 123–124, 128, 145, 156, 163–164, 167–168, 197, 278, 281
  see biomass
  see population characteristics
  see stock assessment

action plans, 5, 15, 21–24, 27, 30, 59, 62, 81, 98–99, 132, 144, 157, 162, 195, 221–222, 225–227, 261, 263, 281, 289

Agenda 21, 13–15, 18, 20, 53, 68, 116, 144, 200

Agreement between Iceland, Norway and Russia Concerning Certain Aspects of Cooperation in the Area of Fisheries (Barents Sea Loop Hole Agreement), 44, 56, 204

Agreement on the International Dolphin Conservation Program (AIDCP), 154, 159–161, 229–230, 284–285
  observer program, 160–161, 229–230, 285
  see also International Dolphin Conservation Program

alien species, 29, 154

allocation systems, 54, 187, 190–191, 199–200, 204, 210, 273, 281
  see also quota regulations

anadromous species, 42, 220, 231, 279

Antarctic, 5, 15, 30, 41, 85–86, 88–89, 115, 122, 144, 149, 151, 154–157, 190, 223, 234, 263, 269, 276–277, 290
  Antarctic Treaty, 30, 154–155, 157, 276–277
  see also Convention for the Conservation of Antarctic Marine Living Resources
  see also Convention for the Conservation of Antarctic Seals
  see also Southern Ocean

Antigua Convention, 13, 15, 21, 59, 65, 85, 121, 124, 151, 162, 191, 193, 196, 205, 284–285
  see Convention for the Strengthening of the Inter-American Tropical Tuna Commission

arbitration, 30, 209, 275, 277
  see also dispute settlement

Argentina, 95, 276

artisanal fisheries, 88

Asia Pacific Fishery Commission (APFIC), 57

assessment, 16, 22–23, 77, 79–81, 85, 87–88, 90, 92–95, 97, 101, 115–116, 118, 121, 123–125, 128, 131–132, 151–153, 155–157, 164–165, 168, 196–198, 226, 228, 244, 252, 262–263, 275–276, 280–284, 286–288, 290–291
  see risk management, assessment
  see stock assessment

associated species, 3, 8, 21–22, 28, 42, 45–46, 54–55, 77–78, 80, 84, 100–102, 115, 119, 122, 144, 148, 151, 153–154, 156–157, 161, 167, 169, 172, 187, 197, 274, 288, 290

Atlantic Ocean, 5, 15–16, 45, 55–56, 60, 85–87, 90, 93–95, 121, 126, 129, 153, 205, 231, 269–271, 274, 280

Australia, 28, 30, 43–44, 62, 127–128, 168, 198, 204–205, 261–262, 276, 286, 288, 290

Bali Plan of Action, 27, 59

Baltic Sea, 55

Barents Sea Loop Hole Agreement, 44, 56, 204
  see Agreement between Iceland, Norway and Russia Concerning Certain Aspects of Cooperation in the Area of Fisheries

Belize, 28, 159, 226, 272, 280, 290

Bergen Ministerial Declaration on Sustainable Development in the ECE Region, 116

Bering Sea Donut Hole Convention
  see Convention on the Conservation and Management of Pollock Resources in the Central Bering Sea

best scientific information available, 22, 25, 77, 116–118, 130–131, 151, 189, 195–196, 205
  see fishery management
  see role of science

biodiversity, 3, 5, 7, 15–16, 21, 29, 31, 59–62, 68, 77–78, 82–83, 100–101, 125, 132, 143–144, 147, 150–151, 153, 166, 169–171, 195, 262, 271, 273–274

management, 3, 5, 7–9, 11, 14–31, 41–42, 44–47, 53–68, 77–78, 80–81, 83–85, 87–88, 90–91, 93–102, 115–134, 143–158, 160–172, 187–210, 219–222, 225–231, 234–236, 238, 241–245, 247–248, 250, 252, 261–264, 270–291
  see also Convention on Biological Diversity

biomass, 44, 46, 80–82, 84–97, 100–102, 118, 122, 125–129, 143, 151, 155, 198, 270, 277–279, 284, 288
  harvest control or decision rule, 85, 101–102, 127, 155, 198, 279
  spawning stock, 82, 85, 90, 93–94, 118, 126, 128–129, 153, 155, 163, 288–289

BirdLife International, 156

Black Sea, 165, 282–283

Bolivia, 160, 227, 281

bycatch, 8, 22, 25–26, 44, 46, 77, 83, 87, 89–92, 94–95, 100–101, 123–124, 143–144, 148, 150–151, 153–157, 159–169, 171, 222–223, 226, 230, 269–272, 274, 276–278, 280–282, 284–291
  dolphins, 90, 151, 153–154, 159–162, 230, 282, 284
  mitigation measures, 22, 123, 143, 153–155, 157, 161, 167–169, 270, 276–278, 282, 286–289, 291
  sea turtles, 160–161, 163, 167, 169, 274, 280, 285–286, 290
  seabirds, 22, 89, 92, 144, 151, 153–157, 159, 162–163, 167–169, 274, 276–277, 280–281, 284–290
  sharks, 22, 89–90, 92, 144, 148, 151, 153–154, 161–164, 166–169, 270, 272, 274, 280–282, 284–288, 290
  see also fishery management

Cambodia, 226

Canada, 19, 60, 93, 124–125, 204, 231–232, 237, 249, 262–263, 270–272, 276, 280, 284, 286

capacity-building, 81, 264
  see education and training

Caribbean Sea, 151

catadromous species, 42

catch documentation schemes, 19, 29, 65–66, 157, 222, 226, 228, 250, 269, 271, 273, 275–277, 279, 281, 283, 285, 287, 289, 291
  see data and information, trade and catch documentation programs
  see World Trade Organization, documentation schemes

catch limit, 79–80, 83, 85–86, 122–124, 238, 282, 284–285, 288
  see quota regulations

catch statistics, 66, 80, 122, 124, 222, 226–228, 247, 275, 277, 279–281, 283, 289, 291
see also stock assessment

catching methods, 83, 226
bottom trawls, 29, 46, 123, 153, 157, 171, 270, 272, 274, 276, 282–283
dredgers, 46, 165, 282–283
driftnets , 25–26, 116, 245, 280
entangling nets, 272
fish aggregating devices (FADs), 90, 98, 124, 161, 163, 167, 169, 280, 282–283, 286
gill nets, 270, 272, 282, 290
handlining, 286, 290
longlining, 22, 90–91, 93, 144, 153–154, 156–157, 161–162, 167, 223, 228, 230, 272, 280–286, 288, 290
midwater trawls, 157, 270, 278
pole-line fishing, 284, 286
seine nets, 160, 272, 282, 290
trap fishing, 93, 274
trawling, 29, 46, 123, 125, 153, 157, 171, 270, 276
trolling , 284, 286, 290

CCAMLR, 5, 14–15, 28–30, 41, 65–66, 85–86, 88–89, 95, 102, 115–116, 121–124, 144–145, 148–149, 151, 153–158, 167, 169–171, 190–191, 193, 196, 198–200, 202, 204–205, 208–209, 223–230, 232–234, 240–241, 244, 246, 248, 251, 263–264, 269, 271, 273, 275–277, 281, 285, 287, 289
see Commission for the Conservation of Antarctic Marine Living Resources

cetaceans, 45, 47, 274
see also dolphins
see also marine mammals

Chile, 30, 42, 44, 56, 61, 208, 261, 276, 284

Chilean seabass, 86
see fish, Patagonian toothfish

China, People's Republic, 56, 65, 276, 278, 280, 284, 286, 290–291

Chinese Taipei, 65, 227, 251, 280–281, 284, 286
see Taiwan

climate change, 53, 78, 82, 116, 132

coastal communities, 31, 64, 78, 83, 100, 133, 206–207

cod fisheries, 44, 86–87, 89, 166, 270, 276
see demersal fisheries, 46, 272

Code of Conduct for Responsible Fishing, 3, 5, 15, 20–21, 53, 172, 221
precautionary approach, 5, 15, 21
UNFSA, 21, 53
see also Food and Agricultural Organization of the United Nations

collapse, 46, 54, 62, 77–78, 82, 84, 87, 89, 94, 96–97, 100, 102, 118, 127, 133, 168, 199, 270
see depleted stocks

Colombia, 61, 159, 285

Commission for the Conservation of Antarctic Marine Living Resources (CCAMLR), 5, 14–15, 28–30, 41, 65–66, 85–86, 88–89, 95, 102, 115–116, 121–124, 144–145, 148–149, 151, 153–158, 167, 169–171, 190–191, 193, 196, 198–200, 202, 204–205, 208–209, 223–230, 232–234, 240–241, 244, 246, 248, 251, 263–264, 269, 271, 273, 275–277, 281, 285, 287, 289
ecosystem approach, 5, 15, 85, 102, 115, 121–122, 144–145, 148–149, 151, 153–156, 167, 169–171, 230, 269, 276
precautionary approach, 5, 15, 30, 85, 95, 102, 115–116, 121–124, 148–149, 151, 153–154, 156, 169–171, 269, 276
boarding and inspection, 234, 240–241, 244, 246, 251, 273, 275, 281, 287, 289
catch documentation scheme, 29, 65–66, 157, 226, 271, 273, 276, 287, 289
Conservation Measures, 15, 28–30, 65–66, 88, 102, 122–123, 144, 149, 153–158, 169–170, 190, 193, 196, 202, 204–205, 208, 227–230, 234, 240–241, 244, 248, 251, 263–264, 271, 273, 275–277, 281, 287, 289
consistency principle, 202
Ecosystem Monitoring Program (CEMP), 155–158, 170, 276–277
inter-RFMO cooperation, 149, 208, 269
observer program, 148, 158, 228–230, 269, 271, 273, 275, 277, 281, 285, 287, 289
participation, 5, 15, 65–66, 149, 190, 200, 234, 269, 271, 275–276, 281, 285, 289
port State measures, 158, 227, 248, 263, 275, 281, 287
Scheme of International Scientific Observation, 157, 229, 277
Scientific Committee, 89, 122–123, 148–149, 153, 155, 157, 167, 169, 190, 196, 276, 281
transshipment, 29, 66, 124, 224, 226–230, 232, 248, 251, 263–264, 271, 277, 281, 287
see also Convention for the Conservation of Antarctic Marine Living Resources

Commission for the Conservation of Southern Bluefin Tuna (CCSBT), 5, 29, 45, 56, 65–66, 90–92, 97–98, 121, 127–129, 148–149, 153–155, 167–168, 190–191, 193, 196–198, 200, 203, 205, 207–209, 224–228, 230, 232, 236, 246, 263–264, 269, 276–277, 281, 287–291
ecosystem approach, 5, 121, 129, 148–149, 153–155, 167–168, 230, 269, 276, 288, 290
ICES, 121, 148, 153, 196, 277, 281
precautionary approach, 5, 121, 127–128, 148–149, 153–154, 269, 276, 288, 290
compatibility provisions, 205, 207, 264
dispute settlement, 29, 128, 149, 193, 207–209, 246, 269, 277, 281, 287, 289, 291
Ecologically Related Species Working Group, 153, 167, 288
Extended Commission, 65, 190, 196, 288–289
observer program, 168, 228, 230, 277, 281, 287, 289, 291
Scientific Research Program, 153, 168
common property, 7, 54
see also freedom of the seas doctrine

Comoros, 261, 290

compatibility principle, 18, 21, 44, 59, 202–203, 206

Compliance Agreement, 3–5, 7, 9, 11, 13, 15–16, 18–22, 24–25, 30–31, 53, 59, 62, 131, 189, 191, 209, 219–220, 229, 242, 261, 266, 270, 272, 274, 276, 278, 280–281, 283–284, 286, 288–290
dispute resolution, 9, 11, 189, 209, 242, 281, 289
flag State, 16, 18–19, 22, 25, 30, 62, 189, 220, 229, 281, 289
see also Food and Agricultural Organization of the United Nations

conciliation, 194, 209, 275, 277, 291
see also dispute resolution

Conference on the Governance of High Seas Fisheries and the UN Fish Agreement, 27, 58
Ministerial Declaration (St. John's Declaration), 16, 27, 57–60

conservation, 3, 5, 7–9, 11, 14–23, 25–31, 41, 44–46, 53, 55–68, 77–78, 80–81, 83, 85, 88, 90, 93–94, 96–98, 100–102, 115–124, 127–133, 144, 146–151, 153–161, 163, 166–171, 187–190, 192–197,

199, 201–210, 219–221, 223–224, 227–230, 234–236, 238, 240–245, 247–248, 250–252, 261–264, 269–282, 284, 286–290
see nature conservation

continental shelf, 14, 17, 42, 87

Convention for the Conservation of Antarctic Marine Living Resources (CCAMLR), 5, 15, 28–30, 41, 65–66, 85–86, 88–89, 95, 102, 115–116, 121–124, 144–145, 148–149, 151, 153–158, 167, 169–171, 190–191, 193, 196, 198–200, 202, 204–205, 208–209, 223–230, 232–234, 240–241, 244, 246, 248, 251, 263–264, 269, 271, 273, 275–277, 281, 285, 287, 289
Madrid Protocol, 155, 158, 281
see also Antarctic

Convention for the Conservation of Antarctic Seals, 149, 155
see also Antarctic

Convention for the Strengthening of the Inter-American Tropical Tuna Commission (Antigua Convention), 13, 15, 21, 59, 65, 85, 121, 124, 151, 162, 191, 193, 196, 205, 284–285
ecosystem approach, 13, 15, 21, 59, 85, 121, 151, 162, 284
precautionary approach, 13, 15, 21, 59, 85, 121, 124, 151, 284
compatibility provisions, 205
see also Inter-American Tropical Tuna Commission

Convention on Biological Diversity (CBD), 15–16, 24, 28–30, 53, 101–102, 116, 132, 144, 171
genetic resources, 29
implementation, 16, 29–30, 132, 144, 171
Jakarta Mandate on Marine and Coastal Biological Diversity, 29, 132, 144
see also biodiversity

Convention on Climate Change, 53, 116
see United Nations Framework Convention on Climate Change

Convention on Fishing and Conservation of Living Resources of the High Seas, 55

Convention on International Trade in Endangered Species of Wild Flora and Fauna (CITES), 15, 28, 30, 94, 99, 101, 128, 144, 157, 163, 239, 275–276
RFMOs, 15, 28, 30, 94, 275–276

Convention on the Conservation of Migratory Species of Wild Animals (Bonn Convention), 30

Convention on the Conservation and Management of Highly Migratory Fish Stocks in the Western and Central Pacific Ocean (WCPF Convention), 5, 25, 29, 45, 60–61, 65–66, 68, 85, 90–92, 98, 121, 129, 148–149, 152–155, 169, 188, 190–191, 193–194, 196–197, 199–200, 203, 205, 207–209, 224, 228–232, 236, 241, 248, 263, 269, 285–287, 289–290
see also Western and Central Pacific Fisheries Commission

Convention on the Conservation and Management of Pollock Resources in the Central Bering Sea (Donut Hole Convention), 41, 56, 86, 102, 127, 153, 168, 197–199, 205, 209, 230, 278–279
compatibility provisions, 205
control measures, 5, 190, 269

Convention on the Conservation and Management of Fishery Resources in the South East Atlantic Ocean (SEAFO Convention), 15, 41, 46, 61, 65, 85, 121, 129, 153, 166, 168, 194, 196, 199–201, 205, 209, 230, 232, 236, 248, 274–275
see also South East Atlantic Fisheries Organisation

Convention on the High Seas, 3, 17, 21, 55, 64, 77, 85, 205, 207, 232, 262, 285–286

Convention on the Law of the Sea, 11, 17, 21, 41, 53, 55, 77, 117, 189, 219
see United Nations Convention on the Law of the Sea

Convention on Wetlands of International Importance (Ramsar), 144, 411, 418

coral, 123, 157, 263

Costa Rica, 159–160, 284

crab fishery, 86, 89, 122, 274, 276
red, 274
see also crustacean

crustacean, 154
see also crab fishery

data and information, 25, 30, 46, 65, 79–81, 115, 117, 130–131, 143, 148, 151, 153, 156, 162–164, 168–169, 222, 225–226, 228, 232–233, 273–275, 277–279, 283–284, 287, 289
collection, 25, 80–81, 115, 117, 131, 143, 148, 153, 162–164, 168–169, 222, 225–226, 273, 275, 277–279, 284
knowledge gaps, 131, 133, 150
reporting, 25, 80, 131, 222, 225–226, 228, 232–233, 273, 275, 277, 279, 283, 287, 289
sharing, 25, 148, 151, 163
trade and catch documentation programs, 222, 227–228, 269–271, 273–275, 277–279, 281, 283, 285–287, 289–291
see also fishery data

deep-sea fisheries, 46, 86, 89, 95, 126, 165–167, 272–274
pelagic fisheries, 25–26, 42, 44, 99–100, 116, 145, 164, 167, 272, 278, 280–281
see also demersal fisheries

demersal fisheries, 46, 272
Alaskan pollock, 86, 89, 99, 168
Aleutian Basin pollock, 127, 198, 278–279
American plaice, 45, 89, 126, 270
Antarctic rock cod, 86, 89, 276
Antarctic toothfish, 86, 88–89, 122, 154, 276
blackfin icefish, 154
black scabbord fish, 89, 272
blue whiting, 272
cod, 3, 44–45, 77, 86–87, 89, 99–100, 126, 148, 166, 204, 270, 276
deepwater hake, 61, 89
Greenland halibut, 86–87, 89, 94, 127, 270–271
ling, 272
marbled rockcod, 154
orange roughy, 46, 272
oreo dory, 61, 89
Patagonian toothfish, 28, 86, 88–89, 122–123, 154, 156–158, 198, 208, 226, 274, 276–277
redfish, 272
Rockall haddock, 272
tusk, 166
witch flounder, 89, 126, 270
wreckfish, 61, 89
yellowtail flounder, 86, 89, 94, 126–127, 270–271
see also deep-sea fisheries

dependent species, 3, 21–22, 45, 77–78, 80, 84, 90, 100–102, 115, 119, 122–123, 144, 148–151, 153, 155–157, 169, 172, 187, 197, 274, 276–277, 290
see associated species

depleted stocks, 3, 9, 24, 62, 79, 81–82, 85–86, 89–90, 95–96, 99, 102, 118–119, 128, 130, 132, 144, 153, 155, 160, 210, 245, 252, 276–277, 281
collapse, 82, 96, 118
serial depletion, 82, 96, 102
see also overfishing
see also stocks

developing States, 23–26, 62, 67, 78, 83, 98, 120, 130, 133, 147, 151, 170–171, 192, 199–200, 207, 231, 234, 261–262, 264, 271, 275, 287
education and training, 24
financial assistance, 133, 264
technical support, 116, 190, 283, 288
see also sustainable development
developing coastal and island States, 24, 78, 83, 98, 130, 133, 147, 151, 171, 200, 205, 207, 274–275, 287
discrete high seas fish stocks, 7, 26, 41–47, 57, 62, 89
see high seas fisheries
dispute settlement, 9, 17, 29–30, 128, 130, 149–150, 187, 193, 199, 207–210, 242, 244, 246, 262, 269, 271, 273, 275, 277, 279, 281, 283, 285, 287, 289, 291
conciliation, 275, 277, 291
International Court of Justice, 277, 283
International Tribunal for the Law of the Sea, 17, 208
see also arbitration
distant water fisheries, 17
see high seas fisheries
dolphins, 28, 153, 159–160, 284
marine mammals, 159
see also cetaceans
Donut Hole Convention, 41, 56, 86, 102, 127, 153, 168, 197–199, 205, 209, 230, 278–279
see Convention on the Conservation and Management of Pollock Resources in the Central Bering Sea (Donut Hole Convention)
ecosystem, 5, 7–9, 13, 15, 21–27, 29, 31, 41, 44, 53, 58–61, 77–78, 80–82, 85, 87, 94, 100–102, 115, 118–123, 127, 129, 132–133, 143–151, 153–156, 158–160, 162–172, 187, 195–196, 205, 209, 222, 230, 245, 252, 261–262, 269–270, 272, 274–276, 278, 280, 282, 284–286, 288, 290
change, 44, 53, 78, 82, 101, 118, 120–122, 132–133, 145, 149, 155, 165, 171, 261
health, 15, 78, 101, 144, 146, 156, 165, 170
integrity, 100, 133, 147, 169
large marine (LME), 143, 145, 148, 155, 159–160, 168–169, 205, 275, 282
marine, 5, 7–8, 15, 22–24, 26–27, 29, 31, 41, 44, 59, 61, 77–78, 80–82, 85, 94, 100–102, 115, 119–120, 122–123, 132–133, 143–151, 153–156, 158–160, 162, 164–165, 167–171, 195, 205, 209, 252, 262, 269, 272, 274–276, 278, 282
modeling, 153, 156, 160, 276
vulnerable marine (VME), 29, 101, 262
ecosystem approach, 5, 7–9, 13, 15, 21–22, 24, 26–27, 53, 58–61, 77, 85, 102, 115, 118–119, 121–122, 129, 143–151, 153–156, 162–172, 187, 195, 230, 269–270, 272, 274, 276, 278, 280, 282, 284, 286, 288, 290
precautionary approach, 5, 7–9, 13, 15, 21–22, 27, 58–61, 85, 102, 115, 118–119, 121–122, 143, 146–149, 151, 153–154, 156, 166, 169–171, 187, 195, 269–270, 272, 274, 276, 278, 280, 282, 284, 286, 288, 290
capacity reduction, 27, 102, 129, 150, 162, 284
dispute resolution, 9, 187
fisheries (EAF), 115, 119, 143–151, 153–156, 162, 164–172, 195, 280, 286
implementation, 5, 8–9, 21, 24, 26–27, 53, 58–59, 61, 121, 129, 143–147, 149–151, 153, 162–165, 169–172, 187, 269, 276, 282, 286, 288, 290
indicators, 118, 129, 148, 150–151, 155–156, 164–165, 170, 282
role of science, 195
see also fishery management
see also ecosystem approach to fisheries management
see also ecosystem-based fisheries management
Ecuador, 42, 46, 61, 159–160, 284–285
education and training, 24
public awareness, 30
EEZ, 3, 14, 17, 41–47, 56, 91, 94, 102, 127–128, 168, 171, 189, 198, 202–207, 224, 231–232, 234–235, 274, 278, 285–286, 288–289
see exclusive economic zone
El Niño, 94
El Salvador, 160, 284
endangered species, 16, 27–28, 83, 92, 94, 101–102, 144, 146–147, 156, 160
see rare species
environmental impact assessment (EIA), 150, 155, 171
environmental protection, 15–16, 53, 155, 277
see marine environmental protection
equity, 8, 15, 53, 146, 151, 199
intergenerational, 53
intragenerational, 15, 53
European Union, 30, 56, 65, 88, 94, 116, 159, 204, 219
European Commission Communication on the Precautionary Principle, 117
European Community, 18–19, 62, 87, 126, 237, 261
exclusive economic zone, 3, 14, 42, 56, 91, 127, 198, 202–203, 224
see also maritime boundaries
experimental fishing, 122–123, 127
exploratory fishing, 46, 95, 123
see also stock assessment
Faroe Islands, 270, 272
finfish fishery, 123, 157, 277
fish stocks, 3, 5–9, 11–12, 14–18, 20–28, 30–31, 41–47, 53–68, 77–78, 80–81, 83–90, 92, 94–97, 99–102, 113, 115–117, 120–121, 125, 130–133, 143–144, 147–149, 151, 153–155, 165–166, 169–172, 187–192, 195, 199, 201–210, 219–221, 229, 236, 245, 248, 251–252, 261, 263, 269, 274, 280, 282, 284, 286–287
see stocks
Fish Stocks Agreement, 3, 5, 7, 11, 15, 21, 29, 41, 45, 53, 56, 58–59, 61, 65, 68, 100, 117, 131, 149, 187, 201–203, 205, 207–208, 219, 251, 261
see United Nations Fish Stocks Agreement
fishery biology, 80
fishery data, 45, 56, 80–81, 88, 90, 122–123, 126, 132, 157, 159, 161, 167–168, 222, 225–226, 230, 274–275, 277–281, 283
collection, 80–81, 90, 123, 132, 157, 161, 167–168, 222, 225–226, 275, 277–281
reporting, 80, 88, 222, 225–226, 230, 275, 277, 279–281, 283
see also data and information
see also monitoring
see also surveillance and enforcement
fishery industry, 171, 412
fishery management, 5, 18, 20–21, 24–25, 42, 45–46, 55–58, 61–63, 65, 80–81, 88, 90, 96, 101–102, 116, 118–119, 122–126, 129, 131–134, 144, 146, 148–151, 153–155, 161–162, 168, 170, 189, 192–193, 195, 198–199, 206, 208, 221–222, 226, 270, 274, 277–279, 283, 290
marine protected areas, 16, 24, 29, 151, 157

capacity, 21, 25, 62, 81, 88, 90, 96, 102, 124, 129, 132–133, 148, 150–151, 155, 162, 170, 192, 199, 222, 283, 290
closed areas, 46, 90, 126, 132, 151, 153, 168, 199, 283
co-operative, 5, 55–56, 80, 149, 151, 193, 199, 221, 277, 279, 283
decision rules, 80, 102, 118–119, 122–124, 133, 149–150, 155, 193, 195, 198, 208, 277, 279
documentation schemes, 65, 222, 226, 277, 279, 283
ecosystem approach to fisheries management (EAFM), 146
ecosystem-based fisheries management (EBFM), 146, 150
fishing rights, 14–15, 17–18, 22, 41, 43, 67, 147, 198–199, 202, 207, 264, 276
gear, 18, 21, 46, 81, 90, 96, 122, 148–150, 153–155, 161, 168, 221, 270, 274, 278, 290
objection procedures, 61, 189, 193, 277, 283
reference points, 81, 96, 116, 118–119, 122–126, 129, 131–133, 144, 151, 193, 195, 198, 277–278
role of science, 195, 209
single vessel phenomenon, 99–100
subsidies, 24, 96, 274
trade measures, 281
"traffic light" approach, 125–126
see also bycatch
see also ecosystem approach, fisheries (EAF)
see also fishing vessels
see also moratoria
see also potential yield
see also precautionary approach
see also quota regulations
see also regional cooperation
see also regional fisheries management organization

fishery products, 20, 90
trade measures, 227

fishery protection, 56, 81, 159
documentation schemes, 19, 29, 65–66, 93–94, 157, 222, 226, 228, 250, 269, 271, 273, 275–277, 279, 281, 283, 285, 287, 289, 291
flag State measures, 18–19, 221, 227–228, 235–236, 238–240, 247–251, 275, 277, 281, 285
port State measures, 95, 221, 228, 236, 247–248, 251–252, 263, 275, 287
trade measures, 19, 29–31, 66, 99, 159, 188, 226–228, 250–252, 263–264, 269, 277, 281, 285, 287, 289, 291
see also illegal fishing
see also surveillance and enforcement
see also unregulated fishing

fishery sciences, 56, 80, 91, 195, 285

fishery survey, 88, 276, 279, 283, 413, 419
see also stock assessment

Fishery Resources Monitoring System (FIRMS), 80, 148

fishing effort, 8, 15, 21, 31, 46, 58, 64, 78, 82–84, 87, 93–99, 123–124, 126, 128, 148, 153, 162, 165–166, 190, 199, 222, 224–226, 233, 251, 272, 275, 279–283, 287

fishing entity, 56, 59, 65–66, 127, 162, 168, 196, 225, 280, 286, 288–291

fishing mortality, 81, 83–85, 87, 93, 98–99, 101–102, 115, 118, 125–127, 130, 144, 148, 151, 155–157, 160, 162–163, 168, 198, 229, 262, 276, 282, 288

fishing vessels, 3, 5, 15–16, 18–23, 25–26, 31, 41, 46, 62–63, 65–66, 82, 87, 95–96, 98–100, 102, 123, 159, 163, 165, 172, 199, 201–202, 206, 219–239, 241, 243, 245, 247–252, 261, 263–264, 270–271, 273–275, 277–279, 281–285, 287–289, 291
data collection, 9, 25, 90, 123, 148, 163, 225–226, 229, 235, 271, 273, 275, 277–281, 284–285, 288, 291
research, 23, 54, 98, 123, 148, 159, 163, 165, 224, 234, 241, 270–271, 275, 277–278, 280, 284–285
flag State, 16, 18–19, 22, 25, 62, 202, 220–221, 223–233, 235–239, 241, 243, 245, 247–251, 271, 273, 275, 277, 279, 281, 285, 289
"genuine link", 16, 25, 224
reflagging, 16, 18–20, 25, 224–225
vessel registers, 16, 19, 31, 57, 66, 220, 224–225, 241, 263, 281, 283, 285
see also fishery management

flags of convenience, 23, 64, 90, 248, 273
see fishing vessels, flag State

flags of noncompliance, 18–19, 188–189, 220, 223–224, 230, 236, 249–251, 263, 271
see surveillance and enforcement

flounder fisheries, 86, 89, 94, 126–127, 270
see demersal fisheries

flag State measures, 18–19, 221, 227–228, 235–236, 238–240, 247–251, 275, 277, 281, 285
see surveillance and enforcement

food chain, 95, 122, 164

food web, 82, 145, 154, 156, 159

Food and Agricultural Organization of the United Nations (FAO), 3–5, 7–9, 11, 13–16, 18–28, 30–31, 42, 45–46, 53–57, 59–62, 65–67, 77–85, 88, 91, 93, 96–97, 99–102, 115–121, 124, 126, 128–132, 143–147, 149–151, 160–163, 165, 167, 169, 172, 189, 195–196, 208, 219–221, 225, 247–248, 251–252, 263, 266, 270–278, 280–284, 286–288, 290–291
Committee on Fisheries (COFI), 14, 57, 149, 247
FishCode Programme, 14
fisheries bodies, 5, 11, 14, 24–25, 46, 53, 56–57, 60–61, 66–67, 80, 121, 149, 151, 162, 189, 208, 282
Fisheries Global Information System (FIGIS), 25, 79
Fishery Committee for the Eastern Central Atlantic (CECAF), 56, 60
Guidelines to Reduce Sea Turtle Mortality in Fishing Operations, 161, 163, 167, 169, 280
International Plan of Action for the Management of Fishing Capacity (IPOA-Fishing Capacity), 23, 27
International Plan of Action for Reducing Incidental Catch of Seabirds in Longline Fisheries (IPOA-Seabirds), 22, 144
International Plan of Action for Conservation and Management of Sharks (IPOA-Sharks), 22, 144
International Plan of Action to Prevent, Deter, and Eliminate Illegal, Unreported and Unregulated Fishing (IPOA-IUU), 23, 62, 221
Model Scheme on Port State Measures, 247–248, 263, 287
Permanent Commission for the South Pacific (CPPS), 46
Regional Fishery Body Network, 46, 208
Rome Consensus on World Fisheries, 26, 77
Rome Declaration on Illegal, Unreported and Unregulated Fishing, 27
Rome Declaration on the Implementation of the Code of Conduct for Responsible Fisheries, 26
South Western Indian Ocean Fisheries Commission (SWIOFC), 46, 56–57, 61, 208
Strategy for Improving Information on Status and Trends of Capture Fisheries, 21, 25, 81
Technical Consultation on the Precautionary Approach to Capture Fisheries, 117

Technical Guidelines on the Ecosystem Approach, 15, 282–283
Technical Guidelines on IUU Fishing, 24, 172, 283
Technical Guidelines on the Precautionary Approach, 15, 21, 116, 118–120, 143, 147, 195, 282
Western Central Atlantic Fishery Commission (WECAFC), 56, 60
see also Code of Conduct for Responsible Fisheries

Forum Fisheries Agency (FFA), 129, 169–170, 224, 231, 277, 287

Framework Agreement for the Conservation of the Living Marine Resources of the High Seas of the South Pacific (Galapagos Agreement), 41, 46, 60–62

France, 159, 202, 204–205, 261, 270, 276, 280, 283–284, 286, 290

freedom of the seas doctrine, 14, 81, 117, 133

high seas, 14, 81
see also common property
see also international law

Galapagos Agreement, 41, 46, 60–62
see Framework Agreement for the Conservation of the Living Marine Resources of the High Seas of the South Pacific

General Fisheries Commission for the Mediterranean (GFCM), 5, 45–46, 55, 57, 65, 85, 90–92, 121, 125, 153, 164–165, 191, 193, 196, 201, 207, 209, 223–224, 228, 230, 232, 236, 269, 280–283
ecosystem approach, 5, 85, 121, 153, 164–165, 230, 269, 280, 282
ICCAT, 5, 45, 55, 65, 85, 90–92, 125, 153, 164, 191, 193, 196, 207, 209, 223–224, 228, 230, 232, 236, 269, 280–283
precautionary approach, 5, 46, 85, 121, 125, 153, 269, 280, 282
control measures, 5, 269, 280
observer program, 228, 230, 281, 283
reform process, 59, 262
Scientific Committee on the Marine Environment and Ecosystem, 164–165, 282
Sub-Committee for Economics and Social Sciences, 125
Sub-Committee on Stock Assessment, 125, 280
geographical distribution, 46, 57
see also migrations

Georgia, 88–89, 122, 155, 157, 205, 227, 276, 281

governance, 3, 5, 7, 9, 11, 15–16, 18–19, 25, 27, 46, 53–54, 57–58, 60, 62–63, 78, 96, 101, 115, 117, 119–120, 130, 144, 147, 149–150, 171–172, 187, 189, 192, 198, 200–201, 209–210, 219, 261–262, 264, 269, 271, 273, 275, 277, 279, 281, 283, 285, 287, 289, 291
fisheries, 3, 5, 7, 9, 11, 15–16, 18–19, 25, 27, 46, 53–54, 57–58, 60, 62–63, 78, 96, 101, 115, 117, 119–120, 130, 144, 147, 149–150, 171–172, 187, 189, 198, 200–201, 209–210, 219, 261–262, 269, 271, 273, 275, 277, 279, 283, 287, 291
high seas, 3, 5, 7, 11, 15–16, 18–19, 25, 27, 46, 54, 57–58, 60, 62–63, 78, 96, 171–172, 189, 201, 210, 219, 261–262, 264, 277, 281, 285, 287, 291
ocean, 5, 15–16, 46, 57–58, 60, 62, 78, 96, 101, 130, 144, 147, 149–150, 261–262, 269, 271, 273, 277, 291

Guatemala, 160, 280, 284
Guinea, Gulf of, 280

habitat, 3, 8, 22, 46, 78, 80, 82, 84, 94, 101, 124, 126, 143–144, 150–151, 164–165, 168, 170–171, 192, 271, 274, 282–284
degradation, 78, 82, 144
restoration, 78, 94, 101
haddock fisheries, 126, 166, 272
see demersal fisheries

hake fisheries, 270, 282
see demersal fisheries

halibut fisheries, 86–87, 89, 127, 270, 278
see demersal fisheries

harvest control or decision rules, 102
see biomass

harvest control, 85, 101–102, 124, 130, 150
fishing effort, 8, 15, 21, 31, 46, 58, 64, 78, 82–84, 87, 93–99, 123–124, 126, 128, 148, 153, 162, 165–166, 190, 199, 222, 224–226, 233, 251, 272, 275, 279–283, 287
see catching methods

high seas, 3, 5, 7, 11, 13–23, 25–27, 30–31, 41–47, 54–58, 60–67, 77–78, 81, 83, 85–87, 89, 94–96, 100, 102, 116, 121, 143, 151, 169–172, 189, 201–208, 210, 219–221, 223–225, 228, 232, 234–236, 244–246, 249, 252, 261–264, 270, 272, 274, 277–278, 281, 285–287, 291

high seas fisheries, 3, 5, 7, 11, 13–16, 18–22, 25–27, 31, 41–43, 46–47, 54–55, 57–58, 61, 63–64, 66–67, 78, 81–83, 85, 87, 102, 116, 121, 151, 172, 203–204, 206, 208, 220, 223, 225, 235, 249, 261–262, 270, 272, 274
discrete high seas stocks, 7, 22, 26, 41–43, 46–47, 57, 87
"purely" high seas stocks, 7, 41, 43

High Seas Task Force, 5, 16, 57, 81, 102, 223–224, 262
see Ministerially-led High Seas Task Force on Illegal, Unreported and Unregulated Fishing

highly migratory fish stocks, 3, 5–9, 11–12, 14–18, 21–22, 24–31, 41–45, 47, 53–59, 61–65, 68, 77–78, 80, 85, 88, 90, 92, 94–95, 97, 99, 113, 115–116, 121, 130–133, 143–144, 147, 149, 153, 169, 171–172, 187, 189–192, 195, 201–203, 205, 207, 209–210, 219–221, 229, 236, 252, 261, 263, 269, 280, 282, 284, 286–287
see migratory species
United Nations Fish Stocks Agreement, 11, 21, 41, 53, 187, 261
United Nations Convention on the Law of the Sea, 17, 21, 41, 53, 55, 77, 219
historical fisheries, 53, 80, 82, 88, 91, 93–94, 128, 199, 280
Honduras, 28, 159–160, 226, 280, 284

Iceland, 23, 56, 88, 94, 146, 204, 263, 270, 272, 274, 280

illegal fishing, 5, 16, 23, 27, 57, 62, 67, 81, 94, 100, 122, 154, 192, 219, 221, 224, 231, 245, 249, 262–263
see also fishery protection

illegal, unreported and unregulated (IUU) fishing, 16, 22–25, 27, 57, 59, 62–63, 66, 81–82, 86, 88, 90, 95, 97–100, 102, 122–124, 133, 154, 156–159, 171–172, 192, 199, 205, 219, 221–229, 232, 234, 242, 244–245, 248–252, 263–264, 270–271, 273, 275, 277, 281, 283, 285, 287, 289, 291
see illegal fishing
see also unregulated fishing

Illex, 95
see squid

IMO, 13, 149
see International Maritime Organization

Indian Ocean, 5, 15, 22, 41, 45–46, 56–57, 61–62, 79, 85, 90–95, 121, 130, 149, 153, 156, 167–168, 190, 205, 223, 231, 261–263, 269, 277, 280, 290–291

Indian Ocean Tuna Commission (IOTC), 5, 22, 85, 121, 129, 149, 153–155, 166–167, 230, 269, 290
precautionary approach, 5, 22, 85, 121, 149, 153–154, 166, 269, 290
control measures, 5, 22, 190, 247, 263, 269, 289, 291
observer program, 228, 230, 277, 287, 289, 291
transshipment, 29, 66, 166, 224, 226–228, 230, 232, 247, 263–264, 277, 287, 291
indicators, 156, 192, 197

reference points, 84, 118, 125, 129, 151, 197, 282
sustainable development, 25, 118, 129, 150–151, 170
Indo-Pacific Tuna Program, 92, 290
see Indian Ocean Tuna Commission

Indonesia, 91, 127, 286, 288, 290

Informal Consultation of State Parties (ICSP), 26, 264
see United Nations Fish Stocks Agreement

Inter-American Tropical Tuna Commission (IATTC), 5, 13, 15, 21–23, 29, 45, 55, 59, 65–66, 83, 85, 90–93, 97–98, 121, 124, 148–149, 151, 153–154, 158–162, 166–167, 190–191, 193, 196–197, 199–200, 203, 205, 207, 209, 223–232, 236, 263–264, 269, 277, 284–285, 287
ecosystem approach, 5, 13, 15, 21–22, 59, 85, 121, 148–149, 151, 153–154, 162, 166–167, 230, 269, 284
precautionary approach, 5, 13, 15, 21–22, 59, 83, 85, 93, 121, 124, 148–149, 151, 153–154, 166, 269, 284
bycatch management, 90–91, 151, 153–154, 160–162, 166–167, 226, 277, 287
compatibility provisions, 205, 207, 264
control measures, 5, 22, 158, 190, 263, 269
data collection, 90, 124, 148, 153, 161–162, 167, 197, 225–226, 229, 269, 277, 284–285
observer program, 59, 158–161, 228–231, 277, 285, 287
transshipment, 29, 59, 66, 124, 166, 224, 226–232, 263–264, 277, 287
Tuna-Dolphin Program, 154, 159–160, 284
see also Convention for the Strengthening of the Inter-American Tropical Tuna Commission (Antigua Convention)

intergenerational equity, 53
see equity

intergovernmental organization (IGO), 149

international environmental law, 5, 7–8, 14, 29, 116

international fisheries law, 5, 7–8, 11–17, 29, 31, 41–42, 53–54, 56, 58, 77, 81, 96, 101–102, 115, 128, 133, 143–144, 190, 208, 222, 235

international law, 5, 7–8, 11–18, 28–29, 31, 41–42, 53–54, 56, 58, 64, 67, 77, 81, 96, 101–102, 115–116, 127–128, 133, 143–144, 159, 188–190, 192, 201–202, 208, 220–222, 224, 227–228, 234–235, 239–240, 244–245, 284
coastal State rights, 17, 202
compliance, 5, 7, 11, 13–16, 18, 31, 53, 64, 67, 77, 159, 188–190, 220–222, 227–228, 234–235, 240, 244–245, 284
consistency, 202, 245
customary, 11, 13–14, 17, 31, 64, 67, 189, 201, 220, 235, 245
duty to cooperate, 13, 16–18, 29, 31, 53, 56, 64, 67, 188–189, 220
flag State, 16–18, 28, 189, 202, 220–221, 224, 227–228, 235, 239–240, 244–245
"hard law", 7, 11, 13, 15, 17, 31
pacta sunt servanda, 188
range State, 28
regime, 5, 7–8, 11, 14–17, 29, 31, 53–54, 58, 64, 67, 77, 96, 102, 115, 133, 143, 159, 188–189, 192, 202, 208, 220, 234–235, 245, 284
"soft law", 7, 11, 13, 15, 17–18, 31, 116, 144, 188, 190
sovereignty, 8, 17, 41, 56, 188–190, 192
use of force, 16–17, 224, 235, 244–245
see also freedom of the seas doctrine
see also maritime boundaries

international trade law, 7, 15, 29, 159, 222, 227–228

International Commission for the Northwest Atlantic Fisheries (ICNAF), 14, 55, 83, 86, 165, 236
see Northwest Atlantic Fisheries Organization

International Commission for the Conservation of Atlantic Tunas (ICCAT), 5, 15, 22, 27–29, 45, 55, 59, 65–66, 83, 85, 90–95, 97–99, 124–125, 148, 153–155, 162–164, 166–167, 188, 190–191, 193, 196–197, 199–200, 203, 205, 207, 209, 223–224, 226–228, 230, 232, 236, 246–247, 251, 263–264, 269, 275, 277, 280–283, 289
Ad Hoc Working Group on the Precautionary Approach, 124
allocation systems, 190–191, 199–200, 281
ecosystem approach, 5, 15, 22, 27, 59, 85, 148, 153–155, 162–164, 166–167, 230, 269, 280, 282
GFCM, 5, 45, 55, 65, 85, 90–92, 125, 153, 164, 191, 193, 196, 207, 209, 223–224, 228, 230, 232, 236, 269, 280–283
precautionary approach, 5, 15, 22, 27, 59, 83, 85, 93, 95, 124–125, 148, 153–154, 166, 269, 280, 282
compatibility provisions, 205, 207, 264
control measures, 5, 22, 27, 95, 190, 247, 251, 263, 269, 275, 280, 289
observer program, 59, 228, 230, 275, 277, 281, 283, 289
participation, 5, 15, 65–66, 98, 190, 200, 269, 275, 280–281, 283, 289
recovery plans, 83, 90, 92–94, 99, 125, 280
reform process, 27, 59
Standing Committee on Research and Statistics, 98, 124, 162, 280
trade measures, 29, 66, 99, 188, 226–228, 251, 263–264, 269, 277, 281, 289
transshipment, 66, 124, 166, 224, 227–228, 230, 247, 251, 264, 277, 281, 283
vessel chartering, 59

International Convention for the Regulation of Whaling, 27, 155

International Council for the Exploration of the Sea (ICES), 55, 81, 85, 87–89, 94, 115, 118, 120–121, 125–126, 147–148, 153, 166, 172, 196, 271–273, 277, 281
Advisory Committee on Fishery Management (ACFM), 125
ecosystem approach, 85, 115, 118, 121, 147–148, 153, 166, 172, 272
precautionary approach, 85, 115, 118, 120–121, 125–126, 147–148, 153, 166, 272
see also Convention for the International Council for the Exploration of the Sea

International Court of Justice (ICJ), 291
see also dispute settlement

International Dolphin Conservation Program (IDCP), 159–160
see also Agreement on the International Dolphin Conservation Program

International Labour Organization, 102

International Maritime Organization, 13, 102, 149

International Tribunal for the Law of the Sea (ITLOS), 17, 127–128, 208, 224, 245, 289
Southern Bluefin Tuna Case(s), 127, 208
see also dispute settlement

International Whaling Commission (IWC), 27, 149, 155, 276–277

invertebrate fisheries
krill fishery, 86, 89, 102, 122, 144, 151, 154–156, 198, 276–277
shrimp fishery, 86–87, 166, 274, 282
squid fishery, 86, 89, 276
islands, 61, 156, 205, 234, 274, 287
IUCN, 93, 101, 275–277, 283
see crab fishery
see World Conservation Union

IUU fishing, 16, 22–25, 27, 57, 59, 62–63, 66, 81–82, 86, 88, 90, 95, 97–100, 102, 122–124, 154, 157–158, 171–172, 192, 199,

205, 219, 221–226, 228, 232, 234, 242, 244–245, 248–252, 263–264, 270–271, 273, 275, 277, 281, 283, 285, 287, 289, 291
see illegal fishing
see also unregulated fishing

Japan, 19, 56, 60–61, 93, 98, 124, 127–128, 162, 208, 226, 235, 262, 270, 272, 276, 278, 280, 283–284, 286, 288–291

Japanese Experimental Fishing Program, 127

joint stocks, 7, 41–42, 46–47, 57, 91, 164, 196, 203, 262
see United Nations Fish Stocks Agreement, transboundary fish stocks

Kenya, 61, 261, 290

Korea, Republic of, 19, 56, 60–61, 90, 127, 270, 274, 276, 278–280, 284, 286, 288, 290

krill fishery, 86, 89, 102, 122, 144, 151, 154–156, 198, 276–277

La Jolla Agreement (Agreement on the Conservation of Dolphins), 159–160, 283–284
see Agreement on the International Dolphin Conservation Program

logbooks, 79, 232, 238, 273
see also data and information

London Conference on Overfishing, 77

Loop Hole Agreement, 44, 56
see Agreement between Iceland, Norway and Russia Concerning Certain Aspects of Cooperation in the Area of Fisheries

LOS Convention, 68
see United Nations Convention on the Law of the Sea

mackerel fisheries, 44, 89, 122, 126, 156, 272, 276
see pelagic fisheries

mare liberum, 14, 54
see freedom of the seas doctrine

marine birds, 154, 156
albatross, 156–157
Antarctic skua, 156
bycatch, 44, 63, 66, 78–80, 85–86, 88, 98, 122, 124, 127, 148, 157, 159, 162–163, 167, 199, 208, 210, 221–222, 225–229, 232–234, 238, 243, 247, 249, 251, 271, 274–277, 280–281, 283, 285–289
penguins, 154, 156
petrels, 30, 154, 156–157
shearwaters, 156

marine environmental protection, 15, 155
see also precautionary approach

marine mammals, 27, 80, 92, 144, 149, 156, 159, 168–169, 277
seals, 149, 156
sea lions, 168
whales, 27, 149, 156
see also cetaceans
see also dolphins
see also whaling

marine pollution, 78, 147, 155, 157–158, 161, 168
debris, 157–158

marine protected area (MPA), 157, 276–277

marine reserve, 132
see marine protected area (MPA)

marine stewardship, 88

maritime boundaries, 3, 7, 14, 41–42, 44, 56, 149, 279
international law, 7, 14, 41–42, 56
sovereignty, 41, 56
territorial sea, 14
maximum economic yield (MEY), 82, 95–96
maximum sustainable yield (MSY), 8, 15, 22, 24, 77–79, 81–85, 90, 92–97, 99, 101–102, 117–119, 121, 124–127, 130, 150, 155, 195, 278, 280, 284–285, 290
see models, bioeconomic
see potential yield
see also exclusive economic zone

Mediterranean Sea, 55, 164–165, 269, 282–283

Mexico, 19–20, 30, 93, 159–160, 280, 284–285

migrations, 3, 5–9, 11–12, 14–19, 21–22, 24–31, 41–47, 53–65, 68, 77–78, 80, 85, 87–88, 90, 92, 94–95, 97, 99, 113, 115–116, 121, 124, 130–133, 143–144, 147, 149, 153, 163–164, 169, 171–172, 187, 189–192, 195, 201–203, 205, 207–210, 219–221, 229, 236, 252, 261–263, 269, 274, 280, 282, 284, 286–288, 290
spawning, 85, 87, 90, 94, 153, 163, 288
see also geographical distribution
see also seasonal distribution

migratory species, 3, 19, 22, 27–28, 30, 42, 44–47, 58, 61, 77–78, 85, 88, 90, 92, 94–95, 99, 143–144, 153, 164, 202–203, 252, 280, 282, 284, 286, 288, 290

Ministerially-led High Seas Task Force on Illegal, Unreported and Unregulated Fishing, 5, 16, 22–25, 27, 57, 59, 62–63, 66, 81–82, 86, 88, 90, 95, 97–100, 102, 122–124, 133, 154, 156–159, 171–172, 192, 199, 205, 219, 221–229, 232, 234, 242, 244–245, 248–252, 262–264, 270–271, 273, 275, 277, 281, 283, 285, 287, 289, 291

models, 53, 80, 82, 84–85, 95–96, 118–120, 122–124, 131, 153, 156, 160, 164–165, 169–170, 172, 196–198, 225–226, 231, 233, 247–248, 252, 262–263, 273, 276, 284, 287, 289
bio-economic, 96

monitoring, 7, 9, 14, 16, 18–19, 23–28, 46, 59–60, 62, 80, 115, 119, 131–132, 148–150, 155–157, 165, 169–170, 172, 189, 199, 210, 219–223, 225, 229–234, 236, 243–245, 250, 261, 263, 269, 271, 273, 275–283, 285, 287, 289, 291
see also data and information
see also fishery data

monitoring, control and surveillance (MCS), 9, 115, 124, 131–133, 210, 219–220, 222–225, 228–235, 242, 246, 252, 269, 275, 281, 287, 291
see monitoring
see surveillance and enforcement

moratoria, 85–87, 89, 94, 100, 126, 128, 153, 163, 165–166, 230, 239, 270–271, 280
see also fishery management

Mozambique, 261, 416
nature conservation, 55, 93, 97, 115, 163, 207, 219
negative lists, 59, 66, 223–224, 251, 264
see surveillance and enforcement, vessel registers

neritic species, 42, 290, 416

New Zealand, 43–44, 46, 62, 95, 127–128, 157, 168, 198, 208, 261, 272, 276, 286, 288

Nicaragua, 160, 280, 284

non-governmental organization (NGO), 153, 201, 250, 270–272, 279

North Atlantic Oscillation, 94

North Pacific Anadromous Fish Commission (NPAFC), 231, 245–246

North Pacific Fur Seal Commission, 83

North Sea, 55, 81, 99, 168, 269, 279, 286

North East Atlantic Fisheries Commission (NEAFC), 5, 14, 41–42, 46, 55, 60, 65–66, 81, 85–87, 89, 94–95, 120–121, 125–126, 151, 153, 166–167, 188, 190–191, 193–194, 196, 200, 202–204, 207–209, 223–225, 228–229, 231–233, 236, 242–244, 246, 248, 251, 262–264, 269, 271–273, 275, 277
 allocation systems, 273
 biodiversity, 5, 60, 125, 151, 153, 166, 262, 271, 273
 ecosystem approach, 5, 60, 85, 121, 151, 153, 166–167, 269, 272
 ICES, 55, 81, 85, 87, 89, 94, 120–121, 125–126, 153, 166, 196, 271–273, 277
 precautionary approach, 5, 14, 46, 60, 85, 95, 120–121, 125–126, 151, 153, 166, 194, 269, 272
 boarding and inspection, 236, 242, 244, 246, 251, 273, 275
 consistency principle, 202–203
 control measures, 5, 95, 190, 248, 251, 263, 269, 275
 reform process, 60, 262
 transshipment, 66, 166, 224, 228–229, 231, 248, 251, 264, 271, 277

Northwest Atlantic Fisheries Organization (NAFO), 5, 14, 22, 24, 27, 41–42, 45–46, 55–56, 59–60, 65–66, 83, 85–87, 89, 94–95, 100–102, 121, 125–127, 148, 153–154, 165–166, 190–191, 193, 196–204, 207, 209, 223–233, 236–240, 242–246, 248–249, 251, 262–264, 269–271, 273, 275, 277
 allocation systems, 273
 ecosystem approach, 5, 22, 24, 27, 59–60, 85, 102, 121, 148, 153–154, 165–166, 230, 269–270
 precautionary approach, 5, 14, 22, 27, 46, 59–60, 83, 85, 95, 102, 121, 125–127, 148, 153–154, 166, 269–270
 boarding and inspection, 236–237, 239–240, 242, 244–246, 251, 273, 275
 consistency principle, 202–203
 control measures, 5, 22, 27, 95, 190, 248, 251, 263, 269, 275
 observer program, 59, 228–231, 237, 271, 273, 275, 277
 reform process, 27, 59–60, 262
 Standing Committee on Fisheries Environment (STACFEN), 270
 Standing Committee on International Control (STACTIC), 60, 190, 239
 transshipment, 66, 166, 224, 227–231, 248, 251, 264, 270–271, 277

Norway, 19, 44, 56, 88, 270, 272, 274, 276, 280, 282

Norwegian-Russian Fisheries Commission, 120, 204

Ocean Conservancy, 161

ocean governance, 5, 16, 58, 60, 101, 144, 147, 149
 see governance

ocean management, 5, 15, 17, 22, 41, 46, 55, 58, 60–62, 67, 78, 85, 87, 90, 95, 129, 133, 144–145, 147, 149–151, 153–155, 158, 161, 169, 190, 202, 205, 252, 262, 271–274, 286
 adaptive, 53–54, 80, 120, 124, 131, 150–151, 162, 284
 competing uses, 58, 144, 147, 149
 integrated, 15, 60, 144, 147, 149–151, 153, 252, 271

oceanic species, 78, 89, 163, 169, 417

Okhotsk, Sea of, 41, 56, 86, 96, 100, 269
 "Peanut Hole", 41, 56, 86, 96, 269

opting out, 102, 289
 see fishery management,objection procedures

Organization for Economic Co-operation and Development (OECD), 251

overexploitation, 3, 5, 7–8, 28, 54–55, 62–63, 77–79, 82–83, 86, 88, 96–97, 100, 102, 116, 118, 120, 122, 130, 133, 144, 168, 172, 187, 195, 198
 see overfishing

overfishing, 7–9, 16, 20–21, 23, 26, 30–31, 44, 67–68, 77–78, 81–86, 90, 92–94, 96, 99–102, 115, 118–119, 125–126, 130, 132–133, 144–145, 147, 150–151, 166, 168, 192, 245, 252, 262, 271, 277, 280–281, 283, 285, 290–291
 definitions, 192
 ecosystem overfishing, 7, 9, 21, 31, 44, 78, 82, 101–102, 115, 118–119, 144, 245, 252, 262
 "species at risk", 78, 101–102, 252
 see also depleted stocks
 potential yield, 82

Pacific Island States, 205

Pacific Islands Forum Fisheries Agency, 224, 287
 see Forum Fisheries Agency

Pacific Ocean, 5, 16, 46, 57, 60–62, 65, 79, 85, 88, 90–91, 93, 95, 124, 158–161, 169, 245, 262–263, 269, 284, 286

Panama, 28, 159–160, 224, 280, 284

pelagic fisheries, 25–26, 42, 44, 99–100, 116, 145, 164, 167, 272, 278, 280–281
 armourhead, 61, 89, 274
 Atlanto-Scandian spring spawn herring, 272
 capelin, 89, 126, 204, 270
 Chilean horse mackerel, 42
 Chilean jack mackerel, 44
 lanternfish, 154, 156
 lightfish, 86
 mackerel, 3, 42, 44, 86, 89, 92, 122, 126, 156, 207, 272–273, 276, 290
 mackerel icefish, 89, 122, 156, 276
 see also deep sea fisheries
 tuna fisheries, 99–100, 164, 167, 280–281

Peru, 42, 61, 160, 208, 276, 284

Philippines, 128, 152, 280, 286, 288, 290

Poland, 56, 272, 276, 278

pollock fisheries, 43, 77, 86, 168, 278–279
 see pelagic fisheries

pollution, 8, 78, 94, 146–147, 155, 157–158, 161, 168
 see marine pollution

population characteristics
 biomass (in weight), 80, 93–94, 96–97, 100, 102, 122, 125, 127–128, 143, 155
 genetic testing, 41, 163
 population dynamics, 148, 164
 population number (in specific number), 42
 population structure, 96, 163
 recruitment, 81–82, 84, 88–89, 93–94, 118, 126, 128, 154–155, 164, 276
 spawning success, 128
 tagging, 41, 45, 163, 168, 277, 281, 285, 289–291
 see also stock assessment
 see also stocks

port State measures, 95, 221, 228, 236, 247–248, 251–252, 263, 287
 see surveillance and enforcement

positive lists, 59, 66, 223–224, 251
 vessel registers and vessel records, 66, 224
 see surveillance and enforcement

potential yield, 82, 413, 416–418, 420
MSY reformulation, 15, 118
overfishing, 7–9, 16, 20–21, 23, 26, 30–31, 44, 67–68, 77–78, 81–86, 90, 92–94, 96, 99–102, 115, 118–119, 125–126, 130, 132–133, 144–145, 147, 150–151, 166, 168, 192, 245, 252, 262, 271, 277, 280–281, 283, 285, 290–291
total allowable catch, 15, 44, 83
quota regulations, 93
see also fishery management

precautionary approach, 5, 7–9, 13–15, 21–22, 27, 30, 46, 58–61, 83, 85, 93, 95, 102, 115–128, 133, 143, 146–151, 153–154, 156, 166, 169–171, 187, 189, 194–195, 252, 269–270, 272, 274, 276, 278, 280, 282, 284, 286, 288, 290
adaptive management, 124, 151, 284
ecosystem approach, 5, 7–9, 13, 15, 21–22, 27, 58–61, 85, 102, 115, 118–119, 121–122, 143, 146–149, 151, 153–154, 156, 166, 169–171, 187, 195, 269–270, 272, 274, 276, 278, 280, 282, 284, 286, 288, 290
burden of proof, 115–117, 121, 123, 128
decision rules, 8, 102, 115, 117–124, 133, 156, 171, 187, 195, 276
implementation, 5, 8–9, 14, 21, 27, 30, 58–59, 61, 83, 116, 120–121, 123–125, 127, 133, 143, 146–147, 149, 151, 153, 169–171, 187, 252, 269, 276, 282, 286, 288, 290
reference points, 8, 22, 83, 85, 115–127, 133, 151, 195, 272, 278, 282, 284, 286
socioeconomic conditions, 14, 119, 169
see also fishery management
see also marine environmental protection

predator prey interactions, 154

quota regulations, 66–67, 102, 166, 249, 270, 279
fishery management, 5, 18, 20–21, 24–25, 42, 45–46, 55–58, 61–63, 65, 80–81, 88, 90, 96, 101–102, 116, 118–119, 122–126, 129, 131–134, 144, 146, 148–151, 153–155, 161–162, 168, 170, 189, 192–193, 195, 198–199, 206, 208, 221–222, 226, 270, 274, 277–279, 283, 290
total allowable catch, 15, 44, 83
see also allocation systems

Ramsar Convention, 144
see Convention on Wetlands of International Importance

rare species, 28, 144
see also species extinction

recruitment, 81–82, 84, 88–89, 93–94, 118, 126, 128, 154–155, 164, 276
see population characteristics

redfish fisheries, 89, 126, 270–273
see demersal fisheries

regional cooperation, 13, 17, 24, 28, 54–55, 57, 149, 159, 205, 208, 219, 234, 264, 277, 287
see also fishery management

regional fisheries management arrangement, 47, 56, 63

regional fisheries management organization, 219

regional economic integration organization (REIO), 56, 65–66, 276, 284

regional fisheries body, 46, 61
see also regional fisheries management organization

regional fisheries management arrangement (RFMA), 56–57, 59–61, 243
participation, 57
see also regional cooperation
see also regional fisheries management organization

regional fisheries management organization (RFMO), 5–9, 11, 13–16, 19–20, 22–31, 41–42, 44–47, 53–60, 62–68, 77–78, 80–81, 83, 85–86, 88–90, 92, 94–95, 97–98, 100, 102, 113, 115–116, 118–122, 124, 126–127, 129–133, 143, 146–151, 153–156, 166, 170–172, 187–210, 219–226, 228–236, 241–252, 261–264, 269–291
allocation (criteria) systems, 187, 273
ecosystem approach, 5, 7–9, 13, 15, 22, 24, 26–27, 53, 58–60, 77, 85, 102, 115, 118–119, 121–122, 129, 143, 146–151, 153–156, 166, 170–172, 187, 195, 230, 269–270, 272, 274, 276, 278, 280, 282, 284, 286, 288, 290
precautionary approach, 5, 7–9, 13–15, 22, 27, 30, 46, 58–60, 83, 85, 95, 102, 115–116, 118–122, 124, 126–127, 133, 143, 146–149, 151, 153–154, 156, 166, 170–171, 187, 189, 194–195, 252, 269–270, 272, 274, 276, 278, 280, 282, 284, 286, 288, 290
data reporting, 9, 25, 80, 88, 131, 225–226, 228, 230, 232–233, 243, 269, 271, 273, 275, 277, 279, 281, 283, 285, 287, 289, 291
decision-making procedures, 9, 187–189, 191–194, 198, 208–209, 219
decision rules, 8, 80, 102, 115, 118–122, 124, 130, 133, 150, 156, 171, 187, 193, 195, 197–198, 208–209, 276–277, 289
dispute resolution, 9, 11, 187, 189, 192–193, 198, 200, 204, 208–210, 242, 244, 262, 277, 281, 289, 291
enforcement measures, 9, 19, 22, 59, 190, 219–223, 228, 235–236, 245–246, 248–252, 263–264, 269, 271, 275, 277, 279, 283, 285
free riders, 64
functions, 9, 22, 53, 56–57, 68, 80, 143, 149, 156, 170, 192, 204, 229, 232–233, 283
geographic area of application, 56
institutional arrangements, 53, 58, 187, 190
inter-RFMO cooperation, 42, 58, 149, 207–208, 269
membership, 5, 62–67, 200, 241, 269–270, 272, 274, 276, 278, 280–281, 283–284, 286, 288, 290
non-Parties, 228, 284
objection procedures, 9, 187, 189, 193–194, 209–210, 219, 236
reform process, 27, 59–60, 262
role of science, 195, 209
"vehicle of good governance", 7
vessel registers, 16, 31, 66, 220, 224–225
see also fishery management
see also regional cooperation
see also regional fisheries body
see also regional fisheries management arrangement
see also surveillance and enforcement

regional fisheries organization, 19, 41, 55–56, 61, 63, 159, 219, 273
see regional fisheries body

Regional Commission for Fisheries (RECOFI), 57

remote sensing, 234
global positioning system (GPS), 233
vessel detection system (VDS), 234
vessel monitoring system (VMS), 234

reproduction, 81

research, 23, 54–56, 58, 79–80, 84, 98, 120, 123–125, 132–133, 146–149, 151, 153, 155–156, 159–171, 196–197, 224, 234, 241, 270–271, 275–278, 280, 284–286, 290

reservations, 188, 244
see fishery management, objection procedures

resilience, 84, 94, 101, 146

Reykjavik Declaration on Responsible Fisheries in the Marine Ecosystem, 23, 144

RFMO, 5, 8–9, 14, 16, 22, 27–29, 31, 41–42, 44, 46–47, 54, 56–57, 59–60, 62–67, 81, 89, 92, 95, 100, 113, 119, 121, 126, 131, 133, 146, 148–149, 151, 153–155, 166, 170–171, 187–203, 205–210, 219–221, 223–226, 228–236, 241–246, 248–252, 261–264, 269–291
  see regional fisheries management organization

Rio Declaration, 13, 15, 18, 20, 53, 116, 200
  see United Nations Conference on Environment and Development (UNCED)

risk management, 8, 22, 84, 97, 99, 102, 116–117, 119–123, 126, 131–132, 146, 169, 194–195, 204, 252
  assessment, 22, 97, 116, 121, 123, 131–132, 252

Rome Technical Conference on the Conservation of the Living Resources of the Sea, 55

Russian Federation, 41, 60, 270, 272, 276, 278–280, 283

Saint Vincent and the Grenadines, 226, 245

scientific research, 55–56, 80, 132, 148–149, 151, 153, 155, 159, 162, 165, 167–168, 170–171, 196–197, 270, 276–278, 280, 284–285
  see research

sea turtles, 161, 163, 167, 169, 280, 285–286, 290
  bycatch, 90, 124

seabirds, 22, 153–157, 162–163, 167–169, 276–277, 281, 284–286, 288–290
  see marine birds

seamounts, 61

season regulations, 153

seasonal distribution, 42
  see also geographical distribution
  see also migrations

security issues, 3, 78
  coast guards, 223, 279
  inter-agency coordination, 26
  ecological security, 78, 151
  food security, 3, 78
  human security, 78
  navies, 15, 100, 224
  see also surveillance and enforcement

sedentary species, 42, 46, 61
  see also migratory species

Seychelles, 261, 276, 290

shared fish stocks, 7, 41, 46–47, 209
  see UN Convention on the Law of the Sea, transboundary stocks
  see United Nations Fish Stocks Agreement, transboundary stocks

shark fisheries, 22, 161, 163, 166–167, 262, 270, 290
  basking, 167
  blue, 163, 290
  bycatch, 90, 124
  finning, 163, 166, 262, 270
  rays, 89, 92, 157, 161–162, 168, 219, 248, 276, 288
  skates, 89, 95, 157, 166, 270–271, 276

shrimp fishery, 86–87, 166, 274, 282

skate fishery, 95
  see shark fisheries

South Africa, 24, 90, 202, 204, 274, 276, 280, 288, 290

South East Atlantic Fisheries Organisation (SEAFO), 5, 15, 25, 41–42, 45–46, 60–61, 65, 85–86, 89, 98, 121, 129, 153, 155, 166, 168–169, 188, 190–191, 193–194, 196–197, 199–201, 205, 208–209, 230, 232, 236, 248, 262–264, 269, 274–275
  allocation systems, 190–191, 199–200
  ecosystem approach, 5, 15, 60–61, 85, 121, 129, 153, 155, 166, 168–169, 230, 269, 274
  precautionary approach, 5, 15, 46, 60–61, 85, 121, 153, 166, 169, 194, 269, 274
  compatibility provisions, 201, 205, 264
  control measures, 5, 190, 248, 263, 269, 275
  dispute resolution, 193, 200, 208–209, 262
  observer program, 168, 230, 275
  Scientific Committee, 89, 98, 129, 155, 168–169, 190, 196–197, 274
  see also Convention on the Conservation and Management of Fishery Resources in the South East Atlantic Ocean

South Pacific RFMO (SPRFMO), 262

South Tasman Rise Agreement, 41, 56

Southern Indian Ocean Fisheries Agreement, 41, 56, 61, 130, 153, 190

Southern Indian Ocean Fisheries Arrangement (SIOFA), 15, 41, 46, 56–57, 60–62, 85, 95, 98, 129–130, 153, 166, 170, 190, 205, 208, 261, 269, 277
  ecosystem approach, 15, 60–61, 85, 129, 153, 166, 170, 269
  compatibility provisions, 205

Southern Ocean, 15, 41, 45–46, 56, 61, 85, 88, 90–93, 95, 122, 130, 144–145, 149, 153–155, 157–158, 168, 190, 205, 224, 263, 269, 276–277
  see also Antarctic

sovereignty, 8, 17, 41, 56, 187–190, 192–193, 204, 209–210, 219
  see international law

Soviet Union
  see Russian Federation

Spain, 159, 162, 272, 276, 280, 283–284

spawning grounds, 87, 90–91, 125, 128, 163, 165, 167, 283, 288–289

species at risk, 78, 86, 89, 96, 99–102, 156, 252, 272
  see rare species

species extinction, 28, 101, 145
  see also rare species

sponges, 123, 157

squid fishery, 86, 89, 276

St. John's Declaration, 16, 27, 57–60
  see Conference on the Governance of High Seas Fisheries and the UN Fish Agreement, Ministerial Declaration

stock assessment, 22, 79–81, 85, 87–88, 92–95, 101, 115–116, 118, 123–125, 128, 131, 151–153, 155, 165, 168, 196–198, 226, 252, 275–276, 280, 282, 284, 286, 288, 290–291
  exploratory fishing, 22, 79–81, 85, 87–88, 92–95, 101, 115–116, 118, 123–125, 131, 151, 153, 155, 165, 196–198, 252, 275–276, 280, 282, 284, 286, 288, 290
  fishery survey, 22, 79–81, 85, 87–88, 92–95, 101, 115–116, 118, 123–125, 131, 151, 153, 155, 165, 196–198, 252, 275–276, 280, 282, 284, 286, 288, 290
  population characteristics, 22, 79–81, 85, 87–88, 92–95, 101, 115–116, 118, 123–125, 131, 151, 153, 155, 165, 196–198, 252, 275–276, 280, 282, 284, 286, 288, 290

stocks, 22, 79–81, 85, 87–88, 92–95, 101, 115–116, 118, 123–125, 131, 151, 153, 155, 165, 196–198, 252, 275–276, 280, 282, 284, 286, 288, 290
stock restoration, 87, 94
see fishery management
see also catch statistics

stock status, 41, 44, 47, 78–81, 86, 89–92, 98, 102, 118–119, 123–129, 132–133, 151, 165, 167, 209, 222, 226, 269–270, 272–276, 278, 280, 282, 284, 286, 288, 290
see stock assessment

Stockholm Declaration, 20
see United Nations Conference on the Environment

stocks, 3, 5–9, 11–12, 14–18, 20–28, 30–31, 41–47, 53–68, 77–102, 113, 115–127, 129–133, 143–144, 147–149, 151, 153–155, 157–160, 163–167, 169–172, 187–193, 195–210, 219–221, 229, 236, 239, 243, 245, 248, 251–252, 261–263, 269–288, 290
shared, 7, 41–42, 46–47, 54, 57, 125, 209, 252, 270, 275, 279
stock mixing, 45, 90, 163, 207
see also depleted stocks
see also population characteristics
see also stock assessment

straddling fish stocks, 3, 5–9, 11–12, 14–18, 21–22, 24–26, 28, 30–31, 41–47, 53–64, 68, 77–78, 80, 85–86, 88–89, 94–95, 97, 99–100, 113, 115–116, 121, 130–133, 143–144, 147, 149, 153, 171–172, 187, 189–192, 195, 201–210, 219–221, 229, 236, 252, 261, 263, 269
see United Nations Fish Stocks Agreement
see United Nations Convention on the Law of the Sea

subsidiarity, 53

surveillance and enforcement, 7, 9, 16, 115, 189, 210, 219, 221, 223, 231, 235, 245, 261, 263, 269–271, 273–275, 277–279, 281, 283, 285–287, 289–291
air surveillance, 231, 271
boarding and inspection, 9, 16, 221, 235, 245, 273, 275, 279, 281, 283, 287, 289, 291
compliance measures, 9, 219, 221, 223, 235, 245, 263, 271, 277, 281, 283, 285, 289
data requirements, 9, 115, 269, 275, 281, 285
electronic log books, 231, 271, 273
flag State measures, 18–19, 221, 227–228, 235–236, 238–240, 247–251, 275, 277, 281, 285
flags of non-compliance, 223
inter-RMFO cooperation, 285
observer programs, 9, 231, 269, 271, 273, 275, 277, 279, 281, 283, 285, 287, 289, 291
port State measures, 95, 221, 228, 236, 247–248, 251–252, 263, 287
trade measures, 19, 29–31, 66, 99, 159, 188, 226–228, 250–252, 263–264, 269, 277, 281, 285, 287, 289, 291
transshipment, 9, 231, 271, 277, 281, 283, 287, 291
vessel detection systems (VDS), 9
vessel monitoring system (VMS), 9, 16, 231, 263, 269, 271, 273, 275, 277, 279, 281, 283, 285, 287, 289, 291
vessel record, 19, 26, 66, 223–225, 287, 291
vessel registers, 16, 263, 281, 283, 285
see also fishery data
see also fishery protection
see also regional fisheries management organization
see also security issues

sustainable development, 3, 5, 11, 13–16, 18, 20, 24–26, 30–31, 53, 57–58, 60, 67, 77–78, 81, 83, 85, 115–118, 120, 129–130, 132–133, 144, 146–147, 149–151, 169–170, 172, 195, 201, 219, 252, 273, 290
see also developing States
see also developing coastal and island States

Taiwan (Taiwan, Province of China), 65
see also Chinese Taipei

territorial sea, 14, 17, 55, 278
see also maritime boundaries

toothfish fisheries, 123, 154, 156, 274, 276–277
see demersal fisheries

total allowable catch (TAC), 44, 66, 82–84, 87–89, 95, 122, 125, 127–128, 132, 157, 167, 197–200, 202, 204–205, 270–271, 273, 276, 288–289
see also potential yield
see also quota regulations

trade documentation schemes, 19, 29, 66, 94, 222, 228, 250, 269, 271, 273, 275, 277, 279, 281, 283, 285, 287, 289, 291
see data and information, trade and catch documentation programs
see fishery protection, documentation schemes

trade information schemes, 222, 250, 275, 277, 279, 287, 289
see data and information, trade and catch documentation programs
see fishery protection, documentation schemes

trade measures, 29–31, 66, 99, 159, 188, 226–228, 250–252, 263–264, 269, 277, 281, 285, 287, 289, 291
see fishery products
see fishery protection

tragedy of the commons, 7, 53–54, 62, 96, 192
see common property

transboundary fish stocks, 7, 41–43, 46–47, 57–58, 282
see United Nations Convention on the Law of the Sea
see United Nations Fish Stocks Agreement

transparency, 9, 15, 18, 23, 25, 29, 53, 77, 80, 102, 131–132, 148–149, 187, 190, 194–195, 197, 199–201, 250–251, 262, 269–272, 274, 276, 278–281, 283–286, 288, 290

transshipment, 9, 66, 124, 166, 220, 224, 227–231, 247–248, 250–251, 264, 270–271, 277–278, 281, 283, 287, 291
see surveillance and enforcement

Treaty of European Union, 116

trophic relationships, 148

tuna fisheries, 45, 57, 88, 90–91, 93, 98, 124, 127, 129, 149, 151, 153–154, 159–161, 167–169, 200, 246, 280–281, 284, 286, 288, 290–291
billfish, 161, 284, 286, 290
dolphins, 153, 159–160, 284
albacore, 90–91, 98, 100, 125, 280, 282, 284, 286, 290
bigeye, 90–91, 93, 98, 100, 124, 129, 160, 162–163, 280–281, 284, 286, 290–291
black marlin, 90, 284, 290
blackfin, 154
blue marlin, 90, 93, 125, 163, 280, 284, 290
bluefin, 5, 45, 90–91, 93, 98, 100, 121, 125, 127, 129, 148, 153, 162–164, 167–168, 198, 200, 208, 224, 246, 269, 280–284, 288, 290
eastern Atlantic bluefin, 45, 82, 90, 92–93, 125, 163, 207, 282
frigate mackerel, 290
little, 57, 169, 246
marlin, 90, 93, 125, 163, 280–281, 284, 290
Mediterranean albacore, 90, 92, 125, 280

Mediterranean swordfish, 90, 92, 125, 280
north Atlantic shortfin mako, 92, 125, 280
north Pacific albacore, 90, 286
northern albacore, 90, 98, 284
northern bluefin, 45, 90, 99–100, 153
ranching/farming, 90, 164, 228, 280–283
sailfish, 90, 125, 280, 284, 290
sauries, 61
skipjack, 90–91, 93, 125, 129, 162, 280, 284, 286, 290
south Atlantic swordfish, 100, 125, 200, 280
south Pacific albacore, 90, 286
southern albacore, 90–91, 284, 290
southern bluefin, 5, 45, 90–91, 93, 98, 100, 121, 127, 129, 148, 153, 167–168, 198, 208, 224, 246, 269, 284, 288, 290
striped marlin, 90, 284, 290
swordfish, 90, 93, 100, 125, 163, 200, 246, 280–282, 284, 290
western bluefin, 45, 90, 93, 121, 125, 148, 163, 200, 224, 269
western Atlantic bluefin, 45, 90, 93, 125, 148, 163, 200, 269
white marlin, 90, 93, 125, 163, 280
yellowfin, 90–91, 93, 151, 154, 159–163, 280, 284, 286, 290
see also pelagic fisheries

UNCED, 3, 15, 18–21, 24, 30–31, 53–54, 60, 68, 115–116, 144, 200
see United Nations Conference on Environment and Development

UNCLOS, 55–56, 84
see United Nations Conferences on the Law of the Sea

UNEP, 28, 30, 99, 102, 149, 283
see United Nations Environment Programme

UNFA, 3
see United Nations Fish Stocks Agreement

UNFSA, 3, 5, 7–9, 11, 13, 15–27, 29–31, 41–42, 44–47, 53–68, 77–81, 83–85, 93–97, 100–102, 115–133, 143–144, 147–151, 154, 160, 162–163, 168–169, 171–172, 187–190, 192–210, 219–221, 223, 225–226, 228–229, 231, 233–236, 241–252, 261–262, 264, 266, 270–274, 276, 278, 280, 283–288, 290
see United Nations Fish Stocks Agreement

United Kingdom, 123, 157, 204–205, 274, 276, 280, 290

United Nations, 3, 11, 15, 17–19, 21, 25–26, 28, 41, 43, 53, 55, 77, 84, 97, 99, 116–117, 143, 149, 187, 219, 244, 261–263
Charter, 244
Commission on Sustainable Development (UNCSD), 26, 53
Secretary-General, 26
see also United Nations General Assembly

United Nations Agreement on Straddling Fish Stocks and Highly Migratory Fish Stocks
see United Nations Fish Stocks Agreement

United Nations Commission on Global Governance, 53

United Nations Conference on Environment and Development (UNCED), 3, 15, 18–21, 24, 30–31, 53–54, 60, 68, 115–116, 144, 200
Rio Declaration, 15, 18, 53

United Nations Conference on the Environment, 3, 15, 18, 53
Stockholm Declaration, 20

United Nations Conferences on the Law of the Sea, 55, 84, 421

United Nations Convention on the Law of the Sea (LOSC), 3, 7–9, 11, 13–18, 25, 27–31, 41–42, 45–47, 53–56, 58–59, 61–64, 77, 83, 85, 117–118, 128, 131, 144, 149, 153, 172, 189, 194–196, 201–203, 206–210, 219–220, 245, 251, 286
compatibility principle, 21
conservation, 17, 21, 41, 53, 55, 77, 117, 219
continental shelf, 17
dispute resolution, 9, 11, 128, 187, 189, 192–193, 198, 200, 204, 208–210, 242, 244, 262, 277, 281, 285, 289, 291
duty to cooperate, 17, 53, 219
environmental protection, 15–16, 53, 155, 277
exclusive economic zone, 3, 14, 41–42, 56, 91, 127, 151, 198, 202–203, 224, 262
fisheries, 3, 5, 7–9, 11–27, 29–31, 41–47, 53–68, 77–88, 90–91, 93–102, 115–133, 143–157, 159–172, 187–191, 194–201, 203–210, 219–226, 229–235, 239, 241, 246–249, 252, 261–263, 269–280, 282–284, 286–287, 290–291
flag State, 16–19, 22, 25, 28, 30, 43, 62, 189, 202–203, 210, 220–221, 223–233, 235–241, 243–245, 247–251, 271, 273, 275, 277, 279, 281, 285, 289
high seas, 3, 5, 7, 11, 13–23, 25–27, 30–31, 41–47, 54–58, 60–67, 77–78, 81, 83, 85–87, 89, 94–96, 100, 102, 116, 121, 143, 151, 169–172, 189, 201–208, 210, 219–221, 223–225, 228, 232, 234–236, 244–246, 249, 252, 261–264, 270, 272, 274, 277–278, 281, 285–287, 291
highly migratory fish stocks, 3, 5–9, 11–12, 14–18, 21–22, 24–31, 41–45, 47, 53–59, 61–65, 68, 77–78, 80, 85, 88, 90, 92, 94–95, 97, 99, 113, 115–116, 121, 130–133, 143–144, 147, 149, 153, 169, 171–172, 187, 189–192, 195, 201–203, 205, 207, 209–210, 219–221, 229, 236, 252, 261, 263, 269, 280, 282, 284, 286–287
integrated approach, 23, 30, 53, 60, 144, 146–147, 153
marine mammals, 27, 80, 92, 144, 149, 156, 159, 168–169, 277
peaceful use of the sea, 9, 53, 208
sedentary species, 42, 46, 61
straddling fish stocks, 3, 5–9, 11–12, 14–18, 21–22, 24–26, 28, 30–31, 41–47, 53–64, 68, 77–78, 80, 85–86, 88–89, 94–95, 97, 99–100, 113, 115–116, 121, 130–133, 143–144, 147, 149, 153, 171–172, 187, 189–192, 195, 201–210, 219–221, 229, 236, 252, 261, 263, 269
transboundary fish stocks, 7, 41–43, 46–47, 57–58, 282

United Nations Environment Programme, 28, 149

United Nations Fish Stocks Agreement (UNFSA), 11, 21, 41, 53, 187, 261
ecosystem approach, 5, 7–9, 13, 15, 21–22, 24, 26–27, 53, 58–61, 77, 85, 102, 115, 118–119, 121–122, 129, 143–151, 153–156, 162–172, 187, 195, 230, 269–270, 272, 274, 276, 278, 280, 282, 284, 286, 288, 290
non-parties, 228, 284
other instruments, 7, 11, 13, 16, 20, 24, 27, 58, 78, 189, 197, 209, 245
precautionary approach, 21, 187
RFMOs, 5–9, 11, 13–16, 19–20, 22–31, 41–42, 44–45, 47, 53–60, 62–68, 77–78, 80–81, 83, 85–86, 88–90, 92, 94–95, 97–98, 100, 102, 115–116, 118–122, 124, 127, 129–133, 143, 147–151, 153–154, 156, 170–172, 187–193, 195–210, 219–226, 228–236, 241–242, 244–252, 261–264, 269, 271, 273, 275–276, 281, 283, 285–289
allocation systems, 187, 273
boarding and inspection, 9, 16, 221–222, 234–237, 239–242, 244–246, 250–251, 273, 275, 279, 281, 283, 287, 289, 291
compatibility principle, 18, 21, 44, 59, 202–203, 206
control measures, 5, 18–19, 22, 27, 31, 77, 95, 123, 190, 219, 221, 247–248, 251, 280
[flag and port State] data collection, 25, 225–226, 229, 235, 271, 273, 275, 281, 285
developing States, 23–26, 62, 67, 78, 83, 120, 130, 133, 147, 151, 171, 192, 199–200, 207, 231, 234, 261–262, 264, 271, 287
dispute resolution, 9, 11, 128, 187, 189, 192–193, 198, 200, 204, 208–210, 242, 244, 262

duty to cooperate, 13, 16–18, 22, 29, 31, 53, 56–57, 63–64, 67, 188–189, 194, 219–220, 241–242
enclaves, 41, 43–44, 100, 204
flag State, 3, 5, 13, 16–20, 22–23, 25, 28, 30–31, 43, 54–55, 62, 64, 67, 131, 188–189, 194, 201–203, 206, 210, 220–221, 223–233, 235–252, 262–263, 271, 273–275, 277, 279–281, 285, 289
highly migratory fish stocks, 3, 5, 7–9, 11, 15–18, 21–22, 24–27, 30–31, 41–42, 44–45, 47, 53–59, 61–65, 68, 77–78, 80, 85, 94–95, 97, 115–116, 121, 130–133, 143–144, 147, 149, 169, 171–172, 187, 189–190, 192, 195, 201–203, 205, 207, 209–210, 219–221, 229, 236, 252, 261, 280, 284, 286–287
Informal Consultation of State Parties, 26
minimum standards, 3, 7, 13, 19–20, 30–31, 77, 83, 85, 115, 120, 129, 133, 150, 172, 189–190, 192–195, 197, 206–207, 209, 252, 261
observer programs, 9, 67, 148, 220, 229, 231, 273, 284–285, 287
port State measures, 95, 221, 228, 236, 247–248, 251–252, 287
"real interest", 5, 57, 63–64, 210, 219, 234, 283
review conference, 5, 9, 11, 16, 23, 26, 30, 42, 59, 93, 261–262, 264
role of science, 195, 209
straddling fish stocks, 3, 5, 7–9, 11, 15–18, 21–22, 24–26, 30–31, 41–42, 44–47, 53–64, 68, 77–78, 80, 85, 94–95, 97, 100, 115–116, 121, 130–133, 143–144, 147, 149, 171–172, 187, 189–190, 192, 195, 201–210, 219–221, 229, 236, 252, 261
transboundary fish stocks, 7, 41–42, 46–47, 57–58
vessel registers, 16, 31, 66, 220, 225
vessel monitoring systems, 9, 16, 19, 26, 220, 231, 278
zonal attachment, 44, 203

United Nations Framework Convention on Climate Change, 116, 411, 422

United Nations General Assembly, 25–26, 116, 262–263
precautionary approach, 116
Ad Hoc Open-Ended Informal Working Group, 26
resolution on fisheries, 25–26
resolution on oceans, 26
resolution on vulnerable marine ecosystems, 262
see also United Nations

United Nations Open-ended Informal Consultative Process on Oceans and the Law of the Sea (UNICPOLOS), 26

United Nations Programme of Action for Sustainable Development
see Agenda 21

United States, 18–19, 21, 26, 30, 43, 55, 60, 93, 99, 116, 124, 146, 159–160, 205, 219, 227, 245, 262, 276
Western Pacific Fisheries Management Council, 161
unregulated fishing, 5, 16, 18, 23, 27, 44, 46, 57, 62, 65, 67, 81, 95, 122, 154, 192, 219, 221, 262–263, 272, 277
see also fishery protection

Uruguay, 29, 224, 276, 280

use conflicts, 67, 244, 422
see ocean management, competing uses

Vanuatu, 159–160, 276, 280, 284–286, 290

Venezuela, 159–160, 280, 284–285

vessel monitoring system (VMS), 16, 19, 220, 222, 225, 231–235, 239, 243, 246, 269, 271, 273, 275, 277, 279, 281, 283–285, 287, 289, 291
see remote sensing
see surveillance and enforcement
see United Nations Fish Stocks Agreement

vessel record, 19, 26, 66, 223–225, 287, 291
see surveillance and enforcement

vessel registers, 16, 19, 31, 57, 66, 220, 222, 224–225, 241, 263, 281, 283, 285
see surveillance and enforcement

Vienna Convention on the Law of Treaties, 189

vulnerability, 97, 144

Western and Central Pacific Tuna Commission, 5, 60, 85, 121, 148, 224, 263, 269
see Western and Central Pacific Fisheries Commission

Western and Central Pacific Fisheries Commission (WCPFC), 5, 25, 29, 45, 60–61, 65–66, 68, 85, 90–92, 98, 121, 129, 148–149, 152–155, 169, 188, 190–191, 193–194, 196–197, 199–200, 203, 205, 207–209, 224, 228–232, 236, 241, 248, 263, 269, 285–287, 289–290
allocation systems, 190–191, 199–200
ecosystem approach, 5, 60–61, 85, 121, 129, 148–149, 153–155, 169, 230, 269, 286, 290
precautionary approach, 5, 60–61, 85, 121, 148–149, 153–154, 169, 194, 269, 286, 290
boarding and inspection, 236, 241, 287, 289
compatibility provisions, 205, 207
control measures, 5, 190, 248, 263, 269, 289
dispute resolution, 193, 200, 208–209, 289
observer program, 228–231, 285–287, 289
Scientific Committee, 98, 129, 155, 169, 190, 196–197, 286, 290
transshipment, 66, 224, 228–231, 248, 287
see also Convention on the Conservation and Management of Highly Migratory Fish Stocks in the Western and Central Pacific Ocean

whales, 27–28, 122, 149, 154, 156
see marine mammals

whaling, 27, 149, 155
see also marine mammals

World Conservation Union (IUCN), 93, 101, 275–277, 283

World Summit on Sustainable Development (WSSD), 5, 13, 16, 21, 24, 77, 81, 83, 94–95, 132, 144–145, 195
Plan of Implementation, 5, 16, 21, 24, 83, 94, 132, 144

World Trade Organization (WTO), 24, 29–30, 99, 102
Agreement on Subsidies and Countervailing Measures, 99
documentation schemes, 29
ecolabelling, 30
subsidies, 24, 30, 99

World Wildlife Fund, 161

zonal attachment, 44, 203
see United Nations Fish Stocks Agreement